T0138629

ESSENTIALS OF
PHYSICAL CHEMISTRY

ESSENTIALS OF PHYSICAL CHEMISTRY

Don Shillady

CRC Press
Taylor & Francis Group
Boca Raton   London   New York

CRC Press is an imprint of the
Taylor & Francis Group, an **informa** business

CRC Press
Taylor & Francis Group
6000 Broken Sound Parkway NW, Suite 300
Boca Raton, FL 33487-2742

© 2012 by Taylor and Francis Group, LLC
CRC Press is an imprint of Taylor & Francis Group, an Informa business

No claim to original U.S. Government works

Printed in the United States of America on acid-free paper
10 9 8 7 6 5 4 3 2 1

International Standard Book Number: 978-1-4398-4097-9 (Hardback)

This book contains information obtained from authentic and highly regarded sources. Reasonable efforts have been made to publish reliable data and information, but the author and publisher cannot assume responsibility for the validity of all materials or the consequences of their use. The authors and publishers have attempted to trace the copyright holders of all material reproduced in this publication and apologize to copyright holders if permission to publish in this form has not been obtained. If any copyright material has not been acknowledged please write and let us know so we may rectify in any future reprint.

Except as permitted under U.S. Copyright Law, no part of this book may be reprinted, reproduced, transmitted, or utilized in any form by any electronic, mechanical, or other means, now known or hereafter invented, including photocopying, microfilming, and recording, or in any information storage or retrieval system, without written permission from the publishers.

For permission to photocopy or use material electronically from this work, please access www.copyright.com (http://www.copyright.com/) or contact the Copyright Clearance Center, Inc. (CCC), 222 Rosewood Drive, Danvers, MA 01923, 978-750-8400. CCC is a not-for-profit organization that provides licenses and registration for a variety of users. For organizations that have been granted a photocopy license by the CCC, a separate system of payment has been arranged.

Trademark Notice: Product or corporate names may be trademarks or registered trademarks, and are used only for identification and explanation without intent to infringe.

Library of Congress Cataloging-in-Publication Data

Shillady, Donald.
   Essentials of physical chemistry / by Donald Shillady.
      p. cm.
   Includes bibliographical references and index.
   ISBN 978-1-4398-4097-9 (pbk.)
   1. Chemistry, Physical and theoretical--Textbooks. I. Title.

QD453.3.S53 2011
541--dc22
                                                       2011005868

Visit the Taylor & Francis Web site at
http://www.taylorandfrancis.com

and the CRC Press Web site at
http://www.crcpress.com

Contents

Preface

This text is really two books in one, perhaps three. The first 9 chapters are intended to meet the new American Chemical Society requirement of one semester of physical chemistry. By good fortune, I have taught such a course five times at Virginia Commonwealth University with special emphasis on ways to make important material understandable to students who may only have had one semester of calculus. This sequence has also been taught once as an experiment at Randolph Macon College using McQuarrie's *Quantum Chemistry*, Second Edition, with a smaller class as a second semester elective. Key topics are included in thermodynamics and kinetics with mathematics that has been tuned to a level that just slightly stretches the student's math ability. These students are used to memorizing vast amounts of material in biology and organic chemistry so we "prime the pump" of their intellect by encouraging memorization of key topics such as the critical point of a van der Waals gas, the Michaelis–Menten equation, and the entropy of mixing. Along the way, we introduce problems related to forensic chemistry as well as a mixture of practical units likely to be encountered in health sciences. Sample tests are included in the text so students can see what is expected of them. The tests are intended to measure knowledge of the formulas and how to use them, as well as to display successful memorization/learning of key concepts.

Spectroscopy is limited in a one semester course; we have tried to use the relatively simple Bohr atom for examples of orbital screening and x-ray emission analysis to get as much meaning from the simple formula as possible. The treatment of chemical kinetics is split into a fundamental Chapter 7 and a more advanced Chapter 8 so that if time is running out in a one semester course, Chapter 8 can be skipped and delayed until the beginning of an elective second semester. A one semester sequence might be "Introduction: Mathematics and Physics Review," Chapters 1 through 7 and 9 with an elective second semester as Chapters 8, 10 through 15 with parts of 16, 17, 18, or 19.

I think it is important to keep the material of an optional second semester visible in the same text with many pictures of successful scientists to coax students into further study. Actually, taking just one semester of physical chemistry really limits a student for further study. I certainly do not agree with the ACS Committee on Professional Training in this matter. Physical chemistry is the axis of science where all the sciences come together along with mathematical models that have been developed over the past 200 years. If perchance my colleagues in physical chemistry have developed a reputation for being difficult, I call on them to make physical chemistry "fun" to train students in this central science. I know from teaching two semesters of physical chemistry in an intensive nine-week "Summer P. Chem." course for over 30 years that by selecting key topics, explaining them well, and teaching the necessary mathematics leads to great morale in a class while serious learning is going on! I often receive messages from students who complete this course saying they are using P. Chem. notes in upper level courses in biochemistry and inorganic chemistry. Topic selection is also a result of my teaching physical chemistry laboratory for many years using successive editions of the Shoemaker, Garland, and Nibler text; thus, material here should form a foundation for that course as well.

This text is being completed just a week before the beginning of another nine-week "Summer P. Chem." course, and it must be considered in view of a series of fast, intense lectures. As usual, there will be students in this class with only one semester of calculus (having never dealt with an integral) and biology majors who are pre-medical students as well as forensic science majors who are taking almost as much chemistry as the formally designated chemistry majors. They all will have recently passed organic chemistry, but probably not have used much mathematics since a prior year in freshman chemistry. We have put the Introduction chapter first because that is the way we have found we can get the students to review whatever math skills they previously used. Putting this chapter first is debatable, but I have received a number of testimonials from past students who comment that the "Math Review" was really important for them. Note that equations in the text are not numbered so as to encourage linking verbal ideas with logic and develop the ability to "read mathematics," a skill this author considers very helpful in science.

I have read the reviews of many other texts in physical chemistry and found that they are seldom positive, although I think a number of those texts are really good. Thus, I anticipate criticism that I should have done more of this or that, but I know from developing this course over 30 years that you don't have time to do everything even in two semesters; thus, the one semester limit requires selection of only a few topics. There should perhaps be more problems, but I know that most teachers are fiercely independent and prefer their own emphasis; thus, the strategy here is to focus on the basics and let the teachers provide their own preferred special topics. In truth, there are already more problems presented here than there is time for in a nine-week summer course. The strategy here is (1) tell the students what you want them to know, (2) tell them what you will test them on, (3) give them a large number of past tests to practice on, and then (4) test them using slightly different questions with at least one new question to separate thinkers from memorizers. This will allow the memorizers to hopefully learn some of what they memorized! We know this process can be successful in bringing a group of students to the national average and above as measured by standardized tests of the American Chemical Society. I will of course be attentive to any outright error in the text, but this material is basically the notes developed over a 30-year period for the students mentioned earlier using many different texts and with many successful outcomes. This material has been taught as a one semester course, where usually the course extended only to Chapter 9, with or without Chapter 8, and in the summer a few students only take the "Introduction: Mathematics and Physics Review" and Chapters 1 through 9.

The "second book" (Chapters 10 through 16) builds on the math skills developed in the first semester, Introduction and Chapters 1 through 9. The emphasis in the second semester is on various forms of spectroscopy built on a foundation of quantum chemistry, and more mathematical details are given as well as some detailed examples that might be useful in Physical Chemistry Laboratory courses. Note that the detailed mathematics is there for interested students in several appendices, but the emphasis on the tests is to learn the bottom line conclusions. Chapter 13 is clearly the most difficult chapter, but hopefully using the "$m = 0$ shortcut" will make it easier. There is use of a small quantum chemistry program (PCLOBE) in Chapters 11 and 16 as explained in Chapter 17. We have found that you can use molecular orbitals in lecture with PCLOBE shown on a video screen often in the second semester. There usually will be three or four students (in a class of 40) who really want to learn this material. Probably not all of the later chapters can be treated in a two semester course, and it is admitted that the nine-week summer course usually stops at Chapter 16 or 18. Chapter 19 is a daring venture into what is usually treated in organic chemistry, but modern pulse-NMR is really an exercise in how to use physical principles of magnetism to make NMR spectroscopy easier to interpret with the COSY technique as a gateway to more advanced methods. While the variational treatment of the H1s orbital in Chapter 16 is *always* treated in this course, the second semester usually ends after Chapter 15. Thus, the "third book" consists of Chapters 17 through 19, which are special topics suitable for senior projects or for prospective graduate students.

Other teachers may want more problems to select from but in the fast nine-week course it seems better to give 5 or 10 problems of good quality a week than a lot of busy work. Some students want more problems rather than derivations while others prefer memorization. Thus, the sample examinations in the text show a balance of some memorization of derivations and other tough numerical problems. After the first few weeks in a fast course, students are motivated to prepare for a looming final examination, and we have found an unorthodox way of coupling their desire to prepare with presentation of "essential" topics. The latter part of the Summer P. Chem. course semester is studying and assimilating old examinations including a number of derivations to memorize/learn. On the last class day of each semester, students go to the board and "explain" a problem from an old test to the class. Teaching to tests? Yes. Does it work as shown by results on standardized tests? Yes! In some universities there are fraternity/sorority files of old tests given by various faculty members. Why not give the whole class the same advantage, particularly if they actually learn from the old tests!

It is not unusual now for a student to bring a laptop to lectures; we have thus included many references to Internet sites with interesting educational supplements. In particular, I owe a debt of gratitude to several authors of interesting Web sites who have given permission to include some of their educational material in this text. Their contributions are noted by name and web address in the captions to their work. Of course, it is still a good idea to have a textbook to lug around, but there are treasures of knowledge on Internet sites. By using the Internet as a supplement, we can combine the results of many excellent tutorial sites to form a richer set of examples. Thus, in the "second book" (Chapters 10 through 16), we gradually extend quantum chemistry concepts to spectroscopic applications. Some other spectroscopic topics are in Chapters 17 through 19 where teachers can choose to emphasize their specialty, and we regret not having time/space for subjects like electrochemistry or discussion of nuclear power. However, we are sure that if the second semester establishes some basic quantum chemistry, a teacher can use the time in a second semester to choose special topics. We have included SCF calculations, Raman spectroscopy, electrospray mass spectrometry, and nuclear magnetic resonance as advanced topics that can be mixed according to the wishes of a given teacher and class, but the main idea is to present more quantitative discussion of the merged science of physical chemistry.

Finally, it is important to acknowledge two colleagues who have brought constructive criticism to bear on my research as well as this text. Professor James F. Harrison of the Michigan State University Chemistry Department has long been a friend and classmate as an undergraduate and in graduate school. His keen intellect has kept me on the right path for a long time and his contributions to Chapters 16 and 17 should be noted. Also, my "style mentor" has been Professor Carl Trindle at the University of Virginia. I cannot match his graceful style in prose and have lapsed back into the conversational style that works for me in lectures. However, Carl has inspired me to use shorter sentences. In addition, my wife Nancy says this is my "last book" and I thank her for her patience in my putting my entire focus on this text for over 9 months and the 30 years before that when most of the notes presented here were developed. My son Douglas has been a valuable computer consultant during my conversion to Word 2007 and Windows 7; he has been of considerable help in preparing and formatting the manuscript. Additional thanks are due to the staff at the Quantum Theory Project at the University of Florida at Gainesville for providing selected photographs from past Sanibel meetings. Special mention should also be made of the cooperation from both the Chemical Heritage Foundation and the Segrè Collection of the American Institute of Physics for assistance with research and selection of photographs. The suggestions and support of my editor, Lance Wobus, is gratefully acknowledged as well. Thanks to the team of project manager Vinithan Sethumadhavan of SPi Technologies, and project editor Richard Tressider and production coordinator David Fausel of CRC Press, who were extremely helpful in what turned out to be a task more complex than expected. Special thanks are also due to the 2010 class of "Summer P. Chem.," who helped me rid the text of typographical errors and check end-of-chapter problems. The second printing includes checking and selected corrections to the quantum mechanics chapters and Appendix A by Prof. Cameron Reed of Alma College. In addition, Prof. Carl Trindle of the University of Virginia assisted in improving some of the Excel graphs.

Donald D. Shillady
Ashland, Virginia

Author

Donald D. Shillady is a native of Montgomery County, Pennsylvania, and an emeritus professor of chemistry at Virginia Commonwealth University (VCU). He has also been an adjunct professor of chemistry at Randolph Macon College. He has taught physical chemistry, physical chemistry laboratory, and quantum chemistry at VCU since 1970 and still teaches "Summer P. Chem." at VCU. A graduate of Drexel University with an MS from Princeton University and industrial experience in electrochemistry, he has been interested in optical activity and magnetically induced optical activity since his PhD thesis at the University of Virginia in 1969. He later expanded his research interests to the larger field of quantum chemistry.

Prof. Shillady has thoroughly enjoyed teaching physical chemistry and physical chemistry laboratory and has carried out experiments in the spectroscopy of optical activity as well as the computation of optical rotatory dispersion (ORD), circular dichroism (CD), magnetic circular dichroism (MCD), complete active space configuration interaction (CASCI) and multi-configurational self-consistent-field (MCSCF) results. To date he has published 77 research papers, edited three books, and completed a monograph with Professor Carl Trindle titled *Electronic Structure Modeling: Connections between Theory and Software*. His other research interests include the biological effects of electromagnetic waves, determination of the absolute configuration of large optically active molecules, integral transform wave functions, properties of metal atom clusters, and the relationship between valence bond and configuration interaction methods in chemistry. The PCLOBE program accompanying this text is a subset of his programming efforts over the years as presented in the simple Gaussian lobe basis representation. This text is his response to the reduction in requirements in physical chemistry by the American Chemical Society at a time when U.S. students have demonstrated lower performance in mathematics and science compared to students of other nations.

Prof. Shillady wishes to promote physical chemistry as fun, using informal language and various mnemonics alongside basic calculus. He presents basic, "essential" physical chemistry in the first 10 chapters with additional material suitable for modern spectroscopy in an attempt to convey the joy of learning clear relationships in molecular behavior. He adamantly regards the central nature of physical chemistry as the keystone in the arch of biology, chemistry, and physics and offers his best material used in a succinct summer course for over 30 years at VCU. This text offers the minimum basic material and provides additional interdisciplinary material in support of his enthusiastic promotion of physical chemistry as a central science.

List of Constants

Since it is unlikely students will learn all the significant figures in the CRC tables, we offer pragmatic values rounded to four significant figures.

90th CRC Value   **≈ Student Value**

Speed of light	: 2.99792458×10^8 m/s	$\approx 2.998 \times 10^8$ m/s
Planck's h	: $6.62606896 \times 10^{-34}$ J s	$\approx 6.626 \times 10^{-34}$ J s
Planck's h-bar	: $1.054571628 \times 10^{-34}$ J s	$\approx 1.055 \times 10^{-34}$ J s
Electron mass	: $9.10938215 \times 10^{-31}$ kg	$\approx 9.11 \times 10^{-31}$ kg
Electron charge	: $1.602176487 \times 10^{-19}$ C	$\approx 1.602 \times 10^{-19}$ C
Proton mass	: $1.672621637 \times 10^{-27}$ kg	$\approx 1.673 \times 10^{-27}$ kg
Avogadro's no.	: $6.02214179 \times 10^{23}$/mol	$\approx 6.022 \times 10^{23}$/mol
1 cal	: 4.184 J	≈ 4.184 J
1 Volt	: 1 Joule/Coulomb	
1 eV	: 1.602177×10^{-19} J	$\approx 1.602 \times 10^{-19}$ J
1 J	: 1 kg m^2/s^2	
1 erg	: 1 g cm^2/s^2	
1 J	: 10^7 erg	
1 N	: 1 kg m/s^2	
1 dyne	: 1 g cm/s^2	
1 N	: 10^5 dyne	
1 gal (U.S.)	: 3.785412 L	≈ 3.785 L
1 quart (U.S.)	: 946.3529 cm^3	≈ 946 cm^3
1 in.	: 2.54 cm (exact)	
R (gas constant)	: 8.314472 Pa m^2/(°K mol)	≈ 8.314 Pa m^2/(°K mol)
R (gas constant)	: 0.08314472 L bar/(°K mol)	≈ 0.08314 L bar/(°K mol)
R (gas constant)	: 0.08205746 L atm/(°K mol)	≈ 0.0821 L atm/(°K mol)
R (gas constant)	: ≈ 1.987 cal/(°K mol) ($PV =$ Energy!)	
1 atm (Earth)	: 1.01325×10^5 Pa $= 1.01325 \times 10^6$ dyne/cm^2	
1 atm (Earth)	: 1.01325 bar $= 101325$ Pa	
1 Rydberg	: 10973731.568527/m	≈ 10973732/m
1 Rydberg	: 13.60569193 eV	≈ 13.606 eV
1 Hartree $= 2$ Rydberg $= 27.21138386$ eV		≈ 27.21 eV

Periodic Table of the Elements

In the table on the next page, the new IUPAC format numbers the groups from 1 to 18. The previous IUPAC numbering system and the system used by Chemical Abstracts Service (CAS) are also shown. For radioactive elements that do not occur in nature, the mass number of the most stable isotope is given in parentheses. Elements 113–116 and 118 have been reported but not yet confirmed.

REFERENCES

G. J. Leigh, Ed, *Nomenclature of Inorganic Chemistry*, Blackwell Scientific Publications, Oxford, U.K., 1990.
Chemical and Engineering News, **63**(5), 27, 1985.
Atomic weights of the elements, 2007, *Pure and Appl. Chem.*, **81**, 2131, 2009.

Periodic Table of the Elements

Key to chart

50	+2 +4
Sn	
118.710	
-18-18-4	

Atomic number → 50
Symbol → Sn
2001 atomic weight → 118.710
Oxidation states → +2 +4
Electron configuration → -18-18-4

New notation (top) — Previous IUPAC form — CAS version

Legend: Metallic solids · Non-metallic solids · Liquids · Gases

1 IA	2 IIA	3 IIIB	4 IVB	5 VB	6 VIB	7 VIIB	8 VIII	9 VIII	10	11 IB	12 IIB	13 IIIA	14 IVA	15 VA	16 VIA	17 VIIA	18 VIIIA	Shell
1 H +1 -1 1.00794 1																	2 He 0 4.002602 2	K
3 Li +1 6.941 2-1	4 Be +2 9.012182 2-2											5 B +3 10.811 2-3	6 C +2 +4 -4 12.0107 2-4	7 N +1+2+3+4+5 -1-2-3 14.0067 2-5	8 O -2 15.9994 2-6	9 F -1 18.9984032 2-7	10 Ne 0 20.1797 2-8	K-L
11 Na +1 22.989770 2-8-1	12 Mg +2 24.3050 2-8-2											13 Al +3 26.981538 2-8-3	14 Si +2 +4 -4 28.0855 2-8-4	15 P +3 +5 -3 30.973761 2-8-5	16 S +4 +6 -2 32.065 2-8-6	17 Cl +1+5+7 -1 35.453 2-8-7	18 Ar 0 39.948 2-8-8	K-L-M
19 K +1 39.0983 -8-8-1	20 Ca +2 40.078 -8-8-2	21 Sc +3 44.955910 -8-9-2	22 Ti +2+3+4 47.867 -8-10-2	23 V +2+3+4+5 50.9415 -8-11-2	24 Cr +2+3+6 51.9961 -8-13-1	25 Mn +2+3+4+7 54.938049 -8-13-2	26 Fe +2+3 55.845 -8-14-2	27 Co +2+3 58.933200 -8-15-2	28 Ni +2+3 58.6934 -8-16-2	29 Cu +1+2 63.546 -8-18-1	30 Zn +2 65.409 -8-18-2	31 Ga +3 69.723 -8-18-3	32 Ge +2+4 72.64 -8-18-4	33 As +3+5-3 74.92160 -8-18-5	34 Se +4+6-2 78.96 -8-18-6	35 Br +1+5-1 79.904 -8-18-7	36 Kr 0 83.798 -8-18-8	-L-M-N
37 Rb +1 85.4678 -18-8-1	38 Sr +2 87.62 -18-8-2	39 Y +3 88.90585 -18-9-2	40 Zr +4 91.224 -18-10-2	41 Nb +3+5 92.90638 -18-12-1	42 Mo +6 95.94 -18-13-1	43 Tc (98) -18-13-2	44 Ru +3 101.07 -18-15-1	45 Rh +3 102.90550 -18-16-1	46 Pd +2+4 106.42 -18-18-0	47 Ag +1 107.8682 -18-18-1	48 Cd +2 112.411 -18-18-2	49 In +3 114.818 -18-18-3	50 Sn +2+4 118.710 -18-18-4	51 Sb +3+5-3 121.760 -18-18-5	52 Te +4+6-2 127.60 -18-18-6	53 I +1+5+7-1 126.90447 -18-18-7	54 Xe 0 131.293 -18-18-8	-M-N-O
55 Cs +1 132.90545 -18-8-1	56 Ba +2 137.327 -18-8-2	57* La +3 138.9055 -18-9-2	72 Hf +4 178.49 -32-10-2	73 Ta +5 180.9479 -32-11-2	74 W +6 183.84 -32-12-2	75 Re +4+6+7 186.207 -32-13-2	76 Os +3+4+6+8 190.23 -32-14-2	77 Ir +3+4 192.217 -32-15-2	78 Pt +2+4 195.078 -32-17-1	79 Au +1+3 196.96655 -32-18-1	80 Hg +1+2 200.59 -32-18-2	81 Tl +1+3 204.3833 -32-18-3	82 Pb +2+4 207.2 -32-18-4	83 Bi +3+5 208.98038 -32-18-5	84 Po +2+4 (209) -32-18-6	85 At +3+5-1 (210) -32-18-7	86 Rn 0 (222) -32-18-8	-N-O-P
87 Fr +1 (223) -18-8-1	88 Ra +2 (226) -18-8-2	89** Ac +3 (227) -18-9-2	104 Rf +4 (261) -32-10-2	105 Db (262) -32-11-2	106 Sg (266) -32-12-2	107 Bh (272) -32-13-2	108 Hs (277) -32-14-2	109 Mt (276) -32-15-2	110 Ds (281) -32-16-2	111 Rg (280)	112 Cn (285)	113 Uut (284)	114 Uuq (289)	115 Uup (288)	116 Uuh (292)		118 Uuo (294)	-O-P-Q

* Lanthanides

58 Ce +3+4 140.116 -19-9-2	59 Pr +3+4 140.90765 -21-8-2	60 Nd +3 144.24 -22-8-2	61 Pm +3 (145) -23-8-2	62 Sm +2+3 150.36 -24-8-2	63 Eu +2+3 151.964 -25-8-2	64 Gd +3 157.25 -25-9-2	65 Tb +3 158.92534 -27-8-2	66 Dy +3 162.500 -28-8-2	67 Ho +3 164.93032 -29-8-2	68 Er +3 167.259 -30-8-2	69 Tm +2+3 168.93421 -31-8-2	70 Yb +2+3 173.04 -32-8-2	71 Lu +3 174.967 -32-9-2

Shell: -N-O-P

** Actinides

90 Th +4 232.0381 -18-10-2	91 Pa +4+5 231.03588 -20-9-2	92 U +3+4+5+6 238.02891 -21-9-2	93 Np +3+4+5+6 (237) -22-9-2	94 Pu +3+4+5+6 (244) -24-8-2	95 Am +3+4+5+6 (243) -25-8-2	96 Cm +3 (247) -25-9-2	97 Bk +3+4 (247) -27-8-2	98 Cf +3 (251) -28-8-2	99 Es (252) -29-8-2	100 Fm (257) -30-8-2	101 Md +2+3 (258) -31-8-2	102 No +2+3 (259) -32-8-2	103 Lr +3 (262) -32-9-2

Shell: -O-P-Q

Introduction: Mathematics and Physics Review

Welcome to "Essentials of Physical Chemistry"! We really have a challenge here! According to the new requirements of the American Chemical Society, only one semester of physical chemistry is required for the BS in chemistry, although a second semester is an acceptable elective. Within chemistry education there has long been a division between students who shy away from mathematics and those who appreciate the quantitative nature of the bridge between chemistry and physics. This is not only unfortunate in closing doors to students who might go further in research but is an unnecessary result of a long tradition of physical chemists reveling in esoteric difficulties. Our position here is that physical chemistry faculty should strive more to popularize the field of physical chemistry without degrading the level of treatment by using enthusiastic joyful teaching methods! There are many applications of physical chemistry where the mathematical modeling has been developed to a point where clean solutions are available and can be compared to experimental results with a sense of wonder at the level of understanding they give to the molecular world that is truly amazing! Here we attempt to provide mnemonics and other analogies to bring chemical principles in focus with calculus-level treatment for the "essential" topics in physical chemistry in one semester, along with enough material to continue further treatment of other topics in an elective semester. It is a "Mission Impossible" to present essential topics in physical chemistry in one semester, so our attitude is "Just the facts, Jack!" If there are four ways to explain something, we will use only one way that we think makes the point, but we will fill in most of the mathematical steps to make that one way as clear as possible. This leads to abbreviated treatments but with sufficient mathematical depth that interested students will still be able to advance to further topics with a good foundation.

Although we need to treat a broader range of topics today, we aim to reach the level of clarity for undergraduates as given in what we consider a pedagogical masterpiece by G. S. Rushbrooke as *Introduction to Statistical Mechanics*, Oxford Press, London, 1962; just the necessary facts, key equations, and only a few problems at the end of each chapter. Our hope is that any student who follows the text with pencil and paper will do well on the Graduate Record Examination or other advanced courses. We cannot cover every topic in one semester but we can give a good foundation with active student participation. After the absolutely essential topics in the first semester, we include further topics for an elective second semester that fill in what should be normally included in a two-semester treatment of physical chemistry, in greater detail at a more leisurely pace. However, we still have to pack some spectroscopy into the first semester because a student may not elect the second semester and yet modern topics are heavily dependent on spectroscopy! Did I say we have a challenge? By the way, some sensitive souls have told me that I should not use capital letters to emphasize words since in this age of texting that denotes shouting, BUT in my lectures I routinely emphasize some words more loudly than others to wake up the back row of the class! Thus no offense is intended, it is just a style mechanism to maintain class attention!

REVIEW OF NECESSARY MATHEMATICS AND PHYSICS

If you have flipped through some pages of this text you may be concerned about the amount and level of mathematics herein. You need to give the text a chance since it is based on over 30 years of teaching students just like yourself who may have sold your calculus text and wondered why you had to take any calculus at all. There is a high probability that you have recently passed a course or two in organic chemistry, which involved little mathematics but massive amounts of learning. The approach of this text is to use your mental synthesis skills learning massive amounts of chemistry to "prime the pump" of your mind using a few examples at a level of two semesters of calculus even though you may have only had one semester! We are going to show the mathematical details (calculus nuggets) of selected cases to provide depth in a few areas but you have to do your part! The author earned a grade of D- in the first quarter of organic chemistry taken in summer school by lounging in a hammock and just moving his eyes over class notes. The second and third quarters of organic chemistry led to grades of A because the present author wrote each reaction over 20 times! Thus it is very important that you use a pencil and paper and copy over the examples and proofs given in this text. Would you expect to learn to swim by reading a book or learn to shoot basketball foul shots by reading about it? This author firmly believes that hand-eye activity in writing equations is a valid way to study physical chemistry as well as organic chemistry. If you expect to just read this text or highlight key passages as if it were a history book, you are already in trouble. However, if you follow along with pencil and paper you will be amazed that priming your mental pump with detailed equations will lead to increased confidence at what you can do with a minimum of mathematics! This approach has worked with timid young ladies as well as overconfident male athletes and the result has always been great morale in the class to the point of actually having "fun" in physical chemistry!

The overwhelming obstacle to the study of physical chemistry is a lack of skill in mathematics whether in calculus or just basic competence in careful addition. Overall, the language of all the physical sciences is mathematics! It is amazing that if you check on journal articles in Chinese, Russian, Turkish, French, German, or English, the equations and tables of numbers are the same! In fact, we will start slowly with calculus examples but the real language of this course is "Calculus"! The author has taught this material to hundreds of students from all over the world whose use of English is a second or third language. This is not a course on the history of science; we seek to form mathematical relationships in your mind. The common language really is Calculus at a level of two semesters of that topic. The prerequisite for this text is at least one semester of calculus and presumably some exposure to trigonometry. That means you know the basic ideas of calculus and have some experience with derivatives but perhaps have not worked with integrals or certainly not integrals over more than one variable. That can be a problem for this course so this "Introduction" should be studied or presented in class before the first "real" chapter. Hopefully your teacher will spend a lecture or two in the first week on this material, but if not it will be in your best interest to study this chapter on your own. We just have to get over the initial barrier to review derivatives and then explore analytical integration (the reverse of taking a derivative) in a brief way that will get students into the game with at least a fighting chance of winning that A grade in physical chemistry! Along the way we might as well review some key topics from sophomore physics, which is another basic prerequisite, and introduce the use of partial derivatives, which we will use a LOT in thermodynamics!

Let us start with a review of derivatives. *It is important to understand for future use that both dx and dy are individually small quantities that may be manipulated using algebra and the derivative is the limiting case of the ratio of the two quantities in the limit of infinitesimal size!*

$$\frac{dy}{dx} \equiv \lim_{\Delta x \to 0} \left(\frac{\Delta y}{\Delta x} \right) = \lim_{h \to 0} \left(\frac{y(x+h) - y(x)}{h} \right).$$

Note the use of the "super equal" sign with three lines, which indicates a definition of a term rather than equal values of two expressions.

Example: Let $y = x^2$, then we find (as expected)

$$\frac{dy}{dx} = \lim_{h \to 0} \left[\frac{(x+h)^2 - x^2}{h} \right] = \lim_{h \to 0} \left[\frac{x^2 + 2xh + h^2 - x^2}{h} \right] = \lim_{h \to 0} \left[\frac{2xh + h^2}{h} \right] = 2x.$$

This can be generalized using the binomial expansion of $(x+h)^n$.

$(x+h)^n = x^n + nhx^{(n-1)} + \cdots + h^n$ so that all the terms of h^2 or higher will go to zero and in the derivative of any variable to an integer power n, THE n FALLS OFF THE ROOF, LEAVING $(n-1)$ as the exponent! That is all we need to know for derivatives of simple polynomials! So we have a simple but powerfully general formula: given $y = x^n$, we can immediately write an easily remembered formula as

$$\left(\frac{dx^n}{dx}\right) = nx^{(n-1)}.$$

This may seem a very informal way to remember this process but in the interest of mental efficiency, it will get the job done and we will seek other correct shortcuts for mathematical operations. (Life is short and we have many equations yet to learn!)

The next special derivative we will use many times involves the derivatives of exponential functions of base "e." The number "e" was implied by the mathematician Napier in 1618, developed further by Bernoulli, and called "e" by Euler in 1727. It is an irrational number like π, but for our purposes we can use a good approximation on our calculator as

$$e \cong 2.7\ 1828\ 1828\ 45\ 90\ 45\ 23536\ldots$$

For those interested in science facts, e can be remembered as roughly 2.7 followed by the digits 1828, 1828 again, and then 45, 90, and 45 but usually you can just enter e^1 on your calculator to verify the approximation. Bernoulli attempted to evaluate the constant using the formula

$$e = \lim_{n \to \infty} \left(1 + \frac{1}{n}\right)^n,$$

and you can easily check the first few digits with your calculator. Napier determined that only one base leads to an exponential function whose functional graph has the property that ALL the derivatives of that function (especially the first derivative) have the same value at any point on the graph as the value of the function itself!

$$\left(\frac{de^x}{dx}\right) = e^x, \quad \frac{d}{dx}\left(\frac{de^x}{dx}\right) = e^x, \quad \frac{d}{dx}\left[\frac{d}{dx}\left(\frac{de^x}{dx}\right)\right] = e^x, \quad \text{and} \quad \left(\frac{d}{dx}\right)^n e^x = e^x \text{ for any } n.$$

That in itself is merely a profound fact but most importantly for us is the similarity to the polynomial formula above when the exponent "falls off the roof."

$$\frac{d}{dx}(e^{ax}) = ae^{ax}.$$

The proof of this derivative is slightly complicated but for our purposes we can memorize it. It is instructive here to introduce the concept of the CHAIN RULE for derivatives of functions, which are in turn functions of variables as for instance when $u = ax$ in $e^u = e^{ax}$. Then we have to take the derivative with respect to u and the chain derivative of u with respect to x so we have:

$$\frac{de^u}{dx} = \frac{de^{ax}}{dx} = \frac{de^u}{du}\frac{du}{dx} = ae^{ax}.$$

Note that the use of the chain rule leads to the ratio $\left(\frac{d}{du}\right)\left(\frac{du}{dx}\right) = \frac{d}{dx}.$

Now consider the case where we can introduce another formula that you may or may not have seen in one semester of calculus for the Taylor expansion of any function of (x) in terms of all its derivatives as applied in this case to e^x. It is worth mentioning the Taylor expansion in this beginning chapter since at a later time we may use it for an approximation when x is small. The general formula for the Taylor expansion is

$$f(x) = f(0) + \sum_{n=1}^{\infty} \frac{x^n}{n!} \left[\left(\frac{d}{dx} \right)^n f(x) \right]_{x=0}, \quad \text{so } e^x \cong 1 + x + \left(\frac{x^2}{2} \right) + \left(\frac{x^3}{3!} \right) + \left(\frac{x^4}{4!} \right) + \cdots$$

Note for future reference, *in any limiting case involving e^x compared to any single function of x^n, the behavior of e^x will dominate since e^x contains every power of x and more higher powers besides; e^x will always "beat" any single power of x in a limiting case.* That is an example of a tedious proof of something you can safely memorize in that when x is small it can be seen from the series expansion that only the first two terms are needed so that we can often use: $e^x \cong 1 + x$, when $(x \ll 1)$.

Along the way we used $n!$ which we will also find very useful later on; it is called "*n*-factorial" and is defined as $n! \equiv 1 \cdot 2 \cdot 3 \cdot 4 \cdot 5 \cdot 6, \ldots n$, that is, the product of successive integers up to n. The use of factorial notation will be common later in the text and it should be clear that it is a convenient form to represent what may be a very large product string of integers. Many of the inexpensive calculators have a special key for $n!$ so try to find out the limit of your calculator and probably you will get an overflow message at about 69! Be prepared to stretch your imagination later on to use the factorial of a number as large as Avogadro's number! We will see later there is a good approximation for factorials of large numbers and Stirling's approximation will be discussed in later chapters.

The next thing to review is that we will often have to take a derivative of a function, which is the *product* of several functions so we need to review the derivative of a product.

Given: $y = U(x) \, V(x)$, then $\dfrac{dy}{dx} = \dfrac{d}{dx}(UV) = U \left(\dfrac{dV}{dx} \right) + V \left(\dfrac{dU}{dx} \right)$.

This principle can be extended to multiple products as well.

Next, we need to consider the case of a *quotient*. I prefer to use the "fall off the roof" idea as applied to a negative power.

Given: $y = \left(\dfrac{U(x)}{V(x)} \right)$, then $\dfrac{d}{dx} \left(\dfrac{U(x)}{V(x)} \right) = \dfrac{d}{dx}(UV^{-1}) = (V^{-1}) \left(\dfrac{dU}{dx} \right) + U(-1)(V^{-2}) \left(\dfrac{dV}{dx} \right)$.

This is equivalent to the form usually given in calculus books if the last term is multiplied by (V/V) to obtain the form as $\dfrac{d}{dx} \left(\dfrac{U}{V} \right) = \left[\dfrac{V(dU/dx) - U(dV/dx)}{V^2} \right]$.

This exercise is shown to illustrate another example:

$$\frac{d}{dx} \left[\frac{1}{x^n} \right] = \frac{d}{dx}[x^{-n}] = (-n)\left[x^{-(n+1)} \right] \quad \text{so that } \frac{d}{dx}[1/x^3] = (-3)(x^{-4}) = -3/x^4.$$

Now we need to venture into what may be new to you in the form of integration. Basically, integration is the reverse of taking a derivative so that when you do the step mentally, you say to yourself, what is the function whose first derivative is the function in the integrand, the function under the integral sign? There are two basic types of integrals, a *definite integral*, which is evaluated between two certain limit values of the argument and results in a definite numerical value. A more general type of integral is an *indefinite integral*, which produces a function whose derivative is the integrand function in the integration formula, *but* (!) since the derivative of a constant is zero, there might have been some constant number associated with the integrated function, which might need some sort of additional information such as a *boundary condition* to evaluate.

Definite integral: $\displaystyle\int_a^b f(x)dx = g(b) - g(a)$, where $\dfrac{dg(x)}{dx} = f(x)$ and we note the order of the limits

implies $g(upper) - g(lower)$. Also note that in one variable, the integral is an *area* which is the product of a sliding value of the function with a tiny slice dx. In some cases, a definite integral can be approximated by plotting the integrand function versus x, cutting out and weighing the paper area under the curve between the limits and comparing to the weight of a known rectangle area of graph paper. Since chemists have access to very accurate balances, it is relatively easy to achieve 1% accuracy by this graphical-weighing method.

Indefinite integral: $\int f(x)dx = g(x) + C$, where $\dfrac{dg(x)}{dx} = f(x)$ and C is an unknown constant. For homework problems, an indefinite integral is incorrect unless the $+C$ term is given.

Examples: $\int x^n dx = \dfrac{x^{(n+1)}}{(n+1)} + C$ and $\int e^{ax}dx = \dfrac{e^{ax}}{a} + C$.

We can treat an important but more complicated case that we will use to justify formulas we merely memorize later but we need to be aware of a process called *integration by parts*, which makes use of definite integration. Consider the integral that is the reverse process of the product derivative rule. Simply put, we wrap a definite integration process around the product rule formula, that is, we perform the definite integration term by term on both sides of the product equation and use the same limits on all the terms. Note that $d(UV) = U\,dV + V\,dU$ does not have the dx denominator. This form of a derivative of just the numerator is called the "differential" and is valid for whatever variable is in the denominator. This concept is often used in thermodynamics.

$$\int_a^b d(UV) = \int_a^b U\,dV + \int_a^b V\,dU, \quad \text{then} \quad \int_a^b U\,dV = \int_a^b d(UV) - \int_a^b V\,dU$$

This simple formal trick can be used in a tedious way to evaluate definite integrals, which cannot be integrated in a single step. This process can seem confusing at first but can be learned. However, there is a problem in that if you arbitrarily choose U and V by separating the UV product in a correct but not efficient way, the process will actually make the integration step more complicated. The strategy is to choose U and V in such a way that the next step is simpler so that multiple applications will lead to a form that is easy to integrate; usually that direction is one that reduces a power in the integrand of the original integral to a lower form.

Example:

$$\int_0^\infty xe^{-ax}dx = \left[\frac{-xe^{-ax}}{a}\right]_0^\infty - \int_0^\infty \left(\frac{-e^{-ax}}{a}\right)dx = (0-0) + \left(\frac{1}{a}\right)\int_0^\infty e^{-ax}dx$$

$$= \left(\frac{-1}{a^2}\right)[e^{-ax}]_0^\infty = \left(\frac{1}{a^2}\right)$$

Now that is a very tedious process (however, it is used in research if necessary) so we only show this once and then generalize to a key formula we can memorize. This is a well-established procedure where tables of "integral formulas" fill several volumes. Fortunately for us we can get around this in this first chapter and simply memorize the final formula.

Key formula: $\displaystyle\int_0^\infty x^n e^{-ax}dx = \dfrac{n!}{a^{n+1}}$

Although we are packing a lot of information into this first chapter, students have absorbed this in my class for many years and I usually spend two or three of the first classes on this material. In my experience, this review is essential for mental comfort in all the later chapter topics. So far all the material mentioned earlier may have been treated in one or two semesters of calculus but to go ahead in thermodynamics we have to understand how to treat more than one variable in the presence of the others. A simple example is the set of basic physical variables (P, V, T, n). While this topic would normally be treated in a course in multivariant calculus, it is unlikely that biology majors will have taken such a course, so we have to introduce the basics here. It really is very simple in that *partial derivatives use the same rules as one-dimensional derivatives except that the other variables are held constant!*

Consider the ideal gas law, which should be familiar from freshman chemistry:

$$PV = nRT, \quad P = \frac{nRT}{V}, \quad V = \frac{nRT}{P}, \quad T = \frac{PV}{nR}, \quad \text{and} \quad n = \frac{PV}{RT}.$$

Although we know R is a constant, there are four variables that have an effect on the overall *state* of an ideal gas.

There are actually 12 possible partial derivatives for the state of a sample of gas but we will only show 3 here, the others are easy to understand considering each variable in turn. Let us treat the possible partial derivatives of P.

$$\left(\frac{\partial P}{\partial T}\right)_{n,V} = \frac{nR}{V}, \quad \left(\frac{\partial P}{\partial V}\right)_{n,T} = -\frac{nRT}{V^2}, \quad \text{and} \quad \left(\frac{\partial P}{\partial n}\right)_{V,T} = \frac{RT}{V}$$

Thus we see the pressure of an ideal gas depends separately on three other variables and the partial derivatives could be rendered into a numerical value by using the values of the other variables. Strictly speaking, the lower right subscripts should be included to indicate all the other variables are being held constant during the derivative evaluations, but if the moles are understood to be constant in a given problem, n is often neglected. Even so any lower subscript can be used to remind us what conditions are held constant. *It is important to learn to verbalize the partial derivatives to give them physical meaning.* For instance $\left(\frac{\partial P}{\partial n}\right)_{V,T}$ could be read as "the change in P with a change in moles while holding volume and temperature constant." A student should find more meaning if the partial derivatives are verbalized but the strategy is to use quantities that are measurable and then use mathematics to manipulate them in ways that lead to new information.

There are two physical quantities that can be measured in the laboratory, which can be tabulated in various handbooks and used in a number of ways to simplify other equations.

Isobaric thermal expansion coefficient: $\alpha \equiv \left(\frac{1}{V}\right)\left(\frac{\partial V}{\partial T}\right)_P$. Note "isobaric" means P is constant.

Here we note this represents how the volume changes as the temperature changes while holding the pressure constant. To make the measurement independent of the amount we divide by the total volume, and this is a positive number. For solids and liquids, this is usually a small numerical value (with units), but it can be large for gases. For an ideal gas we have

$$\alpha = \left(\frac{1}{V}\right)\left(\frac{nR}{P}\right) = \left(\frac{1}{T}\right).$$

Isothermal compressibility coefficient: $\beta \equiv \left(\frac{-1}{V}\right)\left(\frac{\partial V}{\partial P}\right)_T$. Note "isothermal" means constant T.

The β value needs some thought because most substances are compressed when the pressure is increased so the amount by which the volume changes when the pressure increases is negative. Thus the definition includes a minus sign so that the tabulated values will be positive. Nevertheless, the physical phenomenon is that most materials will be compressed to smaller volume as the pressure increases.

Recall that in nature (P, V, T) variables are (seemingly) *independent* of each other for a given quantity of gas n. Thus we could plot the three variables on an (x, y, z) grid. A useful *cyclic rule* can be derived with some thought and a trick using the differential quantities. Although we can plot independent variables on a grid, there may actually be some *state function* that relates them in some dependent way. In the case of a given amount of ideal gas (P, V, T), the ideal gas equation relates the variables, but we can be more general using just verbal description of the (x, y, z) coordinates.

$$dz = \left(\frac{\partial z}{\partial x}\right)_y dx + \left(\frac{\partial z}{\partial y}\right)_x dy.$$

This is a very general differential that can be read as "the change in z equals how much z changes when x changes, holding y constant, multiplied by the amount of change in x plus how much z changes when y changes, holding x constant, multiplied by the amount of change in y." A student should try to read the meaning of the partial derivatives and the total differential before blindly performing correct algebraic manipulations, although that manipulation may be desirable to lead to a new result. Now consider that z is held constant so that $dz = 0$. That leads to a new result that may be useful later and is given here to illustrate how partial derivatives can be manipulated; after all, the derivatives are really ratios of small numbers.

$$\left[0 = \left(\frac{\partial z}{\partial x}\right)_y dx + \left(\frac{\partial z}{\partial y}\right)_x dy\right]_z \quad \text{holding } z \text{ constant.}$$

Next we can use algebra to form $\left(\frac{\partial x}{\partial y}\right)_z$ from the individual differentials and noting the condition that z is constant. Then $\left(\frac{\partial z}{\partial x}\right)_y \left(\frac{dx}{dy}\right)_z = -\left(\frac{\partial z}{\partial y}\right)_x \left(\frac{dy}{dy}\right)_z$ by dividing through by dy. Note that at this point d becomes ∂ so $\left(\frac{dx}{dy}\right)_z = \left(\frac{\partial x}{\partial y}\right)_z$ and of course $\left(\frac{dy}{dy}\right)_z = 1.$

Now multiply both sides by $\left(\frac{\partial y}{\partial z}\right)_x$ and rearranging into alphabetical order (x, y, z) (the order in the product can be permuted) to find $\left(\frac{\partial x}{\partial y}\right)_z \left(\frac{\partial y}{\partial z}\right)_x \left(\frac{\partial z}{\partial x}\right)_y = -1.$

If we plot the independent variables (P, V, T) on the (x, y, z) axis and assume there really is some sort of state function that connects them somehow, then we can generalize the discussion earlier to a similar cyclic relationship in terms of (P, V, T) as

$$\left(\frac{\partial P}{\partial T}\right)_V \left(\frac{\partial T}{\partial V}\right)_P \left(\frac{\partial V}{\partial P}\right)_T = -1.$$

This expression can also be rearranged to use α and β. Thus we can rearrange the cyclic rule and multiply by the factor of $\left(\frac{-1/V}{-1/V}\right)$, which completes the definitions of α and β to obtain

$$\left(\frac{\partial P}{\partial T}\right)_V = \frac{-\left(\frac{\partial V}{\partial T}\right)_P \left(\frac{-1}{V}\right)}{\left(\frac{\partial V}{\partial P}\right)_T \left(\frac{-1}{V}\right)} = \frac{\alpha}{\beta}.$$

This illustrates the sort of algebraic manipulations that are common in thermodynamics. Note that in this case, we have assumed there is some overarching connection in the form of a state function that relates P, V, and T (for a fixed value of moles), but formally it is not necessary to specify the state function as long as the various partial derivatives exist.

While we are stretching your mind with the basics of mathematics that you will need for the remainder of the text, these examples should rapidly introduce you to what we will need without spending two or three more semesters in mathematics courses. All that can be said here is that this method of initial rapid introduction to necessary mathematics has worked in classes for over 30 years. Actually, we have no choice if you are limited to only one semester of calculus and we need to do a thorough treatment of physical chemistry in one basic semester course. By showing the applications of the mathematics in the laboratory topics we want to treat, it has been found that students accept the mathematics, and this is an adequate method of learning, although taking the mathematics in separate courses would be desirable.

Now there is another mathematics topic we need to demonstrate, that of spherical polar coordinates, which occur frequently in physical chemistry. We will often need to reason in three dimensions and use spherical polar coordinates.

Suppose we want to calculate the volume of a rectangular box with dimensions $(L \times W \times H)$, we could set up an integral for the volume as

$$V = \int_0^L dx \int_0^W dy \int_0^H dz = [x]_0^L [y]_0^W [z]_0^H = (L - 0)(W - 0)(H - 0) = LWH, \text{ as expected.}$$

However, when we need to treat a spherical system, we have to convert (x, y, z) to (r, θ, ϕ) coordinates. It is easily seen that $z = r\cos(\theta)$, $x = r\sin(\theta)\cos(\phi)$, and $y = r\sin(\theta)\sin(\phi)$ if you draw the projections of the vector r on the (x, y, z) axis system. That is easy but it is more difficult to convert the volume element to polar coordinates (Figure I.1). This is difficult to visualize but you should be able to see a quasi-cube with dimension $(r\,d\theta)$ as a short arc caused by a small change in θ with fixed r as if the vector acts like a crane arm moving up or down with the base of the crane at the origin. Then another small arc can be generated if the crane arm swings in the ϕ angle but the effective length of the crane arm in the (x, y) plane (the dotted line) is $r\sin(\theta)$ and when there is a small change in ϕ, the arc length is the product (radius)(arc) or $r\sin(\theta)d\phi$. Finally, the thickness of this small quasi-cube is dr and so in the limit of infinitesimals we obtain $r^2\sin(\theta)drd\theta d\phi$ in place of $dxdydz$. Note that the range of the ϕ angle is 0 to 2π, so the crane arm can spin all the way around the z-axis, but the θ angle only needs to vary between 0 and π when coordinated with the ϕ angle to reach any coordinate in the original (x, y, z) coordinate system.

Consider the volume of a sphere of radius a in the spherical polar coordinate system as a key example.

$$V = \int_0^a r^2 dr \int_0^{2\pi} d\phi \int_0^{\pi} \sin(\theta)d\theta = [r^3/3]_0^a [\phi]_0^{2\pi} \int_0^{\pi} \sin(\theta)d\theta = (2\pi a^3/3) \int_0^{\pi} \sin(\theta)d\theta.$$

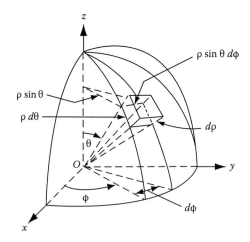

FIGURE I.1 Volume element in spherical polar coordinates. (Adapted from Thomas, G.B., *Calculus and Analytic Geometry*, Addison-Wesley Publishing Co., Reading, MA, 1953, p. 549.)

Now consider the somewhat tricky integral over θ. We see that $\int_0^\pi \sin(\theta)d\theta = [-\cos(\theta)]_0^\pi =$ $[-(-1)-(-1)] = 2$ so we obtain $V = 4\pi\left(\dfrac{a^3}{3}\right)$, which is the familiar formula for the volume of a sphere of radius a. The key point here is the product of integrals over θ and ϕ. Please note for future reference the integral over the angles is 4π. *For the remainder of this text, we do not want to repeat this derivation and when we say "sphere" you should automatically think "4π" over angles.* For students who have not encountered three-dimensional integrals, we appeal to a picture for the volume element and a shortcut for future treatment of spherical systems.

[Three friends, a biologist, a statistician and a physical chemist, went to a horse race and wanted to bet on the winner. The biologist said he wanted to know what the horses ate for breakfast and the statistician wanted to know how the various horses finished in previous races but the physical chemist said nothing and was in deep thought for a long while. Finally the biologist asked what he was thinking and the physical chemist said, "First you assume a spherical horse..."!]

By now you may need more humor beyond the "spherical horse joke" so there is one more integral that we will need which is related to the Napier–Bernoulli–Euler constant e. Assuming you have found your calculus book by now you should look up the natural logarithm formulas and find a key integral as: $\int \dfrac{dx}{x} = \ln(x) + C$.

The mnemonic to remember this integral is to set up the variable as $x = cabin$ so that we can obtain an easily remembered formula:

$$\int \frac{d(cabin)}{cabin} = \ln(cabin) + \text{"}sea\text{"} = \text{A boat?} = \text{Noah's Ark?}$$

Maybe that sort of humorous memory device will help you later on when we encounter this integral many times?

SHORT REVIEW OF VECTOR ALGEBRA/CALCULUS

It may be important to review some basic facts about vectors since this topic comes up in the "dot product" of a velocity vector. The unit vectors $(\hat{i}, \hat{j}, \hat{k})$ point in the (x, y, z) directions and have a length of 1. Then according to the definition $\vec{a} \cdot \vec{b} = |a||b| \cos(\theta_{ab})$, we have $\hat{i} \cdot \hat{i} = \hat{j} \cdot \hat{j} = \hat{k} \cdot \hat{k} = 1$ (normalized) as well as $\hat{i} \cdot \hat{j} = \hat{i} \cdot \hat{k} = \hat{j} \cdot \hat{k} = 0$ (mutually orthogonal). The $(\hat{i}, \hat{j}, \hat{k})$ are the building blocks for vectors in three-dimensional space and form an *orthonomal basis set* for (x, y, z) space. We will need the idea of a "basis set" later on as a way to use components to build a linear combination and "orthonormality" can lead to several time-saving steps later. We especially need the idea of *projecting* a component out of a linear combination. Suppose we have $\vec{v} = a\hat{i} + b\hat{j} + c\hat{k}$ and we want to know what part or how much of that vector is the $\hat{j}$ part. We can take the dot product of the component we want with the linear combination to find

$$\hat{j} \cdot \vec{v} = \hat{j} \cdot (a\hat{i} + b\hat{j} + c\hat{k}) = 0 + b + 0$$

using the orthonormality of $(\hat{i}, \hat{j}, \hat{k})$.

That concept is extremely important in several applications later in this text to greatly simplify tedious spectroscopic concepts involving the Fourier transform. That step allows us to use the idea of projection of an orthonormal component without doing a complicated integral! There are also derivatives of vectors. A vector is a length with an associated direction. The speedometer reading of an automobile gives a scalar speed (40 mph), but you need a dashboard compass to use that speed as a directional vector as (40 mph, west). We can also define a directional "gradient vector" as

$$\vec{\nabla}v \equiv \left(\hat{i}\frac{\partial v}{\partial x} + \hat{j}\frac{\partial v}{\partial y} + \hat{k}\frac{\partial v}{\partial z}\right).$$

Then we can define a second derivative as a dot product of the gradient with itself as

$$\vec{\nabla} \cdot \vec{\nabla} v \equiv \left(\hat{i}\frac{\partial}{\partial x} + \hat{j}\frac{\partial}{\partial y} + \hat{k}\frac{\partial}{\partial z}\right) \cdot \left(\hat{i}\frac{\partial}{\partial x} + \hat{j}\frac{\partial}{\partial y} + \hat{k}\frac{\partial}{\partial z}\right)\hat{v} = \left(\frac{\partial^2 v}{\partial x^2} + \frac{\partial^2 v}{\partial y^2} + \frac{\partial^2 v}{\partial z^2}\right).$$

Note the result is a scalar second derivative in three variables. There is also another type of vector product one may encounter when dealing with magnetic fields called a cross product.

$$\vec{a} \times \vec{b} \equiv \begin{vmatrix} \hat{i} & \hat{j} & \hat{k} \\ a_x & a_y & a_z \\ b_x & b_y & b_z \end{vmatrix} = \hat{i}(a_y b_z - b_y a_z) - \hat{j}(a_x b_z - b_x a_z) + \hat{k}(a_x b_y - b_x a_y).$$

A similar directional cross product called the "curl" exists as $\vec{\nabla} \times \vec{v}$ but we will not need it for this course.

REVIEW OF CLASSICAL PHYSICS

Although our review of calculus should be adequate for the remainder of the text, we still need to review *physics* for that part of physical chemistry. Newton's second law is used heavily in the form: $F = ma$, but we need to remind ourselves that this is a vector equation.

$$\vec{F} = m\vec{a}, \quad \text{where } F_x = ma_x, \ F_y = ma_y, \quad \text{and} \quad \vec{a} = \frac{d\vec{v}}{dt} = \frac{d^2\vec{r}}{dt^2}$$

In units: Force = Newtons in mks; dynes in cgs.
1 Newton $\equiv$ 1 kg m/s^2, 1 dyne $\equiv$ 1 g cm/s^2
In units: Energy $\equiv$ 1 joule = 1 Newton meter, 1 erg $\equiv$ 1 dyne cm (Figure I.2).
Simple conversions: 1 Newton = 10^5 dynes, 1 joule = 10^7 ergs

Recall that mass (kg or g) is an intrinsic property of matter but weight is a force on planet Earth (Anyone here planning a vacation on another planet?) using the acceleration of gravity measured on Earth (which is an average value since Earth is not exactly spherical) so instead of $F = ma$ we use $w = mg$ (small w for weight and capital W for work) with the average value of g usually given as

$$g = 980 \text{ cm/s}^2 = 9.80 \text{ m/s}^2$$

An important concept is Work $\equiv$ (Force)(distance) or $W = Fd$. In mks units work is given in joules and in cgs units in ergs.

1 joule $\equiv 10^7$ ergs
["Erg!" is the last word of a "dying centipede," that is, a dyne-cm]
Chemists often define heat energy in calories.
1 calorie $\equiv$ the heat required to raise the temperature of 1 cm^3 of water (1 c.c.) 1°C (from 4°C
    to 5°C). Later, we will discuss an experiment by Joule to convert heat energy to mechanical
    energy; he found
1 calorie = 4.184 joules

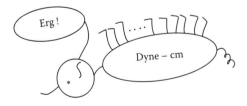

FIGURE I.2   Humorous cartoon to remember the units of energy in the cgs system.

Here we introduce another silly but useful mnemonic. The number 4.184 is easy to remember but which unit is larger? "A calorie is worth many jewels!" Using the rhyming word "jewels" in place of "joules" conveys greater numerical value on the calorie unit so we can remember that in some absolute amount of energy a single calorie is "worth" (larger than) many (4.184) joules. This silly process is important for a student to develop a feeling for a physical quantity, for instance that 1 cm^3 is about the size of a boullion cube of condensed soup and 1°C is one hundredth of the temperature difference between ice water and boiling water (at 1 atmosphere). It is important to reason out the physical units in a calculation rather than to use numbers blindly on a calculator.

$$dW = F_x\, dx + F_y\, dy + F_z\, dz \text{ so we can integrate this to } \int dW = \int F(x)dx$$

A related concept is "Power": Power $\equiv$ (Work/time) but that unit is more useful for electrical measurements than in physical chemistry.

Energy can be "kinetic" or "potential." Note that work is energy:

$$W = \int F_x dx = \int ma_x dx = \int m\left(\frac{dv_x}{dt}\right)dx = \int m(dv_x)\left(\frac{dx}{dt}\right) = \int_0^{v_x} mv_x dv_x = m\left(\frac{v_x^2}{2}\right) \equiv T_x$$

so that work can be in the form of T, which is the usual symbol for kinetic energy (although you may have used "K" in sophomore physics)

Another useful concept in much of physical chemistry is the *conservation of energy*. The main associated concept is that a *force is the negative derivative of some potential*. ($\Rightarrow$ means "implies")

$$V(x, y, z) \Rightarrow F_x = \frac{-\partial V}{\partial x}, \ F_y = \frac{-\partial V}{\partial y}, \text{ and } F_z = \frac{-\partial V}{\partial z}.$$

For simplicity, let us just consider the x-component of the energy.

$$W_x = \int mv_x dv_x = m\int_{v_1}^{v_2} v_x dv_x = \left[\frac{mv_x^2}{2}\right]_{v_1}^{v_2} = T_2 - T_1.$$

But we also know that $W_x = -\int\left(\frac{\partial V}{\partial x}\right)dx = -\int_{V_1}^{V_2} dV = -(V_2 - V_1) = (V_1 - V_2)$.

Then since we have computed W_x two different ways, we can equate the kinetic and potential forms of work $W_x = T_2 - T_1 = V_1 - V_2$.

And so as long as there are potentials whose negative derivatives in space form forces (energy goes down hill!) we have the equality: $T_2 + V_2 = T_1 + V_1$.

This important result tells us that energy can change form but the total value remains the same, that is, *conservation of energy!*

Some important units we will need as well seem random but they are necessary:

1 hour = 60 minutes = 3600 seconds
1 mile = 5,280 feet = 1.6093 kilometers = 1.6093×10^5 cm
1 inch = 2.54 cm
1 pound = 453.6 grams (only on Earth, a pound is a force while a gram is a mass)
1 pound (Avoirdupois) = 16 ounces
F (degrees) = (9/5)C (degrees) + 32 (degrees)
1 calorie = 4.184 joules

Another important point for fractions or units is that $\left(\dfrac{1}{1/3}\right) = 3$; *that is the denominator of the denominator can "flip up" to the numerator* and that can be very useful.

NEWTON–RAPHSON ROOT FINDER

A numerical technique that sometimes is useful is the Newton–Raphson root finding method (Figure I.3). The basic idea is that if the slope of a line/curve $y = f(x)$ is the first derivative (dy/dx), maybe one can extrapolate the slope back to where the function crossed $y = 0$. In order to use this idea, you have to rearrange your equation so that $y(x) = 0$. Then the key equation is

$$x_{n+1} = x_n - \frac{y(x_n)}{\left(\dfrac{dy(x_n)}{dx}\right)}$$

Let us try it on a quadratic for which we know the roots: $y(x) = x^2 - x - 6 = (x + 2)(x - 3) = 0$
Here we can see the roots are $x = -2$ and $x = +3$. Let us try an initial guess of $x = 3.5$.

$$y(x) = x^2 - x - 6 = 0 \quad \text{so} \quad \frac{dy}{dx} = 2x - 1 \quad \text{and} \quad \text{we} \quad \text{have} \quad x_1 = 3.5 - \frac{(3.5)^2 - 3.5 - 6}{2(3.5) - 1} =$$

$3.5 - \dfrac{2.65}{6} = 3.0583$ and then $x_2 = 3.0583 - \dfrac{(3.0583)^2 - 3.0583 - 6}{2(3.0583) - 1} = 3.0583 - \dfrac{0.29489889}{5.1166} =$

3.000664287 leading to

$$x_3 = 3.00066 - \frac{(3.00066)^2 - 3.00066 - 6}{2(3.00066) - 1} = 3.00066 - \frac{3.3004356 \times 10^{-3}}{5.00132} = 3.000059287.$$

Is that close enough? Well what if we make an initial guess of $x = 2.5$?

$$x_1 = 2.5 - \frac{(2.5)^2 - 2.5 - 6}{2(2.5) - 1} = 2.5 - \frac{(-2.25)}{4} = 3.0635,$$

which we know will converge to 3.0.
How about a guess of $x = -1$?

$$x_1 = -1. - \frac{(-1.0)^2 - (-1.0) - 6}{2(-1.) - 1} = -1. - \frac{(-4)}{(-3)} = -2.333333 \quad \text{so we continue to iterate as}$$

$$x_1 = -2.333333 - \frac{(-2.333333)^2 - (-2.333333) - 6}{2(-2.333333) - 1} = -2.333333 - \frac{(1.777776)}{(-5.666666)} = -2.019608$$

and it would seem that the process will converge to -2.0.

The message here is that the Newton–Raphson method will converge but probably to the nearest root if there is more than one root. Even so, this method is quite valuable for finding roots of complicated functions. Of course with a programmable calculator, the iteration can be automated and will converge rapidly!

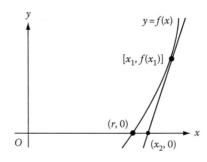

FIGURE I.3 A graphical illustration of the Newton root-finding method, the root is r. (Adapted from Thomas, G.B., *Calculus and Analytic Geometry*, Addison-Wesley Publishing Co., Reading, MA, 1953, p. 226.)

PREVIEW OF COMING ATTRACTIONS WITH REASSURANCE!

That concludes our brief but intensive review of the *mathematics and physics* needed for the remainder of the text, although we may need some special formulas from time to time. In our experience, this initial review is the secret to being able to make a full treatment of selected topics without diluting the coverage, although we will still make use of the simplest correct derivation of other relationships. The student should realize we are attempting to do a "university level" treatment of physical chemistry while being helpful to students with only one semester of calculus. A concurrent second semester of calculus or other mathematics elective is encouraged in the second semester. The student should be aware that even with this helpful approach, the use of scratch paper and working examples is highly encouraged! What follows is a set of homework problems and our experience has shown that a flurry of activity to review and extend calculus skill at the beginning of the course really pays off in student confidence for later topics.

While it is difficult to give broad treatment of many topics in one semester, we can do selected topics at a university level to form a foundation for other science fields such as physics, forensic science, biochemistry, inorganic chemistry, and a possible second semester with some quantum chemistry. Students should be reassured that this text is based on a course that has been presented to a group where on average about one-third of the class only has one semester of calculus and yet the success rate is very high! The secret is that we give examples in the text of specific problems and the homework problems are like the examples. The organization of this text in the mind of the author is to link each chapter topic to a "calculus nugget" as a way to master details of that topic. A hint to understanding each chapter is to look for the "calculus nugget." It is hoped that we can develop calculus skill as we go along so that the reader/student will be able to treat new topics as they may be encountered. We have avoided the encyclopedic coverage of every possible topic and selected a lesser group of "essential" topics. Old tests are provided to show what is expected and a few key topics are treated in detail.

In a possible second semester, more modern material is covered from the twentieth century. Our strategy in the second semester is to instruct students to assimilate as much of the difficult derivations, but *pay special attention to the conclusions*, the final formulas! Yes, we cover some very complicated details for the few who might be interested but a wise student will *focus on the conclusions!* Again a few key topics are treated in detail but old tests are provided to indicate what is expected. The second semester may end at Chapter 15 or 16 but additional material is provided for special projects or by the preference of the teacher. If this material is studied with a positive attitude and use of pencil and paper, a very high rate of success can be expected in terms of real learning by students. Now let's get going!

CALCULUS REVIEW PROBLEMS

1. $\dfrac{d}{dx}\left(4x^3 + 2x + \left(\dfrac{3}{x^2}\right)\right) = ?$

2. $\dfrac{d}{dx}\left(2e^{-4x} + e^{2x^2} + x^2 e^{-3x}\right) = ?$

3. $\dfrac{d}{dx}\left(\dfrac{x^4}{(x^3 + 2x)}\right) = ?$

4. $\dfrac{d}{dx}\left(x^2 e^{-ax^2}\right) = ?$

5. $\dfrac{d^2}{dx^2}\left(\sin\left(\dfrac{n\pi x}{\lambda}\right)\right) = ?$

6. $PV = nRT, \quad \left(\dfrac{\partial P}{\partial T}\right)_{V,n} = ?$

7. $PV = nRT$, $\left(\dfrac{\partial V}{\partial P}\right)_{T,n} = ?$

8. $\displaystyle\int (x^4 + 2x^3 + 3x)dx = ?$

9. $\displaystyle\int_0^\infty 2xe^{-x^2} dx = ?$

10. $\displaystyle\int_0^\infty x^3 e^{-ax} dx = ?$

11. $\displaystyle\int_0^\infty r^2 dr \int_0^{2\pi} d\phi \int_0^\pi d\theta \left[\left(\sqrt{\dfrac{\alpha^3}{\pi}}\right)e^{-\alpha r}\right]\left[\left(\sqrt{\dfrac{\alpha^3}{\pi}}\right)e^{-\alpha r}\right] \sin(\theta) = ?$

12. Calculate the kinetic energy of a 5 ounce sphere (baseball) traveling at 100 mi/h in joules, ergs, and calories.

13. Calculate the square of the vector $\vec{r} = 3\hat{i} + 2\hat{j} + 4\hat{k}$ as $\vec{r} \cdot \vec{r}$.

14. Use the vector in Problem 13 to calculate $\vec{r} \times \vec{r}$.

15. Project the $\hat{j}$ component from the vector in Problem 13.

BIBLIOGRAPHY

Ayres, F. and Mendelson, E., *Schaum's Outline of Calculus*, 5th Edn., McGraw-Hill, New York, NY, 2008.
Thomas, G. B., *Calculus and Analytic Geometry*, Addison-Wesley Publishing Co., Reading, MA, 1956.
Widder, D. V., *Advanced Calculus*, 2nd Edn., Prentice-Hall, Inc., Englewood Cliffs, NJ, 1961.

1 Ideal and Real Gas Behavior

INTRODUCTION TO THE "FIRST ENCOUNTER WITH PHYSICAL CHEMISTRY"

Welcome to "essential physical chemistry"! We will explore parts of this field that work out cleanly in a beautiful way to provide a sense of confidence in your understanding of this interdisciplinary science. At the end you will hopefully learn to appreciate the beauty of concepts that have been discovered since the 1600s by intellectual giants and gain a sense of amazement for the far-reaching effects of those discoveries into all aspects of molecular science. Students who have completed this course often send this author messages telling how this material has helped them later in biochemistry, inorganic chemistry, and physical chemistry laboratory courses. To get the most from this course it is highly recommended that you personally study the material in *Introduction: Mathematics and Physics Review* even if the course lectures start in Chapter 1. It provides a good preparation for the mathematics and physics concepts we will encounter in the rest of the book. We could have put review at the end of the book, but many students have reinforced the view of this author that it needs to be right up front at the beginning and it really makes the understanding of the later material easier.

Historically, the modern age of science and technology begins roughly in the 1600s and scientific knowledge has bloomed exponentially since then, although some concepts originated in ancient Greece and much of clever Roman engineering was lost between 500 and 1600. Thus, we start with the studies of gases by Sir Robert Boyle (Figure 1.1). We arbitrarily assume that applied mathematics is what distinguishes modern science from medieval engineering so you should be aware that in every case we will attempt to unify a concept with some equation, often involving calculus. Thus, you need to pay attention to the worked examples in the chapters, do the assigned problems, and then try the tests at the end of chapters about where a midterm or final examination usually occurs. You should pay attention to the time limits given for those tests and practice those problems until you have that material down cold within the time allowed! Of course, your teacher will give different questions on the tests in your course but if you can do the sample tests you should be ready for almost any variation of that type question: What equipment is needed here? You need a calculator with special functions but a simple $9 solar-powered calculator is adequate if it has $\exp(x)$, $\sin(x)$, $\cos(x)$, $\log(x)$, and $\ln(x)$ with at least eight significant figures and scientific exponential notation. The next requirement is a human brain and an attitude that *you can do this* (!) provided you put some time and effort into the work. So let us get started!

PHENOMENOLOGICAL DERIVATION OF THE IDEAL GAS EQUATION

While mathematical theory often runs roughly parallel to physical science, sometimes faster and other times slower, a key strategy is a process called the "phenomenological approach." In this method a process is studied to determine the variables on which it depends and then an equation is developed, which matches the results of the problem. Often it requires a number of data points to determine whether the result is linear, quadratic, or some higher order in a given variable but the case of the ideal gas law is an excellent starting point to illustrate this method and at the same time enter the important domain of thermodynamics.

Although we could digress to mention early concepts of science by ancient Greek philosophers, we will begin with the first attempts at quantitative studies that can be related with mathematical

FIGURE 1.1 Sir Robert Boyle (1627–1691). An English natural philosopher and chemist, known mainly for careful experiments on gases. He is recognized as the first modern chemist. He was the 14th child of Richard of Cork, an Englishman, but Robert was sent to live with an Irish family at an early age and learned the separate Irish language as a child. He later owned land in Ireland but came to Oxford University for his research. He was a founding member of the Royal Society of London. (Courtesy of the Chemical Heritage Foundation. The Shannon Portrait of the Hon. Robert Boyle, F. R. S. oil on canvas by Johann Kerseboom, British, 1689. Photograph by Will Brown.)

equations. In 1660, Sir Robert Boyle studied gases in Ireland. Much of early scientific research was conducted only by wealthy individuals who had time and funds to conduct special experiments, many of which seem simple today but were of considerable innovation at the time. Consider that today we can routinely purchase glass tubing of uniform bore from several chemical supply companies. However, Boyle's experiments depended on using the height difference of mercury in a glass "J-tube" so that the volume "V" he wanted to study was related to the height of the mercury column "h" by the equation

$$V = \pi r^2 h$$

Here we see in Figure 1.2 that by measuring "h" and assuming a uniform radius for the tubing, the data can be related to the volume "V" of the trapped gas sample. Note that the essential r^2 part of the formula would provide cm^2 while "h" provides "cm" to produce a volume unit of cm^3. Thus, measuring the one-dimensional height difference can give a volume only if there is a proportionality with a uniform bore of the tubing whether the units are in cubic inches or cubic centimeters. In Figure 1.2, the same ruler can be used to measure the volume of the trapped gas (assuming a flat top of the gas chamber) and the difference in height of the mercury levels. While this is only a schematic diagram of Boyle's apparatus, we have captured the key features.

To properly appreciate this seemingly simple device can you imagine how you would melt sand to make glass and draw it into a piece of tubing with uniform bore? By adding more and more liquid mercury to the open end of the right side of the J-tube, the air trapped on the closed left side of the apparatus was gradually compressed. Because of the uniform bore of the tubing, the volume V of the air in the closed end can be calculated by the height of the air bubble space while the difference in the height of the mercury in the left and right sides gives the pressure P applied to the gas expressed in inches of mercury.

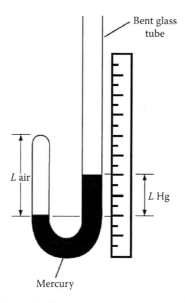

FIGURE 1.2 A schematic of the J-tube Sir Robert Boyle used to study the *PV* relationship.

Boyle's "J-tube" is a very ingenious device, which is simple and accurate. Today this would be called a manometer, which is a device to measure pressures in a laboratory. In Figure 1.2, the difference between the two heights of mercury is given in "inches of mercury" but modern measurements including blood pressure measurements from an arm cuff are usually given in "mm of mercury" and 1 atm of air pressure on planet Earth is standardized as 760 mm of mercury at sea level (since the pressure varies with height above sea level). Please note that *1 in. is exactly 2.54 cm.* (If you ever work on a car or machine with metric dimensions you will find the 13 mm wrench works great on a ½ in. bolt but few others are interchangeable. A Ford Pinto 2000 engine has SAE (Society of Automotive Engineers) bolts externally but internal metric dimensions as built in Germany. This modern example shows that we need to be prepared to convert data in inches as evident in Boyle's data to scientific centimeter units. If we standardize on 760 mmHg as 1 atm (earth) we can also relate the height of 29 11/16 in. of Hg to yet another set of units as

$$29\frac{11}{16} \text{ in.} (25.4 \text{ mm/in.}) = 754.0625 \text{ mm}$$

so

$$754.0625 \text{ mmHg}/(760 \text{ mmHg/atm}) = 0.9921875 \text{ atm.}$$

This brings up a problem having to do with experimental measurements in terms of how accurate we can carry out calculations with data that are only good within 1/16 in. or at best ±1/32 in. (±0.079375 cm) according to Boyle's (subjective) eyes. In fact all experimental data will have some uncertainty. We will postpone the treatment of experimental uncertainty for a few pages so that we do not lose sight of the problem at hand, but we will say here that quoting the value in atmospheres to seven significant figures is artificially precise when the data are good to only the nearest 1/16 in. (0.156875 cm). Here we show some of the actual data from Boyle's 1662 book "*New Experiments Physico-Mechanical, Touching the Spring of Air, and its Effects....*" The data in Table 1.1 lead to an important phenomenological observation in that the product of the pressure (*P*) and volume (*V*) is essentially constant! Although there is some variation in the *PV* product, a wise observer can see the values are nearly constant allowing

TABLE 1.1
Boyle's Data on the Dependence of the Volume
of a Gas on the Pressure of the Gas

Volume	Pressure	$P \times V$
48	$29 \, ^2/_{16}$	1398
46	$30 \, ^9/_{16}$	1406
44	$31 \, ^{15}/_{16}$	1405
42	$33 \, ^8/_{16}$	1407
40	$35 \, ^5/_{16}$	1413
38	37	1406
36	$39 \, ^4/_{16}$	1413
34	$41 \, ^{10}/_{16}$	1415
32	$44 \, ^3/_{16}$	1414
30	$47 \, ^1/_{16}$	1412
28	$50 \, ^5/_{16}$	1409
26	$54 \, ^5/_{16}$	1412
24	$58 \, ^{13}/_{16}$	1412
22	$64 \, ^1/_{16}$	1409
20	$70 \, ^{11}/_{16}$	1414
18	$77 \, ^{14}/_{16}$	1402
16	$87 \, ^{14}/_{16}$	1406
14	$100 \, ^7/_{16}$	1406
12	$117 \, ^9/_{16}$	1411

for the uncertainty of the data to the nearest 1/16 in., so Boyle assumed that the product is constant. This leads to the simplest form of "Boyle's law":

$$PV = C_1 = P_1 V_1 = C_1 = P_2 V_2 \Rightarrow P_1 V_1 = P_2 V_2$$

so we have

$$\frac{P_1}{P_2} = \frac{V_2}{V_1}.$$

Here C_1 is the first constant in this study. We should note that Boyle's experiments were carried out at essentially constant room temperature but do we know that temperature has an effect? Not from this data.

Historically, there was a competition between England and France, which had the effect in science of development of English units still used in the United States and the metric system developed in France and preferred in modern science. Thus, it is interesting that early work by Charles on the temperature dependence was extended by Gay Lussac several years later and is now known as the Charles–Gay Lussac law.

CHARLES' (JACQUES-ALEXANDRE-CÉSAR CHARLES) LAW

While hot air balloons are familiar today, the first documented demonstration occurred on June 5, 1783, when Joseph Montgolfier used a fire to inflate a spherical "hot air" balloon about 30 ft in diameter that traveled about a mile and one-half before it returned to earth. The news of this caused

Jacques-Alexandre-César Charles to try to duplicate this phenomenon. As a result Charles observed that the volume of a gas is directly proportional to its temperature.

$$V \propto T \quad \text{or} \quad \frac{V}{T} = C_2$$

Here, C_2 is a different constant than the C_1 used in Boyle's data. The principle behind the hot air balloon is that hot air expands so the density of the air is lighter in the balloon than the outside air to the extent that the whole "air ship" is lighter than the surrounding air and it rises.

A simple experiment capable of precise date is given on the Purdue "History of Chemistry" Internet site using a 30 mL syringe to give a precise measurement of the total volume of the gas sample as the temperature is raised and measured with the thermometer in the gas. The volume of a 100 mL erlenmeyer flask with a one-hole rubber stopper and thermometer can be measured by weighing the flask, stopper, and thermometer "empty" and then reweighing when filled with water with the stopper and thermometer in place. The difference in mass divided by the density of the water at room temperature gives the volume of the interior of the flask. When the flask has been emptied and dried, a hypodermic needle attached to a 30 mL precision syringe is pushed through the rubber stopper with the syringe set near zero. Then the flask is placed deep into an ice bath, thermally equilibrated, and the stopper is placed in the neck of the flask to the same position as before. The flask is removed from the ice bath and the total volume (flask plus syringe) is noted as the apparatus warms up. A hair dryer can be used to provide heat for a reading at 40°C. The use of the precise syringe is a clever way to improve an experiment with simple equipment. The data obtained are given on the Purdue site as in Table 1.2.

If we plot this and use a modern mathematical way of fitting the "best" line to the data that minimize the square of the deviations between the actual position of n data points and the fitted line, we can use $R^2 \equiv 1 - \dfrac{\sum_i^n (y_i - f_i)^2}{\sum_i^n (y_i - \bar{y})^2}$ and $\bar{y} \equiv \dfrac{\sum_i^n y_i}{n}$ to evaluate the "goodness of fit" to the line of the function $f(x)$ (Figure 1.3). The numerical value of R^2 is routinely available using the Excel program for personal computers and $R^2 \to 1$ for a perfect fit to the line. Using Excel for a least-squares fit of a line to the nine data points we obtain an equation $V = 0.3843(T°C) + 108.85$ with

---

TABLE 1.2
Charles' Law Data

Temperature, °C	Gas Volume, mL
0	107.9
5	109.7
10	111.7
15	113.6
20	115.5
25	117.5
30	119.4
35	121.3
40	123.2

Source: Purdue University, History of Chemistry, http://chemed.chem.purdue.edu/genchem/history/boyle.html

---

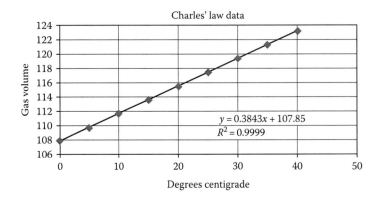

FIGURE 1.3 A plot of the raw data from the "volume-of-gas-flask" experiment. The R^2 value is quite good.

an R^2 value of 0.9999, which indicates a very good fit to a straight line. Supposing the volume could go to zero we find that $V = 0$ at $-280.6°C$.

$$V = 0.3843(T°C) + 107.85 = \frac{-107.85}{0.3843} = -280.6°C$$

Extrapolation across such a long distance magnifies slight errors in the room temperature data and later careful measurements result in a value of $-273.15°C$. To check this we insert a new point into the data set as (-273.0) and replot the data. This time we get a slightly better R^2 value of exactly 1. Thus, we have experimental evidence that the volume of a gas is directly proportional to the Kelvin temperature, that is, Charles' law (Figure 1.4). An associated result is a new "absolute temperature scale" in Kelvin degrees:

$$°K = °C + 273.15°$$

For purists, the symbol for Kelvin degrees should be just K without a degree symbol, but later this will be in conflict with the symbol for equilibrium constants and reaction rate constants so we take the liberty here to designate Kelvin temperatures with a degree symbol as $°K$. Frequent questions on this topic indicate that it is important to state here that differences in centigrade temperature have the

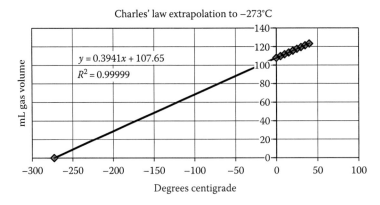

FIGURE 1.4 Charles' law plot of volume versus temperature from the same data but with the added point of zero volume at $-273°C$. This enlightened guess improves R^2 to 0.99999. (Data from http://chemed.chem. purdue.edu/genchem/history/charleslaw.html)

same units as differences in Kelvin temperature. If a chain rule derivative should occur involving temperature differences, $\Delta°K = \Delta°C$; they are the same size unit! While we are establishing simple facts, we note that 1°C is defined as 1/100th of the temperature between that of ice water and boiling water (at 1 atm) as established on the metric (French) scale while the British standardized the Fahrenheit degree as 1/180th of the same temperature gap but then embellished it with the notion that colder temperatures are possible below ice water. Using concentrated salt solutions (as in making homemade ice cream) the original Fahrenheit scale was set to the lowest temperature of water saturated with salt some 32° lower even though we now know far lower temperatures are possible. Thus, we have

$$°F = \left(\frac{180}{100}\right)°C + 32° = \left(\frac{9}{5}\right)°C + 32°$$

It is easy to show that the two temperature scales give the same value at −40° by substituting one for the other either way. In Fahrenheit we see that absolute zero is

$$°F = \left(\frac{9}{5}\right)(-273.15°) + 32° = -459.67\,°F.$$

The Fahrenheit scale is still used in engineering but in chemistry and physics the centigrade scale is used along with the absolute temperature °K.

Now let us combine Boyle's law with Charles' law to improve the overall phenomenological description (for a fixed mass of gas). Consider a two-step process that first changes the pressure of the gas followed by a change in the temperature.

$$P_1T_1V_1 \rightarrow P_2T_1V_x \rightarrow P_2T_2V_2$$

Then by Boyle's law at constant temperature $V_x = \left(\frac{P_1}{P_2}\right)V_1$ and then use Charles' law at constant pressure as

$$V_2 = V_x\left(\frac{T_2}{T_1}\right) = \left(\frac{P_1}{P_2}\right)V_1\left(\frac{T_2}{T_1}\right) \Rightarrow \frac{V_2}{V_1} = \left(\frac{P_1}{P_2}\right)\left(\frac{T_2}{T_1}\right).$$

This leads to a very useful equation *for a fixed mass of gas* as

$$\frac{P_1V_1}{T_1} = \frac{P_2V_2}{T_2}$$

This equation implies the existence of another constant C_3.

$$\frac{PV}{T} = C_3 = nR$$

Now we are close to a general phenomenological equation that takes into account P, V, T, and the moles (n) of the gas, which fixes the mass of the gas sample. Usually this equation is given (in high school chemistry!) as

$$PV = nRT$$

So what is R? And now we come to a key development in 1876.

Avogadro's hypothesis: Equal volumes of different gases at the same temperature and pressure have the same number of molecules.

Amedeo Avogadro (1776–1856) was an Italian physicist who published a basic argument in 1811 that related atomic and molecular weights to definite proportions in compounds but that era was a period of intellectual groping by chemists regarding the meaning of the concepts of atoms and molecules. In 1856 Stanislao Cannizzaro (1826–1910) published a unifying set of course notes that clarified Avogadro's hypothesis and led to the general acceptance of the idea that became the mole concept chemists use today. It is interesting that Avogadro did not know the number that is named for him but by expressing his hypothesis in terms of the number of atoms/molecules he avoided the problem that different gases have different molecular weights. Thus, if we convert the mass of a gas to its mole quantity by dividing by the gram molecular weight M to obtain $n = w/M$ for what is called the mole quantity today and based on the standard of 12.000 g of ^{12}C we now know *Avogadro's number* $= 6.02214179 \times 10^{23}$ as given by a least-squares refinement of modern values of physical constants tabulated in the 90th edn. of the Chemical Rubber handbook. For our purposes we will often use the shorter form as 6.022×10^{23}. This number is a pure number and can refer to one "mole" of anything, ping-pong balls, H atoms, N_2 molecules, etc.

Now we are ready to determine R with the added fact that 1 g molecular weight of many gases has (nearly) the same volume. Actually there are some very slight differences in the molar volume of different gases but the average value at 1 atm and 0°C is 22.414 liters (L). Then

$$R = \frac{PV}{nT} = \frac{(1\,atm)(22.414\,L)}{(1\,mol)(273.15°K)} = 0.082057\,(L\,atm/°K\,mol)$$

Based on the values in the formula the least number of significant figures is five so we need to round off R to five significant figures as 0.082057 (L atm/°K mol). We note that 22.414 L is about the size of a 5 gal solvent can and we need to tabulate some key unit facts in this first chapter. At this point it is easy to introduce the SI equivalent of the gas constant since the only difference is that the pressure is measured in bars where 1 atm = 1.01325 bar. (Note that a "barometer" measures bars.) However, at the lower pressure the molar volume will be larger at about 22.711 L:

$$R = \frac{PV}{nT} = \frac{(1\,bar)(22.711\,L)}{(1\,mol)(273.15°K)} = 0.08314\,(L\,bar/°K\,mol)$$

We will attempt to use the SI units throughout this text and a new generation of students may have only seen SI units in previous texts and hopefully they can "think" in SI units. However, the conversion of the older pressure units in atmospheres or mm of mercury will persist in older literature and some equipment so each student needs to work in their own personal way of dealing with these conversions. In bygone days when slide rules were used instead of calculators, three significant figures were the norm and often answers were only good to two. Under those circumstances we would say $R = 0.082\,(L\,atm/°K\,mol)$ or $R = 0.083\,(L\,bar/°K\,mol)$, easily remembered numbers.

USEFUL UNITS

The density of water is defined to be 1.000000 g/cm^3 at 4°C.
1 pound (avoirdupois) = 453.6 g = 0.4536 kg (on earth)
1 in. = 2.54 cm (exact)
1 mile = 5280 ft = 1.609344×10^3 m = 1.609344 km
1 quart (U.S.) = 946 cc = 0.946 L (check a quart oil can)
1 gallon (U.S.) = 4 quarts (U.S.) = 3.784 L
1 erg = 1 g cm^2/s^2

1 pascal $= 1 \text{ N/m}^2 = 1 \text{ J/m}^3 = 10^7 \text{ erg}/10^6 \text{ cm}^3$
1 pascal $= 10^7 \text{ erg}/10^6 \text{ cm}^3 = 10 \text{ dyne/cm}^2 = 1 \text{ Pa} = 1 \text{ pascal}$
1 newton $= 1 \text{ kg m/s}^2 = 10^5 \text{ g cm/s}^2 = 10^5 \text{ dyne}$
1 joule $= 1 \text{ N m} = 1 \text{ kg m}^2/\text{s}^2 = 10^7 \text{ dyne cm} = 10^7 \text{ erg}$

We need to convert the standard pressure of 1 atm to other units. Imagine freezing metallic mercury and machining 76 nice shiny cubes exactly $(1 \text{ cm}) \times (1 \text{ cm}) \times (1 \text{ cm}) = 1 \text{ cm}^3$ with the density of 13.596 g/cm^3 and then stacking 76 them in a column to evaluate the force on a 1 cm^2 area to yield a pressure as (pressure = force/area).

$$1 \text{ atm} = \rho g h/\text{cm}^2 = (13.6 \text{ g/cm}^3)(980 \text{ cm/s}^2)(76 \text{ cm}) \cong 1.013 \times 10^6 \text{ dyne/cm}^2$$

This is approximate to help visualize the unit but when more accurate values are used the standard conversion is

$$1 \text{ atm} = 1.01325 \times 10^6 \text{ dyne/cm}^2 = 1.01325 \times 10^5 \text{ Pa} = 1.01325 \text{ bar.}$$

Note also that

1 dm $= 10 \text{ cm} = 0.1 \text{ m}$
1 L $= 1 \text{ dm}^3 = 1 \times 10^{-3} \text{ m}^3 = 1 \text{ L}$
1 cm^3 $= 1 \ (1 \times 10^{-2} \text{ m})^3 = 1 \times 10^{-6} \text{ m}^3$
1 atm $= 1.01325 \times 10^5 \text{ pascal where } 1 \text{ pascal} = 1 \text{ Pa} = 1 \text{ N/m}^2$
1 atm $= 1.01325 \times 10^6 \text{ dyne/cm}^2$
1 bar $= 1.00000 \times 10^5 \text{ Pa}$
1 Pa $= 10 \text{ dyne/cm}^2$

It will be convenient for some applications to use cgs units so we need to remember what a dyne is.

1 dyne $= 1 \text{ g cm/s}^2$
1 erg $= 1 \text{ dyne cm} = 1 \text{ g cm}^2/\text{s}^2$

Then

$$R = \frac{(1 \text{ atm})(1.01325 \times 10^5 \text{ N/m}^2 \text{ atm})(22.414 \times 10^{-3} \text{ m}^3/\text{mol})}{(1 \text{ mol})(273.15°\text{K})} \cong 8.314 \text{ J/°K mol}$$

1 cal $= 4.184$ J.
So we have several alternative values for R in different units:

$$R = 0.08206 \text{ L atm/°K mol}$$

$$R = 0.08314 \text{ L bar/°K mol}$$

$$R = 82.06 \text{ cm}^3 \text{ atm/°K mol}$$

$$R = 8.314 \text{ J/°K mol}$$

$$R = 1.987 \text{ cal/°K mol}$$

One of the most common errors students make is to use the incorrect value of R because they forget to check the units. We could standardized the book on SI units, but in the real world there is a lot of equipment out there using all kinds of units and you will be better prepared if you learn to cope with different unit systems including the SI units.

It is very important at this point to note that the product of pressure and volume is always *energy*: $PV = (\text{force/area}) \times (\text{volume}) = \text{force} \times \text{distance} = \text{energy}$. One way to remember this is to chant the rhyme "$PV = \text{energy}!$ $PV = \text{energy}!$ $PV = \text{energy}!$ etc.".

The most common practical unit of pressure is (pounds/inch^2) abbreviated as psi (pounds per square inch). In these units we get

$$\frac{1.01325 \times 10^6 \, \text{dyne/cm}^2}{\left(\frac{453.6 \, \text{g/lb}}{(2.54)^2 \, \text{cm}^2/\text{in}^2}\right)(980 \, \text{cm/s}^2)} = 14.70 \, \text{lb/in.}^2 = 14.7 \, \text{psi}$$

Note here that the unit lb (pound) is in the denominator of the denominator and so flips up to the numerator in the answer.

Finally, a unit that occurs when reading the instructions for inflating the tires of a British or European racing bicycle is that of a "bar." Apparently someone noticed that the 1.01325 constant is close to 1.0 so why not define a "bar" as a clean unit.

$$1 \, \text{bar} = 1.0 \times 10^6 \, \text{dyne/cm}^2 = (1.0/1.01325)(760 \, \text{mmHg}) = 750.06 \, \text{mmHg}$$

To a good approximation 1 bar is 750 mmHg while 1 atm = 760 mmHg. The name "bar" is appropriate because pressure is what a "barometer" measures. To use a common service station air pump to inflate tires in bars just use the simple conversion $1 \, \text{bar} = \left(\frac{750}{760}\right) 14.7 \, \text{psi} = 14.5 \, \text{psi}$ since a bar is a smaller unit than an atm.

MOLECULAR WEIGHT FROM GAS DENSITY (THE DUMAS BULB METHOD)

In bygone days chemists used a simple method of first weighing an empty container and weighing that known volume again with gas inside at a known temperature and pressure. The internal volume could be obtained by the difference in mass of the container empty and filled with water using the density of water. However, when it came time to weigh the container with the unknown gas in it the weight of the empty container (often a glass bulb of about 400 mL volume) the difference in weight (mass) was usually very small compared to the weight (mass) of the empty container. Thus, this method has a very large uncertainty but it can be used with rounded estimates from assumed molecular structures. The method works fairly well for high-molecular weight gases but would be very uncertain for He or H_2. The key idea is to use the mole concept with the ideal gas law.

$$PV = nRT = \left(\frac{w}{M}\right)RT$$

so

$$M = \frac{wRT}{PV} = \left(\frac{w}{V}\right)\frac{RT}{P} = \rho\frac{RT}{P}$$

Here
   "w" is the weight of the gas
   ρ the density
   M is the unknown molecular weight

This method is not very accurate and really only works well for a few gases that are of high molecular weight (CCl_4, $CHCl_3$, etc.).

DALTON'S LAW OF PARTIAL PRESSURES

One curious physical phenomenon associated with gases is the fact that when there is a mixture of gases in a given volume they behave independently so that their pressures are additive. In fact this raises the issue of what we mean by "pressure." Common sense may lead us to expect that volumes are additive as indeed they are for macroscopic objects such as bricks. Thus, it is somewhat thought provoking that several gases can be easily confined in the same volume. This same sort of question also arises for mixtures of liquids to a much less extent as discussed later in Chapter 6. These considerations go to the very heart of the concept of the size of atoms and molecules and how much space is between them in a liquid or gas. As we will soon see, the space between gas molecules is about 100 times their size at 1 atm so there is plenty of space for other molecules. In addition, it will soon become evident that pressure is (force/area) caused by many collisions of gas molecules with the wall of the container. Cavendish in 1781 and Dalton in 1810 contributed to the concept now known as "Dalton's law."

The total pressure exerted by a mixture of gases is equal to the sum of the pressures that each component would exert if placed separately into the container.

Thus, $P_{tot} = P_1 + P_2 + P_3 + \cdots = \sum_i P_i$ but if the gases act as ideal gases, we have $P_{tot} = n_1\left(\dfrac{RT}{V}\right) + n_2\left(\dfrac{RT}{V}\right) + n_3\left(\dfrac{RT}{V}\right) + \cdots = \left(\dfrac{RT}{V}\right)\sum_i n_i = n_{tot}\left(\dfrac{RT}{V}\right)$. Now consider the mole fractions $(n_i/n_{tot}) \equiv \chi_i$. We see that the ratios of the partial pressures to the total pressure are equal to the mole fractions $\left(\dfrac{P_i}{P_{tot}}\right) = \dfrac{n_i\left(\frac{RT}{V}\right)}{n_{tot}\left(\frac{RT}{V}\right)} = \dfrac{n_i}{n_{tot}} = \chi_i = \left(\dfrac{P_i}{P_{tot}}\right)$. As a corollary we note additional conclusions as

$$\frac{n_1}{n_{tot}} + \frac{n_2}{n_{tot}} + \frac{n_3}{n_{tot}} + \cdots = 1, \quad \frac{P_1}{P_{tot}} + \frac{P_2}{P_{tot}} + \frac{P_3}{P_{tot}} + \cdots = 1, \quad \text{and} \quad \chi_1 + \chi_2 + \chi_3 + \cdots = 1.$$

The most common use of Dalton's law is when gases are measured "over water," that is, when a pressure is measured in the presence of moisture, which produces a partial pressure of water vapor as a gas, which in turn contributes a small pressure to the total pressure. This can occur when a reaction produces a gas and the gas is trapped in a container inverted over water. Tables of the vapor pressure of water are readily available in handbooks. Water is often in natural settings where gas pressure is measured in the presence of dew or a layer of water as may occur in "wet" forensic samples. In the distant past chemists often isolated gases as the product of a reaction and allowed the gas to bubble through a water trap, thereby introducing water vapor pressure into the total pressure.

Example: Given a small amount of benzenediazonium chloride that is warmed gently in a closed container to form chlorobenzene and N_2. A tube is attached to the top of the container and inserted under an inverted flask initially filled with water. As the reaction proceeds, most of the N_2 flows through the tube and bubbles up under the water in the flask. After a while an estimated 450 mL of the gas is trapped in the bubble and the pressure at the surface of the water is 750 mmHg and the water temperature is 23°C. Assuming we trapped most of the released N_2 (some is left in the original flask and the tube) and there is a 1:1 stoichiometry of the moles of gas produced to the moles of chlorobenzene, we can estimate the moles of chlorobenzene formed to be equal or greater than the moles of N_2. In a handbook we find that the vapor pressure of water is 2.8104 kPa at 23°C because all modern handbooks report data in SI units but if we have a manometer attached to the reaction flask we can read the pressure in mmHg.

$$P_{H_2O} = \left(\frac{2.8104 \times 10^3 \, \text{Pa}}{1.01325 \times 10^5 \, \text{Pa/atm}}\right)(760 \, \text{mm/atm}) \cong 21.08 \, \text{mmHg}$$

By Dalton's law of partial pressures the total pressure of 750 mmHg of the gas trapped over the water is the sum of the pressure of the water vapor and that of the N_2 so we have

$$P_{N_2} = 750\,\text{mmHg} - P_{H_2O} = (750 - 21.06)\,\text{mmHg} = 728.92\,\text{mmHg}.$$

Then we can calculate the moles of N_2 from the ideal gas equation.

$$n = \left(\frac{PV}{RT}\right) = \left[\frac{\left(\dfrac{728.92\,\text{mmHg}}{760\,\text{mmHg/atm}}\right)(0.450\,\text{L})}{0.08206\,\text{L atm/}^\circ\text{K mol}\,(273.15 + 23)^\circ\text{K}}\right] \cong 0.0178\,\text{mol}\,N_2.$$

This hypothetical problem is given here to illustrate the use of Dalton's law of partial pressures and to discuss the process of *uncertainty analysis*, which may be important if a forensic laboratory technician has to testify in a court case or a chemistry student has to write a report for a laboratory course.

There are more elaborate methods of uncertainty analysis but what we show here [2] is remarkably general and useful not only to give an overall measure of uncertainty in a given result but also permits determination of the cause of most of the uncertainty (Figure 1.5). First we should ask how we are going to merge data in several different units. The answer is to convert the uncertainties to percentages. First, the volume of the gas is 450 mL but uncertain due to the volume of the tube between the water trap and the reaction flask. Let us estimate that the volume is uncertain by 10 mL so in percentage the various units cancel and we have

$$\lambda\%\,(V) = \left(\frac{10\,\text{cm}^3}{450\,\text{cm}^3}\right) \times 100 \cong 2.2222\%$$

$$\lambda\%\,(P) = \left(\frac{1\,\text{mmHg}}{750\,\text{mmHg}}\right) \times 100 \cong 0.1333\%$$

$$\lambda\%\,(T) = \left(\frac{1^\circ\text{K}}{296.15^\circ\text{K}}\right) \times 100 \cong 0.3376\%.$$

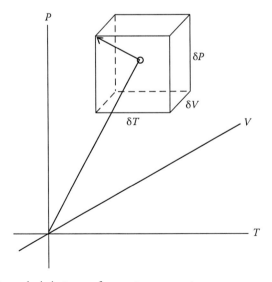

FIGURE 1.5 Uncertainty analysis in terms of percent components.

We use the total pressure of the two gases in the pressure uncertainty since that is what the manometer is measuring with an uncertainty of 1 mmHg (according to our subjective eye) and even though we measure the temperature to the nearest degree (low-precision thermometer) we need to remember we are using the Kelvin temperature in the calculation. Now all the variables are in the same units of percent. In this case we have three variables so we can plot them on a (P, V, T) axis and display the value calculated for the moles as a vector. The relative uncertainties can be displayed as $+/-$ deviations about the tip of the calculated vector result.

Then we use the 3D Pythagoras' theorem to estimate the worst case vector displacement of the tip of the calculated vector result. The result will be in percentage units as

$$\Lambda\% \cong \sqrt{(0.1333)^2 + (2.2222)^2 + (0.3376)^2} \cong 2.2515\%$$

Finally, we can quote the final answer with an estimated uncertainty using the "square root of the sum of the squares of the percent uncertainties in the contributing variables" [2] to find

$$\text{mol N}_2 = 0.0178 \pm 2.2515\% = 0.0178 \pm 4.0076 \times 10^{-4}\,\text{mol}$$

This method is subjective relative to how precise a given observer can measure each variable but standard precision glassware should be estimated to be good to about 0.1% and included in the sum of squares for each item of glassware used so this approximate method can be extended to many variables. We see that according to the uncertainty analysis the calculated result is only good to the fourth decimal place as given. Some texts express error analysis/uncertainty in terms of partial derivatives but it is not easy or clear to assign a partial derivative to a response from a given device in terms of a calculus formula in some cases. For that reason, this author favors the simple formula in terms of percent uncertainties, which can be (subjectively) estimated numerically.

NONIDEAL GAS BEHAVIOR

While the ideal gas law works well for pressures up to about 10 atm and higher temperatures above 25°C, many common processes (air conditioning, refrigeration) involve higher pressures and lower temperatures. If the ideal gas law is truly universal we could define the "compressibility factor" as

$$Z = \left(\frac{PV}{nRT}\right) = 1$$

and expect that if we plot Z against the pressure we should get a flat line (Figure 1.6). When such graphs are plotted for real data, there are large deviations, particularly at low temperatures and/or high pressures.

There are other ways to plot these data to exaggerate the deviations from $Z = 1$, but on the other hand we can see that over a fairly large range of temperatures and pressures the ideal gas law is approximately correct. What are the reasons for the deviations from the ideal? Let us try to patch the ideal gas law for a more detailed treatment. We start by setting up the basic PV behavior and allow for corrections.

$$(P + ?_1)(V + ?_2) = nRT$$

Consider a correction to the pressure, P. If indeed the pressure we measure is due to molecular impacts with a surface in a manometer or a diaphragm in a pressure gauge, is that the actual pressure within the gas? We are creeping up on a new concept that models a gas as a collection of small

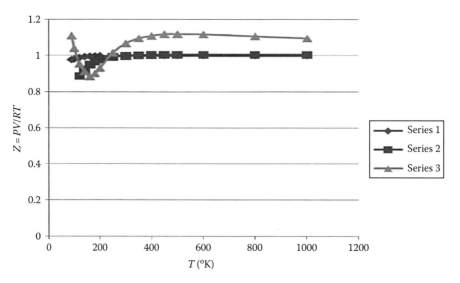

FIGURE 1.6 Compressibility factor of air (a mixture of N_2, O_2, CO_2, Ar, etc.) at different pressures. Series 1 (diamonds) = 1 bar pressure, Series 2 (rectangles) = 10 bar pressure, and Series 3 (triangles) = 250 bar pressure. (Data from *Perry's Chemical Engineers' Handbook*, 6th Edn., McGraw-Hill, 1984. table 3-162. Z-values are calculated from values of pressure, volume (or density), and temperature in Vassernan, Kazavchinskii, and Rabinovich, Thermophysical properties of air and air components, Moscow, Nauka, 1966, and NBS-NSF Trans. TT 70-50095, 1971 and Vassernan and Rabinovich, Thermophysical properties of liquid air and its component, Moscow, 1968, and NBS-NSF Trans. 69-55092, 1970. Courtesy of Mr. Ian C. Roman of Air Liquide, Delaware Research and Technology Center, Newark, DE.)

molecules flying around with a lot of space between them (recall Dalton's law). That idea should include collisions of molecules within the volume. Consider a collision of an auto with a fixed wall compared to a head-on collision with another similar auto. There will be a change of momentum force in the collision with the wall but the force of the head-on collision will be double that with a fixed wall! Thus, the real pressure within the gas volume is actually *higher* than the pressure we measure, although the collisions are relatively infrequent due to the large space between molecules of the gas. An additional consideration is that there may be some weak short-range attractive forces (now known as van der Waals interactions) between the gas molecules that causes them to approach each other more forcefully at short range; thus, leading us to expect a plus sign for this small correction. In fact we can use the idea of concentrations as (moles/volume) for the number of gas molecules in the volume. The pressure correction is small but proportional to the square of the concentration. Of course this idea includes collisions of some molecules with themselves but N molecules can only collide with $(N-1)$ other molecules. However, when N is of the order of 10^{23} then $(N-1)$ is essentially the same as N. We are reminded that when a proportionality is observed phenomenologically we can use a "proportionality constant" to create an equation with a constant as in

$$?_1 \propto \left(\frac{n}{V}\right)\left(\frac{n}{V}\right) = a\frac{n^2}{V^2}.$$

Another factor included in this term is any bimolecular electronic interactions. Thus, the "a" parameter absorbs a number of interaction terms as well as an amount of the bimolecular collision pressure.

Next we need to consider that while the molecules are very small, their volume is not really zero; they have a finite volume and when you have a mole of molecules at very low temperatures tending to condense into a liquid or even a solid, the volume reduces to a smaller but finite volume. This small volume will be proportional to the number of moles of gas as in

$$?_2 \propto n = bn.$$

Thus, a modified gas law was proposed by a Dutch physicist Johannes Diderik van der Waals (1837–1923) in 1873 in his doctoral thesis as a way to simulate the condensation of gases to liquids. He received the Nobel Prize for this work in 1910.

$$\left(P + a \left(\frac{n^2}{V^2} \right) \right)(V - nb) = nRT$$

Chemical engineering students in this class will be aware of more accurate and complicated equations of state, but for this text we will be content to use the van der Waals equation as a useful treatment of real gases. Thus, parameters are available in terms of values of "a" and "b" for a number of gases and each gas has separate parameters, see Table 1.3. We may find some tables of these parameters in older units but lists are available in SI units as well.

In the original 1873 doctoral thesis of van der Waals, the goal was to explain the process of condensation of gases to liquids in a smooth way. The ideal gas equation that predicts the PV product at any fixed temperature should be one branch of a hyperbola. Such "isotherm" curves are indeed found on a plot of pressure versus volume of a fixed amount of gas at constant temperature, for low pressures above the boiling point of a given material. However, as one lowers the temperature a small bump in the isotherm will be observed, which is at first an "inflection point" and then at still lower temperatures enters a region where there is a fog or mist of liquid condensate droplets. It is easy to show that the van der Waals equation is cubic in V by multiplying through by V^2.

$$V^2 \left(P + a \left(\frac{n^2}{V^2} \right) \right)(V - nb) = nRTV^2 \quad \text{or} \quad PV^3 - n(bP + RT)V^2 + n^2aV - n^3ab = 0.$$

TABLE 1.3
van der Waals Constants for Common Gases

Gas	a (L^2 bar/mol^2)	b (L/mol)
He	0.0346	0.0238
Ne	0.208	0.0167
H_2	0.2452	0.0265
Ar	1.355	0.0320
O_2	1.382	0.0319
N_2	1.370	0.0387
CO	1.472	0.0395
CH_4	2.303	0.0431
CO_2	3.658	0.0429
NH_3	4.225	0.0371

Source: Lide, D.R. Ed., *CRC Handbook of Chemistry and Physics*, 87th Edn., CRC Press, Boca Raton, FL, 2006–2007, p. 6–34.

Although the van der Waals gas equation is generally more accurate than the ideal gas law it has a problem in that it is not easy to solve a cubic equation to obtain the volume. It is easy to get the pressure:

$$P = \frac{nRT}{(V - nb)} - \frac{n^2 a}{V^2}$$

but we need a trick to get the volume. By rearranging the equation to a polynomial in V set to zero, we can use the Newton–Raphson root finding method. Thomas [3] shows this method will converge to the nearest root from some initial guess using an iterative process as

$$x_{\text{new}} = x_{\text{guess}} - \left[\frac{f(x_{\text{guess}})}{df(x_{\text{guess}})/dx} \right].$$

This method requires that the function of the polynomial be set to zero and a good initial guess for the root. Ultimately you may have to sketch the curve to see where the nearest root is but if the temperature is relatively high and the pressure is relatively low (with respect to the critical point) we can make an estimate using the ideal gas equation. First we need the derivative with respect to V.

$$3PV^2 - 2n(bP + RT)V + n^2 a = 0.$$

Thus, we have a method to obtain V from the van der Waals equation provided we can make a good initial guess, possibly from the ideal gas equation or from a sketch of the polynomial in V as to where the curve crosses the horizontal axis. That is why we set the equation to zero so that the curve will go through zero at the roots of the equation:

$$V_{\text{new}} = V_{\text{guess}} - \left[\frac{PV_g^3 - n(bP + RT)V_g^2 + n^2 a V_g - n^3 ab}{3PV_g^2 - 2n(bP + RT)V_g + n^2 a} \right]_{V_{\text{guess}}}.$$

Obviously this is a very complicated procedure but it could be coded as an equation in BASIC on a PC to run in a millisecond. You need to set up an iterative loop so that each new value is used for the next guess and stop iterating when the desired number of decimal places in the answer is reached. Because this method is so tedious it is very satisfying to see the answer converge and usually five or less iterations are adequate if the initial guess is good. One of the homework problems will suffice to carry out this procedure to see that it can converge and that an automated way is needed to make it practical.

In Figure 1.7 the schematic diagram plots $V = (V/V\text{-critical})$ and $P = (P/P\text{-critical})$ to yield an equation independent of the (a, b) parameters as we will soon show. The main point is that the isotherms at higher temperature appear as one branch of a hyperbola as expected but at the point marked as the "critical point" the curve has an inflection point and then at still lower temperatures the curve shows its behavior as a cubic curve in V. Careful measurements within the conditions at temperatures lower than the critical point actually confirm the presence of liquid droplets. This leads to a *definition of the critical temperature as that temperature above which a gas cannot be compressed into a liquid phase at any pressure*. At temperatures lower than the critical temperature of a material the gas can be compressed (squeezed) into the liquid form by applying higher pressure.

The inflection point of the van der Waals equation at the critical point is very helpful in a mathematical sense since not only is the first-derivative zero at that point but the first-derivative (slope) changes sign on either side of the critical point so the second derivative (curvature) must also go through zero. We now embark on a series of mathematical manipulations to determine the (a, b) parameters of a given gas from the experimental critical point parameters (P_c, V_c, T_c).

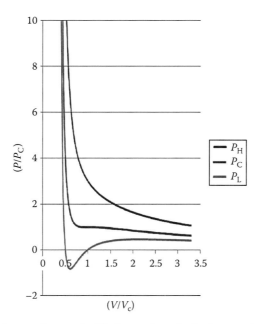

FIGURE 1.7 Selected isotherms of a van der Waals gas. P_C = the isotherm at the critical pressure, P_H = the isotherm above the critical pressure, and P_L = the isotherm below the critical pressure. (Courtesy of Prof. Carl Trindle, Chemistry Department of the University of Virginia.)

First solve the van der Waals equation for P and assume $n = 1$, then take the first and second derivatives of the pressure with respect to the volume at constant $T = T_c$.

$$1.\ P = \frac{RT}{(V - b)} - \frac{a}{V^2}$$

$$2.\ \left(\frac{\partial P}{\partial V}\right)_T = \frac{-RT_c}{(V_c - b)^2} + \frac{2a}{V_c^3} = 0$$

$$3.\ \left(\frac{\partial^2 P}{\partial V^2}\right)_T = \frac{2RT_c}{(V_c - b)^3} - \frac{6a}{V_c^4} = 0$$

Note we have added "c" subscripts in the derivative equations because they are only true (=0) at the critical point. Rearrange Equation (2) and substitute it into Equation (3).

$$\frac{RT_c}{(V_c - b)^2} = \frac{2a}{V_c^3} \quad \text{which} \quad \Rightarrow \quad \frac{2}{(V_c - b)}\left(\frac{2a}{V_c^3}\right) = \frac{6a}{V_c^4}$$

so

$$\frac{4aV_c}{(V_c - b)} = 6a \quad \text{and} \quad 4V_c = 6(V_c - b).$$

Then $4V_c = 6V_c - 6b$ and $6V_c - 4V_c = 6b = 2V_c$ so finally $V_c = 3b$ and $b = V_c/3$. This important result means that the "b" parameter of the van der Waals equation can be obtained as $(1/3)$ of the volume V on the V-axis directly under the critical point. Now let us look for "a." So far we have used Equation (3) so now go back to (2) and use the "b" value.

$\frac{RT_c}{(3b-b)^2} = \frac{2a}{27b^3}$ so $\frac{RT_c}{4b^2} = \frac{2a}{27b^3}$ which means that $T_c = \frac{8a}{27Rb}$. For those who like Arbie roast beef sandwiches, there is a silly way to remember this result as "If you ate $(8a)$ over $(/)$ $27Rb$ (Arbie) sandwiches you will reach a critical temperature (T_c) fever." While it is unlikely you could eat 27 sandwiches, the mnemonic helps you remember the formula. Next we go back to Equation (1) to find "a."

$$P_c = \frac{R\left(\frac{8a}{27Rb}\right)}{(3b-b)} - \frac{a}{9b^2} = \frac{8a}{54b^2} - \frac{a}{9b^2} \times \frac{6}{6} = \frac{2a}{54b^2} = \frac{2a}{6V_c^2} = \frac{a}{3V_c^2}$$

so that $a = 3V_c^2 P_c$.

Let us also see what R looks like for the van der Waals gas:

$$R = \frac{8a}{27T_c b} = \frac{8\left(3P_c V_c^2\right)}{27T_c\left(\frac{V_c}{3}\right)} = \frac{8}{3}\left(\frac{P_c V_c}{T_c}\right).$$

In summary, $b = V_c/3$, $a = 3V_c^2 P_c$, $T_c = \frac{8a}{27Rb}$, and $R = \frac{8}{3}\left(\frac{P_c V_c}{T_c}\right)$. In Figure 1.7 the parameters are the "reduced variables" which can be defined as

$$P_r = \left(\frac{P}{P_c}\right), \quad V_r = \left(\frac{V}{V_c}\right), \quad \text{and} \quad T_r = \left(\frac{T}{T_c}\right).$$

Therefore, we can insert the definitions of (P_r, V_r, T_r) and obtain

$$\left(P_r + \frac{3}{V_r^2}\right)\left(V_r - \frac{1}{3}\right) = \frac{8}{3}T_r.$$

That is a demonstration of the "law of corresponding states" for the van der Waals equation in terms of the reduced variables. Other equations of state should also pass the test of freedom from parameters when expressed in terms of the reduced variables. Learning and/or memorization of the derivation of the van der Waals critical point parameters will simultaneously improve your math skills as well as emphasize the importance of the critical point. The formulas for the critical temperature and critical pressure will be used in later discussion. How fast can you do this derivation?

SUPERCRITICAL FLUID CHROMATOGRAPHY

What good is the critical point? Recently a new form of chromatography has been developed that may be of considerable use in forensic applications in that it does not destroy the sample (depending on the detector) and is capable of resolving some very sticky, gummy materials that normally would be useless goo! Further the equipment is relatively simple and occupies about the space of a normal desk. The technique of supercritical fluid (SCF) chromatography is an analytical method for treatment of materials that are normally difficult to resolve into separate components. While other materials can be used as a mobile phase, the main idea is to use carbon dioxide at a pressure and temperature range that is slightly above the critical temperature as an excellent solvent for high-molecular weight organic materials (Figure 1.8). While detection and identification of various materials are of main interest, the sample can be recovered after passing though the column as long as the detector does not use flame ionization.

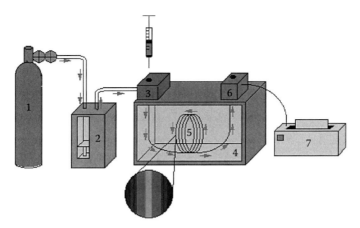

FIGURE 1.8 Schematic diagram of a table top supercritical fluid chromatograph. The numbers in the figure are explained in the Internet site listed where this is an animated GIF picture which simulates the passage of a sample through the instrument. (Karey O'Leary and Advisor Prof. Andrea Detrich of Virginia Tech University at http://www.cee.vt.edu/ewr/environmental/teach/smprimer/sfc/sfc.html)

FLUIDS

The term "fluid" includes not only liquids but also anything that "flows" such as powdered coal or sand slurries in water and of course airflow. Experimentally, the critical temperature for carbon dioxide is 30.98°C, the critical pressure is 73.75 bar, and the critical volume is 94 cm^3/mol [1]. Above the critical temperature, the fluid is called supercritical fluid. We can compare that to the value estimated from the van der Waals equation using data from Table 1.3. The result is within 0.25° of the experimental value.

$$T_c = \frac{8a}{27Rb} = \frac{8(3.658\,\text{L}^2\,\text{bar}/\text{mol}^2)}{27(0.08314\,\text{L bar}/°\text{K mol})(0.0429\,\text{L}/\text{mol})} = 303.88\,°\text{K} = 30.73\,°\text{C}$$

However, we can see that the van der Waals equation is less accurate for the critical volume and the critical pressure

$$V_c = 3b = 3(0.0429\,\text{L}/\text{mol}) = 128.7\,\text{cm}^3/\text{mol}$$

instead of 94 cm^3/mol and

$$P_c = a/3V_c^2 = \frac{3.658\,\text{L}^2\,\text{bar}/\text{mol}^2}{3(128.7\,\text{cm}^3/\text{mol})^2}\left(\frac{1000\,\text{cm}^3}{\text{L}}\right)^2 = 73.61\,\text{bar}$$

which is close to experiment.

Apparently the van der Waals equation gives good estimates of the critical temperature and critical pressure that are of practical interest but does poorly for the critical volume so there are other equations of state chemical engineers use when more accuracy is required!

This critical temperature of CO_2 is only a few degrees above room temperature, which makes it safe to treat organic compounds with little danger of thermal degradation. Although carbon dioxide is an excellent solvent for nonpolar compounds it is less good for polar compounds.

It has been found that for nonpolar materials CO_2 is a very good solvent but to keep the mobile phase (gas) flowing, a column temperature slightly above the critical temperature is used so that the temperature is "super" critical and the pressure is regulated near the critical pressure to keep the mobile phase nearly a liquid (Figure 1.9). Under these conditions SCF chromatography has

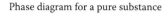

Phase diagram for a pure substance

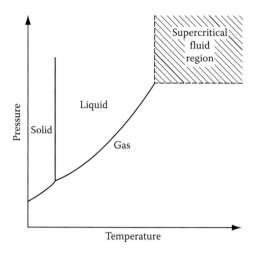

FIGURE 1.9   Diagram of the supercritical phase.

characteristics of both gas chromatography (GC) and liquid chromatography since the mobile phase is gas almost ready to condense to a liquid so that it has both good solvating power and high diffusivity. The proof of these claims can be seen in the examples of the chromatograms of high-molecular weight materials and similar compounds discussed in the following.

A further advantage of SCF chromatography is that the solvating power can be "tuned" by adjusting the pressure and temperature of the mobile phase (usually carbon dioxide) so the density of the mobile phase more nearly matches the density of the sample. These changes can be programmed to enhance and control the elution of various sample components and provide both the temperature programming of GC and the use of a solvent gradient in high-pressure liquid chromatography (HPLC). If separation is more important than detection, SCF can also be used for bulk preparative separations and has been used in industry to remove caffeine from coffee and nicotine from tobacco. Thus, we see that the critical point of a gas is an important topic of considerable practical value.

SUPERCRITICAL FLUID INSTRUMENTATION

From the schematic in Figure 1.7 we see that a SCF chromatograph has great similarity to the type of HPLC where a tank of compressed gas is used as the pressure source but a reciprocating pump is also present to maintain control of the pressure. Similar detectors can be used, although a flame-ionization detector will destroy the eluted materials. There are a number of optical/spectroscopic detectors that are nondestructive and they can be used to monitor the elution process in the case of preparative elution or where a sample is forensic evidence not to be destroyed. Even the columns are similar, although SCF, GC, and HPLC columns are designed as optimum for the intended purpose. Considering the high pressures involved, the SCF columns are more likely to be similar to HPLC columns in cross section but in a longer coil as with GC applications rather than the short (about 1 m) HPLC columns and built to withstand high pressure as for HPLC. One key difference in SCF chromatography is that the mobile-phase temperature and pressure must be adjusted in the flow line to reach the desired supercritical (T, P) condition before the sample is injected and the oven temperature needs to maintain the temperature while the pump maintains the pressure. A restrictor is necessary at the end of the column or after the detector to maintain the pressure in the column. As a practical matter the restrictor may need to be cleaned frequently, but this is similar to routine maintenance of a HPLC apparatus. Variable restrictors are available.

SUPERCRITICAL MOBILE PHASE

While it is possible to use a number of volatile solvents as the mobile phase for SCF chromatography, the most commonly used mobile phase is carbon dioxide. However, carbon dioxide is not a good solvent for polar compounds so it is common to add a small amount of some additional polar organic liquid such as an alcohol or even water as a "modifier." However, the modifier needs to be miscible with carbon dioxide. Much of the other technology associated with either GC or HPLC in terms of sample inlets and types of pumps are adapted to specific applications but the key attribute of the SCF-type chromatography is the maintenance of (T, P) conditions near the critical point of the mobile phase. A selection of columns is available just as for GC or HPLC.

SAMPLE SCF SEPARATIONS

The output results of an SCF chromatogram are presented on a strip chart recorder showing the detector response on the vertical axis and the elution time on the horizontal axis. We show two examples as presented by Karey O'Leary at Virginia Tech University in 1995 (results shown are by permission from http://www.cee.vt.edu/ewr/environmental/teach/smprimer/sfc/sfc.html).

It is evident from these two examples that the resolving power of these SCF chromatograms is excellent under the definition of resolving power as the ability to separate peaks at half height. To be able to resolve different components of pump oil is suggestive that forensic analysis of oils is entirely feasible using SCF chromatography and analysis of environmental samples of pesticides (Figure 1.10) can be carried out at temperatures that are less harsh on compounds that might decompose under destructive gas chromatography-mass spectrometry (GC-MS) testing. Overall SCF chromatography for both detection and preparation is a practical example of the importance of the critical point properties of real gases (Figure 1.11).

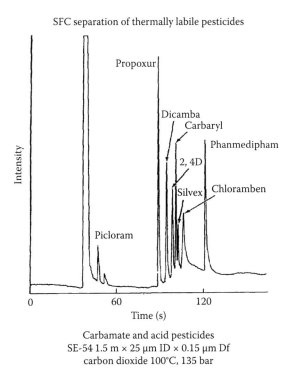

FIGURE 1.10 Example of supercritical fluid chromatography of pesticides.

SFC separation of polymer samples

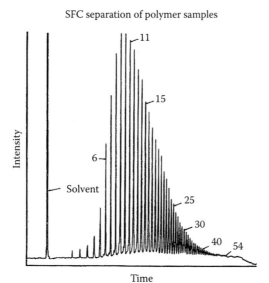

DC silicone fluid separation
SE-54 10 m × 50 μm ID × 0.25 μm Df
carbon dioxide 100°C, 100 bar

FIGURE 1.11   Resolution of silicone polymers using supercritical fluid chromatography.

SUMMARY

This chapter helps us to understand that while the ideal gas law ($PV = nRT$) is generally useful in the range of temperatures above 0°C and pressures up to about 100 atm, extreme conditions can require corrections. Dalton's law leads us to begin to wonder about the behavior of individual gas molecules and the space between them. We also saw that the density of a gas can be related to the molecular weight of the gas. A calculation of the moles of gas based on the Avogadro hypothesis led us to a simple way to consider uncertainties in experimental methods according to a formula based on percent uncertainties in the variable measurements as

$$\Lambda(\%)_{\text{total}} \cong \sqrt{\sum_i [\lambda_i(\%)^2]}$$

While the PV curve for an ideal gas should be the positive branch of a hyperbola at a given temperature (isotherm), experimental data reveal the critical point phenomenon where a gas can condense into a liquid. The work by van der Waals corrects the ideal gas law with small terms and two parameters to formulate

$$\left(P + \frac{n^2 a}{V^2}\right)(V - nb) = nRT$$

which is more accurate than the ideal gas law. Analysis of the van der Waals critical point was based on setting both $(\partial P/\partial V)_{n,T} = 0$ and $(\partial^2 P/\partial V^2)_{n,T} = 0$ at the critical temperature because the PV isotherm curve goes through an inflection point there. Using this calculus condition, the values of the van der Waals (a, b) parameters can be found for a number of real gases. Insertion of the critical parameters into the van der Waals equation leads to an equation that is independent of the parameters and obeys a law of corresponding states in which the equation is expressed in terms of the reduced parameters: $P_R = (P/P_C)$, $V_R = (V/V_C)$, and $T_R = (T/T_C)$, which yields a universal van der Waals equation

$$\left(P_R + \frac{3}{V_R^2}\right)\left(V_R - \frac{1}{3}\right) = \frac{8}{3}T_R.$$

Although the van der Waals gas equation is generally more accurate than the ideal gas law, the equation is cubic in V which leads to a practical difficulty in solving for V. An iterative Newton–Raphson technique is suggested for use with some sort of computer-aided way to evaluate V. The idea of the critical point is applied to introduce an analytical technique termed supercritical fluid chromatography, which is shown to be able to separate and clearly resolve high-molecular weight materials. Discussion followed as to potential applications of SCF chromatography for nondestructive analysis of forensic evidence. The SCF behavior of carbon dioxide was described as a good "solvent" near its critical point. The exercise in calculus to find the van der Waals parameters offers a chance to motivate improvement in mathematical skills when it is seen that the formulas can be used to find the (a, b) parameters for a number of real gases.

PROBLEMS

1.1 Calculate the volume in liters of 2 mol of He gas at 65°F and 740 mmHg pressure.

1.2 Calculate the pressure in a tire inflated to 30 psi in winter at a temperature of 10°F if the tire has the same volume in July when the temperature is 90°F.

1.3 Calculate the pressure of 2 mol of H_2 contained in a 10 L container at 30°C using the van der Waals equation.

1.4 Calculate the volume of 1 mol of CO_2 gas at 30°C when the pressure is 20 bars using the van der Waals equation. (Hint: Estimate V from the ideal gas equation and use the Newton–Raphson method for the van der Waals gas starting from that value.)

1.5 Calculate P_c, V_c, and T_c for NH_3 gas using the van der Waals parameters (a, b) and compare to the values in a handbook or the Internet.

1.6 A glass light bulb shell is sealed by a glassblower and one small tip is pulled out into a long narrow point with an entrance hole of about 1/8 in. diameter. The open bulb is weighed and found to weigh 46.345 g. Then the bulb is completely filled with water using an eye dropper with a thin tip and the filled bulb is found to have a mass of 237.93 g. Use the density of water at 20°C of 0.99821 g/cm^3 to find the internal volume of the bulb. The bulb is then aspirated to remove the water and dried in an oven. When dry and cool the bulb is reweighed and found to still be 46.345 g prior to filling it with about 5 mL of a volatile liquid. The bulb with this small amount of liquid is placed carefully into a 250 mL beaker of water, which is heated to a boil at exactly 100°C. After reaching equilibrium at this temperature and all evidence of liquid is gone the glass tip is sealed with a tiny drop of glue, the bulb is removed from the boiling water, carefully dried, and found to weigh 47.309 g. Assuming the pressure was 1 atm when the bulb was sealed, what is the estimated molecular weight and a possible compound that has a volatile liquid close to this molecular weight?

1.7 Calculate the moles of gas collected over water in a 600 mL container at 756 mmHg pressure at 20°C given that the vapor pressure of water is 2.3388 kPa at 20°C.

1.8 Calculate the uncertainty in the mole answer in Problem 1.7 if the uncertainty is 5 mL in volume, 1 mmHg in pressure, and 0.3°C in temperature. Give the value in % and in moles.

1.9 Calculate the volume of a 9.39 in. diameter basketball in liters. (See *Introduction: Math and Physics Review*.)

1.10 Using $PV = nRT$, $n = 1$, calculate $\alpha = \left(\frac{1}{V}\right)\left(\frac{\partial V}{\partial T}\right)_p$ and $\beta = \left(\frac{-1}{V}\right)\left(\frac{\partial V}{\partial P}\right)_T$. (*Introduction: Math and Physics Review*.)

REFERENCES

1. D. R. Lide, Ed., *CRC Handbook of Chemistry and Physics*, 87th Edn., CRC Press, Boca Raton, FL, 2006–2007, pp. 6–34.
2. R. Livingston, *Physico Chemical Experiments*, 3rd Edn., Macmillan Co., New York, 1957, pp. 22ff.
3. G. B. Thomas, *Calculus and Analytic Geometry*, 2nd Edn., Addison-Wesley, Reading, MA, 1956, p. 226.

2 Viscosity of Laminar Flow

INTRODUCTION

Continuing our appeal to phenomenological derivations, we come to an experimental technique that is very simple to use and has a clean calculus derivation. Despite the simplicity of the measurement of viscosity, it is very useful in several areas of chemistry (polymers), aerodynamics (airplane wing design), hydrodynamics (boat hull design), pharmaceutical delivery (oral delivery in syrups), biophysics (blood flow), and material science (polymers). We are mainly motivated by a need to support the revolutionary Boltzmann's kinetic molecular theory of gases (KMTG; in Chapter 3) with some experimental method. The Boltzmann KMTG can be treated in a cyclical set of self-fulfilling equations (perhaps because it is true!), but a skeptic would require some sort of measurement, mainly because it assumes the existence of very small atoms/molecules never seen individually. Even today there are only a few "pictures" of fat Au atoms on surfaces and x-ray diffraction structures of molecules in crystals. The preponderance of evidence for the size and structure of molecules is firm but indirect. Here, we want to discuss Poiseuille's (Pwaz-e-ay's) law of viscosity for laminar flow [1,2] because it offers several useful applications, but primarily it will be a way to verify Boltzmann's KMTG.

Another modern application is an extension of Einstein's thesis work [1,3,4] on viscous flow of sugar-water solutions applied to the determination of the molecular weight of giant molecules now called polymers. Polymers are typically the result of organic compounds which have both "head" and "tail" reactive groups that can react repeatedly to form large chains, sheets, or bulk materials, which are really a single large molecule. Following WWII, there was a worldwide surge of research in how to make, characterize, and develop application of polymers. This effort continues today in a now mature branch of chemical research, and some amazing properties of specialized polymers have been developed such as high-temperature stability approaching that of metals (polyimides), polymers that change color with temperature, and dry lubricants such as perfluorocarbons. Let us not forget the ubiquitous polystyrene coffee cup. When new polymeric materials are developed, one of the foremost characteristics is the *intrinsic viscosity* of the polymer, and this is measured in a simple way with a pipette, a viscometer, and a stopwatch!

A third motivation is that physical chemistry enters into some aspects of biomedical science, and blood viscosity is a minor diagnostic parameter related to blood-thinning treatment of stroke prevention. Poiseuille's law for laminar flow is a beautiful example of the clean application of calculus to a phenomenological equation which supports Boltzmann's KMTG and is an important method used in polymer science, but we have searched out some biological applications as well. Here, we want to give a foundation to the experimental ideal of laminar flow of fluids, which can be modeled using calculus incremental layers sliding over one another (Figure 2.1). We will see that once we can relate gas viscosity to KMTG, we gain a number of important concepts related to gas-phase chemical reactions, such as binary collision number and the mean free path. We just need some physical data to tie the theory to laboratory reality!

Consider a model of two parallel sheets with one sliding over the other. Common sense tells us that there is some sort of "friction" opposing the sliding motion. *Viscosity is a drag*, literally!

We can develop the idea of a laminar "coefficient of viscosity" from common experience. First, the force required to move one sheet over the other is proportional to the area A of contact between the sheets. Second, more force will be required to move the upper sheet faster. Third, the actual contact between the sheets depends inversely on the contact distance between the sheets since all

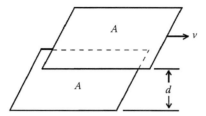

FIGURE 2.1   Sliding layers to derive Poiseuille's law.

materials will have some sort of rough hills and valleys on the surfaces. We can also expect that adding weight to the upper sheet will increase the friction, but really that only makes the surfaces squeeze together more tightly and we have already included the inverse dependence on mean distance d between the layers. We can anticipate that the friction will depend on the applied load on the top sheet, but that will not affect the unit analysis of the friction:

$$f = ma = g\,cm/s^2 = \eta\left(\frac{Av}{d}\right) = \eta\,\frac{cm^2\left(\frac{cm}{s}\right)}{cm} = \eta\,\frac{cm^2}{s}.$$

This leads to the phenomenological units of the coefficient of viscosity in the cgs system as

$$\eta = g/cm\,s \equiv 1\,poise.$$

While this unit is easy to derive using reasoning from everyday experience, the poise (pwaz) is an ancient unit and viscosity is now measured in (pascal seconds), so that 1 poise = 0.1 Pa s in SI units:

$$0.1\,Pa\,s = [(10\,dyne/cm^2)/10]\,s = 1\left(\frac{g\,cm/s^2}{cm^2}\right)s = 1\left(\frac{g}{cm\,s}\right) = 1\,P.$$

More properly called the Hagen–Poiseuille law, it was developed independently by Gotthilf Heinrich Ludwig Hagen (1797–1884) and Jean Louis Marie Poiseuille. Poiseuille's law was experimentally derived in 1838 and formulated and published in 1840 and 1846 by Jean Louis Marie Poiseuille (1797–1869). Hagen also carried out experiments in 1839. While there are a number of derivations, we follow a simple one here from *Physical Chemistry* by Castellan [5].

Consider a pipe with some fluid forced through it by a pressure difference $(P_1 - P_2)$ where $P_1 > P_2$. Although we will eventually consider the phenomenon from a molecular view, we stress the power of calculus here to represent a macroscopic effect in terms of infinitesimals. Assume the fluid (gas, liquid, or slurry) is flowing down, but there is some sort of friction between thin layers as cylinders sliding within each other like concentric rings of pipe or tree rings (Figure 2.2). We can see that in the limit as one goes out to the outer wall, the velocity of the layers must be zero while the velocity is greatest in the center of the tube. Note the total area of the friction is the surface of the outer shell of a cylinder whose radius varies from zero at the center to R at the wall of the pipe, and the variation of the velocity can be described as a derivative (dv/dr), so we can write the frictional force on any given cylinder as

$$f = \eta A\left(\frac{dv}{dr}\right).$$

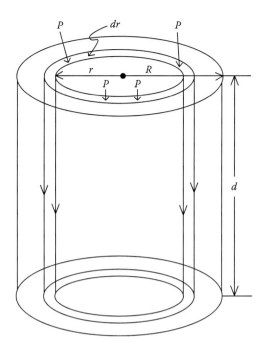

FIGURE 2.2   Calculus diagram of "sliding sleeves" of fluid flowing in a cylinder.

Note this maintains the units described above for the sliding layers since derivative (dv/dr) has units of velocity/length and the total expression is a drag force. However, we need a sign reversal to account for the fact that the velocity decreases as the radius increases. (Let us use $d=l$ to avoid conflict with d/dr.):

$$f = \eta A\left(\frac{-dv}{dr}\right) = \eta(2\pi rl)\left(\frac{-dv}{dr}\right).$$

What is the force driving the fluid? We know from above that it is a *difference in pressure* but for the time being let us just call $P = (P_1 - P_2)$ and note that a pressure is a (force/area), so we need to multiply the pressure by the cross-sectional area of the tube:

$$f = \eta(2\pi rl)\left(\frac{-dv}{dr}\right) = P(\pi r^2).$$

By canceling πr and splitting the derivative into separate differentials, we obtain the variation of velocity in terms of the radius as

$$dv = \left(\frac{-P}{2\eta l}\right)r\,dr,$$

which can be integrated as $\displaystyle\int_0^v dv = \int_R^r \left(\frac{-P}{2\eta l}\right)r\,dr = \left(\frac{-P}{2\eta l}\right)\int_R^r r\,dr.$

Here, we use $v=0$ at $r=R$ because the velocity is zero at the outer wall of the pipe. With these limits, the integrals are easy to do and we get

$$v(r) = \left(\frac{-P}{2\eta l}\right)\left(\frac{r^2 - R^2}{2}\right) = \left(\frac{P}{4\eta l}\right)(R^2 - r^2).$$

In itself this is only a formula that tells us the velocity of the fluid at a specific radius r in terms of the pressure P, the viscosity coefficient η, the length of the tube l, and the outer radius R of the pipe. In order to obtain a formula that can be used to measure something, we ask how much volume flows through the tube in a given time, the bulk flow rate as (volume/time) by integrating the function of r from zero to the outer radius R:

$$\left(\frac{V}{t}\right) = \int_0^R (v)(2\pi r\, dr) = \int_0^R \frac{P(R^2 - r^2)}{4\eta l}(2\pi r\, dr) = \int_0^R \left[\frac{P\pi R^2 r}{2\eta l} - \frac{P\pi r^3}{2\eta l}\right] dr.$$

So, we have

$$\left(\frac{V}{t}\right) = \frac{\pi P R^2}{2\eta l}\left[\frac{r^2}{2} - \frac{r^4}{4R^2}\right]_0^R = \frac{\pi P R^2}{2\eta l}\left[\frac{R^2}{2} - \frac{R^2}{4}\right] = \frac{\pi P R^4}{8 l \eta}.$$

While this derivation is in two parts and requires some careful thinking, we have shown the power of using differential calculus to obtain a relatively simple final formula. The method illustrates the ideas of using small differential quantities with physical reasoning and then integrating to obtain the macroscopic formula. Note especially that the bulk rate of flow depends on the fourth power of the radius of the tube! Doubling the radius of the tube will increase the flow by a factor of 16! Finally, we obtain a way to measure the viscosity coefficient of a fluid by measuring the time it takes for a fixed volume to flow through a pipe of known dimensions:

$$\eta = \frac{\pi P R^4 t}{8 l V}.$$

We are supposed to be learning an essential form of physical chemistry, and once you are aware of laminar viscosity you may notice it in a number of applications. Note that this leads to a simple relationship if we have a standard fluid/liquid with a standardized viscosity such as water. Then, the time for a standard liquid can be compared with the flow time for the same amount of an unknown fluid in the same device to obtain the viscosity of the unknown liquid: $\dfrac{\eta_x}{\eta_{std}} = \left[\dfrac{\left(\frac{\pi P R^4}{8 l V}\right) t_x}{\left(\frac{\pi P R^4}{8 l V}\right) t_{std}}\right] = \dfrac{t_x}{t_{std}}$, so

that $\eta_x = \eta_{std}\left(\dfrac{t_x}{t_{std}}\right)$.

Pure water can be used as a standard for viscosity measurements, although a correction should be applied for changes in density with the temperature:

$$\eta_x = \eta_{std}\left(\frac{\rho_x t_x}{\rho_{std} t_{std}}\right).$$

For H_2O, $\rho \cong 1\,\mathrm{g/cm}^3$.

MEASUREMENT OF VISCOSITY

There are several types of devices which measure viscosity but we will only show the most common type here, the Ostwald viscometer (Figure 2.3).

The Ostwald type can be purchased from a variety of supply houses with different diameter bores. Small bores are used for a low-viscosity range near that of water and larger bores for more viscous liquids. In order to measure the viscosity coefficient of liquids such as motor oil or molasses,

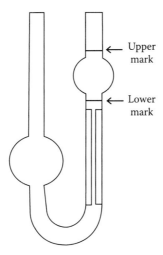

FIGURE 2.3   A typical Ostwald viscometer. (From Gohel, M., Parikh, R., Popat, A., Mohapatra, A., Barot, B., Patel, C., Joshi, H., Sarvaiya, K., Baldaniya, L., Mistry, P., Parejiya, P., Parmar, R., Nagori, S., and Patel, T., Pharmaceutical suspensions: A review, Pharmainfo.net, http://www.pharmainfo.net/free-books/pharmaceutical-suspensionsa-review).

larger bore viscometers are necessary in order to obtain convenient run times (private communication from Prof. MacKnight, Figure 2.4). One merely uses a pipette to standardize the amount of the liquid used and it is drained into a clean viscometer. Then a suction bulb is used to pull the liquid into the reservoir bulb at the top of one arm of the tube and the time it takes for the reservoir to drain past the lower mark on that arm is recorded. Multiple runs can be averaged to improve precision for liquids of low viscosity but for very viscous liquids it is not unusual for run times over 150 s to be reproducible to within 0.1 s, so precision is very good for liquids of higher viscosity (Table 2.1) [6].

FIGURE 2.4   William J. MacKnight is an American polymer scientist at the University of Massachusetts at Amherst. He is a member of the National Academy of Engineering and has received the Ford Prize in High Polymer Physics, the American Chemical Society Award in Polymer Chemistry (Mobil Award), the Distinguished Service Award in Advancement of Polymer Science from the Society of Polymer Science, Japan, and the Herman F. Mark Award from the Division of Polymer Chemistry of the American Chemical Society.

TABLE 2.1
Viscosity of Water at Varied Temperatures

Temperature, °C	Viscosity Coefficient, μPa s	ρ, kg/m^3
0.01	1791.1	999.84
10	1305.9	999.70
20	1001.6	998.21
25	890.02	997.05
30	797.22	995.65
40	653.73	992.22
50	547.52	988.03
60	466.03	983.20
70	403.55	977.76
80	354.05	971.79
90	314.17	965.31
99.606	282.75	958.63
100	12.234	0.58967

Source: Lide, D.R., *CRC Handbook of Chemistry and Physics*, 90th Edn., CRC Press, Boca Raton, FL, 2009–2010. pp. 6-1.

VISCOSITY OF BLOOD

Viscosity can be measured for blood as an auxiliary diagnostic test for diseases in which there are abnormally high levels of proteins or to monitor the effect of blood-thinning agents as a treatment for stroke prevention. The normal range of viscosity of human blood is from 0.99 to 1.55 centipoise and the units are in poise. Blood with viscosity higher than 4.0 centipoise (1 poise = 0.1 Pa s so 0.04 poise = 0.004 Pa s) is considered abnormal and may signal potential circulatory problems. Now for health science students, it may be of interest to consider the average blood flow in an adult human or for forensic students to know possible blood flow in a given time. One complication is that it is known that arteries have elastic walls and flex (bulge) during the high pressure pulse of blood flow, but we will approximate an aorta as a pipe with a fixed diameter. Another substantial problem is that with a pulsating heartbeat, the pressure is not constant; part of the time, there is a higher pressure pulse (beat) while between beats the pressure can be much lower. We can solve this problem with an adjustable factor we can call a "duty factor," which represents the fraction of the time the pressure of the heartbeat pulse is high. Another interesting consideration is that medical measurements still use the high and low pressures measured for blood pressure with an arm cuff in the units of mmHg! Suppose we assume the duty factor is 0.1 as representing the pressure spike of the human heartbeat and then correct the duty factor to a measured value.

Given $\eta = 0.013$ poise, an aorta 6 in. long with an inner diameter of 1/4 in. and blood pressure of 140/80 using a duty cycle of 0.1, calculate the volume of blood flow in gallons/minute:

$$\left(\frac{V}{t}\right) = \frac{\pi\left(\frac{140-80}{760}\right)\text{atm}\left(1.01325 \times 10^6\ \frac{\text{dyne}}{\text{cm}^2\,\text{atm}}\right)\left(\frac{2.54\,\text{cm}}{8}\right)^4 (0.1)}{8(6\,\text{in.})(2.54\,\text{cm/in.})(0.013\,\text{g/cm s})}$$
$$\times \left(\frac{1\,\text{gal}}{4\,\text{qt}}\right)\left(\frac{1\,\text{qt}}{946\,\text{cm}^3}\right)\left(\frac{60\,\text{s}}{1\,\text{min}}\right).$$

So, we find $\left(\dfrac{V}{t}\right) = 2.5548\,\text{gal/min}$. This is too high even with the duty factor of 0.1, so let us standardize the effective duty factor for an average adult human of about 4900 mL/min. Then we have

$$\frac{4900\,\text{mL}}{(4\,\text{qt/gal})(946\,\text{mL/qt})} \cong 1.29493\,\text{gal/min}.$$

Thus, if we wish to use the formula from a rigid pipe for a flexible aortic wall we need to adjust the duty factor:

$$\left(\frac{\text{Duty}}{\text{Cycle}}\right) = (2.5548\,\text{gal/min})\left(\frac{x}{0.1}\right) = 1.29493\,\text{gal/min},$$

so, $x \cong 0.0507$ and we have an effective formula which takes into account the elastic wall of a large artery like an aorta and the pulse nature of the heartbeat as

$$\left(\frac{V}{t}\right) = \frac{\pi\left(\dfrac{P_h - P_l}{760}\right)(1.01325 \times 10^6\,\text{dyne/cm}^2\,\text{atm})\left(\dfrac{2.54\,\text{cm}}{R\,\text{in.}}\right)^4 (0.0507)\left(\dfrac{1\,\text{gal}}{3784\,\text{cm}^3}\right)\left(\dfrac{60\,\text{s}}{\text{min}}\right)}{8(L\,\text{in.})(2.54\,\text{cm/in.})(\eta\,\text{g/cm s})}.$$

We note that in some cases, we could rearrange the equation to calculate the effective radius of the pipe and use the fourth root (square root of the square root) of the rearranged formula if we know the (V/t) bulk flow rate. This formula also teaches us a lot about using different units.

STAUDINGER'S RULE FOR POLYMER MOLECULAR WEIGHT

Although Albert Einstein is most well known for his work in the theory of relativity and for his analysis of the photoelectron effect, he also developed a foundation [3,4] for the theory of solution viscosity. His initial work was on solutions of colloidal spheres and sugar solutions and that work was limited in application. However, as a result of plastics and synthetic rubber being developed during WWII, the field of Polymer Science emerged with great significance to chemical industry.

While chemists developed new synthetic methods for formation of polymers, the inevitable question arose as to the values of molecular weights. The question is complicated by the fact that often the products of polymerizations are "polydisperse," that is, there is a mixture of various molecular weights (n-mers) of similar compounds after the reaction. We can only give a glimpse of polymer science here, but the measurement of viscosity is now a standard technique in determining average molecular weight of large polymer molecules.

Initially, the application of viscosity measurements to polymer solutions extended the relationships derived by Einstein for colloid solutions. While colloids might be assumed to be roughly spherical, polymer molecules can be flexible, rod-like, or plate-like, so adjustments had to be made. Einstein [4] defined some useful terms. Let η_0 be the viscosity of the solvent alone and η be the viscosity of the solution in question. As the solution is diluted, the viscosity will approach the η_0 value, but at other concentrations we can define the relative viscosity as $\eta_r \equiv \left(\dfrac{\eta}{\eta_0}\right)$. A further quantity was defined as the "specific viscosity" as the amount by which the viscosity of a given solution differed from that of the solvent as $\eta_{sp} \equiv \eta_r - 1$. Finally, yet another quantity was defined as the "intrinsic viscosity," which has an interesting graphic property and is believed to be an intrinsic property of the polymer solute. In Einstein's original work [3,4], the intrinsic viscosity for hard spheres is 2.5, but we expect lower values for quasi-linear polymers. Theoretically, the intrinsic viscosity is a pure number but it is sometimes reported as a reciprocal concentration such as deciliters/gram due to the denominator of the definition:

$$[\eta] = \left[\frac{\eta_{sp}}{c}\right]_{c \to 0} \quad \text{and} \quad [\eta] = \left(\frac{\ln \eta_r}{c}\right)_{c \to 0}.$$

In 1930 Staudinger [8] proposed to adapt Einstein's formalism to solutes of polymers, which may or may not be spherical, even rod-like, or plate-like in a simple formula known as Staudinger's rule:

$$[\eta] = KM.$$

The interesting thing about this technique of relating viscosity to molecular weight is that if one can measure the viscosity of various concentrations of polymer in the solvent and plot, both the values of $[\eta] = \left[\frac{\eta_{sp}}{c}\right]$ and $[\eta] = \left(\frac{\ln \eta_r}{c}\right)_{c \to 0}$ on the y-axis of a graph and the concentration $c = $ grams/100 ml on the x-axis (100 mL $= 1$ dL $= 0.1$ L) the two lines should/will meet at the same value of the intrinsic viscosity $[\eta]$. Thus, both the methods of plotting the graph yield the same intrinsic viscosity value. Sometimes other units are reported such as $\left(\frac{g}{dL}\right)\left(\frac{10}{10}\right) = \left(\frac{10\,g}{L}\right)\left(\frac{1000}{1000}\right) = \left(\frac{10\,kg}{m^3}\right)$.

This type of work requires careful laboratory technique, but it is very satisfying to see both lines have the same intercept. Often other types of viscometers are used for this work but the Ostwald viscometer can be used for dilute solutions. Further work by Staudinger and his associates was carried out to find the value of K for various types of polymers, and the relationship was later refined to use two parameters (K, a) as in

$$[\eta] = KM^a.$$

In each case, there had to be calibrations of the molecular weight using other techniques for absolute molecular weights such as melting point methods, osmotic pressure, and light scattering, but the ease of using the viscosity measurements then allows the determination of the molecular weight of an unknown. Today, this is a standard laboratory procedure in industries where polymer properties are measured. Thus, while "viscosity is a drag," its measurement is of great practical importance in industry and of some use as a diagnosis technique in hematology, the study and science of blood.

Example

Castellan [5] gives three data points for polystyrene dissolved in benzene at 25°C as $[C(kg/m^3), \eta(mPa\,s)]$: (21.4, 1.35), (10.7, 0.932), and (5.35, 0.757). Plot these data using both definitions of the intrinsic viscosity and extrapolate to zero concentration. η_0 for benzene is 0.606 mPa s (Figure 2.5). The two intercepts should be close but take the average of the two intercepts as the best value of the intrinsic viscosity. Castellan suggests using the expression $[\eta] = KM^a$ with $K = 1.71 \times 10^{-3}\,m^3/kg$ and $a = 0.74$ (slightly different from the 30°C data in Table 2.2) along with the viscosity to calculate M, the effective molecular weight of this sample:

$$[\eta] = KM^a, \quad \ln([\eta]) = \ln K + a \ln M, \quad \ln M = \frac{\ln([\eta]) - \ln K}{a} = \left(\frac{1}{a}\right)\ln\left(\frac{[\eta]}{K}\right) = \ln\left(\frac{[\eta]}{K}\right)^{\left(\frac{1}{a}\right)},$$

so that $M = \left(\frac{[\eta]}{K}\right)^{\left(\frac{1}{a}\right)}$, now set up tables of $[\eta] = \left[\frac{\eta_{sp}}{c}\right]$ and $[\eta] = \left(\frac{\ln \eta_r}{c}\right)_{c \to 0}$ for $[C(kg/m^3), \eta\,(mPa\,s)]$: (21.4, 1.35), (10.7, 0.932), and (5.35, 0.757).

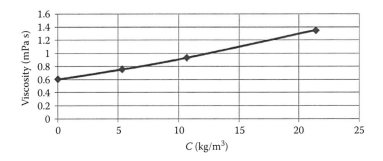

FIGURE 2.5 Raw data of viscosity measurement of polystyrene in benzene with the viscosity of the pure benzene solvent as 0.606 mPa s. Note slight nonlinearity.

TABLE 2.2
Staudinger Constants for Selected Polymer Molecular Weights from Intrinsic Viscosity

Polymer	Solvent	°C	Molecular Weight Range	$K \times 10^4$	a	Reference
Cellulose acetate	Acetone	25	11,000–130,000	0.19	1.03	[7]
Nylon	90% Formic acid	25	5,000–25,000	11.0	0.72	[7]
Polystyrene	Benzene	30	10,000–600,000	1.71	0.72	[7]
Neoprene	Toluene	25	40,000–1,500,000	5.0	0.615	[7]

Note that the K values are in (g/dm) here.

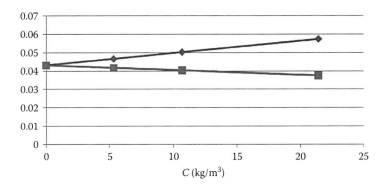

FIGURE 2.6 Plot of (η_{sp}/c) (small diamond points) and $(\ln \eta_r/c)$ (large square points) for polystyrene in benzene. This is not the raw data but is recalculated from the measured viscosity in mPa s.

Use $\eta_r \equiv \left(\dfrac{\eta}{\eta_0}\right)$ and $\eta_{sp} \equiv \eta_r - 1$ to set up tables versus the concentration. Use $\eta_0 = 0.606$:

$m = $ slope for $(\eta_{sp}/c) = (0.05737 - 0.04657)/(21.4 - 5.35) = 6.7290\mathrm{E}{-4}$,
$m = $ slope for $= (\ln \eta_r/c) = (0.03743 - 0.04156)/(21.4 - 5.35) = -2.5732\mathrm{E}{-4}$.

Without doing the graph, we can ask "what is y when $x = 21.4$," within the significant figure of the data (Figure 2.6).

TABLE 2.3
Raw Viscosity Data for Polystyrene in Benzene
with Calculated Limiting Intercept

C	η	η_r	η_{sp}	(η_{sp}/c)	$(\ln \eta_r/c)$
21.4	1.35	2.228	1.228	0.05737	0.03743
10.7	0.932	1.538	0.538	0.05028	0.04023
5.35	0.757	1.249	0.249	0.04657	0.04156

For (η_{sp}/c), we have $0.05737 = (21.4)(6.7290\text{E}{-}4) + b$, so $b = 0.04297$.
For $(\ln \eta_r/c)$, we have $0.03743 = (21.4)(-2.5732\text{E}{-}4) + b$, so $b = 0.04294$.

The average of these two intercepts is 0.042955 or 0.04296 with four significant figures: $[\eta] \cong 0.04296$. On the graph, one line slopes up and the other line slopes down, but both have the same limiting intercept (Table 2.3). Then, we can calculate the limiting molecular weight:

$$M = \left(\frac{[\eta]}{K}\right)^{\left(\frac{1}{a}\right)} = \left(\frac{0.04296}{1.71 \times 10^{-3}\,\text{m}^3/\text{kg}}\right)^{(1.351)} = 77.8919 \cong 78\,\text{kg/m}^3.$$

This seemingly strange molecular weight tells us that the limiting case represents a tangled web of polymer that may be hopelessly knotted with itself and/or actually cross-linked chemically. In the case of synthetic rubber tires, cross-linking may be so great that one can say the whole tire is one molecule! Even so, the intrinsic viscosity is used by polymer scientists to compare different polymers as a measurable property. Although this is but one example, this method is a mainstay procedure in polymer chemistry. The discussion here should prepare you for a very instructive experiment in the text by Shoemaker, Garland, and Nibler [9] where a similar treatment is applied to polyvinyl alcohol (experiment no. 28). While the units of ($78\,\text{kg/m}^3$) are unfamiliar to us as a type of molecular weight, it may be useful to compare the same units for the familiar substance of water. The density of pure water is 997.05 (kg/m^3).

SUMMARY

This short chapter has been a further exercise in the merger of various practical units from several fields to gain experience in unit conversions and at the same time form a foundation for the notion of the viscosity of laminar flow. We should emphasize that the equations we have derived only apply to laminar flow, the kind of flow that occurs when "still waters run deep" over a deep place in a river rather than shallow turbulent flow over rocks. Turbulent flow is difficult to treat mathematically but as we have seen, laminar flow can be treated using the calculus idea of thin layers. The science of laminar flow extends to aerodynamics and is important in the study of airflow over the surface of an airplane wing as well as the entire surface of the aircraft. In addition, the same equations can be applied at lower speed and higher density for the design of boat hulls to maintain laminar flow of water over the hull. Thus, a basic understanding of laminar flow at the macroscopic level has wide applications and we are now prepared to see how gas viscosity can provide experimental verification of the Boltzmann KMTG in the next chapter. The example of a worked problem for the intrinsic viscosity of polystyrene is included because someday viscosity measurement may be a routine task in your job as a chemical scientist. Of course, the "calculus nugget" in this chapter is the derivation of Poiseuille's law, and while most teachers would not expect a student to reproduce that derivation on a test, it has been done by a few students seeking extra credit, so the sequence of steps in the derivation can be "learned." For a few occasional chemical engineering students in the class, the main message of Poiseuille's law is that the flow rate through a pipe is proportional to the fourth power of the internal radius of the pipe, a principle well worth knowing when dealing with "plumbing."

PROBLEMS

2.1 Calculate the bulk volume flow of blood with viscosity 0.02 poise through a 6 inch long aorta of inner diameter 3/8 in. due to a blood pressure of 125/80 mmHg in gallons/min assuming that the pressure is constant (duty factor = 1). Then, multiply the answer by a duty factor of 0.05 to correct for the duration of the heartbeat pulse (and the fact that we are treating the aorta wall as rigid).

2.2 Water can be used as a standard to measure the viscosity coefficient η of an unknown liquid if the temperature is held constant, the same volume of liquid and the same apparatus is used. Given that $\eta = 0.010038$ poise for water at 20°C, calculate the viscosity of an unknown liquid at 20°C if 10 mL of distilled water took 17 s for 10 mL to flow between two marks in an Ostwald (J-tube) viscometer and the unknown liquid took 19 s for 10 mL to flow under the same conditions.

2.3 To show how the R^4 dependence of the Poiseuille law affects flow rate, calculate the bulk flow through a 12 in. long fire hose nozzle with an inner diameter of 2 in. delivering water with $\eta = 0.01 = 0.01$ poise from a pressure of 100 psi and exiting to a pressure of 14.7 psi. Assume there is a pump which can provide the necessary volume and give the answer in gallons/min.

2.4 Calculate the viscosity coefficient η in poise and in Pa s, if 5 gal/min flow (laminar) through a 6 in. long tube 1 in. inner diameter due to a pressure of 18 psi and exit at 14.7 psi.

2.5 Estimate the inner diameter of Lance Armstrong's aorta assuming it is 7 in. long, his blood pressure is 140/60 mmHg and that his heart pumps (as rumored) 9 gal/min. Use $\eta = 0.02 = 0.02$ poise and integrated pulse factor = 0.05.

REFERENCES

1. Neuenschwander, D. E., Albert Einstein's Dissertation, http://www.sigmapisigma.org/radiations/2005/ecp_spring05.pdf
2. Pfitzner, J., Poiseuille and his law. *Anaesthesia* **31**, 273 (1976).
3. Einstein, A., Eine neue Bestimmung der Moleküldimensionen. *Ann. Phys.*, **19**, 289 (1906).
4. Einstein, A., Berichtigung zu meiner Arbeit: Eine neue Bestimmung der Moleküldimensionen. *Ann. Phys.*, **34**, 591 (1911).
5. Castellan, G. W., *Physical Chemistry*, 3rd Edn., Addison-Wesley, London, 1983, p. 942.
6. Lide, D. R., *CRC Handbook of Chemistry and Physics*, 90th Edn., CRC Press, Boca Raton, FL, 2009–2010, pp. 6-1.
7. Mark, H. and A. V. Tobolsky, *Physical Chemistry of High Polymeric Systems*, Interscience, New York, 1950, p. 290.
8. Staudinger, H. and R. Nodzu, Über hochpolymere Verbindungen, 36. Mitteil.: Viscositäts-Untersuchungen an Paraffin-Lösungen. *Berichte* **63**, 721 (1930).
9. Shoemaker, D. P., C. W. Garland, and J. W. Nibler, *Experiments in Physical Chemistry*, 6th Edn., The McGraw Hill Book Co., Inc., New York, 1996.

3 The Kinetic Molecular Theory of Gases

INTRODUCTION

We remind ourselves we are trying to present the essential aspects of physical chemistry and we consider this one of the most essential topics. In our treatment of the van der Waals gas, we have already mentioned the ideas of the collisions of small atoms, which have a lot of space between them as in Dalton's law. Here, we go into further detail regarding the behavior of gas molecules using the ideas of Ludwig Boltzmann (1844–1906), who was one of the intellectual giants of the late nineteenth century and whose "Boltzmann principle" of energy distribution is one of the pillars of modern science. The breakthrough here was due mainly to Boltzmann's PhD thesis on the theory of gases. Here, we will first review the freshman chemistry derivation of part of kinetic molecular theory of gases (KMTG) and then introduce Boltzmann's amazing energy principle.

KINETIC ASSUMPTIONS OF THE THEORY OF GASES

1. A gas is made up of a large number of particles (molecules or atoms) that are small in comparison with both the distance between them and the size of the container.
2. The molecules/atoms are in continuous *random* motion.
3. Collisions between the molecules/atoms themselves and between the molecules/atoms and the walls of the container are *perfectly elastic*.

Let us consider the idea that gas pressure is caused by impacts of atoms/molecules with the wall of a container (or the diaphragm of a pressure gauge). We know a gas will fill any shaped container, but to make the derivation simpler, we assume a cubical container of dimension $L \times L \times L$ where each side is of length L (Figure 3.1). Thus, each inner face of the container has area $A = L \times L$. Looking ahead to the idea that pressure is force/area, we put just one atom in an empty cubical box and analyze the force on one face of the box. Since force is a change in momentum, let us consider the left face of the box in the y–z plane and assume the atom is moving only in the negative x-direction. This simplifies the elastic bounce back into the positive x-coordinate, although in general, the assumed random direction of a molecule would produce a series of random paths throughout the 3D volume of the box. Our thinking is also constrained by a convention in thermodynamics that "*change = after − before*," so we arbitrarily choose the molecule initially traveling in the negative x-direction so that $(\vec{v}_{\text{after}} - \vec{v}_{\text{before}})$ is positive in the derivation. Similarly, the box is cubical to make $V = L^3$, but the container could be of any shape. Since we assumed the atom is in continuous motion and all collisions are elastic, it will bounce off the wall and go in the reverse direction until it reaches the other wall and bounces back again and so on. Since a change in momentum is a force, the collision with the wall causes the (force/area) pressure. Since the collisions are perfectly elastic, the atom will bounce back and forth rapidly. "Perfectly elastic" means that no momentum is lost in the collision, which is an approximation since a hot gas will cool and lose energy, but it is a good approximation over a short period of time. Thus, for convenience, we show the particle moving in the negative x-direction and then reversing:

$$\left.\frac{dp}{dt}\right|_A = (mv_x)_{\text{after}} - (-mv_x)_{\text{before}} = 2mv_x.$$

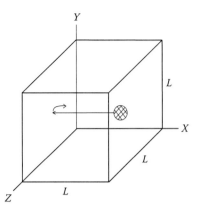

FIGURE 3.1 A single particle in a cubical box undergoing elastic collision with a wall.

Even though collisions will occur on more than one face, we can calculate how long it takes before another collision will occur at this face in terms of the velocity and the distance. This is based on the familiar equation (distance) = (rate)(time), so that the time is the distance/rate in the denominator, which flips up as the inverse, $\dfrac{\#\text{collisions}}{\text{time}} = \dfrac{v_x}{2L}$. Then, the total number of collisions per unit time on area A yields force/area = pressure. So far, so good. The next step

$$\left(\frac{dp/dt}{A}\right) = \left(\frac{v_x}{2L}\right)\left(\frac{2mv_x}{L^2}\right) = \frac{mv_x^2}{L^3} = \frac{mv_x^2}{V} = \frac{\text{force}}{\text{area}}$$

is to multiply by the number of atoms to obtain the pressure as

$$N\left(\frac{f_x}{L^2}\right) = P = \frac{Nmv_x^2}{V}.$$

The following step is an approximation usually used in a freshman presentation in that we note that the motion of the atom is *random* and the velocity is a vector with three components in general. The motion is random so all three components are favored equally, but at the moment we only need v_x^2. If we use the dot product, we can get the square of the velocity containing all three equally weighted components as

$$v^2 = (v_x\hat{i} + v_y\hat{j} + v_z\hat{k}) \cdot (v_x\hat{i} + v_y\hat{j} + v_z\hat{k}) = v_x^2 + v_y^2 + v_z^2.$$

Then, we say that if the motion is really random $v_x^2 = v_y^2 = v_z^2$ and implies that $v^2 = 3v_x^2$ and then, we finally have the pressure as $P = \dfrac{Nm(v^2/3)}{V}$ or interestingly $PV = (1/3)Nmv^2$.

This is a lot like the ideal gas law except for the right side of the equation. Next, we recall from physics that the average kinetic energy of an atom/molecule can be written as ke $= mv^2/2$, so we can write $PV = \left(\dfrac{2}{3}\right)N\left(\dfrac{mv^2}{2}\right) = \dfrac{2}{3}N\,(\text{ke})$. We can also let $N = nN_{Av}$, so the arbitrary number of atoms/molecules, N, can be rewritten as a number of moles n times Avogadro's number N_{Av}. Then, if we define a molar kinetic energy as $N_{Av}\,(\text{ke}) = (\text{KE})$, we have the molar expression as $PV = \left(\dfrac{2}{3}\right)n(\text{KE}) = \left(\dfrac{2}{3}\right)n\left(\dfrac{N_{Av}m\bar{v}^2}{2}\right) = \left(\dfrac{2}{3}\right)n\left(\dfrac{M\bar{v}^2}{2}\right)$. Note, we have introduced an average square of the velocity as $\bar{v}^2$ and the molar mass as M. At this point, we make an assumption using the phenomenological ideal gas law, $PV = \left(\dfrac{2}{3}\right)n(\text{KE}) = nRT$. Thus, equating some experimental data to physical reasoning will be true if $\text{KE} = \dfrac{3RT}{2} = \dfrac{M\bar{v}^2}{2}$. So, we have carried out a path of reasoning "that is probably true" based on the ideal gas law, which we know is only true at low pressure and high temperature, but let us see where this assumption takes us. First, solving this equation for the velocity gives the result

$$v = \sqrt{\frac{3RT}{M}} = \sqrt{\bar{v}^2}.$$

This derivation is not very satisfying because we used a velocity that was not really averaged over all orientations and the result depends on the phenomenological ideal gas law. We also note that the resulting form of "v" is a scalar as the square root of a vector squared using the dot product and we call it "v_{rms}." It is good that we obtain a scalar, but what does it mean to have a "root-mean-square" speed? However, we can use it to estimate the speed as for N_2 gas at 25°C to get some idea of the KMTG velocities. Why not calculate this apparent speed in miles per hour (mph)? Note this is a diatomic molecule and also the formula has no dependence on pressure, just temperature dependence. We better check the units for R in this calculation.

$$v = \left(\sqrt{\frac{3(8.314 \times 10^7\,\text{g cm}^2/\text{s}^2/°\text{K mol})(298.15°\text{K})}{2 \times 14.007\,\text{g/mol}}}\right)\left(\frac{3600\,\text{s/h}}{1.6093 \times 10^5\,\text{cm/mile}}\right) = 1152.6\,\text{mph}.$$

Some things to note are that we need to use either the cgs or mks value of R, remember to double the atomic weight of nitrogen, add 273.15 to the °C, and recall that 1 mile $= 1.6093$ km. Let us do it again in mks.

$$v = \left(\sqrt{\frac{3(8.314\,\text{kg m}^2/\text{s}^2/°\text{K mol})(298.15°\text{K})}{2(0.014007 \times 10^{-3})\,\text{kg/mol}}}\right)\left(\frac{3600\,\text{s/h}}{1.6093 \times 10^3\,\text{m/mile}}\right) = 1152.6\,\text{mph}.$$

That is a very high speed, $v = (1153\,\text{miles/h})\left(\dfrac{5280\,\text{ft/mile}}{3600\,\text{s/h}}\right) = 1691\,\text{ft/s}$, which is faster than a small bore rifle bullet. The .22 LR Stinger is rated at 1435 ft/s. Two obvious questions arise. First, if the molecules are that fast, why is the speed of sound much less at about 1125 ft/s or 768 mph? The answer is that there are a lot of collisions between the gas atoms/molecules and some of the recoil trajectories have backward components, thus slowing the average speed. The second has to do with possible injury from "bullet molecules." Fortunately, the actual mass of the "atomic bullets" is less than 10^{-20} g, so it takes a lot of collisions to make the pressure variations we call "sound" that we sense with our ears.

WEIGHTED AVERAGING: A VERY IMPORTANT CONCEPT

At this point, we would like to proceed to apply the KMTG to experimentally measurable quantities, but we need a firmer foundation for the velocities and speeds of atoms/molecules in the gas phase. The velocity based on the phenomenological ideal gas law is suspect because we know it may not apply to high pressure and/or low temperature, so we need a more rigorous method. The concept/principle of *weighted averaging* occurs in kinetics, statistical thermodynamics, and in quantum mechanics, so we think this is more than just a "math interlude"; it is a unifying principle.

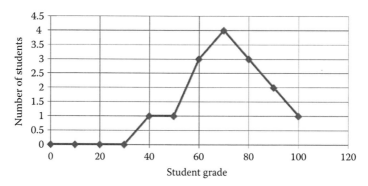

FIGURE 3.2 Hypothetical grade distribution in a class of 15 students.

In this author's experience, this concept needs to be identified in every place it occurs to help students understand what is being averaged. First, let us consider a simple example of weighted averaging. Suppose, we administer a midterm examination to a class of 15 students and record the grades on a graph from the data (Figure 3.2). Although the example is a discrete distribution, we have plotted the data so you can see a "distribution function" line, and we anticipate that if we had grades in increments of 1 point and a class of 250 freshmen, the graph would be a smoother "curve" but still based on a discrete set of points.

We now come to a key idea, which the student should make sure he or she understands since we will use it over and over in later applications. We introduce the symbol $\langle \ \rangle$ to denote an averaging process, in this case a "weighted average" as shown in Table 3.1.

$$\langle G \rangle = \frac{1(100) + 2(90) + 3(80) + 4(70) + 3(60) + 1(50) + 1(40)}{1 + 2 + 3 + 4 + 3 + 1 + 1} = \frac{\sum_i n_i G_i}{\sum_i n_i} = 71.33.$$

Note that the symbol $\sum_i$ indicates a discrete summation over specific values. Also make sure you note that we are weighting the G_i value by the number of times it occurs or is "weighted" in the summation. To gain perspective, we could look at the graph distribution and use the weighting of a given grade, say $n_i/15$ to estimate the *probability* that a student would get a certain grade G_i. So far that is sort of obvious, but the interesting point is that we have to divide "by the number of students in the class" and in effect this "normalizes" the process to the average grade for just "one average" grade of a hypothetical single student. This "normalizing" process will be a key idea in several applications in quantum chemistry as well as here for Boltzmann averaging. Next, we need to take a side trip to the amazing discovery of Boltzmann weighted averaging and ask the question

TABLE 3.1
Weighted Average Grade of Class

No. of Students	Grade
1	100
2	90
3	80
4	70
3	60
1	50
1	40
Weighted average	$(1070/15) = 71.33$

"What would the weighting factor be if we are averaging a sample of a mole (6.022×10^{23}) values for molecules and the range of values of some molecular property is (nearly) continuous?"

We believe that Boltzmann probability weighting is one of the most "essential" concepts in this whole text. We saw above that we can itemize the values of the thing we want to average, but we also need to weight that value and the weighting can be considered as a probability. Boltzmann may have considered a number of mathematical relationships before he "discovered" the formula, now given his name. The basic consideration is that for 3D quantities such as kinetic energy, one needs to merge the contributions in the various (x, y, z) directions in a way that will represent the total *probability*. This follows the common sense understanding of probabilities in that if an event is $1/3$ in x, $1/3$ in y, and $1/3$ in z, then the probability of the event simultaneously in x, y, and z is $1/27$. The idea that a probability *product must also add energies* leads to putting the energy expression in an exponent where a product will add exponents. The properties of the natural logarithm base "e" were known for over 100 years when Boltzmann wrote his PhD thesis, so he chose that base, realizing that separate probabilities in different coordinates would add in the exponent of a common base. We should emphasize that the weighting number n_i above is really the *probability* of the value being weighted when you divide by the number of values. Boltzmann knew that "energy runs down hill" so that higher energies are less probable. On the other hand, he also knew that higher temperatures favor higher energies. Add those factors to your thinking and perhaps you will come up with the same result as he did:

$$\text{Boltzmann probability} = e^{-\varepsilon/kT} = \exp\left(\frac{-\varepsilon}{kT}\right).$$

This amazing relationship came from Boltzmann's consideration of all the above factors. We do not actually know how Boltzmann thought of this relationship, and it was sheer genius at the time, but we need to appreciate the several aspects of this very important mathematical expression. If the temperature is constant, then a given probability is favored less as the energy ε increases and makes the exponential more negative. If the energy is constant, then a higher temperature makes it more probable, while a lower temperature makes it less probable. Finally, if we have different forms of energy such as $(\varepsilon_x, \varepsilon_y, \varepsilon_z)$, the probabilities will add in the exponent to a common base, namely, e. We should also appreciate that ε is a microscopic energy on the atomic scale and that $k = R/N_{Av}$ is the ideal gas constant divided by Avogadro's number N_{Av} so $k = R/N_{Av} = 1.38 \times 10^{-16}$ erg/$^{\circ}$K $= 1.38 \times 10^{-23}$ J/$^{\circ}$K.

Our goal here is to simplify some topics but still treat the full details of the difficult material. Instead of a cubical box, we can shift our attention to a spherical system and consider the motion of atoms/molecules in any direction with a scalar speed v. Velocity is a vector quantity, which requires both a scalar value and a direction. An automobile speedometer indicates only scalar speed and needs to be accompanied by a compass reading to determine the direction needed to specify the velocity of the automobile. Here, we can integrate over all angles (θ, ϕ) with the factor of (4π) shown in the Mathematics Review and only consider the scalar speed v. We can follow the example above of the average grade to set up the average kinetic energy of atoms/molecules in a spherical system. Here, we clearly see that the thing we are averaging is $(mv^2/2)$ and the weighting probability factor is the Boltzmann factor with $\varepsilon = (mv^2/2)$, assuming the only form of energy the particles have is kinetic energy with essentially no potential energy. Actually, there should be an expression of some form of potential energy when the atoms/molecules get very close, but we will see that most of the time they are very far apart. This is called the "hard sphere" approximation. Note that the denominator is needed to divide by all the possible values of the probability just as we divided the weighted grades by the number of students in the class. Later, we will see that the denominator is always the same in Boltzmann averaging these quantities over a spherical volume and can be inverted (flipped) as a factor in the numerator. Here, we want to emphasize the idea of the weighted average compared to the example of the class average grade.

$$\left\langle \frac{mv^2}{2} \right\rangle = \frac{4\pi \int_0^\infty e^{-\frac{mv^2}{2kT}} \left(\frac{mv^2}{2}\right) v^2 \, dv}{4\pi \int_0^\infty e^{-\frac{mv^2}{2kT}} v^2 \, dv} = \frac{4\pi \left(\frac{m}{2}\right) \int_0^\infty e^{-\frac{mv^2}{2kT}} v^4 \, dv}{4\pi \int_0^\infty e^{-\frac{mv^2}{2kT}} v^2 \, dv}.$$

At this point, we have to remind the reader of our agreement; the text will explain difficult topics in simple ways, but the student still has to use scratch paper and write the notes over. Thus, we come to two new integral forms, which need to be learned and used. These two formulas can be derived using integration by parts but that is very time consuming in some cases, so it is acceptable to look up various forms of integrals and use the results in tables. Some tables of integrals have thousands of cases, but in this text, we really only need to learn three definite integrals, the two below and

$$\int_0^\infty x^{2p} e^{-a^2 x^2} \, dx = \frac{1 \cdot 3 \cdot 5 \cdots (2p-1)\sqrt{\pi}}{2^{p+1} a^{(2p+1)}}, \quad \int_0^\infty x^{(2p+1)} e^{-a^2 x^2} \, dx = \frac{p!}{2a^{(2p+2)}}$$

one other, which will be useful in quantum chemistry, namely, the integral $\int_0^\infty x^n e^{-ax} \, dx = \frac{n!}{a^{(n+1)}}$.

We will use these formulas many times in this text so that you will become accustomed to using them. In the case at hand, we first note that the integrands have squared exponents to the base "e," but the power of x in the integrand can be odd or even. Note also that in the exponent of the Boltzmann factor, the value of a is squared relative to the value in the final answer. The formula for the "odd case" $(2p+1)$ can be checked if desired using integration-by-parts for a low value of the exponent, but the "even case" $(2p)$ is more difficult and one can wonder, where the factor of $\sqrt{\pi}$ comes from. Consider the basic integral $I = \int_0^\infty e^{-x^2} \, dx$. Then, we have for the square

$$I^2 = \left(\int_0^\infty e^{-x^2} \, dx\right)\left(\int_0^\infty e^{-y^2} \, dy\right) = \int_0^\infty \int_0^\infty e^{-(x^2+y^2)} \, dx \, dy = \int_0^\infty \int_0^{\pi/2} e^{-r^2} r \, d\theta \, dr,$$ so that we can

integrate over the first quadrant in polar coordinates to obtain

$$I^2 = \left(\frac{\pi}{2}\right) \int_0^\infty e^{-r^2} r \, dr = \left(\frac{\pi}{2}\right) \left[\frac{-e^{-r^2}}{2}\right]_0^\infty = \left(\frac{\pi}{2}\right) \left[\frac{0-(-1)}{2}\right] = \frac{\pi}{4}$$

and then $I = \sqrt{\pi}/2$. When the even case is reduced to simpler form using integration-by-parts, the final step will lead to the $(\sqrt{\pi}/2)$ factor. There will be problems at the end of this chapter to build your skill in applying these formulas, but now we can apply the formulas to obtain the average kinetic energy in the spherical system. Note that the 3D form of this integral will have $v^2 \sin(\theta) dv \, d\theta \, d\phi$ and after integrating over (θ, ϕ), there will still be $v^2 \, dv$ as the differential over v.

Then, we have $\left(\text{noting } \left(\frac{m}{2kT}\right)^{-\frac{5}{2}} = \left(\frac{2kT}{m}\right)^{5/2} \text{ etc.}\right)$

$$\left\langle \frac{mv^2}{2} \right\rangle = \left(\frac{m}{2}\right) \frac{\left(1 \cdot 3 \cdot \frac{\sqrt{\pi}}{2^3} \left(\frac{2kT}{m}\right)^{5/2}\right)}{\left(1 \cdot \frac{\sqrt{\pi}}{2^2} \left(\frac{2kT}{m}\right)^{3/2}\right)} = \left(\frac{3m}{4}\right) \left(\frac{2kT}{m}\right) = \frac{3kT}{2}.$$

This result is comforting in that the direct application of Boltzmann averaging process produces the same result as our previous phenomenological derivation from the ideal gas law. Further, we could perform the same sort of averaging to find the average value of $\langle v^2 \rangle$, but we can obtain the same value by rearranging the average of the kinetic energy to obtain the "root-mean-square" speed, $\sqrt{\langle v^2 \rangle} = v_{rms}$. A more useful form is obtained by multiplying both numerator and denominator by Avogadro's number, so as to use the molar values of R and the gram molecular weight M.

$$\left\langle \frac{mv^2}{2} \right\rangle = \frac{3kT}{2} = \frac{m}{2}\langle v^2 \rangle \text{ so then } \sqrt{\langle v^2 \rangle} = v_{rms} = \sqrt{\frac{3kT}{m}} = \sqrt{\frac{3kT(N_{Av})}{m(N_{Av})}} = \sqrt{\frac{3RT}{M}}.$$

An observant student should also note that if the average kinetic energy of particles in three dimensions is $(3kT/2)$, then one might expect that the kinetic energy is $(kT/2)$ per dimensional degree of freedom. This is called *the law of equipartition of energy* and postulates that each degree of freedom in a given system will lead to an average energy of $(kT/2)$. However, we will eventually learn that it requires extra explanation in the case of vibrational energy at low temperature.

With that exercise in hand, we proceed to the main goal of finding the average speed v instead of "the square-root of the square." Again, we can use Avogadro's number to obtain a formula in terms of mole quantities.

$$\langle v \rangle = \frac{4\pi \int_0^\infty e^{-\frac{mv^2}{2kT}}(v)v^2\,dv}{4\pi \int_0^\infty e^{-\frac{mv^2}{2kT}}(v^2)\,dv} = \frac{4\pi \left[\frac{1}{2}\left(\frac{2kT}{m}\right)^{4/2}\right]}{4\pi \left[1 \cdot \frac{\sqrt{\pi}}{2^2}\left(\frac{2kT}{m}\right)^{3/2}\right]} = \frac{2}{\sqrt{\pi}}\sqrt{\frac{2kT}{m}} = \sqrt{\frac{8kT}{\pi m}}.$$

Thus, we obtain $\langle v \rangle = \sqrt{\dfrac{8kT}{\pi m}} = \sqrt{\dfrac{8kTN_{Av}}{\pi m N_{Av}}} = \sqrt{\dfrac{8RT}{\pi M}}$ as the true average speed of the gas particles, which is similar to the root-mean-square speed but not exactly the same, so we can compare

$$\langle v \rangle = \sqrt{\frac{8RT}{\pi M}} \quad \text{to} \quad \sqrt{\langle v^2 \rangle} = v_{rms} = \sqrt{\frac{3RT}{M}}.$$

By now, you should realize that the denominator of the averaging process used above is the same in each case and refers to the number of particles being averaged. Consider the number N:

$$N = 4\pi \int_0^\infty e^{-\frac{mv^2}{2kT}}v^2\,dv = 4\pi\frac{\sqrt{\pi}}{2^2\left(\frac{m}{2kT}\right)^{3/2}} = \left(\frac{2\pi kT}{m}\right)^{3/2}.$$

Here, 4π will disappear into the factor, but the average numerator will also require a 4π factor.

Thus, we can consider this Maxwell–Boltzmann "distribution function," which can be written as

$$f(v) = 4\pi\left(\frac{m}{2\pi kT}\right)^{3/2} e^{-\frac{mv^2}{2kT}}v^2.$$

You can sketch this function and note that at small values of v, the curve goes up rapidly as a parabola in v, but at some point, the negative exponent starts to decline asymptotically back down to zero. Thus, there must be a maximum in the distribution curve, which refers to the "most probable" speed. In agreement with other texts, we will call this speed α. We can determine this speed by setting the first derivative of $f(v) = 4\pi\left(\dfrac{m}{2\pi kT}\right)^{3/2} e^{-\frac{mv^2}{2kT}}v^2$ to zero to find the maximum where the slope must be zero. Note here that this will be a derivative of a triple product, but the derivative of the constants will do nothing since they are indeed constants, and the derivative of a constant is zero.

$$\frac{df(v)}{dv} = 0 + 4\pi \left(\frac{m}{2\pi kT}\right)^{3/2} \left[\left(\frac{-2mv}{2kT}\right) e^{-\frac{mv^2}{2kT}} v^2 + 2ve^{-\frac{mv^2}{2kT}}\right] = 0.$$

The first zero on the right side is due to the derivative of the constants and the other terms are due to the product of $v^2 e^{-\frac{mv^2}{2kT}}$. Strictly speaking, we see that this derivative can indeed be zero when v is zero and again when v becomes very large in the negative exponent, but we are interested in the flat spot at the top of the peak. Thus, we cancel out $ve^{-\frac{mv^2}{2kT}}$ to leave only $\left(\frac{-m}{kT}v^2 + 2\right) = 0$, and so we find the most probable speed as $\alpha = \sqrt{\frac{2kT}{m}} = \sqrt{\frac{2kTN_{Av}}{mN_{Av}}} = \sqrt{\frac{2RT}{M}}$.

It is difficult to graph $f(v) = 4\pi \left(\frac{m}{2\pi kT}\right)^{3/2} e^{-\frac{mv^2}{2kT}} v^2$ when you insert the tiny mass "m" and use the high velocities that occur. However, we can consider a simpler function that has the same variable dependence by using $f(x) = e^{-x^2} x^2$ as shown in Figure 3.3. The derivative $\frac{df}{dx} = -2x^3 e^{-x^2} + 2xe^{-x^2} = 0 \Rightarrow x_{max} = 1$. We see on the graph that the maximum does indeed occur at $x = 1$, but the shape of the curve is asymmetrical and any weighted average using this distribution will favor higher values of x. By analogy, Figure 3.4 shows that the shape of the Boltzmann distribution is not symmetric and extends out to the higher speeds. However, the most important result from the Boltzmann analysis is that now we know the true average speed:

$$V_{max} = \sqrt{\frac{2RT}{M}} \cong 1.414\sqrt{\frac{RT}{M}}, \quad \langle V \rangle = V_{ave} = \sqrt{\frac{8RT}{\pi M}} \cong 1.596\sqrt{\frac{RT}{M}}, \quad \text{and}$$

$$V_{rms} = \sqrt{\frac{3RT}{M}} \cong 1.732\sqrt{\frac{RT}{M}}.$$

FIGURE 3.3 Ludwig Eduard Boltzmann (1844–1906) was an Austrian physicist, who founded the fields of statistical mechanics and statistical thermodynamics. (Image courtesy of Chemical Heritage Foundation Collections.)

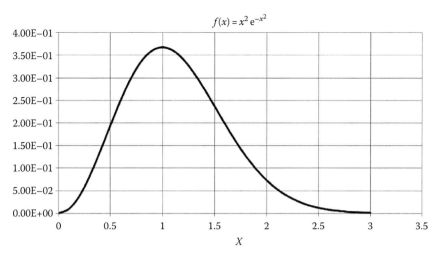

FIGURE 3.4 An asymmetric function with the same functional dependence as the Maxwell–Boltzmann molecular speed distribution. With this simple function, we can see that the maximum is exactly at 1 and the overall distribution is asymmetric to the high values.

We can graph the distribution if we define $F(v) \equiv f(v) / \left[4\pi \left(\dfrac{m}{2\pi kT} \right)^{3/2} \right] = \left[v^2 e^{-\left(\frac{mv^2}{kT} \right)} \right]$. For N_2

at 25°C, $F(v) = [v\,(\text{m/s})]^2 \exp \left\{ - \left[\dfrac{(0.028014\,\text{kg/mol})(\text{m/s})^2 v^2}{2(8.314\,\text{J/mol°K})(298.15\,°\text{K})} \right] \right\}$ and we calculate (Figure 3.5)

$$\alpha = \sqrt{\frac{2RT}{M}} = \sqrt{\frac{2(8.314 \times 10^7 \text{erg/mol}\,°\text{K})(298.15°\text{K})}{28.014\,\text{g/mol}}} = 4.20678 \times 10^4\,\text{cm/s} = 420.678\,\text{m/s}.$$

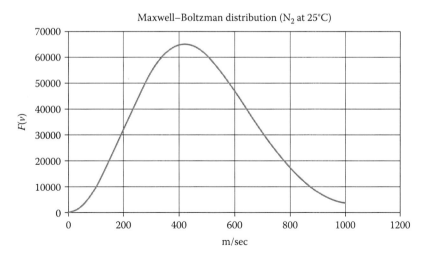

FIGURE 3.5 It is more difficult to identify the maximum using real data, but 420.678 m/s appears to be correct for N_2 at 25°C. This is the distribution function without the factor of $4\pi \left(\dfrac{m}{2\pi kT} \right)^{3/2}$.

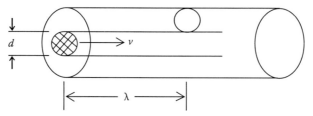

FIGURE 3.6 Swept cylinder during the mean-free-path.

Now that we have the Boltzmann average speed, we can use it for applications that can be measured experimentally to test the theory. Let us consider, gas atoms/molecules as hard spheres of diameter "d." Let us imagine we could ride on an atom in the gas phase as one young lady in my class offered to do based on her enjoyment of "bumper cars" at an amusement park. We can see trouble ahead for her traveling at over 1000 mph.

At first, there is a lot of empty space to travel in, but eventually, there will be a collision with another atom, even if the other atoms are motionless. In Figure 3.6, we imagine a cylinder swept out by an atom/molecule of diameter "d" such that if there is another atom/molecule within that tube, there will be at least a glancing collision or maybe a more direct collision. In 1 s, the number of collisions in the zigzag "collision tube" will be

$$Z_1 = \pi d^2 v n*.$$

Let us define Z_1 as the number of collisions a single sphere will encounter and $n*$ (number of spheres/cm^3) as a unit of concentration. The effective swept area is πd^2 because the minimum radius for contact of two spheres is $R_{\text{contact}} = \left(\dfrac{d}{2} + \dfrac{d}{2}\right) = d$. However, the other spheres are not standing still, they are also moving. We might try to find a way to average over all possible angles but there is a simpler way. We should recall some previous physics experiment or demonstration that any vector can be resolved into Cartesian components so that in principle each approach of two spheres can be resolved into six possibilities: up, down, right, left, forward, and backward. Assuming all the particles have the same average speed (the meaning of "average"), a collision from the rear will be as probable as a head-on collision, so these type of collisions roughly cancel out in terms of their probability. By far, the most common type of collision will be a result of a right-angle collision between two particles with the average speed. Thus, most of the approach speeds will be along the hypotenuse of a triangle with two sides equal to the average speed or $\sqrt{2}\langle v \rangle$ instead of $\langle v \rangle$.

This crude correction produces an improved approximate formula for the collision number Z_1:

$$Z_1 = \left(\sqrt{2}\langle v \rangle\right)\pi d^2 n*.$$

The next question is to wonder how far our young friend can ride a gas particle before a collision occurs. The quantity is called the "mean-free-path" or λ here. If we know the average speed and the number of collisions, then we can calculate λ (Figure 3.7)

$$\lambda = \frac{\langle V \rangle}{Z_1} = \frac{\langle V \rangle}{\sqrt{2}\langle V \rangle \pi d^2 n*} = \frac{1}{\sqrt{2}\pi d^2 n*}.$$

Perhaps, we realize that the idea of collisions is related to the rate of chemical reactions? It turns out that the reaction mechanisms are more complicated than collisions between spheres, but it is certainly true that molecules need to collide in some way in order to react as the necessary first step. If all the

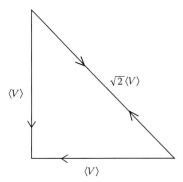

FIGURE 3.7 Assuming two particles (atoms or molecules) are traveling at an average speed of $\langle V \rangle$ and approach each other at right angles, their effective speed of approach is $\sqrt{2}\langle V \rangle$. Random angles of approach can be resolved into right-angle components, so a reasonable approximation of the average rate of approach is $\sqrt{2}\langle V \rangle$ in all directions.

spheres are in motion with the average speed, to get the number of binary collisions, we should multiply the Z_1 collisions by the concentration of targets n^*, but since it takes two spheres to cause one collision, we should divide by 2. This ignores the fact that each sphere should not count a collision with itself but that is a small number, which can be neglected compared with the total number of binary collisions. We also note for future discussion of gas kinetics that the possibility of a three-body collision is so improbable that it can be neglected. Thus, we have for the binary collisions

$$Z_{11} \cong \left(\frac{Z_1 n^*}{2}\right).$$

This is still theoretical, so we need some sort of measurement to confirm these assumptions. Surprisingly, it is possible to construct a gas viscometer similar to the Ostwald viscometer for liquids. If a sufficiently small diameter capillary tube is used, the time to force a given volume of gas through that tiny cylinder will vary with the type of gas and gas viscosity can be measured (Table 3.2). Such viscometers can be standardized using dry air with a relationship derived by the former U.S. National Bureau of Standards [1].

$$\eta = \frac{(145.8 \times 10^{-7})T^{3/2}}{T + 110.4} \text{ poise}.$$

Although, the type of viscometer in Figure 3.8 [2] has been used for years in teaching laboratories, concern regarding the accompanying vapor pressure of mercury has eliminated this apparatus from many laboratories. The main point of interest here is the tiny capillary tube through which the gas sample is forced to flow by the weight of the mercury. The volume between marks "a" and "b" is designed to be exactly 100 mL.

Although, it is interesting to consider the collision properties of gases, the reason we have emphasized the Boltzmann average speed is to get to the measurable quantity of gas viscosity. Consider a rectangular box as in Figure 3.9 that has two slits on the side for inlets from a gas source, which has a temperature gradient in the vertical z-direction, so that the gas entering the upper slit is warmer (faster average speed) than the gas entering the lower slit. Let us assume that the two entrance slits are only two mean free path lengths apart and the temperature gradient along the left side produces a velocity gradient as you increase the z-coordinate (dv/dz).

Assume the right side of the box is open to exhaust the gas. Now, consider the plane between two (1 cm × 1 cm) sheets of gas coming through the slits. With apologies to the IUPAC Committee, we

TABLE 3.2
Viscosities of Selected Gases
at 1 atm and 300°K

Gas	μPa s	Micropoise
Air	18.6	186
Ar	32.9	329
HCl	14.6	146
H_2	9.0	90
D_2	12.6	126
H_2O	10.0	100
D_2O	11.1	111
He	20.0	200
Kr	25.6	256
N_2	17.9	179
O_2	20.8	208
Xe	23.2	232
CO	17.8	178
CO_2	15.0	150
CH_4	11.2	112
HCCH	10.4	104
H_2CCH_2	10.4	104
C_4H_{10}	7.5	75

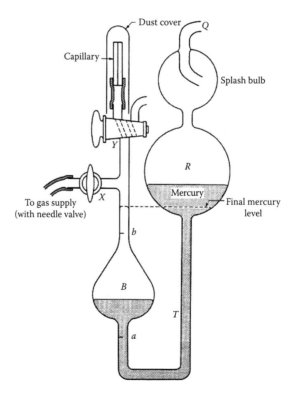

FIGURE 3.8 Gas viscometer. (From Shoemaker, D.P. et al., *Experiments in Physical Chemistry*, 6th Edn., McGraw-Hill Co., New York, 1996, p. 129. With permission.)

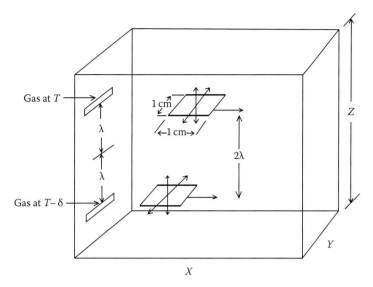

FIGURE 3.9 Schematic showing momentum transfer between layers of gas under a thermal gradient.

use cgs units here because viscosity was derived historically in poise, which is a cgs unit and because we want to emphasize the small domain of the process we are considering. The question is how much vertical interaction is there between the two postage-stamp size sheets of gas as they move to the right? We are only concerned with the few molecules in the upper layer that have components toward the lower layer and the few molecules in the lower layer that have vertical components moving up to the upper layer. The diagram shows that only 1/4 of the possible directions will result in a transfer of molecules with differing velocities from one layer to the other. Since the atoms/molecules have the same mass, the difference in speeds leads to a transfer of momentum and that is a force, which leads to a small viscous drag f_z.

We can itemize this momentum transfer between the two layers, which are separated by 2λ using the formal definition of the speeds due to the vertical gradient as $v_z = (dv/dz)(z + \lambda)$ at the upper slit and $v_z = (dv/dz)(z - \lambda)$ at the lower slit. By assuming 1 cm $\times$ 1 cm layers, we can write the momentum transfer though a (1 cm $\times$ 1 cm) window between the layers. For the moment, ignore the fact that we do not know the speed gradient (dv/dz) and write out the two momenta in the upper and lower layers:

$$mv \downarrow = \left(\frac{1}{4}\right)n^*\langle v\rangle \left(\frac{dv}{dz}\right)(z + \lambda),$$

$$-\left[mv \uparrow = \left(\frac{1}{4}\right)n^*\langle v\rangle \left(\frac{dv}{dz}\right)(z - \lambda)\right],$$

$$m(\Delta v) = \left(\frac{2}{4}\right)n^*\langle v\rangle \left(\frac{dv}{dz}\right)(\lambda) = f_z = \eta_{\text{visc}}x\left(\frac{dv}{dz}\right)(1\ \text{cm}^2).$$

Thus, using our hypothetical "box with slits" we see that we can cancel the unknown gradient (dv/dz) to obtain an expression for the viscosity of the gas as

$$\eta_{\text{visc}} = \left(\frac{1}{2}\right)n^*m\langle v\rangle\lambda = \left(\frac{1}{2}\right)n^*m\langle v\rangle\left(\frac{1}{\sqrt{2}\pi d^2 n^*}\right) = \frac{m\langle v\rangle}{2\sqrt{2}\pi d^2} = \eta_{\text{visc}}.$$

Example

Given the viscosity of CH_4 is 11.2 μPa s (112 micropoise), calculate the effective diameter d, the mean free path λ, and the collision numbers Z_1 and Z_{11} at 1 atm and 27°C.

$$\langle v \rangle = \sqrt{\frac{8RT}{\pi M}}; \quad \eta = \left(\frac{1}{2}\right)\frac{m\langle v \rangle}{\sqrt{2}\pi d^2}; \quad d = \sqrt{\frac{m\langle v \rangle}{2\sqrt{2}\pi \eta}}; \quad M = 12.0108 + 4(1.00785) = 16.0422\,\text{g/mol},$$

$$n^* = \frac{6.022 \times 10^{23}/\text{mol}}{22414\,\text{cm}^3/\text{mol}\left(\frac{300°\text{K}}{273.15°\text{K}}\right)} = 2.4463 \times 10^{19}\,\text{molecules/cm}^3,$$

$$\langle v \rangle = \sqrt{\frac{8(8.314 \times 10^7\,\text{erg}/°\text{K mol})(300°\text{K})}{\pi(16.0422\,\text{g/mol})}} = 6.2922 \times 10^4\,\text{cm/s},$$

$$d = \sqrt{\frac{\left(\frac{16.0422\,\text{g/mol}}{6.022 \times 10^{23}/\text{mol}}\right)(6.2922 \times 10^4\,\text{cm/s})}{2\sqrt{2}\pi(1.12 \times 10^{-4}\,\text{g/cm s})}} = 4.104 \times 10^{-8}\,\text{cm} = 4.104\,\text{Å},$$

$$\lambda = \frac{1}{\sqrt{2}\pi d^2 n^*} = \frac{1}{\sqrt{2}\pi(4.104 \times 10^{-8}\,\text{cm})^2(2.4463 \times 10^{19}/\text{cm}^3)} = 5.463 \times 10^{-6}\,\text{cm} = 546.3\,\text{Å},$$

$$Z_1 = \frac{\langle v \rangle}{\lambda} = \frac{6.2922 \times 10^4\,\text{cm/s}}{5.463 \times 10^{-6}\,\text{cm}} = 1.1518 \times 10^{10}\,\text{collisions/s},$$

$$Z_{11} = \frac{Z_1 n^*}{2} = \frac{(1.1518 \times 10^{10}\,\text{collisions/s})(2.4463 \times 10^{19}/\text{cm}^3)}{2} = 1.4089 \times 10^{29}\,\text{collisions/cm}^3\,\text{s}.$$

We have rounded the numbers to four significant figures because 8.314 J/K mol is only given to four places. Several results are especially worth emphasizing. First, the mean free path is more than 100 times the effective size of the molecule at 1 atm pressure; there is a lot of empty space in a gas at 1 atm. This is typical of conditions at 1 atm. Second, the effective diameter of a little more than 4.1 Å is an effective sphere equivalent shape of the methane molecule. Third, while the Z_1 is high, the Z_{11} value of over 10^{29} collisions/cm^3 s is truly amazing. Is it any wonder that gas-phase reactions are fast? One question, students often ask is how do you calculate n^*? Here, we have used the molar volume of 22.414 L = 22,414 mL at 0°C and 1 atm and corrected it with Charles law for the temperature ratio of 300:273.15.

Finally, a comment is in order regarding the factor of 1/2 in the formula for η. A more complete treatment by Pease [3] integrates over a hemispherical region of the upper and lower layers that produces a factor of $(5\pi/32) = 0.49087$, which is very close to the factor of $1/2 = 0.5$ used here. Thus, we see that we are using a very good approximation.

SUMMARY

This chapter has involved a lot of numerical work as well as some new integral formulas to burrow into the details of KMTG and the use of the Boltzmann principle. Now, we have a more detailed understanding of gases somewhere between the idea of a flowing macroscopic fluid and the inner electronic structure of molecules assuming the atoms/molecules are approximately small spheres. We have learned that there is a lot of action going on inside a gas but there is still a lot of empty space. Perhaps, the most important result of this consideration of the Boltzmann KMTG and gas viscosity is a rough approximate determination of the size of atoms and molecules. In fact, the cgs angstrom unit of 10^{-8} cm is well suited to the size of atoms and molecules and it is common in older texts, but the modern SI unit is the nanometer, easily remembered as "ten-to-the-minus-nine-meter" unit. Since the nanometer is in meters and angstroms are in centimeters, 1 nm = 10 Å. Therefore, a useful outcome of KMTG is a revelation about the actual size of atoms and molecules. Some may question why we needed such an intellectual investment in the derivation of laminar viscosity, but the application to polymer molecular weights is a valuable spin-off and some occasional chemical engineers in the class should find the r^4 dependence of flow rate important. However, that foundation in the theory of laminar viscosity yielded a window into the whole Boltzmann KMTG and that alone justifies the time spent in the derivation of Poiseuille's law.

PROBLEMS

3.1 Compute $\langle v \rangle$, v_{rms}, and the most probable speed α for Ar gas at 25°C and 1 atm pressure in mph.

3.2 Given the viscosity of N_2 is 17.9 µPa s at 27°C and 1 atm pressure, calculate d, λ, Z_1, and Z_{11} for this gas.

3.3 Given the viscosity of He is 20.0 µPa s at 27°C and 1 atm pressure, calculate d, λ, Z_1, and Z_{11} for this gas.

3.4 Evaluate $\displaystyle\int_0^\infty x\, e^{-2x}\, dx$.

3.5 Evaluate $\displaystyle\int_0^\infty x^5\, e^{-3x}\, dx$.

3.6 Evaluate $\displaystyle\int_0^\infty x^6\, e^{-4x^2}\, dx$.

3.7 Evaluate $\displaystyle\int_0^\infty x^5\, e^{-9x^2}\, dx$.

3.8 Evaluate $\displaystyle\int_{-\infty}^\infty x^5\, e^{-9x^2}\, dx$.

3.9 Evaluate $\displaystyle\int_0^{2\pi} d\phi \int_0^\pi \sin(\theta)\,d\theta$.

3.10 Evaluate $\displaystyle\int_0^a r^2\, dr \int_0^{2\pi} d\phi \int_0^\pi \sin(\theta)\, d\theta$.

3.11 Evaluate $\displaystyle\int_{-\infty}^{+\infty} e^{-9x^2} x^4\, dx$ (note limits of integration).

3.12 Evaluate $\displaystyle\int_0^{+\infty} e^{-9x^2} x^5\, dx$ (note limits of integration).

3.13 Evaluate $\displaystyle\int_{-\infty}^{+\infty} e^{-4x^2} x^3\, dx$ (note limits of integration).

3.14 Evaluate $\displaystyle\int_{-\infty}^{+\infty} e^{-4x^2} x^6\, dx$ (note limits of integration).

3.15 Evaluate $\displaystyle\int_0^{+\infty} e^{-4x} x^6\, dx$.

REFERENCES

1. Hilsenrath, J., *Tables of Thermal Properties of Gases*, table 1-B. U.S. *Natl. Bur. Stand. Circ. 564*. U.S. Government Printing Office, Washington, DC (1955).
2. Shoemaker, D. P., C. W. Garland, and J. W. Nibler, *Experiments in Physical Chemistry.*, 6th Edn., McGraw-Hill Co., New York, 1996, p. 129.
3. Pease, R., Kinetic theory of gases. Part II, *J. Chem. Ed.*, **16**, 366 (1939).

4 The First Law of Thermodynamics

INTRODUCTION

In the previous chapter, we sharpened our computational skills and gained an appreciation for the particle model of gases. We now turn our attention to matters of energy and energy flow according to the laws of thermodynamics. In the Math Review chapter, we showed that energy can flow between various forms of kinetic and potential energy but that overall energy is conserved and only the form is changed. Many things can be said about thermodynamics. Mainly, thermodynamics owes more to the steam engine than the steam engine owes to thermodynamics. That means that Watt [1] and other inventors built steam engines and got them to work using raw mechanical reasoning and then thermodynamics was developed/discovered to understand the principles of the engine. We will try to help you gain a foundation of understanding if you will follow along and use pencil and paper to write out some derivations rather than just read the text. It should be understood that while physics majors develop expertise in electromagnetic theory far more than chemistry majors, it is generally true that chemistry majors gain a better understanding of thermodynamics. Chemical engineers use thermodynamics as their main expertise, although augmented by kinetics and transport theory, so the chemistry professionals should take pride that thermodynamics is "their thing," their chance to shine in terms of the scientific method.

Thermodynamics is necessarily more abstract than the study of mechanical devices because it is not always easy to see "heat." You will soon see that thermodynamics is wonderful for providing information about "after-minus-before" processes, but often it tells us little about the mechanism of the process being considered. The good news is that we do not need to know the details of the mechanism of a process, but the bad news is that often thermodynamics does not provide any means to determine the mechanism. This adds to the mystery of thermodynamics since often one can substitute an imaginary process with the same beginning and ending to obtain results without knowing the real mechanism but you will not learn the mechanism. Overall, thermodynamics offers sweeping principles that are important in all the sciences: chemistry, physics, astronomy, and even biology. We will see that there is an ongoing dynamic between the tendency of energy to decrease and a tendency of randomness to increase. Living systems are caught in this dynamic, so basic metabolism is subject to thermodynamics.

HISTORICAL DEVELOPMENT OF THERMODYNAMICS

Although we want to treat thermodynamics from the beginning of quantitative relationships, it may be worth mentioning more primitive ideas. Prior to 1798, one explanation of heat produced by friction was that heat is the release of a substance trapped within materials. The substance was called "caloric." Actually at a very shallow grade school level this idea has some merit but just what is caloric, a liquid, a vapor, some sort of "igneous fluid?" The breakthrough came from an American named Benjamin Thompson (1753–1814) (Figure 4.1) who was born in Woburn, Massachusetts. He became a Major in the British Army at the age of 19 and he left with the British in 1776, after the surrender of Boston. He later entered into the employ of the Bavarian Government and was given the title of Count Rumford. His key experiment was performed in Munich where he was involved in the manufacture of cannon [2]. In the 1700s, cannon were large solid pieces of cast brass or iron

FIGURE 4.1 Count Rumford, Benjamin Thompson (1753–1814). (Courtesy of the Chemical Heritage Foundation.)

with no moving parts. The manufacture merely required boring a cylindrical hole with a large cutting tool. Thompson showed that during the boring with a dull cutting tool, cold water in contact with the cannon could be brought to a boil due to the heat generated by the boring process. He noted that the heat released was proportional to the amount of *work* done on the cannon rather than the amount of material removed, which would limit the amount of caloric released and there was no mass change due to the release of "caloric." A key aspect of his analysis was when he used a very dull cutting tool to continually rub inside an unfinished bore hole for hours without further cutting. It became clear that the work against the resistive friction caused the heat since no cutting was releasing any possible "caloric." On a personal basis Thompson was a person curious about astronomy and other natural philosophy, so he used logic to formulate the connection between heat and work. This change in interpretation brought about a connection between the mysterious nature of heat and the more easily quantified concept of work. Count Rumford was a scientist of his times remaining interested in astronomy, inventing an improved fireplace, and writing further regarding the nature of flames and heat. His work is generally considered a breakthrough in the formulation of thermodynamics because while heat certainly flows like a fluid you cannot capture it in a bottle. Perhaps the mystery of the nature of heat is what still makes thermodynamics challenging for students?

We need to define some terms so pay attention to the verbal meanings and the algebraic sign conventions. While thermodynamics requires thoughtful reasoning, the level of required mathematics is relatively low and the computations are usually easy. Thermodynamics requires a type of reasoning that is light in terms of mathematics but requires considerable use of logic.

DEFINITIONS

"The system" is a region around some mechanical or biological device surrounded by an imaginary boundary, mainly as an attention focusing region or bookkeeping device to define "inner" and "outer." Ultimately, the "system" is the entire universe but thermodynamics can be used to focus on a machine or localized region relative to some thermal boundary, which defines a region of interest. An example would be a thermos bottle or a Dewar flask. The intention of a thermos bottle is to provide a boundary wall against heat flow and experiments could be done within the thermos as isolated from the "outside world," but eventually heat will leak across that boundary relative to conditions in the universe so time is involved to a degree. Again, we will analyze the events in an internal combustion engine where the chemistry of the combustion is much faster than heat can flow through metal walls of a cylinder, so we can model the events within the cylinder as a system thermally isolated from the outside world for a short time interval. Then we can enlarge the domain of the system and consider the energy and heat flow in a region around the engine to consider the whole engine as a unit but isolated from the environment. Lastly, we might want to consider the effect of various thermal devices on the whole planet Earth as an isolated system in space. Thus, the concept of the "system" is largely a matter of choosing a region and a time interval on the part of the person doing the analysis.

$q > 0$, heat energy is absorbed (by the system)
$q < 0$, heat energy is given off (to the environment)
$w > 0$, work done on the system by the environment, especially on a gas
$w < 0$, work done by the system on the environment; especially by a gas

U, internal energy of the system, contains molecular translational, rotational, and vibrational energy relative to a preexisting energy scale of elemental formation in chemistry. Chemistry is "post-creation," the elements are already here with their characteristic electronic energy levels and chemistry studies interactions between elements. Chemical thermodynamics functions at a level of ground state electronic energy levels of the elements. Nuclear chemistry involves a scale of internal energy units relegated to physics.

$\Delta \equiv$ (after – before), this is an order-specific convention definition in all of thermodynamics
$\Delta T = 0$ means "isothermal," same temperature, constant temperature
$\Delta q = 0$ means "adiabatic," constant heat, no heat flow
$\Delta P = 0$ means "isobaric," constant pressure, no change in pressure

Laboratory variables: P, V, T, n, SI units are bars, L, °K, mol but other units do exist.

FIRST LAW OF THERMODYNAMICS

With the definitions and sign conventions above, we can state a simple form of the first law.

$$dU = \not{d}q + \not{d}w$$

In words, this means that the change in internal energy occurs as a change in heat and a change in work. The question in applications is how much change there is in heat and work for a given change in internal energy. Much of the credit for the first law is due to James P. Joule (1818–1889) [3] who was an English physicist who carried out a number of important experiments relating heat and work. An interested student should look up short biographies of Joule on the Internet to appreciate the controversy surrounding his experiments. His several measurements of the relationship between heat in calories and mechanical work (1 cal $= 4.184$ J) eventually triumphed over the "caloric" theory. A warning is needed here regarding the sign of the work term. Many older texts formulate the work term as $-dw$ because almost all the applications involve gases, which expand to push some sort of a piston to do work. In the last 20 years or so almost all texts give the first law with $+dw$ but it all works out fine since compressing a gas is positive work while expanding gas is negative work on the gas but positive work on the environment. This first example shows us that we need to think about the way in which the system and environment act during a given process and worry over using the correct sign of the variables.

Now it is the time to remind ourselves that energy is conserved and that U represents some amount of energy involved in a given process. As such we can represent the change in energy as:

$$\Delta U = U_{\text{after}} - U_{\text{before}}.$$

Conservation of energy is a general principle and U is a *state variable*, which allows us to quantify the amount of an energy change in a given process. This brings up the question as to how the process is carried out. The conservation of energy means that it does not matter how the process is carried out; all that matters is the difference between the end and the beginning of the process. We say that U is a state variable and the change in U is independent of the path. However, we must be careful regarding dq and dw since we know by common sense that surely dw is dependent on the path in terms of what we call "efficiency." A general definition of efficiency used in this text is

$$\text{Efficiency} \equiv \frac{\text{work done}}{\text{energy change used}}.$$

The concept of efficiency can be subjective since it usually implies "useful work" while work done against resistive friction is still work but not "useful" and it generates heat as well. However, the first law says there is a nonsubjective state variable called the internal energy, U, which obeys the conservation principle even if dq and dw vary according to the process used.

Again the notion of efficiency involves the amount of heat energy dq, which might be just sufficient for a given process or wasteful. Thus, dU is an "exact differential" and is independent of path while dq and dw are not exact differentials. Some texts write dq and dw with a slash "/" through the "d" and the "w" but we only need the student to realize that dU is path independent while dq and dw are not. This is important later because we will discover and use other state variables that have the property of path independence nicely explained by Denbigh [4]. Suppose a process involves two steps from (I) → (II) → (III), we can calculate the energy change for (I) → (III) as:

$$\int_I^{II} dU + \int_{II}^{III} dU = (U_{II} - U_I) + (U_{III} - U_{II}) = (U_{III} - U_I).$$

Now suppose (III) is really just a return to (I), then we obtain a special condition for which we have a special symbol indicating integration over a cyclic path with the result of zero,

$$\oint dU = 0.$$

Other cyclic integrals might not be zero but a *state variable* will sum to zero; that is the main characteristic of a state variable. The circle on the integral sign means that the process is carried out over a cyclic process. We can offer a simple insight in that state variables are universal variables and characteristics of the universe while q and w are subjective variables, which depend on how a process is carried out.

A simple example is the idea of gravitational potential energy. If Jack goes up a hill and later tumbles down, we can say two things from common sense. First, if he ended up where he started there is no net change in gravitational potential energy. Second, we cannot tell how much Jack wandered side to side or how much energy he may have expended in other forms, which may not have been recovered in his descent since as far as gravity goes, the only change is between the "after-minus-before" height change. A third observation is that for this process the cycle of gravitational energy satisfies the gravitational cyclic process whose "path integral" (the integrated path) is zero. The path Jack took is typical of a thermodynamic process where we can evaluate the beginning and end of the process but have no information on what happened along the path. If we write the differential of U in terms of T and P we obtain $dU = \left(\frac{\partial U}{\partial T}\right)_P dT + \left(\frac{\partial U}{\partial P}\right)_T dP.$

We will often write differentials in terms of only two laboratory variables assuming a constant sample size in terms of moles n and that there is some unspecified "equation of state," which links P, V, T, n. However, another very important mathematical property is obtained if we use a different path but remember that U does not depend on the path, so that we can realize another relationship,

$$\left(\frac{\partial}{\partial P}\right)_T \left(\frac{\partial U}{\partial T}\right)_P = \left(\frac{\partial}{\partial T}\right)_P \left(\frac{\partial U}{\partial P}\right)_T \quad \text{or} \quad \left(\frac{\partial^2 U}{\partial P \partial T}\right) = \left(\frac{\partial^2 U}{\partial T \partial P}\right).$$

This means that we are measuring the change in U by changing the temperature T and then changing the pressure P, but this will be the same as changing the pressure P first and then changing the temperature T. Check *Introduction: Mathematics and Physics Review* for a review of partial derivatives. Suppose $f(x, y) = 2x e^{3y}$ then $\left(\frac{d^2 f}{dxdy}\right) = \frac{d}{dx}(6x e^{3y}) = \left(\frac{d^2 f}{dydx}\right) = \frac{d}{dy}(2 e^{3y}) = 6 e^{3y}$ because this is a continuous function of both x and y. Most beginning calculus texts teach from a subset of continuous functions so students might expect all second derivatives to have this property but we will find examples in later chapters on quantum mechanics where the order of differentiation does matter in some cases. This property is shared by state variables but within the possibilities of all functions that property is not shared by all cross derivatives. That also means, and is dependent on the fact, that state variables are continuous functions. This relationship between the second derivatives will be very important in later discussions of thermodynamics. We again offer the insight in that state variables are universal variables and characteristics of the universe while q and w are subjective variables that depend on how a process is carried out. In summary, *we observe two important characteristics of state variables, the cyclic path integrals are zero and the mixed derivatives are equal because the state variables are continuous functions.*

ISOTHERMAL PROCESSES

Example
Calculate q, w, and ΔU for the reversible, isothermal expansion of 10 mol of ideal gas from 1 to 0.1 atm at a constant temperature of 0°C.

For the purpose of problem solving in thermodynamics the word "reversible" holds special meaning. It means that PV work is carried out in a way that the internal pressure of the gas is opposed by an external pressure that is only infinitesimally different from the internal pressure by such a small amount that the process could easily be reversed at any time by some small fluctuation. Later we will see that by opposing the internal pressure with the maximum (minus just a tiny amount) external pressure the process will carry out the maximum amount of work. However, for students facing a problem "reversible work" means we can equate the internal and external pressures ($P_{int} = P_{ext}$). In addition, we encounter here the other key word "isothermal," which means $\Delta T = 0$. Recall previous work on gas energy where the Boltzmann average energy of a monatomic gas is ($3RT/2$), which only depends on temperature. Note a monatomic gas will have only three degrees of freedom implying ($RT/2$) energy each for (v_x, v_y, v_z) degrees of freedom. Then for a diatomic gas like N_2 or O_2 (air) there can also be two rotational degrees of freedom, each with ($RT/2$) energy implying an energy of ($5RT/2$). A linear molecule can only rotate two ways but this idea could be extended to polyatomic gases (CH_4), which would lead to ($3RT/2$) for rotational energy and a total of ($3RT$)/mol. We will see later that this idea of ($RT/2$) energy per degree of freedom called the law of energy equipartition is an ideal that is good for translation and rotation but is only realized for vibration at high temperatures. Thus, if the temperature does not change, the energy does not change either. Therefore, we have a simple realization that $\Delta T = 0$ and $\Delta U = 0$ when the temperature is constant. Thus, for this problem we see that $\Delta U = 0 = dq + dw$ so that we also have $q = -w$. Now we come to the only (easy) mathematics in the problem. Note that we have to think about the gas expanding so the work on the gas is negative,

$$w = -\int_{V_1}^{V_2} P_{ext}dV = -\int_{V_1}^{V_2} \left(\frac{nRT}{V}\right)_{int} dV = -nRT \ln\left(\frac{V_2}{V_1}\right) = -nRT \ln\left(\frac{P_1}{P_2}\right), \quad \text{note use of } P_{ext} = P_{int}.$$

Note we have used the inverse relationship between V and P for an ideal gas, which is appropriate because the pressures in the problem are low. Now we can put numbers in the formula.

$$w = -(10\,\text{mol})(1.987\,\text{cal/K mol})(273.15°\text{K})\ln\left(\frac{1\,\text{atm}}{0.1\,\text{atm}}\right) = -12{,}497\,\text{cal} = -52{,}289\,\text{J}.$$

Here we have used 1 cal = 4.184 J (a calorie is worth many jewels!) due to some amazing work by James Prescott Joule. Joule [3] performed a careful experiment measuring the increase in temperature of water caused by a rotating a paddle wheel driven by falling weights in a measurable way. Modern recreations of the experiment produce the value of 4.184 J/cal. Note the work on the gas is negative. Then we use the first law equation to find the heat change.

$$\Delta U = q + w = 0 = q - 12{,}497\,\text{cal} = q - 52{,}289\text{J}.$$

Thus we have $\Delta T = 0, \Delta U = 0, q = 12{,}497\,\text{cal} = 52{,}289\,\text{J} = -w$. From this example, we see that thermodynamics has some key words that have special mathematical implications and that we have to watch the sign conventions carefully. Now let us consider another example where we change one of the key words from "reversible" to "irreversible."

Example
Calculate q, w, and ΔU for the irreversible, isothermal expansion of 10 mol of ideal gas from 1 to 0.1 atm at a constant temperature of 0°C.

Once again we have $\Delta T = 0$ and $\Delta U = 0$ for the same reasons as in the previous example. However, there is a big change in the work because now we have to assume that if the final pressure is only 0.1 atm then the complete expansion was against only that pressure of 0.1 atm. Thus, we set up the first law as before

$$\Delta U = 0 = q + w = q + \left[-\int_{V_1}^{V_2} P_{ext}dV\right] = q - (0.1\,\text{atm})\int_{V_1}^{V_2} dV = q - (0.1\,\text{atm})(V_2 - V_1).$$

Careful reading of the question reveals another key set of words as "ideal gas," so $PV = nRT$.

Thus, $\Delta U = 0 = q + w = q - (0.1\,\text{atm})\left(\frac{nRT}{P_2} - \frac{nRT}{P_1}\right) = q - (0.1\,\text{atm})(10RT)\left[\frac{P_1 - P_2}{P_2 P_1}\right]$ and

so we have a simple expression (easy mathematics!):

$$0 = q - (0.1\,\text{atm})(10\,\text{mol})(1.987\,\text{cal/°K mol})(273.15°\text{K})\left[\frac{0.9\,\text{atm}}{(0.1\,\text{atm})(1\,\text{atm})}\right] = q - 4885\,\text{cal}.$$

Thus, we have $q = 4885\,\text{cal} = 20{,}438\,\text{J}$ and $w = -q$ with $0 = \Delta U$.

Note in particular that while the work is still negative due to the expansion of the gas (negative work on the gas, positive work on the environment), the absolute magnitude is much less than the reversible case and it is generally true that $|w_{rev} = w_{max}|$ as far as the work is calculated. *The maximum work will be obtained by a reversible process.* The sign of the work is determined by whether the system or environment is worked on. So as promised above, the mathematics of thermodynamics is not complicated but more thinking is required as to the sign of the quantities. Thermodynamics encourages thinking more with less automatic use of formulas.

We have mentioned the relationship $U = (3RT/2)$ and $\Delta U = (3R/2)\Delta T$. For a Boltzmann atomic gas, we can take the derivative with respect to temperature while holding the volume constant, recognizing that the laboratory relationships would involve P, V, T, n so maybe at least one other variable is involved assuming that knowledge of two of the three variables would determine the third under some equation of state for a fixed amount (moles) of gas:

$$dU = \left(\frac{\partial U}{\partial T}\right)_V dT + \left(\frac{\partial U}{\partial V}\right)_T dV = C_v dT + \left(\frac{\partial U}{\partial V}\right)_T dV,$$

assuming constant moles.

Thus, we introduce the new quantity $C_V \equiv \left(\frac{\partial U}{\partial T}\right)_V$. This also brings up the realization that for an ideal gas we have $\left(\frac{\partial U}{\partial V}\right)_T = \left(\frac{\partial U}{\partial P}\right)_T = 0$, since the energy of the ideal gas only depends on temperature. In fact we can enlarge the definition of an ideal (monoatomic) gas to mean:

Ideal Gas

1. $PV = nRT$
2. $\left(\frac{\partial U}{\partial V}\right)_T = \left(\frac{\partial U}{\partial P}\right)_T = 0$

We note that these conditions of energy independence from T and P may not be true for real gases or even for the van der Waals gas but they are a reasonable approximation at low pressure and high temperature where the ideal gas equation is good.

ENTHALPY AND HEAT CAPACITIES

Next we come to a practical problem: What happens even in household kitchens when a sealed container is opened? Soft drinks are under some pressure in their containers, so pressure is released when the container is opened. Old fashioned glass milk bottles brought in from the cold would often pop out their cardboard top plugs when warmed in a kitchen environment due to the Charles–Gay-Lussac relationship. Thus, we see opening storage containers involves a small but nonzero pressure push from the interior of a container against the nominal one atmosphere pressure in the environment. This relatively small pressure change occurs in the laboratory as well for every process that is open to the atmosphere and we know that any PV product is energy, even if it is small. Thus, it has proven practical to define another energy quantity called the "enthalpy," which is defined so that the atmospheric pressure and any volume change due to ambient conditions is added to the internal energy as a definition:

$$H \equiv U + PV.$$

Note that H is also a state variable because the PV product is energy and U is already a state variable. This does make some things more complicated even though the new terms may be small in magnitude so that we now have:

$$dH = dU + PdV + VdP.$$

Again we appeal to practical conditions in that the external atmospheric pressure is essentially constant depending only on the weather variations. The atmosphere of the planet is so huge that opening a pressurize vessel or a vacuum chamber will not change the pressure of the atmosphere. Thus, we see that to a good approximation:

$$dH = dU + PdV + 0.$$

But according to the first law $dU = dq - PdV$ allowing for the sign convention of work done by/on gases. Thus, we see that under atmospheric conditions we have

$$dH = dq - PdV + PdV = dq,$$

and we can define a heat capacity under constant pressure conditions as:

$$C_P = \left(\frac{\partial H}{\partial T}\right)_P.$$

In your physics text, there was probably no mention of a difference in the heat capacity whether volume or pressure is held constant and for solids and liquids there is little difference but we will see here that it does make a difference for gases. Students are encouraged to learn this difference and be able to prove it on a test because it is an excellent exercise in the meaning of partial derivatives in thermodynamics. What follows is an example of being able to perform correct algebra but not necessarily reaching the desired simplicity. We need to pay attention to the path through the partial derivatives and gain facility in their manipulation.

$$(C_P - C_V) = \left(\frac{\partial H}{\partial T}\right)_P - \left(\frac{\partial U}{\partial T}\right)_V = \left(\frac{\partial U}{\partial T}\right)_P + P\left(\frac{\partial V}{\partial T}\right)_P + V\left(\frac{\partial P}{\partial T}\right)_P - \left(\frac{\partial U}{\partial T}\right)_V.$$

Please note that $\left(\frac{\partial U}{\partial T}\right)_P \neq \left(\frac{\partial U}{\partial T}\right)_V$, they are not the same. In addition, we can recognize that $\left(\frac{\partial P}{\partial T}\right)_P = 0$ by definition (there is no variation in pressure if the pressure is constant). Next we need to remember that P, V, T are all related by some state function so we need to write

$$dU = \left(\frac{\partial U}{\partial V}\right)_T dV + \left(\frac{\partial U}{\partial T}\right)_V dT$$

and that

$$\left(\frac{\partial U}{\partial T}\right)_P = \left(\frac{\partial U}{\partial V}\right)_T \left(\frac{\partial V}{\partial T}\right)_P + \left(\frac{\partial U}{\partial T}\right)_V \left(\frac{\partial T}{\partial T}\right)_P.$$

With our goal of trying to make things simple we could ask why we expanded dU in terms of T and V, leaving P to be determined by some unspecified state function. The reason is that we want to eliminate $\left(\frac{\partial U}{\partial T}\right)_P$, which is the strange quantity we cannot see any easy way to measure in the laboratory. Thus, we expand dU in terms of the other variables and then impose constant P conditions. While this statement seems abstract here, it is helpful when writing this derivation to understand why we choose to expand dU in terms of T and V.

Of course $\left(\frac{\partial T}{\partial T}\right)_P = 1$ and we can substitute $\left(\frac{\partial U}{\partial T}\right)_P$ into the $(C_P - C_V)$ equation. That leads directly to the new equation as

$$(C_P - C_V) = \left[\left(\frac{\partial U}{\partial V}\right)_T + P\right]\left(\frac{\partial V}{\partial T}\right)_P + \left(\frac{\partial U}{\partial T}\right)_V - \left(\frac{\partial U}{\partial T}\right)_V.$$

So far this is a general equation and we will use it more later, but at this point let us take a short cut and specify that we are using an ideal gas with the extended definition given above.

Then we can use the relationship that $\left(\dfrac{\partial U}{\partial V}\right)_T = 0$ and cancel the last two terms we are left with

$$(C_P - C_V)_{\text{Ideal}} = [0 + P]\left(\frac{\partial V}{\partial T}\right)_P \text{ and } V = \left(\frac{RT}{P}\right)_{n=1} \text{ so } \left(\frac{\partial V}{\partial T}\right)_p = \left(\frac{R}{P}\right).$$

And thus for an ideal gas we have $(C_P - C_V)_{\text{Ideal}} = P\left(\dfrac{\partial V}{\partial T}\right)_p = P\left(\dfrac{R}{P}\right) = R$. This is only

for an ideal gas but it means that if $U = \left(\dfrac{3}{2}\right)RT$ then $C_V = \left(\dfrac{\partial U}{\partial T}\right)_V = \left(\dfrac{3}{2}\right)R$ so that

$C_P = \left(\dfrac{3}{2}\right)R + R = \left(\dfrac{5}{2}\right)R$. Now we can add to the properties of an ideal gas.

1. $PV = nRT$
2. $\left(\dfrac{\partial U}{\partial V}\right)_T = \left(\dfrac{\partial U}{\partial P}\right)_T = 0$
3. $C_P - C_V = R$

ADIABATIC PROCESSES

Next we come to another key word in thermodynamics, "adiabatic." Under some circumstances $q = 0$. One way to achieve this condition is to surround the "system" by a nonconductive layer of insulation such as a vacuum layer (Dewar flask, thermos bottle) or use some sort of glass wool or fiberglass mat around the system as is done with modern refrigerators. Another method is to carry out the process quickly because heat flow tends to be relatively slow. For instance you can put a metal poker into the hottest part of a fireplace for a few moments because the heat will not be conducted up the metal shaft to your hand in the short time it is in the heat. Another more dramatic process is in the combustion chamber of an internal combustion engine. Although such engines appear to be moving rapidly, the explosion of fuel in such an engine is much quicker than the motion of the mechanical parts. For instance an internal combustion engine (ICE) of the four-cycle type (intake, compression, power, exhaust) requires two full revolutions per explosion. Thus, an ICE engine running at a typical 3000 revolutions/minute (rpm) experiences a power stroke duration of about 1/2 of the second revolution of the full four steps, two revolution cycles, so the mechanical time for the explosion is about $\dfrac{(60\,\text{s}/\,\text{min})}{(3000\,\text{rpm})} = 0.020\,\text{s/rev}$ and only about half of every other cycle is the power stroke so the mechanical time for the explosion is about 0.01 s. That is a very long time compared to chemical reactions in the gas phase. We know from the previous discussion of the Boltzmann KMTG that the binary collision number can be of the order of $(10^{28}/\text{cm}^3\,\text{s})$ at 1 atm pressure so in a compressed gas the collision rate will be even higher. Thus, chemical considerations are needed to find a fuel whose burn time more nearly matches the mechanical timing of the engine. A combustion reaction that is too fast for the mechanical parts of the internal combustion engine causes a noticeable sound described as a "ping" or "knock" in the engine and can damage the internal parts of the engine.

For the sake of thermodynamics let us consider a diesel engine. The diesel engine was developed by Rudolf Diesel (1853–1913) who received an initial patent in Europe in 1893 and a U.S. patent in 1898 for an internal combustion engine, which used high compression of air to ignite almost any combustible fuel including crude oil. While he was involved in patent disputes, it certainly seems he was the prime inventor of the documented engines. However, there is evidence that a similar principle has been in use for unrecorded ages by Fiji natives as a fire starter illustrated by a wooden

FIGURE 4.2 A modern "fire piston" based on an ancient device used by natives of the South Pacific to start fires. It is said that Rudolf Diesel was inspired to invent the diesel engine after learning of this device. (The picture is from Wilderness Solutions at http://www.wildersol.com/. With permission of Mr. Jeff Wagner, proprietor of that company.) This picture shows the small hollow in the tip of the piston shaft where tinder is inserted prior to one quick thrust of the plunger into the wooden cylinder. This model shows the original wound string design but other models use a modern rubber O-ring. Mr. Wagner estimates temperatures of about 800°F are achieved in this wooden device.

device using rapid compression of air to ignite tinder shown as a modern version in Figure 4.2. We calculate a typical temperature for such a compression later.

In Figure 4.3 [5] we can see some details of a diesel engine illustrating the heavy design and the lack of a spark plug. It is common for diesel engines to use compression ratio of 22:1 or more and some of the technical design is in the inverted hemispherical combustion chamber shape in the top of the piston. The compression ratio (CR) is the total volume swept out by the movement of the piston from the lowest position to the top position plus the volume of the small combustion volume (in the top of the piston here) divided by the volume of the combustion chamber,

$$CR = \frac{V_{swept} + V_{comb}}{V_{comb}}.$$

Some of the early diesel engines destroyed themselves due to the high compression stresses, which led to stronger, heavier designs. Although there is no electrical spark apparatus it should be pointed out that the small rotating bump labeled "camshaft" is the effective brain of the device in that it regulates the timing of the opening and closing of the spring-loaded valves (there are separate lobes for intake and exhaust valves). In a modern engine, there would also be some mechanical synchronization of the fuel spray some time after the closing of the intake valve. It is likely both valves would be closed when the piston is traveling up in the cylinder to compress the air rapidly and then the fuel will be sprayed into the combustion chamber when the air is at the highest temperature. In modern internal combustion engines, including diesel engines, much of the high technology would be optimized to tight tolerances in the shape of the camshaft lobe (bump). The opening of the exhaust valve by the camshaft also has to be timed to allow exit of exhaust gases after completion of the power stroke of the explosive expansion but then close about the time the intake valve opens to admit a new charge of air. Since the engine may typically be operating at 3000 rpm or faster up to 5000 rpm, considerable engineering research has gone into machining/grinding the shape of the camshaft "bump" to optimize the timing of valve operation in the four-cycle sequence.

The ideal fuel for a diesel engine is cetane (*n*-hexadecane, $C_{16}H_{34}$) with a linear chain structure, which is essentially double the molecular weight of octane (C_8H_{18}) used in gasoline. Other related fuels can be burned in diesel engines but there is an interesting contrast with gasoline engines. Gasoline engines (with spark plugs) need a highly branched octane (2,2,4-trimethyl pentane,

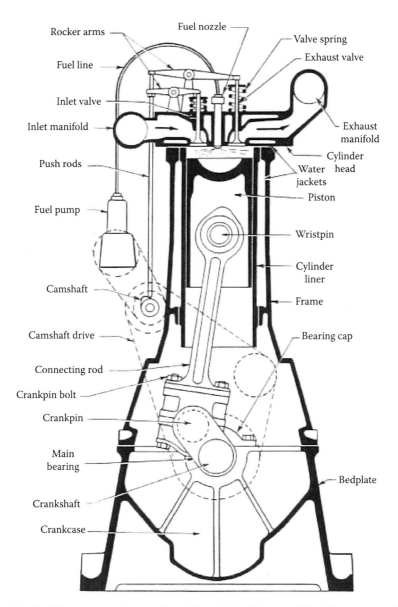

FIGURE 4.3 Detail of diesel engine design. (From Kates, E. J., *Diesel and High-Compression Gas Engines, Fundamentals*, 2nd Edn., American Technical Society, Chicago, IL, 1965, 8th printing 1966, p. 41. With the permission of Power Magazine.)

isooctane) to burn *slower* than the linear *n*-octane while the heavier diesel fuel needs to use *n*-cetane or similar *faster* burning fuels. This is an interesting problem in combustion chemistry to match the rate of the combustion to the mechanical timing of the engine. Early diesel engines characteristically produced a sooty exhaust of incompletely burned fuel but in the last 5 years design modifications have produced diesel engines that have much cleaner exhaust.

The diesel engine is a very good example of a quasi-adiabatic process. Yes, the engine does get hot and requires a system of cooling water, but it is the air that enters the engine on each cycle that we will consider. We only need the first law here to start the analysis:

$$\Delta U = 0 + w.$$

Next we use what we have learned about heat capacities and the sign convention for work done on/by a gas to write

$$dU = nC_V\, dT \quad \text{and} \quad dw = -P\, dV = -\frac{nRT}{V}\, dV.$$

Next we use a trick due to the nature of U as a state variable. While heat and work depend on the way the process is carried out, we should recall that $\Delta U = U_{\text{after}} - U_{\text{before}}$, so we can imagine a process that is idealized and as long as it has the same "before" and "after," we can calculate $\Delta U = U_{\text{after}} - U_{\text{before}}$ for that process and get the same answer as the real process that has the same beginning and ending. We can only do this for state variables as for U and H but we will soon learn there are several state variables that satisfy this "after-minus-before" property. With that information we assume that the compression of the air is carried out reversibly so that we can use the idea that $P_{\text{int}} = P_{\text{ext}}$. Then using the first law we can write

$$nC_V\, dT = \frac{-nRT\, dV}{V},$$

which is a differential equation but we can get rid of the differentials by integrating over the (T,V) changes in the process. (Drag T from the right numerator to the left denominator, cancel n and then integrate.)

$$\left[\int_{T_1}^{T_2} C_V \frac{dT}{T} = -R \int_{V_1}^{V_2} \frac{dV}{V}\right]_{n=1}.$$

Now recall the integral (which we will use often) $\int_x^{x_2} \frac{dx}{x} = \ln\left(\frac{x_2}{x_1}\right)$ and integrate both sides of our equation to obtain $\left(\frac{C_V}{R}\right) \ln\left(\frac{T_2}{T_1}\right) = -\ln\left(\frac{V_2}{V_1}\right) = \ln\left(\frac{V_1}{V_2}\right)$. Now take the anti-ln of the whole equation and raise the T-ratio to the $\left(\frac{C_V}{R}\right)$ power to find $\left(\frac{T_2}{T_1}\right)^{\left(\frac{C_V}{R}\right)} = \left(\frac{V_1}{V_2}\right)$ or perhaps as $V_1 T_1^{\left(\frac{C_V}{R}\right)} = V_2 T_2^{\left(\frac{C_V}{R}\right)}$. Some texts stop here but this is not our favorite form of analysis for a diesel engine. We show an example as an aside to illustrate the important application of adiabatic nozzle expansion for low-temperature spectroscopy. Much of the history of this technique is included in the Nobel Address of J. B. Fenn [6].

ADIABATIC NOZZLE EXPANSION SPECTROSCOPY [7]

A recently discovered way to simplify complex electromagnetic spectra of molecules is to sweep a high-pressure stream of He gas across the sample and carry it to exit into the sample chamber of a spectrometer at a much lower pressure (nominally 1 atm) whereupon the He and the sample will drastically cool and this innovation in spectroscopy has resulted in some remarkably sharp spectral details being resolved in otherwise broad spectral blurs at room temperature. However, the He must be precooled to below 51°K, which is its Joule–Thomson inversion temperature. For simplicity we can consider N_2 gas to illustrate the same point.

Example

Consider the expansion of 100 mL of N_2 gas at 100 atm in a gas bottle at 30°C to 1 atm of pressure through a nozzle so rapidly as to achieve an adiabatic process. What is the temperature of the expanded N_2 if the final volume is 10 L?

N_2 is very close to a diatomic ideal gas, so we use $C_V = \left(\frac{5}{2}\right)R$, $\left(\frac{3R}{2} \text{ translation plus} \frac{2R}{2} \text{ rotation}\right)$ and raise the whole equation to the $\left(\frac{R}{C_V}\right)$ power to find the ratio of the temperatures as $\left[\left(\frac{T_2}{T_1}\right)^{\left(\frac{C_V}{R}\right)} = \left(\frac{V_1}{V_2}\right)\right]^{\left(\frac{R}{C_V}\right)}$ to solve for the ratio of temperatures as

$\left(\frac{T_2}{T_1}\right) = \left(\frac{V_1}{V_2}\right)^{\left(\frac{R}{C_V}\right)}$, so we find that $T_2 = T_1\left(\frac{V_1}{V_2}\right)^{\left(\frac{R}{C_V}\right)} = (303.15°\text{K})\left(\frac{0.100}{10.0}\right)^{\left(\frac{1R}{2.5R}\right)} = 48.046°\text{K}.$

This rapid adiabatic expansion is sufficient to cool the nitrogen to below its boiling point of 77°K, so this is a way to make liquid nitrogen. There is a temperature for each gas called the *Joule–Thomson inversion temperature* and cooling occurs if the initial temperature is below that temperature but the gas heats upon expansion if the initial temperature is above the inversion temperature. At room temperature He is above its inversion temperature and will actually heat up upon expansion. Although there is also a pressure effect, there are absolute temperatures for this effect. For He the temperature is 51°K, for H_2 202°K, for N_2 621°K, and for O_2 it is 764°K (see discussion at http://en.citizendium.org/wiki/Joule-Thomson_effect). Thus, air ($N_2 + O_2$) can be liquefied by adiabatic expansion starting from room temperature and 1 atm, but He and H_2 must be precooled to below their Joule–Thomson inversion temperatures.

DIESEL ENGINE COMPRESSION

Now back to the diesel engine. We start from $V_1 T_1^{\left(\frac{C_V}{R}\right)} = V_2 T_2^{\left(\frac{C_V}{R}\right)}$ and assume air is close to an ideal gas, at least initially, and allow for experimental measurement of C_V and C_P. Although we will see that C_V and C_P have to be measured experimentally since it is a mixture of gases (mostly N_2 and O_2), air does behave like an ideal gas up to about 100 atm. Thus, we can substitute $T = \frac{PV}{nR}$ to obtain

$$V_1\left(\frac{P_1 V_1}{nR}\right)^{\left(\frac{C_V}{R}\right)} = V_2\left(\frac{P_2 V_2}{nR}\right)^{\left(\frac{C_V}{R}\right)}.$$

This equation can be rearranged and raised to the $\left(\frac{R}{C_V}\right)$ power to obtain an equation useful to understand the main principle of the diesel engine. Please note that we will define a new number as $\gamma = \left(\frac{C_P}{C_V}\right)$ and use the ideal gas relationship that $C_P = C_V + R$ and $\frac{C_V}{R} + 1 = \frac{C_V}{R} + \frac{R}{R} = \frac{C_V + R}{R} = \frac{C_P}{R}.$

Thus,

$$V_1^{\left(\frac{C_V}{R}+1\right)} P_1^{\left(\frac{C_V}{R}\right)} = V_2^{\left(\frac{C_V}{R}+1\right)} P_2^{\left(\frac{C_V}{R}\right)},$$

so raise the whole equation to the $\left(\frac{R}{C_V}\right)$ power. Finally we reach $\left[V_1^{\left(\frac{C_P}{R}\right)} P_1^{\left(\frac{C_V}{R}\right)} = V_2^{\left(\frac{C_P}{R}\right)} P_2^{\left(\frac{C_V}{R}\right)}\right]^{\left(\frac{R}{C_V}\right)}$

and $P_1 V_1^{\left(\frac{C_P}{C_V}\right)} = P_2 V_2^{\left(\frac{C_P}{C_V}\right)}$ or $P_1 V_1^{\gamma} = P_2 V_2^{\gamma}.$

This formula is the preferred formula to calculate/estimate the temperature of a gas under adiabatic compression and we will become familiar with it after working out some problems. Suffice it to say that when you hear the word "diesel" you should think of $P_1V_1^\gamma = P_2V_2^\gamma$.

Yes, we have made some approximations and will need to make a few more, mainly in the assumption that air remains an ideal gas up to about 100 atm, but this method is sufficiently accurate to show that the diesel compression can raise the temperature above the flash point (FP) of most fuels. The FP of a liquid is the minimum temperature at which the vapor of the liquid will form an ignitable mixture with air and for cetane/hexadecane, the *CRC Handbook* [8] reports this as 136°C, well below a temperature easily reached in diesel engines.

Example

Calculate the temperature of air compressed adiabatically in a one-cylinder diesel engine from 1040 cc at 25°C to 40 cc. Given that $C_V = (5/2)R$, compute Q, W, ΔH, and ΔU for this compression.

$$\text{"Adiabatic} + \text{diesel" means } P_1V_1^\gamma = P_2V_2^\gamma. \; \gamma = \frac{C_P}{C_V} = \frac{C_V + R}{C_V} = \frac{\left(\frac{5}{2}\right)R + R}{\left(\frac{5}{2}\right)R} = \frac{7}{5} = 1.40.$$

Assume air is an ideal gas.

$P_1 = 1\,\text{atm}$		$P_2 = ?$
$V_1 = 1040\,\text{cc}$	$Q = 0 \rightarrow$	$V_2 = 40\,\text{cc}$
$T_1 = 298.15\,^\circ\text{K}$		$T_2 = ?$

$$P_2 = P_1\left(\frac{V_1}{V_2}\right)^\gamma = (1\,\text{atm})\left(\frac{1040\,\text{cc}}{40\,\text{cc}}\right)^{1.4} = 95.71\,\text{atm,} \quad \text{then} \quad \frac{P_1V_1}{T_1} = \frac{P_2V_2}{T_2} \quad \text{and} \quad T_2 = T_1\frac{P_2V_2}{P_1V_1} =$$

$$(298.15^\circ\text{K})\frac{(95.71\,\text{atm})(40\,\text{cc})}{(1\,\text{atm})(1040\,\text{cc})} = 1097.54^\circ\text{K, which is very hot,} \quad \Delta T = 799.39^\circ\text{K.}$$

Then, even though we assumed an imaginary "reversible" path, the $\Delta U = U_\text{after} - U_\text{before} = nC_V(\Delta T)$, so we need to calculate n, the number of moles.

$$n = \frac{1040\,\text{cc}}{(22,414\,\text{cc/mol})\left(\dfrac{298.15^\circ\text{K}}{273.15^\circ\text{K}}\right)} \cong 0.04251\,\text{mol, so then}$$

$$\Delta U = nC_V\Delta T = (0.04251\,\text{mol})(2.5)(1.987\,\text{cal}/^\circ\text{K mol})(799.39^\circ\text{K}) = +168.81\,\text{cal.}$$

Similar reasoning can be used for $\Delta H = nC_P\Delta T$, which would be the same as for ΔU except for $\left(\dfrac{C_P}{C_V}\right) = \gamma$ so we use the same reasoning with (after–before) and this implies that $\Delta H = \gamma(\Delta U) = (1.4)(168.81)\,\text{cal} = +236.33\,\text{cal.}$ This calculation of ΔH is mysterious because we have not used any detailed information about the process, but we get the answer from ΔU using the powerful (after–before) principle since H is a state variable. Now what about W? We assumed $Q = 0$, so by the first law we have $\Delta U = 0 + W$, so that means that $W = +168.81\,\text{cal.}$ Note that W is positive since we certainly did work ON the air. An added comment is that the effective compression ratio of this engine is $\left(\dfrac{1040\,\text{cc}}{40\,\text{cc}}\right) = 26$. This is a bit higher than most diesel engines, which tend to have compression ratios near 22:1, but the temperature will certainly be high enough to ignite most fuels for any compression ratio greater than about 16:1.

CALORIMETRY AND THERMOCHEMISTRY

Now that we have analyzed some hypothetical applications of the first law of thermodynamics we should ask how these principles apply to chemistry and chemical reactions. The key concept is that elements react to form compounds, presumably to form lower energy situations, but that is not always the case as we will see in the next chapter. Even so, most reactions do result in a lower energy. Since energy is involved, we may think that ΔU is the key to thermochemistry, but we have already mentioned that often pressure and/or volume changes occur during a reaction. Even though these may be small effects, we know that we should work with ΔH. Thus, a question for an experimental science like chemistry is "How can we measure ΔH?" Calorimetry involves mostly simple mathematics but is really the main part of thermochemistry. Careful measurements of ΔH_{comb} are the backbone of thermochemistry.

Chemistry is the science of reactions between elements to form compounds. Physicists can ponder over the creation of the universe and the big bang theory but chemists start with the assumption of the existence of elements. Actually heavy elements are formed from lighter elements such as H and He in the interior of stars and then are distributed throughout space when such stars explode. Every atom (with the possible exception of hydrogen) in your body was once inside a star. The convention assumed in thermochemistry is that "the elements are here and chemistry is the rearrangements of these elements to form compounds so the energy changes are relative to existing elements." In the larger picture of science, we know that nuclear chemistry does occur but for the purpose of chemistry on planet Earth *we assume that the energy of formation of elements in their most abundant form is zero!* (For some elements we use the most abundant form such as graphite for carbon.) We further specify a "standard state" for the element at 1 bar pressure and $\Delta H_f^0(1\,\text{bar}, 298.15°\text{K}) \equiv 0$; older texts used the slightly different value as $\Delta H_f^0(1\,\text{atm}, 298.15°\text{K}) \equiv 0$. Older texts standardize on 1 atm pressure but the newer units specify 1 bar, which is very close to 1 atm (1 atm = 1.01325 bar). The standard state is a very important convenience so that we can tabulate the energy changes of reactions of elements to form compounds.

Another convenience on planet Earth is that our atmosphere has a lot of oxygen that is second only to fluorine in electronegativity; it is very reactive as firemen know all too well. Thus, almost any material will react with oxygen and give off a measurable amount of heat. This type of measurement is called combustion calorimetry and requires a special piece of equipment based on a high-pressure reaction container. One minor consideration in the use of a closed container is that the energy change is measured as $\Delta U = C_V \Delta T$ but we want ΔH. In Figure 4.4 [9], we see the cross section of the total calorimeter device.

The smaller container in the center of the diagram is a heavy stainless steel reaction chamber (Figure 4.5) [9] with thick walls and a screw-type lid with a rubber gasket so that it can be pressurized to about 30 atm of pure O_2 along with a small sample of about 0.5 g (carefully weighed). There is also a simple direct current electrical connection to provide ignition of the sample in the oxygen and many materials will ignite in pure O_2. A carefully measured amount of water (usually 2000 mL) surrounds the combustion vessel and is continuously stirred with a small propeller. In addition, the "water bucket" containing the reaction vessel is within a double-walled fiberglass container providing about 1 in. of insulating air space to further isolate the combustion reaction in a thermal sense. The main data are obtained from a very precise thermometer in the water as to the ΔT for the increase in temperature from the heat of the reaction in the sealed pressure container. Usually, this special thermometer covers the range of about 20°C–35°C in small increments of 0.02°. Note the heat given off by the reaction chamber is absorbed by the 2000 mL of water, so there is a sign change in the heat flow. Next we address the small correction to convert a ΔU value to a ΔH value using the definition $\Delta H = \Delta U + \Delta(PV)$ and $\Delta(PV) \approx (\Delta n_{\text{gas}})RT_{\text{ave}}$, noting that we want to use the apparatus for a small amount of n moles.

$$n\Delta H_{\text{comb}} = -C_V \Delta T + (\Delta n_{\text{gas}}RT_{\text{ave}})n.$$

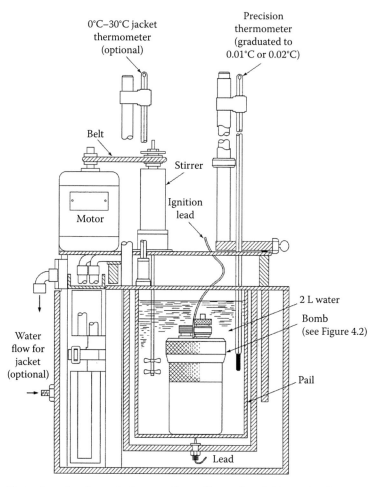

FIGURE 4.4 Heat of combustion calorimeter. (From Shoemaker, D.P. et al., *Experiments in Physical Chemistry*, 6th Edn., McGraw-Hill Companies, Inc. New York, 1996, p. 153, With permission.)

We want to obtain a ΔH value for 1 mol so we have to multiply the equation by the number of moles in the sample, n. The term shown as $-C_V \Delta T$ is the temperature change multiplying the heat capacity C_V of the reaction chamber at constant volume (since we do not want it to actually explode we make the container strong and the sample small to avoid making a hand grenade). The minus sign indicates the heat given out of the reaction chamber is taken in by the 2000 mL of water whose temperature we measure. The whole chamber can be calibrated initially by sending a known current through a metal wire of known resistance inside an empty container and using the formula $\Delta U = i^2 R$ for an electrical heating element. Then an easily purified, stable crystalline material such as benzoic acid can be measured to establish a chemical secondary standard for later routine use.

Example
After calibration with a timed amount of electrical current using $\Delta U = i^2 R$, it is found that a given calorimeter has a heat capacity of $C_V = 2569$ cal/°C = 10748.7 J/°K (the size of 1°C is exactly the same as 1°K). Then using this C_V value exactly 0.600 g of benzene, C_6H_6, is burned in the same calorimeter with an observed temperature rise of 2.332°C at an average temperature of 25°C. Calculate the $\Delta H_{\text{comb}}(25°C)$ for benzene.

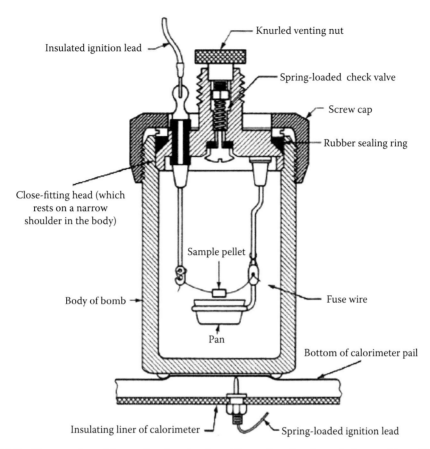

FIGURE 4.5 Cross section of the combustion chamber of heat of combustion calorimeter. (From Shoemaker, D.P. et al., *Experiments in Physical Chemistry*, 6th Edn., McGraw-Hill Companies, Inc. New York, 1996, p. 153. With permission.)

$M = 6(12.0107) + 6(1.00794) = 78.11184$ g/mol, but we do not have a whole mole.
moles benzene $= (0.600$ g$/78.11184$ g/mol$) = 0.00\ 7681294$ mol

We need to make an approximation that even though the temperature changes by 2.332°C on the centigrade scale the energy must be measured in absolute Kelvin degrees so if the initial temperature of the water is carefully started below 25°C at say 24°C and the final temperature is 26.332°C, the average temperature would be $((24 + 26.332)/2) = 25.166$°C or 298.316°K or essentially 25°C for the average temperature outside the reaction container. So if the initial temperature of the water is about 24°C the average temperature will be close to 25°C. While this seems like a crude approximation, the uncertainty is really about $(1/298.15) = 0.00335$. Next we recall the thermodynamic idea that even though the reaction in the combustion temperature may reach some high temperature briefly, the water surrounding the container cools down the gases in the chamber to just 2.351°C above the initial temperature, so the change in energy depends on the change in temperature and not the highest temperature reached. Next we need to correct the ΔU value for the $\Delta(PV)$ term. First we balance the combustion reaction.

$$C_6H_{6(liq)} + \left(\frac{15}{2}\right)O_{2(gas)} \rightarrow 6CO_{2(gas)} + 3H_2O_{liq}; \quad \Delta H_{comb} = ?$$

Recall that the approximate volume of 1 mol of liquid water is slightly more than 18 mL and the molar volume of benzene is about 90 mL but their sum is far less than the 22,414 mL value of a mole of gas so we neglect the volumes of any liquids or solids and estimate the $\Delta(PV)$ term using only the change in the moles of gas, Δn_{gas}. Here we have the net of $(6 - 7.5 \text{ mol}) = -1.5$ mol of gas during the reaction for 1 mol of benzene. Thus, we can compute $\Delta H_{comb}(25°C)$ after dividing through by the number of moles.

$$\Delta H_{comb} = \left[\left(\frac{-C_V \Delta T}{n}\right) + \frac{(\Delta n_{gas})(RT_{ave})n}{n}\right]$$

So inserting the numbers, we obtain

$$\Delta H_{comb} = \frac{(-2569 \text{ cal/deg})(2.332°)}{\left(\dfrac{0.600 \text{ g}}{78.11184 \text{ g/mol}}\right)} + (-1.5 \text{ mol})(1.987 \text{ cal/°K mol})(298°K),$$

and so we find that

$$\Delta H_{comb} = -779,934.7 \text{ cal} - 888.2 \text{ cal} = -780,822.9 \text{ cal} = -3266.9 \text{ kJ}.$$

We can extract additional meaning from this result. First, the reaction is very exothermic, a lot of heat is given off since the number is large and the sign is negative. We easily convert calories to joules using 1 cal = 4.184 J and we have rounded the value of T_{ave} to 298°K since the actual knowledge about T_{ave} is only that it is near 25°C and we see that the whole $\Delta(PV)$ term is much smaller than the main ΔU term. Throughout this text we follow the policy of keeping all the digits on our calculator until rounding to the least number of significant figures at the final answer. The *CRC Handbook* [8] lists the experimental value as -3267.6 kJ.

This method for obtaining heats of combustion has been extended to many thousands of compounds and lists of the values are tabulated (see Table 4.1) in several sources including the *CRC Handbook* [8].

A hint of the usefulness of heats of combustion can be seen in that the heat of combustion for H_2 is the same as the heat of formation of water.

$$H_2 + \frac{1}{2}O_2 \rightarrow H_2O$$

Reactions other than combustion can be treated using a definition for the heat of a reaction:

$$\Delta H_{rxn}^0(298.15°K) \equiv \sum_{product-i} n_i H_{f,i}^0 - \sum_{reactant-j} n_j H_{f,j}^0$$

Now we can see the usefulness of the definition of the heat of formation of elements being zero.

$$\Rightarrow \Delta H_{rxn}^0 = H_f^0(H_2O) - H_f^0(H_2) - \frac{1}{2}H_f^0(O_2) = -285.8 \text{ kJ} - 0 - 0 = -285.8 \text{ kJ}.$$

Note that we have to use the stoichiometric coefficients from the balanced reaction along with the molar H_f^0 values. It turns out that many of the heats of combustion and use of the definition of the heats of formation of elements as zero lead to a whole series of heats of formation of compounds, which have also been tabulated (see Table 4.2) at length in places like the *CRC Handbook* [8] for thousands of compounds. This is possible due to the algebraic summation of enthalpy values. Notice that at this point all reactions are considered to be at 25°C. To summarize, heats of combustion can be used to calculate heats of reactions which lead to standard heats of formation at 25°C.

TABLE 4.1
Selected Values of Heats of Combustion,
ΔH^0_{comb} (1 atm, 298)[a]

Compound	ΔH^0_{comb} (kJ/mol)
C (graphite)	−393.5
CO	−283.0
H_2	−285.8
NH_3	−382.8
H_2NNH_2	−667.1
N_2O	−82.1
CH_4	−890.8
HCCH	−1301.1
H_2CCH_2	−1411.2
H_3CCH_3	−1560.7
C_6H_6	−3267.6
CH_3OH	−726.1
CH_3CH_2OH	−1366.8
CH_3OCH_3	−1460.4
CH_2O	−570.7
HCOOH	−254.6
H_3CCOOH	−874.2
HCN	−671.5
H_3CNO_2	−709.2
H_3CNH_2	−1085.6
$Hg_{(liq)}$	0
$HgO_{(red)}$	−90.79

Source: Lide, D.R., *CRC Handbook of Chemistry and Physics*, 90th Edn.,
CRC Press, Boca Raton, FL, 2009–2010, pp. 5–68.
[a] Note these values are at 1 atm not 1 bar (1 atm = 1.01325 bar).

HESS'S LAW OF HEAT SUMMATION

Now we come to a simple principle that is very important in thermochemistry, Hess's law. Since energy (enthalpy) can be treated quantitatively we can balance the energy of a reaction as well as the mass. Not only that, we can add and subtract the enthalpies of several reactions. Suppose we want to find out the energy change for the hydrogenation of acetylene to form ethane, we can use the heats of combustion to find the enthalpy change for the hydrogenation step.

$$\left[HCCH + \frac{5}{2}O_2 \rightarrow 2CO_2 + H_2O; \quad \Delta H^0_{comb} = -1301.1\,kJ \right] \times (+1)$$

$$\left[H_2 + \frac{1}{2}O_2 \rightarrow H_2O; \quad \Delta H^0_{comb} = -285.8\,kJ \right] \times (+2)$$

$$\left[H_3CCH_3 + \frac{7}{2}O_2 \rightarrow 2CO_2 + 3H_2O; \quad \Delta H^0_{comb} = -1560.7\,kJ \right] \times (-1)$$

$$HCCH + 2H_2 \rightarrow H_3CCH_3; \quad \Delta H^0_{rxn} = [(-1301.1) + 2(-285.8) - (-1560.7)]kJ = -312.0\,kJ$$

TABLE 4.2
Selected Values of Heats of Formation H_f^0 in kJ
at 1 bar and 298.15°K

Compound	H_f^0 (298.15°K, 1 bar) kJ/mol	S_f^0 (298.15°K, 1 bar) J/°K mol
H_2	0	42.55
O_2	0	205.152
CO	−110.53	197.660
CO_2	−393.51	213.785
$C_{graphite}$	0	5.74
HCCH	+227.4	200.9
H_2CCH_2	+52.40	219.3
H_3CCH_3	−84.0	229.2
CH_4	−74.6	186.3
NH_3	−45.94	192.77
HCl	−92.31	186.902
Cl_2	0	233.081
H_2O	−285.830	69.95
H_2CO	−108.6	218.8

This result is actually pretty amazing in that we can use tabulated values of heats of combustion to gain quantitative information on the energy of a hydrogenation reaction. To save space we will not give as many examples of Hess's rule as it deserves but it should be noted that this is a very powerful technique in using tabulated values of a relatively easy measurement of combustion reactions so that many different reactions can be treated. How do we do this? Basically, you first write the reaction you want to treat and balance it. Then you write the combustion reactions for all the species in the reaction of interest and finally multiply the combustion reactions and their energies by factors which produce the mass balance when you add up the reactant and product species. The only new idea here is that you can multiply a combustion reaction by a negative factor to place the combustion reactant on the product side of the mass and energy balance. In effect the chemical "yield arrow →" is treated like a mathematical equal sign because enthalpy is balanced as well as mass. Notice that in the example above the moles of CO_2, H_2O, and O_2 all cancel out. That is the way to check your calculation by making sure all species other than those in the reaction of interest cancel out. Then all you have to do is treat the combustion energies algebraically using the factors you use to balance the reaction of interest.

STANDARD HEATS OF FORMATION AT 298.15°K AND 1 BAR PRESSURE

We have introduced this concept above and the concept is straightforward in the sense that chemists work with elements that already exist and then consider the energy requirements to make compounds. We show below only a few typical values relative to the elements and we expect the energies to be negative but acetylene and ethylene curiously have positive heats of formation. Acetylene and ethylene are stable compounds under most conditions near room temperature and 1 bar pressure but the thermochemistry tells us they are unstable relative to the free elements and we say they are *metastable* with a hint that maybe they are quite reactive under some conditions. Note that for carbon, graphite is the standard form.

TEMPERATURE DEPENDENCE OF REACTION ENTHALPIES

Do all reactions occur at 25°C? Of course not! Thus, we need a way to use H_f^0 values for reactions at other temperatures. The hint of an approach is to note that $C_P = \left(\dfrac{\partial H}{\partial T}\right)_P$, that is, C_P is the amount by which H changes with temperature at constant pressure. In fact we can integrate C_P over a range of temperatures to correct H_f^0 for temperatures other than 298.15°K.

$$\int_{298.15}^{T} C_P \, dT = \int_{298.15}^{T} \left(\frac{\partial H}{\partial T}\right)_P dt = \int_{298.15}^{T} dH = H_T - H_{298.15}.$$

Thus, $H_T = H_{298.15} + \displaystyle\int_{298.15}^{T} C_P \, dT$. However, C_P may depend on the temperature such as $C_P(T)$.

$$H_T^0 = H_{298.15}^0 + \int_{298.15}^{T} C_P(T) dT.$$

Since we may want to use a C_P value not at one of the tabulated temperatures we can fit a polynomial to discrete values of C_P in terms of T. Here "e" is simply the fifth numerical coefficient.

$$C_P \cong a + bT + cT^2 + dT^3 + eT^4$$

While it is possible to get a pretty good fit to most heat capacities with just a polynomial up to T^3, there are some heat capacities for which this is inadequate so we will report data here to T^4 and we will use recent data from the *CRC Handbook* [8] for the data points. As a justification for using the T^4 term we also show the polynomial fit to the CRC data for H_2 (Figure 4.6) where we see that even

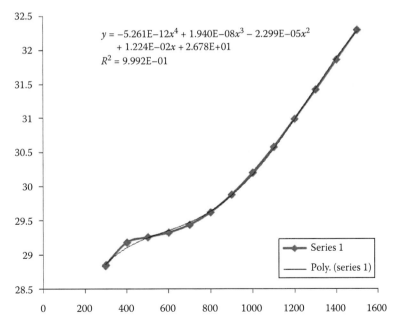

$$y = -5.261E-12x^4 + 1.940E-08x^3 - 2.299E-05x^2 + 1.224E-02x + 2.678E+01$$
$$R^2 = 9.992E-01$$

FIGURE 4.6 A fourth-order polynomial fit to heat capacity data for H_2; data points from the *CRC Handbook* in J/°K mol. Here x is the Kelvin temperature and we can see that even with a T^4 term the fit is not perfect near 400°K and that R^2 is very good but less than 1.000. However, the fit near 1200°K is quite good.

with the T^4 term the fit is not perfect. We need that higher term to improve the smooth interpolation of the heat capacity for any temperature from 300°K to 1500°K. In bygone days, the T^4 term would have caused an extra hardship in the calculation we are about to do if one were limited to a slide rule but today students use calculators and personal computers. In fact we are going to do one example, which could easily be programmed to use a small data library of heat capacity polynomials and H^0_{298} values to automate the calculation of $\Delta H_{rxn}(T)$ values in a few milliseconds on a personal computer. This calculation is the sort of thing that is tedious to do by hand but easy with a computer. However, a general practice in computer programming is to carry out a check of the method using at least one pencil-and-paper calculation.

POLYNOMIAL CURVE FITTING

Most students have heard that "with enough parameters you can draw an elephant," referring to a danger in curve fitting. Parameterization is useful but can lead to nonsense unless applied with care. The numerical value of the "coefficient of determination, R^2" is quoted with the "trend line" fit in Excel as a measure of how good the fit is to the data points. Here $R^2 \equiv 1 - \dfrac{\sum_i^n (y_i - f_i)^2}{\sum_i^n (y_i - \bar{y})^2}$, where $\bar{y} \equiv \dfrac{\sum_i^n y_i}{n}$, f_i are the values of the trend line function at the respective x_i values, and the y_i values are the actual data points from the set of (x_i, y_i) input. We see that the denominator of the second term is a positive number as the square of the deviation of the y_i points from the average value of $(\bar{y})$ and represents the range of the y_i values, but if the computed f_i values of the trend line are all equal to the y_i values, the R^2 value will be 1. Note there is a danger here in that a high-order polynomial can exist, which will pass through every y_i point but oscillate wildly between the points. A second danger is that a tight fit for a set of data points can produce a polynomial, which will diverge greatly from the data set when a value of x is used outside the range of the data set. The polynomial fit should only be used for x values within the range of the data set used for the polynomial. Probably the order of a polynomial fit should not be greater than $(n/2)$ and the best way to fit a curve with a polynomial is to "creep up" on the best fit by slowly increasing the order of the polynomial as R^2 approaches 1 but make sure the order of the polynomial is less than the number of data points. Here we fit a fourth-order polynomial (Figure 4.7) to 13 points.

The specific values of the heat capacities are tabulated in several places but the values used here are from the *CRC Handbook* [8] presented as values for temperatures from 298.15°K to 1500°K.

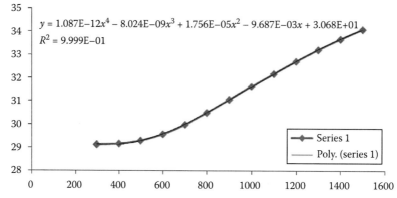

$$y = 1.087\text{E}{-}12x^4 - 8.024\text{E}{-}09x^3 + 1.756\text{E}{-}05x^2 - 9.687\text{E}{-}03x + 3.068\text{E}{+}01$$
$$R^2 = 9.999\text{E}{-}01$$

FIGURE 4.7 C_P heat capacity of HCl from 298.15°K to 1500°K with a fourth-order polynomial fit to specific data points in J/°K mol). The "coefficient of determination, R^2" is very good here with a value of 0.9999, which indicates a near-perfect fit of the polynomial. Here the x value is Kelvin T and T^4 is needed to achieve a near-perfect polynomial fit.

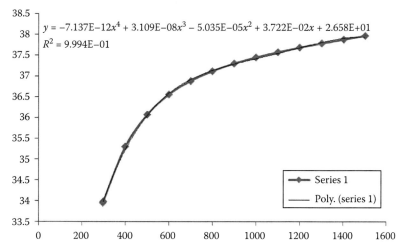

FIGURE 4.8 The polynomial fit to the heat capacity for Cl_2 gas where x in the polynomial is T in °K, the Y-axis is in J/°K mol. The R^2 value of 0.9994 is close to the perfect fit value of 1.0000, which indicates a very good fit.

Few texts show how the polynomials are obtained but today it is easy to use a program such as Microsoft Office Excel to fit a "trend line" polynomial (Figure 4.8) to the modern data given in the *CRC Handbook* [8]. Options within the trend line permit scientific notation extended to four significant figures (tap on the polynomial and then right click) to obtain results shown here.

APPLICATION TO ΔH_{rxn}^0 ($T > 298.15°K$)

Next we show a "once-in-your-life" calculation using the heat capacity polynomials to correct the heat of a reaction for a temperature other than 298.15°K. Obviously this sort of calculation should be programmed for a computer to carry out the detailed steps in future applications but it is educational to do the calculations at least once using just a calculator.

Example
Calculate the $\Delta H_{rxn}(1200°K)$ value for the reaction $1/2H_2 + 1/2Cl_2 \rightarrow HCl$ using H_f^0 values and heat capacity polynomials.

The important concept here is that heat capacities are algebraically subject to the same rules as the H_f^0 values because they are energy quantities. Thus, we have as a formal equation

$$\Delta H_{rxn} = \sum_i^{prod} n_i H_{f,i}^0 - \sum_j^{react} n_j H_{f,j}^0 + \int_{298.15}^{T} \left[\sum_i^{prod} n_i C_{p,i}(T) - \sum_j^{react} n_j C_{p,j}(T) \right] dT.$$

When we use the polynomial heat capacities this leads to

$$\Delta H_{rxn}(T) = \Delta H_{f,298.15}^0 + \int_{298.15}^{T} (\Delta a) dT + \int_{298.15}^{T} (\Delta b) T dT + \int_{298.15}^{T} (\Delta c) T^2 dT + \int_{298.15}^{T} (\Delta d) T^3 dT$$

$$+ \int_{298.15}^{T} (\Delta e) T^4 dT.$$

TABLE 4.3
Polynomial Fits to Selected Heat Capacities, Temperature Range 298.15°K–1500°K

Compound	a	b (E−2)	c (E−5)	d (E−9)	e (E−12)	R^2
H_2	26.78	1.224	−2.299	19.40	−5.261	0.9992
O_2	28.28	−2.347	2.647	−24.15	6.679	0.9997
CO	31.30	−1.699	4.053	−28.34	6.686	1.0000
CO_2	19.29	7.756	−6.818	31.75	−6.123	1.0000
C_{2gas}	53.24	−3.795	1.887	7.421	−5.209	0.9896
$C_{graphite}$	6.227	6.520	−6.046	2.908	−5.881	1.0000
$C_{diamond}$	12.16	8.158	−7.889	40.09	−8.537	0.9999
HCCH	14.29	14.09	−16.54	102.2	−24.20	0.9999
H_2CCH_2	4.279	15.04	−7.623	14.87	0	1.0000
H_3CCH_3	4.084	18.34	−7.623	11.31	0	1.0000
CH_4	26.86	2.817	11.50	−94.51	23.74	1.0000
N_2	31.41	−1.653	3.643	−24.00	5.383	1.0000
NH_3	27.38	2.210	2.487	−24.26	6.197	1.0000
HCl	30.68	−0.9687	1.756	−8.024	1.087	0.9999
Cl_2	26.58	3.722	−5.035	31.09	−7.137	0.9994
H_2O	32.20	−0.4494	2.286	−12.43	2.252	1.0000
H_2CO	27.11	1.052	7.635	−70.77	18.84	0.9999

This is surely a formidable expression but we can simplify it by doing the integrals term by term.

$$\Delta H_{rxn}(T) = \Delta H^0_{f,298.15} + (\Delta a)(T - 298.15) + (\Delta b)\left[\frac{T^2 - (298.15)^2}{2}\right] + (\Delta c)\left[\frac{T^3 - (298.15)^3}{3}\right]$$
$$+ (\Delta d)\left[\frac{T^4 - (298.15)^4}{4}\right] + (\Delta e)\left[\frac{T^5 - (298.15)^5}{5}\right].$$

A common error students make is to use $(T - 298.15)^n$ instead of the correct form as $[T^n - (298.15)^n]$. This sort of problem often appeared on final examinations in the days when students had only slide rules so it should be easier with a calculator. The easy (?) way to set up this problem is to first calculate the Δ terms in the heat capacity polynomial.

Keep in mind the coefficients in the balanced equation $1/2H_2 + 1/2Cl_2 \rightarrow HCl$.

The key formula is $\Delta H_{rxn}(T) = \Delta H^0_{298.15} + \int_{298.15}^{T} (\Delta a + \Delta bT + \Delta cT^2 + \Delta dT^3 + \Delta eT^4)dt$.

Use data from Table 4.3

Compound	a	b (E−2)	c (E−5)	d (E−9)	e (E−12)	R^2
HCl	30.68	−0.9687	1.756	−8.024	1.087	0.9999
H_2	26.78	1.224	−2.299	19.40	−5.261	0.9992
Cl_2	26.58	3.722	−5.035	31.09	−7.137	0.9994

$$\Delta a = a_{HCl} - \left(\frac{a_{H_2}}{2}\right) - \left(\frac{a_{Cl_2}}{2}\right) = \left[30.68 - \frac{(26.78 + 26.58)}{2}\right] = 4.00$$

$$\Delta b = b_{HCl} - \left(\frac{b_{H_2}}{2}\right) - \left(\frac{b_{Cl_2}}{2}\right) = \left[-0.9687 - \frac{(1.224 + 3.722)}{2}\right](10^{-2}) = -0.034417$$

$$\Delta c = c_{HCl} - \left(\frac{c_{H_2}}{2}\right) - \left(\frac{c_{Cl_2}}{2}\right) = \left[1.756 - \frac{(-2.299 - 5.035)}{2}\right](10^{-5}) = 5.423 \times 10^{-5}$$

$$\Delta d = d_{HCl} - \left(\frac{d_{H_2}}{2}\right) - \left(\frac{d_{Cl_2}}{2}\right) = \left[-8.024 - \frac{(19.40 + 31.09)}{2}\right](10^{-9}) = -33.269 \times 10^{-9}$$

$$\Delta e = e_{HCl} - \left(\frac{e_{H_2}}{2}\right) - \left(\frac{e_{Cl_2}}{2}\right) = \left[1.087 - \frac{(-5.261 - 7.137)}{2}\right](10^{-12}) = 7.286 \times 10^{-12}$$

Next we need to compute the main term

$$\Delta H^0_{rxn} = H^0_f(HCl) - \left(\frac{1}{2}\right)H^0_f(H_2) - \left(\frac{1}{2}\right)H^0_f(Cl_2) = -92.31 - (0 + 0)/2 = -92.31\,kJ$$

$$\Delta a \Rightarrow (4.00)(1200 - 298.15) = 3607.4\,J/mol$$

$$\Delta b \Rightarrow (-0.034417)\frac{[(1200)^2 - (298.15)^2]}{2} = -23{,}250.51754\,J/mol$$

$$\Delta c \Rightarrow (5.423 \times 10^{-5})\frac{[(1200)^3 - (298.15)^3]}{3} = 30{,}757.8373\,J/mol$$

$$\Delta d \Rightarrow (-33.269 \times 10^{-9})\frac{[(1200)^4 - (298.15)^4]}{4} = -17{,}180.92635\,J/mol$$

$$\Delta e \Rightarrow (7.286 \times 10^{-12})\frac{[(1200)^5 - (298.15)^5]}{5} = 3622.54675\,J/mol$$

Sum of terms $= -2443.65984\,J/mol = -2.44365984\,kJ/mol \cong -2.44\,kJ/mol$.
Thus, $\Delta H^0_{rxn}(1200°K) = -92.31 - 2.44 = -94.75\,kJ/mol$.

Other texts use more convenient shorter polynomial expansions for this reaction based on older data from 1934–1948 [6] but this result comes from the use of the more recent heat capacity data in the *CRC Handbook* [8] and the R^2 values for the polynomials used here indicate excellent numerical fitting to the experimental data points. As such we believe this result is more accurate than the result using polynomials only up to T^3. This text also shows how the polynomials were determined and provides the R^2 values to evaluate goodness of the fitting procedure. While that is a lot of work for a correction of less than 3%, we admit to a tendency of physical chemists to make extra effort to gain accuracy. Note the interesting alternating signs of the various correction terms that result in the net correction. We carried all the places on a ten place calculator to allow students to follow the computation, but in the end the answer was rounded to only four significant figures. The educational value of this exercise is that we do have a way to correct ΔH^0_{rxn} for temperatures other than 298.15°K. The example also teaches us that when faced with a complicated calculation it is useful to organize the overall process into separate steps. It is tempting to think this problem could be programmed for automation in Basic, f77, or Java to make the whole process less of a chore for humans. Even so the whole process needs to be worked out at least once to check out any automatic program.

OTHER TYPES OF THERMOCHEMISTRY

Here we face the problem of showing the mainstream applications of thermodynamics in one semester, so we neglect other uses of the algebraic additivity of enthalpy values. We only mention here that carefully weighed amounts of minerals can be dissolved in HCl/HF solvents in special solution calorimeters to make accurate estimates of heats of formation of minerals that occur over a period of thousands of years with use of Hess's rule. Other special calorimeters can be used to measure the heat of the process of denaturing (unfolding) proteins in a strong base. Still other calorimeters can be used to measure the heats of sublimation (solid-to-gas), heats of fusion of solids (solid-to-liquid), and heats of vaporization (liquid-to-gas) transitions phase changes but the key idea is the algebraic summation of enthalpy values, which we have treated above mainly for gas phase reactions.

PERSPECTIVE

The student needs to understand that instead of skimming the surface of thermodynamics in a short one-semester treatment we have plowed more deeply into just a few examples of the first law in this chapter. The idea is that we have selected what we think are important illustrations with sufficient detail to prepare an interested student to elect a second semester and yet provide a good foundation for students who stop at just one semester of study and need to apply principles of thermodynamics to other disciplines. We have shown details for partial derivatives and ways to find heats of reactions, but as chemical engineers and chemistry graduate students will tell you, there is a lot more to thermodynamics, except now you have a good preparation for further study.

KEY FORMULAS AND EQUATIONS

For reversible ideal gas processes:

$$w = - \int_{V_1}^{V_2} P\, dV = - \int_{V_1}^{V_2} \left(\frac{nRT}{V}\right) dV = -nRT \ln\left(\frac{V_2}{V_1}\right) = -nRT \ln\left(\frac{P_1}{P_2}\right).$$

For adiabatic ($Q = 0$) temperature changes:

$$V_1 T_1^{\left(\frac{C_V}{R}\right)} = V_2 T_2^{\left(\frac{C_V}{R}\right)}.$$

For adiabatic ($Q = 0$) pressure changes:

$$P_1 V_1^{\left(\frac{C_P}{C_V}\right)} = P_2 V_2^{\left(\frac{C_P}{C_V}\right)}.$$

For constant volume heats of combustion:

$$n\Delta H_{\text{comb}} = -C_V \Delta T + (\Delta n_{\text{gas}} R T_{\text{ave}})n.$$

For heats of reaction at standard conditions:

$$\Delta H_{\text{rxn}}^0(298.15^\circ\text{K}) \equiv \sum_{\text{product}-i} H_{\text{f},i}^0 - \sum_{\text{reactant}-j} H_{\text{f},j}^0.$$

For heats of reaction for $T > 298.15°K$:

$$\Delta H^0_{rxn}(T) = \Delta H^0_{rxn}(298.15) + \int_{298.15}^{T} (\Delta a + \Delta bT + \Delta cT^2 + \Delta dT^3 + \Delta eT^4)dT.$$

PROBLEMS

4.1 Calculate q, w, and ΔU for the reversible, isothermal compression of 10 mol of ideal gas from 1 to 10 atm at a constant temperature of 0°C.

4.2 Calculate the temperature of air compressed adiabatically in a one-cylinder diesel engine from 1060 cc at 25°C to 60 cc. Given that $C_V = (5/2)R$, compute Q, W, ΔH, and ΔU for this compression.

4.3 Use heats of combustion to calculate the $\Delta H^0_{298.15}$ for $3HC \equiv CH \rightarrow C_6H_6$. Although this reaction seems like an improbable stereospecific termolecular collision is required in the gas phase, it does occur on surfaces so much so that high-pressure tanks of acetylene have added impurities to inhibit the formation of benzene.

4.4 Given that 0.500 g of n-heptane (C_7H_{16}) burned in a constant volume combustion calorimeter with $C_V = 1954$ cal/°C causes $\Delta T = 2.934°C$ at $T_{ave} = 25°C$, calculate the molar ΔH^0_{comb} for n = heptane.

4.5 A little known fact is that most gasoline engines will run on "wood smoke" from a smoldering fire of burning wood, paper, mulch, almost any cellulose material in a limited amount of O_2. Although there is an added safety concern regarding having a stove on the same vehicle as an alternate tank of liquid fuel, and the overall power is less than with gasoline, the main reaction is the further combustion of CO. Thus, calculate the heat of the reaction $CO + 1/2O_2 \rightarrow CO_2$ at 1000°K to estimate the molar heat of the reaction in the combustion chamber, $\Delta H^0_{rxn}(1000°K)$.

TESTING, GRADING, AND LEARNING?

Throughout this text we will include actual tests that have been given by the author either in a one semester course (CHEM 305) for forensic majors, CHEM 303–304 for chemistry majors at Virginia Commonwealth University, or CHEM 311–312 at Randolph Macon College. Copies of these tests are provided to all the students along with the answer keys and they are encouraged to add material to their quiz answers that may not be asked (at lower point value). In some courses, only a few students have access to old tests but here we provide old tests to the whole class to give them an "equal study opportunity." Why not, if they actually learn from these old tests?

Physical chemistry 305	Fall 2005 Midterm examination	D. Shillady, Professor, VCU
(Points)	(Attempt all problems)	90 min*

(10) 1. Compute ΔH^0_{298} for the reaction: $C_6H_6 + 3H_{2(g)} \rightarrow C_6H_{12}$, given $\Delta H_{comb}(C_6H_6) = -782.3$ kcal/mol, $\Delta H_{comb}(C_6H_{12}) = -937.8$ kcal/mol, and $\Delta H_{comb}(H_2) = -68.3$ kcal/mol.
(Ans. $\Delta H^0_{298} = 49.4$ kcal)

(15) 2. Given $C_P(CO_2) = 8.87$, $C_P(CO) = 6.97$, and $C_P(O_2) = 7.02$ cal/mol °K and $\Delta H^0_{298} = -67.700$ kcal for $CO_{(g)} + 1/2O_{2(g)} \rightarrow CO_{29(g)}$, estimate the ΔH^0_{350} value at 350°K assuming the C_P values remain constant.
(Ans. $\Delta H^0_{350} = -67.784$ kcal)

---

* Note DS finished this in 61 min. Some of these questions require ideas treated in Chapters 5 and 6. We put this test here because most of it relates to thermochemistry as given in this chapter. The time limit was supposed to be 55 min but since the room was empty after this class, the time limit was extended to 90 min for the students.

(10) 3. Calculate the laminar bulk flow rate in gallons/min for blood with $\eta = 0.014$ poise through an aorta 4 in. long and 1/4 in. inner diameter due to a pressure difference of 130–70 mmHg, and a pulse duty cycle of 0.10. (Ans. 3.558 gallons/min)

(15) 4. The vapor pressure of ethanol is 135.3 mmHg at 40°C and 542.5 mmHg at 70°C. Calculate ΔH_{vap} and the boiling point of ethanol at 760 mmHg.

(Ans. $\Delta H_{vap} = 9883.7$ cal/mol, 78.1°C)

(10) 5. Given $a = 3.610$ atm L^2/mol^2 and $b = 0.0429$ L/mol for the van der Waals equation of CO_2, calculate the pressure (P) of 3 mol of CO_2 in a 5 L container at 50°C. (15.03137 atm)

(15) 6. A 0.500 g sample of n-heptane (C_7H_{16}) burned in a constant volume calorimeter causes $\Delta T = 2.934$°C. If C_V of the calorimeter is 1954 cal/°C and $T_{ave} = 25$°C, calculate ΔH_{comb} of n-heptane at 298°K. (Ans. $\Delta H_{comb} = -1151.3$ kcal/mol)

(15) 7. Using the value of $\Delta H^0_{298} = +17.06$ kcal and $\Delta G^0_{298} = +9.72$ kcal for the reaction $2NOCl_{(g)} \rightleftharpoons 2NO_{(g)} + Cl_{2(g)}$, determine the temperature at which $K_P = 0.600$, assuming ΔH and ΔG remain constant. (Ans. $T = 665.5$°K)

(10) 8. Given $\Delta H_{vap}(H_2O) = 40.71$ kJ/mol, calculate ΔS_{vap} at 100°C. (Ans. $\Delta S_{vap} = 109.1$ J/°K)

REFERENCES

1. Hills, R. L., *James Watt: II The Years of Toil, 1775–1785*, Landmark, Ashbourne, U.K., 2005, pp. 58–65.
2. Thompson, B. (Count Rumford), Experiments upon heat, *Phil. Trans. Rayol Soc.*, 48–80 (1792).
3. Joule, J. P., On the existence of an equivalent relation between heat and the ordinary forms of mechanical power. *Philos. Mag.*, **27**(3), 205 (1845).
4. Denbigh, K., *The Principles of Chemical Equilibrium*, Cambridge University Press, New York, 1961, p. 20.
5. Kates, E. J., *Diesel and High-Compression Gas Engines, Fundamentals*, 2nd Edn., American Technical Society, Chicago, IL., 1965, 8th Printing 1966, p. 41.
6. Castellan, G. W., *Physical Chemistry*, 3rd Edn., Addison-Wesley Publishing Co., 1983, pp. 141–142, based on H. M. Spencer and J. L. Spencer, *J. Am. Chem. Soc.*, **56**, 2311 (1934); Spencer, H. M. and G. N. Flanagen, *J. Am. Chem. Soc.*, **64**, 2513 (1942) and Spencer, H. M., *Ind. Eng. Chem.*, **40**, 2152 (1948).
7. Anderson, J. B. and J. B. Fenn, Velocity distributions in moleculer beams from nozzle sources, *Phys. Fluids* **8**, 780 (1965).
8. Lide, D. R., *CRC Handbook of Chemistry and Physics, 2009–2010*, 90th Edn., 2009–2010, CRC Press, Boca Raton, FL, pp. 5–43.
9. Shoemaker, D. P., C. W. Garland, and J. W. Nibler, *Experiments in Physical Chemistry*, 6th Edn., McGraw-Hill Companies, Inc., New York, 1996, p. 153.

5 The Second and Third Laws of Thermodynamics

INTRODUCTION

While the first law of thermodynamics has some interesting mysteries, which are overcome using the powerful "after-minus-before" principle, energy concepts are familiar to us in science. The next important topic is all around us and taken for granted as a part of life, but most of us do not know it can be made quantitative. It is the principle we sometimes call "Murphy's law" related to the natural tendency of disorder to increase. If you open a brand new deck of cards and drop them from waist height, do you expect they will remain in order? Stack 10 coins heads up and drop them again from waist height; do you expect them to land all heads up? The answer is "no" to both questions, so what is going here? We are hinting strongly that while there is a natural tendency for energy to run "down hill," there is another tendency in nature that tends to increase.

One of the major discoveries of thermodynamics, primarily by Boltzmann, is the way to quantify the phenomenon of disorder beginning with his 1866 PhD thesis. Shortly before that Clausius proposed a second law of thermodynamics in words in 1862. But first, maybe the basic concept was discovered by Sadi Carnot in 1824 (Figure 5.1) [1].

CARNOT CYCLE/ENGINE

The second law of thermodynamics has historically been a mysterious concept, and the basic idea has been verbalized by Clausius, Kelvin, Planck, and others for those who "think in words." One simple statement by Rudolph Clausius (1822–1888) was

> Heat generally cannot flow spontaneously from a material at lower temperature to a material at higher temperature.

The key word here is "spontaneously" because we know that refrigeration can move heat from cold to hot regions, but the natural trend is for heat energy to flow from a hot environment to a cold environment. There are many verbal variations of the second law, and a student can find a great deal of further discussion in other texts but in keeping with the idea of *Essential Physical Chemistry* we choose to pin our explanation on algebraic results from the idealized Carnot cycle (Figure 5.2). Carnot defined a cyclic process in four steps as a process for a hypothetical engine to convert heat to work. The process is usually described with a P, V graph. According to the ideal gas law, the equation for "$PV = $ constant" leads to the positive branch of a hyperbola called an "isotherm." In the previous chapter, we carried out a model analysis of the adiabatic response of a gas to compression and expansion. The shape of the PV curve for an adiabatic step is not an isotherm because we know the temperature changes. Consider an adiabatic expansion of a gas. If $q = 0$ and no heat flows into the gas, the temperature drops, so what would have been an isotherm on the PV graph "sags" to a lower temperature. Similarly, we know from the diesel engine example that adiabatic compression with no heat flow, $q = 0$, leads to a higher temperature, which crosses over isotherms in the upward direction. Thus, we expect isotherms for constant temperature processes but deviations from the perfect hyperbolic shape for adiabatic steps on a P, V graph. There is a dynamic

FIGURE 5.1 Nicholas Léonard Sadi Carnot (1796–1832) was a French physicist and military engineer who published the "Carnot cycle" in 1824. At an early age he attended the Ecole Polytechnique in Paris where there was a very distinguished faculty, the list reads like the very foundations of modern science (Claude-Louis Navier, Gaspard-Gustave Coriolis, Joseph Louis Gay-Lussac, Siméon Denis Poisson, and Andre-Marie Ampere). Carnot died in a cholera epidemic when he was only 36. (Boilly lith., Photographische Gesellschaft, Berlin, courtesy AIP Emilio Segrè Visual Archives, Harvard University Collection.)

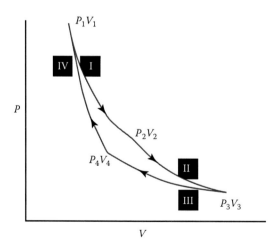

FIGURE 5.2 A qualitative sketch of the Carnot cycle. The actual shape is more of a narrow crescent shape than is often depicted in other sources. Photographs of Carnot's notes show the graph narrow as shown here. An excellent simulation can be found on the Internet at http://demonstrations.wolfram.com/CarnotCycleOnIdealGas/

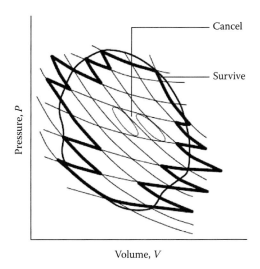

Pressure, *P*

Volume, *V*

Cancel

Survive

FIGURE 5.3 Many Carnot cycles superimposed on an arbitrary *PV* cycle for some other heat engine. Using the calculus idea of breaking up a macroscopic function into small increments, we see that the *PV* energy product will sum up around the edge for the real engine but cancel within the center of the graph. The diagram grid is necessarily coarse here to illustrate the central Carnot cycles but a finer grid could be used to match the real cycle exactly in the limit of very tiny Carnot cycles. The main conclusion is that the Carnot efficiency formula can be applied to any real heat engine.

simulation/demonstration of this at an Internet site by Jacquie Hui Wan Ching, Department of Physics, University of Virginia to be found at http://www.corrosion-doctors.org/Biographies/carnotcycle.htm

Before we give the mathematical details of the Carnot cycle, we should say that there is a good/bad news situation regarding "reality." First, it would appear that there is no technical way to actually construct an exact "Carnot engine"; it is an idealized process that is simplified for mathematical analysis. Second, the good news is that when the *P*, *V* graph of any real engine cycle is plotted on a graph paper using a grid of isotherms and adiabats, the idealized Carnot cycle can be used to flesh out the interior of the real graph in the same way that *dx*, *dy* are used in evaluating an area in calculus, and the outer part of the cycle of the real engine graph will still satisfy the Carnot cycle principles. Thus, this idealized analysis really can be applied to a real heat engine (Figure 5.3).

CARNOT CYCLE

We could specify a theoretical engine of 1 mol gas displacement, 22.414 L, which is huge compared to an automobile V8 engine in the 6 L displacement range, perhaps in the range of size of a steam locomotive engine, but really we only need to specify $n = 1$ in the following discussion. We start at the "hot point" in Figure 5.2 of the cycle with (P, V) values of a nominal fuel explosion or injection of hot gas into some sort of piston arrangement. We know $\Delta U = q + w$ and for gases $w = -\int P \, dv$, so keep track of the signs.

I Isothermal expansion from (P_1, V_1) to (P_2, V_2)
Isothermal $\Rightarrow \Delta T = 0$ so $\Delta U = nC_V \Delta T = 0$ and $q_I = -w_I = +\int_{V_1}^{V_2} P \, dV = RT_h \ln\left(\frac{V_2}{V_1}\right)$
$(\Delta U = 0 = q + w), n = 1$

II Adiabatic expansion from (P_2, V_2) to (P_3, V_3)
Adiabatic $\Rightarrow q_{II} = 0$ so $\Delta U = w_{II} = C_V(T_l - T_h)$ and so $-w_{II} = C_V(T_h - T_l)$.

III Isothermal compression from (P_3, V_3) *to* (P_4, V_4)
Isothermal $\Rightarrow \Delta T = 0$ so $\Delta U = nC_V \Delta T = 0$ and $q_{III} = -w_{III} = +\int_{V_3}^{V_4} P \, dV = RT_1 \ln\left(\dfrac{V_4}{V_3}\right)$
$(\Delta U = 0 = q + w), n = 1$.

IV Adiabatic compression from (P_4, V_4) *to* (P_1, V_1)
Adiabatic $\Rightarrow q_{IV} = 0$ so $\Delta U = w_{IV} = C_V(T_h - T_1)$ and so $-w_{IV} = C_V(T_1 - T_h)$.

Now, we need an important side calculation relating the two adiabatic steps for 1 mol of ideal gas when $q = 0$. (*the key step.*)

$\Delta U = \int C_V \, dT = 0 - \int P \, dV = -RT \int \dfrac{dV}{V}$, which can be rearranged to collect the temperature dependence as $\int C_V \dfrac{dT}{T} = -R \int \dfrac{dV}{V}$. This is easily done algebraically by removing the integration symbols, dragging T under the left side and then reapplying the integral signs. This step is similar to the previous treatment of adiabatic nozzle jet cooling. Here, our goal is to merely derive a key relationship between the volumes and temperatures of the adiabatic steps and note that we kept the minus sign on the work term because we are considering the general case of either an expansion or a compression. Thus, we have

$$C_V \int_{T_1}^{T_2} \frac{dT}{T} = C_V \ln\left(\frac{T_2}{T_1}\right) = -R \int_{V_1}^{V_2} \frac{dV}{V} = -R \ln\left(\frac{V_2}{V_1}\right) = +R \ln\left(\frac{V_1}{V_2}\right).$$

Step II: $C_V \ln\left(\dfrac{T_1}{T_h}\right) = -R \ln\left(\dfrac{V_3}{V_2}\right)$ and Step IV: $C_V \ln\left(\dfrac{T_h}{T_1}\right) = -R \ln\left(\dfrac{V_1}{V_4}\right)$. We can invert the ratio from Step IV (multiply both sides by -1) to obtain

$$C_V \ln\left(\frac{T_1}{T_h}\right) = -R \ln\left(\frac{V_4}{V_1}\right) = -R \ln\left(\frac{V_3}{V_2}\right)$$

and we find $\left(\dfrac{V_4}{V_1}\right) = \left(\dfrac{V_3}{V_2}\right)$ and so $\left(\dfrac{V_4}{V_3}\right) = \left(\dfrac{V_1}{V_2}\right)$.

We do not know how Carnot thought about the equations, but when he decided to sum the quantity, q/T, over all four steps, he found a profound result.

$$\sum\left(\frac{q_{rev}}{T}\right) = \frac{RT_h \ln\left(\frac{V_2}{V_1}\right)}{T_h} + 0 + \frac{RT_1 \ln\left(\frac{V_4}{V_3}\right)}{T_1} + 0 \text{ but } \left(\frac{V_4}{V_3}\right) = \left(\frac{V_1}{V_2}\right) \text{ so the ln terms cancel out.}$$

Note that we used the reversible equation for the adiabatic steps of the PV work, so we must specify that the heat term be noted as q_{rev}. What does this mean? Concisely for the cycle

$$\sum\left(\frac{q_{rev}}{T}\right) = \oint \frac{dq_{rev}}{T} = 0.$$

The importance of this is that a variable that satisfies this matches the same condition as for $\oint dU = 0$ and $\oint dH = 0$, which are what we have called "state variables." Thus, the Carnot cycle is very useful to show the existence of another state variable, which is now called "entropy" with the symbol "S" such that we have

$$\oint\left(\frac{dq_{rev}}{T}\right) = \oint dS = 0.$$

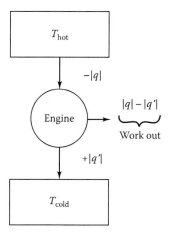

FIGURE 5.4 A schematic showing the entropy changes and efficiency of a heat engine.

Later Boltzmann related entropy to statistical disorder, which we treat later in this chapter. For now, we can see that in the case of a boiling liquid there would be a very large change in disorder when a liquid such as water with a volume of under 19 mL/mol vaporizes to a volume of over 30 L at 373.15°K, since $(22.414 \, \text{L}) \left(\dfrac{373.15°\text{K}}{273.15°\text{K}} \right) \cong 30.62 \, \text{L}$ by Charles' law.

For the case of a boiling liquid, the heat change is called the heat of vaporization, ΔH_{vap}, the temperature is constant at the boiling point T_{bp} and the process is reversible, so we can compute the entropy change as

$$\Delta S = \frac{q_{\text{rev}}}{T} = \frac{\Delta H_{\text{vap}}}{T_{\text{bp}}}.$$

We can also anticipate that if $dH = C_P \, dT$ and $\Delta S = \displaystyle\int \frac{dH}{T} = \int \frac{C_P \, dT}{T} \cong C_P \ln\left(\frac{T_2}{T_1}\right)$, so we see that we will also be able to calculate ΔS for a lot of situations without a phase change but with changing temperature using ΔH and C_P data. Thus, we can add S to U and H in our list (so far) of state variables (Figure 5.4).

CARNOT EFFICIENCY

Now, let us return to the Carnot cycle and consider how much work results for a given amount of heat input q_1, the overall efficiency of the cycle. Engineers would use $\Delta U = q - w_{\text{eng}}$ where "engineering work, w_{eng}" is positive when it is done on the environment, while we use the "IUPAC work, w_{IUPAC}," which is positive when work is done on the system (gas). In the IUPAC interpretation, this difference can be reconciled by using $dw_{\text{IUPAC}} = -P \, dV$ for a gas since an expanding gas affects the environment opposite to work done on the gas. This unfortunate mind-bending difference in the sign of the work seems more sensible under the engineering definition, but mathematically $dw_{\text{IUPAC}} = -P \, dV$ satisfies the IUPAC definition of the first law as $dU = dq + dw_{\text{IUPAC}}$. The sign of the work term is a matter of perspective relative to the system or the environment, but let us try one more explanation. When we write $dw_{\text{IUPAC}} = -P \, dV$, the problem is solved because if the gas is compressed (work done *on* the gas), then dv will be negative and $(-P)(-dv) = +dw$. A student should be warned that there are textbooks with these differing conventions and so we recommend $dw_{\text{IUPAC}} = -P \, dV$ as the solution to the problem. Having said that, we want to evaluate the efficiency, E_{eff}, of a Carnot heat engine in terms of how much work is done on the environment relative to how much energy is input in the first step, so we need to use $-\Sigma w$ and in the previous analysis of the Carnot steps, we have given $-w$ for each step:

$$E_{eff} \equiv \frac{(-\Sigma w)}{q_I} = \frac{\left[RT_h \ln\left(\frac{V_2}{V_1}\right) - C_V(T_l - T_h) + RT_l \ln\left(\frac{V_4}{V_3}\right) - C_V(T_h - T_l) \right]}{RT_h \ln\left(\frac{V_2}{V_1}\right)}.$$

Once again, we use the key relationship from the adiabatic steps as $\left(\frac{V_4}{V_3}\right) = \left(\frac{V_1}{V_2}\right) = \left(\frac{V_2}{V_1}\right)^{-1}$ and substituting that relationship into the ln ( ) terms reverses the sign of one, which allows a lot of cancelation. This results in a very useful and simple formula for the efficiency of a Carnot engine:

$$E_{eff} = \frac{T_h - T_l}{T_h}.$$

This tells us that the efficiency of the hypothetical Carnot engine only depends on the high temperature of the input energy and the low temperature. Let us think about this as $(T_h - T_l)$ representing the heat actually used in the conversion to work, but that the heat released at T_l is wasted. The most important result is that if the (P, V) graph is drawn for any real heat engine, the cycle can be overlaid with a grid of isotherms and adiabats to show that within each tiny calculus differential Carnot cycle all the $\oint ds = 0$ terms cancel internally, and the sum around the outer real engine PV graph also satisfies the same conditions and leads to the same formula.

EFFICIENCY OF REAL HEAT ENGINES

It is important to note that the efficiency formula uses absolute temperatures in °K. On planet earth, ambient temperatures limit the exit temperature T_l of most heat engines to roughly 273°K or higher temperatures. Thus, most heat engines can harness heat energy from hot sources to do work but waste a lot of that heat. A little thought shows that it ought to be possible to select a gas that could operate as a cyclic heat-transport medium between the common temperatures of ice water and steam as in a "steam engine," noting that the gas need not be steam but could be some gas trapped in a cyclic system operating between 273.15°K and 373.15°K. For that semirealistic situation, the efficiency would be

$$E_{eff} = \frac{373.15 - 273.15}{373.15} \cong 0.268 \simeq 27\%.$$

Early in the development of the steam engine, it was realized that if actual water steam is super-heated as a gas well above the boiling point of water to say 800°K and regular liquid is used as a coolant, the engine could be operated between about 800°K and 373°K to improve the efficiency to roughly

$$E_{eff} = \frac{800 - 373}{800} = 0.534 \cong 53.4\%.$$

This is actually very good for a heat engine or any engine for that matter. Compare that to an internal combustion engine at the extreme operating temperatures. Thus, if we assume ideal conditions of about 2300°K for combustion of gasoline and a maximum temperature of the exhaust manifold and valve parts of an iron engine as just below the melting point of iron at about 925°K, we can estimate a maximum efficiency of about 60%:

$$E_{\text{eff}} = \frac{2300 - 925}{2300} = 0.598 \cong 60\%.$$

In practice this is rarely attainable, but even this idealized efficiency shows us that practical heat engines waste a lot of heat energy on the T_1 part of the cycle. Due to internal friction of the moving parts and other design considerations, an actual efficiency of 30% is good. These examples illustrate the power of using the Carnot efficiency formula for any heat engine. All we need to know is the operating range of temperatures in absolute K degrees, but we should realize this is usually an upper limit to the efficiency of a real engine.

ENTROPY AND SPONTANEITY

From life experience, we know that heat tends to flow from hot to cold unless some sort of refrigeration device is used. Although the Carnot cycle established the existence of entropy, let us give some consideration to the sort of simple analysis Clausius would give for a heat engine. Consider a "heat source" at T_{hot} that supplies heat energy to an "engine" so that the heat source provides $|q|$ to the engine, which then discards $|q'|$ heat to the "environment" at T_{cold}. We choose to use the absolute values $|q|$ and $|q'|$ as positive numbers with the algebraic signs to indicate which way the heat flows. The work the engine does is thus the difference in the energy $(|q| - |q'|)$. A simple analysis gives

$$\text{Hot source loses heat } |q|, \quad \text{so } \Delta S_1 = \frac{-|q|}{T_{\text{hot}}},$$

$$\text{Cold source gains heat } |q'|, \quad \text{so } \Delta S_2 = \frac{+|q'|}{T_{\text{cold}}},$$

$$\text{Total } \Delta S = \Delta S_1 + \Delta S_2 = \frac{-|q|}{T_{\text{hot}}} + \frac{|q'|}{T_{\text{cold}}}.$$

Next, we use the idea that "spontaneous heat flow" occurs as long as $\Delta S > 0$. We have to think about this, but we know heat flows from hot to cold and the absolute temperature in the denominator is always positive so the sign of ΔS depends on the heat flow. When the heat flow stops, the ΔS value will be zero. Thus, the heat flow will just stop ($\Delta S = 0$) when $\frac{-|q|}{T_{\text{hot}}} + \frac{|q'|}{T_{\text{cold}}} = 0$, which can be rearranged to $|q'| = |q| \left(\frac{T_{\text{cold}}}{T_{\text{hot}}} \right)$ and $\frac{|q'|}{|q|} = \frac{T_{\text{cold}}}{T_{\text{hot}}}$. If we have to discard $|q'|$ heat energy into the cold sink (exhaust), the maximum energy that can be extracted as work is $|q| - |q'|$. Now define the efficiency of the engine as how much work can be obtained for the input heat absorbed:

$$\text{Efficiency} = \frac{\text{work produced}}{\text{heat absorbed}} = \frac{|q| - |q'|}{|q|} = 1 - \frac{|q'|}{|q|} = 1 - \frac{T_{\text{cold}}}{T_{\text{hot}}} = \frac{T_{\text{hot}} - T_{\text{cold}}}{T_{\text{hot}}}.$$

Thus, we obtain the same formula for the efficiency of a heat engine as from the Carnot cycle. Along the way, we realized that when heat flows from hot to cold, the environment will gain the exhaust heat $|q'|$ so that the environment gains entropy. A profound result of this sort of analysis is that entropy tends to increase in the environment unless there is some other condition and the *overall entropy in the universe tends to increase*. Especially for biology majors and generally for all of us, note that living systems have to expend energy to constantly combat the increase of the randomness in "biological errors." While very clever repair enzymes do amazing things to reduce randomness, the process of aging is a manifestation of increasing entropy.

SUMMARY OF THE SECOND LAW OF THERMODYNAMICS

We have discovered there is a state variable "S" that represents an amount of randomness as in a phase change and is related to the reversible heat change and the temperature at which it occurs as

$$\Delta S = \oint \frac{dq_{\text{rev}}}{T} = \frac{\Delta H}{T} = \int_{T_1}^{T_2} \frac{C_P}{T} \, dT = C_P \ln\left(\frac{T_2}{T_1}\right).$$

We also know that S tends to increase spontaneously; it is a measure of spontaneity in a closed system. However, an important motivation for the Carnot cycle proof is to show the very important relationship: $dS = dq_{\text{rev}}/T$. This leads to a cascade of important relationships that we present rapidly here because we are eager to get to some "essential" equations in our condensed course. Perhaps you have not realized that our process of discovery of S will open a door to more helpful thermodynamic relationships.

EIGHT BASIC EQUATIONS OF THERMODYNAMICS

Now that we know that $dq_{\text{rev}} = T \, dS$ we, can cut a huge time-saving swath through all of thermodynamics and concentrate on the "essentials."

1. The following assumes all work is *reversible PV (gas) work* so that $dw = -P \, dV$.
2. The following assumes *all heat changes are computed reversibly* so that $dq = T \, dS$.

With these two conditions, we can leap forward over 100 years of developments in thermodynamics and that is why it was "essential" to do the Carnot cycle proof.

From the first law: $dU = dq + dw = dq - P \, dv$ so we have $dU = T \, dS - P \, dV$. We have already noted that the energy tends to go down while entropy tends to go up, so in nature there is really a constant trade-off occurring between these two tendencies and what really matters is the difference between the two tendencies. The first treatment of this trade-off was given by H. L. F. von Helmholtz (1821–1894), who defined a new function "A" so that $A \equiv U - TS$ where we have to multiply S by a temperature to get an energy unit. Now, consider the first law again with addition and subtraction of $S \, dT$ to find what is now known as the "Helmholtz free energy, A."

$$dU = T \, dS - P \, dV + S \, dT - S \, dT = T \, dS + S \, dT - P \, dV - S \, dT = d(TS) - P \, dV - S \, dT$$

and we find

$$dU - d(TS) = -S \, dT - P \, dV = d(U - TS) = -S \, dT - P \, dV = dA \quad \text{or} \quad dA = -S \, dT - P \, dV.$$

We see from this that if both T and V are held constant $dT = 0$ and $dV = 0$, $dA = 0$ so that under those conditions $(U - TS) = 0$ and so the Helmholtz energy A indicates an equilibrium; a balanced trade-off between increasing entropy and decreasing energy. That is very interesting and a mathematical truth under reversible heat and work conditions. However, it turns out that it is not very useful in the laboratory, since it implies that pressures must be the only variable if T and V are held constant. Certain experiments can be designed to meet these conditions, but more likely the pressure is constant (at atmospheric pressure) in a laboratory setting unless the system is sealed. Thus, J.W. Gibbs (1839–1903) defined what is now known as the "Gibbs free energy, G," which has proved to be very useful. Gibbs defined $G \equiv H - TS$ and once again all the terms are in terms of energy units. This time we add and subtract $V\, dP$ as well as $S\, dT$ to the first law:

$$dU = T\, dS - P\, dV + S\, dT - S\, dt + V\, dP - V\, dP = d(TS) - S\, dT - d(PV) + V\, dP$$

so that we find

$$dU + d(PV) - d(TS) = -S\, dT + V\, dP = d(U + PV - TS) = d(H - TS) = dG$$

and so we obtain

$$dG = -S\, dT + V\, dP.$$

The equation of $G = H - TS$ is a better indicator of an equilibrium so that when dT and, especially, when dP are zero, then $dG = 0$ indicating an equal trade-off between increasing entropy and decreasing energy. Next, consider the definition of $H \equiv U + PV$ and use the two conditions above to find that that there is a fourth equation for dH as found from

$$dH = dU + P\, dV + V\, dP = (T\, dS - P\, dV) + P\, dV + V\, dP = T\, dS + V\, dP.$$

We have just rushed over more than 100 years of developments in thermodynamics by focusing on key equations, and we can consolidate the "essential" knowledge for this text as

$$dH = T\, dS + V\, dP \quad \text{and} \quad \left(\frac{\partial T}{\partial P}\right)_S = \left(\frac{\partial V}{\partial S}\right)_P,$$

$$dU = T\, dS - P\, dV \quad \text{and} \quad \left(\frac{\partial T}{\partial V}\right)_S = -\left(\frac{\partial P}{\partial S}\right)_V,$$

$$dG = -S\, dT + V\, dP \quad \text{and} \quad -\left(\frac{\partial S}{\partial P}\right)_T = \left(\frac{\partial V}{\partial T}\right)_P,$$

$$dA = -S\, dT - P\, dV \quad \text{and} \quad \left(\frac{\partial S}{\partial V}\right)_T = \left(\frac{\partial P}{\partial T}\right)_V.$$

This is pretty much the jackpot of equations for thermodynamics and now reveals why we chose to prove the Carnot cycle relationship so that we could get to $dq_{\text{rev}} = T\, dS$. Note the alternating pattern of signs and the variables when assembled in the order "$HUGA$," which may help organize the equations in your mind. We could spend a lot more time on the history of the Helmholtz and Gibbs free energy derivations, but in this course on "essentials," we swam through a narrow intellectual cave (Carnot cycle) to reach a beautiful expansive blue grotto of valuable knowledge with these eight equations. The four auxiliary partial derivative equations are called the "Maxwell relationships." They result from the fact that H, U, G, and A are all state variables with "exact differentials." Since it does not matter, which of the two variables change in either order for the basic four "$HUGA$" equations, we can do the second derivatives in either order.

$$dH = T\, dS + V\, dP \quad \text{so} \quad \left(\frac{\partial H}{\partial S}\right)_P = T \quad \text{and then} \quad \left(\frac{\partial^2 H}{\partial P\, \partial S}\right) = \left(\frac{\partial T}{\partial P}\right)_S \quad \text{where we have used the}$$

convention that the most recent derivative is with respect to the lower left variable in the second derivative.

However, we could take the second derivative in the reverse order by taking the first derivative with respect to P first and then with respect to S as shown here:

$$\left(\frac{\partial H}{\partial P}\right)_S = V$$

so that

$$\left(\frac{\partial^2 H}{\partial S \, \partial P}\right) = \left(\frac{\partial V}{\partial S}\right)_P.$$

But a state variable has the property that it does not matter which variable is changed first, so the two second derivatives must be the same.

Thus, we obtain the Maxwell relationship for H as $\left(\frac{\partial T}{\partial P}\right)_S = \left(\frac{\partial V}{\partial S}\right)_P$. You can prove the other three equations easily in the homework problem set.

What is the meaning of these eight equations? They are mathematical truisms, which we can use for problem solving. For instance, while the basic equation for the Helmholtz free energy, A, is not often useful, the fourth Maxwell equation proves to be very useful. For instance, while we see that $\left(\frac{\partial S}{\partial V}\right)_T = \left(\frac{\partial P}{\partial T}\right)_V$ it is very difficult to imagine what $\left(\frac{\partial S}{\partial V}\right)_T$ could possibly mean in terms of something we can measure in the laboratory, but it is very convenient to realize it is equal to the pressure change with respect to temperature when the volume is constant, which is easily measured. Thus, the Maxwell relationships are very valuable in resolving strange dependencies among the four state variables H, U, G, and A as related to laboratory variables (P, V, T). Perhaps we have violated the formal presentation of the historical development of thermodynamics, but for the students we have just saved you countless hours of frustrating reading, and NOW we are ready to do some real thermodynamics (after we get past the third law). It might be a good idea to write the eight equations over a few times till you master the patterns therein. Knowing those eight equations is the key to thermodynamics.

THIRD LAW OF THERMODYNAMICS

So far, we are confident that entropy exists and can be described quantitatively. On the other hand, we have only talked about ΔS so far. The mention of increasing disorder as a liquid vaporizes into a gas or even as a solid melts into a liquid helps us to realize that entropy is related to disorder somehow. This leads up to the idea that there is an absolute value for the entropy of a substance at a given temperature. Once again, the history of the third law is shared by several scientists where credit is given for later consolidation of ideas developed earlier by others. W. Nernst (1864–1941) is generally given the main credit for his work in 1905 called the Nernst heat theorem for which he received the Nobel Prize in 1920. Although Nernst can be said to have founded the field of physical chemistry, his work translated into what is now analytical chemistry. Many students associate his name with the "Nernst equation" of electrochemistry. Much of what is now analytical chemistry was formerly the field of research in physical chemistry, especially in electrochemistry, but a split occurred later in the 1930s when physical chemists were lured into the fields of spectroscopy opened up by the development of quantum mechanics. However, the mathematical basis for the third law was already put in place by Boltzmann in the 1880s and we will mainly use the Boltzmann statistical approach to entropy. In words, we favor a simple statement of the law:

The entropy of a pure, perfectly crystalline substance is zero at 0°K.

Once again, the verbal law is full of hidden meaning. Let us go back to Boltzmann's basic equation (carved into his tombstone) to see the physical implications:

$$S = k \ln W \quad \text{or} \quad S = k \ln \Omega.$$

(Perhaps, we can now see how the letter "W" evolved from the Greek letter "Ω"?)

We know that the natural logarithm of the number 1 is zero. A perfectly crystalline pure substance has a lattice structure that extends in all three dimensions as a perfectly repetitive pattern of atoms or molecules such that one cannot distinguish an imperfection that would aid in defining any list of alternate structures. A perfect crystal only has "one structure," not a list of possibilities where there might be some imperfection here or there. As far as the constant "k" is concerned, Boltzmann applied his statistics at the atom/molecular level and was interested in the gas constant per particle rather than the gas constant per mole so that basically k_B is the gas constant for 1 atom/molecule. It is a precise number based on the gas constant and the best known value of the Avogadro number:

$$k_B = \frac{R}{N_{Av}} = \frac{8.314472 \, \text{J}/^\circ\text{K mol}}{6.0221415 \times 10^{23} \, \text{mol}^{-1}} = 1.3806505 \times 10^{-23} \, \text{J}/^\circ\text{K} \cong 1.38 \times 10^{-16} \, \text{erg}/^\circ\text{K}.$$

So, now maybe we can combine your understanding of the third law. It says that entropy, S, is basically statistical where "W" in Boltzmann's equation refers to "the number of ways the system can exist." In the case of a perfectly crystalline system, W might evolve into a number more than 1 if the atoms swing and sway during vibration, but at 0°K, the vibrations will be minimized, although maybe not completely. There are more pitfalls as we consider this further, but first let us look at the simple interpretation of absolute entropy:

$$S_{tot} = \int_0^{T_{mp}} \frac{C_P(\text{sol}) \, dT}{T} + \left(\frac{\Delta H_{fus}}{T_{mp}}\right) + \int_{T_{mp}}^{T_{bp}} \frac{C_P(\text{liq}) \, dT}{T} + \left(\frac{\Delta H_{vap}}{T_{bp}}\right) + \int_{T_{bp}}^{T} \frac{C_P(\text{gas}) \, dT}{T}.$$

The simple interpretation is that a perfectly crystalline solid increases in randomness with heat at low temperatures until the lattice structure collapses at the melting point temperature where there is a large change in disorder since the liquid is more random than the crystal lattice. Then the liquid warms and becomes more random until the boiling point is reached. Then there is a much larger increase in randomness as the vaporization occurs. After the gas phase is reached, it can still increase its disorder at higher temperatures, so the last term represents the increase in gas entropy with further heating. This is the overall picture, but there has been a lot of research at what happens at very low temperatures near or below 1°K. For practical room temperature thermodynamics, the above equations suffice very well to describe solids, liquids, and gases.

As expected, there are always problems in the details. Let us think about trying to reach absolute 0°K temperature. If we consider highly pure materials, a problem crops up in that many elements have several nuclear isotopes. A really bad case would be to try to crystallize HCl. We would have to somehow isolate either the $^{35}_{17}$Cl or the $^{37}_{17}$Cl isotope as well as sort out the isotopes of $^{0}_{1}$H, $^{1}_{1}$D, and $^{2}_{1}$T. Then we would worry over whether the molecules were vibrating in a net symmetric or asymmetric way (what physicists call quantized "phonon" lattice vibrations). We would also have to worry about whether the nuclear spins are aligned. Experiments have actually been done where a cold sample is put into a strong electromagnet to align the nuclear spins (should we worry about electron spins?) and when the magnetic field is turned off, the spins randomize but absorb heat as they do so; this so-called "adiabatic demagnetization" can be used to obtain very low temperatures of the order of 0.001°K, but the remaining randomness prevents attainment of true 0°K. Thus, for several reasons, it turns out to be impossible to reach absolute $0°K$. Another problem is an isotope effect in molecular structure. Suppose, we somehow freeze DCH_3 to a very low temperature near $1°K$. The basic structure of methane is tetrahedral, but the orientation of the deuterium atom in the molecule could be in any of four positions. Thus, even if we purify the substance and use something like adiabatic nozzle expansion of He followed by the adiabatic demagnetization trick, we will still have a matrix in which there is a randomness of the orientation of the C–D bond that leads to an approximate entropy using Boltzmann's equation of $S \approx k_B \ln(4)$ per molecular unit or on a molar basis $S \approx R \ln(4)$. Similar statistics are easy to see for molecules such as monodeuterobenzene, where we would have something like $S(1°K) \approx R \ln(6)$, $S(1°K) \approx R \ln(3)$ for NDH_2, and so forth just to show the possibilities with deuterium (D) substitution for hydrogen.

The problem of behavior of the heat capacity of solids near $1°K$ has been treated by Einstein and by Debye to include the effects of vibration. For undergraduate treatment, it is sufficient to say that near $1°K$ heat capacities of lattices vary roughly as T^3 [2] so the heat capacity curve increases after $1°K$. Other texts and monographs should be consulted for studies of materials at very low temperatures, but here the "essential" facts are that $S = R \ln(W)$ gives an approximate value for low-temperature entropy due to isotope impurities and that the heat capacity varies as roughly T^3 in the $1°K$ range. There would be a constant "A" characteristic of the material and then $C_P \sim AT^3$. As usual there are alternate verbal descriptions of the third law of thermodynamics but our summary would be

> The entropy of a pure crystalline substance should be zero at $0°K$, but you really cannot get to $0°K$.

We will see in later chapters of this text that there are some strange quantum phenomena, which occur at very low temperatures. In this chapter for what is known as "classical thermodynamics," it is sufficient to say that $S = R \ln(W)$ is a valid formula that predicts an idealized value of zero for the entropy of a substance at $0°$, but there are several reasons why you will not be able to reach absolute zero in a practical way.

ENTROPY OF REACTIONS

Since entropy is a state variable, we can compute the change in entropy in a reaction just as we did for enthalpy:

$$\Delta S^0_{rxn}(298°K) = \sum_i^{prod} n_i S^0_i - \sum_j^{react} n_j S^0_j.$$

Example

$$Hg_{(liq)} + 1/2O_{2(gas)} \rightarrow HgO_{(red)},$$

$$\Delta S^0_{rxn}(298°K) = 70.25 - 75.90 - (205.152/2) = -108.226 \, J/°K \, mol.$$

The entropy change is negative and quite large mainly due to the fact that the random O_2 gas becomes localized in the red solid HgO, which is a drastic reduction in spatial randomness. We note that if ΔS^0_{rxn} is negative, the process is going toward a more ordered state. That is quite against the natural tendency of entropy to increase, but if we compute the heat of the reaction, we see that the reaction is also very exothermic, so we could say in this case the energy released in the reaction makes it possible for the reaction to proceed to a more ordered state (Table 5.1):

$$\Delta H^0_{rxn}(298) = -90.79 - 0 - (0/2) = -90.79 \, kJ/mol.$$

TABLE 5.1
Selected Values of H_f^0 in kJ at 1 bar and 298.15°K
and ΔS_{298}^0 in J/°K

Compound	H_f^0(298.15, 1.000 bar) kJ/mol	S_f^0(298.15, 1.000 bar) J/°K mol
H_2	0	42.55
O_2	0	205.152
CO	−110.53	197.660
CO_2	−393.51	213.785
$C_{graphite}$	0	5.74
HCCH	+227.4	200.9
H_2CCH_2	+52.40	219.3
H_3CCH_3	−84.0	229.2
CH_4	−74.6	186.3
NH_3	−45.94	192.77
HCl	−92.31	186.902
Cl_2	0	233.081
H_2O	−285.830	69.95
H_2CO	−108.6	218.8
$Hg_{(liq)}$	0	75.90
$HgO_{(red)}$	−90.79	70.25

ENTROPY CHANGES AT $T > 298.15°K$

Once again, we need to correct a state variable for temperatures other than the standard state:

$$\frac{\partial \Delta S^0}{\partial T} = \sum_i^{prod} \frac{\partial S_i^0}{\partial T} - \sum_j^{reac} \frac{\partial S_j^0}{\partial T},$$

but we know that

$$\int_{298}^{T} d(\Delta S^0) = \int_{298}^{T} \frac{\Delta C_P \, dT}{T}$$

so we can write

$$\Delta S_{rxn}^0(T) = \Delta S_{rxn}^0(298) + \int_{298}^{T} \left(\frac{\Delta C_P}{T}\right) dT.$$

We know from the previous chapter that we may have to integrate over the various terms of a polynomial heat capacity, but there is a slight difference in the first term in this case. Once again, we can calculate the difference in the C_P polynomial coefficients according to the n_i coefficients in the balanced reaction.

$\Delta a = \sum_i^{prod} n_i a_i - \sum_j^{react} n_j a_j$ and similar expressions for Δb, Δc, Δd, and Δe are obtained using C_P polynomials from Table 4.3. Thus, we need to integrate a slightly different formula for $\Delta S_{rxn}^0(T)$.

$$\Delta S_{rxn}^0(T) = \Delta S_{rxn}^0(298) + \int_{298}^{T} \left(\frac{\Delta a + \Delta bT + \Delta cT^2 + \Delta dT^3 + \Delta eT^4}{T} \right) dT$$

which is easily integrated:

$$\Delta S_{rxn}^0(T) = \Delta S_{rxn}^0(298) + \Delta a \ln\left(\frac{T}{298}\right) + \Delta b(T - 298) + \frac{\Delta c}{2}(T^2 - (298)^2) + \frac{\Delta d}{3}(T^3 - (298)^3)$$

$$+ \frac{\Delta e}{4}(T^4 - (298)^4).$$

The main difference in the entropy integration is in the first natural logarithm term.

TROUTON'S RULE/OBSERVATION

One curiosity of entropy is a relationship between boiling points and heats of vaporization. At an early age, F.T. Trouton (1863–1922) observed a pattern in boiling points and published two papers on a trend now known as "Trouton's rule." This rule falls within the realm of familiar rules in organic chemistry, which often have exceptions but are still useful. Basically Trouton's rule states that

$$\frac{\Delta H_{vap}}{T_{bp}} \approx 10.5\,R \approx 88\,J/^\circ K.$$

In use, this rule requires the boiling point in °K, and it actually works quite well for covalent organic compounds where there is no H-bonding, so that covers many cases in organic chemistry. It is poor for water, carboxylic acids, and alcohols due to the complication of H-bonding but overall, Trouton's rule is useful to estimate the heat of vaporization if the boiling point is known or to estimate the boiling point temperature if the heat of vaporization is known.

We can see in Table 5.2 that there is no trend in the entropies of fusion for melting points of a range of elements or compounds. On the other hand, Table 5.3 shows a near constancy of the entropy of vaporization with notable exceptions for water and acetic acid, which clearly have strong internal H-bonding in the liquid phase. Trouton's rule is more of an "observation" than a derivable equation but still useful for "back-of-the-envelope" estimation of boiling points or heats of vaporization.

TABLE 5.2
Entropy of Fusion for Selected Materials

Element/Compound	T_{mp} (°C)	ΔH_{fus} (kJ/mol)	$(\Delta H_{fus}/T_{mp})$ (J/°K)$_{calc}$
H_2O	0.000	6.01	22.002
$S_{monoclinic}$	115.21	1.721	4.431
Na	97.794	2.60	7.009
K	63.5	2.335	6.936
Mg	650	8.48	9.186
Pb	327.462	4.774	7.949
I_2	113.7	15.52	40.119
C_6H_6	5.49	9.87	35.422
Acetic acid	16.64	11.73	40.478
Naphthalene	80.26	19.01	53.790

TABLE 5.3
Entropy of Vaporization for Selected Materials

Element/Compound	T_{bp} (°C)	ΔH_{vap} (kJ/mol)	$(\Delta H_{vap}/T_{bp})$ (J/°K)$_{calc}$
H_2O	100.0	40.657	108.956
S	1367	154	93.894
Pb	1749	179.5	88.767
I_2	184.4	41.57	90.853
C_6H_6	80.09	30.72	86.966
Acetic acid	117.9	23.70	60.606
Naphthalene	217.9	43.2	87.975
CCl_4	76.8	29.82	85.212
CH_3CH_3	−88.6	14.69	79.599
C_9H_{20}	150.82	37.18	87.695
$C_{10}H_7Br$	259	52.1	97.905
			Average = 88.039

SIMPLE STATISTICAL TREATMENT OF LIQUIDS AND GASES

Let us apply the Boltzmann equation for absolute entropy to the problem of mixing materials. The main point here is to illustrate how entropy tends to lead a process toward randomization even if no energy is involved. The process we are going to show can apply to gases or liquids as an example that merges the third law using the Boltzmann statistical form of entropy with a simple lattice model. We should mention that while the theoretical treatment of gases and solids is worked out in detail, the theory of liquids is still a research frontier, so this simple model is of more interest applied to liquids. Liquids are more ordered than gases but less ordered than crystalline solids. Modern research in computer modeling of liquids can become quite sophisticated, but here we will use a very simple model of an egg carton with positions for 12 eggs. With our understanding of the Boltzmann KMTG and Dalton's law it is easy to imagine how gases mix, so our main interest here is for liquids. We can illustrate the mixing of two liquids (or gases) denoted by red and white poker chips to model and understand what happens when two liquids mix. To maximize meaning in a compact example, we want to show that even when there is no energy component driving the mixing, there is an effect due solely to entropy. Thus, we include the derivation of ΔS_{mix} for binary solutions in this chapter.

In Figure 5.5, we show an egg carton with 12 numbered poker chips placed sequentially. Even with this simple example, we can see that the first poker chip could have been placed in any of the 12 egg wells, but the second chip would only have 11 possibilities, the third chip would only have 10 possible positions, etc. In fact there are $12! = 479,001,600$ ways we could have put the 12 chips into the 12 egg wells, quite a few possibilities!

We could keep track of all those possibilities because there is an implicit order in the orientation of the egg wells and we have placed numbers on the chips to identify each one.

Now, if we flip the numbered chips over so we cannot see the numbers as in Figure 5.6, there is no way to tell which chip was put into what well first, second, third, etc. In fact, we do not even know which chip we picked up first to put in any egg well, so there are 12! possibilities for the order of picking the chips and 12! ways to put them in the egg wells but by just looking at the picture we can see that if the chips are "indistinguishable" there is only one way they can be in the carton, so in the Boltzmann equation, the "W" number comes out to be just 1, the hard way.

$$W = \frac{12!}{12!} = 1$$ so that $S = R \ln(1) = 0$ and as far as the white poker chips are concerned they are "perfectly ordered," so the entropy of this model is zero.

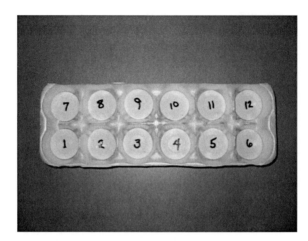

FIGURE 5.5 Twelve numbered white poker chips in a one dozen egg carton.

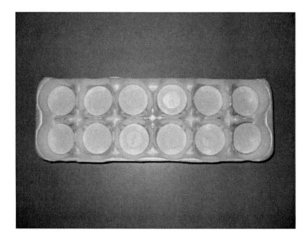

FIGURE 5.6 Twelve "indistinguishable" white chips in 12 egg wells are perfectly ordered.

The concept of "indistinguishability" is important in molecular science because in many cases there is no way to tell one H atom from another as well as many other indistinguishable cases.

Now suppose we want a mixture of red and white chips in the egg carton such that the sum of poker chips is still 12. We could put some red chips in the empty carton in some order as shown in Figure 5.7. Then we could put some numbered white chips into the empty egg wells. Finally, we could turn over the red and white chips to hide the numbering.

Suppose we fill in just six red chips in some random order and hide the labels and then fill the other egg wells with white chips that also have hidden labels, as in Figure 5.6. Now the question is, what is W for this mixture of red and white chips?

$$W = \frac{12!}{(6!)(6!)} = \frac{479,001,600}{518,400} = 924.$$

The 12! in the numerator comes from the fact that there are a total of $6 + 6 = 12$ chips and we could have picked any of the 12 as the first, second, third, etc. to place in the egg carton without noting their color. However, the possible order within the red set of chips could have only been chosen in 6! ways and only 6! ways would have been possible for the order of choosing the white chips.

FIGURE 5.7 Placing only six labeled red chips in the carton shows the many possibilities. Here that corresponds to hiding the labels but the principle should be clear.

Thus, even though there are still 924 possible ways the (six red, six white) arrangement could have been achieved, we now have a way to take care of the fact that all the chips are the same. So, for a binary system of (a, b) species, we have a formula for $W(a, b)$ as

$$W(a, b) = \frac{(N_a + N_b)!}{(N_a!)(N_b!)} \quad \text{and} \quad \text{so } S = k \ln \frac{(N_a + N_b)!}{(N_a!)(N_b!)}.$$

Now, we have to worry over the fact that we have to treat a number of particles of the order of Avogadro's number in the 10^{23} range. Fortunately, two mathematical tricks are available at this point. First, we should realize that we can use the logarithm of the number of particles, which will be a much smaller number while still following the same trends as the number itself. Second, we have available Stirling's approximation for large values of $\ln (n!)$ as found in several handbooks (Table 5.4).

$$\text{Stirling's approximation: } n! \cong (\sqrt{2\pi n})n^n e^{-n}.$$

We can take the natural logarithm of this expression to find a simpler formula:

$$\ln (n!) = n \ln (n) - n + \ln (\sqrt{2\pi n}).$$

TABLE 5.4
Selected Values of Stirling's Approximation for ln (n!)

n	$n!$	$\ln (n!)$	$n \ln (n) - n$	Error	Error (%)
10	3.6288E6	15.1044	13.0258	2.0786	13.7616
20	2.4329E18	42.3356	39.9146	2.4210	5.7186
30	2.6525E32	74.6582	72.0359	2.6223	3.5124
40	8.1592E47	110.3206	107.5552	2.7654	2.4428
50	3.0414E64	148.4777	145.6012	2.8765	1.9373
60	8.3210E81	188.6282	185.6607	2.9675	1.5732

The last term of this expression becomes negligible for very large values of "n," so we arrive at a very useful approximation for the natural logarithm of $n!$ as

$$\ln(n!) \cong n \ln(n) - n.$$

We can see that $n!$ soon exceeds the range of even a 10-place calculator, but we can also see that the percent error decreases as the number gets larger and that when $n = 10^{23}$, the approximation will be very good.

Now consider an egg carton with only red poker chips and another with only white chips. Let red chips be the "A" chips and "B" the white chips. Then in each perfectly ordered carton box we would have (the numbers need not be six and six but could be eight and four etc.)

$$S_A = k \ln\left(\frac{N_A!}{N_A!}\right) = 0 \quad \text{for the red chips} \quad \text{and} \quad S_B = k \ln\left(\frac{N_B!}{N_B!}\right) = 0 \quad \text{for the white chips.}$$

Then after mixing the red and white chips in a larger "two dozen" crate as a simulation of pouring two liquids together, we suppose that A and B are something like n-heptane and n-octane, which are so similar in structure and nonpolar that there is very little energy interaction between them so we make the approximation that $\Delta H_{mix} = 0$. Thus, whatever happens in this mixing is a result of only an entropy effect. Now let us calculate $\Delta S_{mix} = S_{after} - S_{before}$ using the Boltzmann formula:

$$\Delta S_{mix} = k_B \ln\left[\frac{N!}{(N_A!)(N_B!)}\right] - k_B \ln\left(\frac{N_A!}{N_A!}\right) - k_B \ln\left(\frac{N_B!}{N_B!}\right)$$

and then apply Stirling's approximation:

$$\Delta S_{mix} = k_B[N \ln N - N - N_A \ln N_A + N_A - N_B \ln N_B + N_B] - 0 - 0,$$

where

$$N = N_A + N_B.$$

So,

$$\Delta S_{mix} = -k[N_A \ln N_A + N_B \ln N_B - N \ln N].$$

Since $N_A + N_B$ cancels N inside the bracket

$$\Delta S_{mix} = -k_B[N_A \ln N_A + N_B \ln N_B - (N_A + N_B) \ln N].$$

$$\Delta S_{mix} = -k_B\left[N_A \ln\left(\frac{N_A}{N}\right) + N_B \ln\left(\frac{N_B}{N}\right)\right]\left(\frac{N}{N}\right) = -Nk_B\left[\frac{N_A}{N} \ln\left(\frac{N_A}{N}\right) + \frac{N_B}{N} \ln\left(\frac{N_B}{N}\right)\right], \quad \text{where}$$

we have multiplied the bracket by (N/N) and dragged N under both terms. What if N is equal to Avogadro's number? That will convert the expression to a molar basis. Note that N_A/N is a mole fraction.

$$\chi_A = \frac{N_A}{N} \quad \text{and} \quad \chi_B = \frac{N_B}{N}$$ are the respective mole fractions of the A and B species, and we obtain the final expression in terms of mole fractions and if $Nk_B = n_{tot}R$ the final result is

$$\Delta S_{mix} = -nR[\chi_A \ln \chi_A + \chi_B \ln \chi_B],$$

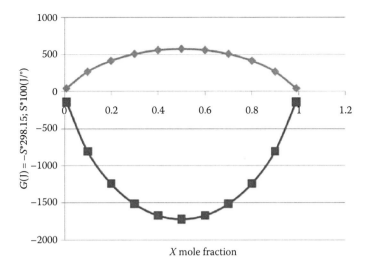

FIGURE 5.8 The entropy of mixing multiplied by 100 (upper line) and the Gibbs free energy (lower line) of mixing at 298.15°K plotted on the same graph. Here $\Delta G_{mix} = 0 - (298.15)(\Delta S_{mix})$ but ΔS_{mix} determines the ΔG_{mix} value at whatever temperature T happens to be. Note ln (0) is undefined, so we plot the edge values as mole fractions of 0.01 and 0.99.

which can be simplified further since for a binary mixture we know that $\chi_B = 1 - \chi_A$, so finally for a binary liquid:

$$\Delta S_{mix} = -nR[\chi_A \ln \chi_A + \chi_B \ln \chi_B] = -nR[\chi \ln \chi + (1 - \chi) \ln (1 - \chi)].$$

We can ask what mole fraction will produce the maximum entropy of mixing? We can do this by taking the derivative of ΔS_{mix} with respect to mole fraction, set the derivative to zero, and then solve for the mole fraction at the maximum. Note that the minus prefix is made positive by the fact that the mole fractions are less than 1, so their logarithms are minus, which leads to a maximum:

$$\frac{\partial \Delta S_{mix}}{\partial \chi} = -nR\left[\chi\left(\frac{1}{\chi}\right) + \ln \chi + (1 - \chi)\left(\frac{-1}{1 - \chi}\right) + (-1) \ln (1 - \chi)\right] = -nR \ln\left(\frac{\chi}{1 - \chi}\right) = 0.$$

The only way this derivative can be zero is if $\left(\dfrac{\chi}{1 - \chi}\right) = 1$ so, we have $\chi = 1 - \chi$ or $2\chi = 1$ and finally, the maximum ΔS_{mix} will occur when $\chi = 1/2$ and $\chi_A = \chi_B$. Well of course, it is also interesting to insert this condition into the Gibbs free energy expression, $\Delta G_{mix} = \Delta H_{mix} - T \Delta S_{mix}$ (Figure 5.8).

We see that $\Delta G_{mix} = 0 + nRT[\chi \ln \chi + (1 - \chi) \ln (1 - \chi)]$ is a *minimum* because the mole fractions are numbers less than 1 so their logarithms are negative and so we can find ΔG_{mix} even when $\Delta H_{mix} = 0$. One last thought is that we can imagine looking at Figure 5.6 when there are more than half red chips and they are indistinguishable. There is less opportunity for variations when the box is nearly all red. Conversely, there is the maximum chance for randomness when the number of red and white chips is equal.

SUMMARY

This chapter has been a survey of the second and third laws of thermodynamics. We have used the Carnot cycle to introduce the concepts of entropy and efficiency, but the most important relationship we found was that $dq_{rev} = T\,dS$. Together with the first law that led to eight very important equations and helped define A and G, two new state variables. The third law was defined and connected to the Boltzmann equation for absolute entropy, $S = k_B \ln W$. Using that equation, we derived the equation for the entropy of mixing in binary solutions. We also discussed the empirical observation of Trouton's rule. Now we can proceed, armed with a basic understanding of the first, second, and third laws of thermodynamics and in possession of eight equations, which are more powerful than we have yet realized.

TESTING, GRADING, AND LEARNING?

We now provide some midterm examinations to assist students in preparing for such a test at about this point in the course. It may be helpful to the students to give this test a bit past the actual midpoint of the course and to give them time to assimilate the material. This author routinely offers the students the option to take their final grade based on the cumulative final examination alone or their cumulative average over the semester, whichever is higher. That policy tends to make the midterm a "practice run" for the final examination, which will include some new questions related to lecture material between the midterm and the final examinations. Providing old tests to the students does focus their study, but they should expect the questions to be modified for the actual examinations. We note here that no student who cut many classes and took only the final examination has ever passed this course.

Now we present an actual examination that was a midterm examination in 2007 for a one-semester presentation in CHEM 305 at Virginia Commonwealth University. After that we show a slightly more difficult midterm examination from the 2009 Summer course. The time allowed for CHEM 303 is longer when a 2 h lecture period is available, but the CHEM 305 time period was only 1 h in the one-semester course.

CHEM 305	Midterm examination, Fall 2007	D. Shillady, Professor
(Points)	(Attempt all problems)	55 min

(15) 1. Calculate the laminar bulk flow rate in gallons/min for blood with $\eta = 0.015$ poise through an aorta 5 in. long and 1/4 in. inner diameter due to a pressure difference of 135–80 mmHg, use a duty cycle factor of 0.05 due to pulsation.   (1.218 gallons/min)

(15) 2. Calculate the temperature of air compressed adiabatically in a one-cylinder diesel engine from 1030 cm^3 at 22°C and 1 atm pressure to 30 cm^3, given $C_V = (5/2)R$. Compute the moles of air assuming 22.414 L/mol of air, Q, W, ΔH, and ΔU for this compression.
$$(T = 1213.7°\text{K}, Q = 0, n = 0.0425, \Delta U = W = +193.95 \text{ cal}, \Delta H = +271.54 \text{ cal})$$

(10) 3. Calculate $\bar{v} = \sqrt{\dfrac{8RT}{\pi M}}$ for He gas molecules in mph at 20°C (He $= 4.0026$ g/mol).
$$(2795.3 \text{ mph})$$

(15) 4. Derive the expression for $\bar{v}$ of a gas molecule using the Boltzmann principle.
(See chapter notes)

(15) 5. A 0.500 g sample of n-heptane (C_7H_{16}) burned in a constant volume calorimeter causes $\Delta T = 2.934°\text{C}$. If C_V of the calorimeter is 1954 cal/°C, calculate ΔH_{comb} of n-heptane at 298°K (use C $= 12.011$ g/mol and H $= 1.008$ g/mol).
$$(-1,151,505 \text{ cal/mol})$$

(10) 6. Compute ΔH_{298}^0 for the reaction $C_6H_6 + 3H_{2(g)} \rightarrow C_6H_{12}$ given the data $\Delta H_{\text{comb}}(C_6H_6) = -782.3$ kcal/mol, $\Delta H_{\text{comb}}(C_6H_{12}) = -937.8$ kcal/mol, and $\Delta H_{\text{comb}}(H_2) = -68.3$ kcal/mol.
$$(\Delta H_{\text{rxn}} = -49.4 \text{ kcal/mol})$$

(10) 7. Given $a = 3.610$ atm L^2/mol^2 and $b = 0.0429$ L/mol for the van der Waals equation of CO_2, calculate the pressure P in atm of 5 mol of CO_2 in a 5 L container at 70°C.

$(P = 25.81$ atm)

(10) 8. Show that $(C_P - C_V) = R$ for an ideal gas.

(See chapter notes)

| Physical chemistry 303 | Summer 2009 | D. Shillady, Professor |
| (Points) | Midterm examination | (Attempt all problems) 120 min |

(15) 1. Using $\eta_{visc} = (1/2)n*m\bar{v}\lambda = 2.08 \times 10^{-4}$ poise at 25°C and 1 atm pressure for O_2, compute d, λ, Z_1, and Z_{11}. (Atomic weight for $O = 15.9994$ g/mol) $(d = 3.573 \times 10^{-8}$ cm, $\lambda = 716 \times 10^{-8}$ cm, $Z_1 = 6.203 \times 10^9$ s^{-1}, $Z_{11} = 7.634 \times 10^{28}$ binary/cm^3 s)

(15) 2. Using the Boltzmann distribution, set up the distribution function for the speed of a gas molecule in any direction; derive formulas for $\bar{v}$, v_{rms}, and the most probable speed α.

(See chapter notes)

(10) 3. Calculate $\bar{v}$ for He (atomic weight 4.002503) gas molecules at 25°C in mph.

(2809.22 mph)

(10) 4. Show $(C_P - C_V) = R$ for an ideal gas.   (See chapter notes)

(10) 5. Compute ΔH^0_{298} for the reaction $3(HC \equiv CH \rightarrow C_6H_{6(l)}$ given the data ΔH^0_{comb} $(HC \equiv CH) = -1301.1$ kJ/mol and $\Delta H^0_{comb} (C_6H_6) = -3267.6$ kJ/mol.

$(\Delta H^0_{298} = -635.7$ kJ)

(15) 6. Find P_c, V_c, and T_c for a van der Waals gas and show the law of corresponding states.

(See chapter notes)

(15) 7. Calculate the temperature of air compressed adiabatically in a one-cylinder diesel engine from 1050 cm^3 at 22°C and 1 atm pressure to 50 cm^3. Given $C_V = (5/2)R$, compute moles air, Q, W, ΔH, and ΔU for this compression.

$(Q = 0$, $W = \Delta U$, $\Delta U = 151.78$ cal $= 636.0$ J, $\Delta H = 212.5$ cal $= 889.1$, $n = 0.0434)$

(10) 8. Calculate the entropy of fusion (ΔS_{fus}) of a compound at 0°C given that its ΔH_{fus} is 39 kJ/mol at 156°C (mp) and the molar C_P values are 28.0 J/°C for the liquid and 20.0 J/°C for the solid.

$(\Delta S_{fus} = 87.263$ J/°K at 0°C)

DS took 80 min

PROBLEMS

5.1 Estimate the absolute entropy of 1-deutero-naphthalene at 1°K using $S = k_B \ln W$.

5.2 Calculate the efficiency of a internal combustion heat engine operating with a heat source at 1000°C and discarding exhaust heat at 700°C.

5.3 Calculate the entropy of mixing for a mixture of n-heptane and n-octane versus mole fraction and sketch a graph showing ΔS_{mix} for mole fractions of n-heptane as 0.01, 0.25, 0.4, 0.5, 0.6, 0.75, and 0.99 on the same graph using a different scale on the Y-axis plot $\Delta G_{mix} = \Delta H_{mix} - T\Delta S_{mix}$ assuming $\Delta H_{mix} = 0$ and $T = 25°C$. Scale $\Delta S_{mix} \times 100$.

5.4 Derive all eight of the basic thermodynamic equations starting from the first law, the definition of H and $dS = dq_{rev}/T$. Derive the four Maxwell relationships using the idea of reversing the order of differentiation.

5.5 Calculate ΔS_{298} for the reaction $H_2 (g) + \frac{1}{2}O_2(g) \rightarrow H_2O$ (liq) at 298.15°K using the data in Table 5.1. Then use that value to correct the value of ΔS_{298} to ΔS_{1000} at 1000°K using the polynomials in Table 4.3.

BIBLIOGRAPHY

Fenn, J. B., *Engines, Energy and Entropy*, Global View Publishing, Pittsburgh, PA, 2003 and W.H. Freeman Co. 1982.

REFERENCES

1. Carnot, S., *Reflection on the Motive Power of Fire*, Dover, New York, 1960.
2. Hill, T. L., *An Introduction to Statistical Thermodynamics*, Addison-Wesley Publishing Co., Reading, MA, 1960, p. 101.

6 Gibbs' Free Energy and Equilibria

INTRODUCTION

In previous chapters, we have stressed that in nature energy tends to decrease while entropy tends to increase. A naive first consideration of any machine or process is that energy is needed to continue operation and we often overlook energy expended on various repair activities that are a form of entropy management. It becomes more obvious that entropy is a factor when one studies chemical processes that "should" occur based on energy considerations but nevertheless require some sort of a catalyst or other special conditions, which imply geometric constraints that overcome the natural tendency of randomness to increase. The value of ΔS is a change in a state variable but the path can be modified by special conditions such as the introduction of a catalytic surface, which allows reactants to meet side-by-side compared to random collisions in the gas phase. Josiah Willard Gibbs (1839–1903) was a foremost U.S. scientist (Figure 6.1) who made important advances in thermo-dynamics applying the new idea of "chemical potential" ($\Delta G/n$) as a free energy per mole of a substance in phase diagrams and applied to equilibria. At the time of his work, few people understood it but it was later developed into the idea of free energy and greatly affected thinking, teaching, and problem solving in chemical engineering. Gibbs' research used what was advanced mathematics in his time but remained at what we call "classical physics" today since he predated quantum mechanics. Gibbs is especially noteworthy in that he carried out research in the United States at a time when the turmoil of the U.S. Civil War and settling in the West were not as conducive to research as was the case in Europe in the late 1800s. However, Gibbs had spent a year each in Paris, Berlin, and Heidelberg and had written contact with foremost scientists in Europe. Gibbs also held the very first PhD in chemical engineering in the United States, awarded in 1863 from Yale University. Other scientists including Albert Einstein regarded Gibbs as a foremost founder of thermodynamics and a true genius. It is indeed humbling to realize that such pure thought by Gibbs, Boltzmann, and others was carried out for the first time without the same sort of support we have now in "the information age," although scientists did study each other's work. Truly we stand on the shoulders of intellectual giants!

For our list of essential topics, we will focus on the main result from Gibbs:

$$\Delta G = \Delta H - T\Delta S$$

Large lists of G_{298}^0 are available [1] as assembled from H_{298}^0 and S_{298}^0 values, so that one can calculate values for chemical reactions as ΔG_{rxn}^0 (298) using balanced reactions and thermodynamic tables. The main usefulness of this process is that one can obtain an equilibrium constant for gas reactions and the concepts for gases can be extended to other phase concentrations. Following Gibbs, we define a concept called the "chemical potential." From the $HUGA$ equations, we have $dG = -S\,dT + V\,dP$ and we specify a new symbol (mu, μ) as $d\mu = \left(\dfrac{dG}{n}\right) = -\overline{S}\,dT + \overline{V}\,dp$ where we specify the equation is for 1 mol and the bar over the entropy and volume indicate values per mole. We look ahead to consideration of an equilibrium at a specific temperature, so when T is constant we have $d\mu = \left(\dfrac{dG}{n}\right) = -\overline{S}\,dT + \overline{V}\,dp = 0 + \left(\dfrac{RT}{P}\right)dP = RTd\ln P$ and we can calculate

FIGURE 6.1 Portrait of Josiah Willard Gibbs (Jr.) (1839–1903) from the Williams Haynes Portrait Collection, Chemical Heritage Foundation Collection. He earned his PhD from Yale (one of the first in the United States) and spent his career there as a professor of mathematical physics.

a change in μ as $\int_1^2 d\mu = RT \int_{P_1}^{P_2} \frac{dP}{P}$, so that we find $\mu_2(T, P_2) - \mu_1(T, P_1) = RT \ln\left(\frac{P_2}{P_1}\right)$. We can use this general formula to reference the chemical potential to standard conditions, such as 298°K and 1 bar.

$$\mu(T, P) = \mu^0 + RT \ln\left(\frac{P}{1}\right) \text{ or using } G_{298}^0 \text{ values } G(T, P) = G_{298}^0(1 \text{ bar}) + RT \ln P. \text{ Many old}$$

texts standardized on 1 atm but the new handbook values are relative to 1 bar (1 atm = 1.01325 bar), so there is not much difference in the numbers and the equation is the same.

Now let us consider a typical equilibrium for gases where we can use pressures according to Dalton's law.

$$a\text{A} + b\text{B} \rightleftharpoons c\text{C} + d\text{D}$$

$$\Delta G = c\mu_\text{C} + d\mu_\text{D} - a\mu_\text{A} - b\mu_\text{B}$$

$$\Delta G = c\mu_\text{C}^0 + d\mu_\text{D}^0 - a\mu_\text{A}^0 - b\mu_\text{B}^0 + RT[c \ln P_\text{C} + d \ln P_\text{D} - a \ln P_\text{A} - b \ln P_\text{B}]$$

but $c \ln P_\text{C} = \ln P_\text{C}^c$, etc.

$$\Delta G = \Delta G_{298}^0 + RT \ln\left\{\frac{P_\text{C}^c P_\text{D}^d}{P_\text{A}^a P_\text{B}^b}\right\} = \Delta G_{298}^0 + RT \ln K_P \text{ where } \left\{\frac{P_\text{C}^c P_\text{D}^d}{P_\text{A}^a P_\text{B}^b}\right\} = K_P. \text{ Note } \Delta G = 0 \text{ at}$$

equilibrium by the very definition of the meaning of equilibrium as a balance of decreasing energy and increasing entropy. Therefore as a result of the equilibrium condition, we have

$$\Delta G_{298}^0 = -RT \ln K_P \text{ and } \ln K_P = -\frac{\Delta G_{298}^0}{RT} \text{ and that leads to } K_P = e^{-\frac{\Delta G^0}{RT}} = \exp\left[\frac{-\Delta G_{298}^0}{RT}\right].$$

We need to make a point here about equilibrium reactions, they are dynamic. A naive idea of an equilibrium is that it oozes in one direction, sets up concentrations, and then coagulates into a sort of semistable pudding. That is far from the truth; the double arrow ($\rightleftharpoons$) is a very active process in a never ceasing reaction in both directions. Modern spectroscopy has shown that even the H atoms in organic compounds can exchange with other H atoms, so a lot of activity is going on in seemingly stable compounds and if there is some sort of reaction it can often reverse itself. Equilibrium reactions are constantly going in both directions even though one direction may be favored. Of course, temperature can affect the extent of the equilibrium in either direction.

Example 1

Suppose we place 0.300 mol of H_2 and 0.100 mol of D_2 in a 2.00 L vessel at 25°C.

$$H_{2(g)} \quad + \quad D_{2(g)} \quad \rightleftharpoons \quad 2HD; \quad \Delta G = -0.700 \, \text{kcal} = -2.929 \, \text{kJ} \quad [2]$$

$$(0.300-x) \qquad (0.100-x) \qquad (2x)$$

$\Delta G_{298}^0 = -RT \ln K_P$, so $\dfrac{+2929 \, \text{J/mol}}{(8.314 \, \text{J/°K mol})(298.15°K)} = \ln K_P \cong 1.1816$. Note that the logarithm comes out to be a pure number after all the units cancel as it must, which is a good way to check the units. The 1 mol value for ΔG_{298}^0 is based on the balanced reaction for 1 mol H_2. This reaction is chosen in our idea of essential physical chemistry to illustrate several points at once, but let us not miss the points. This is an exchange reaction of H and D where only gas phase collisions occur and this implies there is some sort of microscopic mechanism in operation, although as usual the thermodynamics does not give any information on the mechanism. Second, the K_P value is a constant meaning that there is a fixed relationship between the concentrations. Third, the gases are in the same container so we can convert pressures directly to concentrations. Fourth, note that, as written, the ΔG_{298}^0 value is negative. The main idea of ΔG_{298}^0 is that it is negative for a reaction which "tends" to the right as written based on standard state values, but that does not mean it will go to completion. A positive value for ΔG_{298}^0 would mean that the reaction "tends" to go to the left, but not that it would not occur at all. Generally, all reactions that have a negative ΔG_{298}^0 will occur to some extent, but here we are dealing with the case of $\Delta G_{298}^0 = 0$, the condition for an equilibrium.

This example also introduces the idea of "x" as a "molar reaction coordinate, the extent of the reaction"; it is the molar amount of the reaction that occurs on the left side and shows up as moles of product on the right side of the reaction. We will use this same idea when we treat reaction kinetics. Keeping in mind our emphasis on the idea of an equilibrium as a dynamic process, we are actually treating this in a similar way to a kinetics problem but without any mention of time. Let us solve for "x."

$$K_P = e^{1.1816} = 3.2596 = \frac{[P_{HD}^2]}{[P_{H_2}][P_{D_2}]} = \frac{\left[n_{HD}\left(\dfrac{RT}{V}\right)\right]^2}{\left[n_{H_2}\left(\dfrac{RT}{V}\right)\right]\left[n_{D_2}\left(\dfrac{RT}{V}\right)\right]} = \frac{(2x)^2}{(0.3-x)(0.1-x)}$$

Note that the pressures are low enough that the ideal gas law is accurate and the factors of $\left(\dfrac{RT}{V}\right)$ all cancel in the equilibrium expression. It is easy to see that $(3.2596)(0.03 - 0.4x + x^2) = 4x^2$, which leads to $0.2271x^2 + 0.400x - 0.030 = 0$ and can be solved using the quadratic formula: $x = \dfrac{-b \pm \sqrt{b^2 - 4ac}}{2a}$. This leads to $x = \dfrac{-0.400 \pm \sqrt{(0.400)^2 - 4(0.2271)(-0.03)}}{2(0.2271)}$; the positive root is $x \cong 0.072$. Thus, we find at equilibrium: $[H_2] = 0.300 - x = 0.228$, $[D_2] = 0.100 - x = 0.028$, $[HD] = 2x = 0.144$. According to the calculations, almost all of the D_2 has reacted and been converted to HD noting that H_2 can provide two H atoms.

TEMPERATURE DEPENDENCE OF EQUILIBRIUM CONSTANTS

Sometimes it is possible to shift an equilibrium to increase the yield of a desired product. The key equation was given above, which shows temperature dependence through the logarithm.

$\Delta G_{298}^0 = -RT \ln K_P$ and in the example here we have a specific formula: $-\ln K_P = \left(\dfrac{\Delta G_{298}^0}{R}\right)\left(\dfrac{1}{T(K)}\right)$, so that we can show a plot of $\ln (K_P)$ versus $(1/T)$ (Figure 6.2). We will encounter a number of these sorts of plots where the x-axis is a reciprocal temperature, so it is a good idea to carefully consider this graph. If you think about it, the lowest temperature will give the largest value of the x-coordinate, so the right side of the graph refers to the lowest temperature. In the plot shown the y-axis is the negative logarithm of the K_P at that temperature, so the K_P value does indeed change with the inverse temperature in a very linear way. It is perhaps worth noting that this expression is compatible with the Boltzmann principle since

$$K_P = e^{-\frac{\Delta G^0}{RT}} = \exp\left[\frac{-\Delta G_{298}^0}{RT}\right] = e^{-\left(\frac{E}{RT}\right)}.$$

van't HOFF EQUATION

An alternative way to study the effect of temperature on an equilibrium is due to further manipulations by van't Hoff (1852–1911) who was a Dutch physical-organic chemist and the winner of the very first Nobel Prize in 1901 for his research on dilute solutions. Although we have shown a method above, which might be sufficient when ΔG_{298}^0 is available, we show this additional

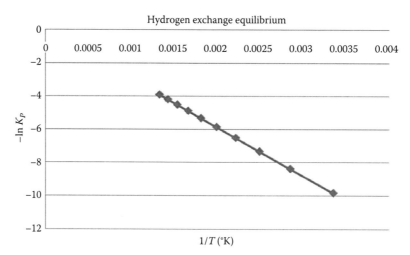

FIGURE 6.2   The plot of $-\ln (K_P)$ versus $(1/T)$ for the $H_2 + D_2 \rightleftharpoons 2HD$ equilibrium.

more complicated method because it produces graphical results similar to the treatment of rate constants in Chapters 7 and 8, and only depends on ΔH^0_{298}. The point of this exercise is to reinforce the idea that *equilibria are dynamic* and are the result of a dynamic trade-off of a forward reaction and a reverse reaction. We start with the same equation as above, $\Delta G^0_{298} = -RT \ln K_P \Rightarrow \ln K_P = -\left(\Delta G^0_{298}/RT\right)$. Then the temperature derivative is

$$\frac{d \ln K_P}{dT} = \frac{-(-\Delta G^0)}{RT^2} - \left(\frac{1}{RT}\right)\frac{d(\Delta G^0)}{dT} = \frac{+\Delta G^0}{RT^2} - \frac{1}{RT}\frac{d(\Delta G^0)}{dT}.$$ But $dG = -S\,dT + V\,dP$, so that

at constant P in a reactor $\left(\frac{\partial G^0}{\partial T}\right)_P = -S^0$. Then at 1 atm $\dfrac{d(\Delta G^0)}{dT} = -\Delta S^0$, so that

$$\frac{d \ln K^0_P}{dT} = \frac{\Delta G^0}{RT^2} - \frac{(-\Delta S^0)}{RT} = \frac{(\Delta H^0 - T\Delta S^0)}{RT^2} + \frac{\Delta S^0}{RT} = \frac{\Delta H^0}{RT^2} = \frac{d \ln K^0_P}{dT}.$$

This equation $\left[\dfrac{d \ln K^0_P}{dT} = \dfrac{\Delta H^0}{RT^2}\right]$ requires only the ΔH^0 value, although a way is still needed to

obtain K^0_P for some initial condition. This equation can be developed further for graphical analysis by integrating the derivative over a range of temperature.

$$\int_{T_1}^{T_2} d \ln K^0_P = \int_{T_1}^{T_2} \frac{\Delta H^0(T)dT}{(RT^2)} \cong \frac{\Delta H^0}{R} \int_{T_1}^{T_2} \frac{dT}{T^2} = \frac{-\Delta H^0}{R}\left(\frac{1}{T_2} - \frac{1}{T_1}\right) = \ln\left[\frac{K^0_P(T_2)}{K^0_P(T_1)}\right].$$

Thus, $\ln K^0_P(T_2) = \ln K^0_P(T_1) - \dfrac{\Delta H^0}{R}\left(\dfrac{1}{T_2} - \dfrac{1}{T_1}\right)$, although this equation assumes that ΔH^0 is constant over a small temperature range.

Example 2

One of the most significant and most highly studied equilibria in the early twentieth century was the Haber ammonia synthesis for which Fritz Haber was awarded the Nobel Prize in 1918. In spite of the fact that the atmosphere of Earth is over 70% nitrogen, it is chemically difficult to use that enormous source of N_2 to produce nitrogen fertilizers because N_2 is quite unreactive. Because of the Haber synthesis, as much as one-third of the world food supply is a result of increased agricultural yield due to nitrogen fertilizers, such as NH_4NO_3, which can be synthesized from NH_3 on a commercial scale. Haber might be considered one of the greatest benefactors to humankind but the original motivation for developing the process was that Germany needed a way to make nitrates for munitions in WWI. In the early 1900s, nitrates were manufactured by acidifying "guano" (bird droppings rich in nitrogen compounds) as found in hundred foot layers on islands off the coast of Chile to obtain nitric acid and then on to nitrates. However, due to various blockades, Germany was cutoff from obtaining guano. Thus, ammonia synthesis was a strategic process, which lengthened WWI but later became very beneficial to agricultural yield. Another consideration is that the Nobel Prize money itself comes from the earnings of the original patent granted to Alfred Nobel for the invention of stabilized tri-nitro-toluene (TNT), so converting atmospheric N_2 to nitrates has great significance to humankind whether for war or peace.

The 90th Edn. of the *CRC Handbook* lists values for NH_3 of $\Delta H^0_{298} = -45.9\,\text{kJ/mol}$ and $\Delta G^0_{298} = -16.4\,\text{kJ/mol}$. Thus, for the equilibrium $N_2 + 3H_2 \rightleftharpoons 2NH_3$ we can calculate $\Delta G^0_{298} - RT \ln K^0_P = 2(-16.4\,\text{kJ/mol}) - 0 - 0 = -32.8\,\text{kJ}$ and so we can find the value of K^0_{298}.

$$\ln K^0_{298} = \frac{-32,000\,\text{J/mol}}{(8.314\,\text{J/mol}^\circ\text{K})(298.15^\circ\text{K})} = 13.2321; \quad K^0_{298} = 5.58 \times 10^5 = \frac{P^2_{NH_3}}{P_{N_2}P^3_{H_2}}.$$

Now since we may want to produce NH_3, we can ask what temperature would shift the equilibrium so that the pressure of the desired product NH_3 is doubled (assuming ΔH_{298}^0 remains approximately constant).

$$\ln\left[\frac{K_P^0(T_x)}{K_P^0(298.15)}\right] = \ln\left(\frac{2}{1}\right) \cong \frac{-\Delta H_{298}^0}{R}\left[\frac{1}{T_x} - \frac{1}{298.15}\right].$$

Thus, $\dfrac{R\ln(2)}{-\Delta H_{298}^0} + \dfrac{1}{298.15} = \dfrac{1}{T_x}$ and so we find that only a modest $11°$ increase is needed.

$\dfrac{(8.314\,\text{J/mol}\,°\text{K})\ln(2)}{(-45{,}900\,\text{J/mol})} + (3.354016435 \times 10^{-3}) = \dfrac{1}{T_x} \Rightarrow T_x = 309.74°\text{K}.$

Actually this process required reaction vessels capable of holding very high pressures and a catalyst is also required. Eventually, a temperature of over $600°$K was found to work best but we can see from this limited example that increasing the temperature shifts the equilibrium toward more NH_3. Here we can also see the first case to comment on the need for care in treating reciprocal temperatures. We recommend carrying all digits available in this sort of calculation and then rounding to the least number of significant figures only at the end of the calculation.

VAPOR PRESSURE OF LIQUIDS

The Gibbs free energy concept can also be used in some cases where ΔG does not appear in the final formula but is still important to analyze the process. An important case is the vapor pressure of liquids. A strange concept that is important in forensic applications is that in principle all solids and liquids have some small vapor pressure. In everyday experience, we can hold items close to our nose and detect faint odors. In some cases like polished granite or other stoneware this slight vapor pressure is negligible to human smell detection but we know dogs and other animals can detect odors perhaps to a sensitivity more than 1000 times that of the human nose. Even then you say granite has no vapor pressure and effectively that is so for the solid but at temperatures where stone becomes molten as in lava there will be a vapor pressure. The point is that when a solid changes into a gas (sublimation) or a liquid changes into a gas (boiling, vaporization) the atoms/molecules of the gas literally "jump" away from the solid or liquid. To motivate this discussion, imagine that when water boils some molecules at the surface of the liquid jump out of the container. However, if we boil water to make tea and turn off the heat and let the water cool, the gas molecules immediately above the surface of the liquid can "crash" into the liquid since their random motion allows some to move toward the liquid. We have already used the idea that phase changes can be considered reversible in our discussion of entropy when we noted $dS = dq_{rev}/T$ and that $\Delta S_{vap} = \Delta H_{vap}/T_{mp}$. Here we extend that idea to treat an equilibrium condition at the surface of a liquid in boiling or at the surface of a subliming solid like CO_2 (dry ice).

Consider the most common case of boiling. At equilibrium $G_{liq} = G_{vap}$ so $dG_{liq} = dG_{vap}$. For the boiling process the most appropriate variables to use are P, T with the assumption that there is some equation of state that will relate P, V, T, and n. Thus, we can relate the general differential of G to one of the *HUGA* equations.

$dG = \left(\dfrac{\partial G}{\partial P}\right)_T dP + \left(\dfrac{\partial G}{\partial T}\right)_P dT = V\,dP - S\,dT$ so we can match the liquid and vapor dG values.

$V_{liq}\,dP - S_{liq}\,dT = V_{vap}\,dP - S_{vap}\,dT$, which leads to $\dfrac{dP}{dT} = \dfrac{(S_{vap} - S_{liq})}{(V_{vap} - V_{liq})}$. Recall $(S_{vap} - S_{liq}) =$

$\dfrac{\Delta H_{vap}}{T_{bp}}$. Thus, $\dfrac{dP}{dT} = \dfrac{\Delta H_{vap}}{T_{bp}(\Delta V_{vap})}$. We make a (good) approximation here due to $V_{liq} \ll V_{vap}$ as for water where the liquid volume is less than 19 mL/mol at $100°$C while the vapor (steam) volume is approximately $(373.15/273.15)\,(22{,}414\ \text{mL/mol}) = 30620\ \text{mL/mol}$ using Charles' law. Thus, we arrive at what is known as the Clapeyron equation, which can be used as is to study transitions in fusion of minerals but

$$\frac{dP}{dT} = \frac{\Delta H_{\text{vap}}}{T_{\text{bp}}(\Delta V_{\text{vap}})} \cong \frac{\Delta H_{\text{vap}}}{T_{\text{bp}} V_{\text{vap}}}$$

is not quite what we need here for liquids. We pause here to note that B. P. E. Clapeyron (1799–1864) was a French engineer who made several contributions to thermodynamics and was actually the person who plotted the Carnot cycle as a PV diagram shown in a previous chapter. Interestingly, Carnot himself actually used the concept of "caloric" in his derivation but Clapeyron put the Carnot cycle into the form we have shown it in this text.

The Clapeyron equation was extended to a more usable form by a German physicist R. G. Clausius (1822–1888) by splitting the $\dfrac{dP}{dT}$ differential and integrating the new form using the ideal gas law for the vapor at low pressure (a very good approximation) to obtain $\dfrac{dP}{dT} = \dfrac{\Delta H_{\text{vap}}}{T\left(\frac{RT}{P}\right)}$, which can be rearranged to $\dfrac{\left(\frac{dP}{P}\right)}{dT} = \dfrac{d\ln P}{dT} = \dfrac{\Delta H_{\text{vap}}}{RT^2}$ and then further to $d\ln P = \dfrac{\Delta H_{\text{vap}}}{R}\left(\dfrac{dT}{T^2}\right) = -\dfrac{\Delta H_{\text{vap}}}{R}\,d\left(\dfrac{1}{T}\right)$ since $d\left(\dfrac{1}{T}\right) = d(T^{-1}) = -T^{-2}\,dT$ (a clever step indeed!).

Graphically, this leads to a plot similar to the case of the temperature dependent equilibrium constant shown above in the section on equilibrium. We see that we can plot a logarithm of the pressure against reciprocal Kelvin temperature and expect to find a straight line with a negative slope.

This can be integrated as $\displaystyle\int_{P_1}^{P_2} d\ln P = \ln\left(\dfrac{P_2}{P_1}\right) = \int_{T_1}^{T_2} \dfrac{\Delta H_{\text{vap}}}{R}\left(\dfrac{dT}{T^2}\right) = -\dfrac{\Delta H_{\text{vap}}}{R}\left[\dfrac{1}{T_2} - \dfrac{1}{T_1}\right]$, so the final working equation becomes the very useful Clausius–Clapeyron equation

$$\ln\left(\frac{P_2}{P_1}\right) = \frac{-\Delta H_{\text{vap}}}{R}\left[\frac{1}{T_2} - \frac{1}{T_1}\right].$$

Note this so-called Clausius–Clapeyron equation has five variables, so that a number of possible problems can be formulated for quiz questions and in a practical sense it is a very useful equation to find ΔH_{vap} from P, T data or a boiling point if one knows ΔH_{vap}.

In Table 6.1, we see values of temperature at which the vapor pressure is at certain values. The final values at 100 kPa are not quite the normal boiling points because 1 atm = 101.325 kPa. This table shows the modern way to represent these types of data, which reveals the vapor pressures of solids at low temperature, but often only provides two data points for P, T in the range of room temperature. Further, the use of only two points can only yield a perfect line when one plots the $\ln(P)$ versus $(1/T)$. In order to show that the approximations made in deriving the Clausius–Clapeyron equation are good but not perfect, we also present in Figure 6.3, the older style data for the vapor pressure of liquid water (H_2O) in mmHg and 1 atm boiling point (see Table 6.2 [2])

$$\ln(P) = \ln(760) - \left(\frac{\Delta H_{\text{vap}}}{R}\right)\left(\frac{1}{T(K)}\right) + \left(\frac{\Delta H_{\text{vap}}}{R}\right)\left(\frac{1}{373.15}\right),$$

TABLE 6.1
Vapor Pressures of Selected Liquids at Various Temperatures (°C)

Vapor Pressure	CCl$_4$	C$_2$H$_5$OH	CH$_3$COOH	H$_2$O	C$_6$H$_6$
1 Pa	−79.4$_{(s)}$	−73$_e$	−42.8$_{(s)}$	−60.7$_{(s)}$	—
10 Pa	−70.8$_{(s)}$	−56$_e$	−26.7$_{(s)}$	−42.2$_{(s)}$	—
100 Pa	−53.5$_{(s)}$	−34$_e$	−8$_{(s)}$	−20.3$_{(s)}$	−40$_{(s)}$
1 kPa	−24.4$_{(s)}$	−7$_e$	14.2$_{(s)}$	7.0	−15.1$_{(s)}$
10 kPa	15.8	29.2	55.9	45.8	20.0
100 kPa (1 bar)	76.2	78.0	117.5	99.6	79.7

Source: Lide, D.R., *CRC Handbook of Chemistry and Physics*, 90th Edn., CRC Press, Boca Raton, FL, 2009–2010, pp. 6–72. With permission.

(s) = solid, e = extrapolated estimate.

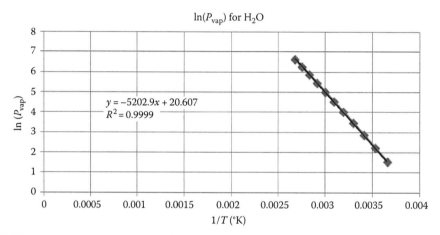

FIGURE 6.3   A plot of ln (P_{vap}) versus ($1/T(K)$) using pressure in mmHg.

TABLE 6.2
Vapor Pressure of Liquids at Various Temperatures in mmHg

Temperature (°C)	CCl$_4$	C$_2$H$_5$OH	CH$_3$COOH	H$_2$O	C$_6$H$_6$
0	32.9	12.7	3.5	4.6	25.3
10	56.0	24.2	6.4	9.2	45.2
20	91.0	44.5	11.8	17.5	75.6
30	142.3	78.5	20.1	31.8	120.2
40	214.8	133.7	34.2	55.3	183.6
50	314.4	219.9	56.3	92.5	271.4
60	447.4	350.2	88.3	149.4	390.1
70	621.1	541.1	137.9	233.7	547.4
80	843.3	812.9	202.3	355.1	753.6
90	1122.0	1187.0	292.7	525.8	1016.1
100	1463.0	1693.0	417.0	760.0	1344.3

Source: Maron, S.H. and Prutton, C.F., *Principles of Physical Chemistry*, The Macmillan Co., New York, 1958, p. 91. With permission.

$$\ln(P) = -\left(\frac{\Delta H_{vap}}{R}\right)\left(\frac{1}{T(K)}\right) + \left\{\left(\frac{\Delta H_{vap}}{R}\right)\left(\frac{1}{373.15}\right) + \ln(760)\right\},$$

or

$$y = mx + b.$$

Table 6.2 tells us nothing about the vapor pressure of these five liquids below 0°C but provides a more detailed view of the vapor pressures in the easily accessible laboratory range between ice water and boiling water. The data for water also give more detail relative to the vapor pressure of water that might be needed for Dalton's law when a gas is collected over water and also give us an appreciation for the relationship between water vapor in the air and a dew point temperature. A direct plot of $\ln(P_{vap})$ versus $(1/T(K))$ yields a slope of $-\left(\frac{\Delta H_{vap}}{R}\right) = -5202.9$ and we find $\Delta H_{vap} = (5202.9)(8.314\,J/mol) = 43256.9\,J/mol$. The value given in the 90th Edn. of the *CRC Handbook* is 43990 J/mol at 25°C. This handbook clearly shows that ΔH_{vap} is not constant but varies slightly with temperature. Let us check that value using two values from Table 6.2

$$\ln\left(\frac{P_2}{P_1}\right) = \frac{-\Delta H_{vap}}{R}\left[\frac{1}{T_2} - \frac{1}{T_1}\right] = \left(\frac{+\Delta H_{vap}}{R}\right)\left[\frac{T_2 - T_1}{T_2 T_1}\right] \text{ or } R\left(\frac{T_2 T_1}{T_2 - T_1}\right)\ln\left(\frac{P_2}{P_1}\right) = +\Delta H_{vap}. \text{ Note}$$

we have used the °C temperature difference in the denominator to remind ourselves that it is a difference of two temperatures and most importantly that the size of 1°K is exactly the same as that of 1°C.

We see from the plot in Figure 6.3 that a least-squares line produces a R^2 value of 0.9999, which indicates a near perfect line. We need to comment on the slope that was found to be -5202.9 from the best line fit. In the Clausius–Clapeyron equation, the log ratio of the pressures would cancel whatever units the pressures are expressed in but the units get mixed up when we separate the pressures and move the ln (760 mm) to the right side of the equation. This is a case where you should cancel the units in the original equation before putting it into the linear form. In addition, you could always use the ratio of the pressure in whatever units to the standard pressure in the same units. The question of units occurs here because we used 760 mmHg/atm. The argument of a logarithm has to be a unitless number.

This value is within the uncertainty of the other values because the pressures in Table 6.1 are only given to three significant figures. It will be found that if you pick any two values for the vapor pressure of water from Table 6.2, you may get slightly different numbers for each pair of vapor pressures that is consistent with the temperature dependent ΔH_{vap} value.

TABLE 6.3
The Vapor Pressure of Solid (Rhombic) I_2 at °C Temperatures for Known Pressure

P	1 Pa	10 Pa	100 Pa	1 kPa	10 kPa	100 kPa
T, °C	$-12.8_{(solid)}$	$9.3_{(solid)}$	$35.9_{(solid)}$	$68.7_{(solid)}$	$108_{(solid)}$	$184.0_{(liquid)}$

Source: Lide, D.R., *CRC Handbook of Chemistry and Physics*, 90th Edn., CRC Press, Boca Raton, FL, 2009–2010, pp. 6–72. With permission.

$I_{2(solid)}mp = 113.7°C = 386.9°K$, $\Delta H_{fus}^0 = 15.52\,kJ/mol = 3.709\,kcal/mol.$
$I_{2(liquid)}bp = 184.4 = 457.6°K$, $\Delta H_{vap}^0 = 41.57\,kJ/mol = 9.936\,kcal/mol.$
$\Delta H_{sub}^0 = \Delta H_{fus}^0 + \Delta H_{vap}^0 = 15.52\,kJ/mol + 41.57\,kJ/mol = 57.09\,kJ/mol = 13.64\,kcal/mol.$

It is worth memorizing some of the numbers related to water because water is so important to our bodies and covers about 70% of the surface of the Earth. A rough estimate would be that ΔH_{vap} (H_2O) $\cong$ 44 kJ/mol $\approx$ 10.52 kcal/mol. Recently, the author met a student from a past class who won a bet for a steak dinner from his boss when he correctly quoted the property of water under discussion in an industrial project; now that is a benefit of education! The property in question was (ΔH_{vap}/g) for water and we see that to a good approximation it is

$$\left(\frac{43990\,\text{J/mol}}{18.01528\,\text{g/mol}}\right)\left(\frac{1\,\text{cal}}{4.184\,\text{J}}\right) = 583.6 \cong 584\,\text{cal/g}.$$

A related fact is the value of the heat of fusion to melt ice per gram as $\left(\dfrac{6010\,\text{J/mol}}{18.01528\,\text{g/mol}}\right)\left(\dfrac{1\,\text{cal}}{4.184\,\text{J}}\right) = 79.73 \cong 80\,\text{cal/g}.$ Since you only need to multiply by 1000 to get calories/kilogram (a mixed set of units) and then multiply by 0.4536 kg/lb (453.6 g/lb) you can convert to calories/pound. While this seems like trivia, chemical engineers worry over such unit conversions all the time and do not have the luxury of having all SI units. Several years ago, a Mars Lander project crashed because part of a computer program was in feet while another part was in meters. We need to know the SI units and other units, which occur in everyday life as well. Consider the British thermal unit, the btu.

1 btu $\equiv$ the heat energy required to raise the temperature of 1 lb (Av.) of water 1°F while 1 cal $\equiv$ the heat energy required to raise the temperature of 1 g of water 1°C.

In calories this would be $(453.6\,\text{g})\left(\dfrac{100°\text{C}}{180°\text{F}}\right)(1°\text{F}) = 252\,\text{cal}.$

Verbally "1 btu = 252" (cal), should be easy to remember. Using another memory device (a calorie is worth many jewels!) we can convert a btu to joules as well:

$$1\,\text{btu} = 252\,\text{cal} = (252\,\text{cal})(4.184\,\text{J/cal}) = 1054.368\,\text{J}.$$

PHASE EQUILIBRIA

One of Gibbs' most enduring contributions was the study of phases. This is very important in metallurgical engineering where complicated phase diagrams are common in the study and formulation of alloys. Other examples occur in the study of as many as eight slightly different phases of ice, all solids but not exactly the same. Other solids have been studied intensively, such as copper–lithium–aluminum alloys for aircraft construction and solid state electronic devices formed from doped silicon. It should be stated here that while a gas phase can only be a gas and liquids easily become homogeneous solution mixtures, solid phases often have more than one crystal structure, which vary with temperatures below the melting point. We leave those complicated cases of alloys to advanced courses and here we want to just discuss the overview of a diagram of three state phases: solid, liquid, and gas. We see in Figure 6.4, a generic phase diagram with boundary lines between the phases and on the left an unusual diagram for water. Water is a very unusual substance in many ways. For most materials the generic shape diagram on the right prevails in which the boundary line between solid and liquid "leans" to the higher temperature side and most materials generally follow this rule. Most materials can be "squeezed" from a liquid into a solid at a given constant temperature by the application of pressure. In the phase diagram for water, we see a very unusual situation where application of pressure to ice at a temperature below the melting point allows the solid to melt into a liquid. Thus, glaciers can "flow" like rivers since their heavy mass provides great pressure on any obstacle in the path of the glacier, pressure melts the ice to a liquid, which runs around the object and then the water refreezes.

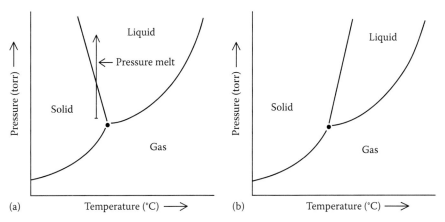

FIGURE 6.4 Schematic illustration of the phase diagram of (a) water compared to (b) most other substances indicating the way in which ice melts under pressure. The scale is exaggerated.

HOW ICE SKATES WORK

While ice skating has been practiced in cold climates for centuries, we can ask how it works. In particular, other slick surfaces would be quickly scratched and scarred if one tries to skate with the usual concave-ground surface on the bottom of the blade with sharp edges. We know that the surface in an ice rink does indeed get scarred but can be easily refinished. However, the key question is how can the skater glide so easily on the surface? In Figure 6.4, we see that at a constant temperature, presumably below the melting point, an increase in pressure will melt the ice. Consider a petite female skater of only 80 lb. Let us assume her skate blade is 1/8 in. wide and 8 in. long to provide 1 in.^{2} of surface. When she places her weight on one foot, that is a considerable applied pressure measured in atmospheres. Evidently, that is enough pressure to melt the ice directly under the blade and provide lubrication for a smooth glide!

$$\frac{(80\,\text{lb})}{(8\,\text{in.})\left(\dfrac{1}{8}\,\text{in.}\right)(14.7\,\text{psi/atm})} \cong 5.44\,\text{atm} = 5.51\,\text{bar}$$

The schematic in Figure 6.4 is only qualitative and there have been more sophisticated studies of the optimum temperature for ice rinks, the shape of concave-ground skate blades and a more complete phase diagram for water, but in the final analysis skates do glide and glaciers do flow. Thus, we have probably oversimplified pressure-melting but it does happen. Figure 6.5 shows a more detailed phase diagram for ice and it is clear that the short line segment of the phase boundary between liquid, water, and Ice I extends downward until at least 200 MPa, although the axes are reversed from Figure 6.4. We note that 200 MPa is equal to 2000 bar so for all practical purposes raising the pressure will "melt" Ice I at a given temperature. How can this be? You need to remember that the water clusters that become a solid are held together mainly by H-bonds, which are weaker than covalent bonds so the pressure is evidently able to disrupt at least some of the H-bonds.

GIBBS PHASE RULE

The Gibbs phase rule is very important in chemical and metallurgical engineering where there can be many phases in the solid state but seldom comes into play in synthetic chemistry. In fact, it is not easy to find a simple example in chemistry. Even so this topic is in our list of essential topics in

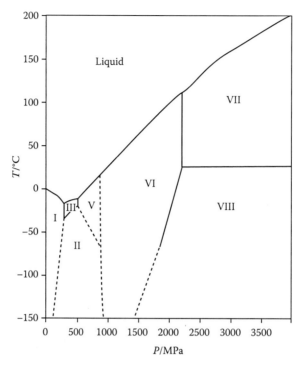

FIGURE 6.5 A more complete phase diagram for water from the 90th Edn. of *CRC Handbook* (by permission). Here temperature (T) is on the vertical axis and pressure (P) is on the horizontal axis. The region of interest is 0–300 MPa which shows the melting temperature decreasing with increasing pressure. Since 1 atm = 1.01325 bar = 1.01325×10^5 Pa, 1 bar = 1.0×10^5 Pa. Thus, 1 MPa = 10 bar and 200 MPa = 2000 bar so a pressure of 5 bar/in.^{2} under an ice skate blade would be on the negative slope of this line. It should be clear that one should consider Ice I and liquid water as the dominant phases up to 200 MPa and down to $-20°C$.

physical chemistry because a basic idea of phase diagrams and phase equilibria is fundamental to understanding the properties of matter. The idea is that there is an overall equation of state that relates (P, V, T, n) for a given system and laboratory constraints fix most of the variables so the question is what flexibility remains within a system in equilibrium. The rule states that when there are "P" phases in a system of "C" components, that results in "F" degrees of freedom to regulate the equilibrium and the equation is

$$F = C - P + 2.$$

Let there be C chemical species in each of P phases, then there will be $P \times C$ mol fractions needed to specify concentrations in each phase. To this we need to add temperature and pressure values so the sum of variables seems to be $PC + 2$, but these variables are not all independent.

In each phase ($\alpha, \beta, \gamma, \delta, \ldots$), the sum of the mole fractions is 1, so that reduces the degrees of freedom by one for each phase or ($-P$) degrees of freedom.

$$\chi_1^\alpha + \chi_2^\alpha + \chi_3^\alpha + \cdots + \chi_C^\alpha = 1$$

$$\chi_1^\beta + \chi_2^\beta + \chi_3^\beta + \cdots + \chi_C^\beta = 1$$

$$\text{--------------------}$$

$$\chi_1^P + \chi_2^P + \chi_3^P + \cdots + \chi_C^P = 1$$

Then we note that the chemical potential of each component is equal within each phase P by virtue of the implicit equilibrium of the C components, so there is also one less degree of freedom relating the chemical potentials within each phase for a reduction of $C(P-1)$ degrees of freedom.

$$\mu_1^\alpha = \mu_1^\beta = \mu_1^\gamma = \mu_1^\delta = \cdots = \mu_1^P$$

$$\mu_2^\alpha = \mu_2^\beta = \mu_2^\gamma = \mu_2^\delta = \cdots = \mu_2^P$$

$$------------------------$$

$$\mu_C^\alpha = \mu_C^\beta = \mu_C^\gamma = \mu_C^\delta = \cdots = \mu_C^P$$

Thus, we can calculate the total number of degrees of freedom in the system as

$$F = (PC) + 2 - P - C(P-1) = C - P + 2.$$

One laboratory experiment that illustrates the Gibbs phase rule is the equilibrium between SO_2 and aniline.

$$SO_{2(gas)} + C_6H_5NH_{2(liquid)} \rightleftharpoons C_6H_5NH_2: SO_{2(solid)}$$

The SO_2 is a noxious gas injurious to mucous membranes, aniline is a slightly tan organic liquid and the complex is a bright orange solid. Thus, $P = 3$ and $C = 2$ so $F = 2 - 3 + 2 = 1$. The author has done this experiment many times. The reaction flask must be tightly sealed because of the noxious SO_2 gas. In preparing the equilibrium, one bubbles the SO_2 gas into about 100 mL of aniline from a small lecture bottle and the orange solid will begin to appear. This must be done in a closed hood. When all three phases are apparent in the reaction flask it is sealed with a fitting for a hose connected to a barometer and the flask is placed into a heating mantle. The equilibrium reaction flask needs to also have a thermometer fitting so that the temperature can be measured. Then there is only one degree of freedom and if the heating mantle determines the temperature of the equilibrium it is only necessary to record the pressure of the system as the remaining degree of freedom to characterize the equilibrium. This experiment also illustrates another convention relating to the "activity" of pure substances. In the case of a solid or a pure liquid, we use the concept of "unit activity" for a pure substance since one can ponder over the question as to what is the "concentration" of a pure liquid or solid? It is not in solution, so no matter how much of a pure solid or liquid is there its concentration is "one." On the other hand the pressure of the gas phase can be measured in moles/liter to relate it to a concentration.

Once the orange complex is formed, it is more convenient to write the equilibrium as the dissociation of the complex.

$$C_6H_5NH_2: SO_2 \rightleftharpoons C_6H_5NH_2 + SO_2, \quad \text{so } K_P = \frac{P_{SO_2}[1_{C_6H_5NH_2}]}{[1_{C_6H_5NH_2:SO_2}]} = P_{SO_2}.$$

Since $\Delta G_{rxn}^0 = -RT \ln K_P = -RT \ln P_{SO_2}$, a plot of the natural log of P_{SO_2} versus $\left(\frac{1}{T(K)}\right)$ should be linear with a slope of $\frac{-\Delta G_{rxn}^0}{R}$. This is a real, but tricky, experiment due to the noxious SO_2 and poses the added problem of disposal of the contents of the reaction flask.

IODINE TRIPLE POINT

One of the oldest forensic techniques is fingerprint analysis. This technique is based on the fact that relatively clean fingertips exude small amounts of fatty acids and other slight secretions from the skin of the fingers. In the case of fresh, sweaty fingerprints on a clean surface, it is sufficient to apply carbon black with a soft brush and then pick up the image of the fingerprint using transparent tape on the sticky side. That is a fortuitous situation but often forensic analysis is faced with smudged partial prints on rough surfaces like cloth or even porous ceramic surfaces. In those cases, any enhancement of the fingerprint image can be helpful in solving a criminal case. The application of iodine vapor is a very old technique where an investigator breaths out through a tube containing calcium carbonate drying agent to remove breath moisture and then the dry breath air flows over iodine crystals as a vapor on to the fingerprint surface. There it is preferentially absorbed into the fingerprint and shows as a dark purple image. This is a very old technique, which fell into disfavor because the image tends to fade away after a few hours. However, recently the use of digital photography has brought this technique back to active use [3]. In particular, the fading aspect is actually an advantage because a fresh iodine image on a rough surface like cloth may be only recognizable as a smudge. The trick is to make a digital photograph of the original image, wait until the iodine fades, and then make another digital photograph without moving the sample. Then simple digital subtraction of the background image leaves only the iodine image. Thus, iodine treatment of fingerprints has undergone renewed interest and use for prints on rough surfaces including cloth with a background pattern.

How does the iodine fingerprint technique work? The basic principle is that solid iodine is one of the few substances that has a significant vapor pressure from the solid and actually sublimes directly from the solid into the gas phase. Although it is common in organic laboratory procedures to purify volatile compounds by sublimation in a vacuum pistol at very low pressure, carbon dioxide and iodine are among only a very few substances that sublime easily at room temperature and pressure. Thus, the technique of sweeping iodine vapor over a surface is a "dry" procedure without any liquid and yet some data is available for the liquid [4].

Now we come to the most interesting type of equilibrium that occurs in the qualitative phase diagrams of Figure 6.4, the place where all three state phase boundaries come together in what is called the "triple point." The triple point is the (P, T) point in the phase diagram where there are three simultaneous equilibria: solid–liquid, solid–gas, and liquid–gas. While it looks like the triple point is the melting point (it is close), it is slightly different. The triple point for water is $273.16^\circ K$ but the melting point is 0.0098° lower, which leads to the standard value of $273.15^\circ K$. Let us use the data from Ref. [5], as shown in Table 6.3. The values of ΔH_{fus}^0 and ΔH_{vap}^0 are given so we can add them to obtain a value for the hypothetical ΔH_{sub}^0 value for the energy required for a whole mole to flash from the solid directly to the vapor, a pure sublimation process without going through the liquid phase. We have all probably seen "dry ice" sublime as solid carbon dioxide goes directly from the solid to the gas and the process here might be called "dry iodine," which is why it is so useful in developing fingerprints without any liquid mess. We should note for future forensic investigators that pure iodine is corrosive and poisonous so one should avoid inhaling it and the use of a squeeze bulb is better than mouth-on-tubing when applying the vapor. We can add the numbers using the "after-minus-before" principle of thermodynamics even though it might be difficult to actually measure this ΔH_{sub}^0 value. With the ΔH_{sub}^0 value we can calculate the hypothetical temperature at which the total sublimation would occur using the Clausius–Clapeyron equation. A word of caution is needed here in that there are a number of $1/T$ values in the calculation and one should not round off the calculation until the end because these reciprocal values are very small and rounding them too soon can lead to large errors. Another problem with the data in Table 6.3 is that the temperatures at which the pressures are very low are probably more uncertain due to the difficulty in measuring such low pressures while the highest vapor pressure of the solid might be contaminated experimentally with interference with some slight melting, so we choose the data point at $68.7^\circ C$ at a pressure of 1 kPa. Then we can solve for the hypothetical T_{sub} temperature.

$$\frac{(8.314\,\text{J/mol}^\circ\text{K}) \ln\left(\frac{101,325\,\text{Pa}}{1000\,\text{Pa}}\right)}{(-57,090\,\text{J/mol})} + \left(\frac{1}{(68.7 + 273.15)^\circ\text{K}}\right) = \left(\frac{1}{T_{sub}}\right), \quad \text{so} \quad T_{sub} = 443.913^\circ\text{K}, \quad \text{the}$$

hypothetical number at which the process of only sublimation would produce 1.01325 bar pressure (1 atm). At the triple point $P_{solid} = P_{liquid}$, so we can write

---

TABLE 6.4
Selected Values of Alpha and Beta for Liquids at Temperatures in °C

Compound	Temperature (°C)	Alpha (10^3/°C)	Beta (10^4/MPa)	Density (g/mL)
H_2O	20	0.206	4.591	0.9982
CH_3OH	20	1.49	12.14	0.7915
CS_2	20	1.12	9.38	1.2632
CH_3CH_2OH	20	1.40	11.19	0.7892
CCl_4	20	1.14	10.50	1.5844 (at 25°C)
C_6H_6	25	1.14	9.66	0.8783
C_8H_{18}	25	1.16	12.82	0.7028

$$\ln\left(\frac{P_{tp}}{101,325}\right) = \left(\frac{-41570\,\text{J/mol}}{8.314\,\text{J/mol}\,^\circ\text{K}}\right)\left[\frac{1}{T_{tp}} - \frac{1}{457.6^\circ\text{K}}\right] = \left(\frac{-57090\,\text{J/mol}}{8.314\,\text{J/mol}\,^\circ\text{K}}\right)\left[\frac{1}{T_{tp}} - \frac{1}{443.913^\circ\text{K}}\right]$$

After canceling the R value, this can be rearranged to isolate the value of T_{tp}.
$T_{tp} = \dfrac{(57,090 - 41,570)}{\left(\frac{57,090}{443.9}\right) - \left(\frac{41,570}{457.6}\right)} = 410.95^\circ\text{K}$. Then we use the liquid equation for the calculation of the

vapor pressure as $\ln\left(\dfrac{P_{tp}}{101,325}\right) = \left(\dfrac{-41,570}{8.314}\right)\left[\left(\dfrac{1}{410.9}\right) - \left(\dfrac{1}{457.6}\right)\right] = -1.2322778$.

This leads to 29549.16 Pa for P_{tp}, which converts to 221.6 mmHg. As calculated, we find the triple point as 411°K, 222 mmHg, 411°K, 0.2916 atm, or 137.8°C, 0.2955 bar. Both the temperature and the vapor pressure we have calculated for the triple point are perhaps higher than expected but the vapor pressure is certainly consistent with the idea that it is easy to obtain a substantial vapor pressure for fingerprint enhancement at room temperature. In research applications, it would be necessary to use 64 bit precision of about 14 significant figures in a computer program to obtain more precise values due to the use of $1/T$ values numerous times and we doubt that the calculated values are within 5% of experimental values because even though the Clausius–Clapeyron equation is accurate, we are cautious rounding reciprocals. The 90th Edn. of the *CRC Handbook* does not give the temperature of the I_2 triple point but this calculation has taught us about the existence of a triple point and provides information about the vapor pressure of I_2 sublimation relative to the renewed use of iodine vapor fingerprint enhancement (see Figure 6.6.)

(C_P–C_V) FOR LIQUIDS AND SOLIDS

While we are discussing solids, liquids, and gases we can consider the difference in heat capacities for solids and liquids. We will now need to use some of the information from the *HUGA* set of equations. Along the way we will repeat the case for an ideal gas and show where the derivation changes for the general case. We start from the definitions of C_P and C_V.

$$C_P - C_V = \left(\frac{\partial H}{\partial T}\right)_P - \left(\frac{\partial U}{\partial T}\right)_V = \left(\frac{\partial U}{\partial T}\right)_P + P\left(\frac{\partial V}{\partial T}\right)_P + V\left(\frac{\partial P}{\partial T}\right)_P - \left(\frac{\partial U}{\partial T}\right)_V, \quad \text{but} \left(\frac{\partial P}{\partial T}\right)_P = 0.$$

FIGURE 6.6 An iodine enhanced fingerprint. (Photo provided courtesy of Forensics Source 2010.) Thanks to Eric Schellhorn, director of marketing, and Floyd Wilson who developed a print on an outside rough surface as a severe demonstration as requested. Close examination reveals clear print lines suitable for computer analysis. See Ref. [3] for additional examples.

Recall

$$\left[\left(dU = \left(\frac{\partial U}{\partial T}\right)_V dT + \left(\frac{\partial U}{\partial V}\right)_T dV\right)\left(\frac{1}{dT}\right)\right]_P \Rightarrow \left(\frac{\partial U}{\partial T}\right)_P = \left(\frac{\partial U}{\partial T}\right)_V + \left(\frac{\partial U}{\partial V}\right)_T \left(\frac{\partial V}{\partial T}\right)_P.$$

Then

$$(C_P - C_V) = \left(\frac{\partial U}{\partial T}\right)_V + \left(\frac{\partial U}{\partial V}\right)_T \left(\frac{\partial V}{\partial T}\right)_P + P\left(\frac{\partial V}{\partial T}\right)_P - \left(\frac{\partial U}{\partial T}\right)_V = \left(\frac{\partial V}{\partial T}\right)_P \left[\left(\frac{\partial U}{\partial V}\right)_T + P\right].$$

Recall that for the ideal gas $\left(\frac{\partial U}{\partial V}\right)_T = 0$ and $\left(\frac{\partial V}{\partial T}\right)_P = \left(\frac{R}{P}\right)_{n=1}$, which lead to $C_P - C_V = R$. However, this time we use one of the *HUGA* equations:

$$dU = T\,dS - P\,dV \quad \text{and} \quad \left(\frac{\partial U}{\partial V}\right)_T = T\left(\frac{\partial S}{\partial V}\right) - P.$$

$$(C_P - C_V) = \left(\frac{\partial V}{\partial T}\right)_P \left[T\left(\frac{\partial S}{\partial V}\right)_T - P + P\right] = \left(\frac{\partial V}{\partial T}\right)_P \left[T\left(\frac{\partial S}{\partial V}\right)_T\right].$$

That was pretty easy, but now we need another equation from the *HUGA* set $dA = -SdT - PdV$ and $\left(\frac{\partial S}{\partial V}\right)_T = \left(\frac{\partial P}{\partial T}\right)_V$, so $(C_P - C_V) = \left(\frac{\partial V}{\partial T}\right)_P T\left(\frac{\partial P}{\partial T}\right)_V$.

We recognize these quantities as things we can measure in a laboratory but not the most convenient expression for routine use. Next, we need a general expression of the volume differential (and set it to zero for a new relationship): $dV = 0$

$$dV = \left(\frac{\partial V}{\partial T}\right)_P dT + \left(\frac{\partial V}{\partial P}\right)_T dP = 0, \quad \left(\frac{\partial P}{\partial T}\right)_V = -\frac{\left(\frac{\partial V}{\partial T}\right)_P}{\left(\frac{\partial V}{\partial P}\right)_T}.$$

Thus, multiplying by (V^2/V^2) we can cast this into easily measurable quantities and we find

$$(C_P - C_V) = \frac{-\left(\frac{\partial V}{\partial T}\right)_P T \left(\frac{\partial V}{\partial T}\right)_P}{\left(\frac{\partial V}{\partial P}\right)_T} \left(\frac{V^2}{V^2}\right) = \frac{TV\alpha^2}{\beta}; \quad \alpha \equiv \frac{\left(\frac{\partial V}{\partial T}\right)_P}{V}, \quad \beta \equiv \frac{-\left(\frac{\partial V}{\partial P}\right)_T}{V}.$$

$$(C_P - C_V) = \frac{TV\alpha^2}{\beta}.$$

This general expression is most useful for liquids since (C_P-C_V) is usually very small for solids.

Example 3

Let us calculate $C_P - C_V$ for water at 20°C. The challenge is to sort out the units. First, we need the molar volume of water at 20°C using the density at 20°C of 0.9982063 g/cm^3 and the molecular weight of 18.01528 g/mol, which yields (18.01528 g/mol/0.9982063 g/cm^3) = 18.04765 cm^3/mol = 18.04765×10^{-6} m^3/mol. Use α and β from Table 6.4.

Note that 1 cm^3 = (0.01 m)^3 = 1×10^{-6} m^3 and we need to recall that 1 Pa = 1 N/m^2.

$$(C_P - C_V) = \frac{TV\alpha^2}{\beta} = \frac{(293.15°K)(18.04765 \times 10^{-6}\, m^3/mol)(0.206 \times 10^{-3}/°C)^2}{(4.591 \times 10^{-4}/10^6\, Pa)}$$

Note a common problem in data tables is that the power of 10 is shifted, so here $[10^3\,(°C)^{-1}]$ means that the number in the table has been multiplied by 1000 and the units are reciprocal degrees centigrade. Similarly $[10^4(MPa)^{-1}]$ means that the number in the table has been multiplied by 10^4 and the units are reciprocal megapascals. We also flipped reciprocal pascals into m^2/N in the denominator of the expression. Thus, we find that

$$(C_P - C_V) = \frac{TV\alpha^2}{\beta} = 0.489032\, Nm/mol\,°K = 0.489032\, J/mol\,°K = 0.116882\, cal/mol\,°K.$$

While this is a small value compared to $R = 8.314\, J/mol\,°K = 1.987\, cal/mol\,°K$, it is not a small number meaning that C_P and C_V are noticeably different for liquid water. We did this as an exercise showing the use of two of the *HUGA* equations and an interesting application of units. This is one of the topics usually shown in a graduate course in thermodynamics but in our approach of only covering a few topics well we have shown an application of some of the *HUGA* equations and there will be others. Knowing the *HUGA* equations and how to use them is roughly half of all thermodynamics.

OPEN SYSTEMS: GIBBS–DUHEM EQUATION FOR PARTIAL MOLAL VOLUMES

The relationship we are about to describe is due to the work of Pierre Duhem (1861–1916) a French physicist who translated Gibbs' work into French and was in his own rights a prolific author of thermodynamic studies. So far the applications of thermodynamic (except for the on-stream ammonia synthesis discussed above) have been for what are "closed systems" where it is possible to enclose the "system" with a boundary and separate it from the "environment." Many of the synthetic applications in chemical engineering are carried out with on-stream processing rather than in a batch reactor, a system in which a continuous flow of reactants is processed and continuous product flows out of some sort of reaction chamber. While most laboratory synthesis is carried out in batch fashion, there are also static phenomena, which depend on adding an arbitrary amount of one

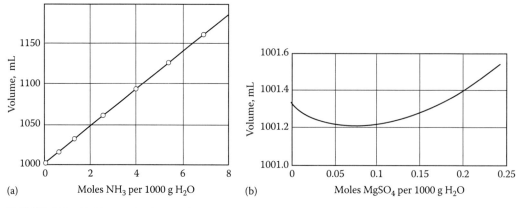

FIGURE 6.7 (a) The total volume of an ammonia–water solution. (b) Volume of a solution of $MgSO_4$ in water. (From Lewis, G.N. and Randall, M., *Thermodynamics*, McGraw-Hill, New York; Barrow, G.A., *Physical Chemistry*, 5th Edn., McGraw-Hill, New York, 1988. With permission.)

reagent which then affects another reagent. In the spirit of our emphasis on essentials, we limit our treatment here to binary solutions (again). We should pause to appreciate that solutions are wondrously homogeneous. Once dissolving occurs, a true solution is essentially immune from settling of the solute due to gravity. If molecules were just like wooden blocks, we could imagine that the total volume of a solution would be the simple sum of the volumes of the two components, the solvent and the solute. In Figure 6.7a [6], we see such a situation where ammonia molecules are dissolved in water. In this case, the total volume is the simple sum of the water molecules and the ammonia molecules. Note that in this case both water and ammonia molecules are electrically neutral, so the volume effect is relatively free of electrical interactions even though there is a chemical interaction in the form of an equilibrium,

$$NH_3 + H_2O \rightleftharpoons NH_4OH \rightleftharpoons (NH_4)^+ + (OH)^-.$$

We know from other courses in chemistry that

$$K_i = \frac{[NH_4^+][OH^-]}{[NH_4OH]} \cong 1.8 \times 10^{-5},$$

so in fact there are only small amounts of charged ions in this weakly basic solution. Although both water and ammonia molecules are polar, even the dipole moments are similar with $\mu_{H_2O} \cong 1.85\,D$ and $\mu_{NH_3} \cong 1.47\,D$, so the sum-of-blocks works very well. A dipole of 1 Debye unit (1 d) is defined to be an electrical vector due to a charge separation of $(+1)\,(-1)$ electron units by a distance of 1 Å. Thus 1 $d = 10^{-18}$ statcoul cm or 10^{-10} esu angstroms, a non-SI unit. We will discuss dipole moments further in a later chapter but you may have encountered dipole moments in organic chemistry? In Figure 6.7b [6], we see a slightly different and mysterious situation when $MgSO_4$ is dissolved in water. The solution volume actually *contracts* for dilute solutions. This is believed to be due to the attraction of the water dipole moments to the charged species Mg^{2+} and $[SO_4]^{2-}$, which causes attractive packing of the ions, at least in the primary solvent shell. Eventually the volume increases as more and more $MgSO_4$ is added to the fixed amount of water (1000 g H_2O).

In Figure 6.8 [6], we see an amazing situation where the ethanol molecules dissolved in water have an aliphatic, hydrophobic alkane portion of the molecule as well as the polar –OH end of the molecule. Thus, ethanol can simultaneously repel water molecules from the aliphatic part of the molecule while at the same time participating in the H-bonding interactions with the water molecules. Thus, we see that in an "open system" where various amounts of solute are added to a

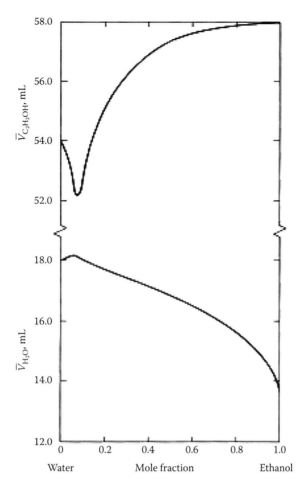

FIGURE 6.8 Partial molal volumes of water and ethanol. (From Lewis, G.N. and Randall, M., *Thermo-dynamics*, McGraw-Hill, New York; Barrow, G.A., *Physical Chemistry*, 5th Edn., McGraw-Hill, New York, 1988. With permission.)

fixed amount of water some complicated interactions can and do occur. The study of these concentration effects defines a slightly new term "molal (m)" where the concentration of moles of solute are dissolved in exactly 1000 g of solvent (often water). In practice, we usually want to know how much solute is in a liter of a solution, ignoring the overall large amount of solvent (water), so molar (M) concentrations are very useful. When we are studying solvent interactions the total volume of the solution can vary with the amount of solute, so the molal (m) concentration is relative to a fixed amount of solvent (1000 g). 1 M $\equiv$ mol solute in 1 L of solution but 1 m $\equiv$ mol solute in 1000 g of solvent.

Let us examine the naive assumption that the total volume of a solution is the sum of the volumes of the solvent and the solute. In terms of what we expect on the basis of common sense addition of wooden blocks we expect that $V_{tot} = V_1 n_1 + V_2 n_2 = \left(\dfrac{\partial V_1}{\partial n_1}\right) n_1 + \left(\dfrac{\partial V_2}{\partial n_2}\right) n_2$. This means that we assume that $dV = \left(\dfrac{\partial V_1}{\partial n_1}\right) dn_1 + \left(\dfrac{\partial V_2}{\partial n_2}\right) dn_2$, but $dV = V_1 dn_1 + n_1 dV_1 + V_2 dn_2 + n_2 dV_2$. According-ing to the Gibbs–Duhem relationship, this can only be true if $n_1 \, dV_1 + n_2 \, dV_2 = 0$. Although this is only a limited example of a volume effect for open systems, the Gibbs–Duhem idea can be applied to other properties such as the chemical potential. What does it mean? As a student, the author did not understand this but gradually came to appreciate it in the following sense.

> *If you want to believe that a chemical property is the simple sum of its parts you have to allow that property to adjust to the presence of the other components!*

This seemingly strange departure from ordinary experience is due to the microscopic nature of the molecular world just as Dalton's law of partial pressures needs to be interpreted in terms of very small atoms/molecules, which can intermingle and seem to occupy the same volume. By the way, just what is the molarity (M) of water when we have 1 L of just water? In most laboratory situations, the concentration of the solute is in the range of 0.01 M or even less so we tend to ignore the solvent. However, by definition of molarity (M) we can write:

$$(1000 \, g \, H_2O/L)/(18.01528 \, g/mol) = 55.508 \, M.$$

This is usually ignored when dealing with water solutions of inorganic ions but if you think about it even simple salt solutions involve at least one solvent shell around each ion as an "aqueous ion" but this is a small amount of the water, which still leaves a large concentration of water capable of dissolving more/other solutes.

$$NaCl + H_2O \rightleftharpoons Na^+_{aqueous} + Cl^-_{aqueous}$$

Example 4

Although the ammonia solution has a volume we expect from common sense, let us consider the next more complicated case of NaCl dissolved in water. Starting with 1000 g of water we can dissolve varying amounts of NaCl in the solvent and now we have to be aware of the fact that water is quite "polar" with a dipole moment of 1.8546 debyes [5] so the negative (O lone pairs) end of the water molecules will be attracted to the Na^+ ions and the positive end of the water dipoles (H-atoms) will be attracted to the Cl^- ions. There may also be an effect that short range order is induced in the H-bonding of the second layer of water molecules around a solvated ion. In addition, there may be some long range repulsion between ions with the same charge, but this will be a very small effect in dilute solutions. Thus, we come to realize that there are electrical effects within the solution as well as the usual consideration of H-bonding in aqueous solutions. Suppose careful laboratory measurements of the total volume of 1000 g of water and various moles (m) of NaCl solute can be fitted to a polynomial in m of NaCl [7].

$$V_{tot} = 1003.00 + 16.62m + 1.77m^{\frac{3}{2}} + 0.12m^2.$$

Then it is easy to calculate the "partial molal volume" of NaCl in terms of the molal concentration as:

$$\overline{V}_{NaCl} \equiv \left[\frac{\partial V_{tot}}{\partial m}\right] = \left[16.62 + \left(\frac{3}{2}\right)(1.77)m^{\frac{1}{2}} + 2(0.12)m\right](mL/mol).$$

Then we can obtain the differential of the NaCl concentration as:

$$d\overline{V}_{NaCl} = \left[\left(\frac{1}{2}\right)\left(\frac{3}{2}\right)(1.77)m^{\frac{-1}{2}} + 2(0.12)\right]dm.$$

But what about the partial molal volume of the water as affected by the charged ions?

The Gibbs–Duhem equation gives us another relationship to allow the partial molal volume of the water to actually vary with the concentration of the NaCl using $n_{H_2O} \, d\overline{V}_{H_2O} + n_{NaCl} \, d\overline{V}_{NaCl} = 0$.

We can use the density of water at 25°C of 0.9970480 g/cm^3 and the molecular weight of 18.01528 g/mol to calculate the molar volume of water at 25°C as:

$$(18.01528 \, g/mol)/(0.9970480 \, g/cm^3) = 18.06862 \, cm^3/mol = \overline{V}^0_{H_2O}.$$

Then we can integrate the Gibbs–Duhem equations to find the partial molal volume of water.

$$\int_{\overline{V}^0}^{\overline{V}_1} d\overline{V}_1 = \left(\frac{-1}{n_1}\right) \int_0^m n_2 \, d\overline{V}_2, \quad so \quad \int_{18.06862}^{\overline{V}_{H_2O}} d\overline{V}_{H_2O} = \left(\frac{-1}{n_{H_2O}}\right) \int_0^m m \left[\left(\frac{3}{4}\right)(1.77)m^{\frac{-1}{2}} + 2(0.12)\right] dm.$$

This looks complicated but just do the integral term by term and insert the numbers.

$$\overline{V}_{H_2O} - 18.06862 \, mL/mol = \left. \frac{\left(\left(\frac{3}{4}\right)(1.77)\left(\frac{2}{3}\right)m^{\frac{3}{2}} + 2(0.12)m^2\left(\frac{1}{2}\right)\right)}{-\left(\frac{1000 \, g}{18.01528 \, g/mol}\right)} \right|_0^m, \quad that \ is \ the \ right \ side \ of \ the$$

equation is to be evaluated between the limits of 0 and m for the general treatment of any molal concentration of NaCl. Moving the volume of pure water to the right side and carefully evaluating the other formula at the upper limit of m and the lower limit of 0 leads to

$$\overline{V}_{H_2O} = 18.06862 - \left[\frac{\left(\frac{1.77}{2}\right)m^{\frac{3}{2}} + 0.12m^2}{55.50844}\right] \quad (as \ cm^3/mol).$$ Note that the partial molal volume of the

water actually contracts with increasing amounts of NaCl. Combining these results and rounding the numbers to fewer places considering the limited significant figures in the original polynomial we find (finally):

$$\overline{V}_{H_2O} = 18.069 - 0.0159m^{\frac{3}{2}} - 0.00216m^2 \quad and \quad \overline{V}_{NaCl} = 16.62 + 2.66m^{\frac{1}{2}} + 0.24m^2.$$

Aside from some esoteric calculation, what does this mean? First, it means that electrical interactions within the solution cause departures from the usual sum-of-blocks total volume of the solution. This is a warning that on the molecular scale electric effects become important in liquids. Second, notice that the partial molal volume of the NaCl solute gets larger as "m" increases since that brings in the charged particles to the solution. On the other hand, the partial molal volume of the water decreases as more charged ions are in solution; the water "shrinks" on a molar basis as the NaCl is added. Third, we can calculate the total volume of the solution using the moles of water and the moles of NaCl by multiplying the moles by the formulas we have obtained for a given value of "m." Use of the Gibbs–Duhem equation now allows us to maintain our naive idea of just multiplying the molar volume formulas by the number of moles and adding to get the total. This sort of study of solution behavior is currently an area of considerable research, although the approach uses large and complicated computer simulations rather than the relatively simple example we have shown for a binary solution.

CHEMICAL POTENTIAL FOR OPEN SYSTEMS

Historically, one of the main contributions of J. W. Gibbs to the development of thermodynamics was his extension of $G = H - TS$ to open systems. This is an important consideration for on-stream processes encountered by chemical engineers. We have already introduced the concept of the *chemical potential*, $\mu_i = (G_i/n_i)$, in two previous applications in this chapter; first in the treatment of gas species $\mu(T, P) = \mu^0 + RT \ln P$ and then again in the discussion of the Gibbs phase rule. So far the treatments referred to closed systems and it seemed that μ_i is just another way to express the G_i value per mole. However, in an open system we can include the change in the moles of a given substance in the change in G_i in a partial molal sense using a more general formula.

$$dG = -S\,dT + V\,dP + \left(\frac{\partial G}{\partial n_1}\right)_{T,P,n_{i\neq1}} dn_1 + \left(\frac{\partial G}{\partial n_2}\right)_{T,P,n_{i\neq2}} dn_2 + \left(\frac{\partial G}{\partial n_3}\right)_{T,P,n_{i\neq3}} dn_3 + \cdots$$

This shows that the change in G depends on the changing amount of moles of each component.

Thus, we have $dG = -S\,dT + V\,dP + \sum_i \left(\dfrac{\partial G}{\partial n_i}\right)_{T,P,n_{j\neq i}} dn_i$ for an open system.

This can have far reaching effects in the analysis of an on-stream process but we will just show the implications here and leave further treatment to engineering texts. Recall that

$$A \equiv U - TS = U + (PV - PV) - TS = H - TS - PV = G - PV$$

Thus $dA = dG - P\,dV - V\,dP = \left[-S\,dT + V\,dP + \sum_i \left(\dfrac{\partial G}{\partial n_i}\right)_{T,P,n_{j\neq i}} dn_i\right] - P\,dV - V\,dP$

Then $dA = -S\,dT - P\,dV + \sum_i \left(\dfrac{\partial G}{\partial n_i}\right)_{T,P,n_{j\neq i}} dn_i$. Also using $H = G + TS$ we find that

$dH = T\,dS + V\,dP + \sum_i \left(\dfrac{\partial G}{\partial n_i}\right)_{T,P,n_{j\neq i}} dn_i$ and along with the use of $U = H - PV$ we find that

$dU = T\,dS - P\,dV + \sum_i \left(\dfrac{\partial G}{\partial n_i}\right)_{T,P,n_{j\neq i}} dn_i$. In each case, the $\left[\sum_i \left(\dfrac{\partial G}{\partial n_i}\right)_{T,P,n_{j\neq i}} dn_i\right]$ terms accompany the state variables in an open system. This can be reinterpreted to realize that these terms are part of the differentials of each of the state variables. However, the amazing thing is that

$\mu_i = \left(\dfrac{\partial G}{\partial n_i}\right)_{T,P,n_{j\neq i}}$ in each case! We can also write the general case of the total differential in each case for H, U, G and A as follows:

$$dH = T\,dS + V\,dP + \sum_i \left(\frac{\partial H}{\partial n_i}\right)_{S,P,n_{j\neq i}} dn_i \quad dU = T\,dS - P\,dV + \sum_i \left(\frac{\partial U}{\partial n_i}\right)_{S,V,n_{j\neq i}} dn_i$$

$$dG = -S\,dT + V\,dP + \sum_i \left(\frac{\partial G}{\partial n_i}\right)_{T,P,n_{j\neq i}} dn_i \quad \text{and} \quad dA = -S\,dT - P\,dV + \sum_i \left(\frac{\partial A}{\partial n_i}\right)_{T,V,n_{j\neq i}} dn_i.$$

Comparing the total differentials to the derived equations in terms of $\sum_i \mu_i\,dn_i$ we find that

$\mu_i = \left(\dfrac{\partial G}{\partial n_i}\right)_{T,P,n_{j\neq i}} = \left(\dfrac{\partial A}{\partial n_i}\right)_{T,V,n_{j\neq i}} = \left(\dfrac{\partial H}{\partial n_i}\right)_{S,P,n_{j\neq i}} = \left(\dfrac{\partial U}{\partial n_i}\right)_{S,V,n_{j\neq i}}$. Now the basic equations for open systems are modified to allow for the change in moles of species involved as

$$dH = T\,dS + V\,dP + \sum_i \mu_i dn_i,$$

$$dU = T\,dS - P\,dV + \sum_i \mu_i dn_i,$$

$$dG = -S\,dT + V\,dP + \sum_i \mu_i dn_i,$$

$$dA = -S\,dT - P\,dV + \sum_i \mu_i dn_i.$$

To reinforce the idea, we say again that in every case we have $\mu_i = \left(\dfrac{\partial G}{\partial n_i}\right)_{T, P, n_{j \neq i}}$, which shows how important the chemical potential is to analysis of open systems and the significance of Gibbs' contribution. However, we leave further applications of the chemical potential to engineering texts. We restrict the "essential" applications to closed batch processes bench chemists use.

This concludes our survey of just about everything with Gibbs' name on it and includes an introduction to some concepts of what goes on "inside a solution." More modern treatments require extensive quantum calculations [8] and the use of statistical thermodynamics that we will introduce in a later chapter. However, in the hands of a gifted scientist [9] the simpler cellular automaton method also provides valuable insights.

MODELING LIQUIDS

While there are some recent [10] characterizations of liquid water resulting from intensive quantum mechanical calculations, we can briefly look at a simple but powerful application of the cellular automata approach. In principle, one should treat the quantum mechanics of the interior of nuclei, the electronic structure and long range interactions on a statistical scale averaging over mole quantities but there are several intermediate treatments. One method is called molecular dynamics, which uses rigid "tinker toy" models of molecules and classical dynamics ($f = ma$) with a pseudo-random number generator to move the molecular models. This approach has been successful in simulating some aspects of complicated biochemical reactions and is widely used in biological simulations [11]. An even simpler method is called the method of cellular automata (CA), which uses a flat grid of cells wound onto the surface of a large torus (donut) to simulate a quasi-infinite region of liquid. The method then uses simple probability rules to make decisions about molecular movement with a pseudo-random number generator providing the direction. The science of how to make a truly random sequence of numbers is a problem in itself. A seemingly foolproof method (which is not truly random and actually repeats eventually) is to pick an arbitrary eight digit integer, square it and pick the middle digits from the answer (digits 4, 5, 6, 7, 8, 9, 10, and 11). Then divide that by 100,000,000 to find a decimal value between 0 and 1 to use as a probability between "$0 = $ no" and "$1 = $ yes." There is actually a specialty field of mathematics for ways to generate a truly random sequence of numbers. You can see the fallacy in the simple example we have given in that the answer depends on the initial "seed" number we start with and if we start two sequences with the same seed we will get the same sequence, but it is approximately random.

In the model developed by Kier and Chang [5,12–14], a square grid is wrapped on to a donut form and probabilities are considered for only two parameters (Figure 6.9). First, there is the *"Joining Parameter," J(A,B), which defines the tendency of movement of species A toward or away from species B when the two are separated by a vacant cell. J > 1 simulates an attraction and J < 1 simulates a repulsion. For J = 0 A cannot move at all. Second, there is the Breaking Probability, P_{AB}, which assigns a probability that a species A next to species B will break apart from B; it is a measure of the "stickiness" of the two species. If $P_{AB} = 0$ the species will not separate while if $P_{AB} = 1$ there is no tendency for A to stay next to B.* When species A, B, C, and D can potentially be next to each other the "stickiness probability" is given by the product of the pair probabilities as P_{AB}, P_{AC}, and P_{AD}. If A is surrounded by four species there is no movement. While these rules seem simple and arbitrary they incorporate chemical knowledge gained from other sources and are actually the result of careful considerations. Kier and Chang [5] have developed a combined rule for simulating water molecules and the square grid with four possible neighbors seems to represent in a probabilistic way the possibility of four bonds of water to neighbor atoms in a flickering, quasi-diamond lattice that can be formed using H-bonding. The combined rule is

$$\log J = -1.5 P_B + 0.6 \quad \text{with } T = 100 P_B$$

FIGURE 6.9 Prof. Lemont Kier, senior fellow of the Center for the Study of Biological Complexity at Virginia Commonwealth University, speaking to the Hanover Master Gardener Association, which he founded in 1988. Medicinal chemists have great respect for and interest in natural compounds from plants. Professor Kier has been the author of seven books and 278 scientific papers to date and was an early pioneer in the use of molecular orbital theory in pharmaceutical chemistry. (See *Molecular Orbital Theory in Drug Research*, Academic Press, New York, 1971.)

The final clever approximation is to define the temperature as an energy analog related to the make/break probability. It takes scientific creativity to postulate these relationships but they are based on experience and physical reasoning in an analog way. What are the results?

In Figure 6.10, we see a set of sensible results for what we know about water. At low temperature the water molecules aggregate to form large clusters, presumably to solidify into a solid mass at $0°C$ and at higher temperatures there are a few water molecules that have no nearest neighbors, anticipating the gaseous state of steam. The value of this study is whether it gives a good representation of what we know about water and that is where some amazing correlations can be drawn from this simple model. First, f_0 and f_1 are essentially linear in temperature so it is not surprising that they correlate with the temperature with a good fit as:

$$T\,(°C) = (-490.28)f_0 + (622.60)f_1 + 4.46; \quad R^2 = 0.996$$

That is just a matter of fitting linear data to the temperature. The background of Professor Kier is research in chemical pharmacology where it is a valid strategy to look through large amounts of biological data and search for correlations and the main goal is to find relationships. This strategy is more phenomenological than deriving an equation from Newton's laws of mechanics and the first order of business is to find relationships through correlations. This is similar to major component analysis in certain analytical chemistry methods. However, the relationships may not be linearly independent and further experiments may need to be carried out to determine physical constants for equations describing those correlations. There is such a theoretical framework in the work of Jhon et al. [15] called the significant structure theory of liquids. Perhaps the most important meaning of the work shown here is that a number of properties of water have been shown to have strong correlations to the

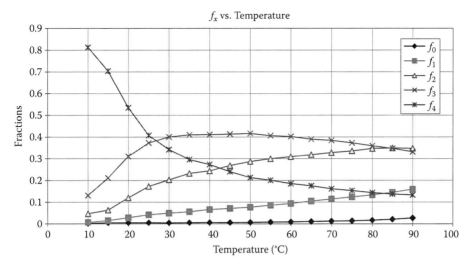

FIGURE 6.10 Results of simulating the populations of clusters of water molecules using the Kier–Chang cellular automata model as a function of temperature with the rules given in the text. Series f_4 starting at 81% for low temperature (10°C) is the fraction of clusters of molecules with four neighbors, Series f_3 is the fraction with three neighbors, Series f_2 the fraction with two neighbors, Series f_1 the fraction with one neighbor and finally Series f_0 is the fraction of molecules with no nearest neighbors. (Redrawn from the data given in Kier, L.B., *Chem. Biodivers.*, 4, 2540, 2007. With permission.)

fractional amounts of the various clusters. Thus, these results offer rough estimates of the fractional amounts of the various clusters and could be used as information for a more elaborate statistical thermodynamic analysis in terms of fractional amounts of clusters, but it should be understood that these clusters are not static species but rather average structures.

A more complex phenomenon is the heat capacity of liquid water and this can be fitted with small amounts of three of the fractions as (Table 6.5):

$$C_P \text{ (cal/g)} = 1.0478 - 0.0488f_2 - 0.0620f_3 - 0.0446f_4; \quad R^2 = 0.995.$$

Next, Kier [12] matched the liquid surface tension to f_1 as the pair interaction most likely important in holding the surface together in the reduced dimensionality compared to the bulk and found

$$\sigma \text{ (dyne/cm)} = -93.72f_1 + 75.33; \quad R^2 = 0.996.$$

TABLE 6.5
Calculated and Observed Heat Capacity of Water in (cal/g) at °C

C_P (Obs.)	C_P (Calc.)
1.0013	1.0011
0.9999	1.0002
0.9988	0.9988
0.9983	0.9981
0.9980	0.9978
0.9980	0.9978
0.9980	0.9982
0.9983	0.9983
0.9985	0.9985

Then the vapor pressure is most likely related to the fraction of molecules with no nearest neighbors (gas-like) and perhaps related to the fraction of dimers and free molecules; Kier [12] found

$$\log P \, (\text{mmHg}) = -24.30f_0 + 15.64f_1; \quad R^2 = 0.997.$$

Again, for the isothermal compressibility they found a relationship in terms of the cluster fractions.

$$\beta = \kappa = \frac{-1}{V}\left(\frac{\partial V}{\partial P}\right)_T (10^{-6}/\text{bar}) = 79.60 - 43.61f_2 - 39.57f_3 - 30.32f_4; \quad R^2 = 0.991.$$

The dielectric constant of water can also be fitted to

$$\varepsilon = -178.88f_1 + 55.84; \quad R^2 = 0.994,$$

and finally it is also possible to fit the viscosity of water using only the f_4 fraction as with

$$\eta \, (cP) = 1.439f_4 + 0.202; \quad R^2 = 0.965.$$

On one hand it should be clear that the coefficients of these excellent correlations offer little interpretation in terms of fundamental constants but on the other hand the sum of these several correlations offers credibility to the overall CA parameters used here. We can see that some properties depend strongly on f_1, which suggests dependence on dimer structure dominance while the vapor pressure does sensibly depend on molecules with no neighbors, which is this model's way of indicating a gas. At the very least, this body of fitted parameters forms a calibrated basis to study dilute aqueous solutions. There have been other treatments of liquid water [8,15,16], which also depend on the assumption of hydrogen-bonded clusters of larger size and 3D structure which are successful in predicting properties, so the concept of molecular clusters in liquids has support from other research. As pointed out by Kier [12], this model fulfills the hypothesis of Haggis [16] regarding the properties of water being due to fractional amounts of clusters of various size. A conclusion for undergraduate students is that relatively simple models of liquids can lead to results which correlate with physical properties. However, the dynamic nature of molecules in the liquid state does not lead to simple formulas. Dynamic models can provide a better understanding of the nature of liquid structures.

SUMMARY

In this chapter, we have briefly treated chemical equilibria in a quantitative way using Gibbs' free energy concept $G = H - TS$ and the concept of chemical potential for closed systems as $\mu(T, P) = \mu^0 + RT \ln P$. Probably the most important aspect of the chapter was the derivation and use of the Clausius–Clapeyron equation $\ln\left(\dfrac{P_2}{P_1}\right) = \dfrac{-\Delta H_{\text{vap}}}{R}\left[\dfrac{1}{T_2} - \dfrac{1}{T_1}\right]$ applied to the boiling of liquids and to the triple point of a phase diagram for iodine relative to fingerprint enhancement. We then developed the general case of $(C_p - C_V) = \dfrac{TV\alpha^2}{\beta}$ showing it is small but nonzero and less than R for liquids. Open systems were encountered in which partial molal volumes of materials adjust their volumes in the presence of one another. Open systems also need added terms for the chemical potential $\mu_i = \left(\dfrac{\partial G}{\partial n_i}\right)_{T,P,n_{j\neq i}}$ but we consider only the "essential" closed systems. An example of the use of the method of Cellular Automata (CA) modeling to simulate liquids was given to show that the behavior of liquids is still an area of research.

PROBLEMS

6.1 Calculate the moles of H_2, D_2, and HD for the equilibrium in Example 1 at 0°C and again at 50°C and predict whether the yield of HD is increased or decreased at higher temperatures.

6.2 Calculate ΔS_{vap} for ethanol using the vapor pressures of 24.2 mmHg at 10°C and 541.1 mmHg at 70°C. Report the values you find for ΔH_{vap} and the normal boiling point.

6.3 Calculate the boiling point of benzene using the data in Table 6.2 and the vapor pressure at 20°C.

6.4 Calculate ΔH_{vap}^0 for CCl_4 using two well spaced temperatures from Table 6.1 or 6.2 and then find the "normal boiling point" (at 1 atm pressure) of CCl_4.

6.5 Calculate the pressure on a single ice skate blade 10 in. long by (0.125 in.) wide supporting the weight of an adult male skater weighing 180 lb. Give the answer in atm, bar, psi and pascals.

6.6 Calculate the vapor pressure of I_2 crystals using the ΔH_{sub}^0 value due to the temperature of 98°F from the breath of a forensic investigator.

6.7 Calculate $C_P - C_V$ for CS_2 at 20°C.

6.8 Suppose you are employed by a laboratory that has synthesized and patented a new liquid detergent called "Brand X" and they want to know its partial molal volume in water, so they can calculate how much to mix with water to achieve a given bottle volume at a given concentration of the detergent. By performing a number of total volume measurements you obtain points related to the moles of the detergent and fit a polynomial to the data using a computer program. Your supervisor wants to know the partial molal volume formulas as a function of "m" moles of detergent from the polynomial you have fitted to the total volume data which is $V_{tot}\,(cm^3) = 1002.83 + 12.4635m + 0.9834m^2$ where "m" is the moles of the detergent. Calculate $\overline{V}_{H_2O}$ and $\overline{V}_{detergent}$ for this solution.

6.9 Derive the expression $(C_P - C_V) = \dfrac{TV\alpha^2}{\beta}$. If this question occurred on an examination, how long would it take for you to do the derivation: 5 min, 3 min? This exercise is an excellent review of the manipulation of partial derivatives.

6.10 Given the value of $\Delta H_{298}^0 = +17.06\,kcal$ and $\Delta G_{298}^0 = +9.72\,kcal$ for the equilibrium $2NOCl_{(g)} \rightleftharpoons 2NO_{(g)} + Cl_{2(g)}$, estimate the temperature at which $K_P = 0.500$ assuming the values of ΔH and ΔG remain constant.

TESTING, GRADING, AND LEARNING?

Here we provide another actual midterm examination from the intense Summer P. Chem. course at VCU in 2008. If students are told they are responsible for derivations like the van der Waals critical point and the Carnot cycle they will have a chance to learn them, but it is unlikely students will "learn" such derivations without fair warning. Since students can learn massive amounts of encyclopedic information in organic chemistry and biochemistry, there is no reason not to expect them to learn key multistep derivations. Learning these key derivations improves the level of the math skill in the class.

Physical chemistry 303   Midterm, Summer 2008   D. Shillady, Professor
(Points)   (120 min. Attempt all problems.)

(15) 1. Show that $\sum \left(\dfrac{q_{rev}}{T}\right) = 0$ for a Carnot cycle; derive the efficiency formula.

(Answer in Chapter 4, how fast can you do the derivation?)

(10) 2. Calculate the laminar bulk flow rate in gallons/min for blood with $\eta = 0.015$ poise through an aorta 5 in. long and (1/4 in.) inner diameter due to a pressure difference of (120–60) mmHg. Multiply by a "duty factor" of 0.05 to compensate for pulsation.

$[(V/t) = 1.3285 \text{ gallons/min}]$

(15) 3. Find P_c, V_c, and T_c for a van der Waals gas and show the law of corresponding states.

(Answer in Chapter 1, how fast can you do the derivation?)

(10) 4. Derive the expressions for $\overline{V}$ and $\sqrt{\overline{V^2}}$ of a gas molecule using the Boltzmann principle. (Answer in Chapter 3, how fast can you do the derivation?)

(10) 5. Compute ΔH^0_{298} for the reaction: $C_6H_6 + 3H_{2(g)} \rightarrow C_6H_{12}$ given the data $\Delta H_{comb}(C_6H_6) = -782.3\,\text{kcal/mol}$, $\Delta H_{comb}(C_6H_{12}) = -937.8\,\text{kcal/mol}$, and $\Delta H_{comb}(H_2) = -68.3\,\text{kcal/mol}$. ($\Delta H^0_{298} = -49.4$ kcal, use Hess's rule of summation)

(15) 6. Calculate the temperature of air compressed adiabatically in a one-cylinder diesel engine from 1035 cm^3 at 25°C to 35 cm^3. Given $C_V = (5/2)R$, compute moles of air, Q, W, ΔU, and ΔH for this compression if the initial pressure is 1 atm. ($T_2 = 1155$°K, $P_2 = 114.61$ atm, mol $= 0.0423$, $Q = 0$, $W = \Delta U = +180$ cal, $\Delta H = +252$ cal)

(15) 7. Derive the expression for $\left(\dfrac{\partial U}{\partial T}\right)_T$ of a van der Waals gas and use it to compute ΔU for the isothermal expansion of 7 mol of Ne gas from 50 to 500 L at constant 25°C given "a" $= 0.21$ L^2 atm/mol^2. (Hint: Use $dU = T\left(\dfrac{\partial P}{\partial T}\right)_V - P$)

$$\left(\Delta U = n^2 a \int_{50\,L}^{500\,L} \frac{dV}{V^2} = n^2 a \left[\frac{-1}{V}\right]_{50\,L}^{500\,L} = +4.485\,\text{cal} = +18.765\,\text{J, note sign}\right)$$

(5) 8. Show that $(C_P - C_V) = R$ for an ideal gas. (Answer in Chapter 4, how fast can you do it?)

(5) 9. Calculate $\overline{V}$ for He gas in mph at 25°C (He $\Rightarrow 4.002602$ g/mol)

$$\left(\overline{V} = \sqrt{\frac{8RT}{\pi M}} = \sqrt{\frac{8(8.314 \times 10^7 \,\text{erg/K mol})(298°\text{K})}{\pi(4.002602)}}\left[\frac{3600\,\text{s/h}}{1.6093 \times 10^5\,\text{cm/mile}}\right] = 2809.4\,\text{mph}\right)$$

BIBLIOGRAPHY

Hill, T. L., *An Introduction to Statistical Thermodynamics*, Addison-Wesley, Reading, MA, 1960, p. 186.

REFERENCES

1. Lide, D. R., *CRC Handbook of Chemistry and Physics*, 90th Edn., CRC Press, Boca Raton, FL, 2009–2010, pp. 6–72.
2. Maron, S. H. and Prutton, C. F., *Principles of Physical Chemistry*, The Macmillan Co., New York, 1958, p. 91.
3. Petersen, S. L., S. L. Naccarato, and G. John, December 2007/January 2008, Forensic Magazine. See the article at http://www.csigizmos.com/products/latentdevelopment/enhancinglatent.html
4. Lide, D. R., *CRC Handbook of Chemistry and Physics*, 87th Edn., CRC Press, Boca Raton, FL, 2007, p. 9–48.
5. Kier, L. B. and C.-K. Cheng, A cellular automata model of water, *J. Chem. Inf. Compu. Sci.*, **34**, 647 (1994).
6. Lewis, G. N. and Randall, M.: Revised by Pitzer, K.S. and L. Brewer (1961), *Thermodynamics*, 2nd Edn., McGraw-Hill, New York, 1961.
7. Barrow, G. A., *Physical Chemistry*, 6th Edn., The McGraw-Hill Companies, Inc., New York, 1996, p. 305.
8. Bukowski, R., K. Szalewicz, G. C. Groenenboom, and Ad van der Avoird, Predictions of the properties of water from first principles, *Science*, **315**, 1249 (2007).
9. Lide, D. R., *CRC Handbook of Chemistry and Physics*, 90th Edn., CRC Press, Boca Raton, FL, 2009–2010, p. 9–52.
10. Wall, F. T., *Chemical Thermodynamics*, 3rd Edn., W. H. Freeman and Co., San Francisco, CA, 1974, p. 463.
11. Goodman, J. M., *Chemical Applications of Molecular Modeling*, The Royal Society of Chemistry, Cambridge, U.K., 1998.
12. Kier, L. B., A cellular automata model of bulk water, *Chem. Biodivers.*, **4**, 2540 (2007).

13. Kier, L. B., C.-K. Cheng, and P. G. Seybold, Cellular automata models of aqueous solution systems, *Rev. Comput. Chem.*, **17**, 205 (2001).
14. Kier, L. B., P. G. Seybold, and C.-K. Cheng, *Modeling Chemical Systems Using Cellular Automata*, Springer, Dordrecht, The Netherlands, 2005.
15. Jhon, M. S., J. Grosh, T. Ree, and H. Eyring, The significant-structure theory applied to water and heavy water, *J. Chem. Phys.*, **44**, 1465 (1966).
16. Haggis, G. H., J. B. Hasted, and T. J. Buchanan, The dielectric properties of water in solutions, *J. Chem. Phys.*, **20**, 1452 (1952).

7 Basic Chemical Kinetics

INTRODUCTION

We have taken a quick trip through what we consider essential thermodynamics and for those students who are only going to take one semester of physical chemistry we have to make sure we treat the basics of chemical kinetics. A glance at the table of contents' chapter headings will reveal that we can continue to do more with kinetics in an additional chapter or skip to some basic spectroscopy and then return to more kinetics in the second semester with an emphasis on the molecular level. Here, we want to make sure we establish the mathematical basis for the time dependence of chemical reactions at a macroscopic level. Once again we are giving what we believe are the essential aspects of kinetics here and then visit more advanced kinetics in the second semester.

The main concept we need to develop is the "Extent of Reaction." *The extent of the reaction is related to the mole quantity change during the reaction and is based on 1 mol so that we can relate to whatever the coefficients are in the balanced reaction.* On a macroscopic scale as used in kinetics we consider the extent of the reaction related to the process of "moles in and moles out." So, even if there is a detailed treatment of a reaction mechanism, the extent of reaction is a mole quantity related to progress of the reaction toward the product. Consider the reaction

$$A + 2B \rightarrow C + 3D$$

Let "x" be the extent of reaction. We are interested in the rate of the overall reaction but what do we mean by "rate"? We can measure the rate by the appearance of species C or D or we can measure the rate of disappearance of A or B.

$$\text{Rate} = +\frac{d[C]}{dt} = +\left(\frac{1}{3}\right)\frac{d[D]}{dt} = -\frac{d[A]}{dt} = -\left(\frac{1}{2}\right)\frac{d[B]}{dt}.$$

When it comes to measuring a rate, we can use a variety of techniques but they must be quantitative. Note that we can measure the rate as an appearance of a product or as a disappearance of a reactant relative to the 1 mol extent of reaction. A rate can be measured by a count of events per unit time as appearance/disappearance of individual molecules or converted to moles using Avogadro's number. As a practical matter, a student should be alert to any physical variable which changes in time during a reaction and can be related quantitatively to moles of the reacting species. Another clue to how to proceed is the fact that we show the rate as a derivative which will lead to the need to solve differential equations by integrating the rate equation.

FIRST-ORDER REACTIONS

The most basic type of rate equation is the first-order decay and we will give complete details of the mathematics here. There are a number of spontaneous reactions in nuclear chemistry and organic chemistry. A basic characteristic of any reaction is that the more reactant there is, the more the reaction will proceed but as the amount of reactant decreases the reaction will be slower. Thus, the rate of the reaction is proportional to the concentration of the reactant.

$$A \rightarrow B, \quad \frac{-d[A]}{dt} = k_1[A]$$

The method we want to teach here is to identify the extent of reaction "x" relative to the initial concentrations. An increase of B corresponds to the appearance of "x." Let "a" be [A] at time $t = 0$ and for simplicity, let [B] $= 0$ at $t = 0$. Then, k_1 has units of $(1/t)$ and where "x" is now the concentration of [B] for times greater than zero (zero is whenever you start your clock!).

$$\frac{+dx}{dt} = k_1(a - x) \quad \text{and} \quad x = 0 \quad \text{at } t = 0.$$

Perhaps it is a good idea to write the variables under the chemical reaction as follows:

$$A \quad \rightarrow \quad B$$
$$(a - x) \qquad x; \quad t > 0$$

Here, we use the key concept that if there is a "proportionality" such as $C \propto D$ we can immediately write $C = kD$ using the basic idea that a proportionality symbol "$\propto$" can be replaced by "$= k$," where k is called the proportionality constant. In kinetics, the proportionality constant is called the rate constant. Here we add a subscript to the rate constant to indicate the order of the reaction as k_1 for a first-order reaction. We will try to do this for higher orders but eventually in complicated cases we may abandon this simpler convention. Next, we can rearrange the kinetic equation to separate x and t variables.

$$\frac{dx}{(a - x)} = k_1 \, dt,$$

so we can integrate this to

$$\int \frac{dx}{(a - x)} = \int k_1 \, dt.$$

If we recall our joke in the math review chapter $\int \frac{d(\text{cabin})}{\text{cabin}} = \ln(\text{cabin}) + C$, we can integrate the rate equation and apply the boundary conditions to the indefinite expression. If we use cabin $= (a - x)$, then $d(\text{cabin}) = -dx$ so we need a minus sign in the answer. We find

$$-\ln(a - x) = k_1 t + C,$$

then at $t = 0$, $x = 0$, so we have $-\ln(a) = C$.

Note that both sides of the indefinite equation would have a constant of integration but since we do not know either of them we combine the constants into one value, C. Then we have $-\ln(a - x) = k_1 t - \ln(a)$ which can be rearranged to $\ln\left(\frac{a}{a - x}\right) = kt$ and then we take the anti-ln of the whole equation to find the final result in what should be a familiar form $\left(\frac{a}{a - x}\right) = e^{k_1 t}$.

A first-order reaction is usually treated as a decay of the original concentration (invert the equation), so, finally, we have

$$(a - x) = a \, e^{-k_1 t} \quad \text{or} \quad a(t) = a_0 \, e^{-k_1 t}.$$

Although that is the solution, in nuclear chemistry it is common to refer to the "half-life" of an original amount. Let us see what the equation looks like when $(a - x) = a/2$, which occurs at a time $t_{1/2}$. $\left(\dfrac{a}{2}\right) = a\,e^{-k_1 t_{1/2}}$ so, canceling "a" we find $(1/2) = e^{-k_1 t_{1/2}}$ and when we take the natural log of the reciprocal of the whole equation we find that $\ln(2) = k_1 t_{1/2} = 0.69314718$. Most of the time this is rounded to $(k_1)(t_{1/2}) = 0.693$, so we also have $k_1 = \dfrac{0.693}{t_{1/2}}$ as well as $t_{1/2} = \dfrac{0.693}{k_1}$. The rounding to only 0.693 is left over from the days when that was sufficiently accurate for use with a slide rule but you can use the more accurate value if you wish.

PROMETHIUM: AN INTRODUCTION TO NUCLEAR CHEMISTRY

We consider that one of the "essential" parts of physical chemistry is some awareness of nuclear chemistry. While the periodic chart poses as a list of stable elements, there are hints of irregularity by the absence of elements no. 43 (technetium, Tc) and no. 61 (promethium, Pm). Modern students are also aware of unstable elements beyond no. 92 (uranium, U). Since nuclear reactions seem to follow a sequence of first-order reactions, we take some time to mention a few of the mechanisms that are occurring in the first-order processes.

Pm was long searched for, but not discovered until nuclear chemistry became a laboratory science and small samples can now be prepared. Pm is elusive for several reasons. Although it has some 44 known isotopes, not one of them is stable. However, it illustrates several aspects of nuclear decay, even though all the processes are first order. The isotopes range from $^{128}_{61}\text{Pm}$ to $^{163}_{61}\text{Pm}$ and the three most stable isotopes are $^{145}_{61}\text{Pm}$ ($t_{1/2} = 17.7$ years), $^{146}_{61}\text{Pm}$ ($t_{1/2} = 5.53$ years), and $^{147}_{61}\text{Pm}$ ($t_{1/2} = 2.623$ years; data from Ref. [1]). We have ventured into this topic because of the simplicity of first-order kinetics, which is so typical of the several forms of nuclear decay and because historically the research of the Curies (Marie, Pierre, and their daughter, Irene) goes across the modern boundaries of chemistry and physics and deserves mention in a "physical chemistry" text. However, even basic nuclear physics is beyond the scope of this text except to refer to Ref. [2], which gives an overview of the main principles. The isotopes of Pm, all have sufficiently short half-lives so that any that was present in primordial Earth has long since decayed but small amounts are present due to decay of any one of several isotopes of Nd, which undergo beta decay (a neutron decays to a proton and ejects an electron, thus increasing the atomic number by 1).

$$^{147}_{60}\text{Nd} \rightarrow\ ^{147}_{61}\text{Pm} + \beta^-;\quad t_{1/2} = 10.98\,\text{days}.$$

The main point here is that the case of Pm shows that while nuclear decay is usually thought of as decreasing to a lower atomic number, some reactions (beta decay, a form of electron emission) actually increase the atomic number of the elemental species. Note that in either case, we can use first-order kinetics for a given single step.

In the 1800s, there were several other gaps in the known periodic chart such as for promethium and technetium but while Pm is radioactive it is so scarce that it was not detected until 1944 at the Oak Ridge National Laboratory by Jacob A. Marinsky, Lawrence E. Glendenin, and Charles D. Coryell. Other radioactive elements are more plentiful in the crust of the Earth and Marie Curie was interested in isolating such elements perhaps more because of their radioactivity than to describe a new element. Thus, Madame Curie's work went far beyond isolating two new elements. This research opened up a new realization that a number of heavier elements have radioactive isotopes. Marie Curie and her husband Pierre carried out a (laborious) chemical separation of two new elements (polonium and radium) that are radioactive and you cannot separate the chemical significance from the physical significance. Since the work of the Curies, we now know that there are many possible isotopes, particularly of the heavier elements. The purpose of this short section is to make undergraduates aware that natural nuclear chemistry is going on all the time. Nuclear reactions are probably the source of heat within the core of the Earth and stars form heavier elements by fusion reactions starting with hydrogen.

Let us consider the decay of Pm by two mechanisms: (1) electron capture where a 1s electron is pulled into the nucleus to make a neutron from one of the protons and (2) beta emission where an electron is ejected and a neutron decays into a proton. Neutrons are not stable and when they are out of a nucleus they have a half-life of 10.3 min but what determines whether a proton becomes a neutron or a neutron becomes a proton depends on the other protons and neutrons in a given nucleus.

$$n \rightleftharpoons p^+ + e^-.$$

$$^{145}_{61}\text{Pm} + e^-_{1s} \rightarrow ^{145}_{60}\text{Nd}; \quad t_{1/2} = 17.7 \text{ years}.$$

$$\left[\begin{array}{l} ^{146}_{61}\text{Pm} + e^-_{1s} \rightarrow ^{146}_{60}\text{Nd} \\ ^{146}_{61}\text{Pm} \rightarrow ^{146}_{62}\text{Sm} + e^-_{\beta} \end{array} \right]; \quad t_{1/2} = 5.52 \text{ years}.$$

$$^{147}_{61}\text{Pm} \rightarrow ^{147}_{62}\text{Sm} + e^-_{\beta}; \quad t_{1/2} = 2.623 \text{ years}.$$

With that very brief exposure to the complexity of nuclear chemistry and perhaps a better appreciation of the significance of the research of the Curies, let us consider the simplest case:

$$^{147}_{61}\text{Pm} \rightarrow ^{147}_{62}\text{Sm} \quad \text{with } t_{1/2} = 2.623 \text{ years}.$$

How much of a 10 g sample will remain after 5 years (Figure 7.1)? We simply insert the data into the equation,

$$a = (10 \text{ g})e^{-k_1(5 \text{ years})} = (10 \text{ g}) \exp\left[-\left(\frac{0.693}{2.623 \text{ years}} \right)(5 \text{ years}) \right] = 2.669 \text{ g}.$$

Thus, we see that the first-order kinetic equation is much simpler than the consideration of the decay mechanism. From a geological point of view, this shows that any primal amount of Pm would have rapidly decayed in the lifetime of the Earth of over 4 billion years. In fact the age of the Earth has been estimated by the presence or absence of radioactive isotopes in the crust of the Earth related to their isotopes, although the original amounts have to be estimated. Nuclear decay is a prime example of first-order kinetics and as with the case of any kinetic problem, the decay is totally dependent on

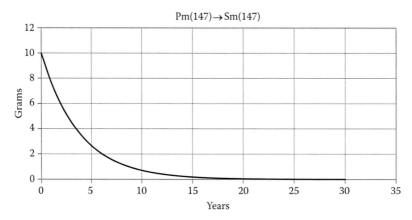

FIGURE 7.1 The radioactive decay of $^{147}_{61}\text{Pm} \rightarrow ^{147}_{62}\text{Sm}$ based on an initial 10 g sample.

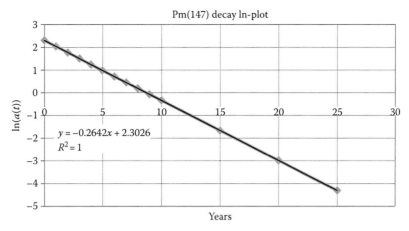

FIGURE 7.2   A plot of ln (value of remaining Pm sample) vs. time in years.

the amount of the material which can decay. In many cases, the half-life is known more accurately than here but Pm decays so fast that the small amounts that have been made limit the precision.

If the amounts of (a, t) are known for a number of points k_1 can be determined by plotting $\ln\left(\dfrac{a(t)}{a_0}\right) = -k_1 t$ or $\ln[a(t)] = -k_1 t + \ln(a_0)$ where the slope is $(-k_1)$. Here, the R^2 value is a value of 1 indicating a perfect fit. The slope is -0.2642 and we know that $k_1 = \dfrac{0.693}{t_{1/2}}$ so we find from the plot that $t_{1/2} = \dfrac{0.693}{0.3013} = 2.623$ years, which was the value we were given. Thus, if you know data in the form of $\{a(t), t\}$ values, you can find the value for k_1 and $t_{1/2}$ for a first-order rate process. For those who are careful about units, we see that when concentration units of $[x]$ are in mol/L, and then $\dfrac{d[x]}{dt} = k_t[a - x]$ means that k_t has units of reciprocal time (Figure 7.2).

MADAME CURIE AND RADIOACTIVITY

While Pm was never found in nature, it has been made in modern nuclear reactors. However, there were other holes in the periodic chart in the 1800s and Marie Sklodowska was a bright young woman who came from Warsaw to study in Paris with Henri Becherel at the University of Paris. There she met and married Pierre Curie who was studying magnetism. However, during isolation of radioactive uranium from pitchblende ore, Marie was sure she had discovered another element in the ore, which was much more radioactive than the uranium. Pierre was intrigued and joined Marie in the laborious separation and isolation of polonium ($Z = 84$) and then a second element, radium ($Z = 88$). Students of physical chemistry should note the combination of chemistry and physics in this discovery. Radium was relatively easy to isolate but polonium is similar to bismuth in its properties and there was also bismuth in the pitchblende ore, so a lengthy process of fractional crystallization was carried out to separate polonium salts from bismuth salts.

The world press was initially captivated by the beautiful young Polish woman who shared the 1903 Nobel Prize in physics with Henri Becherel and Pierre Curie and she distributed the prize money to acquaintances and students. She and Pierre did not realize the effect that the radioactivity had on their health and in 1906 a weakened Pierre walked in front of a horse carriage and was killed. The Curies had two daughters before Pierre died and one daughter, Irene Joliot-Curie, was also awarded a Nobel Prize in chemistry in 1935 for discovering that aluminum became radioactive when bombarded with alpha particles. Note that the decay of radium can provide energetic

FIGURE 7.3   Marie (Sklodowska) Curie (1867–1934) was a Polish-French scientist who shared a Nobel Prize in Physics in 1903 for the discovery of radium and polonium and received another Nobel Prize in chemistry in 1911 for isolating and characterizing radium. She was the first woman to win a Nobel Prize and the only one of two to win in two different fields (Pauling won in chemistry and peace). She also pioneered use of portable x-ray units for battle wounds in WWI.

alpha particles. Prior to Marie's second Nobel Prize in 1911 there was a controversy in the press about her private life and only a few close friends stood by her but she was awarded the 1911 Prize in chemistry. The entire story of her career and up-and-down relationship with the French press is detailed in a balanced way in Ref. [3]. In Figure 7.3, we see her without eye protection which may be due to a photographic setting (although eye protection for laboratory chemists was optional until recently). However, the old style wash bottle in her hand is the type where you put your mouth on the input pipe to blow out wash water and that is hardly what one would use in a laboratory today when working with almost any chemicals let alone radioactive materials. Thus, it is likely she was not only exposed to radiation but also ingested small amounts of radioactive material. Many of these early discoveries in physics and chemistry were considered wondrous but later proved to be hazardous. Some personal anecdotes are that she hired a Polish-speaking nurse for her daughters so they would learn Polish as well as French and her (radioactive) laboratory notebooks when last examined showed patterns to make diapers. Because she was the first woman to win a Nobel Prize and was beautiful, many people adored her, so a student will find many Internet sites with her history. Perhaps because the discoveries were made in France and her daughter continued research in radioactivity, France has a long history of using nuclear power for generation of electricity and has one of the highest dependencies on nuclear power generation in the world (>70%). In 1995, Francois Mitterrand, President of France, officiated as Marie and Pierre were reburied in the Pantheon, the mausoleum of France's most illustrious citizens.

RADIUM

There are many isotopes of Radium from $^{201}_{88}$Ra to $^{234}_{88}$Ra but the one with the longest half-life is $^{226}_{88}$Ra with a half-life of 1599 years and decays by emitting an alpha particle (He^{2+}) with an energy of 4.870 million electron volts (MeV).

Example 1

In their original fractional crystallization separation method, the Curies eventually obtained about 0.1 g of radium chloride. How much of a 0.1 g sample of $^{226}_{88}$RaCl$_2$ isolated in 1911 would remain today (2011)? Thinking like a chemist, if you assume that radium decays to Rn by emitting $^4_2$He^{2+} to form $^{222}_{86}$Rn, the Rn would escape as a gas. Then the Cl$_2$ originally associated with the radium cation might leave as a gas along with the radon but what matters is that the radium is gone.

$$a(t) = a(0)\,e^{-\left(\frac{0.693}{1599\,\text{years}}\right)(2011-1911)\,\text{years}} = (100\,\text{mg})(0.958) = 95.8\,\text{mg}.$$

So, we see as a laboratory source of alpha particles the supply would be pretty constant over a long period of time. Another consideration is that radium is in the same column of the periodic chart as Ca and so biologically it might have similar chemistry to Ca and become trapped in bone tissue where it would be radioactive for a long time. Thus, this interlude regarding the fact that first-order decay is a useful model for nuclear processes has provided an opportunity to discuss some aspects of nuclear chemistry. Considering the crossover of physics and chemistry in the work of the Curies (Marie, Pierre, and Irene) and information in the popular domain regarding nuclear chemistry, we think this brief discussion is justified as an essential part of physical chemistry.

SECOND-ORDER RATE PROCESSES: [A] = [B]

In some texts, the [A] = [B] case is the only one shown for second-order rate processes because it is easy to derive, but actually it is not very general and applies only to cases of reagents which dimerize or to cases where the concentrations have been carefully prepared so that in fact [A] = [B] by initial preparation. We will see below that the general case of [A] $\neq$ [B] is not very difficult and much more general. Nevertheless, let us derive this case and see how we could treat the data graphically.

A + A → B with [B] = x and initially we have [A] = a and [B] = 0 at t = 0 so, we can write $+\dfrac{dx}{dt} = k_2(a-x)(a-x) = k_2(a-x)^2$. We now see why the rate is second order because the sum of the exponents of the concentration terms is two. Once again we separate the variables and integrate the two sides of the equation separately but with a combined integration constant.

$\displaystyle\int \frac{dx}{(a-x)^2} = \int k_2\,dt$ so, we find $\dfrac{1}{(a-x)} = k_2 t + C$. At $t = 0$, $\dfrac{1}{a} = 0 + C$ so we can write $\dfrac{1}{(a-x)} - \dfrac{1}{a} = k_2 t$ or $\dfrac{a-(a-x)}{a(a-x)} = k_2 t$ which is the same as $\dfrac{x}{a(a-x)} = k_2 t$. Note that for the second-order case we see that here that $\dfrac{d[x]}{dt} = k_2[a-x]^2$ so k_2 has (1/time concentration) units and if the concentrations are in (mol/L) the units of $k_2 = $ (L/mol time). The value of the rate constant can be obtained by plotting $y = mx + b$ using $\left(\dfrac{1}{a-x}\right) = k_2 t + \left(\dfrac{1}{a}\right)$, the slope $= k_2$.

SECOND-ORDER RATE PROCESSES: [A] $\neq$ [B]

This is the more general case of a second-order rate process where the concentrations of the two species are not the same. In some texts, this is avoided because of the problem with integrating the rate expression but we show here a special theorem [4] that permits solution of this problem "by inspection." Here the reaction is $A + B \rightarrow C$ where $[C] = x$ with $[A] = (a - x)$ and $[B] = (b - x)$ for $t > 0$ but at $t = 0$ we initially have $[A] = a$, $[B] = b$ and $[C] = 0$. We can set this up as before for the first-order case to keep track of the concentrations as

$$\underset{(a-x)}{A} + \underset{(b-x)}{B} \rightarrow \underset{x}{C}$$

Then $+\dfrac{dx}{dt} = k_2(a - x)(b - x)$ which leads to $\displaystyle\int \dfrac{dx}{(a - x)(b - x)} = \int k_2\, dt$. At first glance this looks like a difficult integral on the left side and maybe some of you will dig out your calculus text to look up "the method of partial fractions" which is a tedious way to split the integral into two simpler integrals, but there is an easy way!

Theorem:

Let $p(x)$ and $q(x)$ denote polynomials in the variable "x" with no factor common (cancel all such beforehand) and let the degree of $p(x)$ be less than $q(x)$. Consider the case in which $q(x)$ has a linear factor $(x - a)$, not repeated. Then, provided $p(x)$, $q(x)$, $f(x)$, and $g(x)$ are all continuous, we can determine the partial fraction coefficient $c = c(a)$ for the term $c/(x - a)$.

Proof: Let $g(x) = (x - a)f(x)$ so that we have $f(x) = \dfrac{p(x)}{q(x)} = \dfrac{g(x)}{(x - a)} = \dfrac{c}{(x - a)} + h(x)$. Here $h(x)$ is the left over part of $f(x)$ after we separate the $\dfrac{c}{(x - a)}$ part and c is the number we seek. Then we can rearrange $\dfrac{g(x)}{(x - a)} = \dfrac{c}{(x - a)} + h(x)$ to $g(x) = c + (x - a)h(x)$. Now take the limit as $x \rightarrow a$. $\displaystyle\lim_{x \to a} [(x - a)f(x)] = \lim_{x \to a} g(x) = \lim_{x \to a} [c + (x - a)h(x)] = g(a) = c$, Q.E.D.

We hope this little proof has not made the process more mysterious than it is in an actual application. The point is that mentally we can separate a factor of $\dfrac{c}{(x - a)}$ from a function with $(x - a)$ in the denominator and determine the coefficient "c" simply by letting $x = a$ everywhere in the remainder of the function. We need a simple example. Consider

$$\frac{1}{(x - a)(x - b)} = \frac{c_1}{(x - a)} + \frac{c_2}{(x - b)} = \frac{1}{(x - a)(a - b)} + \frac{1}{(x - b)(b - a)} = \frac{(x - b) - (x - a)}{(x - a)(x - b)(a - b)}.$$

So, using the theorem above we have shown that $\dfrac{1}{(x - a)(x - b)} = \dfrac{1}{(x - a)(x - b)}$. You will see that when we use this theorem what we do mentally is factor out the $(x - a)$ part of the denominator and just substitute $x = a$ in the remainder of the expression and then move on to the next factor of $(x - b)$ in the denominator and substitute $x = b$ in the rest of the function, including the $(x - a)$ part. With practice on a few examples this can be performed mentally by inspection and can even be used in some other forms of kinetic reactions [4] including the rarely used third-order case. Thus, we can proceed to solve the $A + B \rightarrow C$ case posed above.

$$\int \frac{dx}{(a - x)(b - x)} = \int \frac{dx}{(a - x)(b - a)} + \int \frac{dx}{(b - x)(a - b)} = \int k_2 dt, \quad \text{which is much easier to integrate.}$$

At $t = 0$, $x = 0$, so we can solve for the value of C, the combined constant of integration.

$$\int \frac{dx}{(a-x)(b-a)} + \int \frac{dx}{(b-x)(a-b)} = \frac{-\ln(a-x)}{(b-a)} + \frac{-\ln(b-x)}{(a-b)} = \frac{\ln\left[\frac{(a-x)}{(b-x)}\right]}{(a-b)} = k_2 t + C.$$

$$\frac{\ln\left[\frac{(a)}{(b)}\right]}{(a-b)} = 0 + C, \text{ so we have } \frac{\ln\left[\frac{(a-x)}{(b-x)}\right]}{(a-b)} = k_2 t + \frac{\ln\left[\frac{a}{b}\right]}{(a-b)} \text{ or rearranged to } \frac{\ln\left[\frac{b(a-x)}{a(b-x)}\right]}{(a-b)} = k_2 t.$$

This is not the most general form because the coefficients of A and B are both 1 but we are ready for an example. Students could look up such integrals in tables or use a computer program to solve such problems but here we offer a way to use a simple calculus trick which is general to many other cases and only requires knowing $\int \frac{d(\text{cabin})}{\text{cabin}} = \ln(\text{cabin}) + C$ followed by algebra.

Example 2

Data is given in a paper by W. J. Svirbely and J. F. Roth [5] for the kinetics of the reaction between propionaldehyde and hydrocyanic acid in aqueous solution at 25°C. This is excellent data for teaching second-order kinetics because the concentrations of both reactants are given at different times as follows:

Time, min	2.78	5.33	8.17	15.23	19.80	∞
[HCN]	0.0990	0.0906	0.0830	0.0706	0.0653	0.0424
[C$_3$H$_6$O]	0.0566	0.0482	0.0406	0.0282	0.0229	0.0000

First, we notice that at $t = \infty$ there is still some HCN remaining (0.0424 mol/L) and it does no harm to remind the reader that HCN is extremely poisonous. Perhaps you have encountered this type of reaction before and recognize that the propionaldehyde is the "limiting reagent" but here we mainly need to observe that in our model system [A] $\neq$ [B]. Let us "reset our clock," so that time zero is measured from the first time when we know the concentrations of both reactants, namely at 2.78 min. Another good practice is to select data points which are far apart in time to get the best overall picture of the rate process, namely at 19.80 min to reduce short term fluctuations in the data. The next question is whether the process is really second order? The simple answer is that it is a good guess that when two species react, the rate process is second order. Nevertheless, we will test this assumption and if the rate constant (k_2) is not constant over the span of time in the data then our assumption is incorrect. Here, we are only giving the essentials of the most common type of kinetics but an excellent book on this subject is available in the text by Moore and Pearson [6]. Thus, we assume second-order kinetics and see if the data fits that assumption. Let $[A] = a = [HCN]$ and $[B] = b = [Pr]$ at $t = 2.78$ min. Then, we can use the formula we derived above and insert the data values relative to the concentrations at 19.80 min.

$$\frac{\ln\left[\frac{b(a-x)}{a(b-x)}\right]}{(a-b)t} = k_2 = \frac{\ln\left[\frac{0.0566(0.0653)}{0.0990(0.0229)}\right]}{(0.0424 \, \text{mol/L})(19.80 - 2.78 \, \text{min})} = 0.677261 \, \text{L/mol min}.$$

Note that the concentrations $[a - x]$ and $[b - x]$ are the actual values in the table of data, while the values for $[a]$ and $[b]$ are the values of the data at 2.78 min. In the denominator, the difference in the two concentrations will always be 0.0424 mol/L, but the time is measured starting at 2.78 min, so we have to subtract that from the 19.80 min. Now let us check to see if the data supports the idea of a second-order process using the data at 8.17 min.

$$\frac{\ln\left[\frac{(0.0566)(0.0830)}{(0.0990)(0.0406)}\right]}{(0.0424\,\text{mol/L})(8.17 - 2.78\,\text{min})} = 0.682438\,\text{L/mol min.}$$

That is close enough to the previous value to declare this process is second order and we would ordinarily expect that the earlier data would be less precise than the later data since the reaction is occurring faster in the beginning leading to more uncertainty in the data. Thus we can round the two values (or even compute an average of all four values possible from the given data) and report that the reaction is indeed second order with a rate constant of $k_2 = 0.68\,\text{L/mol min.}$

Example 3
In some cases, a reaction may seem to be first order even though we know two species are involved. These rates are called pseudo-first-order reactions and can occur when the concentration of one of the reactants is in great excess. Consider the following data from Ref. [7] (with permission). A reaction flask is set up containing methyl acetate and 1 M HCl at 25°C. The almost unspoken second reactant is the water in the HCl solution which is of the order of 55 M in H_2O. Actually the HCl is a catalyst and is not consumed during the reaction but aids the hydrolysis.

$$\underset{(a-x)}{CH_3COOCH_3} + \underset{(b-x)}{H_2O} + HCl \rightarrow \underset{x}{CH_3OH} + \underset{x}{CH_3COOH} + HCl$$

It should be obvious that a "hydrolysis" reaction involves water and in fact some water is used up during the reaction. However, if the concentration of the methyl acetate is less than 1M the loss of 1 mol of water out of a concentration of 55 M will hardly be missed and further 1 mol of water is only about 18 mL. Thus, on the basis of moles and volume, the amount of water in the 1 M HCl solution seems almost constant. The rate constant really should be

$$+\frac{dx}{dt} = k_2[H_2O]\,[CH_3COOCH_3] = k_2[55]\,[CH_3COOCH_3] = k'[CH_3COOCH_3]$$

but since the water concentration changes very little during the reaction we can just absorb the water concentration into an effective pseudo-first-order constant $k' = (k_1[55])$. Consider the data carefully.

Time, s	339	1242	2745	4546	∞
Volume, mL	26.34	27.80	29.70	31.81	39.81

You need to understand the way the experiment is set up. Suppose a 2 L flask is set up about half full with a solution of 1 M HCl and a clock is started when an unknown amount of methyl acetate is added to the solution. Then at times which are noted by the clock reading, a pipette of unknown size (but the same quantitative volume for each aliquot) is used to draw out an aliquot which is drained into a 250 Erlenmeyer flask containing about 50 mL of ice and water to "quench" (i.e., slow down) the hydrolysis reaction at a lower temperature. Then the acid in the flask is titrated with carefully standardized 0.100 M NaOH to a phenolphthalein pink end point and the volume of the titration is recorded in the table above. There is uncertainty in the titrations caused by the "creeping end point" which makes it important to obtain a lot of titration points in the hope that the error will average out. Usually it will be helpful to have many more points than given here. First, we are getting a "handle" on this reaction rate using a titration of the acid in this solution as a function of time but it was already 1 M in HCl before the additional amount of acetic acid is produced by the hydrolysis. Thus, the titration volumes in the table are mainly a titration of the 1 M HCl with a slowly increasing amount of acetic acid. Note the final titration, which is usually done by using the same pipette for an aliquot from the reaction solution several days after the start of the reaction, reaches a limiting value at $t = \infty$. How can we get a quantitative rate constant, even if it is a pseudo-first-order rate constant, from such ill-defined data? The key in this case and in many other reaction rates is to ignore a lot of unnecessary information and *focus on the "handle" of the data which connects the extent of the reaction to time.* In this case, the "handle" is the "amount yet to go" in the reaction. Reset your clock to $t = 0$ with the data at 339 s, then the "amount yet to react" is

$$[(V_{HAc} + V_{HCl})_{t=\infty} - (V_{HAc} + V_{HCl})_{t=339\,s}] = a_0 = V_\infty - V_{339}.$$

Then we can use the usual first-order equation $a(t) = a_0\,e^{-k't}$ and $\ln\left(\dfrac{a(t)}{a_0}\right) = -k't$ $a(t) = V_\infty - V_{t-339}$ so we write $\left(\dfrac{-1}{t}\right)\ln\left[\dfrac{(V_\infty - V_{t-339})}{(V_\infty - V_{339})}\right] = k'$ and we can make a table of k' values:

$$\left(\frac{-1}{(1242 - 339)\,s}\right)\ln\left[\frac{39.81 - 27.80}{39.81 - 26.34}\right] = \left(\frac{-1}{903\,s}\right)\ln\left[\frac{(12.01)}{(13.47)}\right] = 1.270491 \times 10^{-4}\,s^{-1},$$

$$\left(\frac{-1}{(2745 - 339)\,s}\right)\ln\left[\frac{39.81 - 29.70}{39.81 - 26.34}\right] = \left(\frac{-1}{2406\,s}\right)\ln\left[\frac{(10.11)}{(13.47)}\right] = 1.192602 \times 10^{-4}\,s^{-1},$$

$$\left(\frac{-1}{(4546 - 339)\,s}\right)\ln\left[\frac{39.81 - 31.81}{39.81 - 26.34}\right] = \left(\frac{-1}{4207\,s}\right)\ln\left[\frac{(8.00)}{(13.47)}\right] = 1.238468 \times 10^{-4}\,s^{-1}.$$

$$\text{Average } k' = 1.2325 \times 10^{-4}\,s^{-1}$$

We see that the calculated value of k' fluctuates. The main reason for this is that the titrations are subject to a large uncertainty due to the fact that even in the cold ice water of the titration flask the reaction is still proceeding. Ideally, the end point of the titration should be the first visual pink of the titration, but since the reaction is still proceeding at the lower temperature, the pink will fade as more acetic acid is produced and then the titrator will add more base and then that will fade, etc., so the end point of the titration is subject to considerable uncertainty. Even so this experiment has been performed by hundreds of students and the results are remarkably reproducible, although with a large uncertainty.

Time, s	1242	2745	4546
$(k'/s) \times 10^{-4}$	1.270491	1.192602	1.238468

Example 4
Now we come to a more general treatment of a second-order reaction in which the coefficients in the balanced reaction are not 1:1. This is a set of data which is reported in Ref. [7] (with permission). This is the sort of quantitative data one needs for precise work and should be familiar from the use of oxidation–reduction titrations in quantitative analysis. Although only the $[(S_2O_3)^{2-}]$ needs be reported, we assume that the sodium salt is used for aqueous solubility.

$$H_2O_2 + 2Na_2(S_2O_3) + 2HCl \rightarrow 2H_2O + Na_2(S_4O_6) + 2NaCl.$$

As usual, the oxidizing agent $[H_2O_2]$ is reduced and the reducing agent $[(S_2O_3)^{2-}]$ is oxidized. The important part of this reaction for the kinetic study is that the coefficients on the peroxide and the thiosulfate are in 1:2 ratio. Thus, the reaction is of the type $A + 2B \rightarrow C$. The data is given as

Time, min	16	36	43	52
$[(S_2O_3)^{2-}]$	0.01030	0.00518	0.00416	0.00313

At $t = 0$, $[H_2O_2] = 0.03680$ and $[(S_2O_3)^{2-}] = 0.02040$ at pH 5.0.

This time, we are given the initial concentrations and the intermediate concentrations of only one reactant. We do not know the order of the reactions but we will guess it is second order because there are two reacting species and if the rate constant checks out to be a constant that will confirm the process is second order. Let $[H_2O_2] = a$ and $[(S_2O_3)^{2-}] = b$, then we can write $\dfrac{+dx}{dt} = k_2(a - x)(b - 2x)$ and that leads to an integrated expression as

$$\int \frac{dx}{(a-x)(b-2x)} = \int k_2\, dt = \int \frac{dx}{(a-x)(b-2a)} + \int \frac{dx}{(b-2x)\left(a - \dfrac{b}{2}\right)} \text{ using the theorem above.}$$

Then we find $\dfrac{-\ln(a-x)}{(b-2a)} + \dfrac{-\ln(b-2x)}{2\left(a - \dfrac{b}{2}\right)} = \dfrac{\ln\left[\dfrac{(a-x)}{(b-2x)}\right]}{(2a-b)} = k_2 t + C$ and if $x = 0$ when $t = 0$ we

have $\dfrac{\ln\left[\dfrac{(a)}{(b)}\right]}{(2a-b)} = 0 + C$ so we find $\dfrac{\ln\left[\dfrac{(a-x)}{(b-2x)}\right]}{(2a-b)} = k_2 t + \dfrac{\ln\left[\dfrac{a}{b}\right]}{(2a-b)}$ and $\dfrac{\ln\left[\dfrac{b(a-x)}{a(b-2x)}\right]}{(2a-b)t} = k_2$.

This can be generalized for any reaction of the type $A + nB \rightarrow C$ with the value of "n" as $\dfrac{\ln\left[\dfrac{b(a-x)}{a(b-nx)}\right]}{(na-b)t} = k_2$ but here $n = 2$ so we can use the data for a second-order rate process.

This problem is made more difficult than the previous example by the fact that we need to use quantitative reasoning to obtain the concentrations. We have concentrations for what we are calling "b" here and according to the coefficients in the balanced equation "a" will disappear only half as fast as "b" is being used up. Once again we choose the longest time period at $t = 52$ min to get the best overall view of the process. At $t = 52$ min, the concentration of the $[H_2O_2]$ will be given by

$$[H_2O_2] = 0.03680 - \{0.02040 - [(S_2O_3)^{2-}]\}/2$$

That is the key to the whole problem. We can now fill in the table of values for the $[H_2O_2]$.

Time, min	16	36	43	52
$[H_2O_2]$	0.03175	0.02910	0.02868	0.028165

$$\frac{\ln\left[\dfrac{b(a-x)}{a(b-2x)}\right]}{(2a-b)t} = k_2 = \frac{\ln\left[\dfrac{(0.02040)(0.028165)}{(0.03680)(0.00313)}\right]}{[2(0.03680) - 0.02040\,\text{mol/L}](52\,\text{min})} = 0.580929\,\text{L/mol min.}$$

Let us check this with the data at 36 min.

$$\frac{\ln\left[\frac{b(a-x)}{a(b-2x)}\right]}{(2a-b)t} = k_2 = \frac{\ln\left[\frac{(0.02040)(0.02910)}{(0.03680)(0.00518)}\right]}{[2(0.03680) - 0.02040\,\text{mol/L}](36\,\text{min})} = 0.593134\,\text{L/mol min}$$

The average of these two values is 0.587 L/mol min and once again there perhaps should be more confidence in the value from the longer time period of 52 min but the two values are close enough to support the assignment of a second-order rate process.

There is still a caution in assigning the order of a rate equation. Although, we have taken the precaution of calculating the rate constant using the assumed order over several data points in the examples above, it will be necessary in general to test that assumption over a wide range of time intervals if the rate equation is not a "tight fit" with variability in the calculated rate constant. The key comparison will be to compare the error range for alternate models to determine the "best" order. In the examples above, we did see variation in the values of the calculated rate constant but in some cases it may be due to experimental error. The examples we have shown are fairly clear cut determinations of the order but the case of the pseudo-first order hydrolysis was included to show how a true second-order reaction can seem like a first-order reaction and in that case the drift in the order is also complicated by the problem that the ice quenching of the reaction samples is not complete, so there is an added analytical problem in the data.

ARRHENIUS ACTIVATION ENERGY

In the example above with the pseudo-first order hydrolysis reaction, the procedure tried to "quench" the reaction by adding the aliquot sample to ice water. This implies there is a temperature effect on the reaction rate. The simplest treatment of this effect was by Svante A. Arrhenius (1859–1927), a Swedish physicist who formulated the first explanation of the temperature dependence of reaction rate constants. Most organic chemistry texts show a single potential energy barrier to a reaction along a reaction coordinate. Note that a *"reaction coordinate" is a distance along a path of progressive geometrical distortion of the reacting species which results in the formation of product.* The "extent of reaction" is a quantitative mole concept while the "reaction coordinate" is a geometrical concept. A little thought leads one to the realization that there must be more than one reaction coordinate since most organic reactions lead to more than one product molecule, but each reaction coordinate will have a barrier that needs to be overcome before the overall ΔH^0_{rxn} is realized. Although we do not know how Arrhenius thought of this, the formula incorporates aspects of the Boltzmann principle. In order for a molecule to reach the energy to go over the barrier there would be a Boltzmann probability of having that energy.

$$k = A\,e^{-\left(\frac{E^*}{RT}\right)}$$

Since the exponent is a unitless number, the value of A has the units of the rate constant k. Here, A is the "Arrhenius constant" and is believed to be related to the number of binary collisions Z_{11} we encountered in the kinetic theory of gases. In the experience of this author, a few calculations for reactions of small molecules using collisions augmented with a "steric factor" do give qualitative agreement with experiment for gas phase reactions. However, A becomes merely a large number when fitted to data for reactions in solution. Even so, we can take the natural log of the equation and gain an appreciation for how the rate constants change with temperature and that leads to an experimental value for what is called the "activation energy, E^*." The activation energy is related to the amount of energy required to pass over the barrier in the reaction coordinate. Because of the simplicity of the Arrhenius formula and the ability to fit experimental data with only two parameters (A, E^*), the Arrhenius formula is extensively used in computer programs which model gas phase reactions that turn out to be far more complicated than one might think. For instance, it takes well over 100 reactions with Arrhenius factors to model the details of all the free radicals and intermediate species in the simple combustion of the methane flame [8].

$$CH_4 + 2O_2 \rightarrow CO_2 + 2H_2O$$

The details of such complex mechanisms require a simple model capable of parameter fitting, so the Arrhenius model is still used for studies of combustion of jet fuel and various rocket engine fuels. A later chapter will explore more detailed treatments but here we can use the equation in natural logarithm form. If $K_{rate} = A\,e^{-\left(\frac{E^*}{RT}\right)}$, we have $\ln(K(T)) = \ln A - \left[\frac{E^*}{R}\right]\left(\frac{1}{T}\right)$ and this leads to another graph with $(1/T)$ along the x-axis. A side issue in notation is that we will have to use K_{rate} for the rate constant to distinguish it from the Boltzmann constant "k" and of course we will use "K" for Kelvin temperature so this author favors using °K for the temperature when all three variations of (k, K_{rate}, °K) are used in the same equation.

Example 5

An example from Ref. [7] will be treated here with the simpler Arrhenius equation and will be used again later to compare this method to more sophisticated treatments. The data consists of measurement of the solvolysis of an alkyl halide (1-chloro, 1-methyl cycloheptane) in a solution of 80% ethanol for what is nominally an SN1 replacement of the chloride ion. In a later chapter, we will examine this reaction in more detail as a model of a steric hindrance in a carbocation substitution reaction but here we are just interested in the measured reaction rates at various temperatures.

Temp, °C	0	25	35	45
K_{rate}, s^{-1}	1.06×10^{-5}	3.19×10^{-4}	9.86×10^{-4}	2.92×10^{-3}
$1/T$ (°K)	3.660992×10^{-3}	3.354016×10^{-3}	3.245173×10^{-3}	3.143172×10^{-3}
$\ln K_{rate}$	-11.454656	-8.050319	-6.921854	-5.836172

Clearly the reaction speeds up as the temperature increases, so the question is whether the data agrees with the equation proposed by Arrhenius. Thus, we plot the data versus $(1/T)$ as

$$\ln(K_{rate}) = -\left(\frac{E^*}{R}\right)\left(\frac{1}{T}\right) + \ln(A) \text{ in linear } y = mx + b \text{ form.}$$

The best-fit line yields a value of R^2 which indicates a very good linear fit. The variable "x" on the graph refers to the x-axis which is $1/T$ (°K) here and the slope has a value of -10872. Using $m = -\left(\frac{E^*}{R}\right)$ we can calculate the activation energy according to Arrhenius as $-\left(\frac{E^*}{R}\right) = -10,872$ and we have to recognize that the $(1/T)$ cancels the temperature units in R.

$$(10,872)R = E^* = (10872)(8.314\,J/mol) = 90.3898\,kJ/mol = 21.604\,kcal/mol.$$

Next, we can ask what value is predicted for the A parameter of the Arrhenius plot. Since "y" on the best-fit line refers to $\ln(K_{rate})$, the "b" term is really $\ln(A) = 28.365$, so we can find A by taking the anti-ln to obtain $A = \exp(28.365)/s = 2.08335 \times 10^{12}\,s^{-1}$, which is a very large number. In the rate equation, the actual rate could be $2.08335 \times 10^{12}\,s^{-1}$ (L/mol). Rough calculations with actual

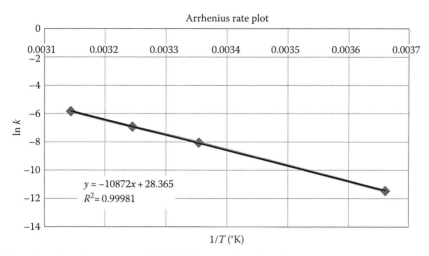

FIGURE 7.4   Plot of ln (K_{rate}) versus [$1/T$ (°K)] for an alkyl chloride solvolysis.

concentrations yield numbers slightly lower than KMTG Z_{11} binary collision numbers, typically (10^{28}/cm^3/s) in the gas phase, due to geometrical requirements in the reaction mechanism. However, in the liquid phase the reactants must find each other in a crowded liquid. Unfortunately, the lack of precise descriptions of most liquids makes the comparison difficult. More importantly, the steric requirements for solution reactions can vary widely. Thus, it seems best to describe A as "related" to the number of effective collisions of the reactants. We can say that for gas phase reactions, which are dependent on thermal velocities for the collision number, we might expect $A \propto \sqrt{T}$, since $\bar{v} = \sqrt{\dfrac{8RT}{\pi M}}$, but the Arrhenius A value is not a function of the temperature, it is just a large constant.

The plot of the data in Figure 7.4 yields a value for E^* which is typical of a number of organic reactions which often have an activation energy of about 20 kcal/mol. It may be of general use to ask what is the effect of a 10°C increase of temperature on a typical reaction rate and we can use this reaction to give an answer for a change from 25°C to 35°C.

$$\frac{k_{308}}{k_{298}} = \frac{A\,e^{-\frac{E^*}{R(308)}}}{A\,e^{-\frac{E^*}{R(298)}}} = e^{+\left(\frac{E^*}{R}\right)\left[\frac{308.15 - 298.15}{(298.15)(308.15)}\right]} = e^{\left(\frac{20000\,\text{kcal/mol}}{1.987\,\text{cal/mol}^\circ}\right)\left[\frac{10^\circ}{(298.15)(308.15)^{\circ 2}}\right]} = e^{1.0955574} \cong 2.99085.$$

So as a rule of thumb, a reaction with an activation energy (E^*) of about 20 kcal/mol or 83.68 kJ/mol will speed up by a factor of 3 for an increase of 10°C. In fact, we see from the data that we have values for the rate at 25°C and 35°C which give the ratio as 3.09. *Note in that calculation the value of A did not matter since it cancels in the ratio* and for that reason the Arrhenius formula was useful even if the value of A was not known.

THE CLASSIC A → B → C CONSECUTIVE FIRST-ORDER REACTION

The next example is a classic problem in both nuclear chemistry as well as chemical engineering. (By the way, a student who complained that he would never see this problem in "real life" was sitting in a seminar the very next day when another student was presenting the results of his PhD research showing a time-dependent series of NMR peaks. In the data, a certain peak (A) decreased to form a second peak (B) and that peak reached a maximum but then decreased to form a final peak (C). The PhD candidate then proceeded to use this solution to analyze the kinetics of his data!) The idea is obvious for nuclear processes because nuclear decay follows successive step-by-step transformations from one isotope to

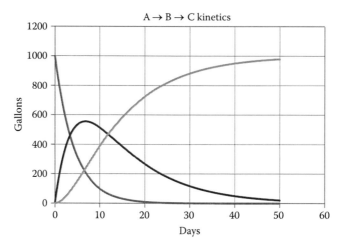

FIGURE 7.5 Contents of tanks A, B, and C in the A → B → C example.

another. Although nuclear reactions often pass through many steps, we have emphasized that all are spontaneous first-order processes. This example can easily be adapted to successive nuclear decay steps by substituting the word "atoms" for "gallons" in what follows. Engineering students will likely find this example more familiar if we use the analogy of flowing liquid, so, let us consider three large tanks for liquids, tank A, tank B, and tank C situated on a hill with A highest and C lowest (Figure 7.5). All three tanks can hold up to 1100 gal of solution but the pipe connecting tanks A and B allows tank A to drain with a $t_{1/2}$ of 3 days, while the pipe connecting tanks B and C is smaller allowing tank B to drain with $t_{1/2}$ of 8 days. For the sake of the boundary conditions, assume exactly 1000 gal of water is stored in tank A at time $t = 0$ and the valve between tanks A and B is closed with tanks B and C empty. Let us suppose that tank A is just for water storage while tank B is used to add NH_4NO_3 to the water to make a liquid fertilizer and tank C is just a holding tank for slow distribution of the fertilizer liquid from the last valve out of tank C to an irrigation system. Assume that the NH_4NO_3 solution is dilute and ignore any volume change of the water. The question is if both valves (A → B) and (B → C) are opened at the same time ($t = 0$) and the outlet of tank C is closed, at what time will the maximum volume be in tank B? Just to be complete, what will the maximum volume of tank B be at that time?

Let N_0 represent the initial gallons of water in tank A, which is 1000 gal here but might be a different number in another situation. Let N_A be the amount of solution (gallons) in tank A at any later time with the similar meaning for N_B and N_C. We can save effort since we know tank A will drain in a first-order way with $t_{1/2} = 3$ days, so we can immediately write $N_A = N_0 e^{-k_a t}$.

The next step is the tricky part because as soon as some water drains into tank B from tank A, some of it will start to drain into tank C, although at a slower rate. Thus, we write

$$\frac{dN_B}{dt} = +k_a N_A - k_b N_B = +k_a N_0 e^{-k_a t} - k_b N_B$$

where we have used k_a and k_b for the two first-order rate constants and we emphasize water coming into tank B with a "+" sign and water/solution leaving tank B with a "−" sign. Please take a minute to understand what is happening here because the next step is tricky and we want to make sure you have the overall picture of water coming into tank B from tank A but immediately starting to drain into tank C. Next, we come to what is usually a whole chapter in a textbook on differential equations. Collect the two terms involving N_B and multiply the whole equation by an exponential "integrating factor." Write the operation and then we will explain it.

$$\left[\frac{dN_B}{dt} + k_b N_B = k_a N_0 e^{-k_a t}\right] e^{k_b t} \Rightarrow e^{k_b t}\left(\frac{dN_B}{dt}\right) + k_b e^{k_b t} N_B = \frac{d}{dt}\left(e^{k_b t} N_B\right) = k_a N_0 e^{(k_b - k_a)t}.$$

This is really an amazing mathematical trick because we should see that the two terms in N_B would be part of a derivative of a product if only there was a factor of $e^{k_b t}$, so why not multiply the whole equation by that factor? Now we can integrate both sides of the equation:

$$\int d(e^{k_b t} N_B) = \int k_a N_0\, e^{(k_b - k_a)t}\, dt \Rightarrow e^{k_b t} N_B = (k_a N_0)\frac{e^{(k_b - k_a)t}}{(k_b - k_a)} + C \text{ but } N_B = 0 \text{ at } t = 0, \text{ so we}$$

have $C = \dfrac{-k_a N_0}{(k_b - k_a)}$ and $e^{k_b t} N_B = \dfrac{k_a N_0\, e^{(k_a - k_b)t}}{(k_b - k_a)} - \dfrac{k_a N_0}{(k_b - k_a)}$. Then collecting common factors we

have $e^{k_b t} N_B = \dfrac{k_a N_0\, e^{(k_a - k_b)t}}{(k_b - k_a)} - \dfrac{k_a N_0}{(k_b - k_a)} = \dfrac{k_a N_0}{(k_b - k_a)}\left[e^{(k_b - k_a)t} - 1\right]$ and after we divide both sides

by $e^{k_b t}$, we finally reach

$$N_B = \frac{k_a N_0}{(k_b - k_a)}\left[e^{-k_a t} - e^{-k_b t}\right].$$

That gives us a formula for N_B at any time $t > 0$. When will the contents of tank B reach a maximum? We need to set $\left(\dfrac{dN_B}{dt}\right) = 0$ and solve for $t_{\max N_B}$. Thus, $\left(\dfrac{dN_B}{dt}\right) = \left[\dfrac{k_a N_0}{(k_b - k_a)}\right]\left(-k_a e^{-k_a t} + k_b e^{-k_b t}\right) = 0$. Therefore $k_a e^{-k_a t} = k_b e^{-k_b t}$; now take the natural log of the whole equation, so $\ln k_a - k_a t = \ln k_b - k_b t$ or $\ln\left(\dfrac{k_a}{k_b}\right) = (k_a - k_b)t$. Thus, $t_{\max N_B} = \dfrac{\ln\left(\frac{k_a}{k_b}\right)}{(k_a - k_b)}$. Now find $t_{\max N_B}$ using the numerical values,

$$t_{\max N_B} = \frac{\ln\left(\frac{k_a}{k_b}\right)}{(k_a - k_b)} = \frac{\ln\left(\dfrac{\frac{0.693}{3\,\text{days}}}{\frac{0.693}{8\,\text{days}}}\right)}{\left(\dfrac{0.693}{3\,\text{days}}\right) - \left(\dfrac{0.693}{8\,\text{days}}\right)} = \frac{\ln\left(\frac{8}{3}\right)}{0.693\left(\dfrac{8-3}{24\,\text{days}}\right)} = 6.7936\,\text{days}.$$

With this information we can find out how much water/solution is in tank B at that time:

$$N_B = \frac{k_a N_0}{(k_b - k_a)}\left[e^{-k_a t} - e^{-k_b t}\right] = \frac{\left(\dfrac{0.693}{3\,\text{days}}\right)(1000\,\text{gal})}{\left(\dfrac{0.693}{8\,\text{days}}\right) - \left(\dfrac{0.693}{3\,\text{days}}\right)}\left[e^{-\left(\frac{0.693}{3}\right)(6.7936)} - e^{-\left(\frac{0.693}{8}\right)(6.7936)}\right],$$

and so $N_B(\max) = (-1600\,\text{gal})(-0.346975) = 555.16\,\text{gal}$. We might as well calculate the contents of tank A and tank C at this time, assuming the outlet valve of tank C is closed.

$N_A = (1000\,\text{gal})e^{-\left(\frac{0.693}{3\,\text{days}}\right)(6.7936\,\text{days})} = 208.19\,\text{gal}$, so assuming conservation of the initial 1000 gal (neglecting evaporation) we can calculate the volume in tank C as whatever is not in tank A or tank B. The volume in tank $C = (1000 - 208.19 - 555.16) = 236.65$ gal at $t = 6.7936$ days. We also realize that eventually all the solution will end up in tank C with tanks A and B completely empty. If this was a problem in nuclear decay, we might have so few individual atoms starting from a number like 1000 atoms that we would have to round the values of N_A, N_B, and N_C to integer values

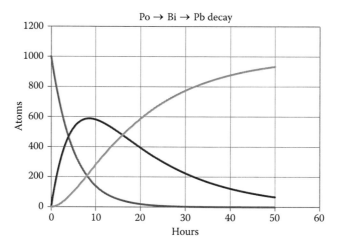

FIGURE 7.6 A plot of the successive decay of $^{204}_{84}Po \rightarrow {}^{204}_{83}Bi$ with $t_{1/2} = 3.53\,h$ followed by $^{204}_{83}Bi \rightarrow {}^{204}_{82}Pb$ with $t_{1/2} = 11.2\,h$ based on an initial 1000 atoms of $^{204}_{84}Po$. Thanks to Prof. Steven Yates of the University of Kentucky for suggesting this example.

but if mole quantities are involved in a chemical reaction we might have to use more significant figures in the half-life values than just 0.693. Even so we have solved the problem and you can judge whether more or less significant figures are warranted for a particular case. Note that we have solved the numerical example for the case where $k_a > k_b$ and the temporary maximum in tank B will occur under that situation. If $k_a < k_b$, the temporary buildup in tank B will not occur.

A → B → C Water Tank Summary:

$$N_0 = 1000\,\text{gal}; \quad t_{1/2}(A) = 3\,\text{days}; \quad t_{1/2}(B) = 8\,\text{days}; \quad t_{\max\,N_B} = 6.7936\,\text{days}$$
$$N_A = 208.19\,\text{gal at } t_{\max\,N_B}; \quad N_B = 555.16\,\text{gal at } t_{\max\,N_B}; \quad N_C = 236.65\,\text{gal at } t_{\max\,N_B}$$

Now that we have some understanding of the way in which the decay scheme works, we can look back and understand some of the difficulty Marie Curie had in isolating radioactive polonium (named for her home country of Poland). Using modern data from Ref. [1], we find $t_{1/2} = 3.53\,h$ for the electron capture of $^{204}_{84}Po \rightarrow {}^{204}_{83}Bi$ followed by another electron capture by $^{204}_{83}Bi \rightarrow {}^{204}_{82}Pb$ with $t_{1/2} = 11.2\,h$ to form stable $^{204}_{82}Pb$. Although other isotopes are involved, this scheme shows the difficulty in isolating Po from Bi while the decay process is going on. Note that the time scale in Figure 7.6 is in hours.

Considering the many applications of this type of problem such as nuclear decay and various forms of time-dependent spectroscopy (NMR, UV–VIS, etc.) there is sufficient detail to the solution presented above to allow it to be used in a number of situations and it is certainly one of the "essential" aspects of basic kinetics in physical chemistry.

SPLITTING THE ATOM

After the work by Marie Curie and her daughter Irene established a new field of research in the radioactivity of elements, others carried out similar experiments to begin to understand the internal structure of the nucleus and a period of increased research occurred in the 1930s. At the Kaiser Wilhelm Institute in Berlin and the Niels Bohr Institute in Stockholm, a drama unfolded in 1938 that ushered in the atomic age. Lise Meitner was a petite, shy Austrian girl who made friends with a

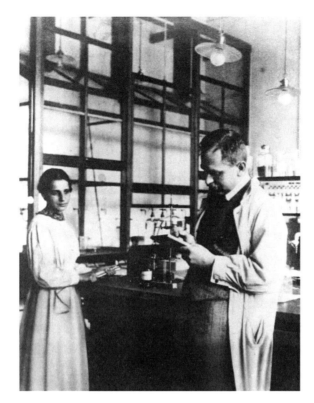

FIGURE 7.7 Lise Meitner (1878–1968) was an Austrian physicist who first described the splitting of a uranium nucleus as "fission" and Otto Hahn (1879–1968) who was awarded the 1944 Nobel Prize in chemistry for analyzing the elemental fragments of uranium fission. He missed the Nobel ceremony because at the time he was a prisoner of war in a British camp. Meitner was later recognized for her key role in the interpretation of Hahn's data when the United States awarded her the Fermi Award jointly with Hahn and his assistant Fritz Strassmann. Judging by her youthful appearance this was probably taken at the Kaiser Wilhelm Institute in Berlin in 1913. Details are given by David Bodanis in the historical novel "$E = mc^2$."

charming young German named Otto Hahn when they met in college and at first she worked for him and later he worked for her (Figure 7.7). She was much better at theory but he was a very good chemist and they were working on trying to make a new element by aiming a beam of *slow* neutrons at uranium hoping to make a new isotope but the uranium kept disappearing and quantities of barium showed up in the beam target.

When Germany annexed Austria in 1938, Meitner who was formerly an Austrian citizen became a German citizen. Then she could not remain at the Kaiser Wilhelm Institute because her parents were Jewish (although she was baptized as a Christian in 1908). As a result, she went to the Niels Bohr Institute in Stockholm, although she could not speak Swedish. She continued to correspond with Hahn about their experiments and he apparently admitted he did not understand what was happening to the uranium. This event and all the human drama is described by David Bodanis [9] in the historical novel "$E = mc^2$." While walking in the snow on a Christmas eve in 1938, in Sweden, Lise Meitner and her nephew Robert Frisch added up the apparent atomic masses of the residual elements found by Hahn and came to the astounding conclusion that about 1/4 of a proton mass was missing and possibly converted into 200 MeV of energy in the form of flying particle debris. Hahn had split the atom but Meitner interpreted the experiment. Hahn quickly published the results ignoring her contribution and he was awarded the Nobel Prize in Physics in 1944. The oversight of her slightly later paper, published in *Nature* (February 11, 1939) using the Bohr "Liquid Drop" model of the nucleus, was partially rectified when the United States awarded her the U.S. Fermi Prize in 1966 jointly with Hahn and his assistant Fritz Strassmann. Meitner and her nephew coined the term "fission" to describe the splitting of the uranium nucleus and today she is known as the "Mother of Fission." Meitner refused to participate in the Manhattan Project because she did not want to work on a bomb. Since that time the kinetic scheme we have described as A → B → C has been extended to many cases of successive first-order decay by radioactive isotopes. Many of the heavier elements are in fact radioactive but with long half-lives and they eventually decay to some stable isotope of Pb in many cases. In retrospect, what Hahn was observing was probably a small amount of a nuclear reaction:

$$^{235}_{92}\text{U} + {}^1_0\text{n} \rightarrow {}^{141}_{56}\text{Ba} + {}^{92}_{36}\text{Kr} + 3{}^1_0\text{n}.$$

Since Kr is a gas, only Ba would show up in the analysis of the solids in the U target. It is important to note that this reaction produces more neutrons than are needed to start the reaction so a chain reaction is definitely possible in the presence of more $^{235}_{92}\text{U}$.

In the late 1950s and early 1960s, undergraduate science education in the United States often included a course in "radiochemistry." This author took such a course which included a set of laboratory experiments using radioactivity counters and careful chemistry experiments with radioactive isotopes. Today such courses seldom occur in chemistry curricula due to the dangers of radioactivity. However, this author believes physical chemistry is possibly the last opportunity for some undergraduate science majors to gain a minimal understanding of nuclear chemistry which is still important in the present age. In the absence of a potentially dangerous laboratory course, we can at least include some paper-and-pencil examples here. In a wonderful general chemistry text by G. Sasin and R. Sasin [10] (reprinted as a soft cover course-pack later by this author [11]), there is a very short chapter on "nuclear chemistry." This material was of great interest in the 1950s. Although Irene Joliot-Curie (daughter of Marie Curie) had won a Nobel Prize in 1935 (in chemistry) for making aluminum radioactive due to bombardment with alpha particles, public interest peaked following WWII. The text by Sasin and Sasin is usefully succinct, so we include some of their material to give a sense of the way in which the alchemist dream of changing one element to another can occur with modern facilities in nuclear reactors and particle accelerators.

In the examples of nuclear reactions which follow we will focus on *conservation of particles* even though we know from Einstein's formula $E = mc^2$ that small amounts of mass can be converted into enormous amounts of energy and we will use integers for particle numbers. In this way, we have to balance the isotope superscripts as well as the subscripts. Note in the fission of $^{235}_{92}\text{U}$, the number of (protons + neutrons) in the superscript adds up to 236 on both sides of the reaction and the number of protons in the subscripts adds up to 92 on both sides of the reaction. Here are a few nuclear reactions given in the Sasin and Sasin text [10,11] (by permission):

A. α-Particle bombardment
   1. Emission of a proton, a process common to many elements from C to K

$$^{27}_{13}\text{Al} + {}^4_2\text{He} \rightarrow {}^{30}_{14}\text{Si} + {}^1_1\text{H}.$$

   2. Emission of a neutron

$$^{11}_5\text{B} + {}^4_2\text{He} \rightarrow {}^{14}_7\text{N} + {}^1_0\text{n}.$$

B. Proton bombardment

$$^7_3\text{Li} + {}^1_1\text{H} \rightarrow 2{}^4_2\text{He}.$$

C. Deuteron bombardment

$$\textstyle^{7}_{3}Li + {}^{2}_{1}D \rightarrow {}^{8}_{4}Be + {}^{1}_{0}n.$$

D. Neutron bombardment

$$\textstyle^{28}_{14}Si + {}^{1}_{0}n \rightarrow {}^{28}_{13}Al + {}^{1}_{1}H.$$

E. Electron bombardment

$$\textstyle^{9}_{4}B + {}^{0}_{-1}e \rightarrow {}^{8}_{3}Li + {}^{1}_{0}n \text{ followed by } {}^{8}_{3}Li \rightarrow {}^{8}_{4}Be + {}^{0}_{-1}e.$$

We can see that an α-particle is a nucleus of He as the dication He^{2+}. When electrons are given off from a nuclear reaction they are called "β-particles" and electrons can be absorbed or emitted. In some heavy elements, 1s electrons can sometimes be captured into a nucleus and combine with a proton to form an additional neutron in the nucleus which reduces the number of protons and changes the elements atomic number (Z) to ($Z-1$) as $^{0}_{-1}e + {}^{1}_{1}H \rightarrow {}^{1}_{0}n$. Thus, for the purpose of chemistry we can regard a neutron as the combination of an electron and a proton. While this has been controversial in the past it is known that neutrons are unstable outside a nucleus and they decay into an electron and a proton as $^{1}_{0}n \rightarrow {}^{0}_{-1}e + {}^{1}_{1}H$ with $t_{1/2} = 10.3$ min.

Of course we have not balanced the energy by this sort of analysis and in actual nuclear reactions gamma rays are often emitted (particularly in neutron bombardment) to balance the energy and the energy of fast emitted particles is part of the energy balance. Finally, as deduced by Lise Meitner, the sum of the masses may not add up and missing mass has been converted into tremendous amounts of energy often measured in "millions of electron volts" as MeV. The missing amounts of mass are usually small enough that we can count the particles as integers in our simple "particle balance." Today nuclear physics is a maturing science but there are still some subatomic particles to be detected experimentally. Even so, Lise Meitner is the "Mother of Fission" and the research discoveries by the Curies (Marie, Pierre, and Irene) led to modern nuclear chemistry. An astute student should note the connections between physics and chemistry in these discoveries.

PROBLEMS

7.1 An unstable isotope that can occur in fallout from a nuclear blast is $^{90}_{38}Sr$, which has a half-life of 29.1 years and has similar chemistry to Ca, so that when it falls on vegetation (grass), herbivores (cows) can ingest it and it can enter the human food chain through milk. Since it is radioactive and children need the nutrition and Ca in milk, $^{90}_{38}Sr$ is considered a long duration health hazard from nuclear fallout and was one of the main reasons for an international ban on above-ground testing of nuclear weapons. Calculate how much of a 1 g sample of $^{90}_{38}Sr$ will remain after 50 years.

7.2 If a 500 gal tank has a valve which allows it to drain from 500 to 250 gal in 2 h, how much of 300 gal in the same tank will drain in 1 h?

7.3 Using the data in Example 2, determine the time when the concentration of propionaldehyde will be exactly 0.020 M.

7.4 Using the data of Example 3, determine the time when the volume of the aliquot titration will be 35.00 mL using the same pipette and the same NaOH titration concentration.

7.5 Using the data of Example 4, determine the time at which the thiosulfate ion concentration will be 0.0035.

7.6 It is sometimes said that many organic reactions double their rate for an increase of only 10°C. Assuming that is true, use the Arrhenius dependence of reaction rates on temperature to determine the approximate value of E^* that would lead to a doubling of an organic reaction rate when heated from 25°C to 35°C.

7.7 Consider the decay of free neutrons $^1_0n \rightarrow {}^{\ 0}_{-1}e + {}^1_1H$ with $t_{1/2} = 10.3$ min. How long will it take for 1% of 1 mol of free neutrons to decay?

7.8 To make sure you understand the mathematical steps, repeat the derivation of the formula for N_B in the $A \rightarrow B \rightarrow C$ reaction sequence and calculate the time of the maximum number of $^{204}_{83}Bi$ atoms if $^{204}_{84}Po = 1000$ atoms at $t = 0$ given the half-life of $^{204}_{84}Po$ is 3.53 h and the half-life of $^{204}_{83}Bi$ is 11.2 h. Calculate the maximum number of $^{204}_{83}Bi$ atoms and the number of $^{204}_{84}Po$ and $^{204}_{82}Pb$ atoms, when $^{204}_{83}Bi$ is at a maximum. Check your results against Figure 7.6.

7.9 Predict the products of the following nuclear reactions based only on *integer* mass units, ignore energy balance, $\Delta E = (\Delta m)c^2$ and emission of gamma rays. Some possibilities are ^1_0n, ^1_1p, $^4_2\alpha = {}^4_2He^+$, and $^{\ 0}_{-1}e = {}^{\ 0}_{-1}\beta$.

   (a) $^{24}_{12}Mg + {}^1_0n \rightarrow {}^{24}_{11}Na + ?$ (b) $^{14}_7N + {}^4_2He \rightarrow {}^{17}_8O + ?$

   (c) $^9_4Be + {}^4_2He \rightarrow {}^{12}_6C + ?$ (d) $^{27}_{13}Al + {}^4_2He \rightarrow {}^{30}_{15}P + ?$

   (e) $^{14}_7N + {}^1_0n \rightarrow {}^{11}_5B + ?$ (f) $^{31}_{15}P + {}^1_0n \rightarrow {}^{28}_{13}Al + ?$

   (g) $^{24}_{11}Na \rightarrow {}^{24}_{12}Mg + ?$

REFERENCES

1. Lide, D. R., *CRC Handbook of Chemistry and Physics, 2009–2010,* 90th Edn., CRC Press, Boca Raton, FL, p. 11–155.
2. Anonymous physicist, http://universe-review.ca/F14-nucleus.htm#shell
3. Fromen, N., Marie and Pierre Curie and the discovery of polonium and radium, Nobelprize.org, The official web site of the Nobel Prize, http://nobelprize.org/nobel_prizes/physics/articles/curie/
4. Shillady, D. D., A theorem to simplify the derivation of certain rate equations, *J. Chem. Educ.*, **49**, 347 (1972).
5. Svirbely, S. J. and J. F. Roth, Carbonyl reactions. I. The kinetics of cyanohydrin formation in aqueous solution, *J. Am. Chem. Soc.*, **75**, 3106 (1953).
6. Moore, J. W. and R. G. Pearson, *Kinetics and Mechanism*, 3rd Edn., John Wiley and Sons, New York, 1981.
7. Daniels, F. and R. A. Alberty, *Physical Chemistry*, 3rd Edn., John Wiley and Sons, New York, 1967, pp. 371–372.
8. Markatou, P., L. Pfefferle, and M. D. Smooke, A computational study of methane-air combustion over heated catalytic and non-catalytic surfaces, *Comb. and Flame*, **93**, 185 (1993).
9. Bodanis, D., $E = mc^2$, Berkley Books, New York, 2000, Chap. 9.
10. Sasin, G. and R. Sasin, *Theory and Problems in General Chemistry*, Drexel Institute of Technology, 1952. Private Communication from R. Sasin, Emeritus Dean, Franklin and Marshall College.
11. Shillady, D., R. Sasin, and M. Hobbs, Chemistry in the news (coursepack for CHEM 112 at Virginia Commonwealth University) 1998. Available from UpTown Copying Service, Richmond Va.

8 More Kinetics and Some Mechanisms

INTRODUCTION

In Chapter 7, we tried to form a good foundation for the study of reaction rates using quantitative measurements. The topic of kinetics deserves a full-semester course, and the classic text is *Kinetics and Mechanism* [1] initially by Frost and Pearson with an updated third edition by Moore and Pearson. Here, we go beyond the straightforward first- or second-order reactions to a few complicated multistep reactions. The main theme of this chapter is the use of the "steady-state approximation," which is a pencil-and-paper method to treat reactions, which include transient intermediate species in the overall reaction. Today, complex reactions are studied using computer modeling, but the pencil-and-paper steady-state treatment still has educational value in explaining the principles of transition-state intermediates (Eyring model), chain reactions, and enzyme kinetics. One goal of this chapter is to learn how to treat reactions according to the Eyring transition-state model to report entropy changes as well as energy changes in the transition state. Another goal is to appreciate how complicated chain reactions can be by studying the solvable scheme for the reaction between H_2 and Br_2. Finally, the important case of enzyme kinetics is treated by deriving the Michaelis–Menten equation with and without a competitive inhibitor. These cases and a few others all depend on some form of the steady-state concept. This admittedly short list of applications was selected as the "essential" topics needed by students in prehealth science, forensic science, and chemistry. Informal interviews of students from this course now in industry or graduate studies have helped form this small list over a number of years. This author always asks graduates of this course, "Did you use the topics we learned in physical chemistry?" and the list of topics has been adjusted several times. Thus, the topics here are the result of that selection process.

BEYOND ARRHENIUS TO THE EYRING TRANSITION STATE

Before the invention of copying machines, scientists distributed their work through journal articles as they do today. But there was no easy way to copy an article then; so when an article appeared, the author would order a hundred or so "reprints" to be mailed to interested parties upon written request. There were even special postcards issued by departments to be used by faculty to request reprints. The reprints themselves were often stapled into very nice covers. As a graduate student assigned to an inventory task, the author discovered boxes and boxes of nicely bound reprints from the 15 years of Prof. Henry Eyring's tenure at Princeton University before he moved to the University of Utah. Henry Eyring's name was eventually on over 685 publications, and many books covering a wide variety of topics in physical chemistry but he is best known for his work on absolute rate theory. We might add that his lectures were usually very enthusiastic and highly animated, entertaining as well as full of special insight (Figure 8.1).

Here, we can show Eyring's genius in reinterpreting the Arrhenius formula. According to Eyring's theory, there is in almost every reaction a key "transition state." Rather than just use the "extent of reaction" in moles to treat the overall reaction turnover, the Eyring treatment imagines some molecular distortion of the internal coordinates of the combined "activated complex," which

Henry Eyring

February 20, 1901–December 26, 1981

FIGURE 8.1 Henry Eyring developed the theory of the kinetic transition state. Prof. Eyring published more than 680 research papers and was an enthusiastic lecturer. Perhaps, his most important contribution to physical chemistry was the formulation of the "transition-state" concept in chemical reactions. (Courtesy of the University of Utah, see also http://www.nap.edu/html/biomems/heyring.html).

then leads to a dissociation to form product(s). It required Eyring's genius to provide a mathematical treatment to modify the Arrhenius concept to describe the transition-state process. In what follows, we may not use exactly the reasoning that Eyring used but still attempt to impart the same clever process. Schematically, we have

$$K_{\text{rate}} = A \, e^{-\left(\frac{E^*}{RT}\right)} \quad \text{from Arrhenius}$$

$$A + B \rightleftharpoons [AB]^{\ddagger} \rightarrow C + D$$

where we have used the "$\ddagger$" symbol to indicate the activated complex. Eyring generated this new symbol in chemistry to indicate this transient species, which is the key to the rate-determining step of a reaction. As such, the decay of the activated complex is essentially a first-order process no matter how it came to be.

Let us carefully consider the "$\rightleftharpoons$" part of this process. It means that many times, the activated complex may be formed but then fall apart to go back to the starting materials or sometimes proceed to the product. This leads to a concept called "the steady-state approximation," which is a key idea in several of the examples which follow. The idea is that the reactants (perhaps assisted by solvent molecules) collide or get near one another long enough to form the activated complex but it may be an unstable bottleneck in the process. Like any truly dynamic equilibrium, the process rapidly proceeds in both directions while setting up a steady-state concentration of the activated complex. Here is a statement of the steady-state idea:

$$[AB]^{\ddagger} \ll 1 \text{ but } [AB]^{\ddagger} \neq 0, \text{ and since it is so small then } \frac{d}{dt}[AB]^{\ddagger} \cong 0 \text{ or } [AB]^{\ddagger} \cong \text{constant.}$$

Eyring's transition-state theory was developed in the 1930s and strained every computational capability at that time. Now it is possible to use modern computer programs to study the rearrangement of the reactants along a "reaction coordinate," which is different from the "extent of reaction" used in Chapter 7. Early calculations by Eyring, Gershinowitz, and Sun [2] as well as by Hirschfelder, Eyring, and Topley [3] explored the simplest reaction of

$$H + H_2 \rightarrow H_2 + H.$$

Their results and others were reviewed later in 1976 by D. G. Truhlar [4] who also investigated that reaction with more modern computer programs only to find that the early work by Eyring et al. was mainly qualitative. Even so, Eyring had an understanding of the basic concepts but not the means to do accurate calculations.

The "reaction coordinate" is often a tortuous rearrangement of the atoms into a contorted shape of the activated complex, but the path to the complex can be "stepped" using computer simulations. Such a path can be shown for reactions such at H_3 mentioned above or the rearrangement of

$$HOCN + H \rightarrow H + OCNH \quad or \quad OCNH \rightarrow HOCN.$$

Such reactions may have relatively simple "reaction coordinates" in a line (H_3) or on some relatively simple arc of one H in a plane relative to a fixed linear group like (OCN^-). However, in general, reactions of larger polyatomic molecules involve many simultaneous adjustments in the atom coordinates to reach the transition-state geometry. We choose to apply the Eyring analysis to the solvolysis data that we have previously treated using the Arrhenius method. That will allow us to make a comparison of the Arrhenius results with the Eyring results and we want to apply the method to a typical organic chemistry reaction. We want to incorporate physical chemistry principles into organic chemistry.

Usually, it takes some increase in energy to rearrange the atoms into the geometry of the activated complex, and in particular, the shape requires a change in entropy to go from separate reactant molecules to a (temporarily) more ordered shape as the activated complex. Thus, Eyring's first daring step was to propose that the Arrhenius activation energy is really $\Delta G^{\ddagger}$. It may seem that we should use $E^* = U = H - PV$ but Eyring proposed $E^* = \Delta G^{\ddagger}$:

$$E^* \Rightarrow \Delta G^{\ddagger} = \Delta H^{\ddagger} - T\Delta S^{\ddagger}.$$

The first modification Eyring made was to rewrite the Arrhenius equation in terms of $\Delta G^{\dagger}$ as,

$$k = (f)e^{-\frac{\Delta G^{\ddagger}}{RT}} = (f)e^{-\left[\frac{(\Delta H^{\ddagger} - T\Delta S^{\ddagger})}{RT}\right]} = \left[(f)e^{+\left(\frac{\Delta S^{\ddagger}}{R}\right)}\right]e^{-\left(\frac{\Delta H^{\ddagger}}{RT}\right)} = A\,e^{-\left(\frac{E^*}{RT}\right)},$$

So, Eyring implies

$$\left[(f)e^{+\left(\frac{\Delta S^{\ddagger}}{R}\right)}\right] = A.$$

Since the exponential has no units, "f" has units of inverse time.

As we will see, that yields much more information about the activated complex as well as satisfies the need to interpret the meaning of the Arrhenius A value. The value and meaning of the new factor f is some yet-to-be-determined function with inverse time units as befitting a first-order rate constant. In addition, the usual values of E^* and $\Delta H^{\ddagger}$ are not very different, so the general idea of the activation energy remains the same.

The next step is daring genius on the part of Eyring, the comments here can only speculate at how he arrived at his final formula. He addressed the question of how the activated complex evolved into the product. We will study molecular vibrations in a later chapter but we can imagine that the activated complex is a group of atoms with rubber bands for bonds. In the process which leads to the product molecule(s), Eyring supposed that the weakest bond in the activated complex would break and lead to the product(s). There is a point of controversy here. The Eyring transition-state model assumes that the transition state exists briefly but for a finite time as a unit. More detailed calculations show that the reaction pathway usually travels through a molecular cluster that has at least one unstable vibration with an imaginary vibrational frequency due to a "saddle point" in the energy surface. Other experimental evidence shows that the reaction rate is quantized in steps according to the positive part of the saddle point with the inverted potential providing the "pass over the mountain," so it may be that the Eyring vibrational frequency refers to the quantized energy of the usual positive potential at the saddle point and not the imaginary frequency part of the saddle point in the energy surface (Figure 8.2).

Recall that the average energy in the Boltzmann treatment of gases led to the average kinetic energy of $(3/2)RT$ per mole or $(3/2)k_BT$ per molecule, implying $(1/2)k_BT$ energy per degree of freedom in each molecule. Then, consider a vibrating harmonic oscillator as a model for a molecular vibration. The unusual thing about such an oscillator is that when the vibrational coordinate passes over the minimum, it has all kinetic energy and no potential energy but when it is at either limit (turning point) it stops momentarily with no kinetic energy and all potential energy before reversing direction. *Thus, a vibration has both kinetic and potential energy* characteristics, and it will be shown in a later chapter that the energy is proportional to $(2/2)k_BT$ since two forms of energy are present. So, Eyring used the energy of the vibration as $E_{vib} = h\nu = 2k_BT/2 = k_BT$, which is equivalent to saying

$$\nu = k_BT/h,$$

which has inverse time units.

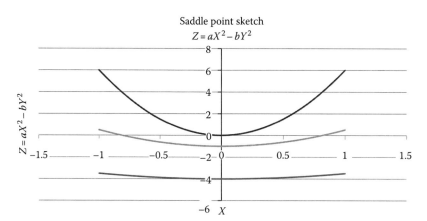

Saddle point sketch
$Z = aX^2 - bY^2$

FIGURE 8.2 A simple sketch of a saddle point potential in the X–Z plane. In the Y–Z plane as viewed "from the side," there is an inverted parabola. As viewed in the plane of the paper here the lower, most shallow, parabola is in the foreground and as you proceed "up" to the $(0, 0, 0)$ point, the positive parabola becomes more narrow. The formula for this surface is $Z = aX^2 - bY^2$. The Eyring reaction coordinate would proceed up over the activation energy hill in the Y–Z plane but still be in a parabolic vibrational potential in the X–Z plane. Assuming the positive parabolic potential leads to vibrational levels (Chapter 12), the frequency of the positive parabola perpendicular to the reaction coordinate is the "gate keeper" condition to allow passage over the saddle point. Thus, a vibrational frequency is involved but the analogy of the words "frequently" and "frequency" is mathematical and semantic. Once we substitute $h\nu = kT$, only the energy is specified and no direct information about the key vibrational mode is needed.

Now, are you ready for the clever step? In your mind mull over the idea that however "frequently" an event occurs that can be described as its "frequency." Thus, the Eyring method uses a frequency factor for the occurrence of the reaction event:

$$f = \nu = \frac{k_B T}{h},$$

which leads to the combined expression for the Eyring rate constant as

$$K_{\text{rate}} = \left[\left(\frac{k_B T}{h} \right) e^{+\left(\frac{\Delta S^{\ddagger}}{R} \right)} \right] e^{-\left(\frac{\Delta H^{\ddagger}}{RT} \right)}.$$

Now we can apply Eyring's equation for the rate constant to the solvolysis reaction we are considering. There are at least two slightly different ways to solve this problem for the values of $\Delta H^{\ddagger}$ and $\Delta S^{\ddagger}$. We expect that $\Delta H^{\ddagger}$ will be close in value to the Arrhenius value of E^* but we might not know what to expect for $\Delta S^{\ddagger}$, although it might be negative since the separate reactant species are combined to form a more ordered transition state. Thus, this system can be solved as a problem with two equations in two unknowns if we use two values of the experimental rate constant. Let us try the simplest method first using the method of substitution.

EXAMPLE

Given $K = 3.19 \times 10^{-4}$/s at 25°C and $K = 2.92 \times 10^{-3}$/s at 45°C, we choose data points reasonably far apart to avoid small irregularities in closely spaced points yet avoid the data at 0°C because we want to use one of the equations to get results for 25°C. We then rearrange the natural logarithm of the equation, so we can cast it into two equations in two unknowns:

$$R \ln \left[\frac{K_{\text{rate}} h}{k_B T} \right] = \Delta S^{\ddagger} - \frac{\Delta H^{\ddagger}}{T},$$

so, we find for the two temperatures:

$$(8.314) \ln \left[\frac{(3.19 \times 10^{-4})(6.6260693 \times 10^{-34})}{(1.3806505 \times 10^{-23})(298.15)} \right] = -311.8406407 = \Delta S^{\ddagger} - \frac{\Delta H^{\ddagger}}{298.15}$$

$$\left[(8.314) \ln \left[\frac{(2.92 \times 10^{-3})(6.6260693 \times 10^{-34})}{(1.3806505 \times 10^{-23})(318.15)} \right] = -293.9720128 = \Delta S^{\ddagger} - \frac{\Delta H^{\ddagger}}{318.15} \right] (-1)$$

$$-17.8686279 = \Delta H^{\ddagger} \left(\frac{1}{318.15} - \frac{1}{298.15} \right)$$

We subtracted the second equation from the first equation to eliminate $\Delta S^{\ddagger}$ temporarily and find $\Delta H^{\ddagger}$.

So, we find that

$$\Delta H^{\ddagger} = \frac{-17.8686279}{\dfrac{(298.15 - 318.15)}{(318.15)(298.15)}} = 84.74770588 \, \text{kJ/mol} = 20.25518783 \, \text{kcal/mol}.$$

Mathematical note: A common student error can occur at this point if the differences of the reciprocals are rounded, subtracted, and then the final reciprocal is rounded again:

$$\frac{1}{\left(\dfrac{1}{308.15}\right) - \left(\dfrac{1}{298.15}\right)} \cong \frac{1}{(3.25 \times 10^{-3}) - (3.35 \times 10^{-3})} \cong \frac{1}{-0.10 \times 10^{-3}} \cong -10,000$$

vs.

$$\frac{1}{\left(\dfrac{1}{308.15}\right) - \left(\dfrac{1}{298.15}\right)} = \frac{(308.15)(298.15)}{-10} = -9187.49225,$$

which is easier and more accurate.

Recall that the E^* value from the Arrhenius fit as

$$E^* = 90.390 \, \text{kJ/mol} = 21.604 \, \text{kcal/mol}.$$

Next, we come to a questionable step with this substitution method. What temperature should we use to calculate $\Delta S^{\ddagger}$? We choose 25°C but only because that is a standard for other values.

Then at 25°C we find $\Delta S^{\ddagger} = -27.59544238 \, \text{J/deg} = -6.59546902 \, \text{cal/deg}$.

We have shown the ten significant figures because the numbers vary over many powers of ten and we want to be careful in finding $\Delta S^{\ddagger}$, which is a small number. This is also one of those tricky calculations where you separate units when you take the natural logarithm, so that you need to make sure your units are correct before you rearrange the logarithmic equation. Now let us check the $\Delta S^{\ddagger}$ value at 45°C:

$$-293.9720128 = \Delta S^{\ddagger} - \frac{84747.70588}{318.15} = \Delta S^{\ddagger} - 266.3765704$$

so then we find that $\Delta S^{\ddagger} = -27.59544238 \, \text{J/deg} = -6.59546902 \, \text{cal/deg}$.

While it is very comforting to find the same value for $\Delta S^{\ddagger}$ at both temperatures as a result of solving two equations in two unknowns, you might expect that if you do not keep all the places on your eight or ten significant figure calculator you might get a different answer at different temperatures. You may have noted with annoyance that we have been using the full number of significant figure values for the physical constants and this is one time that you do have to be precise. Having said that, the main thing is that $\Delta S^{\ddagger}$ negative. This means that in this reaction, the change in entropy probably is caused by some sort of solvent ordering around the immanent cation site. Let us think about that for a moment. The cycloheptane ring is not flat or in a chair form, so the presence of the added methyl group prevents easy attack by the solvent on the chloro substituent from one side. The only plausible way the SN1 departure of the Cl$^-$ can be encouraged to leave the ring is by some fairly ordered arrangement of solvent molecules, so the negative $\Delta S^{\ddagger}$ value comes from local ordering of solvent near the chloride ion. Essentially, all reactions have a negative $\Delta S^{\ddagger}$ value in the Eyring model due to the need for a fairly specific geometry in the activated complex.

GRAPHICAL–ANALYTICAL METHOD FOR $\Delta H^{\ddagger}$ AND $\Delta S^{\ddagger}$

The method above is useful for estimating the Eyring parameters using only two data points in a test situation (hint), but for research we would want to use all the available data. In Figure 8.3, we plot the natural logarithms of the rate constants versus $[1/T(K)]$ and get a very good fit to a straight line just as we did for the Arrhenius plot but here we will analyze the data in a different way (it is the same plot).

Here, we can develop some useful relationships using the slope of the graph and the fact that $\Delta G^{\ddagger} = \Delta H^{\ddagger} - T\Delta S^{\ddagger}$. Let us start with the definition of the Eyring rate constant:

$$K_{\text{rate}} = \left(\frac{k_B T}{h}\right)\left(e^{+\frac{\Delta S^{\ddagger}}{R}}\right)\left(e^{-\frac{\Delta H^{\ddagger}}{RT}}\right) = \left(\frac{k_B T}{h}\right)e^{-\frac{\Delta G^{\ddagger}}{RT}},$$

so then

$$\ln(K_{\text{rate}}) = \ln\left(\frac{k_B T}{h}\right) + \left(\frac{\Delta S^{\ddagger}}{R}\right) - \left(\frac{\Delta H^{\ddagger}}{RT}\right).$$

To compare with the slope of the graph, we need

$$\frac{d\ln(K_{\text{rate}})}{d\left(\frac{1}{T}\right)} = \frac{\left(\frac{k_B}{h}\right)\frac{dT}{d\left(\frac{1}{T}\right)}}{\left(\frac{k_B T}{h}\right)} + 0 - \left(\frac{\Delta H^{\ddagger}}{R}\right).$$

Aside we also need

$$\frac{dT}{d\left(\frac{1}{T}\right)} = \frac{d}{d\left(\frac{1}{T}\right)}(T) = \frac{d}{d\left(\frac{1}{T}\right)}\left[\frac{1}{\left(\frac{1}{T}\right)}\right] = \frac{d}{d\left(\frac{1}{T}\right)}\left(\frac{1}{T}\right)^{-1} = (-1)\left(\frac{1}{T}\right)^{-2} = -T^2.$$

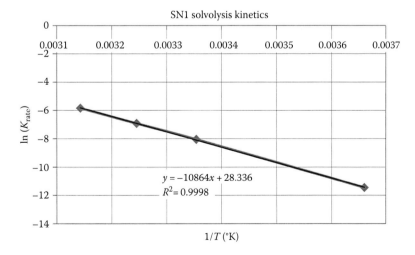

FIGURE 8.3 Plot of $\ln(K)$ versus $(1/T)$ for the solvolysis reaction.

So

$$\frac{d \ln(K_{rate})}{d\left(\frac{1}{T}\right)} = \left(\frac{-T^2}{T}\right) + 0 - \left(\frac{\Delta H^{\ddagger}}{R}\right)$$

and

$$\left\{ \left[\frac{d \ln(K_{rate})}{d\left(\frac{1}{T}\right)} \right] + T \right\} (-R) = \Delta H^{\ddagger}.$$

Next, we see that

$$\Delta G^{\ddagger} = -RT \ln\left(\frac{K_{rate}h}{k_B T}\right) = \Delta H^{\ddagger} - T\Delta S^{\ddagger} \quad \text{and} \quad \Delta S^{\ddagger} = \frac{-(\Delta G^{\ddagger} - \Delta H^{\ddagger})}{T} = \frac{\Delta H^{\ddagger} - \Delta G^{\ddagger}}{T}.$$

Thus,

$$\Delta S^{\ddagger} = \left(\frac{\Delta H^{\ddagger}}{T}\right) - \left(\frac{\Delta G^{\ddagger}}{T}\right) \quad \text{and} \quad \frac{\Delta G^{\ddagger}}{T} = -R \ln\left(\frac{K_{rate}h}{k_B T}\right).$$

So finally we have,

$$\Delta S^{\ddagger}(T) = R\left[\ln\left(\frac{K_{rate}h}{k_B T}\right) - \frac{\left(\frac{d \ln(K_{(rate)})}{d\left(\frac{1}{T}\right)}\right)}{T} - 1 \right] = R\left[\ln\left(\frac{K_{rate}h}{k_B T}\right) - \left(\frac{slope}{T}\right) - 1 \right],$$

$$\Delta H^{\ddagger}(T) = (-R)\left[\left(\frac{d \ln(K_{rate})}{d\left(\frac{1}{T}\right)} \right) + T \right] = (-R)[(slope) + T],$$

and

$$\Delta G^{\ddagger}(T) = -RT \ln\left(\frac{K_{rate}h}{k_B T}\right).$$

Now, we see that all three activation values depend on the temperature. Let us evaluate them and compare with the two-point values obtained above. From the graph, we see the slope of $\frac{d \ln(K_{rate})}{d\left(\frac{1}{T}\right)} = -10,864 = slope$, so we can insert this number into the formulas. The slope can also be obtained from two known rate values if careful arithmetic is used for $\frac{\Delta \ln K}{\Delta(1/T)}$, as here

$$\frac{\ln(9.86 \times 10^{-4}) - \ln(3.19 \times 10^{-4})}{\left(\frac{1}{308.15}\right) - \left(\frac{1}{298.15}\right)} = \ln\left(\frac{9.86 \times 10^{-4}}{3.19 \times 10^{-4}}\right)\left[\frac{(308.15)(298.15)}{-10}\right] = -10,367.77,$$

but it is better to use the slope from a graph fitted to several points as shown above $(-10,864)$ or use two rates far apart in temperature. The best two-point slope would be over the full range of the data that spans from 0°C to 45°C here, and we would obtain a slope from those two points as

$$\frac{\ln(2.92 \times 10^{-3}) - \ln(1.06 \times 10^{-5})}{\left(\dfrac{1}{318.15}\right) - \left(\dfrac{1}{273.15}\right)} = \ln\left(\frac{2.93 \times 10^{-3}}{1.06 \times 10^{-5}}\right)\left[\frac{(318.15)(273.15)}{-45}\right] = -10,856.85.$$

We will use the slope of the graph here $(-10,864)$ since it is available from a least-squares fit:

$$\Delta H^{\ddagger}(25°C) = (-R)[(\text{slope}) + T] = (-8.314\,\text{J/mol}°\text{K})[-10,864 + 298.15],$$

so we find

$$\Delta H^{\ddagger} = +87.844\,\text{kJ/mol} = +20.995\,\text{kcal/mol},$$

$$\Delta S^{\ddagger}(T) = R\left[\ln\left(\frac{K_{\text{rate}}h}{k_B T}\right) - \left(\frac{\text{slope}}{T}\right) - 1\right],$$

$$\Delta S^{\ddagger}(25°C) = (8.314\,\text{J/mol}°\text{K})\left\{\ln\left[\frac{(3.19 \times 10^{-4})(6.6260693 \times 10^{-34})}{(1.3806505 \times 10^{-23})(298.15)}\right] - \left(\frac{-10,864}{298.15}\right) - 1\right\},$$

$$\Delta S^{\ddagger}(25°C) = -17.208822\,\text{J/deg mol} = -4.1130072\,\text{cal/deg mol}.$$

We can calculate $\Delta G^{\ddagger}$ in two ways. First,

$$\Delta G^{\ddagger}(25°C) = -(8.314\,\text{J/mol}°\text{K})(298.15°\text{K})\ln\left[\frac{(3.19 \times 10^{-4})(6.6260693 \times 10^{-34})}{(1.3806505 \times 10^{-23})(298.15)}\right]$$

or

$$\Delta G^{\ddagger}(25°C) = +92.975\,\text{kJ/mol}.$$

And so finally, $\Delta G^{\ddagger} = \Delta H^{\ddagger} - T\Delta S^{\ddagger} = +87,844 - (298.15)(-17.208822) = +92.975\,\text{kJ/mol}.$

While it is comforting that we get the same answer for $\Delta G^{\ddagger}$ in two different ways and that is a check on the calculations for $\Delta H^{\ddagger}$ and $\Delta S^{\ddagger}$, the opportunities for error in these calculations are many. Assuming you might be tested on these equations (hint), it would be best to "practice" doing these calculations several times when you are rested with a clear head.

SUMMARY OF GRAPHICAL METHOD RESULTS AT $T = 25°C$

$$\Delta H^{\ddagger}(25°C) = +87.844\,\text{kJ/mol} = +20.995\,\text{kcal/mol},$$

$$\Delta S^{\ddagger}(25°C) = -17.208822\,\text{J/}°\text{mol} = -4.1130072\,\text{cal/}°\text{mol}, \quad \text{and}$$

$$\Delta G^{\ddagger}(25°C) = +92.975\,\text{kJ/mol} = +22.222\,\text{kcal/mol}.$$

One conclusion from this study of the simple form of Eyring's absolute reaction rate theory is that we can obtain $\Delta H^{\ddagger}$ and $\Delta S^{\ddagger}$ using two equations in two unknowns over a limited temperature range. However, the more general analytical formulas show that the values are definitely dependent on the temperature. Considering the many ways to make errors in these calculations, it might occur to a person using these equations for research that the appropriate constants could be stored in a simple computer program in BASIC or JAVA and all that needs to be given to the program as input data is the desired temperature and the slope of the ln (K) plot.

FURTHER CONSIDERATION OF SN1 SOLVOLYSIS

The problem from Brown and Borkowski [5] we have been considering above is rich in possible details for further study. At this point, we only want to make two comments. First, we want to expand your understanding of the activation barrier. In Freshman texts and even some organic chemistry texts, the activation energy is usually presented in the Arrhenius form with a single energy barrier and an exothermal release of ΔH_{rxn}.

In Figure 8.4, we show at least three possible reactions as a result of the initial (rate-determining step) formation of the SN1 carbocation. The point is that beginning students often talk about "the" reaction product as if there is only one product. Particularly, in organic chemistry and to a lesser extent in general, there are often several product paths away from the activated complex. However, it may well be that the same activation energy barrier applies to initial transition-state complex. It should be noted that in organic chemistry, the yields of desired products are usually less than 100% due to "side reactions." Here, the solvent is 80% ethanol and 20% water, perhaps to aid in solubility of the aliphatic precursor, so we expect the major product will be a result of the polar water molecules reacting with the carbocation but there is a lot of ethanol, so we should expect some formation of the ethyl ether product as well. Then, a third product is possible from elimination of a H from the cation to form an internal double bond. That serves to remind us that while the overall rate-determining step is formation of the cation, there are several optional reaction pathways for the cation in such a mixed solvent. Now that we know about the importance of $\Delta S^{\ddagger}$, we should have a more sophisticated idea of what happens on the molecular scale. In later work, Eyring extended the details of the simple scheme we have shown here to multiple pathways and detailed treatment of the energy levels of the transition state, but we have only shown the key concepts here.

The second point is that there is a secondary time dependence built into a lot of kinetic pathways. This leads to the very useful concept of the "rate-determining step" (RDS) in which the slowest step in a complicated sequence of many steps controls the overall rate of a sequence. Although we have only shown a few detailed examples in Chapter 7, the good news is that we usually only need to examine the kinetics of the slowest time-bottleneck in a complicated sequence and then that step can usually be treated with a first or second order analysis. In the case considered above, nothing happens until the carbocation is formed and then what happens later is fast so the Eyring transition-state analysis is appropriate to the overall rate.

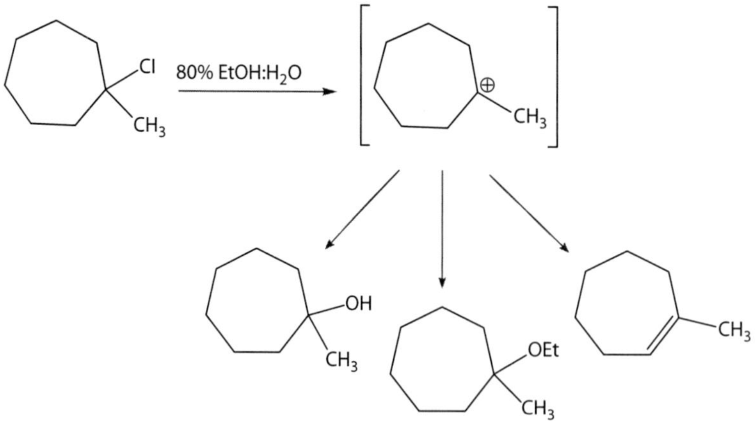

FIGURE 8.4 Side reactions in solvolysis. (Drawings courtesy of Prof. Suzanne Ruder, Virginia Commonwealth University.)

CHAIN REACTIONS AND THE STEADY STATE

So far here and in Chapter 7, we have considered straightforward analysis of reaction rates measured by "the extent of the reaction" based on mole turnover rate, but there are many reactions which follow more complicated rate equations. The classic problem of this more complicated type was first treated by Bodenstein and Lind [6] and later by Christiansen [7], Herzfeld [8], and Polanyi [9]. The first experimental data for the reaction of hydrogen and bromine were fitted accurately to an expression, which is far from what might be expected:

$$H_2 + Br_2 \rightarrow 2HBr; \quad \frac{d[HBr]}{dt} = \frac{k[H_2][Br_2]^{\frac{1}{2}}}{1 + \left(\frac{k'[HBr]}{[Br_2]}\right)}.$$

The value of k' was 0.10, and the overall rate constant was fitted to an Arrhenius plot with a value for E^* of 175 kJ/mol. This mystery was successfully explained later by Herzfeld and by Polanyi as due to a series of intermediate reactions. While this overall reaction is largely due to the ease with which H_2 and Br_2 can be broken apart to free radical atomic species, it is now known that there are many similar reactions which can be treated by the same mathematical approximation called "the steady-state approximation":

1. Specifically, write down the postulated intermediate reactions with their rate constants and assign an identifying label to each reaction.
2. Identify transient, short-lived species, typically free radicals.
3. Write the equations for creation and annihilation of the transient species and set the time derivative to zero to approximate the idea that the concentration of these transient species reaches some steady value, but it is constant so the time derivative is zero.
4. Use the steady-state equations to solve for the expressions of the transient species.
5. Perform "clever algebra" to consolidate and simplify the steady-state equations in terms of the concentrations of the reactants and products. In some cases, one or more of the steady-state equations may lead nowhere, which probably means that you postulated a nonexistent reaction or wrote an incomplete description of the creation and annihilation of that transient.
6. Substitute the steady-state concentrations into the basic rate reaction either for disappearance of reactants or for appearance of product(s).

Before we show examples of this procedure, it may be worth mentioning that this approach is a very good "pencil-and-paper" way to test postulated mechanisms. Even today, with sophisticated spectroscopic equipment and the latest quantum chemistry computer programs, one should try to test a mechanism for a research problem with pencil, paper, and the steady-state method; it is that useful.

Steady-State Example No. 1: $H_2 + Br_2 \rightarrow 2HBr$

Now we are ready to treat the Bodenstein–Lind rate equation with the steady-state method. We write the individual steps with roman numbers for better manipulation:

$$(I) \quad Br_2 \xrightarrow{k_1} 2Br\bullet$$
$$(II) \quad Br\bullet + H_2 \xrightarrow{k_2} HBr + H\bullet$$
$$(III) \quad H\bullet + Br_2 \xrightarrow{k_3} HBr + Br\bullet$$
$$(IV) \quad H\bullet + HBr \xrightarrow{k_4} H_2 + Br\bullet$$
$$(V) \quad Br\bullet + Br\bullet \xrightarrow{k_5} Br_2$$

(Add the equations.)

$$0 \cong \frac{d[\text{Br}\bullet]}{dt} = k_1[\text{Br}_2] - k_2[\text{Br}\bullet][\text{H}_2] + k_3[\text{H}\bullet][\text{Br}_2] + k_4[\text{H}\bullet][\text{HBr}] - k_5[\text{Br}\bullet]^2$$

$$0 \cong \frac{d[\text{H}\bullet]}{dt} = k_2[\text{Br}\bullet][\text{H}_2] - k_3[\text{H}\bullet][\text{Br}_2] - k_4[\text{H}\bullet][\text{HBr}]$$

$$0 \cong k_1[\text{Br}_2] - k_5[\text{Br}\bullet]^2$$

This example is very good for teaching because it illustrates the method but has an easy solution because of a fortuitous substitution. So, substitute the steady-state equation for [H•] into the equation for the [Br•] and note the cancelation due to $k_2[\text{Br}\bullet][\text{H}_2] = k_3[\text{H}\bullet][\text{Br}_2] + k_4[\text{H}\bullet][\text{HBr}]$. Thus we find $0 \cong k_1[\text{Br}_2] - k_5[\text{Br}\bullet]^2$, which can be solved for

$$[\text{Br}\bullet]_{ss} \cong \sqrt{\left(\frac{k_1}{k_5}\right)[\text{Br}_2]}.$$

Now substitute the expression for $[\text{Br}\bullet]_{ss}$ into the [H•] equation to find

$$k_2[\text{H}_2]\left(\frac{k_1}{k_5}\right)^{\frac{1}{2}}[\text{Br}_2]^{\frac{1}{2}} = (k_3[\text{Br}_2] + k_4[\text{HBr}])[\text{H}\bullet]_{ss},$$

and we can solve for $[\text{H}\bullet]_{ss}$ as follows:

$$[\text{H}\bullet]_{ss} = \frac{\left(\frac{k_1}{k_5}\right)^{\frac{1}{2}} k_2[\text{H}_2][\text{Br}_2]^{\frac{1}{2}}}{(k_3[\text{Br}_2] + k_4[\text{HBr}])}.$$

Now we are ready to write the overall rate equation as the appearance of HBr, and we substitute the steady-state concentrations we have found.

$$\frac{d[\text{HBr}]}{dt} = k_2[\text{Br}\bullet]_{ss}[\text{H}_2] + k_3[\text{H}\bullet]_{ss}[\text{Br}_2] - k_4[\text{H}\bullet]_{ss}[\text{HBr}].$$

(The next step is amazing!)

$$\frac{d[\text{HBr}]}{dt} = \left(k_2[\text{H}_2]\left(\frac{k_1}{k_5}\right)^{\frac{1}{2}}[\text{Br}_2]^{\frac{1}{2}}\right) + \left[\frac{\left(\frac{k_1}{k_5}\right)^{\frac{1}{2}} k_2[\text{H}_2][\text{Br}_2]^{\frac{1}{2}}(k_3[\text{Br}_2] - k_4[\text{HBr}])}{(k_3[\text{Br}_2] + k_4[\text{HBr}])}\right].$$

Put this expression over a common denominator and the numerator terms in k_4 cancel:

$$\frac{d[\text{HBr}]}{dt} = \frac{2k_2k_3\left(\frac{k_1}{k_5}\right)^{\frac{1}{2}}[\text{Br}_2]^{\frac{3}{2}}[\text{H}_2]}{(k_3[\text{Br}_2] + k_4[\text{HBr}])}.$$

Now divide the numerator and denominator by $(k_3[Br_2])$, and we obtain Bodenstein–Lind formula:

$$\frac{d[HBr]}{dt} = \frac{2k_2 \left(\frac{k_1}{k_5}\right)^{\frac{1}{2}} [H_2][Br_2]^{\frac{1}{2}}}{\left[1 + \left(\frac{k_4}{k_3}\right)\left(\frac{[HBr]}{[Br_2]}\right)\right]} .$$

While this is an amazing explanation of the Bodenstein–Lind rate equation, it is not an isolated incident. There are many other complex reactions, which can be solved with this method.

STEADY-STATE EXAMPLE NO. 2: THERMAL CRACKING OF ACETALDEHYDE [10]

This next example is not quite perfect because it gives a solution with a leftover radical unaccounted for. However, it is shown here as an example of what to expect in research. Suppose we want to understand the thermal decomposition of acetaldehyde. Rice and Herzfeld [10] studied the thermal "cracking" of hydrocarbons as part of a very important study related to petroleum processing. Here, we present the thermal cracking of acetaldehyde. Consider the following scheme for the thermal decomposition reaction [11]:

$$CH_3CHO \xrightarrow{\Delta} CH_4 + CO + CH_3CH_3.$$

(I) $CH_3CHO \xrightarrow{k_1} CH_3\bullet + \bullet CHO \quad E^* = 76.0\,\text{kcal/mol}$

(II) $CH_3\bullet + CH_3CHO \xrightarrow{k_2} CH_4 + \bullet CH_2CHO \quad E^* = 10.0\,\text{kcal/mol}$

(III) $\bullet CH_2CHO \xrightarrow{k_3} CO + \bullet CH_3 \quad E^* = 18.0\,\text{kcal/mol}$

(IV) $\bullet CH_3 + \bullet CH_3 \xrightarrow{k_4} CH_3CH_3 \quad E^* = 0\,\text{kcal/mol}$

$$0 \cong \frac{d[CH_3\bullet]}{dt} = k_1[CH_3CHO] - k_2[\bullet CH_3][CH_3CHO] + k_3[\bullet CH_2CHO] - k_4[\bullet CH_3]^2$$

$$0 \cong \frac{d[\bullet CH_2CHO]}{dt} = k_2[\bullet CH_3][CH_3CHO] - k_3[\bullet CH_2CHO]$$

$$0 \cong \frac{d[\bullet CHO]}{dt} = k_1[CH_3CHO]$$

The third steady-state species ($\bullet$CHO) is apparently constant and only depends on the amount of acetaldehyde, so that is a dead end as far as the steady-state substitution process goes. There may be some other steps leading to H_2CO but here $\bullet$CHO does not contribute to the production of CH_4. However, from the second equation, we find that

$$[\bullet CH_2CHO]_{ss} = \left(\frac{k_2}{k_3}\right)[\bullet CH_3][CH_3CHO].$$

Then from the first steady-state equation, we have by substitution of the $[\bullet CH_2CHO]_{ss}$ expression:

$$k_1[CH_3CHO] - k_2[\bullet CH_3][CH_3CHO] + \left(\frac{k_3 k_2}{k_3}\right)[\bullet CH_3][CH_3CHO] - k_4[\bullet CH_3]^2 \cong 0,$$

and the middle terms cancel leaving

$$[\bullet CH_3]_{ss} = \left(\frac{k_1}{k_4}\right)^{\frac{1}{2}} [CH_3CHO]^{\frac{1}{2}}$$

for the steady-state concentration of the methyl radical. As usual we have to choose something which we can measure to follow the "extent of the reaction" and so, we choose the appearance of methane, CH_4. Then the rate is

$$\frac{d[CH_4]}{dt} = k_2[\bullet CH_3] [CH_3CHO] = \left(\frac{k_1}{k_4}\right)^{\frac{1}{2}} k_2[CH_3CHO]^{\frac{3}{2}},$$

with perhaps unexpected dependence on the acetaldehyde concentration to the (3/2) power. The original paper by Rice and Herzfeld gives approximate E^* values for the various reactions in the overall scheme (in kcal of course, since the paper was written in 1934). We can estimate the overall activation energy from the final rate expression by combining the Arrhenius forms of the several rate constants:

$$k_2\left(\frac{k_1}{k_4}\right)^{\frac{1}{2}} = A_2\left(\frac{A_1}{A_4}\right)^{\frac{1}{2}} e^{-\frac{\left[10.0+\frac{(76.0+0)}{2}\right]kcal}{RT}} = A_2\left(\frac{A_1}{A_4}\right)^{\frac{1}{2}} e^{-\left(\frac{48,000\,cal}{RT}\right)} = A'e^{-\left(\frac{48,000\,cal}{RT}\right)}.$$

Thus, even though we are not given the values of the A constants, we can estimate the E^* energy. There are many other such free radical mechanisms in thermal cracking of hydrocarbons and these have been in use in designing petroleum refining operations for many years.

STEADY-STATE EXAMPLE NO. 3: THE LINDEMANN MECHANISM

An important use of the steady-state concept is the application to nominal unimolecular reactions that have high activation energies. This was first formulated by Frederick A. Lindemann (1886–1957), an English physicist. Some examples are isomerizations, such as $CH_3NC \rightarrow CH_3CN$ or decompositions like $CH_3CH_2Cl \rightarrow CH_2 = CH_2 + HCl$. For simplicity, let us consider a unimolecular isomerization, such as what seems to be $A \rightarrow B$ in the gas phase. Having specified that this is a gas-phase reaction, we need to remember what we learned about the kinetic theory of gases and also Dalton's law of partial pressures. From Dalton's law, consider that there might be another gas present, M, but at the end of this discussion we will be free to let $M = A$ as well. Then, we have activation by collisions. Remember that $\bar{v} = \sqrt{\frac{8RT}{\pi M}}$, but that is only the average velocity. The Boltzmann distribution shows there are some molecules with much higher velocities. Collisions with other higher energy molecules can transfer energy to the A molecules and raise their energy above the activation energy to react to form A*.

$A + M \underset{k_{-1}}{\overset{k_1}{\rightleftharpoons}} A^* + M$ and then $A^* \xrightarrow{k_2} B$ but we can consider A* as a transient steady-state intermediate, so

$$\frac{d[A^*]}{dt} = k_1[A] [M] - k_{-1}[A^*] [M] - k_2[A^*] \cong 0.$$

We can solve this for $[A*]$ as $[A*] = \dfrac{k_1[M][A]}{k_2 + k_{-1}[M]}$ and then $\dfrac{d[B]}{dt} = k_2[A*] = \dfrac{k_2 k_1[M][A]}{k_2 + k_{-1}[M]} = \text{rate}$. Now suppose that $[M] \to \infty$ meaning that there is some other gas present that is at a much greater concentration than $[A]$. In that case, we see that $\dfrac{d[B]}{dt} \to \left(\dfrac{k_2 k_1}{k_{-1}}\right)[A] = k'[A]$, which is what we thought initially. This can happen whether M is some other gas in great abundance or whether M = A at a high concentration (pressure) of A itself. Now consider that the whole system is a low pressure. Then we have $\dfrac{d[B]}{dt} = \left\{\dfrac{k_1 k_2[M]}{k_2 + k_{-1}[M]}\right\}[A] = k_{obs}[A]$ as the general solution, but what if there is no $[M]$ and the only other gas for collisions is when $[M] = [A]$? Thus, at relatively low pressure and only A present we have:

$$\frac{d[B]}{dt} = \frac{k_1 k_2[A]^2}{k_2 + k_{-1}[A]} \simeq k_1[A]^2,$$

assuming $k_{-1} \ll k_2$. Thus, the reaction will appear to be first order at high pressure and second order at low pressure. We also can see that adding an inert gas such as He or Ar (not totally inert) could be used to change the order of such a reaction depending on the pressure of the inert gas. The mathematics of the steady-state treatment is easy here but the thought process of this type of problem involves careful reading of the gas pressures and numerical values of the individual rate constants.

ENZYME KINETICS

Although enzymology is a specialty field in biochemical/pharmaceutical chemistry, it is so important to all health sciences that we need to include a basic part of it in our list of "essential" topics in physical chemistry. Enzymes are very special in many ways. First, they are usually large linear polypeptides with specific sequences of amino acid "residues" which have the amazing ability to *coil up in solution from a single strand into a globular shape* with a special "active site" and remain water soluble. Think back to experiments in an organic chemistry laboratory course, how many organic reactions were carried out in water? Then again consider that the human body is approximately 70% water and that planet Earth is about 70% covered by water. Water is a polar solvent and most organic compounds are not very soluble in water. *We should ponder how nature enables complicated organic reactions in an aqueous medium. The answer is that enzymes are little floating organic laboratories which can carry out highly specific (often stereospecific) organic reactions while in an aqueous medium.* This situation is accomplished by wrapping a special catalytic arrangement of the side chains inside the polypeptide with a hydrophobic shell, a concept formulated by Prof. Walter Kauzmann in Figure 8.5. A crude analogy would be an oil drop in water where the oil drop includes a special active site in the interior. Considering that enzymes have to carry around their laboratory building in water, they turn out to be amazingly efficient and often the turnover rate for an enzymatic reaction is many thousand times faster than the reaction would be if the reactants were refluxed in an organic chemistry pot. Finally, enzymes can often do organic reactions in a single step which would require many, many steps using organic synthesis techniques. One of the reasons for successful specificity and efficiency of enzymes is that often the pocket of the active site gently assists the "substrate" molecule into the conformation of the activated complex for that reaction. Enzymatic oxidation–reduction reactions often have metal atoms in the active site, although not all enzymes have metal atoms. The fact that Copper can be either +1 or +2 and the small difference in the energy of Cu as $(Ar)3d^9 4s^2$ or Cu as $(Ar)4s^1 3d^{10}$ makes it ideal for oxidation–reduction reactions. The Cu–O–Cu active site in catechol oxidase is special in an electrochemical sense as well as having some geometric specificity (Figure 8.6).

Previous examples have introduced the very useful concept of the "steady state" and that applies here as well. In addition, the rapid forward and backward reactions in an equilibrium still apply for

FIGURE 8.5 Prof. Walter J. Kauzmann (1916–2009) was an American physical chemist whose research spanned thermodynamics (Kauzmann's paradox of supercooled liquids), quantum chemistry (1957 text), and biochemistry (the hydrophobic effect in enzymes). He was the Chair of Chemistry at Princeton University from 1964 to 1968 and the Chair of the Department of Biochemistry from 1980 to 1981. He is probably best known for his work on the thermodynamics and optical activity of proteins. (From Princeton University Department of Chemistry. With permission.)

enzyme reactions in the sense that a substrate molecule may enter an active site but wander out before a reaction occurs. Since enzymes are very efficient, the in-out part of the equilibrium may favor fewer unreacted exits, but the treatment includes that possibility.

Enzymes can be very difficult to isolate and purify but they are usually so efficient that one can purchase milligram quantities of purified enzyme from biochemical companies for several hundred dollars per milligram but then only use microgram quantities for a given experiment. It is important to know that only a small amount of enzyme is necessary to carry out reactions. This leads to a quite different type of laboratory requiring refrigerator storage and even a large walk-in room, which is refrigerated to isolate enzymes and carry out reaction studies.

We have left this discussion until after description of the Eyring transition-state theory because there are many similarities. The main difference is that usually the active site pocket of an enzyme has two main geometrical attributes. First, there is the "lock and key" analogy, which notes that usually the entrance to the active site is stereospecific to a particular substrate molecule or a class of molecules. We see that in Figure 8.6 with catechol oxidase for substituted phenols. Much has been made of the lock-and-key concept in pharmaceutical research since that is how substrate specificity is achieved and many medicinal drug molecules are designed with a specific shape and

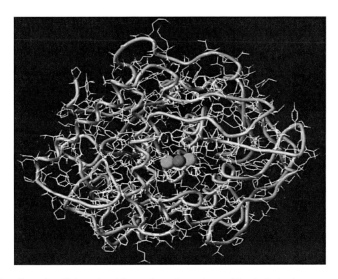

FIGURE 8.6 A drawing of *o*-diphenol oxidase where the polypeptide chain is shown as a heavy strand. The active site of the enzyme is shielded but accessible in the center with a Cu–O–Cu moiety shown as large spheres. (From the Brookhaven Protein Data Bank as 2P3X.pdb. Thanks to Prof. Glen E. Kellogg of the Virginia Commonwealth University, Medical College of Virginia.)

with charged ion sites in specific regions. However, getting the key into the lock is only the first step, it is necessary to turn the key. Thus, a second geometrical feature of the active site is that it often encourages the conformation of the substrate molecule to assume the geometry of the transition state. No wonder enzyme reactions can be both stereospecific and efficient. We see that with catechol oxidase, although the specificity is lower than with some other enzymes but it is tuned to oxidize an *o*-diphenol. This particular enzyme was isolated from grapes but catechol oxidase is common in many fruits and vegetables. It is apparently part of a protective mechanism in that whenever there is a physical injury to an apple, banana, or potato, catechol is released and the enzyme converts it to benzoquinone, which is an antiseptic to bacteria and fungal infections. Later the benzoquinone reacts further with oxygen to produce black spots. On the one hand, the plant is protecting itself but on the other hand most people consider blackened fruit as spoiled, so two strategies are used to keep the fruit from blackening: cooling to slow the reaction and packing in a nitrogen atmosphere to keep oxygen away from the produce.

BASIC MICHAELIS–MENTEN EQUATION

The treatment of enzyme kinetics was given by Leonor Michaelis (1875–1940), a German biochemist and physician, and Maud Menten (1879–1960), a Canadian biochemist and physician, and is called the Michaelis–Menten [12] equation today.

The model for the reaction is $E + S \underset{k_{-1}}{\overset{k_1}{\rightleftharpoons}} (E \cdot S) \xrightarrow{k_2} P + E$ (E = enzyme, S = substrate, and P = product), but a special consideration is given to the concentration of the free enzyme $[E_{free}]$ compared to the total amount of the enzyme in the solution $[E_{tot}]$, so that we have $[E_{free}] = [E_{tot}] - (E \cdot S)$. We proceed to the rate step, which is susceptible to the steady-state analysis:

$$\frac{d[E \cdot S]}{dt} = k_1[E_{free}][S] - k_{-1}[E \cdot S] - k_2[E \cdot S] \cong 0,$$

balancing creation/loss of the $(E \cdot S)$ species.

So $k_1([E_{tot}] - [E \cdot S])[S] = [E \cdot S](k_{-1} + k_2)$ and so $k_1[E_{tot}][S] = [E \cdot S]\{(k_{-1} + k_2) + k_1[S]\}$. Now we can solve for the steady-state concentration:

$$[E \cdot S]_{ss} = \frac{k_1[E_{tot}][S]}{(k_{-1} + k_2) + k_1[S]}.$$

$$\text{Rate} = \frac{d[P]}{dt} = V = k_2[E \cdot S]_{ss} = \frac{k_2 k_1[E_{tot}][S]}{(k_{-1} + k_2) + k_1[S]}.$$

This expression can be simplified by dividing the numerator and denominator by k_1 to obtain a new combined constant:

$$\left(\frac{k_{-1} + k_2}{k_1}\right) \equiv K_M,$$

$$V = \frac{k_2 k_1[E_{tot}][S]}{(k_{-1} + k_2) + k_1[S]} = \frac{k_2[E_{tot}][S]}{\left(\dfrac{k_{-1} + k_2}{k_1}\right) + [S]} = \frac{k_2[E_{tot}][S]}{K_M + [S]} = \frac{V_{max}[S]}{K_M + [S]} = V.$$

Next, we have the clever step of inverting the equation $\dfrac{1}{V} = \left(\dfrac{K_M}{V_{max}}\right)\left(\dfrac{1}{[S]}\right) + \left(\dfrac{1}{V_{max}}\right)$. Inverting the whole equation puts it in the form of $y = mx + b$ to fit a straight line. Without inverting the equation, you can see that when [S] is low the rate increases rapidly with substrate but as [S] increases in the denominator as well as the numerator, the rate will reach a plateau at some value we can call V_{max}. This plateau can be reached by adding a very large amount of substrate:

$$V_{max} = \lim_{[S] \to \infty} V = k_2[E_{tot}].$$

We can use this to provide meaning for K_M. Consider the rate when V is $(V_{max}/2)$. $\dfrac{1}{(V_{max}/2)} = \dfrac{2}{V_{max}} = \left(\dfrac{K_M}{V_{max}}\right)\left(\dfrac{1}{[S_{1/2}]}\right) + \left(\dfrac{1}{V_{max}}\right)$ or $2 = \left(\dfrac{K_M}{[S_{1/2}]}\right) + 1$ by canceling V_{max}.

Thus, we can finally say that $K_M = [S_{1/2}]$, that is the substrate concentration when $V = V_{max}/2$. In words, "*K_M is the substrate concentration when the rate is half the maximum rate.*"

EXAMPLE: A HYPOTHETICAL ENZYME

Let us consider synthetic data for "Enzyme-X" which is similar to the actual data for pancreatic carboxypeptidase [13,14] (note the name of an enzyme ends in "-ase"). We use synthetic data, so we can insert key points into an Excel plot. It should be clear that at low substrate concentration, the rate increases rapidly as more substrate is added. However, in spite of the efficiency of an enzyme, there is only a small amount in solution. Thus, the rate approaches a limit as more and more substrate saturates the active sites of the enzymes in solution and eventually reaches a limit, V_{max} as shown in Figure 8.7. Special points have been inserted into the data so that you can see the limiting rate of 0.090 mM/s and see that at half that rate, 0.045 mM/s, the substrate concentration is 0.0065 mM which is the value of K_M. The value of K_M is not easily seen on the first plot but when we show the double reciprocal plot in Figure 8.8 we get a more precise value of K_M value in two ways. There are really two intercepts in the double-reciprocal plot, which apparently was the innovation of Lineweaver and Burk [15] in 1934. We can use the third and fourth columns of Table 8.1 to make such a plot for our "Enzyme-X" data in Figure 8.8.

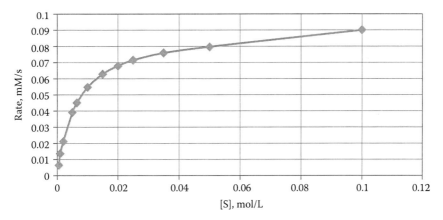

FIGURE 8.7 Typical data for enzyme kinetics, similar to data for pancreatic carboxypeptidase (see Ref. [13]) using synthetic values to show $V_{max} = 0.090$ mM/s and $K_M = 6.50$ mM at $[S] = 0.045$ mM.

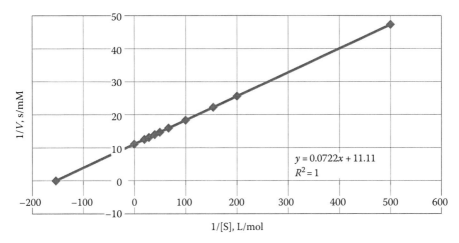

FIGURE 8.8 Double reciprocal plot for Enzyme-X data similar to pancreatic carboxypeptidase (see Ref. [13]). The $x = 0$ and $y = 0$ points have been inserted into the data to show the intercepts $(0, 1/V_{max})$ and $(-1/K_M, 0)$. The slope $= (K_M/V_{max}) = 0.0722$ mol/s and the y-intercept $= 1/11.11$ s/mM $= (1/V_{max})$, so $V_{max} = 0.090$ mM/s and $K_M = 6.5$ mM.

We have added two calculated points after the best line was fitted to the data, so that the x- and y-intercepts show up as points on the graph using the Michaelis–Menten equation:

$$\frac{1}{V} = \left(\frac{K_M}{V_{max}}\right)\left(\frac{1}{[S]}\right) + \left(\frac{1}{V_{max}}\right),$$

so, from the graph we see that

$$\left(\frac{K_M}{V_{max}}\right) = 0.0722$$

and

$$\left(\frac{1}{V_{max}}\right) = 11.11,$$

TABLE 8.1
Typical Data for "Enzyme-X"

S, mM/L	Rate, V, mM/s	1/[S], L/mol	(1/V), s/mM
0.5	0.0064	2000.0	155.56
1.0	0.0136	1000.0	73.33
2.0	0.0212	500.0	47.22
5.0	0.0391	200.0	25.56
6.5[a]	0.0450[a]	153.85[a]	22.22[a]
10.0	0.0546	100.0	18.33
15.0	0.0628	66.7	15.93
20.0	0.0679	50.0	14.72
25.0	0.0714	40.0	14.00
35.0	0.0759	28.57	13.17
50.0	0.0797	20.0	12.55
100.0[a]	0.0900[a]	10.0[a]	11.11[a]

[a] Points added by calculation to show the V_{max} limit and $K_M = 0.0065$ mM.

so $K_M = (0.0722)/(11.11) = 0.00649865 \cong 0.0065$ mM and $V_{max} = 1/11.11 = 0.090009 \cong$ 0.090 mM/s. So, with the Lineweaver–Burk plot, we see that when $y = 0$, we have $x = -153.85 = -1/K_M$, which checks the value of K_M as 0.0065 mM. Then, when $x = 0$, $y = 1/V_{max}$ and $V_{max} = 0.090$ mM/s again.

Thus, we see that enzyme kinetics uses the steady-state approximation and the double-reciprocal plot to provide a robust approach to study the reaction and K_M is a useful concept that tells us what concentration of the substrate will give one half of the maximum rate.

MICHAELIS–MENTEN WITH COMPETITIVE INHIBITOR

There are actually several other cases of enzyme reactions, but keeping to our list of "essential" physical chemistry we will only treat the important case of a competitive inhibitor of the normal substrate since this is at the heart of much pharmaceutical research. At the simplest level, one can use the "lock-and-key" concept to imagine that there are other molecules slightly different from the natural substrate molecule. Suppose the natural substrate has a methyl group exposed in a certain place. There could be a similar molecule that is the same but without the methyl group and it can probably fit into the same active site cavity but might not do the same reaction. Another molecule might be the same as the natural substrate but have an ammonium ion instead of the natural methyl group. The ammonium group will probably fit in the same space as the methyl group but the charge on the ammonium ion may severely change the chemistry in the active site. Other possibilities exist but the point is that there are other molecules, which can compete with the natural substrate but which do not do the same chemistry.

The concept of the active site in a floating, mobile, water-soluble enzyme can be extended to biological "receptors" that are fixed in cell membranes and a similar analysis can be applied to competitors to natural substrates. This means that it is important to have an analysis procedure for competitive inhibition. Consider the Michaelis–Menten equations with an inhibitor "I":

$$E + S \underset{k_{-1}}{\overset{k_1}{\rightleftharpoons}} (E \cdot S) \xrightarrow{k_2} E + P$$

and

$$E + I \underset{}{\overset{K_1}{\rightleftharpoons}} (E \cdot I); \quad K_I = \frac{[E \cdot I]}{[E_{free}][I]}$$

$$[E_{free}] = [E_{tot}] - [E \cdot S] - [E \cdot I] = [E_{tot}] - [E \cdot S] - K_I[E_{free}][I],$$

so we rearrange to find $[E_{free}]$:

$$[E_{free}] = \frac{([E_{tot}] - [E \cdot S])}{(1 + K_I[I])}.$$

Now we proceed with the familiar Michaelis–Menten derivation for the usual steady-state approximation:

$$\frac{d[E \cdot S]}{dt} = k_1[E_{free}][S] - k_{-1}[E \cdot S] - k_2[E \cdot S] \cong 0.$$

Insert the formula for $[E_{free}]$ into the steady state, so

$$\frac{k_1[S]([E_{tot}] - [E \cdot S])}{(1 + K_I[I])} = (k_{-1} + k_2)[E \cdot S].$$

But we need to isolate $[E \cdot S]$ and you should follow this step with pencil and paper:

$$\frac{k_1[S][E_{tot}]}{(1 + K_I[I])} = \frac{[E \cdot S]k_1[S]}{(1 + K_I[I])} + (k_{-1} + k_2)[E \cdot S] = [E \cdot S]\left\{ \frac{k_1[S] + (1 + K_I[I])(k_{-1} + k_2)}{(1 + K_I[I])} \right\}.$$

Cancel the $(1 + K_I[I])$ denominator on both sides of the equation and we obtain

$$[E \cdot S]_{ss} = \frac{k_1[S][E_{tot}]}{(1 + K_I[I])(k_{-1} + k_2) + k_1[S]}.$$

As before $V = k_2[E \cdot S]_{ss}$ and $V_{max} = k_2[E_{tot}]$, and these values would be measured using the Michaelis–Menten measurements without I. Thus, we have

$$V = \frac{k_2 k_1[S][E_{tot}]}{(1 + K_I[I])(k_{-1} + k_2) + k_1[S]},$$

now use $k_1 = \frac{1}{(1/k_1)}$ and we see $K_M = \frac{(k_{-1} + k_2)}{k_1}$ appear.

$$V = \frac{k_2[S][E_{tot}]}{(1 + K_I[I])K_M + [S]} = \frac{V_{max}[S]}{(1 + K_I[I])K_M + [S]},$$

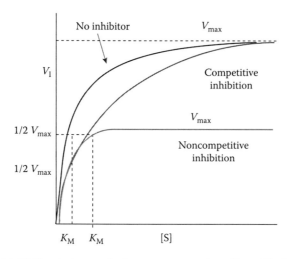

FIGURE 8.9 Types of inhibition in terms of substrate concentration. (From Kimball, J., Enzyme kinetics, http://users.rcn.com/jkimball.ma.ultranet/BiologyPages/E/EnzymeKinetics.html. With permission.)

so,

$$\left(\frac{1}{V}\right) = \left\{ \frac{(1 + K_I[I])K_M}{V_{max}} \right\} \left(\frac{1}{[S]}\right) + \left(\frac{1}{V_{max}}\right).$$

This is just like the Michaelis–Menten equation with a modified slope but the same y-intercept. Note that when $\left(\frac{1}{[S]}\right) = 0$, we have the same y-intercept as before, $\left(\frac{1}{V}\right) = \left(\frac{1}{V_{max}}\right)$. However, when $\left(\frac{1}{V}\right) = 0$, the x-intercept changes if $K_I[I] > 0$. Therefore, one can add a suspected inhibitor molecule to the enzyme solution and carry out the rate measurement and do the Michaelis–Menten analysis. If the slope of the Lineweaver–Burk line changes, that is proof of competitive inhibition by species "I" since $\left\{ \frac{(1 + K_I[I])K_M}{V_{max}} \right\} =$ slope.

An excellent example of the effect of inhibition is given on the Internet [16] at http://users.rcn. com/jkimball.ma.ultranet/BiologyPages/E/EnzymeKinetics.html with data for *o*-diphenol oxidase shown in Figures 8.9–8.11. The enzyme is prepared from the supernatant liquid of homogenized apples. This enzyme is responsible for the fact that apples turn brown when the skin is removed. A Lineweaver–Burk plot is easily obtained using a spectrometer to measure the reaction as an increase in optical density at 540 nm with time. A similar set of data can be obtained with the presence of para-hydroxybenzoic acid which is shown to be an inhibitor and a third set of data is obtained showing the effect of phenylthiourea as a noncompetitive inhibitor. There are many other useful descriptions of enzyme kinetics on the Internet. Our function here is to provide the algebraic derivation which is not often given.

MICHAELIS–MENTEN SUMMARY

Enzymes are very efficient with high reaction rates when the substrate concentration is low but they can become saturated when the substrate concentration is high [16]. Competitive inhibition results

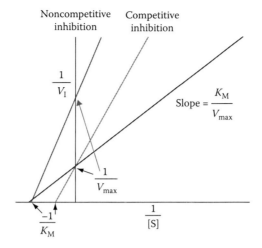

FIGURE 8.10 Lineweaver–Burk plots of inhibition.

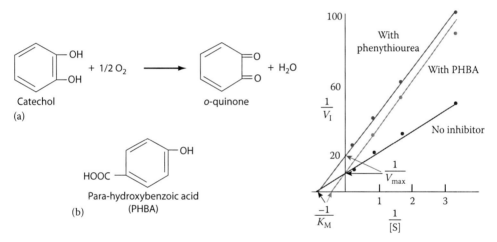

FIGURE 8.11 A study of o-diphenol oxidase with catechol substrate including competitive (PHBA) and noncompetitive (phenylthiourea) inhibitors. (From Kimball, J., Enzyme kinetics, http://users.rcn.com/jkimball.ma.ultranet/BiologyPages/E/EnzymeKinetics.html. With permission.)

in blockage of the limited enzyme sites, so it takes a higher substrate concentration to reach V_{max} and so K_M is higher in the presence of a competitive inhibitor. A non-competitive inhibitor may bind to the enzyme at some other place than directly in the active site but still cause a loss of efficiency by the enzyme. The analysis of a noncompetitive inhibitor requires more knowledge of the actual binding mechanism and is less amenable to easy algebraic treatment. A noncompetitive inhibitor may actually change the enzyme in some small way that reduces the efficiency of the enzyme. A third type of inhibitor is a "suicide substrate" that enters the active site and reacts chemically with the interior of the active site. This has use in some medical research where the effect of blocking an enzyme is being studied. In the case of a suicide enzyme, the animal host may die or sometimes reach a dormant state while new enzyme is biosynthesized and the animal recovers when new enzyme is available.

KINETICS CONCLUSIONS

While we are still self-constrained to limit our treatment to what we believe is essential to physical chemistry, we have added further examples to the Chapter 7 treatment of reaction kinetics, which include some aspects of multistep mechanisms and introduced the steady-state approximation. The steady-state concept was then extended to the Eyring transition-state concept and used again for the critical step in the Michaelis–Menten treatment of enzyme kinetics. This has been a fast tour of some complicated algebra but in our experience students who "learn" the derivations have a deeper appreciation for the concepts. Casual interviews of students from past classes have revealed that the Michaelis–Menten derivations have been the most useful aspect of this chapter.

PROBLEMS

8.1 Use the data in Example 5 of Chapter 7 at 0°C and 45°C to compute the Arrhenius "A" and E^* values.

8.2 Use your answers from problem 8.1 above to calculate the temperature at which the reaction rate would be twice the rate at 0°C.

8.3 Use the data in Example 5 of Chapter 7 with the "two-point" method on page 159 to calculate $\Delta H^{\dagger}$ and $\Delta S^{\dagger}$ at 35°C using data from 35°C to 45°C. This is likely the method of choice in a test situation, so you need to practice this "two-point method." This method is less reliable when the points are close together.

8.4 Use the data in Problem 8.1 with the "graphical method" on pages 162–163 to calculate $\Delta H^{\dagger}$ and $\Delta S^{\dagger}$ at 35°C instead of 25°C.

8.5 Rederive the Bodenstein–Lind rate law for the reaction of $H_2 + Br_2$ and review what you were taught in organic chemistry about the reactivity of free radicals. Look up the dissociation energy of $Br_2 \rightarrow 2Br$ in the chemical rubber handbook and compare it with the energy of dissociation of $H_2 \rightarrow 2H$.

8.6 Given the table of data below for the reaction of o-diphenol oxidase with catechol, draw the Lineweaver–Burk plot and find the slope of the graph and use the value of the y-intercept $(1/V_{max})$ to obtain the K_M value. Compare that value to the reciprocal of the x-intercept where $(1/V) = 0$. Submit your graph along with your calculations. We will need three significant figures to compare with the data in the next problem to see the effect of an inhibitor.

[S]	4.8 mM	1.2 mM	0.6 mM	0.3 mM
1/[S]	0.21	0.83	1.67	3.33
ΔOD_{540} (V_i)	0.081	0.048	0.035	0.020
$1/V_i$	12.3	20.8	28.6	50.0

The values for $(1/V_i)$ have been rounded to three significant figures for a smoother fit to the V_i values, which are only given to two significant figures. However, Prof. Kimball rounds $(1/V_1)$ further to only two significant figures. In either event, biological data are often more scattered than physical data but some of the $(1/S)$ data are rounded to three significant figures; so we carry three significant figures in $(1/V_1)$. When using biological data, it is essential to use least-squares fits.

8.7 Given the table of data below for the reaction of o-diphenol oxidase with catechol in the presence of parahydroxy benzoic acid (a competitive inhibitor), draw the Lineweaver–Burk plot and find the slope of the graph and use the value of the y-intercept $(1/V_{max})$ to obtain the modified K_M value. Calculate the value of K_M from the slope and compare it with the reciprocal of the x-intercept where $(1/V) = 0$. Compare the value of the new K_M value with the K_M from the problem. Submit your graph along with your calculations.

[S]	4.8 mM	1.2 mM	0.6 mM	0.3 mM
1/[S]	0.21	0.83	1.67	3.33
ΔOD_{540} (V_i)	0.060	0.032	0.019	0.011
$1/V_i$	16.7	31.3	52.6	90.9

TESTING, GRADING, AND LEARNING?

Next, we present a final examination from the 2008 Summer course in CHEM 303. The time limit was 3 h including 110 min. for an ACS standardized test and the students knew that the grade on this test would be their final grade if the score was higher than their average including the score on the final.

Physical chemistry 303   Final examination   Summer 2008   D. Shillady, Professor
(points)   (Attempt all problems)

(20) 1. Using $\eta_{\text{visc}} = \left(\dfrac{1}{2}\right) n^* m \bar{v} \lambda = 2.08 \times 10^{-4}$ g/cm s at 25°C and 1 atm for O_2, compute σ, λ, Z_1, and Z_{11}.   (atomic weight $O = 15.9994$ g/mol, $\sigma = 3.57$ Å, $\lambda = 716$ Å, $Z_{11} = 7.63 \times 10^{28}$)

(20) 2. Show that $\sum \left(\dfrac{q_{\text{rev}}}{T}\right) = 0$ for a Carnot cycle, derive the Carnot efficiency formula.

(20) 3. Calculate the temperature of air compressed adiabatically in a one-cylinder diesel engine from 1040 cm³ at 25°C and 1 atm to 40 cm³. Given $C_V = (5/2)R$ compute Q, moles of air, W, ΔU, and ΔH for this compression.
   $(Q = 0, n = 0.0425, T = 1097°\text{K}, W = \Delta U = 168.8 \text{ cal}, \Delta H = 236.3 \text{ cal}, P = 95.71 \text{ atm})$

(20) 4. Derive the expressions for $\bar{v}$, $\sqrt{\overline{v^2}}$, and $\sqrt[3]{\overline{v^3}}$ of a gas using the Boltzmann principle.

(20) 5. Given the normal boiling point of CH_3CH_2OH is 78.45°C and the vapor pressure is 78.5 mmHg at 30°C, compute ΔH_{vap} for CH_3CH_2OH and compute the vapor pressure at 70°C.
   $(\Delta H_{\text{vap}} = 9924 \text{ cal/mol}, P = 535.7 \text{ mmHg at } 70°\text{C})$

(20) 6. Find P_c, V_c, and T_c for a van der Waals gas and show the law of corresponding states.

(20) 7. Derive the expression for $\left(\dfrac{\partial U}{\partial V}\right)_T$ of a van der Waals gas and use it to compute ΔU for the isothermal compression of 10 mol of Ne gas from 500 to 50 L at 25°C $\left(\text{``}a\text{''} = 0.21 \right.$

   L^2 atm/mol²; use $dU = T\left(\dfrac{\partial P}{\partial T}\right)_V - P \Bigg)$ ($\Delta U = -9.15$ cal, note the sign)

(20) 8. If $A - k_1 \rightarrow B - k_2 \rightarrow C$ with $t_{1/2}(A) = 3$ h and $t_{1/2}(B) = 8$ h, calculate the time when [B] is a maximum given $[A] = 100$ and $[B] = [C] = 0$ at $t = 0$.   (Answer: 6.794 h)

(10) 9. $^{198}_{79}$Au is radioactive with $t_{1/2} = 2.70$ days. How long will it take for 69% of 250 g to decay?   (Answer: 4.563 days)

(20) 10. Derive the formula for ΔS_{mix} of a binary liquid, assuming $\Delta H_{\text{mix}} = 0$. Show that ΔG_{mix} is a minimum and ΔS_{mix} is a maximum when the mole fractions are 0.5.

(20) 11. Consider data for the reaction: $H_2O_2 + Na_2S_2O_3 - [H^+] \Rightarrow 2H_2O + Na_2S_4O_6$

Time (min)	16	36	43	52	at $T = 0$, $[H_2O_2] = 0.0368$
$[(S_2O_3)^{2-}]$	0.01030	0.00518	0.00416	0.00313	and $[(S_2O_3)^{2-}] = 0.0204$

How long will it take until $[(S_2O_3)^{2-}] = 0.0030$?   (Answer: 53.3 min)

BIBLIOGRAPHY

Houston, P. L., *Chemical Kinetics and Reaction Dynamics*, McGraw Hill, Boston, MA, 2001.
Lesk, A. M., *Introduction to Protein Architecture*, Oxford University Press, London, 2000.

REFERENCES

1. Moore, J. W. and R. G. Pearson, *Kinetics and Mechanism*, 3rd Edn., John Wiley and Sons, New York, 1981.
2. Eyring, H., H. Gershinowitz, and C. E. Sun, Potential energy surface for linear H_3 (and why the axes are not at 90°), *J. Chem. Phys.*, **3**, 786 (1935).
3. Hirschfelder, J., H. Eyring, and B. Topley, Reactions involving hydrogen molecules and atoms, *J. Chem. Phys.*, **4**, 170 (1936).
4. Truhlar, D. G. and R. E. Wyatt, History of H_3 Kinetics, *Ann. Rev. Phys. Chem.*, **27**, 1 (1976).
5. Brown, H. C. and M. Borkowski, The effect of ring size on the rate of solvolysis of the 1-Chloro-1-methylcycloalkanes, *J. Am. Chem Soc.*, **74**, 1894 (1952).
6. Bodenstein, M. and S. C. Lind., Geschwindigkeit der Bildung des Bromwasserstoffs aus seinen Elementen, *Z. Phys. Chem.*, **57**, 168 (1906).
7. Christiansen, J. A., On the reaction between hydrogen and bromine, *K. Dan, Vidensk. Selsk. Mat.-Fys. Medd.*, **1**, 14 (1919).
8. Herzfeld, K. F., Zur Theorie der Reaktionsgeschwindigkeiten in Gasen, *Ann. Phys.*, **59**, 635 (1919).
9. Polanyi, M., Causes of forces of adsorption. *Z. Electrochem.*, **26**, 49 (1920).
10. Rice, F. O. and K. F. Herzfeld, The thermal decomposition of organic compounds from the standpoint of free radicals. VI. The mechanism of some chain reactions, *J. Am. Chem. Soc.*, **56**, 284 (1934).
11. Castellan, G. W., *Physical Chemistry*, 3rd Edn., Addison-Wesley Publishing Co., Reading, MA, 1983, p. 844, problem 32.32.
12. Michaelis, L. and M. L. Menten, *Biochemische Zeitschrift*, **49**, 333 (1913).
13. Lumry, R., E. L. Smith, and R. R. Glantz, Kinetics of carboxypeptidase action. I. Effect of various extrinsic factors on kinetic parameters, *J. Am. Chem. Soc.*, **73**, 4330 (1951).
14. Tinoco, I., K. Sauer, and J. C. Wang, *Physical Chemistry, Principles and Applications in Biological Sciences*, 3rd Edn., Prentice Hall, Upper Saddle River, NJ, 1995, p. 429.
15. Lineweaver, H. and D. Burk, The determination of enzyme dissociation constants, *J. Am. Chem. Soc.*, **56**, 658 (1934).
16. Kimball, J., Enzyme kinetics, http://users.rcn.com/jkimball.ma.ultranet/BiologyPages/E/EnzymeKinetics.html Special thanks to Prof. John Kimball of Harvard for contributing the data and graphs related to the reaction of o-diphenol oxidase (catechol oxidase) with catechol. More is available from his site at Kimball's Biology Pages, http://biology-pages.info

9 Basic Spectroscopy

INTRODUCTION

An essential concept for physical chemistry in the twenty-first century is spectroscopy that inter-twines developments in astronomy, physics, and technical innovations in instrumentation as now applied to chemistry. While we focus on chemical applications, we can briefly mention historical highlights in physics because the key concept is the *quantization of energy*. Energy quantization is a revolutionary concept that took most of the early twentieth century to discover, prove, and describe but it is now the backbone of spectroscopy. *Spectroscopy measures various forms of light energy that are absorbed or emitted only at specific wavelengths.* That is due to the fundamental concept that at the level of atoms and molecules, energy occurs in "quantum chunks" that are so small that in everyday life, we think energy is continuous but it is not. The analogy we offer to students is the difference between smooth peanut butter and chunky style peanut butter, because if you examine smooth peanut butter with a simple lens you can see tiny chunks. Thus, smooth or chunky is a matter of size in peanut butter and also in energy.

In this chapter, we will attempt to provide an overview of several forms of spectroscopy to set the scene for more detailed descriptions in Chapter 10. Thus, we will have to compress the historical development to focus on the important case of the hydrogen spectrum and the Bohr model of the quantized levels within the H atom. We will revisit the details of the early twentieth century discoveries in chemistry and physics in Chapter 10. Since this may be the end of a one-semester course, we will stretch the Bohr model to treat x-rays but then show the need for more modern methods. We do this to introduce several spectroscopic techniques within a simple mathematical model and to create interest for a second semester of physical chemistry with the question "If energy is quantized, what does this mean?"

PLANCK'S DISCOVERY

The father of quantization was really Max Planck who received the Nobel Prize in 1918 for his work on blackbody radiation in 1901. However, there were earlier signs of a strange new concept regarding energy in the work of others. For instance, science historians might go all the way back to 1814 when German optician Josef von Fraunhofer observed individual lines in the spectra of stars and the sun. Later Gustav Kirchhoff, a German physicist, made known his work with Robert Bunsen on observation of dark (and light) lines in the spectrum of the sun, and this work was developed further by several amateur astronomers, most notably by William Huggins (1824–1910) who sold his silk exchange business and set up a private observatory just outside of London. Thus, even in the late 1800s, scientific research was still carried out by individuals at their own expense as in our discussion of Sir Robert Boyle in the 1600s. In the late 1800s, a breakthrough in astronomy was to attach a spectroscope (a prism) to a telescope and separate the image of stars and the sun into separate lines/images of different color. As the science progressed, it became possible to measure the wavelengths of the various colors of light. Measurements by Huggins and later by H. C. Vogel (1841–1907) were then studied by Johann Balmer (1825–1898) who published an analysis of a few visible lines from the spectrum of hydrogen, which is abundant in space and in the spectra of stars. An excellent modern history of these developments is given by Becker [1], which gives a glimpse of the importance of astronomy to the development of physics and chemistry.

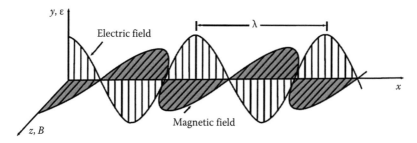

FIGURE 9.1 The classical understanding of light as an electromagnetic wave. Make special note of an oscillating magnetic field as well as an oscillating electric field. In this representation, the fields are restricted to a single plane and would be said to be "plane polarized."

```
log (λ) cm  <---  +----+----+----+----+----+----+----+----+----+----+----+----+----+--->
                   4    3    2    1    0   −1   −2   −3   −4   −5   −6   −7   −8   −9  −10 −11
Range                   Radio           Microwave  Infrared    Visible Ultraviolet  X-ray     γ-ray
```

Other discoveries were being made in Physics in the late 1800s which seemed unrelated to spectra at first. James Clerk Maxwell (1831–1879), a Scottish physicist and mathematician, published several revolutionary papers uniting the theory of electric and magnetic fields in 1861–1862. Maxwell noted that light might be an electromagnetic wave (Figure 9.1). Soon thereafter, Heinrich Rudolf Hertz (1847–1894) was the first investigator to send and receive radio waves over a short distance at Karlsruhe Polytechnic in Germany in 1865. Hertz demonstrated that radio waves could be diffracted just as can light waves and generally confirmed Maxwell's prediction that light is an electromagnetic wave. For our purpose, we need to understand that the visible range of light is only a narrow range of wavelengths and that other forms of radiation differ only in their wavelength. Maxwell's equations and Hertz's experiments led to a wave explanation with a very fast, but finite, propagation speed of close to 3×10^{10} cm/s. Today this speed, "c," is known more accurately than many of the other physical constants. In fact, the other constants are measured in terms of "c" that is now *defined* as 299,792,458 m/s $= 2.99792458 \times 10^{10}$ cm/s. The wavelength of a wave is the length of the wave for one full cycle and the frequency of light is related to the wavelength by the constant speed of their product.

$$c = \lambda \nu$$

So,

$$\nu = \frac{c}{\lambda}$$

and

$$\lambda = \frac{c}{\nu}$$

or

$$\frac{1}{\lambda} = \bar{\nu} = \frac{\nu}{c}$$

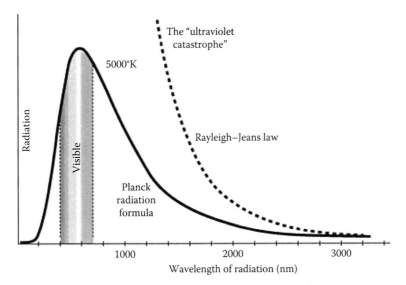

FIGURE 9.2 The blackbody curve in wavelength representation showing the failure of the Rayleigh–Jeans law for short wavelengths and the tantalizing agreement at long wavelengths. The Planck equation is essentially an exact fit to the experimental data. (From Nave, C.R., Blackbody radiation, Georgia State Univ. http://hyperphysics. phyastr.gsu.edu/hbase/mod6.html With permission. See also other parts of the Hyper-Physics site.)

Maxwell's electromagnetic theory predicted that the energy of a light wave was proportional to the sum of the squares of the field components ($E^2 + B^2$). However, in 1901, Max Planck solved a mystery in fitting the shape of the spectrum from a hot object, the so-called blackbody radiation curve, by assuming that light energy occurs in quantum chunks. In his derivation, he assumed the energy of the light was proportional to the frequency and he had to fit an empirical parameter "h" to $\varepsilon = h\nu$. Planck's derivation was not accepted at first, but in 1905, Albert Einstein interpreted another experiment, called the photoelectric effect, in which he used $\varepsilon = h\nu$. In the photoelectric effect, light is used to eject an electron from the surface of a metal (in a vacuum) and an opposing voltage is used to "stop" the flight of the electron so that the "stopping voltage" is a measure of the energy of the ejected electron. When stopping voltage data from this experiment is plotted against the frequency of the exciting light, the graph is a straight line and the slope is the same value Planck had assumed to fit the blackbody radiation curve. Not only that but Planck's formula fits the blackbody radiation curve with essentially an exact fit. There was a competing theory from the British group of Rayleigh and Jeans which only fit the experimental curve at long wavelengths (Figure 9.2). The failure of the Rayleigh–Jeans treatment at short wavelengths was called the "Ultraviolet Catastrophe" but while Planck solved the problem, his solution was not accepted immediately because of the empirical parameter "h." Thus, it took over five years for the scientific community to digest the idea that energy occurs in quantum chunks but the two experiments put the concept of quantization on a firm foundation. The modern proportionality constant is called Planck's Constant "h," where $h = 6.6260693 \times 10^{-34}$ J s $= 6.6260693 \times 10^{-27}$ erg s. Both M. Planck and A. Einstein received a separate Nobel Prize for their work.

RADIO WAVES

At present, amplitude modulation (AM) radio stations in the United States are limited to 50,000 W transmitters. The watt is a unit of power: 1 W $= 1$ J/s. The older radio AM technology uses a carrier wave of constant frequency but varies the amplitude of the wave. Newer FM (frequency modulation) stations maintain a constant amplitude but vary the frequency over a narrow range.

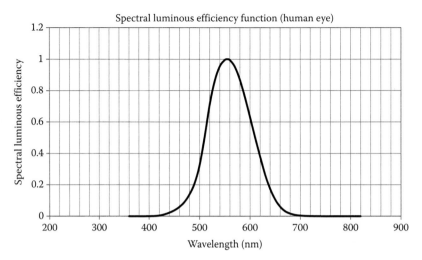

FIGURE 9.3 The average response of the human eye defines the term "visible." Note that the red laser light visible with bar code cash registers has a wavelength of 633 nm while a "black light" Hg bulb has most of its intensity at 365 nm. The maximum sensitivity is at 555 nm (green) which is useful if survival depends on visual acuity in a jungle or forest. (From Lide, D.R., *CRC Handbook of Chemistry and Physics*, 90th Edn., CRC Press, Boca Raton, FL, 2009. With permission. Data sources: (a) The basis for physical photometry, CIE Publication 18.2, 1983; (b) CIE standard colorimetric observers, ISO/CIE No. 10527, 1991; (c) Kaye, G.W.C. and Laby, T.H., *Kaye and Laby Tables of Physical Constants*, 16th Edn., Longman Group Ltd., Harlow, Essex, 1995.)

The AM frequencies are generally in the kilohertz (kHz) range while the FM frequencies are in the megahertz (MHz) range, where 1 Hz = 1 cycle/s. Thus, an AM station emitting waves with a frequency of 1140 kHz (WRVA, Richmond, USA) is using a carrier wave frequency of 1.140×10^6 cycles/s so we can calculate the wavelength:

$$\lambda = \frac{c}{\nu} = \frac{2.99792458 \times 10^8 \, \text{m/s}}{1.140 \times 10^6 \text{s}^{-1}} = 26.2976 \, \text{m}.$$

That is a long wavelength but it is still a form of light.

Why cannot we see the top of a radio transmitter at night if it is sending out the energy of a 50,000 W light bulb? The answer is that the human eye is only sensitive to a narrow range of wavelengths roughly between 3600 Å (deep violet) and 8200 Å (red) or in the range 3.6×10^{-5} cm to 8.2×10^{-5} cm since 1 Å is 1×10^{-8} cm. The actual range where the average human eye can perceive a scale of brightness is less and the average spectral luminous efficiency curve is given in Figure 9.3. The maximum sensitivity is at about 555 nm (green). Visual sensitivity to green in a forest or jungle environment may be a survival attribute. Another thing to remember is that the energy of the electromagnetic wave is proportional to the frequency according to Planck's discovery of $\varepsilon = h\nu$, so *blue light has more energy than red light*. Since $c = \lambda\nu$, shorter wavelength means a higher frequency and vice versa. In spectroscopy, it is common to refer to the high-energy end of the frequency scale as the "blue edge" while the low-energy end of the scale is the "red edge." The verbal history is compressed here but we will give mathematical details in Chapter 10.

BALMER'S INTEGER FORMULA

Our story regarding atomic and molecular spectra really starts with the work of a Swiss mathematician Johann Balmer (1825–1898), in 1885, when he was successful in fitting a formula to the available wavelengths of the H spectrum [2]. The main point of Balmer's formula is that it involves *small integers*; ponder that for a moment, integers. There are some excellent demonstrations of the use of prisms to separate the wavelengths of light and examples of the H lines on the Internet (http:// csep10.phys.utk.edu/astr162/lect/light/absorption.html) where you can see the colors of the lines as well as the fact that the lines occur at specific wavelengths. At that time, the most accurate wavelengths for the H spectrum were available from A. J. Angström (1814–1874), a Swedish physicist for whom the wavelength unit is named (1.0×10^{-10} m $= 1.0 \times 10^{-8}$ cm $= 1$ Å). Balmer was a mathematician who spent most of his career teaching at a school for girls, but at the age of sixty in 1885, he succeeded in fitting the wavelengths of the H spectrum to the formula:

$$\lambda = \frac{hm^2}{m^2 - n^2},$$

where he referred to his symbol "*h*" as the "hydrogen constant" but "*m*" and "*n*" as integers (Balmer's "*h*" is not the same as Planck's "*h*"). Using $n = 2$ and $m = 3, 4, 5, 6, \ldots$, he fitted the H spectral lines and even predicted the wavelength of a new line for $m = 7$, which was later observed at 397 nm by Angström. Later other series of lines, not in the visible part of the electromagnetic spectrum, were discovered when $n = 3, 4, \ldots$. While we should appreciate the intellectual accomplishments of these early scientists and the role of amateur astronomy in the development of physics and chemistry, we leave further study of such history to interested students with Ref. [1]. Instead, we have offered this discussion to emphasize the role of integers in Balmer's formula.

A significant extension of Balmer's work occurred in 1888 when the Swedish physicist Johannes Rydberg (1854–1919) developed a similar formula using the reciprocal of the wavelength:

$$\bar{\nu} = \frac{1}{\lambda} = RZ^2 \left(\frac{1}{n_1^2} - \frac{1}{n_2^2} \right); \quad n_1 < n_2.$$

This formula applies only to atoms/ions with just one electron such as H, He^{1+}, Li^{2+}, Be^{3+}, etc., where Z is the number of protons in the nucleus. Although the constant "*c*" has been standardized, the value of R is the most accurately measured number in physical science with an relative uncertainty of only 6.6×10^{-12} in the 90th Edn. of the *CRC Handbook*. The modern value of the Rydberg constant "*R*" is 109737.31568527 cm^{-1} and early measurements could be made to at least 109737 cm^{-1} *before* Bohr derived his formula in 1913. Looking back at Rydberg's work, it is clear that the specific value of his constant is dependent on his choice of $(1/\lambda)$ units, and we will see that this unit is still used in infrared spectroscopy. The use of the reciprocal square of integers is an extension of Balmer's formula.

In a later chapter, we will explore the events in physics between 1885 and 1913, which were tumultuous in the wonder of the discoveries, but we want to focus here on the spectrum of the H atom. Although the work by Planck in 1901 and the interpretation of the photoelectric effect by Einstein in 1905 are very important to the overall development of modern spectroscopy, the next breakthrough for understanding the spectrum of the H atom occurred in 1913 with the theoretical model of Niels H. Bohr (1885–1962), a Danish physicist who received the Nobel Prize for this work in 1922 (Figure 9.4). Bohr was a "pencil-and-paper" theorist who made a major discovery by postulating the quantization of angular momentum. Note that earlier work by Planck in 1901 had postulated that energy exists as small chunks of size $\varepsilon = h\nu$, which was an earth-shaking concept that few people really believed. However, Planck's treatment led to a model of the broad spectrum emitted from a hot body with a fit of his theoretical data points to the experimental data that was essentially "exact" and thus hard to refute. We do not know how Bohr arrived at the idea that angular momentum is quantized but we can note that momentum is embedded in some energy formulas. We can see that if momentum is quantized that will also quantize the

FIGURE 9.4 Niels Henrik David Bohr (1885–1962) was a Danish physicist who unified spectral data from atoms (mainly H) with a theory of quantized energy in 1913 for which he was awarded the Nobel Prize in Physics in 1922. He earned his doctorate in 1911 from the University of Copenhagen and then studied further under Ernest Rutherford in Manchester England who was studying the nature of atomic structure. He hypothesized the multielectron shells which were useful in organizing chemical concepts. Bohr became a professor of physics at the University of Copenhagen in 1916. In 1920, he was named director of the Institute of Theoretical Physics at the University of Copenhagen. He was named a Fellow of the Royal Society of London in 1926 and received the Royal Society Copley Medal in 1938. He fled Denmark during WWII and assisted scientists in the Manhattan project at Los Alamos in the United States but returned to Copenhagen after the war and promoted the peaceful uses of atomic energy.

energy as well. While this is reasonable after the fact, it took the genius of Bohr to extend this to angular momentum as $mvr = n\hbar$:

$$E = \frac{mv^2}{2} = \frac{(mv)^2}{2m} = \frac{p^2}{2m}$$

and momentum $p = mv$.

Thus, if energy is quantized (Planck 1901), momentum is even more fundamental than energy, because energy can be expressed in terms of momentum. The discovery of quantized energy and the fact that energy can be expressed in terms of momentum implies that momentum is probably quantized as well. Then angular momentum in a rotating system should also be quantized and only exist as specific chunks, according to Bohr's hypothesis. By 1913, only a few people really believed that energy is "chunkified," but the combination of Planck's blackbody spectrum in 1901 and Einstein's 1905 explanation of the photoelectric effect had convinced a few scientists that quantization of both energy and momentum does exist. Whereas Planck had used "h" as his proportionality constant for quantized energy, Bohr chose "$\hbar$" (h-bar) as the basic unit of angular momentum in a rotating system where h-bar = Planck's $\dfrac{h}{2\pi}$. We are now going to show Bohr's derivation that is very clever but it is one of those treatments where you can do correct algebra without getting the right answer unless you take a certain path.

Here, we come to another problem in that we will be dealing with electrical units, a field that has developed over time in various laboratories and over several hundred years resulting in as many as five different systems [3]. The modern SI units have tried to unify this situation but at the cost of introducing a new annoyance in the form of a factor of $4\pi\varepsilon_0$ that pops up all over the equations. For the simplicity of explaining this derivation, we will use a system of units that goes forward with the energy in units of "electron volts" that is used by most physicists and nuclear physicists. Let us set out the Bohr hypotheses and develop the formula before we analyze the meaning of the results. Bohr assumed that the electron in the H atom moved in a (flat) circular orbital around a positive ion (Ze^+) in one of the "allowed orbitals" determined by the momentum quantization. For the H atom, $Z=1$, but the derivation can be applied to He^+ ion as $Z=2$, Li^{2+} ion as $Z=3$, Ne^{9+} as $Z=10$, or even U^{91+} as $Z=92$; for any system with just one electron in orbit around a positive nucleus. *The model does not apply to more than one electron.* The electrical interactions are a result of the e^- charge on the electron and a positive charge of Ze^+ on the positive nucleus. Note that at the time Bohr worked on his theory, the Rydberg formula cited above was well known to scientists as providing a very accurate fit to the measured spectral lines of atoms or ions with only one electron. There is a lesson here in how theorists function in using some fragmentary experimental evidence to check a pencil-and-paper theory but we must admit that no one else had an explanation of why/how the Rydberg formula worked or the physical principles behind it.

1. $mvr = n\left(\dfrac{h}{2\pi}\right) = n\hbar$, for $n = 1, 2, 3, \ldots$, that is, quantize the angular momentum of the electron ($mvr = n\hbar$).

2. $\dfrac{mv^2}{r} = \dfrac{Ze^2}{r^2}$, this balances the centripetal force of the electron with the electrostatic attraction. In fact, the electrostatic attraction of the electron by the positive ion continually pulls (accelerates) the path of the electron into a curve just as a rock on a string is pulled into a circular path.

   The key step occurs right here, in that Bohr solved for the velocity in terms of the velocity instead of taking the square root to find "v." Thus, he used an unusual algebra step so that he could insert the quantization of the angular momentum:

$$v = \frac{Ze^2}{mvr} = \frac{Ze^2}{n\hbar}.$$

3. $E_{\text{tot}} = T + V = \dfrac{mv^2}{2} - \dfrac{Ze^2}{r}$, the total energy is the sum of kinetic (T) and potential (V) parts.

4. From 2, we see

$$mv^2 = \frac{Ze^2}{r} = m\left(\frac{Ze^2}{n\hbar}\right)^2 = \frac{mZ^2e^4}{n^2\hbar^2} = \frac{Ze^2}{r}$$

so

$$r = \frac{n^2\hbar^2}{mZe^2} = \left(\frac{n^2}{Z}\right)\left(\frac{\hbar^2}{me^2}\right) = r(n, Z).$$

5. $E_{\text{tot}} = \dfrac{m}{2}\left(\dfrac{Ze^2}{n\hbar}\right)^2 - \dfrac{Ze^2}{\left(\frac{n^2}{Z}\right)\left(\frac{\hbar^2}{me^2}\right)} = \left(\dfrac{mZ^2e^4}{n^2\hbar^2}\right)\left[\dfrac{1}{2} - 1\right] = -\left(\dfrac{Z}{n}\right)^2\left(\dfrac{me^4}{2\hbar^2}\right) = E(n, Z)$

Note that we now have not only a formula for the quantized energy $E(n, Z)$ but a formula for the radius of each quantized orbit. It is quite desirable to investigate the formulas to obtain meaningful units for the energy and radius. At this point, we need to grapple with a few non-SI units. First, the definition of *an "electron volt" is the energy gained by an electron when accelerated through a potential of 1 V.*

1 V $=$ 1 J/C (1 Volt $=$ 1 joule per coulomb)

1 C $=$ 1 A s (1 Coulomb $=$ 1 Ampere second)

1 F $=$ 96,485.3383 C $\cong$ 96,485 C $=$ 1 mol of electrons $=$ 1 Faraday $=$ 1 F

1 F is the amount of A s that will electroplate 1 g atom of $Ag^+ + e^- \rightarrow Ag^0$

1 electron charge $=$ (96,485.3383 C/6.0221415 $\times 10^{23}$) $=$ 1.60217653 $\times 10^{-19}$ C
$\cong$ 1.602 $\times 10^{-19}$ C

1 eV $=$ 1 (J/C)(1.60217653 $\times 10^{-19}$ C) $=$ 1.60217653 $\times 10^{-19}$ J $=$ 1.60217653 $\times 10^{-12}$ erg
$\cong$ 1.602 $\times 10^{-19}$ J

The mass of an electron $= m_e =$ 9.1093826 $\times 10^{-31}$ kg $=$ 9.1093826 $\times 10^{-28}$ g $\cong$ 9.11 $\times 10^{-28}$ g

The mass of a proton $= m_p =$ 1.67262171 $\times 10^{-27}$ kg $=$ 1.67262171 $\times 10^{-24}$ g $\cong$ 1.67 $\times 10^{-24}$ g

1 mol of electron volts $=$ (6.0221415 $\times 10^{23}$/mol)(1.60217653 $\times 10^{-19}$ J)
$=$ 96.485 kJ/mol $=$ 23.061 kcal/mol

In the 1930s, electrochemistry was a major part of physical chemistry and laboratory measurements were related to easily reproducible experiments. Thus plating out 1 mole of silver metal from a solution of $AgNO_3$ was an easy way to measure coulombs with an ammeter to measure current and a clock measuring seconds. The Faraday constant then requires further definitions of an ampere etc., but those constants can be obtained through measurements and calculations from electroplating silver. Today, the modern values are all subjected to a least squares fit of all the known constants with the best experimental data except, as mentioned above, the value of "c" is now fixed and not subject to further measurement. The value of "c" is the kingpin of most of all the other constants.

The units mentioned so far are either SI or accepted for use with SI units but we need an additional value from the electrostatic unit system. In that system, charge is measured in "statcoulombs" as related to the cgs system where Coulomb's law can be written as a force between two charged particles separated by a distance.

$$\text{Force} = ma = \frac{qq}{r^2}$$

so the charge

$$q \propto \sqrt{mar^2} \sim \sqrt{\frac{(\text{g cm})(\text{cm}^2)}{s^2}} \sim \frac{g^{1/2}\, cm^{3/2}}{s} \sim \text{statcoulomb}$$

The numerical conversion from statcoulombs to coulombs is 3.335641 $\times 10^{-10}$, so we can convert the electrochemical coulombs to statcoulombs in cgs units.

$$q_e = \frac{(1.60217653 \times 10^{-19}\, C)}{(3.335641 \times 10^{-10})} = 4.803204 \times 10^{-10} \text{ statcoulomb}$$

Let us consolidate the key values to make formulas that are easy to remember.

$$a_0 = \left(\frac{\hbar^2}{me^2}\right) = \frac{(6.6260693 \times 10^{-27} \text{ erg s})^2}{4\pi^2(9.1093826 \times 10^{-28} \text{ g})(4.803204 \times 10^{-10} \text{ g}^{1/2} \text{ cm}^{3/2}/\text{s})^2}$$

$$= 0.5291772 \times 10^{-8} \text{ cm}$$

$$\left(\frac{me^4}{2\hbar^2}\right) = \frac{4\pi^2(9.1093826 \times 10^{-28} \text{ g})(4.803204 \times 10^{-10} \text{ g}^{1/2} \text{ cm}^{3/2}/\text{s})^4}{2(6.6260693 \times 10^{-27} \text{ erg s})^2(1.60217653 \times 10^{-12} \text{ erg/eV})} = 13.6057 \text{ eV}$$

In the second step, we have used the conversion from erg to eV in the denominator. Since $1 \text{ eV} = 1.60217653 \times 10^{-19}$ J, we just multiply by 1.0×10^{-7} to put the constant into ergs.

Although we have already delved into several unit conversions, we need just one more formula to be on our way to understanding spectroscopy. The next formula is by far and away the most useful formula in spectroscopy. If you attend a research seminar and the speaker gives energies in kilocalories/mole he/she is probably an older chemist, if the speaker uses kilojoule/mole energies he/she is probably a younger chemist, but if the speaker is a physicist you will probably hear all the energy values in electron volts. Thus, even though the *CRC Handbook* and other texts strive to use only SI units, our recommendation is to get used to electron volt energies. In x-ray analysis later in this chapter the units are usually in kiloelectron volt (thousands of electron volts), so there is a large part of the scientific community using electron volts.

With a warning to the students, *we have selected electron volts as the most useful energy unit to relate spectroscopy experiments to theory* in the sense that a student can imagine the physical units. However, the physical constants are revaluated every three years or so which makes past research papers subject to drift in the values of the constants. Around 1960, quantum chemists addressed this problem and chose yet another set of units in which $c = \hbar = m_e = q_e = 1$ to simplify theoretical equations in "atomic units," so that the equations were expressed totally in the basic mathematical units. In these units (used by quantum chemistry computer programs), a person only needs to know the latest value of an energy unit called the hartree and the latest value of the Bohr radius (a_0) to convert computer results back to laboratory results. At present (2010), 1 hartree $= 27.2113845$ eV and $a_0 = 0.52917720859 \times 10^{-8}$ cm. We will not use these units until Chapter 17 but you can see that the Bohr formulas simplify further in these units:

$$E(n, Z) = -\left(\frac{Z}{n}\right)^2\left(\frac{me^4}{2\hbar^2}\right) = -\left(\frac{Z}{n}\right)^2(0.5 \text{ hartree}); \quad r(n, Z) = \left(\frac{n^2}{Z}\right)\left(\frac{\hbar^2}{me^2}\right) = \left(\frac{n^2}{Z}\right)a_0.$$

A VERY USEFUL FORMULA

Here, we present a very simple but powerful formula. This formula is so useful that you can sit in a research seminar or lecture and do mental arithmetic on the spot and then make a very intelligent comment such as "Yes, Professor, but the wavelength for that transition should be . . ." Recall that Planck realized that the energy of a light wave is proportional to the frequency of the wave and evaluated the proportionality constant to be "h." As mentioned above, the same number occurs in the slope of the data for the photoelectric effect as analyzed by A. Einstein in 1905. The modern value is $6.6260693 \times 10^{-27}$ erg s. Assume there are two energy levels in a molecular system such that the difference between the two levels is a quantum with energy $\Delta E = h\nu = \dfrac{hc}{\lambda}$. This situation happens so often in spectroscopy that we can develop a useful shortcut formula if we assume the λ value is in angstroms (1.0×10^{-8} cm) and we always want the energy value in electron volts. Then we have

$$\Delta E \text{ (eV)} = \frac{hc}{\lambda \text{ (Å)}} = \frac{(6.6260693 \times 10^{-34} \text{ J s})(2.99792458 \times 10^8 \text{ m/s})(10^{10} \text{ Å/m})}{(1.60217653 \times 10^{-19} \text{ J/eV})(\lambda \text{ (Å)})}$$

or

$$\Delta E\,(\mathrm{eV}) = \frac{12398.41906}{\lambda\,(\mathring{A})}$$

and more generally useful as

$$\Delta E\,(\mathrm{eV}) \cong \frac{12{,}398}{\lambda\,(\mathring{A})}\,.$$

This was first brought to our attention many years ago [4] and has proved to be very useful for quick estimates of wavelength or energies in electron volts doing mental arithmetic while sipping coffee in a seminar.

PRELIMINARY SUMMARY OF THE BOHR ATOM

$$r(n, Z) = \left(\frac{n^2}{Z}\right)(0.5291772\,\mathring{A}); \quad \Delta E\,(\mathrm{eV}) \cong \frac{12{,}398}{\lambda\,(\mathring{A})}; \quad E(n, Z) = -\left(\frac{Z}{n}\right)^2 (13.6057\,\mathrm{eV})$$

These three formulas can be used for quite a few applications which will be our introduction to spectroscopy. The most obvious application is to compare the energy formula to the experimental wavelengths of the H atom spectrum ($Z = 1$):

$$\Delta E\,(\mathrm{eV}) = E_2 - E_1 = (-1)\left[\frac{1}{n_2^2} - \frac{1}{n_1^2}\right](13.6057\,\mathrm{eV}) = \frac{hc}{\lambda}$$

This has the same type denominator as the Balmer formula and when the other numbers are compared, it is found that the Bohr equation is essentially the same as the Balmer equation. There is only a slight difference due to the fact that the nucleus in the Bohr model is fixed at the center of the atom while the real spectra include the fact that the electron and proton both orbit around the center-of-mass (the see-saw balance point) of the two particles. That is really very close to the position of the proton because it is much more massive than the electron. When this correction is made to the Bohr formula, the agreement with the experimental spectra is essentially exact.

One other unit we may encounter in spectroscopy, particularly in infrared spectroscopy, is the "wave number." Basically, the wave number is just the reciprocal of the wavelength:

$$c = \lambda\nu$$

so

$$\nu = \frac{c}{\lambda}$$

and

$$\bar{\nu} \equiv \left(\frac{1}{\lambda}\right) = \left(\frac{\nu}{c}\right)$$

This unit may have come about due to the way some early experiments were done to measure wavelengths and was important in the derivation of Rydberg's formula but it is now established in infrared spectroscopy. In particular, the combined Balmer–Rydberg formula in wave numbers is

$$\bar{v}_H = -109{,}737.31568525 \left(\frac{1}{\text{cm}}\right)\left[\frac{1}{2^2} - \frac{1}{n^2}\right]; \quad n = 3, 4, 5, \ldots$$

We show the modern value that was initially known only to about six significant figures. However, through the work of Johannes Rydberg (1854–1919), a Swedish physicist, this number is probably the most refined constant in spectroscopy. Since H is a common element in space and much of early spectroscopic data came from astronomy measurements, the Rydberg became a unit of energy. In 1913, Bohr compared his theoretical equation to experiment using the value of the Rydberg:

$$E(n, Z) = -\frac{me^4}{2n^2\hbar^2}$$

for $Z = 1$, so,

$$E_2 - E_n = -\frac{me^4}{2n^2\hbar^2}\left[\frac{1}{2^2} - \frac{1}{n^2}\right] = h\nu = \frac{hc}{\lambda}$$

and then

$$\frac{(E_2 - E_n)}{hc} = -\frac{me^4}{2n^2\hbar^2(hc)}\left[\frac{1}{2^2} - \frac{1}{n^2}\right] = \frac{1}{\lambda} = \bar{v}.$$

We can see that the Bohr's factor of constants should agree with the experimental value of the Rydberg. We have already noted there needs to be a small correction to the value of "m" since a tiny center-of-mass correction should be applied (the electron and proton actually rotate about their mutual center of mass which is very close to the position of the more massive proton).

$$\frac{me^4}{2n^2\hbar^2(hc)} = \frac{me^4(4\pi^2)}{2h^3c} = \frac{(4\pi^2)(9.1093826 \times 10^{-28}\text{ g})(4.803204 \times 10^{-10}\text{ g}^{1/2}\text{ cm}^{3/2}/\text{s})^4}{2(6.6260693 \times 10^{-27}\text{ erg s})^3(2.99792458 \times 10^{10}\text{ cm/s})} = R_H$$

Thus, we calculate $R_H = 109{,}737.2794 \left(\dfrac{1}{\text{cm}}\right)$, and we see that using modern values of the constants without the center-of-mass correction to "m" we get six figure agreement with the experimental value. This was a major triumph for Bohr and the amazing agreement with the rydberg constant made believers in the idea that angular momentum is quantized.

Following the triumph of the Bohr theory, let us consider some limitations of the Bohr model of the atom. First, there are only flat circular orbitals and you have probably seen orbitals in organic chemistry textbooks that have different 3D shapes due to further research since 1913. However, because the ΔE (eV) $= E_2 - E_1$ values are correct, at least for ($n \rightarrow n+1$) transitions, the Bohr model was a breakthrough in understanding the energy levels of atoms. Note also that these flat orbitals do not give insight as to how atoms combine into molecules.

A second philosophical dilemma with the model is that the electrons in orbitals have angular momentum, so they must be moving in some sense of the word and in fact we calculated the formula for their velocity. However, in the theory of radio transmitters, the radiated signal is caused by sending electrons moving back and forth in some sort of antenna. Thus, it is known that moving electrons radiate energy so why/how is it possible that electrons can find stable orbitals that do not radiate energy? Today, we would say that radio antennas radiate because the electronic excitation is causing the electrons in the metal to change energy levels rapidly but the non-excited energy levels are stable and do not radiate. These sort of questions were raised about the Bohr model but until 1926 there was no better model. There is no further accurate agreement with experiment except for spectra of light element ions like He^+, Li^{2+}, etc., one-electron ions of light elements.

SIGNIFICANCE OF THE BOHR QUANTUM NUMBER *n*

Today, students in this class will have been exposed to a form of the periodic chart that incorporates many discoveries since 1913. However, in the early twentieth century the periodic chart was organized mainly by atomic weights with less organization than we enjoy today. At that time, the rare gases were believed to be totally inert but their atomic numbers allowed Bohr to postulate an "electron shell model" of atoms heavier than H with occupancies of 2, 8, 8, 18, 18, ... based largely on the atomic numbers of the rare gases and the differences between them. For instance, Ne ($Z = 10$) has 8 more electrons than He ($Z = 2$) and Ar ($Z = 18$) has 8 more electrons than Ne. Then Kr ($Z = 36$) has 18 more electrons than Ar, and so on. The Bohr orbits were assigned labels as the K, L, M, N, ... shells. Strictly speaking, the Bohr model only has one quantum number, "*n*." Later, Arnold Sommerfeld (1868–1951) postulated elliptical orbitals to introduce additional angular momentum rules into the shell model. Sommerfeld introduced the l-quantum number, the m-quantum number, and even the fourth quantum number for spin based on later developments. For the time being, we will push the Bohr model to its extremes using just the n-quantum number. In spite of several more recent descriptions of atoms since 1913, the Bohr model survives in some terminology related to the inner K, L, and M shells as we will soon see, because x-rays come from transitions between these shells. Although the Bohr model does little to explain chemical bonding in molecules, it remains a concept related to atoms. Even today, the emblem of the International Atomic Energy Agency (IAEA) still shows an atom with the Bohr–Sommerfeld elliptical orbits as a representation of atomic orbitals.

ORBITAL SCREENING

Consider the limitation we have already mentioned in that the Bohr model only applies to atoms/ions with just one electron. Please note that when there is more than one electron, repulsion between the electrons alters the energy situation. However, it is possible to adjust the model for more than one electron, assuming we understand that we are leaving the realm of high accuracy and merely modeling trends. Probably, every student has done the sodium flame test in freshman chemistry laboratory. In more sophisticated models, the fluffy yellow flame is believed to be due to a transition from a 4p orbital back down to a 3s orbital, and the strong yellow "D" line at 589 nm is in fact two separate lines (a doublet) believed due to two possible spin states of the 4p orbital. The Bohr model has none of this detail but if we assume that the 4p orbital has nearly the same energy as the 4s orbital (same shell in the Bohr model), we can treat the Z value as a parameter and fit the model to a system of an "ion core" of charge (Ze^+) with an outer electron in a 4s orbital (assuming it has the same energy as a 4p orbital; not really true but close). Note 589 nm $= 5890$ Å. Thus,

$$\Delta E = E_4 - E_3 = \left(-Z_{eff}^2\right)\left[\frac{1}{4^2} - \frac{1}{3^2}\right](13.6057\,eV) = Z_{eff}^2(0.661388\,eV) = \frac{12{,}398}{5890} = 2.104923\,eV.$$

We can solve this for Z_{eff}. Thus, $Z_{eff} = \sqrt{\dfrac{2.104923\,eV}{0.661388\,eV}} \cong +1.784$; what does this mean? The effective nuclear charge might have been expected to be $+1$ assuming that the inner 10 electrons "covered up" (10/11) of the nuclear charge of the Na nucleus. However, the effective nuclear charge is greater than $+1$ by a considerable amount. This can be interpreted as "imperfect screening" of the nuclear charge by the first 10 electrons (the Ne core). We can see from the formula for the Bohr radius that the radius increases as the square of the quantum number "n," and so, as the orbits of the $n = 2$ shell get larger than the $n = 1$ shell, their ability to be between the outer electron and the nuclear charge of the core is reduced so the outer electron experiences more of the uncovered nuclear charge. Thus, the Bohr model can give a qualitative explanation for the concept of electron screening. We emphasize that this result departs from the precision of six-figure accuracy of the spectroscopic measurements but provides a valuable concept.

X-RAY EMISSION

Here is another way in which the simple Bohr model can be used to give qualitative ideas about electronic structure that is tantalizingly semiquantitative. X-rays were discovered by W. C. Roentgen (1845–1923), a German physicist who carried out the first experiments in 1895 and was awarded the very first Nobel Prize in Physics in 1901. A process called Auger x-ray emission spectroscopy is due to the use of an electron beam to knock electrons out of inner orbitals of atoms whereupon the outer electrons of the formerly neutral atoms "fall" down in a cascade of steps to again fill the lowest orbitals. When this happens, electromagnetic radiation is given off just as the yellow flame emission occurs due to an outer electron of Na. This process was studied by Pierre Victor Auger (1899–1993), a French physicist who found that electrons are emitted (and scattered) as well as x-rays. Since the $n = 1$ level is the lowest level, the energy gap between the $n = 2$ and $n = 1$ levels is large enough to produce the emitted radiation in the x-ray range. X-rays from the ($n = 2 \rightarrow n = 1$) transition are called K_α and those from ($n = 3 \rightarrow n = 1$) K_β.

As one proceeds to heavier elements, the number of electrons increases but the increasing nuclear charge pulls the inner orbitals closer to the nucleus as the Bohr radius decreases. According to the Bohr model of circular shells, the extension to more electrons went as K, L, M, N, ... shells containing 2, 8, 8, 18, 18, ... electrons respectively. The point is that for elements beyond Cd ($Z = 48$), the L shell as well as the innermost K shell are pulled in tightly, so that x-rays can be easily observed from electrons falling into a partially empty L shell from the M and N shells. The Bohr model advocates justified the 2, 8, 8, 18, ... sequence using just the one "n" Bohr quantum number from the atomic numbers of the rare gas atoms and we can use the 2, 8, 8 sequence for the innermost shells. You can also see that for high values of Z, the inner L shell might be partly empty and electrons could cascade from the M and N shells. Thus, transitions from ($n = 3 \rightarrow n = 2$) are called L_α and those from ($n = 4 \rightarrow n = 2$) are called L_β. While K_α transitions can still be created for $Z > 48$, a higher energy electron beam is needed, so it has been found that reducing the excitation beam energy to 20,000 V or even just 10,000 V allows use of the same dispersive system and detectors for the L_α of $Z > 48$ as for the K_α transitions of elements with $Z < 49$.

We are emphasizing x-ray wavelengths to make the connection to the Bohr theory that originated from wavelength data. However, most modern x-ray fluorescence (XRF) data is reported in kilo-electron volt energies from a detector under a high voltage potential (Figure 9.5). Of course, we know $\lambda\,(\mathrm{Å}) = \dfrac{12{,}398}{\Delta E(\mathrm{eV})}$ will give us an accurate conversion between electron volts and wavelength in angstroms, so we can convert from one representation to the other. However, the table of x-ray data was last reported as wavelengths in the *CRC Handbook* in the 1976 edition. Another recent consideration is that in order to keep the range of x-rays within a given order of diffraction and detector limits the longer wavelength M_α and M_β, x-rays can be used for heavier elements. However, the $M_{\alpha,\beta}$ transitions are less easily assigned in terms of clean integers. We will see that even the $L_{\alpha,\beta}$ transitions are subject to non-integer screening of the nuclear charge so the $M_{\alpha,\beta}$ transitions are identified and cataloged as empirical data and we will not attempt an analysis of the

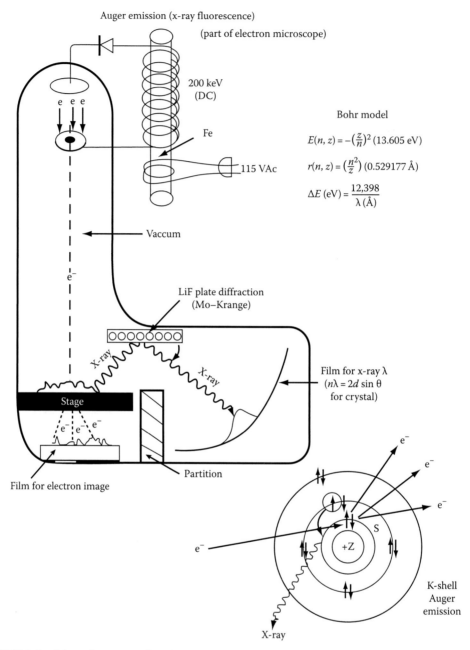

Auger emission (x-ray fluorescence)

(part of electron microscope)

200 keV
(DC)

e e e

Fe

115 VAc

Vaccum

e⁻

LiF plate diffraction
(Mo–Krange)

X-ray

X-ray

Stage

e⁻ e⁻ e⁻

Film for electron image

Partition

Film for x-ray λ
($n\lambda = 2d \sin \theta$
for crystal)

Bohr model

$$E(n, z) = -\left(\frac{z}{n}\right)^2 (13.605 \text{ eV})$$

$$r(n, z) = \left(\frac{n^2}{z}\right) (0.529177 \text{ Å})$$

$$\Delta E \text{ (eV)} = \frac{12{,}398}{\lambda \text{ (Å)}}$$

+Z

S

e⁻

e⁻

e⁻

e⁻

K-shell
Auger
emission

X-ray

FIGURE 9.5 Schematic cartoon of an electron microscope with an XRF attachment.

internal screening of the M shell here. Software that accompanies a specific x-ray spectrometer will include these kiloelectron volt energies, characteristic of elements within the range of the detector and the diffraction plate used within the instrument.

X-rays have been important in medical science for many decades, and x-ray machines usually use a Cu target and the x-rays from the wavelength of the K_α line. The Cu K_α x-ray is near optimum for biological tissue, since its wavelength of 1.541 Å (Table 9.1) is close to the average organic bond length. Consider typical bond lengths between carbon atoms in biological molecules such as

TABLE 9.1
X-Ray Auger/Fluorescence Wavelengths of Selected Elements

Element	λ (nm)	Element	λ (nm)	Element	λ (nm)	Element	λ (nm)
3 Li	K_α 22.8	28 Ni	K_α 0.1658	53 I	L_α 0.3149	78 Pt	L_α 0.1313
4 Be	K_α 11.4	29 Cu	K_α 0.1541	54 Xe	L_α 0.3016	79 Au	L_α 0.1276
5 B	K_α 6.76	30 Zn	K_α 0.1435	55 Cs	L_α 0.2892	80 Hg	L_α 0.1241
6 C	K_α 4.47	31 Ga	K_α 0.1340	56 Ba	L_α 0.2776	81 Tl	L_α 0.1207
7 N	K_α 3.16	32 Ge	K_α 0.1254	57 La	L_α 0.2666	82 Pb	L_α 0.1175
8 O	K_α 2.362	33 As	K_α 0.1176	58 Ce	L_α 0.2562	83 Bi	L_α 0.1144
9 F	K_α 1.832	34 Se	K_α 0.1105	59 Pr	L_α 0.2463	84 Po	L_α 0.1114
10 Ne	K_α 1.445	35 Br	K_α 0.1040	60 Nd	L_α 0.2370	85 At	L_α 0.1085
11 Na	K_α 1.191	36 Kr	K_α 0.09081	61 Pm	L_α 0.2282	86 Rn	L_α 0.1057
12 Mg	K_α 0.989	37 Rb	K_α 0.09256	62 Sm	L_α 0.2200	87 Fr	L_α 0.1030
13 Al	K_α 0.834	38 Sr	K_α 0.08753	63 Eu	L_α 0.2121	88 Ra	L_α 0.1005
14 Si	K_α 0.7125	39 Y	K_α 0.08288	64 Gd	L_α 0.2047	89 Ac	L_α 0.0980
15 P	K_α 0.6157	40 Zr	K_α 0.07859	65 Tb	L_α 0.1977	90 Th	L_α 0.0956
16 S	K_α 0.5372	41 Nb	K_α 0.07462	66 Dy	L_α 0.1909	91 Pa	L_α 0.0933
17 Cl	K_α 0.4728	42 Mo	K_α 0.07093	67 Ho	L_α 0.1845	92 U	L_α 0.0911
18 Ar	K_α 0.4192	43 Tc	K_α 0.06750	68 Er	L_α 0.1784	93 Np	L_α 0.0889
19 K	K_α 0.3741	44 Ru	K_α 0.06431	69 Tm	L_α 0.1727	94 Pu	L_α 0.0868
20 Ca	K_α 0.3358	45 Rh	K_α 0.06133	70 Yb	L_α 0.1672	95 Am	L_α 0.0847[a]
21 Sc	K_α 0.3031	46 Pd	K_α 0.05854	71 Lu	L_α 0.1620	96 Cm	L_α 0.0828[a]
22 Ti	K_α 0.2749	47 Ag	K_α 0.05594	72 Hf	L_α 0.1570	97 Bk	L_α 0.0809[a]
23 V	K_α 0.2504	48 Cd	K_α 0.05350	73 Ta	L_α 0.1522	98 Cf	L_α 0.0791[a]
24 Cr	K_α 0.2290	49 In	L_α 0.3772	74 W	L_α 0.1476	99 Es	L_α 0.0773[a]
25 Mn	K_α 0.2102	50 Sn	L_α 0.3600	75 Re	L_α 0.1433	100 Fm	L_α 0.0756[a]
26 Fe	K_α 0.1936	51 Sb	L_α 0.3439	76 Os	L_α 0.1391	101 Md	L_α 0.0740[a]
27 Co	K_α 0.1789	52 Te	L_α 0.3289	77 Ir	L_α 0.1351	102 No	L_α 0.0724[a]

Sources: Weast, R.C., *CRC Handbook of Chemistry and Physics*, 53rd Edn., CRC Press, Cleveland, OH, 1971, p. E-131; http:// en.wikipedia.org/wiki/X-ray_fluorescence

[a] The values are rounded to four significant figures.

1.54 Å, single; 1.39 Å, double; and 1.31 Å, triple. If the x-ray wavelength is much larger than the space between the atoms, the material will be opaque to the electromagnetic waves. If the wavelength is much smaller than the spacing between the atoms, the electromagnetic wave will pass through the material. Thus, a wavelength is needed in the range of about 1.5 Å to produce a semitransparent image from exposure of the material to these x-rays. For Cu, we have a qualitative estimate for the wavelength of

$$\Delta E = E_1 - E_2 = (29)^2 \left[\frac{1}{1^2} - \frac{1}{2^2} \right] (13.6058\,\text{eV}) = 8581.85835\,\text{eV}$$

and

$$\lambda_{K_\alpha} = \frac{12,398}{8581.8\,\text{eV}} = 1.445\,\text{Å}.$$

So we see that the K_α x-ray emission of a Cu target can be calculated from the simple Bohr model but there is an error of about 6.2% compared to the experimental value of 1.541 Å; not exactly the six figure accuracy of the Rydberg value but close. Note, we have reversed the sign of the energy difference for an emission. The answer is sufficiently close to the experimental value that we are confident that our simple model has captured the main principle of the phenomenon. For further use later, let us ask what the effective nuclear charge is that would produce the correct wavelength:

$$Z_{\text{eff}} \cong \sqrt{\frac{12,398\,\text{eV\,Å}}{(1.541\,\text{Å})(0.75)(13.6057\,\text{eV})}} = 28.0791,$$

not the bare nuclear charge of 29.

That is quite revealing and suggests that the $n = 1$ shell is not completely empty. Using modern information about electron spin and the idea that orbitals can contain two electrons with opposite spin, it appears that there is still one electron in the $n = 1$ orbital and only one is missing. The Bohr theory postulates that the orbital occupancy should be 2, 8, 8, 18, ... with no reason given except that the periodic chart implies that occupancy, so the $n = 1$ orbital should have two electrons and the K_α data implies there is still one electron in the $n = 1$ orbital.

Today, most computer monitors and TV screens are some variant of a flat screen but not so long ago all TV screens and computer monitors were a picture tube in which a beam of electrons moved across the screen and excited phosphors on the inside of an evacuated "cathode ray tube." That meant that you were facing an electron beam hitting the inside of a glass tube with energy of 20,000 V or so. What about K_α x-rays from cathode ray screens? While the heaviest (highest Z) element in glass (SiO_x) would be Si ($Z = 14$), the early green phosphors were a form of ZnO, so Zn was present ($Z = 30$). Later, color TV tubes had lanthanide salts for various colors such as Eu_2O_3 for the red color and so some Eu was present ($Z = 63$). While the impact of the electron beam with Si atoms produced only "soft x-rays," the heavier elements could produce x-rays with shorter wavelengths capable of breaking bonds in biological compounds. Thus, there was a safety issue with cathode ray tubes, especially for young developing children sitting close to the picture tube. Now, flat screen picture screens are not only more convenient but they also have eliminated an x-ray hazard that accompanied the use of cathode ray tubes.

This author is old enough to have purchased new shoes at a time when department stores had "fluoroscopes," which were real-time x-ray machines with inlets for your feet, and you could "show your mother" on a small TV-like screen that the shoes were big enough by wiggling your toes to show the space inside the new shoes. It is now known that high exposure to x-rays can cause sterility, so by now those old fluoroscopes are long gone. Fortunately, shoes tend to last a year or more so children's exposure to x-rays was infrequent. Modern radiology technicians work behind a lead–glass wall and wear a film badge to monitor the extent of exposure on any given day.

FORENSIC/ANALYTICAL USE OF AUGER X-RAYS

One of the most interesting recent forensic developments is alloy analysis applied to bullets and metal shell casings. It has been found that with the sensitivity of modern instrumentation it is possible to analyze the alloys in forensic samples to match bullets to a particular box of cartridges due to slight variations in the alloy composition. A variety of techniques are available for alloy analysis such as various forms of optical emission spectra and mass spectroscopy but to continue our survey of applications of the Bohr equation, we want to discuss XRF [5]. The overall Auger process of aiming a beam of high energy electrons to knock inner electrons out of atoms produces scattered "Auger electrons" as well as x-ray emission due to outer electrons "falling" into inner orbits. We have shown above that certain metal targets (Cu) can be selected to generate a preponderance of K_α x-rays for use in medical imaging although the Bohr model is too simple and results in an error of about 6.2%. That implies that various elements have characteristic K_α and L_α wavelengths that can be used for elemental analysis within our understanding that the model does not include the effect of the repulsion between electrons. Our schematic in Figure 9.5 is oversimplified but shows some features that should be part of the science education of undergraduates. The diagram shows how a beam of electrons can be accelerated to high velocity to form what is called an "electron microscope" to form an image on film or on a fluorescent screen by transmission of the electron beam accelerated through a potential of 75,000 V or even higher. An electron beam can be focused using electrostatic lenses (not shown on the schematic) and the development of electron microscopes has reached a high level of sophistication to visualize bacteria and viruses in biological samples.

The abbreviation SEM indicates a "scanning electron microscope" that can use an x–y scan to form a pixel image due to changes in transmitted intensity of the beam. However, in the last few years it has been realized that the same device can be used with a lower voltage of the electron beam of typically 20,000 V. The lower voltage is still sufficient to cause K_α x-rays for elements up to $Z = 48$ (Cd) but the same LiF diffraction monochrometer and detector can then be used to measure L_α and L_β XRF (x-ray fluorescence) since for elements with $Z > 48$, the L shell is much deeper in energy. We can use the Bohr formula to estimate the L_α wavelength also:

$$\Delta E_{L_\alpha} = E_2 - E_3 = (Z^2)\left[\frac{1}{2^2} - \frac{1}{3^2}\right](13.6057\,\text{eV}) = Z^2\left[\frac{5}{36}\right](13.6057\,\text{eV}) = Z^2(1.88968)\,\text{eV}$$

We can use the experimental wavelengths of the L_α transitions to probe the interior of the heavier elements. We saw above that we could use the integer Z value of the nucleus for the K_α wavelengths of the $(n = 2) \rightarrow (n = 1)$ transition of the Bohr atom and obtain reasonable results.

However, for the case of the effective nuclear charge Z_{eff} for the excited outer electron in the sodium flame test, we found a noninteger value due to incomplete screening of the nuclear charge by the shell of electrons between the $(n = 3, n = 4)$ shells of the atom and what we can call the Neon-core of the inner part of the Na atom. In the case of the L_α energy, the transition is from the $(n = 3) \rightarrow (n = 2)$ level, so the K shell is between the L shell and the bare nucleus for sure. Further, the L shell may still be mostly occupied with perhaps one vacancy so the Ne core may be mostly intact.

For elements heavier than Cd, we expect that the Bohr radius of the $(n = 1)$ is quite small so that the screening (covering up) of the nuclear charge by the K shell should be close to 2 for the two electrons in the K shell. However, the next eight electrons in the L shell may still be mostly there with perhaps only one or two vacancies. Thus, the effective nuclear charge for the heavier elements should be close to $Z_{\text{eff}} = Z - 8$. Consider In where $Z = 49$. From Table 9.1, we see that the wavelength of the L_α transition is 0.3772 nm = 3.772 Å. From the formula above for the L_α transition energy, we can solve for the value of Z_{eff} for the charge that is seen by the electron falling into the L shell as

$$\Delta E_{L_\alpha} = E_2 - E_3 = \frac{12{,}398\,\text{eV\,Å}}{3.772\,\text{Å}} = Z_{\text{eff}}^2(1.88968)\,\text{eV}.$$

Thus, for In ($Z = 49$), we find

$$Z_{\text{eff}} \cong \sqrt{\frac{12{,}398\,\text{eV\,Å}}{(3.772\,\text{Å})(1.88968\,\text{eV})}} = 41.7057.$$

TABLE 9.2
Nuclear Screening of the K Shell and a Partial
L Shell in Selected Elements

Element	Z	L_α (eV)	λ_{L_α} (Å)	Z_{eff}	KL_{screen}
Sn	50	3.444	3.600	42.6904	7.3096
Cs	55	4.286	2.892	47.6302	7.3698
Nd	60	5.230	2.370	52.6147	7.3853
Tb	65	6.2728	1.977	57.6074	7.3926
Yb	70	7.4140	1.672	62.6417	7.3538
Re	75	8.6150	1.433	67.6641	7.3359
Hg	80	9.987	1.241	72.7103	7.2897
Rn	86	11.724	1.057	78.7851	7.3149
Th	90	12.966	0.956	82.8424	7.1576
Pu	94	14.279	0.868	86.9404	7.0560

This can be interpreted as the effective charge out of the bare nuclear charge of 49 that is not covered by the K shell and what there is of the L shell. Thus, $(49 - 41.7057) = 7.2943$ so the clean formula of the one-electron Bohr atom is now muddled up with internal electron–electron repulsion and the idea that as the shell radius gets large the electrons get spread out more and cannot completely cover up the nuclear charge on a 1:1 basis. The same effect is almost constant across the periodic chart for the L_α transitions from atomic number 49 (In) to the data for Pu ($Z = 94$) as seen for selected elements in Table 9.2. This result also indicates that the electron giving off the x-ray energy is not falling into an empty L shell and that may be a function of how hard the atom was hit with the incoming electron, because this effective charge number indicates there are definitely some other electrons in the L shell. Assuming the number of electrons is an integer, the noninteger effective charge means that as the electrons move they cannot be everywhere at once. Thus, even if the electrons are very fast, both the K shell and what there is of the L shell cannot completely cover an integer amount of the nuclear charge. These considerations are useful to increase our appreciation of what is going on inside an atom. Historically, this is probably as far as one can push the Bohr model without including electron–electron repulsion and a better description of the orbitals but it is interesting to see that the K- and L-Auger transitions behave almost like the one-electron Bohr model when the model is adjusted for an effective nuclear charge.

Although our short list implies that the average value of the screening might be 7.3, previous work by Moseley soon after 1913 preferred a value of 7.4 [6]. Mosely was a brilliant British chemist who was tragically killed in action in WWI at the Battle of Gallipoli at the age of 27. Some writers have said that Bohr's shell model was not believed until the work of Mosely; note concurrent dates of discovery.

X-RAY FLUORESCENCE

While Auger processes do lead to x-ray emission, there is a variation in the technique that offers more sensitivity and generality. Overall, x-ray emission techniques are not as sensitive as some other analytical methods but offer simple sample preparation and simultaneous imaging of very small samples and *elemental analysis*. An application combining these advantages is the ability to check the elemental composition of doped microelectronic devices. Forensic samples such as shot or bullets offer an abundant sample size, so the element ratios within a strong signal is the desired information. These examples can use the Auger emission of the SEM beam or reduce the voltage to measure L_α transitions. With the standard SEM method, scattered Auger electrons require a grounding connection to the sample in some cases to bleed off the secondary scattered electrons but the x-rays tend to be generated in a region of space near the sample stage. Some x-rays can be collimated using a slit or hole that faces a crystal-lattice grid such as a plate of LiF to cause diffraction of the x-rays into a sort of "x-ray rainbow." The dispersed spectrum can then be detected using photographic film blackening or with modern electronic detectors. One possible detector is a small block of pure Si with a core region of Li immersed in liquid N_2 and under a voltage potential. When an x-ray hits the Li in the detector, a cascade of ionized electrons results in a "pulse-count." The Li core is the active electrode and the Si provides a nonconducting shell around the Li. The low temperature is to provide a low "dark current thermal ionization" until an energetic x-ray hits the detector and causes a big signal due to ionized electrons. The detector can be moved along a wavelength track or a number of the detectors can be used in fixed positions corresponding to different wavelengths. Students have asked why the device used for the spectrum in Figure 9.9 required liquid N_2. The answer is that the detector needs to be at a low temperature ($77°K$) to reduce spurious thermal signals. A very good description of x-ray detection is given in Ref. [7].

X-RAY DIFFRACTION

At this point, we need to explain an important aspect of almost any form of spectroscopy, which is the need to disperse a spectral rainbow into individual components. For optical spectra in the visible or even the infrared range, one can employ a wedge (prism) of a transparent material such as glass, quartz, or even potassium bromide. Different refractive indices for different colors will fan out (disperse) the rainbow of a beam of light. There is another method of dispersing light. Light can be "diffracted" from a grid of grooved lines on a reflecting surface (a grating) so that light is reflected differently from the crests and troughs of the grooves. Then the electric fields of the light cancel out except at certain angles of reflection based on the basic law of diffraction that we illustrate in Figure 9.6. This phenomena can also be used to diffract light from regular crystal lattice features if we have a sufficiently robust crystal to withstand the energy of x-rays. Here that implies an ionic substance with strong electrostatic lattice forces. Thus, the lattice of LiF can act as a diffraction device. Figure 9.6 shows that the electric field waves will tend to cancel out from adjacent scattering sites except at a certain angle defined by the Bragg scattering angle discovered in 1913 by William Lawrence Bragg (son, 1890–1971) and William Henry Bragg (father, 1862–1942), a father–son team who shared the Nobel Prize in Physics for this work in 1915.

$$n\lambda = 2d \sin(\theta); \quad n = 1, 2, 3, \ldots$$

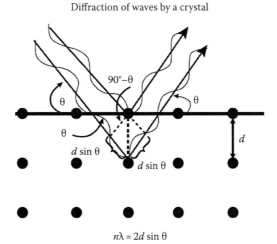

$$n\lambda = 2d \sin \theta$$

FIGURE 9.6 Bragg diffraction mechanism of an electromagnetic wave by a crystal lattice.

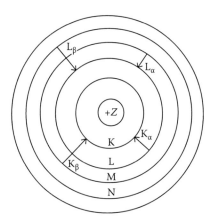

FIGURE 9.7   Bohr orbits K, L, M, and N showing x-ray Auger/fluorescence transitions.

FIGURE 9.8   Ron Jenkins (1932–2002) was an English scientist who lived in the United States. He wrote four books on the analytical use of x-ray spectroscopy and published about 230 scientific papers and 11 book chapters. He taught these topics at the Denver x-ray conference for some 30 years. He was the first recipient of the Ron Jenkins Award given every two years by the organizers of the Denver x-ray conference and was executive director of the International Center for Diffraction Data. He founded the *X-ray Spectrometry—An International Journal* and he is the only person to have received both the Birks Award for x-ray spectrometry and the Barrett Award for x-ray diffraction. (By permission from the International Center for Diffraction Data.)

The main reinforcement of the waves will occur for the $n = 1$ angle, but the intensity of the second order ($n = 2$) angle can be large and often one must use filters or even a tandem second monochromator to purify the first-order wavelength. In the case of x-rays, one can use a lattice of purified LiF to disperse a source of x-rays [7]. In the highly oversimplified schematic of Figure 9.5, we see a naive dispersal of the K_α bands of x-rays from various elements. Actual electron microscopes with XRF attachments are more complicated than Figure 9.5, but students need to know about the basic principles based on the schematic in Figure 9.7 and greatly developed by Dr. Ron Jenkins (Figure 9.8) and others. For instance, the schematic shows the principle of using the ratio of windings in a transformer to create a higher voltage because you need alternating current to increase voltage with a transformer. We also show the use of a simple diode to allow a half-wave rectification of the high voltage alternating current to create a pulsed direct current beam. That is only for an educational schematic, in an actual device there would be full-wave rectification with capacitors in parallel to damp ripples in the direct current to a high degree so as to allow a continuous electron beam to approach the optical properties of a microscope using visible light. The idea of an electron microscope is to use a beam of electrons in place of a light beam. Thus, an electron microscope uses very sophisticated electrostatic lenses to simulate optical lenses so the diagram is very crude but emphasizes the Auger process and the dispersal of K_α or L_α x-rays into a range of characteristic bands for the various elements.

Let us develop a formula from the Bohr model, which shows the way to calculate the K_α for each element. We know the K_α transition is from $n = 2$ to $n = 1$, so we will have the same quantum numbers for the K_α wavelength of all elements (in this simple model), only Z changes.

$$\Delta E = (Z^2)\left(\frac{1}{1^2} - \frac{1}{2^2}\right)(13.6057\,\text{eV}) = \frac{12398}{\lambda_{K_\alpha}} = Z^2\left(\frac{3}{4}\right)(13.6057)$$

so we rearrange for λ_{K_α} as

$$\lambda_{K_\alpha}\,(\text{Å}) = \left(\frac{12,398}{13.6057}\right)\left(\frac{4}{3Z^2}\right) = \frac{1214.981}{Z^2}$$

and we see the wavelength is inversely proportional to the square of the atomic number of the element. Thus, if we want to calculate the wavelength of the K_α wavelength for cadmium we would find for Cd, $Z = 48$, so that the wavelength would be estimated as 0.527335 Å for the x-ray. However, we saw the Bohr model led to a 6.2% error for Cu and Cd has a larger atomic number. The experimental value for the K_α fluorescence wavelength of Cd is 0.5357 Å with only a 1.6% error, so we see our simple Bohr model is good for the heavier elements. In order to eliminate this error it would be necessary to run experimental Auger fluorescence on known samples and create a correction table or graph. If we want to compute just the K_α voltage we can use the expression reduced to a constant times Z^2

$$\Delta E = (Z^2)\left(\frac{1}{1^2} - \frac{1}{2^2}\right)(13.6057\,\text{eV}) = (Z^2)(10.204275\,\text{eV}).$$

In the opinion of this author, it is important for students to learn the basic principles of instrumentation but we need an application for motivation so an actual spectrum is shown in Figure 9.9 as output obtained for a sample of lead shot. It is evident that the lead is in abundance but is not pure and several other elements are present. Perhaps the most interesting feature of the scan is that there are really two peaks for lead, M_α (2.343 keV, 5.076 Å) and M_β (2.442 keV, 5.299 Å) that are the energies emitted when the first and second electrons fall down into the partially empty $n = 3$ orbital. The less intense peak at higher energy is the M_β transition. Note that the entire spectrum falls with a range of (0–10 keV), so the K and L transitions are not reported but M transitions are within the parameters of the instrument. The detection of small peaks for other elements is treated with the aid of a software program that accompanies the instrument. For forensic applications, the relative amounts of the impurities could vary from sample to sample and might be matched uniquely to some unfired shot in a box of cartridges/shells obtained from a suspect in a crime. Thus, this sample XRF spectrum illustrates $M_{\alpha,\beta}$ transitions for a heavy metal of interest to forensic problems along with illustration of elemental impurities in the lead.

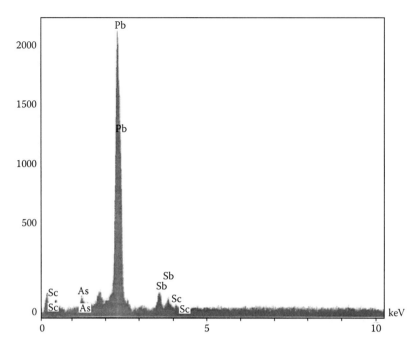

FIGURE 9.9 XRF of nominal lead shot measured at Virginia Commonwealth University using a Kevex Quantex ISI-130 SEM-EDX electron microscope with XRF attachment and a 20 keV excitation beam. The *y*-axis is in counts because the intensity was measured with an internal pulse counter and the *x*-axis is given directly in kiloelectron volt as reported with the software associated with the instrument. The spectrum was run by James Spivey at VCU. The XRF spectrum of the lead shot reveals it is mainly lead but contains other metals as well. Note the presence of poisonous As and Pb. Thanks are due to Rhonda Stroud of the Naval Research Laboratory for interpretation of the spectrum and the assignment of the peak between As and Pb at 1.740 keV as due to a K line from Si.

ELECTRONIC ABSORPTION SPECTROSCOPY/SPECTROPHOTOMETRY

One of the most useful applications of spectroscopy in forensic science and in many biochemical analyses is colorimetric spectrophotometry. We have exhausted the application of the simple Bohr equation but if this is the end of your one semester in physical chemistry, we need to mention the spectroscopy of electronic excitations of molecules. Quite a few relatively simple analyses depend on the absorbance of a "color" at a specific wavelength in the range of 700–200 nm. This range is called the ultraviolet–visible range (UV–Vis). Most UV–Vis spectrometers are not able to record spectra below 210 nm due to solvent absorption and/or oxygen absorption at 180 nm. Gas samples in special gas cells can be studied by flowing N_2 gas though the spectrometer to displace O_2 and reach 175 nm or by evacuating the entire instrument to reach 130 nm. The cuvette sample container will also limit the wavelength range. Glass or plastic cuvettes are opaque below 380 nm while more expensive quartz glass cuvettes will transmit down to 220 nm. Below 200 nm, special CaF_2 lenses are necessary, so in practical reality the spectral range effectively stops at about 380 nm for single wavelength spectrophotometers or 220 nm using quartz cuvettes in an expensive scanning spectrometer.

There are also reagents with colors at longer visible wavelengths as well as useful chemical properties. Aqueous $KMnO_4$ solutions transmit an intense purple color (due to absorbance of yellow) and the compound is a strong oxidizing agent. Thus, $KMnO_4$ is used for a variety of spectrophotometric analyses. We introduce it here because of the general use in various analyses but especially because it illustrates the concept of *electronic absorbance*. There are also characteristic groups in organic compounds that tend to absorb light in specific color ranges; they are called *chromophores*. In organic compounds, chromophores tend to have double bonds with delocalized electrons. Some inorganic compounds also have chromophore qualities, often in the visible range of the spectrum. The Bohr model would say that the $KMnO_4$ transitions are within the $n = 3$ shell of Mn but we are beginning to see the limitations of the Bohr model here. By following the historical development of spectroscopy and using only the simple mathematics of the Bohr model, we cannot invoke a modern explanation of electronic spectroscopy for molecules. Here, we want to mention this important technique that is simple in use and has wide application in this survey of spectroscopy.

Spectrophotometry is usually carried out at a single, optimum, wavelength that is established during calibration with solutions of known concentrations. The procedure is to use a simple 1 cm path glass cuvette of solution to measure an unknown concentration. $KMnO_4$ is a strong oxidizing agent, so a great number of schemes can be used to measure its concentration following some test reaction. An Internet search on the use of $KMnO_4$ in various tests/assays should result in finding many applications of this very useful reagent.

In order to define the type of absorbance we are discussing, we need to define the Beer–Lambert law:

$$T = \frac{I}{I_0} = e^{-\varepsilon_\lambda [C](l)} = \exp\{-\varepsilon_\lambda [C \, mol/L] \, [\text{path length (cm)}]\}; \quad A \equiv \varepsilon_\lambda [C](l).$$

The IUPAC recommended term for $\varepsilon(\lambda)$ is the "molar absorption coefficient" and a student should note the spelling with a "p"; A is "absorbance" spelled with a "b." "T" is the ratio of light "*transmitted* through the sample compared to a blank, not the kinetic energy T" and the absorbance "A" is the logarithm (specify the base) of T as $A = -\ln T = \varepsilon_\lambda [C] l$. Here, "$[C]$" is the concentration in moles/liter, "l" is the sample cell path length in cm, and ε_λ is the "molar absorption coefficient" at a given wavelength λ. The intensity "I" is the amount of light transmitted through cell, solvent, and sample while "I_0" is the "blank" signal of light intensity measured for only the cell and solvent. This is a simple formula that is exploited in a large number of forensic/analytical "color tests," often at a single wavelength selected for maximum sensitivity.

Example

We can use the absorbance scale on the spectrum marked "curve 1" in Figure 9.10 to deduce the concentration of the $KMnO_4$ solution. We estimate the 525 nm peak is at $A = 0.85$, and we assume a 1 cm path so we can solve for $[C]$ as

$$A = \varepsilon_{525}[C](l) = \left(\frac{2455}{M \, cm}\right)[C](1 \, cm) \cong 0.85.$$

and then we find

$$[C] \cong \frac{0.85}{\left(\dfrac{2455}{M \, cm}\right)(1 \, cm)} \cong 3.46 \times 10^{-4} \, M.$$

In the simple example above, we note that the molar absorption coefficient for $KMnO_4$ [8] is quite large at 525 nm and that the Beer–Lambert law can be used for very dilute solutions.

Low-cost spectrophotometers are available for single wavelength measurements but a scanning spectrometer is a more substantial piece of instrumentation that can produce a spectral curve. The scanning option is a decided luxury compared to hand-plotting data for single-point readings.

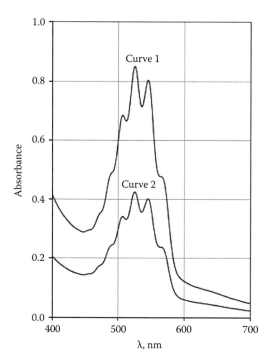

FIGURE 9.10 The visible spectrum of aqueous $KMnO_4$. The wavelength usually chosen for maximum sensitivity and stability is 525 nm. The molar absorption coefficient of aqueous $KMnO_4$ is $\varepsilon_{525} = 2455/M\,cm$ [8]. Note this is an absorbance of green-yellow-orange light and the transmitted light is the remaining red + blue = purple color of whatever white light source is behind the sample. The sample absorbs light but the human eye "sees" transmitted or reflected light. (Spectrum contributed by James Holler of the University of Kentucky.)

Most spectrometers offer the option of plotting the spectrum in terms of absorbance "A," "$\%T$", or "$\varepsilon(\lambda)$." Thus a spectral scan can be recorded with a constant concentration and path length but the $\varepsilon(\lambda)$ will vary with wavelength and a scanned spectrum can be obtained as $\dfrac{A(\lambda)}{[C](l)} = \varepsilon(\lambda)$. UV–Vis spectra can then be related to the energies at which electronic transitions absorb energy in terms of $\varepsilon(\lambda)$. For $KMnO_4$, the 525 nm absorbance corresponds to

$$\Delta E\,(eV) = \frac{12{,}398}{5250\,\text{Å}} \cong 2.36\,eV.$$

INTERPRETING ELECTRONIC SPECTRA

It should be clear that the Bohr model cannot easily explain the electronic spectrum of the permanganate ion, $(MnO_4)^-$. We had to provide a foundation in thermodynamics and kinetics in the early chapters. Much of that material is what this author would call "old but essential" science from before 1913. However, a huge modern area of science is still relatively untouched here so far in the form of molecular quantum mechanics (Chapters 11 through 14). Consider the spectrum of $KMnO_4$. Historically, Bohr orbitals were followed by Erwin Schrödinger's improved solution of the H atom orbitals in 1926 and that was extended to specifically include the effect of electron spins by Paul Dirac in 1928. With these orbitals, a team under D. R. Hartree set up a process called the "self-consistent field" (SCF) method in the 1930s. The Hartree SCF method was improved by V. A. Fock and J. C. Slater in 1930. An oversimplified description of the SCF method is to use 3D-orbital functions ("probability clouds" instead of Bohr rings) to describe electron positions, calculate how the interactions between the electrons change the weighting of the orbital functions to *lower the energy*, then recalculate the electron interactions. This was done iteratively until convergence was achieved. From thermodynamics, we know "energy goes down-hill" while "entropy tends to increase." For the SCF process, there is a powerful theorem shown in Chapter 16 called the Variation Theorem that proves that lower energy approaches the true energy from above as long as the numerical calculations are correct; *even for a function which is a guess*. All of this activity in the 1930s was applied to one-center cases of atoms. Molecules larger than H_2 were beyond treatment. In 1951, a breakthrough by C. C. J. Roothaan (Chapter 17) was made when a procedure was derived as the linear combination of atomic orbitals to treat molecules. Even so, the applications were limited to molecules of light elements, mostly organic compounds.

Rapid improvement in computers played a large role in the development of "quantum chemistry" in the 1960s but still there was a computational barrier against treating metal complexes including all the electrons. However, the early computer programs for the atomic Hartree SCF method were refined and very fast for one-center atomic problems. Thus, the emerging field of quantum chemistry was ripe for an explosive development by K. H. Johnson [9] and his colleagues in 1972. He used the one-center computer programs in what was called the multiple-scattering X-alpha (MS-$X\alpha$) method with a "muffin-tin" treatment of molecules. Each atom was treated as a spherical system set in an outer spherical field where electrons could roam and "exchange" with each other as well as be scattered in the space between the atoms. Since electrons are indistinguishable, accurate calculations need a way to express the possibility of the interchange (exchange) of electrons. A major innovation was to use an approximation invented/derived by J. C. Slater called the "$X\alpha$" exchange formula, an empirical way to calculate the exchange energy that lowers the calculated energy.

$$V_{X\alpha}(\vec{r}) \cong -6\alpha \left[\left(\frac{3}{8\pi} \right) \rho(\vec{r}) \right]^{\frac{1}{3}},$$

where $\rho(\vec{r}) = \psi * \psi(\vec{r})$. With that approximation to allow indistinguishable electrons to exchange positions, the rest of the calculation could use Coulomb's law for electron–electron repulsion, $V = +\dfrac{e_{q1}e_{q2}}{r_{12}}$, and electron attraction to nuclei, $V = -\dfrac{(Ze_q)e_q}{r_{12}}$. It was found necessary to represent the kinetic energy of the electrons as proportional to the second derivative of the electron cloud wave functions but this was all worked out from Schrödinger's 1926 derivation. The Coulomb interactions could be evaluated using numerical integration over a grid of points in a way similar to the trapezoid rule for 1D integration and so we see Johnson's scheme in Figure 9.11. The state-of-the-art at that time was that the method was well established as long as the problem was for only one atom at the center of the coordinate system. Johnson's innovation was to merge these spherical atoms into molecules. Although you will have to read later chapters to fully appreciate the Johnson "muffin-tin potential" and J. C. Slater's contributions to quantum chemistry, we show you the results here to offer a more modern treatment of electrons than the flat Bohr orbitals and to try to explain the electronic spectrum of $KMnO_4$.

In 1972, Johnson used Schrödinger's orbital notation first derived in 1926 and that is probably familiar to you from your freshman chemistry text. In Figure 9.12, we see the energy level scheme for MnO_4^- with the Mn valence shell orbital levels on the left and orbital levels for O on the right. In the middle, we see the calculated levels for the complete MnO_4^- tetrahedral complex (valence shell). Here, we see the use of symmetry labels for a T_d point group, the condensed spatial description of a tetrahedral object. Point group theory is discussed in Chapter 18. This author was at a conference where these results were presented (The Quantum Chemistry Symposium No. 6, 1972, held at Sanibel Island, Florida) and in spite of many experts in attendance, there was a hush over the audience when Figures 9.13 and 9.14 were shown of the contours of electron orbital clouds in the O–Mn–O bonds. At the time, this was truly amazing.

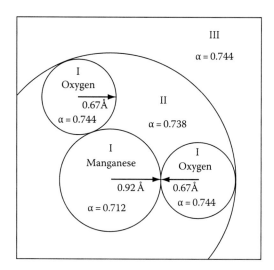

FIGURE 9.11 The Xα "muffin-tin" potential for the tetrahedral MnO_4^- ion showing the values of α used in each (spherical) cup of the muffin-tin. Within each spherical region a numerical Hartree one-center SCF calculation was carried out with Xα exchange and the entire complex was contained in an outer region of exchange. The wave functions were matched at each boundary to maintain single-valued, finite, and continuous functions. Matching the boundaries used mathematics developed for particle-scattering, hence the name "multiple scattering." Although this method certainly depends on using a computer, it is simpler than methods used today (2010) but more sophisticated than the Bohr theory. (With permission from Johnson, K.H. and Smith, F.S., *Phys. Rev. B*, 5, 831, 1972. Copyright 1972 by the American Physical Society.)

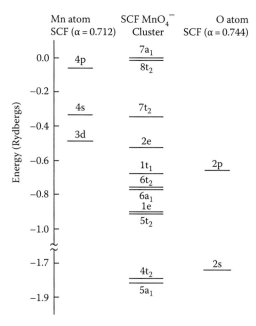

FIGURE 9.12 The energy level scheme showing the valence levels of atomic Mn on the left, atomic O on the right and the results of the Xα calculation for the MnO_4^- complex in the middle. The HOMO is the $1t_1$ and the LUMO is the 2e level using the symmetry labels appropriate to the tetrahedral (T_d) point group. Note the use of Rydberg energy units championed by the MIT group under J. C. Slater. Today we would use 1 rydberg = 0.5 hartrees. (With permission from Johnson, K.H. and Smith, F.S., *Phys. Rev. B*, 5, 831, 1972. Copyright 1972 by the American Physical Society.)

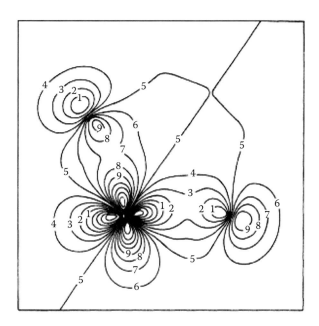

FIGURE 9.13 Contour map of the normalized $5t_2$ "ψ orbital cloud" of the O–Mn–O σ-bonds between the Mn atom and two of the O atoms. The values of the contour lines are $\psi = -0.2$ for line no. 1 and $+0.2$ for line no. 9 with increments of 0.05. That means that line no. 5 is a nodal plane where the electron wave function ψ is 0. This is a plot of ψ, which can be positive or negative, but the electron density $\rho = \psi * \psi$. The "bonds" are the buildup of electron probability between Mn and O. (With permission from Johnson, K.H. and Smith, F.S., *Phys. Rev. B*, 5, 831, 1972. Copyright 1972 by the American Physical Society.)

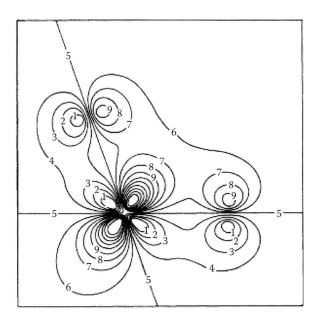

FIGURE 9.14 Contour map of the normalized 1e π-bonding "ψ orbital cloud" for the same O–Mn–O plane. The contour lines have the same values between -0.2 and 0.2 as in Figure 9.13. Here we see the 2p π orbitals extending out from the sides of the O atoms overlapping with a hybrid orbital on the Mn atom. There is a sign change in ψ going across contour 5 but the electron density will all be positive as $\psi * \psi = \rho$. (With permission from Johnson, K.H. and Smith, F.S., *Phys. Rev. B*, 5, 831, 1972. Copyright 1972 by the American Physical Society.)

One of the main changes in interpretation of electron orbits going from the Bohr rings to Schrödinger "ψ clouds" is that the Schrödinger interpretation is more delocalized. A second key difference is that the wave function has to be squared to obtain the actual probability as $\psi * \psi = \rho$. Rather than picturing electrons circling nuclei as in the Bohr model, the more recent interpretation is that the electrons are very light and moving very fast so all we can say about their position is that they are described by a function that is the square root of the probability, a wave function "cloud" rather than a ring. In Figures 9.13 and 9.14, we introduce pictures drawn of such wave functions in the form of contour maps. Contour maps are 2D representations with a third dimension represented as a closed line to show the outline of a "slice." Contour maps are often used to show height in land maps. It was not until the late 1960s and 1970s that computers were capable of easily calculating such functions and making the drawings. The simple conclusion from Figures 9.13 and 9.14 is that MnO_4^- is a tetrahedral anion complex and each of the four O atoms is connected to the Mn atom by two bonds ($\sigma + \pi$) for a total of eight bonds. Developments in chemistry and physics between 1913 and 1972 and on to 2010 are outlined in later chapters. Please note that while the derivations in the later chapters are difficult, *the emphasis is on learning the conclusions from those derivations,* so just "read" the derivations and pay attention to the results and it will not be difficult at all (note the experimental facts in Table 9.3) if we learn to look for the conclusions, the "take home message" of tedious calculations.

We have jumped ahead rapidly in the historical sequence of this chapter, because the Bohr atom model cannot provide an explanation of the $KMnO_4$ spectrum and to encourage you to take the second semester of physical chemistry to learn about more modern developments. But now what does this mean relative to the purple color of $KMnO_4$? One of the limitations of this MS-Xα method was that the only way you can get reasonable values for the transition energies between the calculated orbitals is to use what became known as the "half-electron method." In that method, it was realized that if you excite an electron from the ground state to an upper orbital, there should be some reorganization of the ground state orbital clouds. Thus, 0.5 of an electron was placed into the occupied orbital and 0.5 of an electron was placed in the higher orbital to represent a "transition" for an electronic excitation. At the time, this was accepted because it was an easy calculation and gives reasonable results, but it was clearly the weakest part of the theoretical treatment. Today, there are much more sophisticated ways to find the transition energies that account for the reorganization of the other electrons when one is excited to a higher level. But here is the payoff. Using the calculated HOMO $\rightarrow$ LUMO energy as the $1t_1 \rightarrow 2e$ transition we find that

$$\lambda_{HOMO \rightarrow LUMO} = \frac{12,398}{2.3 \, eV} = 5390 \, \text{Å} = 539 \, nm.$$

Look back at Figure 9.10. While 525 nm is usually chosen as the most stable wavelength for spectrophotometry, it is clear that overall, the peak of the absorbance spectrum is very close to 539 nm. In fact, 539 nm is probably closer to the true transition energy of $\Delta E(1t_1 \rightarrow 2e)$ than 525 nm since

TABLE 9.3
Electronic Transition Energies for MnO_4^-

Transition	Unrelaxed SCF	Half-Electron	Experiment
$1t_1 \rightarrow 2e$	2.1	2.3	2.3
$6t_2 \rightarrow 2e$	3.2	3.3	3.5
$1t_1 \rightarrow 7t_2$	4.5	4.7	4.0
$5t_2 \rightarrow 2e$	5.3	5.3	5.5

Source: Johnson, K.H. and Smith, F.C., *Phys. Rev. B,* 5, 831, 1972.

the lowest energy part of a broad electronic transition starts at the red edge. At the time this calculation was presented, it was sensational and it eventually led many researchers to continue to improve the Xα idea to what is now modern density functional theory (DFT). This type of calculation, although an approximation, encouraged further development of the idea that the electron–electron exchange can be parameterized with more refined functions. Today, the majority of modern quantum chemistry invokes some form of a parametric calculation for electron–electron exchange within a DFT program, although there are competing models that try to avoid any parameters. In Chapter 17, we discuss the fact that even the best DFT exchange potentials still use some form of analytical "Hartree–Fock exchange." The problem of how to represent electron–electron exchange between indistinguishable particles is not fully solved as of 2010 and research continues to search for exact formulas.

GENERAL PRINCIPLES OF SPECTROSCOPY

We hope a few points were made regarding spectroscopy in this brief treatment. This has been a "broad brush" view of spectroscopy using mostly the simple mathematics of the Bohr H atom, so that we can consider several types of spectroscopy in what may be the only opportunity if this is a one-semester treatment. Spectroscopic topics in the latter chapters of this text will be more precise but require a foundation of more detailed mathematics.

1. Spectroscopy involves measurement of electromagnetic radiation as absorbed or emitted from atoms and molecules between definite energy levels and can be characterized by wavelength or frequency. Light energy is proportional to frequency, $\varepsilon = h\nu$, and angular momentum is quantized as well ($mvr = n\hbar$).
2. The Bohr model of the H atom was derived to obtain two key formulas:

$$E(n, Z) = -\left(\frac{Z}{n}\right)^2 (13.6057\,\text{eV})$$

and

$$r(n, Z) = \left(\frac{n^2}{Z}\right)(0.5291772\,\text{Å})$$

3. The electron volt unit of energy and the wave number quantity were encountered in this brief introduction to spectroscopy. A survey of units used in spectroscopy was presented as a preparation for further studies. The valuable formula $\Delta E\,(\text{eV}) = \dfrac{12{,}398}{\lambda\,(\text{Å})}$ was derived.
4. We noted some details of basic electronic circuits and diffraction of electromagnetic radiation. LiF was mentioned as a crystal for use as an x-ray diffraction device.
5. Examples were shown here based mainly on applications of the Bohr model of the atom for the H-atom spectrum and XRF. The Bohr model only applies to systems with a single electron orbiting a single positive ion. The model can be used for systems with more than one electron to treat an outer electron shell as orbiting a spherical inner ion with a noninteger effective nuclear charge. It was necessary to define the molar absorption coefficient $\varepsilon(\lambda)$ to understand molecular electronic spectra and the Beer–Lambert law was illustrated for aqueous $KMnO_4$.
6. Concepts like excitation, emission, quantum numbers, and electron volts have all been introduced in a framework of simple algebra, so in later chapters we can focus on the quantum phenomena with familiarity of concepts and units already in hand.

7. The example of the very useful spectrophotometric reagent $KMnO_4$ is analyzed using ideas that will be presented in the later chapters of this text to illustrate the concept of electron orbitals as "clouds" instead of Bohr rings and an illustration is given for an excellent early calculation of the electronic absorbance of $KMnO_4$.

PROBLEMS

9.1 Given the ionization potential of the Li atom is 5.391719 eV, compute the effective nuclear charge (Z_{eff}) experienced by the outer 2s electron using the Bohr model. Use the value of Z_{eff} to calculate the Bohr radius of the 2s orbital.

9.2 Given the ionization potential of the Na atom is 5.139076 eV, compute the effective nuclear charge (Z_{eff}) experienced by the outer 3s electron using the Bohr model. Use the value of Z_{eff} to calculate the Bohr radius of the 3s orbital.

9.3 Given the ionization potential of the K atom is 4.3406633 eV, compute the effective nuclear charge (Z_{eff}) experienced by the outer 4s electron using the Bohr model. Use the value of Z_{eff} to calculate the Bohr radius of the 4s orbital.

9.4 Given the ionization potential of the Rb atom is 4.177128 eV, compute the effective nuclear charge (Z_{eff}) experienced by the outer 5s electron using the Bohr model. Use the value of Z_{eff} to calculate the Bohr radius of the 5s orbital.

9.5 Compute the Bohr radius and the energy of the H atom for $n = 1, 2, 3, 4$.

9.6 Calculate the energy and radius of the $n = 1$ orbit for H, He^+, Li^{2+}, Fe^{25+}, and U^{91+}.

9.7 Calculate the energy of the $(n = 1)$ to $(n = 2)$ transition of the H atom in electron volts, wave numbers, wavelength, and frequency.

9.8 Make a small table of the K_α values from Table 9.1, the K_α(calc) value and the corrected value K_α^*(calc) along with the λ_{K_α} value from the experimental value and the value calculated from K_α^*(calc) energy for the following elements: B, Na, Cl, Fe, and Ag.

9.9 Estimate the absorbance (A) of $KMnO_4$ at 525 nm in curve 2 of Figure 9.10 to calculate the concentration of the solution.

9.10 Assume that LiF exhibits a first-order Bragg diffraction at an angle of 29.6° for 6.23 keV x-rays. Use the Bragg's law to compute the "d" value from this data. Keep in mind that although LiF has a cubic crystal structure, there are quite a few possible distances that line up in the crystal when viewed from various angles but what is the "d" for this Bragg reflection?

BIBLIOGRAPHY

Beckhoff, B., B. Kanngießer, N. Langhoff, R. Wedell, and H. Wolff, *Handbook of Practical X-Ray Fluorescence Analysis*, Springer, Dorcdrecht, the Netherlands, 2006.

Bertin, E. P., *Principles and Practice of X-Ray Spectrometric Analysis*, Kluwer Academic/Plenum Publishers.

Buhrke, V. E., R. Jenkins, and D. K. Smith, *A Practical Guide for the Preparation of Specimens for XRF and XRD Analysis*, Wiley, New York, 1998.

Jenkins, R., *X-Ray Fluorescence Spectrometry*, 2nd Edn., Wiley, Chichester, 1999.

Jenkins, R. and J. L. De Vries, *Practical X-ray Spectrometry*, Springer-Verlag, New York, 1973.

Jenkins, R., R. W. Gould, and D. Gedcke, *Quantitative X-Ray Spectrometry*, Marcel Dekker, New York, 1981.

Van Grieken, R. E. and A. A. Markowicz, *Handbook of X-Ray Spectrometry*, 2nd Edn., Vol. 29, Marcel Dekker Inc, New York, 2002.

X-ray fluorescence, http://en.wikipedia.org/wiki/X-ray_fluorescence.

REFERENCES

1. Becker, B. J., Celestial spectroscopy: Making reality fit the myth, *Science*, **301**, 1332 (2003).
2. Balmer, J. J., Notiz über die Spectrallinien des Wasserstoffs, *Ann. der Physik und Chemie*, **25**, 80 (1885) [as translated and published by H. A. Boorse and L. Motz in *The World of the Atom*, Vol. 1, Basic Books, New York, 1966; augmented by C. Giunta].
3. Jackson, J. D., *Classical Electrodynamics*, 1st Edn., John Wiley & Sons Inc., New York, 1962, Table 2, page 618.
4. Wehr, M. R. and J. A. Richards, *Physics of the Atom*, Addison-Wesley Publishing Co., Inc, Reading, MA, 1960, page 73.
5. Thomsen, V., Spectrometers for elemental spectrochemical analysis, Part II: X-ray fluorescence spectrometers, *Spectroscopy*, **25**(1), 46 (2010).
6. Mosely, H. G. J., The high frequency spectra of the elements, *Phil. Mag.*, 1024 (1913).
7. Bryndol, S., B. Y. Danon, and R. Block, X-ray imaging with parametric X-rays (PXR) from a lithium fluoride (LiF) crystal, *Nuclear Instruments and Methods in Physics Research A*, **560**, 589 (2006).
8. Kalbus, G. E., V. T. Lieu, and L. H. Kalbus, A spectrophotometric study of the permanganate-oxalate reaction, *J. Chem. Ed.*, **81**, 100 (2004).
9. Johnson, K. H. and F. C. Smith, Chemical bonding of a molecular transition-metal ion in a crystalline environment, *Phys. Rev. B*, **5**, 831 (1972).

10 Early Experiments in Quantum Physics

INTRODUCTION

We tried to introduce the idea of quantization in Chapter 9 as a completion of a one semester course. However, we skipped over some really interesting events in the history of Science between 1900 and 1913 when Bohr derived the quantized energy of the H atom. First we want to carry out the 1901 Planck derivation of the formula for blackbody radiation [1]. Many texts just show the curve, write $\varepsilon = hv$, and move on. As a student, this author found that limited explanation very frustrating since energy quantization is a fundamental concept. Even among graduate texts in quantum mechanics, we are aware of only one that does the complete treatment which we will draw upon for the mathematics [2] but supplement with a narrative that we have found helpful to students over the years. Then in 1905, Albert Einstein (1879–1955), one of the most influential scientists of all time, gave an explanation of the photoelectric effect [3] for which he received the Nobel Prize in 1921. Even reducing our list to essential topics, we need to discuss the Davisson–Germer experiment [4]. The photoelectric effect introduces the idea that light waves can act as particles while the Davisson–Germer experiment showed that particles can act like waves and confirmed the De Broglie equation [5]. However, you can be assured that it will not be as difficult as you might have anticipated and if you can absorb the meaning of just these three key experiments you should be able to begin thinking in terms of quantum mechanics! Despite our slow historical development, this is 2010 and we have to get to the twenty-first century somehow!

STEFAN–BOLTZMANN LAW: RELATING HEAT AND LIGHT—PART I

We do not want you to forget the thermodynamics you learned in earlier chapters but historically there was a shift in science with the idea of energy quantization in 1900. There was awareness of the connection between heat and light before the late 1800s but one of the first quantitative treatments was by Boltzmann and his doctoral mentor Jozef Stefan (1835–1893), an Austrian physicist and Ludwig Boltzmann's PhD thesis advisor. Prior to the concept of quantization, the arguments were thermodynamic in nature. Recall the HUGA equations of thermodynamics and the form of the first law for a closed system as $dU = TdS - PdV$ and the equation for the Helmholtz free energy $dA = -Sdt - PdV$. Equating the second derivatives $\dfrac{\partial^2 A}{\partial T \partial V} = \dfrac{\partial^2 A}{\partial V \partial T} \Rightarrow -\left(\dfrac{\partial S}{\partial V}\right)_T = -\left(\dfrac{\partial P}{\partial T}\right)_V$ which will be useful here and leads to $\left(\dfrac{\partial U}{\partial V}\right)_T = T\left(\dfrac{\partial P}{\partial T}\right)_V - P$. Another fact needed is beyond the level of this text as a relationship that comes from Maxwell's electromagnetic theory in that there is a weak "electromagnetic pressure" $P = \dfrac{\rho}{3}$, often discussed in Astronomy relative to intense light from stars. This feeble pressure is predicted by Maxwell's equations for electromagnetic waves but its measurement is made difficult by thermal gas heating even in a partial vacuum. The common Crookes radiometer (http://www.strangeapparatus.com/Crooke_s_Radiometer.html) actually works via thermal heating of air around small paddle wheel vanes exposed to intense light in a partial vacuum. However, in 1933, Bell and Green [6] improved on earlier experiments and succeeded in measuring this small pressure to rotate some small vanes (glass plates) suspended by delicate quartz fibers in a vacuum of 10^{-6} mmHg. Chemistry students should be pleased that the HUGA equations we learned earlier are also important in astronomy. Suffice it to say there is such pressure as related to the density of the radiation $\rho(T)$, which is a function of the absolute temperature. Then we have $\rho(T) = \dfrac{U}{V}$ so that $U = \rho(T)V$ so the energy is the energy density per unit volume times the volume and $\left(\dfrac{\partial U}{\partial V}\right)_T = \rho$. Then, when we express the pressure in just one of three dimensions, $P = \dfrac{\rho}{3}$, we find $\left(\dfrac{\partial P}{\partial T}\right)_V = \dfrac{1}{3}\dfrac{d\rho}{dT}$. Now the energy expression becomes $\left(\dfrac{\partial U}{\partial V}\right)_T = T\left(\dfrac{\partial P}{\partial T}\right)_V - P$ so that $\Rightarrow \rho(T) = \dfrac{T}{3}\left(\dfrac{d\rho(T)}{dT}\right) - \dfrac{\rho(T)}{3}$. Rearranging we find $\dfrac{4\rho(T)}{T} = \dfrac{d\rho(T)}{dT}$ and further $\int \dfrac{4dT}{T} = \int \dfrac{d\rho(T)}{\rho(T)} \Rightarrow 4 \ln T + C = \ln \rho(T)$. This can be put in final form using experimental measurements of the constant to find that $\alpha = anti \ln(C) \cong 7.5657 \times 10^{-16}$ (J/m^3 K^4) and that

$$\rho(T) = \alpha T^4.$$

This expression is used by astronomers to estimate the temperature of stars, and later in this chapter we will ask if the Planck theory agrees with this macroscopic relationship.

We have shown the Stefan–Boltzmann Law to indicate a boundary between macroscopic thermodynamics and a revolutionary new way of thinking about energy in terms of quantization.

BLACKBODY RADIATION: RELATING HEAT AND LIGHT—PART II

We are now going to show the details of one of the most important advances in science in the early twentieth century and we are going to give all the details so you appreciate there is no philosophical "wiggle-room" around the inescapable conclusion of energy quantization. The derivation is in two parts. The first part has to do with counting how many waves can fit into a cubical volume and is tricky, but correct. The wave-counting part of the problem is just a necessary exercise to compare the final equation to experimental data. The second part of the derivation is the revolutionary concept but the mathematics is much simpler and only involves summing a geometric series. While the "mode counting" part of the derivation is necessary, a student should ponder deeply over the summation of a discrete series to get the average energy per mode! The second step is easily learned and it really is an important part of science education!

One of the unsolved mysteries of the late 1800s was an understanding of the blackbody radiation spectrum. Ideally, a black body is one which is in thermal equilibrium between absorbing and emitting radiation, as such it would appear "black." A real blackbody can be approximated by a box with a pinhole since light going in the hole would be absorbed but what light comes out would be characteristic of the temperature of the interior and the spectrum does not depend on the material used for the box. We offer a very crude schematic in Figure 10.1 to convey the sense of the problem. Common experience teaches us that hot objects can give off a color that changes from red to orange to yellow to white as the temperature increases. In principle, careful measurements can be made of the spectrum of an object using a dispersing element such as a prism or a grating and recording the dispersed light on film or more modern electronic devices (Figure 10.2). Due to the success of the Maxwell equations and the experiments of Hertz, there was an air of confidence among physicists that all the equations were available to explain the blackbody spectrum. If you read some of the histories of this period in *Science* on the Internet, you will better appreciate the mental groping regarding the nature of light. The Rayleigh–Jeans treatment of this problem was carried out from 1900 to 1905 by British scientists and perhaps overlooked the 1901 work by Max Planck (1858–1947), a German scientist who later received the Nobel Prize for his work in 1918 (Figure 10.3).

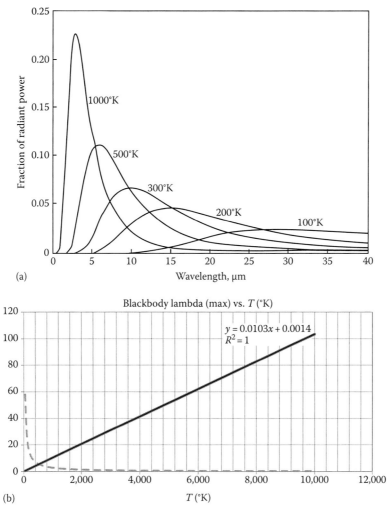

FIGURE 10.1 (a) The radiative power emitted by a blackbody at various temperatures illustrates Wein's Law. (b) The blackbody curve maximum wavelength in μm (10^{-6} m) at different temperatures (dash) clearly showing the shift of the intensity maximum toward the blue (shorter wavelength) as the temperature increases. The shift in wavelength maxima is shown in (b). The straight solid line is the frequency in 10^{13}/s of the maximum of the blackbody radiation. While the smooth curves in (a) show a trend, the straight line of the frequency maxima with temperature is extremely good with $R^2 = 1$, and the dashed line is the relative height of the maxima. (Data from Lide, D.R., *CRC Handbook of Chemistry and Physics*, 90th Edn., CRC Press, Boca Raton, FL, 2010, pp. 10–243. With permission.)

In relation to the introductory thermodynamic equations of the Stefan–Boltzmann law we note that Max Planck had been a professor of physics since 1889 specializing in thermodynamics. There is a very interesting history of Planck's discovery on the Internet at http://www.daviddarling. info/encyclopedia/Q/quantum_theory_origins.html. In fact Planck's interest was initially related to an equation he had tried to find relating Boltzmann's entropy to Wein's law. Wein's law was simply that the color of a hot object shifts with temperature and Planck developed the quantized equation to explain Wein's law. Wein's "law" is just an empirical observation that Planck tried to put on a firm foundation, although Planck approached the problem from a thermodynamic approach. Wien's Law relates the wavelength of the spectral maximum to temperature as [7]

$$\lambda_{max} T = 2.90 \times 10^{-3}\, m \cdot K$$

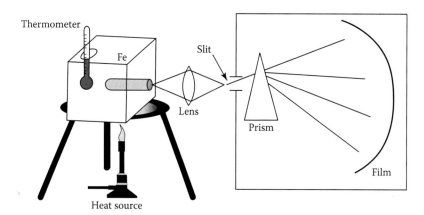

FIGURE 10.2 A schematic of measurement of blackbody radiation from a heated source.

FIGURE 10.3 Max Planck (1858–1947) was a German physicist who solved the problem of the blackbody radiation curve by assuming energy is quantized which is the foundation of quantum theory. He is considered one of the foremost scientists of the twentieth century and a number of research institutes in Germany are named after him. He received the 1918 Nobel Prize in physics for his work. We have mentioned that Albert Einstein encouraged Planck's interpretation of quantization and here Planck is presenting the Planck gold medal to Albert Einstein in 1929.

The Rayleigh–Jeans formula for the blackbody radiation curve failed disastrously for short wavelengths and high frequencies but Planck was perhaps motivated mostly by his interest in Wien's empirical "law." At that time it was believed that there would be many tiny discrete oscillators in the cube of metal shown in Figure 10.2 and these oscillators emitted light as a function of temperature so we will follow the classical Rayleigh–Jeans derivation up to a point. Assume the dimensions of the block are $L \times L \times L = V$. While we are interested mainly in the internal waves in the block, we do need to have an "observation hole" and we might discuss the light waves in the cavity only but here we assume the waves are part of what is going on inside the whole block and just continue across the interior hole. That is a key point because Planck assumed the oscillators in the wall of the observation cavity set up the waves. We will see that Planck derived an exact fit to the blackbody radiation spectrum but he did not believe it himself for several years and others regarded it as a sort of parameterized fit due to the use of h as an adjustable constant.

An important point occurs here in that in 1905, Einstein [2] called attention to the idea of quantized radiation waves "in the cavity," which are now called "photons." This is a very difficult concept but perhaps we can say that if the walls of the cavity are emitting light then the light observed in the hole will be that emitted light. Then if waves are involved, assume that only an integer number of wavelengths can fit into each dimension and that the waves continue throughout the block and across the observation hole (a thinking person will note that automatically introduces integers into the derivation!) we have $\left(\dfrac{L}{\lambda}\right) = n$. Multiplying by 2π we obtain $2\pi\left(\dfrac{L}{\lambda}\right) = n2\pi = \left(\dfrac{2\pi}{\lambda}\right)L$. We can define another characteristic of a wave called the "wave number" as $k_\lambda = \left(\dfrac{2\pi}{\lambda}\right)$ and so we get $k_\lambda L = n2\pi$ and this will apply in each of the (x, y, z) directions. Thus we have three values for the three dimensions $k_{\lambda x}L_x = n_x 2\pi$, $k_{\lambda y}L_y = n_y 2\pi$, and $k_{\lambda z}L_z = n_z 2\pi$. However, we know from polarizing sunglasses that light can be polarized in two planes orthogonal to the direction of propagation so any arbitrary polarization has to be some combination of two possible polarizations and we will have to multiply our count of possible waves by 2 later on. For now we see that our scheme for counting the number of waves in (L_x, L_y, L_z) depends on a triad of values as (n_x, n_y, n_z) including degeneracies (for a cube) such as $(1, 2, 3)$, $(2, 1, 3)$, $(3, 1, 2)$, etc. Thus we need a counting scheme that will be able to treat these degeneracies as well as high values of the indices (n_x, n_y, n_z). Since we know the n values will be very high for a large number of possible waves, we need a "scoreboard" representation and a very clever method is given in [2] (Figure 10.4). The relationship of this scoreboard is similar to the scoreboard in a stadium. The scoreboard is not the game but it is related to the game! Here we see that as the n values get larger and larger we can define another variable $n^2 = n_x^2 + n_y^2 + n_x^2$ as the square of a radius vector in

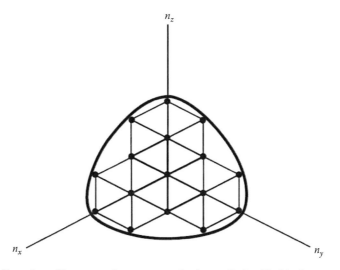

FIGURE 10.4 A "scoreboard" to count the energy modes in a radiating blackbody.

"n-space" and we note that the n values can be negative so the possible values of this degenerate n-space forms a surface that tends to be like the surface of a sphere.

Thus the number of modes can be written as (# modes) $= 2(4\pi)n^2 dn$. Then we use $k = \dfrac{2\pi n}{L}$ to find $dn = \dfrac{Ldk}{2\pi}$ and so we find that the number of nodes per unit volume is $\left(\dfrac{n}{V}\right) = 8\pi\left(\dfrac{kL}{2\pi}\right)^2\left(\dfrac{L}{2\pi}\right)\dfrac{dk}{L^3} = \dfrac{k^2 dk}{\pi^2}$. Finally we convert back to use $c = \lambda\nu$ and $dk = \dfrac{2\pi}{c}d\nu$ to find $\left(\dfrac{n}{V}\right) = \dfrac{k^2 dk}{\pi^2} = \left(\dfrac{1}{\pi^2}\right)\left(\dfrac{2\pi\nu}{c}\right)^2\left(\dfrac{2\pi}{c}\right)d\nu = \dfrac{8\pi\nu^2 d\nu}{c^3}$.

Although this is a rather complicated way to count all the nodes in a given volume, it has been checked over and over and we are telling you the "right way" rather than just making a statement. This difficult task has a reward however. In the late 1800s, the Boltzmann influence would predict an average energy per mode of $(k_B T)$ for each "mode" since they were due to some sort of oscillator. An oscillator intrinsically has $2\left(\dfrac{k_B T}{2}\right) = k_B T$ for the combined kinetic and potential degrees of freedom that cannot be separated, so the number of modes per volume should be multiplied by $k_B T$! So that is simple, after the tedious but correct counting scheme we get the Rayleigh–Jeans formula as $\left(\dfrac{n}{V}\right) = \rho(\nu)d\nu$ (Figure 10.5). Thus we have

$$\text{Rayleigh–Jeans} \Rightarrow \rho(\nu)d\nu = \frac{8\pi k_B T \nu^2 d\nu}{c^3}.$$

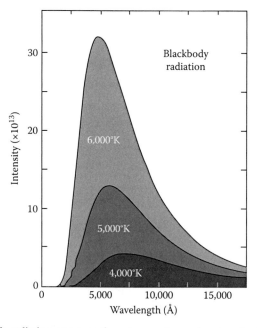

FIGURE 10.5 The Planck radiation curve at three temperatures showing the increase in intensity with temperature. One can also see a slight shift of the peak toward the blue at the temperature increases as is familiar when heated objects glow red, orange, yellow, then white as the temperature increases. (From Prof. Mike Guidry of the University of Tennessee Knoxville Department of Physics as used in their Astronomy 162 course. http://csep10.phys.utk.edu/astr162/lect/light/radiation.html With permission.)

It is easy to see that as the frequency increases as v^2 in the ultraviolet and the x-ray range, the "density of states" $\left(\dfrac{n}{V}\right) = \rho$ will increase to ∞ contrary to experimental data! We also need to alert the reader to an interesting dilemma here. Note that since $c = \lambda v$, $dv = \dfrac{-cd\lambda}{\lambda^2}$. Since *the convention in plotting graphs is to show the x-axis increasing to the right* we will have to agree to suppress the sign of one or the other if we plot the graph in terms of v or λ and we have $\rho(v)dv = \dfrac{8\pi k_B T v^2 dv}{c^3}$ using frequency or $\rho(\lambda)d\lambda = \dfrac{8\pi k_B T}{c^3}\left(\dfrac{c}{\lambda}\right)^2\left(\dfrac{-c}{\lambda^2}\right)d\lambda = -\dfrac{8\pi k_B T d\lambda}{\lambda^4}$ when using wavelength. However, as one goes to shorter wavelengths $(\lambda \to 0)$ the density of states still diverges to ∞ so the Rayleigh–Jeans formula diverges in either frequency or wavelength form! We did the tricky counting scheme in detail so that you know it is correct and that is not the problem. Rayleigh and Jeans depended on the average energy per mode as $\bar{\varepsilon} = \dfrac{\displaystyle\int_0^\infty \varepsilon e^{-\frac{\varepsilon}{k_B T}}d\varepsilon}{\displaystyle\int_0^\infty e^{-\frac{\varepsilon}{k_B T}}d\varepsilon} = \dfrac{(k_B T)^2}{(k_B T)} = k_B T$ where we have used $0! = 1$ and our old friend from Introduction as $\displaystyle\int_0^\infty x^n e^{-ax}dx = \dfrac{n!}{a^{n+1}}$. Surely we do not think Boltzmann was incorrect?

In 1901, Max Planck pondered this dilemma and tried a revolutionary assumption. Maybe the problem is in using $\displaystyle\int_0^\infty d\varepsilon$ that assumes the energy is a smooth continuous variable? That is, maybe the energy is not a smooth continuous variable? *Thus Planck tried using Σ instead of $\int d\varepsilon$.* An integral sign is a form of continuous addition but if energy exists as small chunks, quanta, then maybe you have to use a discrete summation over the states. One can pose the question as to whether smooth peanut butter is really an oil or just some finer version of chunky peanut butter. Close examination shows that smooth peanut butter is just smaller chunks of peanut in an oil base! Thus Planck considered the average energy of discrete "quantum" chunks (Figure 10.6).

$\bar{\varepsilon} = \dfrac{\sum_{n=0}^\infty (n\varepsilon)e^{-\frac{n\varepsilon}{k_B T}}}{\sum_{n=0}^\infty e^{-\frac{\varepsilon}{k_B T}}} = \dfrac{\sum_{n=0}^\infty (n\varepsilon)x^n}{\sum_{n=0}^\infty x^n}$ where we have used the previous analogy of the average

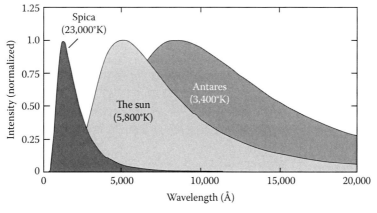

FIGURE 10.6 Normalized blackbody radiation for three stars using the Stefan–Boltzman equation to estimate the temperature of the stars. (From Prof. Mike Guidry of the University of Tennessee Knoxville Department of Physics as used in their Astronomy 162 course. http://csep10.phys.utk.edu/astr162/lect/light/radiation.html With permission.)

grade of students in a class to make sure we divide the thing we are averaging by the number of students in the class, "normalizing" the weighted-average. It helps to simplify the summation if we define $x \equiv \left(e^{-\frac{\varepsilon}{k_B T}} \right)$ and note that Planck defined the discrete energy as $E = \varepsilon, 2\varepsilon, 3\varepsilon, 4\varepsilon \ldots$ You may recall a geometric series from calculus? Let $f(x) = 1 + x + x^2 + x^3 + x^4 + \cdots$ for $x < 1$.

$$f(x) = 1 + x + x^2 + x^3 + x^4 + \cdots$$
$$- [xf(x) = x + x^2 + x^3 + x^4 + x^5 \cdots]$$

$$(1 - x)f(x) = 1$$

As long as $x < 1$ we can do this trick to find the sum of $f(x) = \dfrac{1}{(1 - x)}$. Further, we can take the derivative of $f(x)$ as $\dfrac{df(x)}{dx} = \dfrac{d}{dx}\left[\dfrac{1}{1 - x} \right] = \dfrac{-(-1)}{(1 - x)^2} = 1 + 2x + 3x^2 + 4x^3 + 5x^4 + \cdots$, but recall $\sum_{n=0}^{\infty} (n\varepsilon)x^n = \varepsilon x + 2\varepsilon x^2 + 3\varepsilon x^3 + 4\varepsilon x^4 + \cdots = (\varepsilon x)(1 + 2x + 3x^2 + 4x^3 + \cdots)$ and this means that $\bar{\varepsilon} = \dfrac{[\varepsilon x/(1 - x)^2]}{[1/(1 - x)]} = \dfrac{(\varepsilon x)}{(1 - x)} = \dfrac{\varepsilon e^{-\frac{\varepsilon}{k_B T}}}{\left(1 - e^{-\frac{\varepsilon}{k_B T}}\right)} = \dfrac{\varepsilon}{\left(e^{+\frac{\varepsilon}{k_B T}} - 1\right)}$ in a closed form! Next, Planck made a very famous assumption based on Maxwell's 1865 hypothesis that light is an electromagnetic wave and Hertz' 1886 experiments that radio waves can be transmitted and diffracted. Planck proposed that the energy of light is proportional to frequency, $\varepsilon = h\nu$. Note that at this point Planck did not know h and in fact his idea was so revolutionary (energy chunks indeed!) that one of the criticisms of his calculation was that he had to adjust the value of h by choosing a value that fit the data. In fact if you check the dates, Jeans published additional work in 1905 using the Rayleigh–Jeans formulation after Planck's 1901 work! So now we are ready for the amazing result of Planck's assumption of quantized energy.

$$\text{Planck} \Rightarrow \rho(\nu)d\nu = \frac{8\pi\nu^2}{c^3}\left[\frac{h\nu}{e^{\frac{h\nu}{k_B T}} - 1} \right]d\nu = \frac{8\pi h\nu^3 g\nu}{c^3 \left(e^{+\frac{h\nu}{k_B T}} - 1\right)}$$

Skeptics criticized h as an adjustable parameter, but when Planck chose $h = 6.626 \times 10^{-34}$ J · s he was able to fit the experimental to the experimental data for essentially an exact fit! One of the main critics was Wilhelm Ostwald (1853–1932), a German physical chemist, who did not accept the atomistic theory and believed energy is continuous. While Planck also was skeptical about the existence of atoms, he had to adjust his thinking when his equation produced an exact fit to experiment based on quantization. In 1909, Ostwald was awarded the Nobel Prize for his work with catalysis. From this brief discussion, you can see that even at this late date Boltzmann's 1866 KMTG prediction of tiny gas atoms was not widely accepted. The term "ultraviolet catastrophe" was only used later by Paul Ehrenfest in 1911 and Planck was motivated mostly by the shift in wavelength peak with temperature due to his background in thermodynamics.

Note the long wavelength agreement of the Rayleigh–Jeans function with experiment, does Planck's formula satisfy this condition?

$$\lim_{\nu \to 0} \left(\frac{h\nu}{e^{\frac{h\nu}{k_B T}} - 1} \right) = \lim_{\nu \to 0} \left[\frac{h}{\left(\frac{h}{k_B T}\right)e^{+\frac{h\nu}{k_B T}}} \right] = k_B T \text{ where we have used L'Hopital's rule for the limit. That}$$

is amazing for the low frequency limit but what about the high-frequency limit? Recall from Chapter 0 that $e^x = 1 + x + x^2 + x^3 + x^4 + \cdots$ and that e^x will dominate any integer power of x^n because e^x contains in its series every power of x and one higher power for any given single power of x!

Thus we have $\lim\limits_{v \to \infty} \left[\dfrac{hv}{e^{\frac{hv}{k_B T}} - 1} \right] = \left(\dfrac{\infty}{e^{\infty}} \right) = 0$. So the amazing thing about Planck's average energy expression is that it goes to zero at the high frequency end and is asymptotic to the experimental curve at the low frequency end! We have indulged in your patience with the tedium of the derivation because *it is exact! Energy is quantized!*

One additional amazing confirmation is that we can integrate Planck's formula over all frequencies to obtain the total emissive power to compare to the Stefan–Boltzmann law $\int_0^\infty \rho(v)dv = \dfrac{8\pi h}{c^3} \int_0^\infty \dfrac{v^3 dv}{\left(e^{\frac{hv}{k_B T}} - 1 \right)} = \left(\dfrac{8\pi h}{c^3} \right) \left(\dfrac{k_B T}{h} \right)^4 \int_0^\infty \dfrac{x^3 dx}{(e^x - 1)}$ by substituting $x = \dfrac{hv}{k_B T}$. However, this involves a very difficult integral, which can be solved by an expansion technique and term-by-term integration but it is readily found in a table of integrals as $\int_0^\infty \dfrac{x^3 dx}{(e^x - 1)} = \dfrac{\pi^4}{15}$.

Thus we can derive $\int_0^\infty \rho(v)dv = \left(\dfrac{8\pi h}{c^3} \right) \left(\dfrac{k_B T}{h} \right)^4 \left(\dfrac{\pi^4}{15} \right) = \left(\dfrac{8\pi^5 k_B^4}{15 h^3 c^3} \right) T^4 = \alpha T^4$, which agrees with experiment!

PHOTOELECTRIC EFFECT

Once again we invoke our strategy of not discussing all of the amazing discoveries in the early 1900s but focusing on the "essential" facts. Some biographies of Albert Einstein reveal that perhaps he was the foremost advocate of the idea of tiny atoms in nature, agreeing with Boltzmann. Throughout the 1800s, there were a number of experiments that showed that light could knock electrons out of a metal surface. The strange thing about the effect was that it seemed to depend on the wavelength/color of the light and did not depend on the intensity of the incident light. In Figure 10.7, we show a schematic of an evacuated chamber and a simple direct current circuit used to measure the current of electrons flowing through the vacuum from a metal surface to a collection electrode.

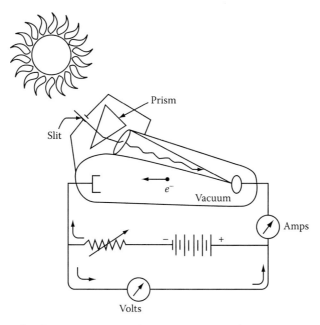

FIGURE 10.7 Schematic of the measurement of the photoelectric effect. Note the opposite polarity of the stopping potential part of the circuit.

The circuit also includes a battery for a voltage source and a variable resistor to adjust the voltage of the battery output, but note especially that the polarity of the battery is opposite to the polarity of the phototube! This arrangement is called a "bucking potential" and it opposes the current flow from the phototube due to a voltage controlled by the variable resistor (the small slash of the battery symbol is the negative plate, the larger slash is the positive plate). Assuming a white light source, a slit and a prism can be used to select the color of the light. It was found the photocurrent ceased as the color reached a limit in the red direction. The experiment was designed to use the adjustable bucking potential to find the voltage on the voltmeter at the bottom of the circuit that would stop the current flow reading on the ammeter. A series of stopping voltages could be measured and related to the wavelength or frequency of the light but that is not the main effect. In 1905, Einstein [3] published a very important paper explaining the main meaning of the experimental data by pointing attention to the red-most cutoff frequency after which the experiment no longer worked.

Einstein postulated that energy from the light was being used to knock electrons out of the surface of the metal and that energy was converted into kinetic energy of the ejected electron. Note this experiment depended on the ability to generate a vacuum inside the tube or else the electron would simple collide with a gas molecule and probably form an anion. Einstein's brilliant conclusion was that the frequency of the red-most color where the photocurrent became zero no matter what the bucking potential was represented the energy required to remove the electron from the metal and he called it the Work Function (W_f) (Figure 10.8). Although it is not used in the basic equation to follow, it is important to know that Einstein postulated that the energy chunks of the light were discrete particles called "photons." This started a long discussion in *Science* over what is called the "wave-particle duality" as to how Maxwell's classical electromagnetic waves could be quantized. The best answer is that there are "wave packet pulses" centered at a given frequency by very slight differences in the main frequency but more mathematics (Fourier analysis) would be needed to show this. First we can plot the "stopping potential" of the bucking voltage that stops the photoelectron flow at a series of wavelengths converted to frequencies as shown in Figure 10.9 for sodium metal. It is clear from the graph that there will be zero photocurrent at ν_0 which is at 2.3591 eV and the slope is 0.4134 eV/10^{14} Hz, but the line stops to the left of 2.3591 eV at 5.7043 10^{14} Hz or $\lambda = 5256$ Å. Plotted against frequency this is means we can equate the kinetic energy of the flying electron to (Table 10.1)

$$\frac{mV^2}{2} = C(\nu - \nu_0).$$

One might ask what is the constant C? Einstein then used Planck's proportionality constant and equated the total energy of the incoming light photon to the kinetic energy and the energy to knock the electron out of the metal, the "work function" W_f.

$$E = h\nu = W_f + \frac{mV^2}{2} \quad \text{or} \quad \frac{mV^2}{2} = h\nu - W_f = h(\nu - \nu_0) = C(\nu - \nu_0) \quad \text{and so} \quad C = h!$$

Given the data is only reliable to the nearest millivolt (0.001 eV), we can convert the slope to joules as $h \cong (0.4134 \text{ eV}/10^{14} \text{ Hz}) (1.60217653 \times 10^{-19} \text{ J/eV}) = 6.6234 \times 10^{-34} \text{ J} \cdot \text{s}$ That is very interesting but maybe we ought to also plot the limiting stopping potentials versus the ν_0 frequencies for a number of metals to see if that has a linear dependence. We use the W_f values from the *CRC Handbook* [8], which only gives three significant figures for the values in eV so we have used the most precise formulas to convert to ν_0 in attempt to get a more precise value for h. It is again clear that the linear fit is very good but the slope is slightly different at 0.4136 (eV/10^{14} Hz) that leads to $h = 6.626602128 \times 10^{-34}$ J $\cdot$ s, which should be rounded to $h = 6.627 \times 10^{-34}$ J $\cdot$ s. A thinking

FIGURE 10.8 Albert Einstein (1879–1955) was a German physicist whose work affected scientific thought more than any scientist since Isaac Newton. His flurry of papers in 1905 on the photoelectric effect, Brownian motion, special relativity and mass–energy equivalence ($E = mc^2$; $m = E/c^2$) established him as a foremost scientist of his time. In 1911, his calculations using relativity predicted that light from a distant star would be bent by the gravity of the Sun, which was confirmed by Sir Arthur Eddington in 1919 using an eclipse of the Sun. Other strange effects related to time dilation have more recently been confirmed during experiments in space flight. Einstein received the 1921 Nobel Prize in Physics for his work interpreting the photoelectric effect because his theory of special relativity was not well understood. He also received the Copley Medal from the Royal Society in 1925, the Royal Astronomical Society Gold Medal in 1926, and the Benjamin Franklin Medal from the Franklin Institute in 1936, as well as the Max Planck Gold Medal from the German Physical Society in 1929 shown in Figure 10.3. Einstein immigrated to the United States in 1933 when Adolf Hitler became Chancellor in Germany and took a position at the Institute for Advance Study in Princeton where he remained until his death at age 76 in 1955. In 1939, he sent a letter to President Franklin Roosevelt of the United States warning of the danger of Germany developing an atomic bomb, and the U.S. Manhattan Project won the race to build an atomic weapon under the leadership of Robert Oppenheimer while the German effort under Werner Heisenberg suffered constant setbacks due to Partisan raids in Norway and Allied bombing of German Industry.

student will realize there is a factor of h included in the formula used as λ (angstroms) $= \dfrac{12398.4}{\Delta E \text{ (eV)}}$ so Figure 10.10 is not a surprise, but the slope in Figure 10.9 is amazing! That is a direct measurement of $\varepsilon = h\nu$! The modern value of h is $6.6260693 \times 10^{-34}$ J · s so the photoelectric effect gives the value of h graphically within the number of significant figures (3) of the eV data! Wait a minute, isn't that the value Planck had to use to "fudge" his blackbody spectrum? A lot of credit should be given to Einstein for his interpretation of the photoelectric effect and reinforcing Planck's work. In fact, Albert Einstein received a Nobel Prize for this work in 1921 (in reserve), actually in 1922, but delivered his Nobel Address in 1923. Although the concept of photons can be mixed with classical electromagnetic waves, we should recognize that Einstein thought of the photoelectric effect in terms of quantized chunks of light energy. We have chosen this experiment as one of our "essential" topics because it confirms Planck's hypothesis and gives a direct way to measure the value of h.

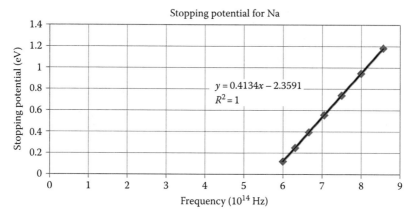

FIGURE 10.9 The stopping potential for photoelectrons from sodium versus frequency of the exciting light (photons). The modern value of the work function of sodium is 2.36 eV. (From Lide, D.R. Ed., *CRC Handbook of Chemistry and Physics*, 90th Edn., CRC Press, Boca Raton, FL, 2009–2010, pp. 12–121.)

TABLE 10.1
Work Function for Selected Metals
versus Photoelectric ν_0 Values

Element	W_f (eV)	λ_0 (Å)	ν_0 (10^{14} Hz)
Cs	1.95	6358.15	4.715087
K	2.29	5414.15	5.537204
Na	2.36	5253.56	5.706464
Li	2.93	4231.54	7.084720
Mg	3.66	3387.54	8.849855
Be	4.98	2489.64	12.041606

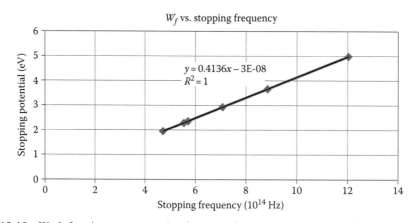

FIGURE 10.10 Work function versus stopping frequency for selected Group I and II metals.

DE BROGLIE MATTER WAVES

We now come to a mysterious concept first put forth by Prince Louis De Broglie in 1923.
Roughly, here is the reasoning due to De Broglie (Figure 10.11).

$$c = \lambda v, \text{Hertz 1888}, \quad E = hv, \text{Planck 1901}, \quad E = mc^2, \text{Einstein 1905 [9]}$$

$$E = hv = \frac{hc}{\lambda} = mc^2 \Rightarrow \lambda_{light} = \left(\frac{h}{mc}\right)_{light} \Rightarrow \lambda_{matter?} = \left(\frac{h}{mv}\right)_{matter?}, \quad \text{De Broglie 1923 [6]}$$

The problem with interpretation of this idea of "matter waves" is that waves should exhibit diffraction which is not commonly observed macroscopically. Consider a typical rifle bullet of 4.2 g traveling at a speed of 965 m/sec. The De Broglie matter wavelength for that bullet would be

$$\lambda = \frac{h}{mv} = \frac{6.62609 \times 10^{-34}\,\text{J} \cdot \text{s}}{(0.0042\,\text{kg})(965\,\text{m/s})} \cong 1.635 \times 10^{-34} \left[\frac{\text{kg(m/s)}^2\text{s}}{\text{kg} \cdot \text{m/s}}\right] = 1.635 \times 10^{-34}\,\text{m which is smal-}$$

ler than any available technique can measure! On the other hand, consider the De Broglie wavelength of an electron with 5 eV of energy as a typical circumstance within a molecule.

FIGURE 10.11 Louis-Victor Pierre Raymond De Broglie (1892–1987) was a French scientist who wrote a revolutionary PhD thesis in 1924 that formed the basis for Schrödinger's wave mechanics in 1926. De Broglie was awarded the Nobel Prize in physics in 1929. He also received the Henri Poincare Medal in 1929 and the Albert I of Monaco prize in 1932. De Broglie had been an officer in the French Army during World War I stationed at radio facilities at the Eiffel Tower and had read about radio waves and some of Einstein's relativity papers. After the war, he enrolled in a doctoral program at the University of Paris and wrote a revolutionary paper in 1923 [6], which was the basis for his PhD thesis in 1924. However, none of his professors understood the idea, and his thesis was rejected until the Chair of the Department of Physics, Prof. Pierre Langevin, sent the thesis to Einstein who approved it and De Broglie was awarded the PhD for his work. Actually, the work was based on Einstein's special relativity and we remind the reader that what is often called "nonrelativistic wave mechanics" is already based on a relativistic principle.

$$E = \frac{mV^2}{2} \Rightarrow V \Rightarrow \sqrt{\frac{2E}{m}} = \sqrt{\frac{2(5.0\,\text{eV})(1.60217653 \times 10^{-19}\,\text{J/eV})}{9.1093826 \times 10^{-31}\,\text{kg}}} = 1.36205154 \times 10^6\,\text{m/s}$$

$$\lambda = \frac{h}{mV} = \frac{6.62609 \times 10^{-34}\,\text{J}\cdot\text{s}}{(9.1093826 \times 10^{-31}\,\text{kg})(1.36205154 \times 10^6\,\text{m/s})} = 5.4847606 \times 10^{-10}\,\text{m or about}$$

$\lambda \cong 5.485 \times 10{-8}\,\text{cm} = 5.485\,\text{Å}$, which is comparable in size to a molecule. Thus, wave mechanics is useless for macroscopic calculations but essential for molecular calculations!

DAVISSON–GERMER EXPERIMENT

Following the 1923 paper by De Broglie [5], a number of experimental specialists tried to verify the wave properties of particles. A group at Bell Laboratories designed an experiment to test the De Broglie idea using low energy electrons. Several papers were published but the most detailed appeared in 1927 by Davisson and Germer [10]. Clinton Davisson (1881–1958) was awarded the Nobel Prize for this work in 1937. The initial experiments consisted of scattering a beam of electrons with a uniform low energy at various angles off of small blocks of platinum and nickel. These experiments gave just broad angular patterns and did not follow De Broglie's equation.

In 1925, there was an accident at the Bell Laboratories site in which a liquid air container exploded and broke the glass enclosure of the experiment while a small block of nickel was being bombarded by an electron beam in a vacuum. When air rushed into the experimental chamber, a layer of oxide formed on the surface of the nickel block, which was very hot from the electron impacts. That was a serendipitous accident because after repairs were made, the block had to be heated for a long time under vacuum to vaporize the oxide layer. An electric heater under the block brought the temperature of the block close to the melting point of the nickel under vacuum for perhaps a month or more. At the end of that time the polycrystalline nickel had large areas of annealed crystalline nickel clearly visible. When the electron beam experiment was restarted, the angular pattern of the scattered electrons showed sharp peaks and eventually some 30 peaks were assigned to scattering off various planes of the face-centered cubic (fcc) nickel crystal surface. If you draw an fcc structure as in Figure 10.10 you can see that you could slice through the lattice in a number of ways, and the $\{1, 1, 1\}$ plane is formed through the corners (a, a, a) opposite to $(0, 0, 0)$ where a is the unit cell dimension. Many planes can be defined in terms of the unit cell dimension such as $\{1, 1, 1\}$, $\{1, 1, 0\}$, and $\{1, 0, 0\}$ where the indices refer to the (x, y, z) dimensions. In their paper, Davisson and Germer say they oriented the surface of the Ni crystals perpendicular to the $\{1, 1, 1\}$ planes as in Figure 10.12 and apparently used a spacing between the planes of 2.18 Å. A modern value of the fcc structure of nickel shows the a parameter to be 3.5238 Å [11], and the formula for the spacing between $\{1, 1, 1\}$ planes is

$$d_{lmn} = \frac{a}{\sqrt{l^2 + m^2 + n^2}} = \frac{3.5238\,\text{Å}}{\sqrt{1 + 1 + 1}} = 2.03447\,\text{Å}$$

The modern value of the Ni (fcc) unit cell makes the angle of the diffracted wave slightly higher than their 50° conclusion but there are uncertainties associated with the angle of the surface to the electron beam and whether the whole surface was pure $\{1, 1, 1\}$ (Figure 10.13). Still the calculated angle agrees with the overall pattern of the scattering peaks shown in their paper. Most of the scattering peaks corresponded to the well-known x-ray scattering patterns that strengthen the wave analogy! Note they specifically state that they used $n\lambda = 1d\sin(\theta)$, which makes sense according to the diagram shown in Figure 10.14 for a perpendicular beam compared to the grazing incidence equation of $n\lambda = 2d\sin(\theta)$.

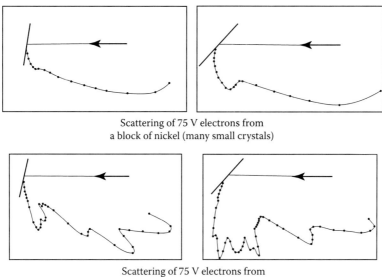

Scattering of 75 V electrons from
a block of nickel (many small crystals)

Scattering of 75 V electrons from
several large nickel crystals

FIGURE 10.12 The Davisson–Germer discovery after annealing Ni crystals. (Reprinted with permission from Davisson, C. and Germer, L.H., *Phys. Rev*, 30, 705, 1927. Copyright 1927 by the American Physical Society.)

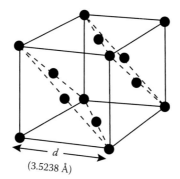

d
(3.5238 Å)

FIGURE 10.13 The Ni (fcc) unit cell showing the {1, 1, 1} planes.

The most pronounced scattering peak occurred for 54 eV electrons and was observed at an angle of about 50°–55° from the incoming beam, which was perpendicular to the metal surface. This is a different orientation from the usual Bragg angle with the angle of incidence equal to the angle of reflection but when the phase of the waves is right for reinforcement such diffraction can occur using the equation of $n\lambda = 1d\sin(\theta)$ instead of $n\lambda = 2d\sin(\theta)$.

$$E = \frac{mV^2}{2} = 54\,\text{eV so } V = \sqrt{\frac{2(54\,eV)(1.60217653 \times 10^{-19}\,\text{J/eV})}{9.1093826 \times 10^{-31}\,\text{kg}}} = 4.358354871 \times 10^6\,\text{m/s}.$$

$$\lambda = \frac{h}{mV} = \frac{6.6260693 \times 10^{-34}\,\text{J}\cdot\text{s}}{(9.1093826 \times 10^{-31}\,\text{kg})(4.358354871 \times 10^6\,\text{m/s})} = 1.668954288 \times 10^{-10}\,\text{m}$$

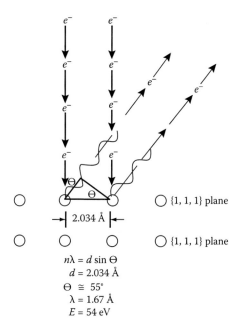

$$n\lambda = d \sin \Theta$$
$$d = 2.034 \text{ Å}$$
$$\Theta \cong 55°$$
$$\lambda = 1.67 \text{ Å}$$
$$E = 54 \text{ eV}$$

FIGURE 10.14 "Half-Bragg" $n\lambda = d \sin(\theta)$ diffraction of 54 eV electrons in a beam perpendicular to the $\{1, 1, 1\}$ plane of Ni observed by Davisson and Germer [10]. It is clear that Davisson and Germer used $n\lambda = 1d \sin(\theta)$ on page 723 of their paper.

so $\theta = \sin^{-1}\left(\dfrac{\lambda}{1d}\right)$. Then $\theta = \sin^{-1}(0.820338608) = 55.1187°$ roughly in agreement with the experimental result using $n\lambda = 1d \sin(\theta)$! Some of the curves shown in the original paper can be interpreted as peaks at angles slightly larger than 50° while others show the peaks at almost exactly 50°. Since modern technology uses such scattering to characterize surfaces, the wave diffraction of electrons has been established beyond any uncertainties in the original experiment (Figure 10.15). Because the raw

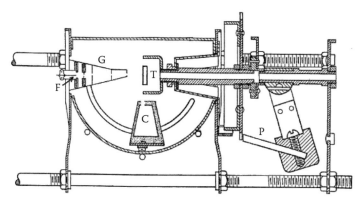

FIGURE 10.15 A cross section view of the original Davisson–Germer apparatus. The Ni crystal is the target "T," "G" is the electron gun and "C" is the collector, which can travel on a curved track to measure the angle of scattered electrons entering the collector. "F" is a heated tungsten filament, which emits thermal electrons as the source. "P" is the electrical potential wire. Note the width of the entrance to the detector cup seems to be at least 2° of arc leading to some uncertainty of the angle by a small amount. (Reprinted with permission from Davisson, C. and Germer, L.H., *Phys. Rev.*, 30, 705, 1927. Copyright 1927 by the American Physical Society.)

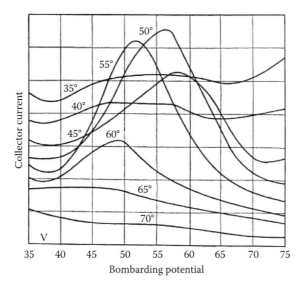

FIGURE 10.16 Raw data from the Davisson–Germer experiment showing detection of scattered electrons at various angles of the collector versus the energy of the electron beam. A 55° angle calculated in the text is very close to the simple protractor measurement compared to the 50° peak shown here for a beam of 54 V. Note the acceptance width of the detector cup surely is wider than 1° and no uncertainty is given for how close the {1, 1, 1} plane was oriented to be perpendicular to the incident beam so the calculation with the modern cell constant for Ni changes the angle slightly but it is still within the variation of the data shown. (Reprinted with permission from Davisson, C. and Germer, L.H., *Phys. Rev.*, 30, 705, 1927. Copyright 1927 by the American Physical Society.)

data was presented in an unusual way, we also show the diagram of the original equipment with the sliding arc track for the detector cup relative to the electron beam from the "gun."

The surface of the nickel block was oriented so that the {1, 1, 1} surface was exposed and a very strong peak was observed using a collector cup with a galvanometer to measure the diffracted electrons and many careful checks were made to test the energy of the scattered electrons compared to random background scattering, and the 1927 paper [10] is an amazing example of thorough scientific work. In Figure 10.14 we sketch the path of electrons incident perpendicular to the surface of the target block with the spacing of 2.034 Å between the Ni atoms arrayed in the {1, 1, 1} plane. Although Davisson and Germer used a {1, 1, 1} spacing of 2.18 Å to arrive at a 50° angle for the 54 V beam compared to a modern value of 55° using a spacing of 2.034 Å (Figure 10.16), the observation of the second large diffraction at 65 V is a very convincing demonstration of diffraction (Figure 10.17). The most interesting thing about the experiment is that the beam is directly at the surface of the block perpendicular to the surface and the fact that the diffraction follows what might be called a "half-Bragg" rule of $n\lambda = 1d \sin(\theta)$.

There are two important results from the Davisson–Germer experiment. Foremost is the agreement with the De Broglie equation and that is the overwhelming immediate payoff of the result. In fact, L. De Broglie was awarded the Nobel Prize not long after in 1929! Clinton Davisson (1881–1958) was awarded the Nobel Prize for this work later in 1937. For our later chapters, this is also the foundation of "wave mechanics," the Schrödinger equation, quantum chemistry, molecular orbital theory, and modern spectroscopy. It really is that important to prove particles have wave properties or at least behave as if there is some sort of common wave behavior in the mathematics. Having said that, one might think the experiment is over and done with, but reading the full 1927 paper shows that a number of additional diffraction peaks were assigned to gas molecules absorbed on the Ni surface. Thus a second more recent use of this effect is the study of surface chemistry with improved methods of simultaneous detection of multiple diffraction spots similar to x-ray diffraction.

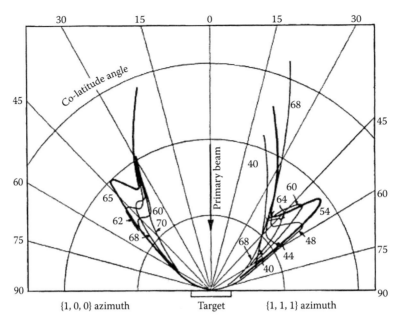

FIGURE 10.17 Schematic of the primary "54 V," first order ($n = 1$) diffraction of electrons at approximately 55° from {1, 1, 1} spacing of 2.034 Å between planes of Ni atoms. This unusual way of graphing the raw data shows the angular variation for two different sets of results of different voltages relative to diffraction from a {1, 0, 0} lattice plane at 66 V on the left and from a {1, 1, 1} lattice plane on the right at 54 V. (Reprinted with permission from Davisson, C. and Germer, L.H., *Phys. Rev.*, 30, 705, 1927. Copyright 1927 by the American Physical Society.)

SUMMARY

This chapter is intended to be the beginning of a second semester with emphasis on molecular quantum mechanics. Here we have emphasized three essential experiments which form the foundation for quantum mechanics and quantum chemistry by actual laboratory experiments and theory, which knit the results together. The conclusions are that energy is indeed quantized (exists as very tiny chunks) and particles can have wave properties as well as light waves behaving like particles! Although this chapter has straightforward formulas, the next few chapters will provide mind-bending details on the differential equations solved by Erwin Schrödinger in 1926 and so we begin here to provide the key facts in this chapter and will provide similar summaries in the more difficult chapters.

1. The radiated light energy from a light source is proportional to T^4. Light waves have a very small, but measureable pressure.

2. $\rho(v)dv = \dfrac{8\pi v^2}{c^3}\left[\dfrac{hv}{e^{\frac{hv}{k_B T}} - 1}\right]dv = \dfrac{8\pi h v^3 dv}{c^3\left(e^{\frac{+hv}{k_B T}} - 1\right)}$ is the formula for the light energy radiated

   from a blackbody source. It depends on energy occurring in discrete chunks or "quanta" where $\varepsilon = hv$, and while Planck derived this in 1901 the equation was not seriously accepted at first because the value of h was considered an adjustable parameter even though the formula fits the data exactly.

3. In 1905, Einstein interpreted the photoelectric effect and showed that a graph of stopping potential versus light frequency had a slope given by the same value of the h Planck used

in the blackbody derivation. This added tremendous credibility to Planck's blackbody equation. Once again, the formula $\lambda\ (\text{A}°) = \dfrac{12,398}{\Delta E\ (\text{eV})}$ proved to be useful.

4. In 1923, De Broglie proposed the seemingly strange concept that there is some sort of mathematical pilot wave that can be used to describe the behavior of particles as given by the formula $\lambda = \dfrac{h}{p} = \dfrac{h}{mv}$. The matter wavelength of electrons was found to be similar in size to chemical bonds leading to the idea that we need to learn "wave mechanics" to understand the behavior of small particles such as electrons and nuclei.

5. In 1927, a key experiment by Davisson and Germer at Bell Laboratories in the United States showed that a beam of electrons (particles) were diffracted from an annealed crystal lattice of Ni atoms. The property of diffraction is a key characteristic of waves. This set the stage to use "wave mechanics" to describe phenomena at the small scale of atoms and molecules. In addition, a new technique of "low-energy electron diffraction" (LEED) was developed to study material surfaces at the atomic level.

PROBLEMS

10.1 Insert the values of the constants into the formula for $\alpha = \dfrac{8\pi^5 k^4}{15h^3 c^3}$ and compare the result to the numerical value given in the text.

10.2 To show how the maximum of the Planck formula shifts with temperature set the first derivative of $\rho(\lambda)$ with respect to λ to zero and rearrange it to $y(x) = e^{-x} + \left(\dfrac{x}{5}\right) - 1 = 0$ where $\left(\dfrac{hc}{\lambda_m kT}\right) = x$. Then $\dfrac{dy}{dx} = -e^{-x} + \dfrac{1}{5}$, which can be used in a Newton–Raphson iteration to find x from guessed values. $x_{better} = x_{guess} - \dfrac{\left[e^{-x_g} + 0.2 x_g - 1\right]}{\left[0.2 - e^{-x_g}\right]}$. Start with a good guess of $x = 5.0$ and in a few iterations you will obtain Wein's constant for $\lambda_{\max} T = \dfrac{hc}{W k_B}$ where W is the converged constant from the iteration. This problem "explains" why the color of a hot object changes with the temperature as the maximum of the blackbody shifts with temperature. This effect and the constant value were known before Planck solved the problem so it is important to show that Planck's law agrees with previously known data.

10.3 On an Internet site called the "Physics Forum" at http://www.physicsforums.com, a student with the identification of "georgeh" and the topic title of "Stopping Voltage" says he has a problem to find Planck's constant and the work function, W_f, from two data points for sodium metal. He cites two points: a stopping potential of 1.85 V at 300 nm and another stopping potential of 0.82 V at 400 nm. Convert the wavelengths to frequencies and plot the stopping potential on the vertical axis and the frequencies on the horizontal axis. Extrapolate the line between the two points to 0.0 Stopping Potential and calculate the W_f value for sodium in electron volts. Calculate the effective value of the slope of the two points and convert the units to $J \cdot s$/frequency to compare the value to Planck's constant.

10.4 Calculate the De Broglie wavelength of a baseball with a mass equal to 5.25 oz avoirdupois traveling at 90 mph. compare that to the De Broglie wavelength of an electron traveling at 10% of the speed of light in vacuum which is comparable to the speed of a 1s electron in a heavy atom.

10.5 Davisson and Germer also reported a strong diffraction intensity peak for 65 eV electrons at an angle of 44° from the incident beam striking the face of their Ni crystal perpendicular to the surface, which they assigned to the {1, 0, 0} plane of the Ni crystal structure.

Calculate the "d" distance this implies between Ni atoms in this plane. Compare your answer to the value of 2.18 Å used by Davisson and Germer for crystalline Ni.

10.6 Look up the Davisson–Germer paper in *Physical Review*, vol. 30, p. 705, (1927) and look at the picture of the Ni crystal in their Figure 5. Then look up the word "anneal" in a dictionary and give your own brief description of what annealing metal means.

REFERENCES

1. Planck, M., On the law of the energy distribution in the normal Spectrum, *Ann. Phys.*, **4**, 553 (1901).
2. Powell, J. L. and B. Craseman, *Quantum Mechanics*, Addison-Wesley Publishing Co., Inc., Reading, MA, 1961.
3. Einstein, A., Concerning an heuristic point of view toward the emission and transformation of light, *Ann. Phys.*, **17**, 132 (1905).
4. Davisson, C. and L. H. Germer. Reflection of electrons by a crystal of nickel. *Nature*, **119**, 558 (1927).
5. De Broglie, L., Radiations. — *Ondes et quanta, Comptes Rendus*, **177**, 507 (1923), see also http://www.davis-inc.com/physics/
6. Bell, M. and S. E. Green, On radiometer action and the pressure of radiation, *Proc. Phys. Soc.*, **45**, 320 (1933), see also Hull, G. F., M. Bell, and S. E. Green, Notes on the pressure of radiation, *Proc. Phys. Soc.*, **46**, 589 (1934).
7. McQuarrie, D. H., *Quantum Chemistry*, University Science Books, Sausalito, CA, 1983, p. 10.
8. Lide, D. R., *CRC Handbook of Chemistry and Physics*, 90th Edition, CRC Press, Boca Raton, FL, 2009–2010, p. 12–121.
9. Einstein, A., Does the inertia of a body depend on its energy content? *Ann. Phys.* (Lpz.), **18**, 639 (1905); see also http://astrol.panet.utoledo.edu/~ljc/world2.html
10. Davisson, C. and L. Germer, Diffraction of electrons by a crystal of nickel, *Phys. Rev.*, **30**, 705 (1927).
11. Lide, D. R., *CRC Handbook of Chemistry and Physics*, 87th Edition, CRC Press, Boca Raton, FL, 2006–2007, p. 4–161.

11 The Schrödinger Wave Equation

INTRODUCTION

This chapter is primarily about "wave mechanics" since that is the most convenient way to introduce undergraduates to quantum mechanics using calculus. An equivalent form called "matrix mechanics" will be discussed briefly in a later chapter. Consider again the 1923 paper by De Broglie and the experiments that validated the particle-wave duality in Chapter 10. One might well ask that if there really is some "wave" that describes the behavior of particles, then is there an equation that the wave obeys? Even today it is difficult to say what the "wave" is, but it may help to find an equation it obeys. Step back a moment to some basic calculus:

$$\left(\frac{d^2}{dx^2}\right)\sin(ax) = \left(\frac{d}{dx}\right)[a\cos(ax)] = (-a^2)\sin(ax).$$

Perhaps you did not notice the pattern before but we can put this into a general form as

$$(Operator)(Eigenfunction) = (Eigenvalue)(Eigenfunction).$$

The word "eigen" in German means "characteristic, unique, peculiar, special..." and only certain functions satisfy this condition called an eigenfunction equation. An analogy that has been successful in explaining this to undergraduates is to consider an apple tree with ripe apples on it. If you hit the branches with a stout stick some apples will fall off the tree but the tree will still be there. The operator is the act or operation of hitting the tree with the stick, the tree is the eigenfunction and the apples are the eigenvalue(s). The eigen word comes from German because this relationship was first linked to the De Broglie wave idea by Erwin Schrödinger in 1926 in a series of papers that are among the most important in modern science [1]. Schrödinger (1887–1961) was an Austrian physicist who received the Nobel Prize for his work in 1933 (Figure 11.1).

We will now present a derivation of the Schrödinger equation that may not be the way he thought of it but that follows from limited use of calculus and simple algebra. Since De Broglie implied there is some sort of invisible, untouchable, mathematical pilot wave accompanying the motion of particles, we assume the general form of such a wave and relate it to its own second derivative. We will use ψ since it is universally used for the wave function.

Let $\psi = A\sin\left(\frac{2\pi x}{\lambda}\right)$. Then $\frac{d^2\psi}{dx^2} = \left(\frac{d^2}{dx^2}\right)A\sin\left(\frac{2\pi x}{\lambda}\right) = \left(\frac{-4\pi^2}{\lambda^2}\right)A\sin\left(\frac{2\pi x}{\lambda}\right)$. That would be

true for any wave but we want to apply it to a "matter wave" using $\lambda = \dfrac{h}{mv}$, so we substitute λ_{DB}.

$$\left(\frac{d^2}{dx^2}\right)A\sin\left(\frac{2\pi x}{\lambda}\right) = \left[\frac{-4\pi^2}{\left(\frac{h}{mv}\right)^2}\right]A\sin\left(\frac{2\pi x}{\lambda}\right) = \left[\frac{-(mv)^2}{\hbar^2}\right]A\sin\left(\frac{2\pi x}{\lambda}\right).$$ Note that $\left(\dfrac{h^2}{4\pi^2}\right) = \hbar^2$.

One more step is needed. Recall $E_{\text{tot}} = \dfrac{mv^2}{2} + V = \dfrac{(mv)^2}{2m} + V = T + V = H_{op}$ where $H_{op} = T + V$ is the total energy operator with T as the conventional symbol for the kinetic energy (often K in

233

FIGURE 11.1 Erwin Rudolf Josef Alexander Schrödinger (1887–1961) was an Austrian theoretical physicist who is famous for his wave equation treatment of quantum mechanics. He was awarded the Nobel Prize in physics in 1933. He became a full professor at the University of Zurich in 1921 and in 1926 wrote four extremely important papers establishing wave mechanics. He left Germany in 1933, although his parents were Christian, and moved to various lecturing positions for several years until in 1940 he was invited to set up an Institute for Advanced Study in Dublin. There he continued to write many papers and a small book *What is Life*, with conjecture that genes are molecular, which inspired Francis Crick (of Watson and Crick) and Max Delbruck among others to study DNA.

sophomore physics texts) and V as the conventional symbol for whatever potential energy there may be. With that understanding, we have $T = \dfrac{mv^2}{2} = \dfrac{(mv)^2}{2m} = (E_{tot} - V) \Rightarrow (mv)^2 = 2m(E_{tot} - V)$ so

that now $\left(\dfrac{d^2}{dx^2}\right) A \sin\left(\dfrac{2\pi x}{\lambda}\right) = \left[\dfrac{-(mv)^2}{\hbar^2}\right] A \sin\left(\dfrac{2\pi x}{\lambda}\right) = \left[\dfrac{-2m(E_{tot} - V)}{\hbar^2}\right] A \sin\left(\dfrac{2\pi x}{\lambda}\right)$ or

$\left(\dfrac{d^2}{dx^2}\right)\psi = \left[\dfrac{-2m(E_{tot} - V)}{\hbar^2}\right]\psi$. This is where it gets interesting in that Schrödinger identified the

energy with a mathematical operator! Thus $\left(\dfrac{-\hbar^2}{2m}\right)\left(\dfrac{d^2}{dx^2}\right)\psi = (E_{tot} - V)\psi$ and further we have

$\left(\dfrac{-\hbar^2}{2m}\right)\left(\dfrac{d^2}{dx^2}\right)\psi + V\psi = E_{tot}\psi$ and finally $\left[\left(\dfrac{-\hbar^2}{2m}\right)\left(\dfrac{d^2}{dx^2}\right) + V\right]\psi = E_{tot}\psi$ or $H_{op}\psi = E_{tot}\psi$.

This has major implications in that there is a mathematical (calculus) operator H_{op}, which represents the total energy of a particle (in only one dimension so far) and a function ψ, which incorporates the De Broglie condition and is an eigenfunction of the total energy operator. Note the left side of the equation must be in energy units since the right side is in terms of E_{tot}.

We can carry the analysis of the energy units further. Classically (sophomore physics), the kinetic energy would be $\left(\dfrac{mv^2}{2}\right) = \left(\dfrac{m^2v^2}{2m}\right) = \left(\dfrac{p^2}{2m}\right) \sim \left(\dfrac{-\hbar^2}{2m}\dfrac{d^2}{dx^2}\right)_{op}$ so maybe, just maybe (?)

$p_{op}^2 \sim -\hbar^2 \dfrac{d^2}{dx^2}$ and there is also a momentum operator $p_{op} \sim \left(\dfrac{\hbar}{i}\right)\dfrac{d}{dx}$ where $i = \sqrt{-1}$? In this case the spatial variable is x but it is known from the physics subfield called mechanics that for every momentum in a given coordinate system there is a corresponding coordinate; for every (p, q) pair there is a q for each momentum in that coordinate as p_q. Not every coordinate system will be (x, y, z) so the general condition is $p_q = \left(\dfrac{h}{i}\dfrac{d}{dq}\right)$. Note that there is no special need to rewrite the potential energy V as anything other than a multiplicative "operator." The main action of the energy operator is in the momentum operator. Maybe it is too soon to oversimplify wave mechanics, but the main principle is indeed simple, just follow a few direct steps.

1. Write the total energy expression in terms of classical momenta and coordinates.
2. Insert the equivalent operator $\left(\dfrac{\hbar}{i}\dfrac{d}{dq}\right)$ wherever momentum occurs.
3. Consider any function of the coordinates such as the potential energy to be a simple multiplicative operator.
4. Form the total energy operator, $H_{op} = T + V$ and write the Schrödinger equation as

$$H_{op}\psi = E_{tot}\psi \quad \text{where } H_{op} = \frac{p_{op}^2}{2m} + V(q) = \left(\frac{-\hbar^2}{2m}\frac{d^2}{dq^2}\right) + V(q).$$

5. Solve the differential equation by whatever means to find ψ and E_{tot}, noting that there may be a set of functions $\{\psi_n\}$ with corresponding eigenvalues $\{E_n\}$.

It would all be so simple if step 5 really is easy to do. We need to simultaneously introduce you to some simple techniques in solving certain types of differential equations while at the same time solving an easy problem that has sufficient application to laboratory measurements to be realistic. It is traditional to use the "particle-in-a-box" (PIB) problem for this purpose. To provide motivation, let us consider the ultraviolet spectrum of all-*trans* polyenes, *trans*-butadiene will suffice. We should know from organic chemistry that there is a principle of "sigma-pi-separability," which alerts us to the idea that the four electrons in 2P π orbitals are oriented in a plane perpendicular to the plane of the atoms $H_2C=CH–CH=CH_2$, which contains all the 2P σ bonds and the C1s orbitals (Figure 11.2). Sigma-Pi separability is a good approximation because the 2P π orbitals are odd functions with a node (sign change) in the plane of the atoms while the 2P σ orbitals are even functions with respect to reflection in the molecular plane and the product of an even and odd function integrates over all space to zero. The sigma and pi orbitals are "orthogonal" in a first approximation, although certainly there is some coulomb repulsion between the electrons and even a strange phenomenon called "exchange" due to the fact that electrons are indistinguishable and can occasionally trade places! Further, the 2P σ orbitals are spatially more compact than the larger, diffuse 2P π orbitals. Thus for several reasons, we consider the sigma bond skeleton of the molecule to be "frozen" and screening all but +4 nuclear charges with the electronically soft 2P π orbitals forming linear combinations to hold the four pi electrons. Thus we consider a path along the trans structure as a square box in which there are four electrons.

Another simplifying assumption is that these electrons suffer no mutual repulsion but do tend to pair into two spin pairs $(\alpha, \beta)_{\pi 1}$ and $(\alpha, \beta)_{\pi 2}$. We rely on previous explanations of spin pairing in

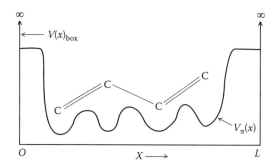

FIGURE 11.2 Schematic one-dimensional potential for the pi-electrons of *trans*-butadiene with the structure of the molecule tilted up in the plane of the figure. With near perfect screening by the C1S electrons and the C2SP sigma core electrons the pi-electrons "see" only smooth minima in the potential over the C atom positions. We neglect H atoms as part of the electronic "core."

other chemistry courses. The proper treatment of electron spin requires inclusion of a discussion of relativity beyond the level of this course. Two comments can be made however. It can be shown using the Dirac model of the H atom that the spin-up α electron has *intrinsic angular momentum* of $+\dfrac{\hbar}{2}$ and the spin-down β electron has $-\dfrac{\hbar}{2}$ angular momentum and they are not quite exactly in the same orbital, but it is sufficient at this level to imagine that there are two electrons with opposite magnetic orientation paired up in the same spatial orbital. Of course we also assume you have been shown in organic chemistry that pi-orbitals are highly delocalized so that the center bond of butadiene has considerable double bond character.

Now we are ready to define the mathematics of the (quantized) PIB. Assume there is a potential V, which is zero along the box but that keeps the electrons in the box by rising straight up at the ends of the box to $+\infty$.

$V = 0$, $0 < x < L$ but $V = \infty$, $x \le 0$ and $V = \infty$, $x \ge L$ where the box is defined by $0 \le x \le L$.

Since the potential is zero in the box, the particle(s) can only have kinetic energy so we form H_{op}.

$$H_{op} = \frac{p_{op}^2}{2m} + 0 = \frac{\left(\frac{\hbar}{i}\frac{d}{dx}\right)^2}{2m} + 0 = \frac{-\hbar^2}{2m}\frac{d^2}{dx^2} + 0,$$ so $H\,\psi = E\,\psi$ and $\dfrac{-\hbar^2}{2m}\dfrac{d^2\psi}{dx^2} = E\psi$. The next step is

to solve the differential equation (after rearrangement) $\dfrac{d^2\psi}{dx^2} + \left(\dfrac{2mE}{\hbar^2}\right)\psi = 0$. This is a second-order differential equation (involves a second derivative) but is a type that is easy to solve by factoring the operator. Define $D \equiv \left(\dfrac{d}{dx}\right)$, then $\left(D + i\sqrt{\dfrac{2mE}{\hbar^2}}\right)\left(D - i\sqrt{\dfrac{2mE}{\hbar^2}}\right)\psi = 0$. This is the simplest type of a second-order differential equation which is usually taught in the early part of a text on differential equations but not all second-order differential equations can be factored in this way. Now we need to think. The equation says to apply two first-order operations in succession so we have cracked apart a second-order equation into the product of two first-order equations that are easier to solve. In particular, we see that if we find ψ that solves the rightmost first-order equation, the result will be zero so it will not matter what the left parenthesis is and we can interchange the order of the two operator parentheses. That means that we can/will get two solutions, and the general solution will be some linear combination of the two. Let us solve the rightmost part of the equation as $\left(D - i\sqrt{\dfrac{2mE}{\hbar^2}}\right)\psi = 0$. We have solved this type equation for the case of first-order kinetics. For simplicity consider $(D - a)\psi = 0$ where a is a constant.

$\left(\dfrac{d}{dx} - a\right)\psi = 0 \Rightarrow \dfrac{d\psi}{dx} = a\psi \Rightarrow \dfrac{d\psi}{\psi} = adx$ and then $\displaystyle\int \dfrac{d\psi}{\psi} = a\int dx$ and so $\ln(\psi) = ax + C$. Now take the antiln of the equation to find $\psi = Ae^{ax}$ where $C = \ln A$. Before we go further, please note that $(-a)$ in the equation became $(+a)$ in the solution so when we encounter this type of differential equation, we can simply write the solution as the constant with opposite sign as a power of base-e! Using that idea we can write down the two solutions of our PIB problem.

$\psi(x) = C_1 e^{-i\left(\sqrt{\frac{2mE}{\hbar^2}}\right)x} + C_2 e^{+i\left(\sqrt{\frac{2mE}{\hbar^2}}\right)x}$ and that is the solution except we do not know C_1 or C_2 and we need to use boundary conditions to find these constants.

Now recall Euler's rule that $e^{i\theta} = \cos(\theta) + i\sin(\theta)$ and apply it to both terms. We do not know C_1 or C_2 but there would be a cos( ) part from both terms and a sin( ) part from each term. Let the coefficients of these two parts be new unknown constants so we have

$\psi(x) = A\cos\left[\left(\sqrt{\dfrac{2mE}{\hbar^2}}\right)x\right] + iB\sin\left[\left(\sqrt{\dfrac{2mE}{\hbar^2}}\right)x\right]$. However, at $x = 0$ the potential energy

goes up to $+\infty$ so physically no particle can be there and the wave function must be zero there; so $\psi(0) = 0$. Since $\cos(0) = 1$, the only way we can have $\psi(0) = 0$ is if $A = 0$. The other constant B is

still unknown but it can be nonzero. Now we have $\psi(x) = iB\sin\left[\left(\sqrt{\dfrac{2mE}{\hbar^2}}\right)x\right]$. The fact that

i appears as a factor may seem strange but we will find a way around that soon.

Next we need to apply the boundary condition to the right side of the box at $x = L$ and again we see that the wave function must go to zero at that wall. This is easily seen by considering

$\psi = \dfrac{\left(\frac{d^2\psi}{dx^2}\right)}{\left(\frac{2m}{\hbar^2}\right)[V - E]}$. As long as $\left(\dfrac{d^2\psi}{dx^2}\right)$ is finite, an infinite denominator makes $\psi = 0$.

Up to this point the value of the energy E could be any value but we are about to see it become

quantized! At $x = L$ we have $\psi(L) = iB\sin\left[\left(\sqrt{\dfrac{2mE}{\hbar^2}}\right)L\right] = 0$. We could set B to zero but then

there is no wave function at all! That is called the trivial solution and corresponds to no particle in the box! Is there any other way this can be zero? Yes, the $\sin(\theta)$ is zero every time $\theta = \pi$, 2π,

$3\pi, \ldots, n\pi$. Thus we find the quantization condition for the energy as $\left(\sqrt{\dfrac{2mE}{\hbar^2}}\right)L = n\pi$, which

means that $\left(\sqrt{\dfrac{2mE}{\hbar^2}}\right) = \dfrac{n\pi}{L}$ and $E_n = \dfrac{n^2\pi^2\hbar^2}{2mL^2} = \dfrac{n^2\pi^2 h^2}{2m(2\pi)^2 L^2} = \dfrac{n^2 h^2}{8mL^2} = E_n$. Now the energy can

have only certain values!

We still need to find the value of B. There was a lot of philosophical groping in the 1920s as scientists tried to interpret the meaning of the De Broglie wave but the physicists were well educated in electromagnetic theory. It was known that while the oscillating electric field of a light wave could be written as $E = Ae^{2\pi i v t}$, with the complex number $i = \sqrt{-1}$ embedded in the formula, only an intensity $I = (Ae^{+2\pi i v t})^*(Ae^{+2\pi i v t}) = A^2$ could be measured in the laboratory as a real number. Further, the intensity is the square of the amplitude of the wave. We have already seen that the complex number $i = \sqrt{-1}$ has occurred in our $\psi(x)$ wave function. Thus it was realized that $\psi^*\psi \propto$ *probability*. We can only measure real numbers in the laboratory but $\psi(x)$ can be complex so we need to measure the probability as the product of the wave function and its complex conjugate where if any part of the formula has $i = \sqrt{-1}$ it must be changed to $-i$ in the complex conjugate. Note if $\psi = a + ib$, then $\psi^* = a - ib$ and $\psi^*\psi = (a - ib)(a + ib) = a^2 + b^2$ which is a totally real number!

The next question is how to assign probability based on $\psi^*\psi$? "Certainty" might have been assigned 100% but instead $\psi^*\psi = 1$ was chosen as certainty with lesser values for other lower probability events. Therefore if we integrate our probability function over all possible values of the variable, the result should be set to 1. Thus we have $1 = \int \psi^* \psi \, dx$. Here $\psi^* = -iB \sin\left(\frac{n\pi x}{L}\right)$ and $(-iB)(iB) = +B^2$, so the "*" makes the result a real number. $\int_0^L B^2 \sin^2\left(\frac{n\pi x}{L}\right) dx = 1$. This condition will allow us to evaluate B! It is common practice to neglect the "i" if it is a direct factor since we have seen that it will disappear whenever an integral for a probability is carried out but it has served here to illustrate the need for the complex conjugate. Thus we have to do the integral and solve for the B coefficient. $\int_0^L B^2 \sin^2\left(\frac{n\pi x}{L}\right) dx = 1 = B^2 \int_0^L \left[\frac{1 - \cos\left(\frac{2n\pi x}{L}\right)}{2}\right] dx$ using the half-angle formula from trigonometry. We will have to split this into two integrals.

$$1 = B^2 \left(\frac{1}{2}\right) \left\{ \int_0^L dx - \int_0^L \cos\left(\frac{2n\pi x}{L}\right) dx \right\} = \left(\frac{B^2}{2}\right) \left\{ L - \left[\frac{\sin\left(\frac{2n\pi x}{L}\right)}{\left(\frac{2n\pi}{L}\right)}\right]_0^L \right\} = \left(\frac{B^2}{2}\right)[L - 0] = \frac{B^2 L}{2}$$

If $\frac{B^2 L}{2} = 1$, then we have found $B = \sqrt{\frac{2}{L}}$ so in conclusion we have the full solution (neglecting i)

$$\psi_n(x) = \left(\sqrt{\frac{2}{L}}\right) \sin\left(\frac{n\pi x}{L}\right) \quad \text{and} \quad E_n = \frac{n^2 h^2}{8mL^2}, \quad n = 1, 2, 3, 4, 5, \ldots, n$$

Note also that the $n = 0$ case is the "no-particle" case and the energy is quadratic in n for $n > 0$.

Let us consider the ultraviolet spectrum of butadiene (Table 11.1). Assuming two electrons loosely spin-paired in each orbital and neglecting the interaction between the electrons, we can fill two energy levels with four electrons. Then the lowest energy excitation will be from the highest occupied molecular orbital (HOMO) $n = 2$ to the lowest unoccupied molecular orbital (LUMO) $n = 3$. We can estimate the wavelength of this transition as a one-electron jump from $n = 2$ to $n = 3$. $\Delta E = E_3 - E_2 = h\nu = h\left(\frac{c}{\lambda}\right) = \frac{(9 - 4)h^2}{8mL^2} \Rightarrow \lambda = \frac{8mL^2 c}{5h}$, but what shall we use for L? We can estimate L as three times the length of the aromatic bond in benzene since the middle bond is part of the pi-electron system as $3 (1.4 \text{ Å}) = 4.2$ Å. Then we can find λ as $\lambda = \frac{8mL^2 c}{5h} = \frac{8(9.10938215 \times 10^{-31} \text{ kg})(4.2 \times 10^{-10} \text{ m})^2(2.99792458 \times 10^8 \text{ m/s})}{5(6.62606896 \times 10^{-34} \text{ J} \cdot \text{s})}$ and so $\lambda \cong 1.1632 \times 10^{-7}$ m $= 1.1632 \times 10^{-5}$ cm $= 1163.2 \times 10^{-8}$ cm $= 1163.2$ Å. This is too far into the vacuum ultraviolet so let us treat L as a variable parameter and find what value of L will fit the experimental wavelength of 2170 Å [2, p. 103].

TABLE 11.1
Absorption Bands of Selected *trans*-Polyenes

Compound	# π Electrons	λ (Å)
Ethylene	2	1625
Butadiene	4	2170
Hexatriene	6	2510
Octatetraene	8	3040

Source: Davis, J.C., *Advanced Physical Chemistry*, The Ronald Press Co., New York, 1965, p. 103.

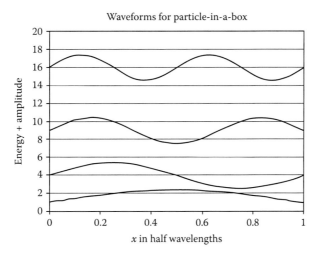

Waveforms for particle-in-a-box

x in half wavelengths

FIGURE 11.3 The PIB wave functions for butadiene spaced vertically approximately according to energy ($\propto n^2$). Note the number of nodes is $(n-1)$. (From Trindle, C. and Shillady, D., *Electronic Structure Modeling: Connections between Theory and Software*, CRC Press, Boca Raton, FL, 2008. With permission.)

$$L = \sqrt{\frac{5h\lambda}{8mc}} = \sqrt{\frac{5(6.62606896 \times 10^{-34}\,\text{J}\cdot\text{s})(2.170 \times 10^{-7}\,\text{m})}{8(9.10938215 \times 10^{-31}\,\text{kg})(2.99792458 \times 10^{8}\,\text{m/s})}} \cong 5.736 \times 10^{-10}\,\text{m}$$ so a

better estimate of the effective length of the box is $L = 5.736$ Å. While we probably did not need to carry so many significant figures in the constants and our estimated length is off by over 1.5 Å, the fitted value is reasonable. The extent of the diffuse 2P π orbitals surely extends out over the H atom region on the ends of the molecule. Thus the model is actually quite good considering the approximations made along the way.

In Figure 11.3, we see the first four solutions of the model and for butadiene the first two waves would be the occupied orbitals with orbital $\psi_3 = \sqrt{\frac{2}{L}}\sin\left(\frac{3\pi x}{L}\right)$ as the LUMO. Note that $\psi_2 = \sqrt{\frac{2}{L}}\sin\left(\frac{2\pi x}{L}\right)$ is the HOMO. We can see that the number of nodes in the wave function increases with the value of n, the quantum number. In Chapters 16 and 17, methods will be developed to express molecular orbitals as linear combinations of atomic orbitals (LCAO) with weighting coefficients for each atomic orbital. Table 11.2 shows that the same sinusoidal pattern and number of nodes occurs for butadiene when higher-level approximations are used and the molecular orbitals are expressed as columns of weighting coefficients.

In Figure 11.4 we see the same box with the probability functions $\psi^*\psi$ and the positive peaks indicate where the particle is most likely to be found. At low energy, the particle is most likely to be found in the middle of the box while at higher energy the particle will be rattling rapidly back and forth so its position will be more likely to be spread out.

Since this simple model is solvable, let us try to learn as much as we can about the solutions. First of all, there are many solutions and they form an "orthonormal set," $\{\psi_n\}$. We have normalized the various $\psi_n = \sqrt{\frac{2}{L}}\sin\left(\frac{n\pi x}{L}\right)$ functions and they can be shown to be orthogonal in the sense that $\int_0^L \psi_n^*\psi_m d\tau = 0$ so they form an orthonormal set. In the familiar Cartesian three-space, we know that a vector can be represented in terms of the $(\hat{i}, \hat{j}, \hat{k})$ unit vectors as $\vec{r} = x\hat{i} + y\hat{j} + z\hat{k}$ where $\hat{i} \cdot \hat{j} = 0$, $\hat{i} \cdot \hat{k} = 0$ and $\hat{j} \cdot \hat{k} = 0$ so that all the unit vectors are mutually at right-angles and hence orthogonal in

TABLE 11.2
Molecular Orbitals for *trans*-Butadiene,
2Pz Coefficients by Column

	π_1	π_2	π_3^*	π_4^*
2Pz(1)	0.376	−0.525	0.618	−0.502
2Pz(2)	0.488	−0.403	−0.462	0.682
2Pz(3)	0.488	0.403	−0.462	−0.682
2Pz(4)	0.376	0.525	0.618	0.502

Note: Weighting coefficients of an STO-3G SCF-MO calculation, see Chapter 17.

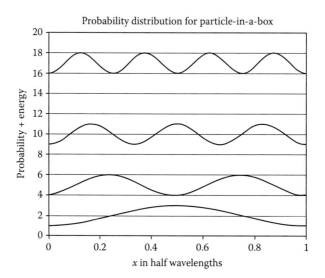

FIGURE 11.4 $\psi^*\psi$ Probabilities for the PIB butadiene wave functions. (From Trindle, C. and Shillady, D., *Electronic Structure Modeling: Connections between Theory and Software*, CRC Press, Boca Raton, FL, 2008. With permission.)

the dot-product sense. The extension to multidimensional "Hilbert space" algebra is necessary in quantum mechanics because there are many possible eigenfunctions but they are mutually orthogonal in the sense that there is an integral over the product instead of $\hat{i} \cdot \hat{j} = |i||j| \cos(\pi/2) = 0$. Students should be aware that there are really only about five or six problems that have known exact solutions and although there is no proof that all the solutions form complete sets, there is no known exception. As far as is known, a quantum mechanical solution to the Schrödinger equation should/will have a set of solutions that form a complete set for any problem with the same boundary conditions and there will be a set of corresponding eigenvalues. The set of orthogonal functions may need to be normalized to the arbitrary certainty of 1 to reach the full condition of orthonormality. This concept of a complete set of eigenfunctions and eigenvalues will be very useful for estimating solutions to problems not yet solved.

Let us use this exactly solvable model to learn more about quantum mechanics. Note that we could apply the Hamiltonian operator more than once and generate the value of the energy raised to a power $H_{op}(H_{op}\psi_n) = H_{op}(E_n\psi_n) = E_n(H_n\psi_n) = E_n^2\psi_n$, in fact $H_{op}^m\psi_n = E_n^m\psi_n$. This is so because ψ_n is an exact eigenfunction of H_{op} and we say E_n is a "good quantum number." However there are other quantities we would like to extract from the wave function. Return to the question of the most probable position "x" the particle will have in the lowest energy state. Although we can treat the coordinate as an operator, $x \to x_{op}$, the function is not an eigenfunction of this operator! We can see that $x\psi_1 = \sqrt{\dfrac{2}{L}} x \sin\left(\dfrac{1\pi x}{L}\right)$ is just another function of x, not an eigenfunction. However, there is an alternate way to obtain the average value of the coordinate x or the expectation of the average value of x as $<O> \equiv \dfrac{\int \psi^* O \psi d\tau}{\int \psi^* \psi d\tau}$ where O is any operator, so $<x> = <n|x|n> \equiv$

$\int_0^L \sqrt{\dfrac{2}{L}} \sin\left(\dfrac{n\pi x}{L}\right)(x)\sqrt{\dfrac{2}{L}} \sin\left(\dfrac{n\pi x}{L}\right) dx$. When we use previously normalized functions we do not need the denominator since it will be 1, but we should recall the analogy to the average grade of a class where we have to divide by the number of students in the class. Thus there is a way to do a weighted average of a quantum mechanical operator with the slight difference of inserting the operator between ψ^* and ψ. In undergraduate slang, we sometimes call this the "sandwich integral" where the operator is sandwiched between ψ^* and ψ, but really it is the "expectation value of the operator".

Now let us evaluate $<x> = \int_0^L \sqrt{\dfrac{2}{L}} \sin\left(\dfrac{\pi x}{L}\right)(x)\sqrt{\dfrac{2}{L}} \sin\left(\dfrac{\pi x}{L}\right) dx = \left(\dfrac{2}{L}\right)\int_0^L x \sin^2\left(\dfrac{\pi x}{L}\right) dx$.

$<x> = \left(\dfrac{2}{L}\right)\int_0^L x \sin^2\left(\dfrac{\pi x}{L}\right) dx = \left(\dfrac{2}{L}\right)\int_0^L x\left[\dfrac{1 - \cos\left(\frac{2\pi x}{L}\right)}{2}\right] dx$ and again do two integrals. $<x> =$

$\left(\dfrac{1}{L}\right)\left[\int_0^L x\, dx - \int_0^L x \cos\left(\dfrac{2\pi x}{L}\right)\right] = \left(\dfrac{1}{L}\right)\left\{\left(\dfrac{x^2}{2}\right)\Big|_0^L - \left[\dfrac{x \sin\left(\frac{2\pi x}{L}\right)}{\left(\frac{2\pi}{L}\right)}\right]_0^L + \int_0^L \dfrac{\sin\left(\frac{2\pi x}{L}\right) dx}{\left(\frac{2\pi}{L}\right)}\right\}$ where we

have used integration by parts for the second integral. To be complete we show details.

$<x> = \left(\dfrac{1}{L}\right)\left\{\left(\dfrac{L^2 - 0}{2}\right) - (0 - 0) - \left[\dfrac{\cos\left(\frac{2\pi x}{L}\right)}{\left(\frac{2\pi}{L}\right)^2}\right]_0^L\right\} = \left(\dfrac{1}{L}\right)\left\{\left(\dfrac{L^2}{2}\right) - 0 - \left[\dfrac{1 - 1}{\left(\frac{2\pi}{L}\right)^2}\right]\right\} = \dfrac{L}{2}$ and

so we find by calculation that the average value of the x coordinate is $<x> = L/2$. In fact this is true for any level n. Now in case you think you can just use common sense to guess properties all of the time let us ask what is the average value of $<x^2>$. Again we can set up the sandwich integral and carry out the integration by parts (twice) and we find that

$<x^2> = \int_0^L \sqrt{\dfrac{2}{L}} \sin\left(\dfrac{n\pi x}{L}\right)(x^2)\sqrt{\dfrac{2}{L}} \sin\left(\dfrac{n\pi x}{L}\right) dx = \left(\dfrac{L^2}{3}\right)\left[1 - \left(\dfrac{3}{2n^2\pi^2}\right)\right]$, leaving the proof to the

homework. Recall that in classical mechanics (sophomore physics) there is a coordinate for every momentum so let us consider $<P_x>$ as the average value of the x-momentum.

$$<P_x> = \int_0^L \sqrt{\dfrac{2}{L}} \sin\left(\dfrac{n\pi x}{L}\right)\left(\dfrac{\hbar}{i}\dfrac{d}{dx}\right)\sqrt{\dfrac{2}{L}} \sin\left(\dfrac{n\pi x}{L}\right) dx = \left(\dfrac{\hbar n\pi}{iL}\right)\int_0^L \sin\left(\dfrac{n\pi x}{L}\right)\cos\left(\dfrac{n\pi x}{L}\right) dx$$

$$<P_x> = \left(\dfrac{\hbar n\pi}{iL}\right)\int_0^L \sin\left(\dfrac{n\pi x}{L}\right)\cos\left(\dfrac{n\pi x}{L}\right) dx = \left(\dfrac{\hbar n\pi}{iL}\right)\left[\dfrac{\sin^2\left(\frac{n\pi x}{L}\right)}{\left(\frac{n\pi}{L}\right)}\right]_0^L = 0, \quad \text{so } <P_x> = 0.$$

That makes sense because momentum is a vector and the average direction of the momentum averages out to be exactly zero as the particle moves back and forth, but, what is $<P_x^2> = 0$?

$$<P_x^2> = \int_0^L \sqrt{\dfrac{2}{L}} \sin\left(\dfrac{n\pi x}{L}\right)\left(\dfrac{\hbar}{i}\dfrac{d}{dx}\right)^2\sqrt{\dfrac{2}{L}} \sin\left(\dfrac{n\pi x}{L}\right) dx = \left(\dfrac{n\pi\hbar}{L}\right)^2 = \dfrac{n^2 h^2}{4L^2}, \quad \text{so } (<P_x>)^2 \neq <P_x^2>.$$

This checks with the quantized energy as $E_n = \dfrac{P_x^2}{2m} \Rightarrow P_x^2 = 2mE_n = (2m)\left(\dfrac{n^2 h^2}{8mL^2}\right) = \left(\dfrac{n^2 h^2}{4L^2}\right)$.

Heisenberg [3] is widely credited with pointing out this problem from a statistical point of view and we can define the "variance" as $\sigma^2(x) = \langle x^2 \rangle - \langle x \rangle^2$ and also $\sigma^2(P_x) = \langle P_x^2 \rangle - \langle P_x \rangle^2$. Then we can consider the fact that x is the corresponding coordinate to P_x and wonder about their mutual effect on each other. Thus consider the product $\sigma_x \sigma_{P_x}$. Here we come to some intricate algebra but it will be worth it to find what is called the *Uncertainty Principle* [2, p. 96].

$$\sigma_x^2 = \left(\frac{L^2}{3}\right)\left[1 - \frac{3}{2n^2\pi^2}\right] - \left(\frac{L^2}{4}\right) = L^2\left(\frac{4-3}{12}\right) - \frac{L^2}{2n^2\pi^2} = \frac{L^2}{12}\left[1 - \frac{6}{n^2\pi^2}\right] = \frac{L^2}{12}\left[\frac{n^2\pi^2 - 6}{n^2\pi^2}\right].$$

Then $\sqrt{\sigma_x^2} = \frac{L}{2\sqrt{3}}\frac{\sqrt{n^2\pi^2 - 6}}{n\pi}$ and $\sqrt{\sigma_{P_x}^2} = \frac{nh}{2L}$ so using a trick [2, p. 96] of $(-2 = -3 + 1)$ we find

$$\sigma_x \sigma_{P_x} = \left(\frac{L}{2\sqrt{3}}\right)\left(\frac{nh}{2L}\right)\frac{\sqrt{n^2\pi^2 - 6}}{n\pi} = \frac{\hbar}{2}\sqrt{\frac{n^2\pi^2 - 6}{3}}$$

$$= \frac{\hbar}{2}\sqrt{\frac{n^2\pi^2}{3} - 3 + 1} = \frac{\hbar}{2}\sqrt{1 + \left(\frac{n^2\pi^2 - 9}{3}\right)} > \frac{\hbar}{2}.$$

We see that even for $n = 1$ the quantity in the square root will be greater than 1 and the uncertainty will increase as the quantum number n increases but the minimum value will be about $\left(\frac{\hbar}{2}\right)$.

This is the Heisenberg Uncertainity Principle: $\quad (\Delta P_x)(\Delta x) \geq \left(\frac{\hbar}{2}\right).$

What does this mean physically? The conclusion is that if we reduce the uncertainty in a coordinate, the momentum uncertainty will increase and vice versa. Further, the limit of uncertainty in both momentum and position cannot be made simultaneously any smaller such that their product of uncertainties is less than $\hbar/2$. While this is very small, it does indicate there is a tiny amount of "slop" in $(\Delta P_x)(\Delta x)$ and hints that unless we are dealing with an exact eigenvalue we will need to use some sort of statistical interpretation of quantum mechanics.

DEFINITION OF A COMMUTATOR

What is the cause of this uncertainty? We noted earlier that x and P_x are related and have a mutual effect. Another way to quantify the effect of position and momentum is by using a "commutator." In real arithmetic and algebra with real numbers, we are used to interchanging the order of factors as in $2 \times 3 = 3 \times 2 = 6$, that is the commutator $[3, 2] = 0$, but when we use calculus operators that interchange of order may not work. Let us define a quantity called the commutator as a bracket that represents *the amount by which interchanging the order of two successive operators makes a difference* (on some arbitrary function $f(x)$). Thus $[P_x, x]f(x) \equiv P_x x f(x) - x P_x f(x) = \left(\frac{\hbar}{i}\frac{d}{dx}\right)x f(x) - x\left(\frac{\hbar}{i}\frac{d}{dx}\right)f(x)$. Note the first term involves a derivative of a product while the second term does not. Thus we find that

$$[P_x, x]f(x) \equiv \left(\frac{\hbar}{i}\frac{d}{dx}\right)x f(x) - x\left(\frac{\hbar}{i}\frac{d}{dx}\right)f(x) = \left(\frac{\hbar}{i}\right)\left\{f(x) + x\frac{df}{dx} - x\frac{df}{dx}\right\} = \left(\frac{\hbar}{i}\right)f(x)$$

and so $[P_x, x] = \hbar/i$ for whatever $f(x)$ the operators are applied to and the order does matter! Thus the uncertainty principle is related to the fact that you really cannot know both the position and momentum exactly in a simultaneous way. Physically this can be put in very simple terms in that when you try to measure the momentum exactly the position becomes uncertain and if you pin down the momentum to a definite value, then the position is blurred. Fortunately for macroscopic measurements $\left(\hbar/2\right)$ is very small but for atoms and molecules this becomes a problem in precision measurements. Note we can sometimes use "Dirac notation" for the sandwich integral "expectation value" expressed as a "bra-ket" where the complex part on the left is the "bra" and the real part on the right is the "ket" for operator "O" in $\int \psi_m^* O \psi_n d\tau = \langle m|O|n \rangle$.

POSTULATES OF QUANTUM MECHANICS

Having solved only one problem in quantum mechanics we allow for the fact that there are only a very few known solved problems and try to write down the postulated rules and then we will apply the rules to another problem to reinforce the concepts.

Postulate I   The state of a quantum-mechanical system is completely specified by a function $\Psi(r, t)$ that depends on the coordinates of the particle and on time. This function, called the wave function or the state function, has the important property that the product of $\Psi^*(r, t)\ \Psi(r, t)\ dxdydz$ is the probability that the particle lies in the volume $dxdydz$ located at $r(x, y, z)$ at time t.

(Note that "completely specified" means that the wave function contains all the information that can be obtained about the system using quantum mechanics! That provides tremendous motivation to solve the Schrödinger equation and find the explicit wave function!)

Corollary   In order for the wave function to be used in the Schrödinger equation, it must have several mathematical properties:

1. It must be finite.
2. It must be continuous and single valued.
3. It must be defined for at least first and second derivatives.

Postulate II   To every observable laboratory measurement in classical mechanics there corresponds an operator in quantum mechanics.

Corollary   Cartesian coordinates (x, y, z), spherical polar coordinates (r, θ, ϕ), or in general any set of coordinates, q, merely become multiplicative operators, while the corresponding momentum operators, P_q, become differential operators such as $\left(\dfrac{\hbar}{i} \dfrac{\partial}{\partial q} \right)$.

Postulate III   In any measurement of the observable associated with the operator A, the only exact values that will ever be observed are the eigenvalues a_j which satisfy the eigenvalue equation

$$A\psi_j = a_j\psi_j.$$

Corollary   If the state function is not an eigenfunction of the operator A, then only an average value can be obtained as from many measurements; see Postulate IV.

Postulate IV   If a system is in a state described by a normalized wave function Ψ, then the average value of the observable corresponding to the operator A is given by

$$<a> = \int\limits_{-\infty}^{+\infty} \Psi^* A\Psi d\tau.$$

Corollary If the wave function is not normalized, then the average value of the observable corresponding to the operator A is given by

$$<a> = \frac{\int_{-\infty}^{+\infty} \Psi * A \Psi d\tau}{\int_{-\infty}^{+\infty} \Psi * \Psi d\tau}.$$

Postulate V The wave function or state function of a system evolves in time according to the time-dependent Schrödinger equation

$$H\Psi(r, t) = i\hbar \frac{\partial \Psi}{\partial t}.$$

While time-dependent processes can be treated, our emphasis will be on time-independent wave functions at the undergraduate level.

PARTICLE ON A RING

There is another problem in quantum mechanics that is exactly solvable and can be applied to chemistry. Let us assume there is a particle constrained to move only on a ring of fixed radius, a. We could consider this to be a tiny glass bead with a hole riding on a lubricated ring of thin wire, but anyone who has recently taken a course in organic chemistry should immediately see the analogy to that delocalized ring drawn on aromatic compounds such as benzene. This is an important problem because it is solvable and because it illustrates the general rules of quantum mechanics we have outlined earlier. First we write the classical momentum and assume the potential energy on the ring is zero, but here we have to use polar coordinates with a fixed radius and only allow the angle θ to vary so the coordinate of interest is $dc = a\,d\theta$ for a circumference c and a radius a. This problem is solvable and is especially important because it is another illustration of the proper way to treat the complex arithmetic involving $i = \sqrt{-1}$. For the sake of simplicity, we assume that the potential energy of the particle on the ring is zero so we only consider the kinetic energy in polar coordinates and form the Hamiltonian operator. $H = T + V = T + 0 = \frac{\left(\frac{\hbar}{i} \frac{d}{a\,d\theta}\right)^2}{2m} + 0 = \frac{-\hbar^2}{2ma^2} \frac{d^2}{d\theta^2}$ and then set up $H\psi = E\psi$ to be solved. Thus $\frac{-\hbar^2}{2ma^2} \frac{d^2\psi}{d\theta^2} = E\psi \Rightarrow \frac{d^2\psi}{d\theta^2} + \frac{2ma^2 E\psi}{\hbar^2} = 0 \Rightarrow$ $\left[D + i\sqrt{\frac{2ma^2 E}{\hbar^2}}\right] \left[D - i\sqrt{\frac{2ma^2 E}{\hbar^2}}\right] \psi = 0$. This is very similar to the PIB problem but there are some important differences. First the differential circumference $dc = a\,d\theta$ has a constant in it and as we will soon see, the boundary conditions are different because the wave function must be continuous and connect smoothly as we go around the ring. However, we do know how to solve this kind of second-order differential equation as $\psi = C_1 e^{-i\sqrt{\frac{2ma^2 E}{\hbar^2}}\theta} + C_2 e^{+i\sqrt{\frac{2ma^2 E}{\hbar^2}}\theta}$. We do not know either C_1 or C_2 but we do know that we must enforce $\psi(2\pi) = \psi(0)$ to join the beginning and end of the wave function smoothly as it goes around the ring (in a standing wave). If there is a sudden jump in the wave function at $\psi(2\pi)$, the second derivative will be undefined at the break point and the whole equation will be meaningless! Since the two solutions only differ in the sign of the exponent, that sign can be absorbed into the value of $\pm\theta$ and there is only one form of the wave function as $\psi = C e^{\pm i\sqrt{\frac{2ma^2 E}{\hbar^2}}\theta} \Rightarrow \psi(0) = C e^0 = \psi(2\pi) = C e^{\pm i 2\pi \sqrt{\frac{2ma^2 E}{\hbar^2}}} \Leftrightarrow$ $2\pi \sqrt{\frac{2ma^2 E}{\hbar^2}} = 0, 2\pi, 4\pi, \ldots, n(2\pi)$ so we find that $\sqrt{\frac{2ma^2 E}{\hbar^2}} = \pm n$ and then we have

$$\sqrt{\frac{2ma^2E}{\hbar^2}} = \pm n \Rightarrow E_n = \frac{n^2\hbar^2}{2ma^2}, \, n = 0, \pm 1, \pm 2, \ldots.$$ Then we also find that $\psi_n = Ce^{\pm in\theta}$, $n = 0, \pm 1,$ $\pm 2, \ldots$. Note there is a solution for $n = 0$ in this case! Also we have what are called degenerate solutions for $n = \pm 1, \pm 2, \ldots$, that is for a given n-level there are two states with the same energy. States with the same energy but different wave functions are said to be degenerate.

The last thing to do is to normalize the wave functions so once again we set the integrated probability to 1.

$$1 = \int_0^{2\pi} (Ce^{in\theta})*(Ce^{in\theta})d\theta = C^2\int_0^{2\pi} e^{i(n-n)\theta}d\theta = C^2\int_0^{2\pi} d\theta = C^2(2\pi - 0) = 2\pi C^2 = 1 \text{ so } C = \frac{1}{\sqrt{2\pi}}.$$

Let us take this opportunity to show the wave functions are orthogonal for any $m \neq n$.

$$\int_0^{2\pi} \left(\frac{e^{in\theta}}{\sqrt{2\pi}}\right)^* \left(\frac{e^{im\theta}}{\sqrt{2\pi}}\right)d\theta = \left(\frac{1}{2\pi}\right)\int_0^{2\pi} e^{i(m-n)\theta}d\theta = \left[\frac{e^{i(m-n)\theta}}{2\pi i(m-n)}\right]_0^{2\pi} = 0, \quad m \neq n.$$

Since any integer multiple of the full 2π range will be the same at the upper and lower limits, they cancel. Thus we have a lowest level of $n = 0$ with zero energy whose wave function is $\psi_0 = \frac{1}{\sqrt{2\pi}}$, not 0. Above that energy there are degenerate energy pairs. Finally we have the complete solution $\psi_n = \frac{e^{in\theta}}{\sqrt{2\pi}}$ and $E_n = \frac{n^2\hbar^2}{2ma^2}, \, n = 0, \pm 1, \pm 2, \ldots$. We also have another interesting relationship here in that the wave functions are also eigenfunctions of the angular momentum.

$$\left(\frac{\hbar}{i}\frac{d}{ad\theta}\right)\frac{e^{\pm in\theta}}{\sqrt{2\pi}} = \left(\frac{\pm n\hbar}{a}\right)\frac{e^{\pm in\theta}}{\sqrt{2\pi}}.$$

The equal value with opposite sign for the n quantum number implies a pair of degenerate energy orbitals but with the particle traveling in opposite directions.

While any exact solution should be appreciated in a mathematical sense, there really is a major conclusion here. Note the pattern of the energy levels in Table 11.3 from a self-consistent-field calculation including "core orbitals" as well as 2Pz π-electrons. The numbers in parentheses are the orbital numbers in the presence of many lower energy core orbitals and the negative energies are for occupied orbitals with positive energies for empty orbitals. There is one non-degenerate level followed by two sets of double degenerate levels. The pattern returns to a non-degenerate level for the benzene molecule because there are only six pi orbitals. *The point is that the particle-on-a-ring (POR) model predicts and explains the (4n+2) rule of aromaticity in organic chemistry!*

Let us use the POR model to estimate the HOMO $\rightarrow$ LUMO transition in benzene.

$$\Delta E = E_2 - E_1 = h\nu = \frac{hc}{\lambda} = \frac{(2^2 - 1^2)\hbar^2}{2ma^2} = \frac{3h^2}{2ma^2(2\pi)^2} \Rightarrow \lambda_{\pi\rightarrow\pi^*} = \frac{8\pi^2ma^2c}{3h}.$$

Let us ignore the H atoms of benzene and note that a hexagon can be made into six isosceles triangles with approximately 1.4 Å sides (actually 1.395), which is a good approximate value for a, and let us assume that the six pi electrons are three spin-pairs in levels $n = 0$, $n = 1$, and $n = -1$. Then the HOMO $\rightarrow$ LUMO transition can be computed using the formula derived earlier.

$$\lambda_{\pi\rightarrow\pi^*} = \frac{8\pi^2ma^2c}{3h} = \frac{8\pi^2(9.10938215 \times 10^{-31} \text{ kg})(1.4 \times 10^{-10} \text{ m})^2(2.99792458 \times 10^8 \text{ m/s})}{3(6.62606896 \times 10^{-34} \text{ J} \cdot \text{s})},$$

TABLE 11.3
STO-3G One-Electron Energies for Pi-Molecular
Orbitals of Benzene (Hartrees)

$E(24) = 0.4940$
$E(23) = 0.2664$
$E(22) = 0.2664$
$E(21) = -0.2699$
$E(20) = -0.2699$
$E(17) = -0.4354$

Note: STO-3G $\Rightarrow$ Slater-type-orbitals fitted with a linear combination of three Gaussian orbitals. See Chapter 17.

$\lambda_{\pi \to \pi^*} \cong 2.12607 \times 10^{-7}$ m $= 2126.07 \times 10^{-10}$ m $\cong 2126$ Å but the experimental transition for benzene is at 262 nm $= 2620$ Å. Let us consider what value of the radius would produce the experimental wavelength.

$$\lambda_{\pi \to \pi^*} = \frac{8\pi^2 m a^2 c}{3h} \Rightarrow a = \sqrt{\frac{3\lambda h}{8\pi^2 mc}} = \sqrt{\frac{3(2620 \times 10^{-10} \text{ m})(6.62606896 \times 10^{-34} \text{ J} \cdot \text{s})}{8\pi^2 (9.10938215 \times 10^{-31} \text{ kg})(2.99792458 \times 10^8 \text{ m/s})}},$$

$\lambda_{\pi \to \pi^*} = \dfrac{8\pi^2 m a^2 c}{3h} \Rightarrow a \cong 1.554 \times 10^{-10}$ m $= 1.554$ Å. Noting that the C–H bonds are about 1 Å the width across the benzene molecule from H to H would be about 4.8 Å for a radius of 2.4 Å. Thus we have had to adjust the effective radius to be slightly larger than directly over the C atoms but not by much and one might imagine that the pi electrons are somewhat attracted to the H atoms. Actually, we have only adjusted the radius to fit the experimental wavelength. However, it is pleasing that such a simple model comes so close to the experimental value! Figure 11.5 shows the coefficients ($\times 1000$) of the pi-orbitals of benzene from an all-electron self-consistent field calculation where it is evident that the lowest energy orbital is totally symmetric with all the coefficients the same, while the next orbitals have nodal patterns in a way that corresponds to the POR wave solutions if we use $e^{i\theta} = \cos(\theta) + i\sin(\theta)$.

Can this model be extended further? With one additional, fairly severe, assumption, the POR model can be extended to other aromatic hydrocarbons [4–6]. The main assumption is that the interior bonds of a poly-aromatic hydrocarbon such as naphthalene are less important than the delocalized pathway on the outer circumference of the ring structure. A crude demonstration can be made using a wire coat hanger that has a specific shape but that can be easily bent into a circle (with a handle). If we break the middle bond in naphthalene and bend the outer bonds into a circular polygon, we can apply the POR model to the ring. There are ten bonds in the outer edge of naphthalene and we can use the approximation that their length is about 1.4 Å which is roughly the bond length in benzene. Thus we can set the circumference of our model ring equal to (10) (1.4 Å) $= 14$ Å $= C = 2\pi a$. That gives us a radius for our POR ring.

$a = \dfrac{14 \text{ Å}}{2\pi} = 2.228$ Å. Then we note there are 10 pi-electrons in naphthalene so we expect the POR model to include 5 pairs of electrons. That means the HOMO $\to$ LUMO transition will be from $n = 2$ to $n = 3$. Then we can calculate the $\pi \to \pi^*$ wavelength as

$$\Delta E = E_3 - E_2 = h\nu = \frac{hc}{\lambda} = \frac{(3^2 - 2^2)\hbar^2}{2ma^2} = \frac{3h^2}{2ma^2(2\pi)^2} \Rightarrow \lambda_{\pi \to \pi^*} = \frac{8\pi^2 m a^2 c}{5h}.$$

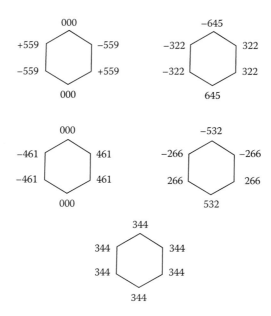

FIGURE 11.5 The 2Pz coefficients of benzene multiplied by 1000, obtained from a STO-3G SCF-MO calculation using the program PCLOBE, see Chapter 17. Only the lowest five orbitals are shown. (From Trindle, C. and Shillady, D., *Electronic Structure Modeling: Connections between Theory and Software*, CRC Press, Boca Raton, FL, 2008. With permission.)

$$\lambda_{\pi \to \pi^*} = \frac{8\pi^2(9.10938215 \times 10^{-31}\,\text{kg})(2.22816 \times 10^{-10}\,\text{m})^2(2.99792458 \times 10^8\,\text{m/s})}{5(6.62606896 \times 10^{-34}\,\text{J} \cdot \text{s})}$$

$$\lambda_{\pi \to \pi^*} = 3.231217 \times 10^{-7}\,\text{m} \cong 3231.217 \times 10^{-10}\,\text{m} = 3231\,\text{Å}.$$

Experimentally the leading edge of the ultraviolet spectrum of naphthalene occurs at about 3150 Å but electronic spectra of large molecules are broadened by vibrational interactions as we will see in a later chapter. Thus we will take the calculated $\pi \to \pi^*$ wavelength as qualitatively correct; quite good considering the simplicity of the model. Overall as the aromatic molecules become larger, the POR model is useful for qualitative reasoning. The POR model was first discussed by Platt [4] in 1949, expanded by Moffit [5] in 1954 and used for interpretation of magnetic circular dichroism spectra by Michl [6] in 1978.

COMPARISON OF PIB AND POR APPLICATIONS

An instructive comparison of the PIB and POR models is given by the diagram in Figure 11.5. The main point for students is to know whether to use the degenerate levels in the POR model or the non-degenerate levels in the PIB model. The next important approximation is to estimate length or circumference using an average of the –C=C– bond length as 1.4 Å. From then on we have to realize these models mainly apply to conjugated hydrocarbons. We should remember that the POR model provides understanding of aromaticity. We have used these simple, but solvable, model systems to illustrate the postulates of quantum mechanics and now we are ready to proceed to more realistic problems.

ADDITIONAL THEOREMS IN QUANTUM MECHANICS

We now have seen two problems with exact solutions in quantum mechanics. Actually there are only a few such exact solutions remaining such as the harmonic oscillator, the rigid rotor, the hydrogen atom, and the forced harmonic oscillator, and we need to know some general principles which may help with approximate solutions to unsolved problems. The first useful concept is really a self-fulfilling definition that helps enforce quantum mechanics agree with laboratory measurements in spite of the complications due to complex arithmetic.

Definition: Define the adjoint of an operator as $<\psi|A|\psi> = <\psi|A^\dagger|\psi>^*$

This can also be stated in words that an operator A usually operates to the right on a function but it can also operate to the left on the complex adjoint $\psi^* = <\psi|$ and the definition above means that if the operator acts to the left it must be the adjoint form of the operator. This is the characteristic of a "Hermitian" operator. This also means that in the matrix mechanics form of quantum mechanics a given "matrix-element" of a Hermitian matrix is related to another element on the other side of the (upper left to lower right) diagonal of the matrix by the relationship $A_{mn} = A_{nm}^*$. The following theorem shows why this definition is useful.

Theorem 1 *The eigenvalues of a Hermitian operator are real numbers.*

Proof: Given $A\psi = a\psi$, form the expectation value $\int \psi^* A\psi d\tau$, then operate to the left and the right.

$$a^* \int \psi^*\psi d\tau = \int \psi^* \overleftrightarrow{A} \psi d\tau = a \int \psi^*\psi d\tau.$$

Start in the middle and operate inside the integral to the right and to the left using the adjoint rule. Now subtract the right side of the equation from the left to obtain a new condition

$$a^* \int \psi^*\psi d\tau - a \int \psi^*\psi d\tau = (a^* - a) \int \psi^*\psi d\tau = 0.$$

Think about that. Assuming $\int \psi^*\psi d\tau \neq 0$, that means that $(a^* - a) = 0$ and that can only be true if $a = b + ic = a^* = b - ic$ and that can only be true if $c = 0$. Thus a is a real number! Q.E.D. This can be generalized to expectation values of Hermitian operators as given elsewhere [7] but here it is sufficient to define an adjoint operator and that the definition guarantees real eigenvalues of Hermitian operators. The next theorem is very useful but is built on the previous theorem.

Theorem 2 *Two different eigenfunctions of a Hermitian operator are orthogonal if their eigenvalues are not equal.*

Proof: Given $A\psi_1 = a\psi_1$ and $A\psi_2 = b\psi_2$ with $a \neq b$, consider an integral between $<\psi_1|A|\psi_2>$.

$$a^* \int \psi_1^*\psi_2 d\tau = \int \psi_1^* \overleftrightarrow{A} \psi_2 d\tau = b \int \psi_1^*\psi_2 d\tau.$$

Once again subtract the right side of the equation from the left side to obtain

$$(a^* - b) \int \psi_1^*\psi_2 d\tau = 0.$$

We know that $a^* = a$ because it is a real eigenvalue and since $a \neq b$ we see that $(a - b) \neq 0$. Then the only conclusion is that $\int \psi_1^*\psi_2 d\tau = 0$, which means ψ_1 and ψ_2 are orthogonal, Q.E.D.

Theorem 3 *A single set of eigenfunctions can exist for two different Hermitian operators if the operators commute.*

Proof: Assume there really are two operators, A and B, which have the same eigenfunction ψ so that $A\psi = a\psi$ and $B\psi = b\psi$ and that $[A, B] = 0$. Again we set up the integral $<\psi|[A, B]|\psi>$.

$$0 = \int \psi^*(0)\psi d\tau = \int \psi^*[A, B]\psi d\tau = \int \psi^*(AB - BA)\psi d\tau = (a^*b^* - ba) \int \psi^*\psi d\tau = 0, \text{ Q.E.D.}$$

This is an "existence" theorem, it means that there can be such a ψ or $\{\psi_n\}$ but even if we show the two operators commute it does not help find the set $\{\psi_n\}$. However, the most important result of this theorem is that if two operators commute (not all do!) then we can "know" their eigenvalues or expectation values simultaneously. For two operators that do not commute we may be able to find the observable for one, while the other is not completely defined or vice versa. We can mention that the angular momentum operators (Lx, Ly, Lz) do not commute and a given set of eigenfunctions for the Lz operator may not give a clear interpretation of the eigenvalues for (Lx, Ly) using the eigenfunctions of Lz. We delay further discussion of this problem until the chapter on the H atom. The main use of this third theorem is to use the eigenfunctions of some particular operator, which commutes with the operator we are interested in to evaluate the expectation value in the known set of eigenfunctions. We will see that there are some blind alleys in quantum mechanics and we often have to use any trick we can think of to evaluate what we want to know, even if it is a roundabout approach and for those cases this third theorem can be useful.

SUMMARY

In this chapter we explored the question of how to use the De Broglie matter waves to describe molecular phenomena.

1. The (time-independent) Schrödinger wave equation $H\psi = E\psi$ was derived from the second derivative of an arbitrary wave function by incorporating $\lambda = \dfrac{h}{mv} = \dfrac{h}{p}$ into the second derivative expression of the wave function. The characteristics of an eigenvalue equation were noted.

2. It was noted that there is a total energy operator, H, called the Hamiltonian operator, which can be set up by writing the kinetic energy operator in terms of momentum and the potential energy without change except to regard it as a multiplicative operator. The key to converting classical physics formulas into quantum mechanical operators is merely to use $p_{op} = \left(\dfrac{\hbar}{i} \dfrac{d}{dq}\right)$ where q is the coordinate corresponding to the momentum and $\hbar = \dfrac{h}{2\pi}$. Then all one has to do is to solve the differential equation $H\psi = E\psi$ to find the wave function ψ and the energy E. In principle, the wave function ψ contains all the available information about a given system and $\psi^*\psi$ can be interpreted as a probability. Physical quantities are described by "operators" as calculus functions, which can be applied to the wave function to obtain the quantized values of physical observables.

3. The problem of noninteracting particles trapped in a one-dimensional "box" with infinitely high walls was solved to find the results $E_n = \dfrac{n^2 h^2}{8mL^2}$ and $\psi_n = \sqrt{\dfrac{2}{L}} \sin\left(\dfrac{n\pi x}{L}\right)$. The $n = 0$ level is a trivial solution for "no particle" because $\sin(0) = 0$ and the energy levels are non-degenerate with quadratically increasing spacing in n. However, the actual spacing between the levels can be quite small as for the quantized translational energy of a gas molecule.

4. The PIB model can be used to estimate the electronic transition wavelengths of linear polyenes by using 1.4 Å per bond in the conjugated chain, although slightly longer values of the total length L give better answers. The simple model with $V = 0$ in the box gives qualitatively useful interpretation of $\pi \rightarrow \pi*$ transitions in organic compounds assuming no electron–electron repulsion or exchange and simple spin pairing of two electrons per orbital.

5. A similar pi-electron model can be solved for the POR where $E_n = \dfrac{n^2 \hbar^2}{2ma^2}$ and $\psi_n = \dfrac{e^{\pm in\theta}}{\sqrt{2\pi}}, n = 0, \pm 1, \pm 2, \pm 3, \ldots$. In this case, the $n = 0$ level is not zero and the levels above $n = 0$ are doubly degenerate; this leads to a correspondence with the $4n + 2$ rule for aromatic character in organic chemistry.

6. The POR model can be extended to other aromatic pi-electron ring systems by ignoring the inner cross-ring bonds and using a ring defined by the radius "a" of a circle whose circumference is obtained as 1.4 Å times the number of bonds in the outer circumference of the pi-electron ring system. This provides a qualitative model for aromatic ring systems assuming perfect spin pairing of two electrons per orbital and neglect of any repulsion or exchange between electrons.

7. A number of theorems are introduced related to the properties of solutions of the Schrödinger equation. It is especially noted that valid wave functions must be (a) finite, (b) continuous, and (c) single-valued.

PROBLEMS

11.1   Estimate the wavelength of the HOMO $\rightarrow$ LUMO $\pi \rightarrow \pi*$ of all-trans octatetraene.

11.2   Calculate the value of the box length to bring the PIB HOMO $\rightarrow$ LUMO $\pi \rightarrow \pi*$ wavelength of octatetraene into agreement with the experimental wavelength of 3040 Å.

11.3   Use integration by parts twice to derive the expression for the average value of $<x^2>$

$$<x^2> = \int_0^L \sqrt{\frac{2}{L}} \sin\left(\frac{\pi x}{L}\right)(x^2)\sqrt{\frac{2}{L}}\sin\left(\frac{\pi x}{L}\right)dx = \left(\frac{L^2}{3}\right)\left[1 - \left(\frac{3}{2n^2\pi^2}\right)\right].$$

11.4   Show that $[P_x, x] = -[x, P_x]$, hint apply the operators to a dummy function $f(x)$.

11.5   Use the uncertainty relationship to estimate the uncertainty in the product $(\Delta x)(\Delta P_x)$ for the $n = 1, 2, 3, 4,$ and 5 levels of the PIB problem. Use factors of $\left(\dfrac{\hbar}{2}\right)$.

11.6   Estimate the HOMO $\rightarrow$ LUMO $\pi \rightarrow \pi*$ wavelength for anthracene ($C_{14}H_{10}$) using the Platt Perimeter extension of the POR model.

11.7   Compare (calculate) the HOMO $\rightarrow$ LUMO $\pi \rightarrow \pi*$ wavelength for azulene ($C_{10}H_8$) using the Platt Perimeter extension of the POR model; compare your result for the example calculation for naphthalene in the text. What does this say about the Perimeter model?

11.8   Evaluate the energy of the $n = 1$ and $n = 2$ levels of an electron-in-a-box for $L = 10$ Å in joules and show the wave functions are orthogonal by direct integration of $\int \psi_1^* \psi_2 d\tau$. Use trigonometry relationships as needed. ($\sin(2\theta) = 2\sin(\theta)\cos(\theta)$ might help).

11.9   Evaluate the angular momentum of the $n = 0, n = 1$ and $n = 2$ levels for a POR where the radius is 1 Å, in h-bar units. Note the POR wave functions are eigenfunctions of the angular momentum as well as the energy!

11.10   Prove the wave functions of the $n = 1$ and $n = 2$ levels of the POR model are orthogonal by direct integration of the product of the normalized functions.

STUDY, TEST, AND LEARN?

Realistically, Chapters 10 and 11 contain some important material that is perhaps the last really "must learn" material in the second semester. There is a considerable amount of basic physics in Chapter 10, and Chapter 11 has the basic concepts of quantum chemistry. From now on, the material will emphasize learning the main conclusions but only using the mathematics as a reference. However, this is a good point to have a quiz over Chapters 10 and 11. On another page we will show some tests from a nearby four year Liberal Arts College, Randolph Macon College where the physical chemistry courses are CHEM 311 and 312 with 312 as an elective.

Once again, the students were encouraged to memorize derivations and to try to answer every question with the additional knowledge that their grade would be based on the grade they achieve on the final examination or their average including the final examination grade, whichever is higher.

Chemistry 312	Randolph Macon College	Spring 2009	D. Shillady, Professor
(Points)	Quiz #1	(Attempt all problems)	55 min

(20) 1. Compute the De Broglie matter wavelength of an electron ejected from Cu ($W_f = 4.45$ eV) by a two-photon absorption of 4880 Å from an intense Ar^+ laser ($\lambda_{DB} = 15.42$ Å).

(20) 2. Derive the quantized energy levels of a "POR" and use the perimeter model to estimate the first $\pi \rightarrow \pi^*$ wavelength of anthracene ($C_{14}H_{10}$) and normalize the wave function. Use ($-C \approx C -$ of 1.4 Å) ($\lambda_{\pi \rightarrow \pi^*} \cong 4533$ Å).

(10) 3. Estimate the Auger wavelength of x-rays emitted when Cu ($Z = 29$) is used as an electron beam target and an $n = 2$ electron "falls" to a 1s Bohr orbital ($\lambda_{x-ray} \cong 1.445$ Å).

(10) 4. If two eigenfunctions of the same Hermitian operator have different eigenvalues, prove the eigenfunctions are orthogonal, after first proving the eigenvalues are real (see notes).

(20) 5. Crystalline NaCl has a simple cubic unit cell with a Na–Cl distance of 2.76 Å. Calculate the angle of the $n = 2$ LEED pattern for the full-Bragg diffraction of 150 eV electrons. ($n\lambda = 2d \sin(\theta)$, $\theta = 21.253°$, angle of incidence = angle of reflection).

(20) 6. Adjust the "length, L" of the PIB model to fit the first $\pi \rightarrow \pi^*$ transition for all-*trans*-hexatriene (C_6H_8) to the experimentally observed wavelength of 2510 Å ($L \cong 7.3$ Å).

REFERENCES

1. Schrödinger, E., Quantizierung als Eigenwertproblem (Erste Mitteilung) **79**, 361 (1926); Schrödinger, E., Quantizierung als Eigenwertproblem (Zweite Mitteilung), *Ann. Phys.* **79**, 489 (1926); Schrödinger, E., Quantisierung als Eigenwertproblem (Dritte Mitteilung), *Ann. Phys.* **80**, 437 (1926); Schrödinger, E., Quantisierung als Eigenwertproblem (Vierte Mitteilung), *Ann. Phys.* **81**, 109 (1926).
2. Davis, J. C., *Advanced Physical Chemistry*, The Ronald Press Co., New York, 1965.
3. Heisenberg, W., Über den anschaulichen Inhalt der quantentheoretischen Kinematik und Mechanik, *Zeitschrift fur Physik*, **43**, 172 (1927).
4. Platt, J. R., Classification of spectra of cata-condensed hydrocarbons, *J. Chem. Phys.*, **17**, 484 (1949).
5. Moffit, W., The electronic spectra of cata-condensed hydrocarbons, *J. Chem. Phys.*, **22**, 320 and 1820 (1954).
6. Michl, J., Magnetic circular dichroism of cyclic .pi.-electron systems. 1. Algebraic solution of the perimeter model for the A and B terms of high-symmetry systems with a $(4N + 2)$-electron [n]annulene perimeter, *J. Am. Chem. Soc.*, **100**, 6801 (1978).
7. Trindle, C. and D. Shillady, *Electronic Structure Modeling: Connections between Theory and Software*, CRC Press, Boca Raton, FL, 2008, p. 15.

12 The Quantized Harmonic Oscillator: Vibrational Spectroscopy

INTRODUCTION

Now that we have learned some of the principles which apply in quantum mechanics, we move on to the next more difficult problem of a quantized harmonic oscillator. The goal here is to provide a rigorous application of the polynomial method of solving differential equations on a relatively simple case and to provide some insight into how the Schrödinger equation was first solved [1]. Then we proceed to application in the form of worked examples. We have tried to give sufficient details of this solution to allow a student to follow the derivation with pencil and paper but do not forget to ponder over the spectroscopic applications! This author would agree that it is more important to absorb the main conclusions of this material than to master the derivations. In fact, the highest recommendation of this author is to always ask "What does this mean?" and absorb the conclusions for a future activity called "thinking" rather than just memorizing facts.

The previous particle-in-a-box (PIB) and particle-on-a-ring (POR) problems both had $V = 0$ and only dealt with the kinetic energy operator. The essence of the harmonic oscillator is a parabolic potential energy $V = \dfrac{kx^2}{2}$ where k is the "spring constant" or "restoring force constant." According to the idea that *a force is the negative derivative of a potential*, we have $f(x) = -\dfrac{dV}{dx} = -\dfrac{2kx}{2} = -kx$ so k is the proportionality factor of the force of a spring stretched or compressed away from $x = 0$. The classical case can be solved easily by equating two forces $f = ma = m\dfrac{d^2x}{dt^2} = -kx$. The solution from sophomore physics is $x = A \sin\left[\left(\sqrt{\dfrac{k}{m}}\right)t\right]$ and this can be shown by direct substitution. $\dfrac{dx}{dt} = A\sqrt{\dfrac{k}{m}}\cos\left[\left(\sqrt{\dfrac{k}{m}}\right)t\right]$ and $\dfrac{d^2x}{dt^2} = -A\left(\dfrac{k}{m}\right)\sin\left[\left(\sqrt{\dfrac{k}{m}}\right)t\right] = -\left(\dfrac{k}{m}\right)x$. So $m\dfrac{d^2x}{dt^2} = -kx$, Q.E.D.

This could have been solved to obtain the same result in more complicated ways but we want to build on your experience from problems in physics and we note that "A = amplitude."

Here we are faced with the *quantum mechanical* treatment of the harmonic oscillator. This potential does not take into account the fact that a chemical bond will eventually break when stretched more than about 10 Å but it does provide a very good approximation to the small vibrations of a bond in low energy states. An interesting fact plays a role here in that on planet Earth at average temperatures around 25°C, almost all molecules are in very low vibrational states. This is fortuitous for the use of a potential which is most accurate at low energies. We see in Figure 12.1 a sketch of a parabolic potential superimposed on a model of a dissociative bond potential. Here the value of the potential is a minimum at $r = r_0$ and the energy is negative for a "bound" state but we will measure the energy as positive above the minimum of the parabolic potential. We see that the parabolic potential makes a good fit to a real potential for a considerable

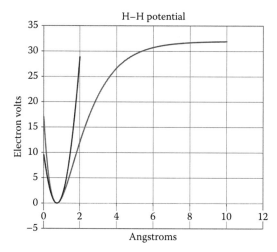

FIGURE 12.1 A parabolic potential superimposed on a Morse potential using actual data for the H_2 molecule. An exact but more complicated solution also exists for the Morse potential. (See Ref. [3])

portion of the bound region but eventually the parabola extends to $+\infty$ while the real potential is asymptotic to a zero energy dissociation limit as the bond is stretched. The region where the real potential departs from the harmonic parabola is called the "anharmonic" region but most of the applications deal with the harmonic region down low in the parabolic well.

In Figure 12.1, we have an opportunity to introduce an approximate function due to Morse [2].

$$V(r) = D_e \left[1 - e^{-a(r-r_0)} \right]^2.$$

This so-called Morse potential has three parameters, D_e, a, and r_0, which can be fitted to experimental data for a variety of diatomic molecules and even some vibrations of polyatomic molecules. The Morse potential has two shortcomings. It starts from zero and remains positive while the real potential is below the axis as a negative number and the short range part of the potential actually crosses the vertical axis (as does the parabolic potential as well) while the real potential would not. However, the lower part of the well is a good approximation to a harmonic well and the Morse potential does dissociate to a limit. An exact solution to the Morse potential exists [3] as one of the few exact solutions in quantum mechanics, but the solution is best left to specialized courses. Other discussions are available on the Internet [4]. To apply this to the H_2 molecule we use data from the 90th Edn. of the *CRC Handbook* [5] noting a similar very accurate quantum mechanical calculation of the energy and bond length [6] (Table 12.1).

TABLE 12.1
Spectroscopic Parameters for H_2

Bond length: $r_0 = 0.74144$ Å
Energy at minimum: $D_e = 1.17445$ hartrees $= 31.95840977$ eV
Force constant: $k = 5.75$ N/cm $= 5.75 \times 10^5$ dyne/cm $= 5.75 \times 10^{-3}$ dynes/Å
$a = \sqrt{k/2D_e} = 0.74932661/$Å

It can be shown by matching the second derivative of the Morse potential to the second derivative of the parabolic potential that $a = \sqrt{k/2D_e}$ [3,4], so we can plot both the harmonic and Morse potential on the same graph. With D_e in electron volts and r_0 in Å along with the calculated value of $a = \sqrt{k/2D_e}$ we obtain data points to plot in Figure 12.1.

The graph shows that the harmonic and Morse potential wells superimpose fairly well at low energy and probably the first few energy levels correspond for both models. That may be all we need for ground states most of the time on planet Earth. The exact potential well can also be constructed from spectroscopic data but it is adequate to introduce the Morse potential to provide a useful model. Both the parabolic and Morse potentials are incorrect in that the short range part crosses the vertical axis which is meaningless so the potentials are only shown for $x > 0$ but the fit for low energy levels is still useful.

Just as with the PIB model, the particle (effective mass of the vibrating parts of the molecule) will have kinetic energy and rattle back and forth in the parabolic well. The commutator idea in the previous chapter could be used to completely solve this problem. However, our goal here is to show how the "essential" quantized energy condition occurs. While the calculus here is usually found in a further course in differential equations (recommended), we think it is better to use functions and derivatives instead of formal operators. If we follow a derivation based on operator commutators, it would add another unfamiliar aspect to the problem so we will present the polynomial method in the assumption that a teacher can put the steps on the board or that a student can follow it on scratch paper. There is just no way around using some form of detailed mathematics to show how this problem was solved! However, here we take the time-honored path used first by Schrödinger [1] and beautifully detailed by Pauling and Wilson in 1935 [7] to solve this problem with a polynomial expansion technique, which uses standard calculus methods. We do this partly so that we can allude to the polynomial expansion method when we solve the H-atom using Schrödinger's method. Note the Schrödinger H-atom treatment is a problem in three dimensions (compared to the flat Bohr model) while the harmonic oscillator is our best chance to give a complete solution in only one dimension. At this point, a student easily frustrated by mathematical details is advised to go to the section marked **Harmonic Oscillator Results** to obtain the key results and then return to check out the rigorous details of the mathematical basis of the facts related to the quantized harmonic oscillator.

HARMONIC OSCILLATOR DETAILS

As is now usual, we start by writing the Hamiltonian operator and attempt to solve the differential equation. The basic strategy is that we recall the idea of a Taylor power series expansion, which can represent any function as a (potentially infinite) power series. Thus we hope we can write $\psi(x) = \sum_{n=0}^{\infty} a_n x^n$ and find some way to evaluate the values of a_n. However, we have to first suffer through a few changes in variable to achieve a "simple" equation! We give more details than most texts at this point so that you can follow the derivation with pencil and paper or the teacher can put these steps on the board for slow appreciation. $H = \left(\dfrac{-\hbar^2}{2m} \dfrac{d^2}{dx^2} \right) + \dfrac{kx^2}{2}$ where $x \equiv (r - r_0)$ and we will define the meaning of the mass later. Thus $\dfrac{-\hbar^2}{2m} \dfrac{d^2\psi}{dx^2} + \dfrac{kx^2}{2} \psi = E\psi$. Now let $x = \beta\xi$ and absorb all the physical units into β so that ξ is without units. Then $\dfrac{-\hbar^2}{2m\beta^2} \dfrac{d^2\psi}{d\xi^2} + \dfrac{k}{2}\beta^2\xi^2\psi = E\psi$ and this is useful for unit analysis related to energy. In units $E \sim \dfrac{\hbar^2}{2m\beta^2} \sim \dfrac{k}{2}\beta^2 \Rightarrow \beta^2 = \dfrac{\hbar}{\sqrt{mk}}$. What follows is a mind-bending sequence of variable changes so we seek as much meaning as possible from the initial unit analysis. We will also find a curiosity in that the harmonic oscillator is the only known case where the classical result for the frequency will be the same as the quantum formula. You may recall from sophomore physics that a periodic frequency is given by $\nu = \dfrac{1}{2\pi}\sqrt{\dfrac{k}{m}}$ so then we substitute that formula into $\dfrac{k}{2}\beta^2 = \dfrac{k}{2}\dfrac{\hbar}{\sqrt{mk}} = \dfrac{\hbar}{2}\sqrt{\dfrac{k}{m}} = \dfrac{h}{2}\left(\dfrac{1}{2\pi}\sqrt{\dfrac{k}{m}}\right) = \dfrac{h\nu}{2}$ and a really shrewd guess might lead us to suspect that the quantized levels will involve something like $E_n \propto \dfrac{h\nu}{2}$? Let us now get past the messy algebraic substitutions. First, multiply the equation by $2m$ and divide by $-\hbar^2$. That produces $\dfrac{d^2\psi}{dx^2} - \dfrac{2mkx^2}{2\hbar^2}\psi = \dfrac{-2mE}{\hbar^2}\psi$, so then $\dfrac{d^2\psi}{dx^2} + \dfrac{2mE}{\hbar^2}\psi - \dfrac{mkx^2}{\hbar^2}\psi = 0$. Next define $\alpha \equiv \dfrac{2m\pi\nu}{\hbar}$ so that $\alpha \equiv \dfrac{2m\pi}{\hbar}\left(\dfrac{1}{2\pi}\sqrt{\dfrac{k}{m}}\right) = \dfrac{\sqrt{mk}}{\hbar}$ and then $\alpha^2 = \dfrac{mk}{\hbar^2}$. Thus if we also define $\lambda \equiv \dfrac{2mE}{\hbar^2}$ we find that $\dfrac{d^2\psi}{dx^2} + (\lambda - \alpha^2 x^2)\psi = 0$. This looks simpler but still has a term proportional to x^2, which prevents use of the factoring technique so helpful in solving the PIB and POR problems. Since λ is a constant we might ask if there is an asymptotic solution when $x \gg \lambda$ so we could neglect λ. Then $\dfrac{d^2\psi}{dx^2} \cong \alpha^2 x^2 \psi$ but that still has no clean solution, although it suggests $\psi = [e^{-\left(\frac{\alpha x^2}{2}\right)}]f(x)$ because a solution with $e^{+\left(\frac{\alpha x^2}{2}\right)}$ would not be finite for $x \to \infty$. Then we have for finite range in x, $\dfrac{d\psi}{dx} = -\dfrac{2\alpha x}{2}e^{-\left(\frac{\alpha x^2}{2}\right)}f(x) + e^{-\left(\frac{\alpha x^2}{2}\right)}\dfrac{df}{dx}$ and the second derivative is $\dfrac{d^2\psi}{dx^2} = \left[-\alpha e^{-\left(\frac{\alpha x^2}{2}\right)}f(x) + (\alpha^2 x^2)e^{-\left(\frac{\alpha x^2}{2}\right)}f(x) - (\alpha x)e^{-\left(\frac{\alpha x^2}{2}\right)}\left(\dfrac{df}{dx}\right) - (\alpha x)e^{-\left(\frac{\alpha x^2}{2}\right)}\left(\dfrac{df}{dx}\right) + e^{-\left(\frac{\alpha x^2}{2}\right)}\left(\dfrac{d^2 f}{dx^2}\right)\right]$.

After the derivatives are combined, we divide through by $e^{-\left(\frac{\alpha x^2}{2}\right)}$ to obtain the short range equation as $\left[\dfrac{\frac{d^2\psi}{dx^2} + (\lambda - \alpha x^2)\psi}{e^{-\left(\frac{\alpha x^2}{2}\right)}}\right] = \dfrac{d^2 f(x)}{dx^2} - 2\alpha x\dfrac{df(x)}{dx} + (\lambda - \alpha)f(x) = 0$. Next we can relate to a known equation solved by a French mathematician Charles Hermite [8,9] (1822–1901) with one more change of variable as $f(x) = H(\sqrt{\alpha}x) = H(\zeta)$. Here we see the mathematics of Schrödinger as filtered through the Pauling and Wilson text so we do not know exactly how Schrödinger thought of this solution. A sensible student will realize that we are really only interested in the final result while giving Schrödinger credit for solving a difficult problem! Now for our final variable change we substitute $\zeta = x\sqrt{\alpha}$ or $x = \dfrac{\zeta}{\sqrt{\alpha}}$. We will need the chain rule here, $d\zeta = \sqrt{\alpha}dx$ so that $\dfrac{d}{dx} = \dfrac{d}{\left(\frac{d\zeta}{\sqrt{\alpha}}\right)} = \dfrac{\sqrt{\alpha}d}{d\zeta}$. Then we come to the final working equation. Make a note of this coordinate scaling by $\sqrt{\alpha}$ for later interpretation because α contains information about the frequency and mass of the oscillator so that x will be scaled differently for different molecules!

$\dfrac{d^2 H(\zeta)}{d\zeta^2}\Big/\alpha - 2\alpha\left(\dfrac{\zeta}{\sqrt{\alpha}}\right)\dfrac{dH(\zeta)}{d\zeta}\Big/\sqrt{\alpha} + (\lambda - \alpha)H(\zeta) = 0$ using the chain rule that simplifies if we "swing up" $\sqrt{\alpha}$ in the derivative terms to find $\dfrac{d^2 H(\zeta)}{d\zeta^2} - 2\zeta\dfrac{dH(\zeta)}{d\zeta} + \left(\dfrac{\lambda}{\alpha} - 1\right)H(\zeta) = 0$. Schrödinger would have been aware of mathematical research in polynomials during 1860–1900 and he would have recognized an equation studied by Hermite [8,9]. Now we are ready for the main polynomial strategy by expanding $H(\zeta) = a_0 + a_1\zeta + a_2\zeta^2 + a_3\zeta^3 + a_4\zeta^4 + \cdots = \sum_{n=0}^{\infty}a_n\zeta^n$. Thus we insert the series into the equation. We use $\frac{d}{d\varsigma}\sum_{n=0}^{\infty} a_n\varsigma^n = \sum_{n=1}^{\infty} na_n\varsigma^{(n-1)}$ and $\frac{d^2}{d\varsigma^2}\sum_{n=0}^{\infty} a_n\varsigma^n = \sum_{n=2}^{\infty} n(n-1)a_n\varsigma^{(n-2)}$ to obtain specifically

$$1 \cdot 2a_2 + 2 \cdot 3a_3\varsigma + 3 \cdot 4a_4\varsigma^2 + 4 \cdot 5a_5\varsigma^3 + \cdots$$

$$-2a_1\varsigma - 2 \cdot 2a_2\varsigma^2 - 2 \cdot 3a_3\varsigma^3 + \cdots$$

$$+\left(\frac{\lambda}{\alpha} - 1\right)a_0 + \left(\frac{\lambda}{\alpha} - 1\right)a_1\varsigma + \left(\frac{\lambda}{\alpha} - 1\right)a_2\varsigma^2 + \left(\frac{\lambda}{\alpha} - 1\right)a_3\varsigma^3 + \cdots = 0.$$

Since each power of ς must be zero separately we obtain a lot of equations!

$$1 \cdot 2a_2 + \left(\frac{\lambda}{\alpha} - 1\right)a_0 = 0$$

$$2 \cdot 3a_3 + \left(\frac{\lambda}{\alpha} - 1 - 2\right)a_1 = 0$$

$$3 \cdot 4a_4 + \left(\frac{\lambda}{\alpha} - 1 - 2 \cdot 2\right)a_2 = 0$$

$$4 \cdot 5a_5 + \left(\frac{\lambda}{\alpha} - 1 - 2 \cdot 3\right)a_3 = 0$$ and so forth as $n \to \infty$. For the moment, it looks like we have a solution but there is still one more problem. When we deal with infinite power series, we have to ask whether they converge or just increase to ∞. Consider a similar power series $e^{\varsigma^2} = 1 + \varsigma^2 + \frac{\varsigma^4}{2} + \frac{\varsigma^6}{3!} + \frac{\varsigma^8}{4!} + \cdots + \frac{\varsigma^n}{\left(\frac{n}{2}\right)!} + \frac{\varsigma^{n+2}}{\left(\frac{n}{2}+1\right)!} + \cdots$ There is a test for convergence called the "ratio test" to see if the terms are gradually getting smaller. $\frac{C_{n+1}}{C_n} < 1$? Here we find $\frac{C_{n+1}}{C_n} = \frac{(n/2)!}{\left(\frac{n}{2}+1\right)!} \to \frac{1}{n}$ and although the terms get smaller, the cumulative sum is not limited so we have a divergent series for e^{ς^2}. Yes, the series converges to e^{ς^2}, but that function goes to ∞ as ς increases. Now consider the ratio of successive terms we see in the preceding text for the Hermite polynomials. $\frac{a_{n+2}}{a_n} = -\frac{\left(\frac{\lambda}{\alpha} - 2n - 1\right)}{(n+1)(n+2)} \to \frac{1}{n}$, so we see that whether we consider the series based on even or odd values of n, the series will diverge! We have already been through a lot of mathematics, which you might not see until a course in differential equations but you should recognize this as a roadblock Schrödinger must have faced. The power series must be truncated to obtain a set of finite polynomials! Note the negative sign in the coefficient ratio. Thus we let the energy embedded in the λ parameter be such that $\left[\left(\frac{\lambda}{\alpha}\right) - 1 = 2n\right]$. Then after a certain number of terms, the energy condition will be reached and *the series will terminate* and prevent divergence! Amazing! Thus we have $\frac{\left(\frac{2mE}{\hbar^2}\right)}{\left(\frac{2m\pi\nu}{\hbar}\right)} - 1 = 2n$ and then we solve for the energy. $E_n = (2n+1)\left(\frac{\pi h\nu}{2\pi}\right) = (2n+1)\frac{h\nu}{2} = \left(n + \frac{1}{2}\right)h\nu = E_n$

In the experience of this author, an interested undergraduate student can carry out the variable changes we have indicated on scratch paper but it is better if the class teacher can put the steps on a board to slowly appreciate the process. Although we follow roughly similar steps as the very clear treatment in the classic text by Pauling and Wilson [7] we wanted to at least show how the equation can be solved with some added comments. This is a key derivation in just one dimension, which pales in difficulty to the problem solved by Adrien-Marie Legendre (1752–1833), another French mathematician whose derivation of the Legendre polynomials in 1785 opened the way for Schrödinger. If you think this derivation was mind-bending, we want to assure you it is very easy compared to the derivation for the key equation of the rigid rotor (solved mainly by Legendre). We are going to spare the students the very difficult steps for the rigid rotor and the H atom solution, which are much more complicated but that involves a similar polynomial analysis.

HARMONIC OSCILLATOR RESULTS

What about the actual form of the wave function? The mathematics requires us to backtrack through all the variable changes that were made. Here we present the final results, which are normalized. Other more advanced texts in quantum mechanics will give further details, and if a student is interested in further work, it would be a good idea to purchase the text by Pauling and Wilson [7] in its original form or the reprinted version from Dover Press.

$$\psi_n = N_n e^{-\frac{\varsigma^2}{2}} H_n(\varsigma),$$

where $N_n = \sqrt{\left(\frac{\alpha}{\pi}\right)^{1/2} \frac{1}{2^n n!}}$ and $E_n = \left(n + \frac{1}{2}\right) h\nu$ with $\nu = \left(\frac{1}{2\pi}\right)\sqrt{\frac{k}{m}}$

$$H_n(\varsigma) = (-1)^n e^{\varsigma^2} \frac{d^n(e^{-\varsigma^2})}{(d\varsigma)^n} = \sum_{k=0} \frac{(-1)^k n! (2\varsigma)^{(n-2k)}}{(n-2k)! k!}.$$

Note there are odd and even solutions (Table 12.2).

Note that the normalization constant $N_n = \sqrt{\left(\frac{\alpha}{\pi}\right)^{1/2} \frac{1}{2^n n!}}$ is a function of $\alpha = \frac{4\pi^2 m\nu}{h}$, which in turn depends on m and ν for a particular molecule so we present the functional form of a few low-energy wave functions with the normalization constant and $\alpha = 1$ just to see their shapes. The graphs we show are the usual presentations that show a number of zero-crossing "nodes" and a special effect that *as the energy levels increase, the amplitude of the waves increase near the sides of the parabolic potential well*. This is evident in the wave functions we show in the following text, most noticeably in $\psi_4^* \psi_4$. This means that the velocity of the particle is greatest when moving over $x = 0$ but since it has to slow to reverse direction at the sides (turning points) of the potential well it is slightly more probable to be near the sides of the well. It is important for undergraduates to

---

TABLE 12.2
Selected Hermite Polynomials

Even	Odd
$H_0(\varsigma) = 1$	$H_1(\varsigma) = 2\varsigma$
$H_2(\varsigma) = 4\varsigma^2 - 2$	$H_3(\varsigma) = 8\varsigma^3 - 12\varsigma$
$H_4(\varsigma) = 16\varsigma^4 - 48\varsigma^2 + 12$	$H_5(\varsigma) = 32\varsigma^5 - 160\varsigma^3 + 120\varsigma$
$H_6(\varsigma) = 64\varsigma^6 - 480\varsigma^4 + 720\varsigma^2 - 120$	$H_7(\varsigma) = 128\varsigma^7 - 1344\varsigma^5 + 3360\varsigma^3 - 1680\varsigma$

Note: $\varsigma = \sqrt{\alpha} x$.

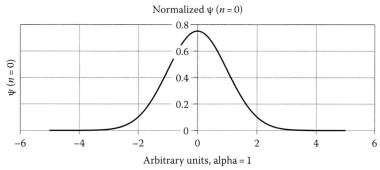

FIGURE 12.2   The normalized oscillator function $e^{-\left(\frac{\varsigma^2}{2}\right)}$.

understand that quantum mechanics does indeed scale-up to macroscopic behavior but when we deal with atoms and molecules, quantum effects occur in the microscopic domain.

Most textbooks illustrate the lowest wave functions of the harmonic oscillator by plotting the normalized functions as $\psi_n = N_n e^{-\left(\frac{\varsigma^2}{2}\right)} H_n(\varsigma)$ as we have done. However, the α variable will stretch the x-coordinate by the amount $\sqrt{\alpha}$ and $\alpha = \dfrac{2m\pi v}{\hbar}$ (Figure 12.2). This means the x-coordinate will be distorted by the mass and frequency for a given case. Here we show the normalized shapes of the first four wave functions and their square, which is the probability according to $\psi^*\psi$. We should not be surprised that the shapes may be totally dominated by the leading terms [10] of the $H_n(\varsigma)$ polynomials depending on a particular set of values for m and v of a certain molecule (Figures 12.3 through 12.11). Here we could have used numerical values for H_2 which is a light molecule in terms of mass but the force constant is by no means the lowest in diatomic molecules as shown in Table 12.3.

$$E_{vib}/hc = \omega_e\left(v+\frac{1}{2}\right) - \omega_e x_e\left(v+\frac{1}{2}\right)^2 + \cdots$$

and

$$E_{\rm rot}/hc = BJ(J+1) - D_v[J(J+1)]^2 + \cdots$$

where $B_v = B_e - \alpha_e\left(v+\dfrac{1}{2}\right) + \cdots$ and $D_e = D_v + \cdots$

Note that N_2 and CO have very high force constants and N_2 has the highest in Table 12.3. Consider that $N{\equiv}N$ clearly has a triple bond and so does $C{\equiv}O$ when you draw the Lewis electron structure!

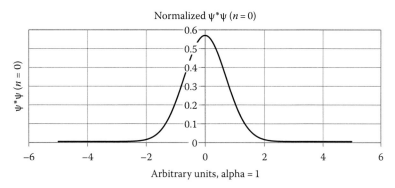

FIGURE 12.3   The normalized $\psi^*\psi$ probability for the $n=0$ oscillator state.

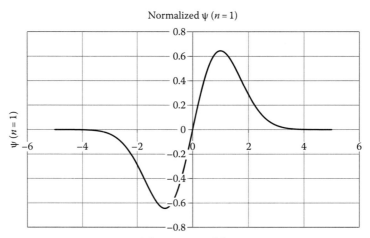

FIGURE 12.4 Normalized $n = 1$ oscillator function $(2\varsigma)e^{-\left(\frac{\varsigma^2}{2}\right)}$.

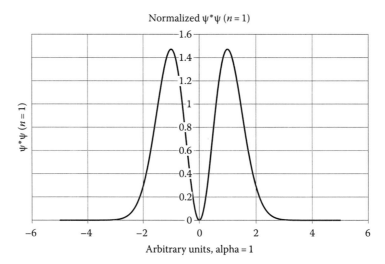

FIGURE 12.5 $\psi^*\psi$ Probability for the $n = 1$ oscillator state.

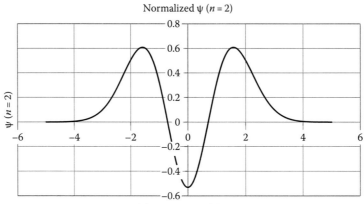

FIGURE 12.6 Normalized $n = 2$ oscillator function $(4\varsigma^2 - 2)e^{-\left(\frac{\varsigma^2}{2}\right)}$.

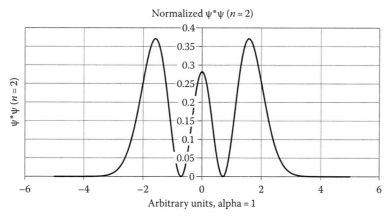

FIGURE 12.7 $\psi^*\psi$ Probability for the $n = 2$ oscillator state.

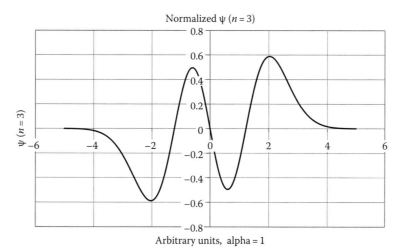

FIGURE 12.8 Normalized $n = 3$ oscillator function $(8\varsigma^3 - 12\varsigma)e^{-\left(\frac{\varsigma^2}{2}\right)}$.

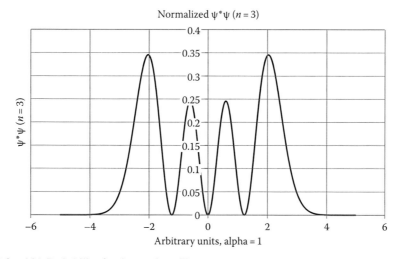

FIGURE 12.9 $\psi^*\psi$ Probability for the $n = 3$ oscillator state.

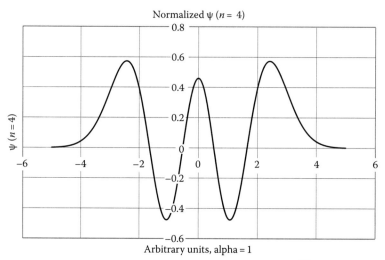

FIGURE 12.10   Normalized $n = 4$ oscillator function $(16\varsigma^4 - 48\varsigma^2 + 12)e^{-\left(\frac{\varsigma^2}{2}\right)}$.

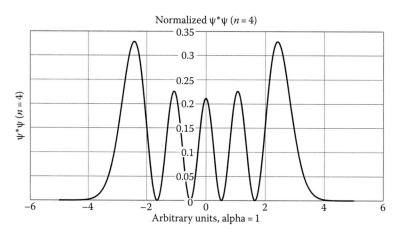

FIGURE 12.11 $\psi^*\psi$ Probability for the $n = 4$ oscillator state.

REDUCED MASS

Consider two masses connected by a rigid bar as a model of a diatomic molecule. The two masses need not be equal as m_1 and m_2 with the length of the bar as r. Now balance the model on a seesaw pivot point as in Figure 12.12 such that the distance from the pivot point to m_1 is r_1 and the distance from the pivot point to m_2 is r_2 such that $r = r_1 + r_2$. By the lever principle, we have at balance $m_1 r_1 = m_2 r_2$. Thus $r_1 = \dfrac{m_2 r_2}{m_1}$ and $r_2 = \dfrac{m_1 r_1}{m_2}$. Then $r = r_1 + \dfrac{m_1 r_1}{m_2} = \left(1 + \dfrac{m_1}{m_2}\right) r_1 = \left(\dfrac{m_2 + m_1}{m_2}\right) r_1$ so that $r_1 = \left(\dfrac{m_2}{m_1 + m_2}\right) r$. By analogy, $r_2 = \left(\dfrac{m_1}{m_1 + m_2}\right) r$ as well. In a rotating system, the role of mass is replaced by the moment of inertia $I = \sum_i m_i r_i^2 = m_a r_1^2 + m_b r_2^2 = \dfrac{m_a m_b^2 r^2}{(m_a + m_b)^2} + \dfrac{m_b m_a^2 r^2}{(m_a + m_b)^2} = \dfrac{(m_b + m_a) m_a m_b}{(m_a + m_b)^2} r^2 = \dfrac{m_a m_b}{(m_a + m_b)} r^2$.

TABLE 12.3

Force Constants and Spectroscopic Constants for Selected Diatomic Molecules

Molecule	k (N/cm)	ω_e (cm^{-1})	$\omega_e x_e$ (cm^{-1})	B_e (cm^{-1})	α_e (cm^{-1})	D_e (10^{-6} cm^{-1})	r_e (Å)
H–H	5.75	4401.21	121.34	60.853	3.062	47100	0.74144
BeH	2.27	1247.36	36.31	10.3164	0.3030	1022.1	1.3426
BeD	~ same	1530.32	20.71	5.6872	0.1225	313.8	1.3419
BH	3.05	2366.9	49.40	12.021	0.412	1242	1.2324
BD	~ same	1703.3	28	6.54	0.17	400	1.2324
HF	9.66	4138.32	89.88	20.9557	0.798	2151	0.91681
HD	~ same	2998.19	45.76	11.0102	0.3017	59	0.91694
H-^{35}Cl	5.16	2990.95	52.82	10.5934	0.30718	531.94	1.27455
D-^{35}Cl	~ same	2145.16	27.18	5.448796	0.113292	140	1.27458
HBr	4.12	2648.97	45.22	8.46488	0.23328	345.8	1.41444
DBr	~ same	1884.75	22.72	4.245596	0.084	88.32	1.4145
HI	3.14	2309.01	39.64	6.426365	0.1689	206.9	1.60916
LiH	1.03	1405.65	23.20	7.51373	0.21665	862	1.59490
LiD	~ same	1054.80	12.94	4.23310	0.09155	276	1.5941
NaH	0.78	1172.2	19.72	4.9033634	0.1370919	343.40	1.88654
NaD	~ same	826.1	?	2.557089	0.051600	93.46	1.88654
KH	0.56	983.6	14.3	3.416400	0.085313	163.55	2.243
KD	~ same	707	7.7	1.754	0.0318	50	2.240
RbH	0.52	936.9	14.21	3.020	0.072	123	2.367
CsH	0.47	891.0	12.9	2.7099	0.0579	113	2.4938
CsD	~ same	619.1	?	1.354	?	20	2.505
C≡O	19.02	2169.81	13.29	1.93128075	0.01750390	6.1216	1.12823
CS	8.49	1285.15	6.50	0.8200434	0.0059182	1.336	1.53482
N≡N	22.95	2358.57	14.32	1.99824	0.017318	5.76	1.09769
Li–Li	0.26	351.43	2.61	0.67264	0.00704	9.87	2.6729
Na–Na	0.17	159.13	0.72	0.154707	0.008736	0.581	3.0789
BeO	7.51	1487.32	11.83	1.6510	0.0190	8.20	1.3309
MgO	3.48	784.78	5.26	0.57470436	0.00532377	1.2328	1.74838
CaO	3.61	723.03	4.83	0.444441	0.003282	0.6541	1.8221

Source: Lide, D. R., *CRC Handbook of Chemistry and Physics*, 90th Edn., CRC Press, Boca Raton, FL, 2009, 9-108–9-112.

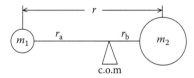

FIGURE 12.12 Two different masses balanced on a see-saw center-of-mass pivot.

This is an important simplification! It says that in a system that rotates around a center-of-mass pivot point the individual masses can be represented by an equivalent single mass with a value of $I = \dfrac{m_a m_b}{(m_a + m_b)} r^2$ and so we can substitute $\mu = \dfrac{m_a m_b}{(m_a + m_b)}$ for the mass in a harmonic oscillator. This implies that a real diatomic molecule can rotate around its center of mass while it vibrates with an equivalent one-body mass, μ. That is why we used $\mu = \dfrac{m_H m_H}{(m_H + m_H)} = \dfrac{m_H}{2}$ in the calculation of α earlier.

ISOTOPE SHIFT IN THE VIBRATIONAL FUNDAMENTAL FREQUENCY

From Table 12.3, we can see a number of interesting trends. The most obvious is the shift in frequency between deuterated hydrides such as HCl and DCl and a number of others. Although the atomic mass of H is given as 1.00794 g/mol that is only an average value for the natural abundance of H on planet Earth, which includes small amounts of deuterium and tritium, so the actual atomic mass of $^1H = 1.007825032$ and the actual mass of $^2D = 2.014101778$ g/mol (0.0115% abundance on Earth). This difference in mass of about 2:1 can have a large effect on the vibrational frequency and is called an *isotope shift*. There can be other isotope shifts such as for ^{35}Cl and ^{37}Cl, which are smaller but the largest isotope shifts due to differences in mass can be observed for substitution of D for H and hydrides are a good place to look for such an effect. Consider the ratio of the fundamental frequencies of the H compound and the D compound. Based on electrostatics of the charges on the nuclei, there is no reason to believe that the electronic bonding is any different in DCl than in HCl so the shape of the potential well should be the same and so $k_{HCl} = k_{DCl}$. The bond length might shorten a slight amount in DCl since the D part of the molecule is "heavier" and less likely to vibrate out to larger bond lengths (examine Table 12.3 to see if deuterated hydrides have shorter average bond lengths than the corresponding hydride) but the bond strength should be the same based on the nuclear charges and the number of electrons. Let us use rough approximations to the molecular weights as 35 for ^{35}Cl, 1 for H, and 2 for D just to show the effect of the masses on the vibrational frequencies using the reduced masses.
$$\frac{\nu_{HCl}}{\nu_{DCl}} = \frac{\left(\frac{1}{2\pi}\right)\sqrt{\frac{k_{HCl}}{\mu_{HCl}}}}{\left(\frac{1}{2\pi}\right)\sqrt{\frac{k_{HCl}}{\mu_{DCl}}}} = \sqrt{\frac{\mu_{DCl}}{\mu_{HCl}}} =$$

$\sqrt{\frac{(2 \cdot 35/37)}{(1 \cdot 35/36)}} \cong \sqrt{2(36/37)} = 1.39497 \cong 1.4$. We have used approximate atomic weights because of the slight factor of the ratio of the inverse total molecular weights but roughly we expect replacement of H by D to reduce the vibrational frequency by a factor of $(1/\sqrt{2}) = 0.707$ or for the H-compound to have a frequency higher than the D-compound by roughly 1.4 From data $\frac{\bar{\nu}_{HCl}}{\bar{\nu}_{DCl}} = \frac{\left(\frac{1}{\lambda}\right)_{HCl}}{\left(\frac{1}{\lambda}\right)_{DCl}} = \frac{(\nu/c)_{HCl}}{(\nu/c)_{DCl}} = \frac{2990.95/cm}{2145.16/cm} \cong 1.3942$.

The same sort of isotope shift is observed in polyatomic molecules as well and we show in Figures 12.13 and 12.14 [13] the corresponding redshifts in the spectrum of water due to substitution

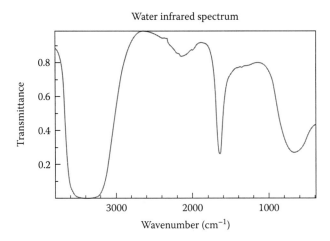

FIGURE 12.13 A low resolution infrared spectrum of liquid water, H_2O. (From *NIST Chemistry WebBook*, http://webbook.nist.gov/chemistry)

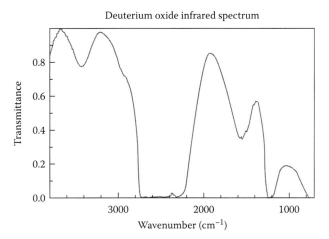

Deuterium oxide infrared spectrum

FIGURE 12.14 A low resolution liquid phase spectrum of D_2O showing the H/D isotope shift of the 3200 cm^{-1} of water to a much lower energy around 2750 cm^{-1}. (From *NIST Chemistry WebBook*, http://webbook. nist.gov/chemistry)

of deuterium for proton-hydrogen. The effect is blurred due to the spectra of liquids but the hydrogen/deuterium isotope shift is the largest observed in vibrational spectroscopy and is easily seen when comparing the infrared spectra of H_2O and D_2O.

HERMITE RECURSION RULE

Next we should note a useful *recurrence relationship* between the Hermite polynomials. $\varsigma H_n(\varsigma) = n H_{n-1}(\varsigma) + \left(\frac{1}{2}\right) H_{n+1}(\varsigma)$, which relates a given n-level to the levels above and below.

For instance $\varsigma H_2(\varsigma) = 2H_1(\varsigma) + (1/2)H_3(\varsigma) = 2(2\varsigma) + \left(\frac{1}{2}\right)(8\varsigma^3 - 12\varsigma) = 4\varsigma^3 + 4\varsigma - 6\varsigma$, so indeed we find that $\varsigma H_2(\varsigma) = \varsigma(4\varsigma^2 - 2)$. We also check the wave function normalization.

Consider whether $\psi_0 = \left(\frac{\alpha}{\pi}\right)^{1/4} e^{-\left(\frac{\varsigma^2}{2}\right)} = \left(\frac{\alpha}{\pi}\right)^{1/4} e^{-\left(\frac{\alpha x^2}{2}\right)}$, the $n = 0$ wave function, is normalized.

$$\int \psi_0^* \psi_0 d\tau = \int_{-\infty}^{+\infty} \left(\left(\frac{\alpha}{\pi}\right)^{1/4} e^{-\left(\frac{\alpha x^2}{2}\right)}\right)^* \left(\left(\frac{\alpha}{\pi}\right)^{1/4} e^{-\left(\frac{\alpha x^2}{2}\right)}\right) dx = 2\int_0^\infty \left(\frac{\alpha}{\pi}\right)^{1/2} e^{-\alpha x^2} dx$$

$$= 2\left(\frac{\alpha}{\pi}\right)^{1/2} \sqrt{\frac{\pi}{\alpha}} \left(\frac{1}{2}\right) = 1.$$

Note we have used $\int_{-\infty}^{+\infty} (even)dx = 2\int_0^\infty (even)\,dx$ along with $\int_0^\infty e^{-ax^2} dx = \frac{1}{2}\sqrt{\frac{\pi}{a}}$.

The wave functions are also orthogonal because the product of an odd and even function integrated over all space is zero and the (even × even) or (odd × odd) products cancel due to the nodal patterns. In summary, $\int \psi_m^* \psi_n d\tau = \delta_{mn}$ where $\delta_{mn} = 1, m = n$ or $\delta_{mn} = 0, m \neq n$. Note δ_{mn} is the so-called Kroneker delta, which is a compact way to indicate an orthonormal set $\{\psi_n\}$.

INFRARED DIPOLE SELECTION RULE

The most important reason we need to study the harmonic oscillator wave functions is to determine how a given molecule will react to the sinusoidal electric field of an impinging light wave. Most molecules have an electric dipole moment due to some displaced electronic charge caused by electronegativity differences between atoms. Much of the original work on measuring dipole moments and defining the concept was carried out by Peter Debye (1884–1966), a Dutch physical chemist and Nobel Laureate who later moved to Cornell University in the United States. Originally, the unit of the Debye was defined as a separation of two opposite sign charges of magnitude q_e separated by 1 Å. This could be related to ionic compounds with bond lengths typically measured in Å. The SI conversion is 1 Debye $= 3.33564 \times 10^{-30}$ C m as measured in Coulombs and meters. Hydrogen chloride (HCl) has a typical value of 1.1086 D [11] and monomeric gas phase formaldehyde (H_2CO) a value of 2.332 D [11]. A dipole moment is usually shown in diagrams as an arrow with the point at the negative end and a plus sign as the tail of the arrow. Dipole moments are vectors and the direction of the dipole moment in *ortho*-dichlorobenzene is not the same as for *meta*-dichlorobenzene.

We will use a dipole interaction explanation for infrared absorption. Although the classical description of a light wave shows an oscillating field, let us consider the wavelength of infrared light, which is typically absorbed by a carbonyl group near 1700 cm^{-1}. Note that the typical wavelength of infrared light is much larger than the size of the molecules as for the example of monomeric formaldehyde which is the prototypical carbonyl compound.

$$\left(\frac{1}{\lambda}\right) = \bar{\nu} = 1700/\text{cm} \Rightarrow \lambda = \left(\frac{1 \text{ cm}}{1700}\right) \cong 5.882 \times 10^{-4} \text{ cm} = 58{,}820 \text{ Å, which is much longer}$$

than the size of small molecules typically less than 10 Å. *Thus the actual situation is that during the absorption process, the dipole of the molecule is in an electric field $\vec{E}$, which is approximately constant.* If the molecular dipole couples with the electric field, the bond may be lengthened or compressed slightly by the light wave leading to more or less vibrational amplitude according the interaction $\vec{E} \cdot \vec{\mu}$. Here $\vec{\mu} = q\vec{r}$ for some internal charge separation q along a distance $\vec{r}$. Thus we ask for the expectation value (average) of the interaction of the dipole moment with a given bond length and factor out the approximately constant charge separation ε along a bond that is causing the dipole moment as $\varepsilon <m|(r - r_e)|n> = \varepsilon <m\left|\frac{\varsigma}{\sqrt{\alpha}}\right|n>$ between any two states. Although it looks like ε and $\sqrt{\alpha}$ are complications, they can be factored out of the integral and we are only interested in what states can be connected by this transition mechanism and not the absolute value of the interaction.

Using $\psi_n = N_n H_n(\varsigma) e^{-\frac{\varsigma^2}{2}}$, we can evaluate the transition dipole interaction between states $|m\rangle$ and $|n\rangle$ as $\left\langle m\left|\frac{\varsigma}{\sqrt{\alpha}}\right|n\right\rangle = \int_{-\infty}^{+\infty} N_m H_m e^{-\frac{\varsigma^2}{2}} \left[\frac{\varsigma}{\sqrt{\alpha}}\right] N_n H_n(\varsigma) e^{-\frac{\varsigma^2}{2}} d\varsigma$ where $\varsigma = \sqrt{\alpha}x = \sqrt{\alpha}(r - r_e)$. Then invoke the recursion relation so $\varsigma H_n(\varsigma) = nH_{n-1}(\varsigma) + (1/2)H_{n+1}(\varsigma)$ inside the integral to find that $\left\langle m\left|\frac{\varsigma}{\sqrt{\alpha}}\right|n\right\rangle = N_m N_n \left(\frac{1}{\sqrt{\alpha}}\right) \int_{-\infty}^{+\infty} e^{-\varsigma^2} H_m(\varsigma)\left[nH_{n-1}(\varsigma) + \left(\frac{1}{2}\right)H_{n+1}(\varsigma)\right] d\varsigma$. Now the N_n coefficient is not correct for the $(n-1)$ and $(n+1)$ states but we can multiply and divide by the needed value in each case to complete the formula and use the Kroneker delta function.

$$\left\langle m\left|\frac{\varsigma}{\sqrt{\alpha}}\right|n\right\rangle = N_m N_n \left(\frac{1}{\sqrt{\alpha}}\right) \int_{-\infty}^{+\infty} e^{-\varsigma^2} H_m(\varsigma)\left[\left(\frac{N_{n-1}}{N_{n-1}}\right)nH_{n-1}(\varsigma) + \left(\frac{1}{2}\right)\left(\frac{N_{n+1}}{N_{n+1}}\right)H_{n+1}(\varsigma)\right] d\varsigma$$

which leads to $\left\langle m\left|\frac{1}{\sqrt{\alpha}}\right|n\right\rangle = \left(\frac{N_m N_n}{\sqrt{\alpha}N_{n-1}}\right)n\delta_{m,n-1} + \left(\frac{N_m N_n}{\sqrt{\alpha}N_{n+1}}\right)\frac{\delta_{m,n+1}}{2}$.

This means that the dipole transition can only be nonzero if $m = n \pm 1$! Thus if the oscillator is in state $|n>$ the transition will only occur to a state $|m>$, which is one level above or one level below $|n>$. That was a lot of complicated reasoning but it was necessary to get a very useful result! It should be noted that the dipole mechanism depends on the fact that there is a permanent nonzero dipole moment in the molecule. In the case of a homonuclear diatomic molecule, such as O_2, N_2, or Li_2, there is no dipole moment and stretching the bond will not produce a dipole so another technique (the Raman effect) must be used to detect transitions that depend on a change in polarizability, which is a second-order effect caused by induced dipole moments. We give a detailed discussion of Raman spectroscopy in Chapter 18.

What does this "selection rule" of $\Delta n = \pm 1$ mean for us? Think about it. We have $E_n = \left(n + \dfrac{1}{2}\right)h\nu$ so we have $\Delta E = \pm h\nu$. That means that since the spacing of the harmonic oscillator energy levels are equal, it does not matter whether the transition is $0 \rightarrow 1, 4 \rightarrow 5, 2 \rightarrow 3$, or $1 \rightarrow 2$, the energy *should* be the same *if the potential is a perfect parabola*! Thus, we expect that infrared spectral bands should be unique narrow peaks, well almost. We should be aware that in Figure 12.1 we have discussed the energy levels that are in the parabolic potential well. The way in which the Morse potential (and the true potential) broadens to the dissociation limit has a result that is documented in the solutions to the exact solution of the Morse potential, which shows the energy levels gradually get closer together. However, those upper states will be less likely to be populated according to the Boltzmann energy principle of $e^{-\left(\frac{n h \nu}{kT}\right)}$. Transitions involving the higher n-states are called "hot bands" because they cannot occur unless the temperature is high enough to populate those upper states. Since the hot bands are less probable and the energy gap between states decreases, the infrared bands tend to be sharp on the blue edge but broaden and taper off on the red edge of the bands. Even so, because on planet Earth most molecules are in the $n=0$ state at room temperature, most infrared bands are narrow and unique for fixed molecules in solid lattices or with some broadening for liquid samples. In polyatomic molecules, there is a complication in that there are sometimes "combination bands" in which two infrared bands are excited simultaneously. Fortunately, these combinations can usually be separated from the fundamental bands using algebraic addition/subtraction of the band energies.

$3N - 6$ OR $3N - 5$ VIBRATIONS?

When we discussed the energy equipartition law in Boltzmann's KMTG discussion, we needed to count the number of vibrations. A polyatomic molecule with N atoms would have 3N degrees of freedom if the atoms were separated. As a unit, a polyatomic molecule has 3 degrees of freedom for translation. However, there is a special case for the rotation of a linear molecule. There is no physical way to observe or measure rotation about the linear axis of a linear molecule so in effect there is no rotation about that axis. Thus a nonlinear molecule will have 3 degrees of rotation but a linear molecule will only have 2. Then the remaining degrees of freedom for internal vibration is either $(3N - 3 - 3) = (3N - 6)$ for nonlinear molecules or $(3N - 3 - 2) = (3N - 5)$ for a linear molecule! We have discussed diatomic molecules here with only one fundamental frequency but in polyatomic molecules there are $3N - 6$ (nonlinear) or $3N - 5$ (linear) fundamental vibrations and they are linearly independent as a result of the way in which the polyatomic multi-vibrational problem is solved. In solids or liquids, our reasoning that there should be one (blurred) main band for each fundamental vibration is observed. However, we should realize that molecules in the gas phase can rotate freely so we can expect more details related to rotational energy levels of molecules in the gas phase.

Characteristic IR bands have been thoroughly documented and used as a sort of "fingerprint" analysis of solids and liquids [11]. In Figure 12.15, we see an idealized vibrational spectrum of formaldehyde with $(3N - 6) = 6$ sharp bands predicted for fingerprint identification. Note the strong band at 1750 cm^{-1}, which is characteristic of many other carbonyl compounds. In Figure 12.16 [13], we see an actual spectrum of formaldehyde where it is clear that the bands are broader and more complicated than the ideal spectrum. However, the 1750 cm^{-1} band is still strong. For a gas-phase spectrum there are additional effects due to molecular rotation and anharmonic deviation from the strict $\Delta n = \pm 1$ rule, but the finger print idea is still approximately valid. In Figure 12.17 [13], we see a characteristic fingerprint infrared spectrum of benzene. Before nuclear magnetic resonance (NMR) became highly developed, infrared bands were used for structural analysis of compounds

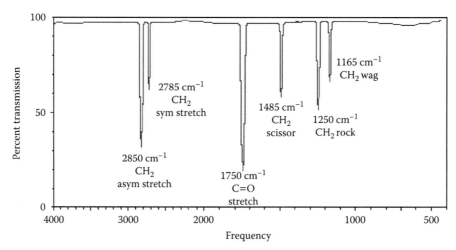

FIGURE 12.15   Normal modes of formaldehyde. (From William Reusch, Chemistry Department, Michigan State University. See his infrared tutorial at http://www.cem.msu.edu/~reusch/VirtualText/Spectrpy/InfraRed/infrared.htm).

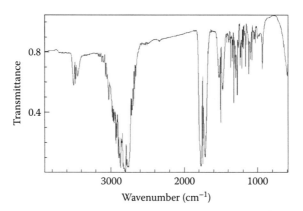

FIGURE 12.16   Low resolution spectrum of formaldehyde, note 1750 cm^{-1} C=O stretch. (From *NIST Chemistry WebBook*, http://webbook.nist.gov/chemistry)

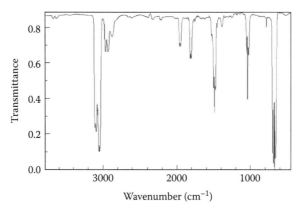

FIGURE 12.17   Low resolution infrared spectrum of benzene vapor showing characteristic fingerprint bands of an aromatic ring. (From *NIST Chemistry WebBook*, http://webbook.nist.gov/chemistry)

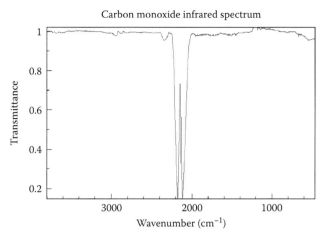

FIGURE 12.18   Low resolution spectrum of the fundamental vibration of CO gas. (From *NIST Chemistry WebBook*, http://webbook.nist.gov/chemistry)

and the strong band at 3030 cm^{-1} due to a –C–H stretch attached to an aromatic ring was a telltale characteristic of an aromatic ring.

In the next chapter, we will see more details of infrared spectra of gases where it will be evident that the dipole selection rule correctly predicts the $\Delta n = \pm 1$ rule. However, modern instrumentation with higher sensitivity reveals lower intensity transitions for $\Delta n = 2, 3, 4, \ldots$. Before NMR spectroscopy became available, IR spectroscopy of solids was a mainstay of structural assignments and identification of chemical compounds. Today, NMR is the first choice for structural analysis of organic compounds but infrared spectroscopy is still useful for studies of gases in the atmosphere. Consider the spectrum of carbon monoxide in Figure 12.18 [13] at low resolution. It would seem there is one main band, but even this low resolution spectrum has a "notch" in the middle which will be explained in the next chapter as a "Q-branch". Clearly, there is more detail to be obtained with higher resolution, but from the dipole selection rule we only expect this one main band. Historically, the infrared spectra of solids and liquids would seem to obey the dipole selection rule very well and we would have to work very hard with the scanning spectrometers of the 1970s to see much more, although in some cases higher resolution was possible.

In recent years, new technology in the form of Fourier transform spectrometers has revolutionized infrared spectroscopy in both higher resolution and much greater sensitivity. A composite of several isotopic species of CO is shown in Figure 12.19 from the HITRAN data base [12]. There are several important points here. First, the possible combination of isotopes is surprising, although these are laboratory data sets where high concentrations of rare isotopes have been prepared. With natural abundance it is likely you would only see the $^{12}C^{16}O$ species. Second, the main point of this figure is that the dipole selection rule is almost useless in this case! We know from the low-resolution spectrum given earlier that the CO fundamental vibration is near 2150 cm^{-1} and yet in the broad spectra in Figure 12.20, we see bands near 4000, 6000 cm^{-1} and even a faint band near 8000 cm^{-1}, in other words we see $\Delta n = 1, 2, 3, 4$! If we go back to the parabola fitted to a Morse curve earlier in this chapter for H$_2$, we see that we can fit the parabola exactly to the curve at the minimum, but the overall fit of the parabola is not perfect. The point is that the dipole selection rule is for a perfect parabola but *the true potential is anharmonic and is not a perfect parabola!* While some real potentials will be more shallow or deeper than what is shown there for H$_2$, it should be clear from Table 12.3 that H$_2$ has a typical force constant. Thus the spectra of the various isotopic species of CO shows not only the (0->1) vibrational transition but (0->2), (0->3), and (faintly) (0->4) as well! Finally, we need to look at the units of the *y*-axis where it becomes evident that the arbitrary intensity scale varies from about 10^{-19} (cm^{-1}/(molecule cm^{-2})) for the (0->1) transition

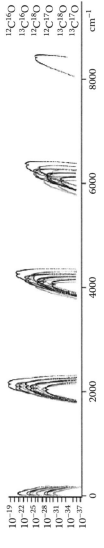

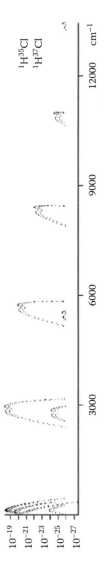

FIGURE 12.19 Overtone spectra of CO for several isotopic species from the HITRAN Library [12]. The left edge shows the far infrared pure rotational spectra while the bands near 2000 cm^{-1} show the nominal vibrational spectra and a notch is evident for the "Q branch" near 2150 cm^{-1}. However, there are also several strong overtones near 4000 cm^{-1}, 6000 cm^{-1} and even a faint overtone near 8000 cm^{-1}. This amazing data has only been available since 2009.

FIGURE 12.20 The infrared bands of isotopic species of HCl from the HITRAN data source [12]. The overtone vibrational bands occur at multiples of 2885 cm^{-1} up to $v = 4$ ($n = 4$, harmonic level) due to anharmonicity. The first set of bands at the extreme left (low energy) of the spectrum is due to pure rotation transitions discussed in the next chapter.

to about 10^{-25} for the (0->4) overtone transition. That means that the (0->4) overtone is observed but it is about one million times less intense! In a liquid sample, it is unlikely such weak transitions would be observed due to solvent interactions, so for liquid samples the dipole selection rule is useful. However, we now know that modern instrumentation is able to detect deviations from the dipole selection rule of the perfect parabolic potential for gas samples.

Just to show that CO is not the only gas that shows the violation of the dipole selection rule, we present similar spectra for HCl isotopic species from the same HITRAN data source [12]. Here we can again see what are apparently vibrational transitions all the way to (0->4) called overtones, and again the intensity decreases from about 10^{-19} to about 10^{-25} for a reduction of about 1 million in the same units. We emphasize again that the leftmost spectrum is a series of lines due to pure rotational transitions (see Chapter 13) while the second set of spectra near 3000 cm^{-1} are rotational–vibrational transitions showing a notch in the middle for a missing "Q-branch." We will treat these spectra in detail and explain the "Q-branch" in the next chapter, but it is useful to note here that the HITRAN spectra show both the low-energy pure rotational lines and the rotation–vibration lines. This author is used to seeing only the $(n=0)$->$(n=1)$ vibrational transition spectra recorded using relatively insensitive scanning spectrometers. Thus seeing the effects of the anharmonic transitions from $(n=0)$->$(n=4)$ is truly amazing and not likely shown in earlier texts!

We do not expect to see such details of anharmonicity at lower resolution. For forensic analysis, one would probably be using infrared bands for chemical identification in what is called the "IR-fingerprint" bands. Often the samples are solids or liquids and only the positions of the fundamental bands are of interest. Of course small samples or low concentrations will only be observable in terms of the most intense bands which are determined by the dipole selection rule. Through documentation, a number of tables of the main infrared bands have been developed and for some small molecules such as formaldehyde, it is possible to actually work out the frequencies and the atom motions in the various vibrations using computer programs. Each vibrational motion in a polyatomic molecule is a linearly independent motion called a "vibrational mode." If a movie were made of a vibrating molecule you would see a blur of complicated jiggling motions but mathematically the normal modes are orthogonal linearly independent motions, each with a specific eigenvalue frequency.

RAMAN SPECTROSCOPY

So far we have discussed infrared absorption of electromagnetic radiation (light), which depends on the sample molecules having a dipole moment. There is another useful form of vibrational spectroscopy which depends on reemission of light. It involves a mechanism in which the electromagnetic wave distorts the electronic structure and induces temporary dipole moments in the sample molecules. Homonuclear molecules such as H_2, N_2, O_2, etc., and certain symmetric vibrational modes of polyatomic molecules do not have dipole moments. However, their infrared spectra can be observed using a second-order effect which depends on "induced dipole moments." We briefly discussed electronic levels of molecules in Chapter 9 so we know transitions between them are usually in the UV-Vis energy range. The PIB and POR problems also discussed electronic energy levels. Raman spectroscopy is a form of vibrational spectroscopy but it uses intense light in the visible range of the spectrum to excite molecules from the ground electronic state to an excited electronic state. When the exciting light is reemitted (scattered) the wavelength of the reemitted light can differ by vibrational quanta as shown in Figure 12.21. This "effect" was discovered in 1928 by C. V. Raman (Figure 12.22).

Depending on the orientation of the light wave and the molecule, an intense light wave can polarize the electron shells of a molecule and create a temporary dipole moment. Let there be a 3×3 *polarizability tensor* $[\alpha]$, which satisfies a relation with the components of the electric field of the light as $[E_x, E_y, E_z]$ to induce dipole components $[\mu_{ind-x}, \mu_{ind-y}, \mu_{ind-z}]$, which can then couple with one or more vibrations. Here it is sufficient to note that the electric field of a light wave can actually induce temporary dipole components in a molecule with a zero dipole moment and that can lead to

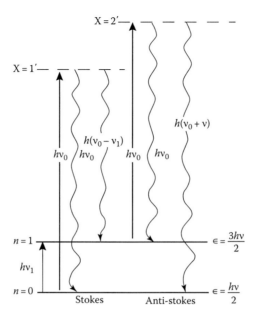

FIGURE 12.21 A diagram of the Raman mechanism. The upper state labeled "X" may not be known but is an electronically excited state which presumably has vibrational levels available for an allowed transition. Most of the light absorbed at the exciting wavelength/frequency is reemitted at the same wavelength but due to the vibrational (or rotational) selection rule some of the "scattered light" is at a wavelength/frequency higher or lower than the exciting light by a vibrational (or rotational) quantum.

FIGURE 12.22 Sir Chandrasekhara Venkata Raman (1888–1970) was an Indian physicist who discovered the inelastic scattering phenomenon which bears his name in 1928 and for it he was awarded the Nobel Prize for physics in 1930. Raman scattering produces scattered photons which differ in frequency from the radiation source which causes it, and the difference is related to vibrational and/or rotational properties of the molecules from which the scattering occurs. It has become more prominent in the years since powerful monochromatic laser sources can provide the scattering power. (Photograph by A. Bortzells Tryckeri, AIP Emilio Segre Visual Archives, W. F. Meggers Gallery of Nobel Laureates.)

vibrational transitions. This mechanism emphasizes the nonzero aspects of the *polarizability tensor* [α] instead of the dipole moment.

$$\vec{\mu} = \alpha\vec{E} = \begin{bmatrix} \alpha_{xx} & \alpha_{xy} & \alpha_{xz} \\ \alpha_{yx} & \alpha_{yy} & \alpha_{yz} \\ \alpha_{zx} & \alpha_{zy} & \alpha_{zz} \end{bmatrix} \begin{bmatrix} E_x \\ E_y \\ E_z \end{bmatrix} = \begin{bmatrix} \mu_x \\ \mu_y \\ \mu_z \end{bmatrix}_{induced}.$$

You can tell from the double index on the elements of [α] such as α_{xy} that this is a second-order effect and generally depends on a more intense light source.

A simple mechanical analogy requiring the concept of a tensor is asking how a force applied at an arbitrary angle to a corner of a rectangular box would be transmitted throughout the box. The force of an arbitrary blow would be transmitted in all three (x, y, z) directions but to different extents depending on the angle of the impact. One set of (x, y, z) vectors would be required to describe the impact while a second set of vectors (x', y', z') would describe the response of the box. A 3×3 tensor could then be used to describe how the arbitrary force would be transmitted.

The Raman spectrum is observed by using an intense monochromatic source applied to a sample and the scattered light is gathered at 90° from the incident beam. There is considerable intensity from the incident beam and the scattered light at the wavelengths of interest on either side of the monochromatic source are much less intense. Thus the exciting wavelength may obscure the Raman wavelengths and care is required to design the optical system. A monochrometer is used to disperse the emitted light and display both the Stokes and the anti-Stokes lines. Further analysis of the spectra involves the use of the polarizability tensor mentioned earlier. Principles of symmetry can also be applied to interpret the spectra as shown in Chapter 18.

In Figure 12.23, we see a Raman spectrum of CCl_4 [14] that shows the weak anti-Stokes bands as well as the more intense Stokes lines. The intensity of the anti-Stokes lines is less because the

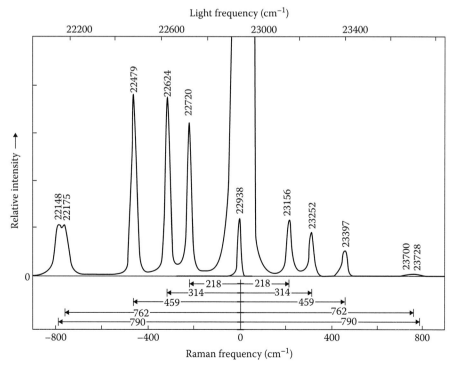

FIGURE 12.23 CCl_4 Raman spectrum excited by 4358 Angstrom light from a Hg arc. (From Tobias, R. S., *J. Chem. Ed.*, 44, 2, 1967. With permission of The Journal of Chemical Education.)

Boltzmann probability of the sample molecule being in a higher energy state is less prior to the electronic excitation. If the corresponding Stokes line is already weak, it may be very difficult to observe the anti-Stokes lines. The spectrum here is from older technology, which used a Hg arc lamp for excitation before powerful lasers were available.

SUMMARY

This chapter introduces the student to the method of seeking a solution to a differential equation in the form of a polynomial, which was probably the method used by Schrödinger in 1926 to solve this problem in terms of Hermite polynomials. An undergraduate with two semesters of calculus should be able to follow this key derivation with pencil and paper, assuming enough time and patience are available, but we still provide most of the detailed calculus steps here. For students with only one semester of calculus, it would be better for the teacher to show the steps on the board in lecture.

1. A series of clever changes in variable enable one to convert the one-dimensional Schrödinger equation with a potential of $V = \dfrac{kx^2}{2}$ into the differential equation previously solved by Hermite for truncated short polynomials multiplied by an exponential factor. The constant k is the force constant of a given bond potential as if it were a "spring." The most important fact emerging from the solution is that $E_n = \left(n + \dfrac{1}{2}\right)h\nu$, which means that even at the $n = 0$ level, there is $E_0 = \dfrac{h\nu}{2}$, the so-called "zero-point energy." The successive levels of the oscillator are equally spaced and a dipole selection rule predicts that $\Delta n = \pm 1$. Thus all transitions should result in the same transition frequency in the infrared region for one characteristic band of each vibration of a molecule. This idealized concept is spoiled for higher n values by the fact that the upper part of the potential for a real molecule is not a harmonic parabola, it is more like a Morse potential. A diatomic bond needs to have a nonzero electric dipole moment in order to absorb infrared radiation according the dipole selection rule of $\Delta n = \pm 1$; a homonuclear diatomic like N_2 or O_2 is thus without vibrational absorption by the electric dipole mechanism. With very sensitive measurements, it can be shown that the $\Delta n = \pm 1$ rule can be violated and observed via weak "overtone" vibrations where $\Delta n = +2, +3, +4, \cdots$.

2. For diatomic molecules, the two masses can be combined into a single "reduced mass" as calculated relative to the center of mass balance point given by $\mu = \dfrac{m_1 m_2}{(m_1 + m_2)}$ using actual masses. If atomic weights are used, the formula must be divided by Avogadro's number. Once μ is available, it is surprising to find the oscillator frequency is given by the classical oscillator formula $\nu = \dfrac{1}{2\pi}\sqrt{\dfrac{k}{\mu}}$. Isotope shifts can be estimated by the ratio of frequencies given by $\dfrac{\nu_1}{\nu_2} = \dfrac{\frac{1}{2\pi}\sqrt{\frac{k}{\mu_1}}}{\frac{1}{2\pi}\sqrt{\frac{k}{\mu_2}}} = \sqrt{\dfrac{\mu_2}{\mu_1}}$ so that substituting a deuteron (D) for a proton (H) in a compound X will shift the former X-H vibration by about 1.4 to the red ($\bar{\nu}_{XD} = \bar{\nu}_{XH}/1.4$).

3. For structural assignments and forensic applications, the main molecular $\Delta n \Rightarrow 0 \to 1$ transitions in the infrared are characteristic of various functional groups in chemical structure. On planet Earth, the average temperature is such that most molecules are in the $n = 0$ level at room temperature so that $\Delta n \Rightarrow 0 \to 1$ is most easily observed with low resolution spectrometers. For molecules, there are $3N - 5$ (linear) or $3N - 6$ (nonlinear) fundamental vibrations. Actual infrared spectra exhibit broadening and fine structure

including a mysterious notch called a "Q-branch," which stimulates our interest in further analysis in the next chapter.

4. Intense monochromatic light sources (lasers) can be used to induce temporary dipole moments in symmetric molecules or symmetric vibrational modes in polyatomic molecules to observe the Raman effect. Raman spectra record vibrational (and rotational) transitions in scattered visible light wavelengths from molecules without dipole moments. Raman spectroscopy has enjoyed renewed interest due to the availability of modern lasers.

PROBLEMS

12.1 Given the harmonic potential $V(x) = \dfrac{kx^2}{2}$ and the Morse potential $V(x) = D_e[1 - e^{-ax}]^2$, where $x = (r - r_0)$, take the second derivative of each potential and substitute $x = 0$ to equate the curvature of the two potentials at the minimum energy and show that $a = \sqrt{\dfrac{k}{2D_e}}$.

12.2 Write out the details of the product $\displaystyle\int_{-\infty}^{+\infty} \psi_0^* \psi_1 dx$ and show why the result is zero.

12.3 Use the force constant given in Table 12.3 for N_2 with $\nu = \dfrac{1}{2\pi}\sqrt{\dfrac{k}{\mu}}$ to calculate the vibrational frequency. Note here μ is just half the atomic mass of one N atom (14.007 g/mol)/2.

12.4 Assuming the bond force constant $k = 4.84 \times 10^5$ dyne/cm is the same for H-^{35}Cl and H-^{37}Cl, calculate as the isotope shift difference in their vibrational frequencies and $\Delta\bar{\nu} = (\Delta\nu/c)$ for the splitting in cm^{-1}. Use M(H) = 1.00794 g/mol, M(^{35}Cl) = 34.96885269 g/mol, and M(^{37}Cl) = 36.96590259 g/mol.

12.5 Predict the number of normal-mode vibrations for the following molecules: CH_4, O=C=C=C=O (linear), benzene (C_6H_6), cyclohexane (C_6H_{12}), and formaldehyde (CH_2O).

REFERENCES

1. Schrödinger, E., Quantizierung als Eigenwertproblem (Zweite Mitteilung), *Ann. Phys.*, **79**, 489 (1926).
2. Morse, P., Diatomic molecules according to the wave wechanics. II. Vibrational levels, *Phys. Rev.*, **34**, 57 (1929).
3. Eyring, H., J. Walter, and G. E. Kimball, *Quantum Chemistry,* John Wiley & Sons, New York, 1944, page 272.
4. Zielinski, T. J., Exploring the Morse Potential, Division of Chemical Education, Inc., American Chemical Society, 1998, http://jchemed.chem.wisc.edu/JCEDLib/SymMath/collection=001/MorsePotential_nb.pdf
5. Lide, D. R., *CRC Handbook of Chemistry and Physics*, 90th Edn., CRC Press, Boca Raton, FL, 2009–2010, p. 9–50, p. 9–99 and pp. 9-103–9-107.
6. Kolos, K. and L. Wolniewicz, Accurate adiabatic treatment of the ground state of the hydrogen molecule, *J. Chem. Phys.*, **41**, 3663 (1964).
7. Pauling, L. and E. B. Wilson, *Introduction to Quantum Mechanics, With Applications to Chemistry*, McGraw-Hill Book Co., New York (1935), Chapter III. The copyright is now owned by Dover Press, New York and is published as an unabridged reprint.
8. Hermite, C., *Compt. Rend. Acad. Sci. Paris,* **58**, 93–100 and 266–273 (1864).
9. Weisstein, E. W., Hermite polynomial, *MathWorld*, Wolfram Research, (2011). http://mathworld.wolfram.com/HermitePolynomial.html
10. Slater, J. C., Atomic shielding constants, *Phys. Rev.*, **36**, 57 (1930).
11. Lide, D. R., *CRC Handbook of Chemistry and Physics*, 90th Edn., CRC Press, Boca Raton, FL, 2009–2010, p. 9-108 to 9-112.
12. Rothman, L. S., I. E. Gordon, A. Barbe, D. ChrisBenner, P. F. Bernath et al., The HITRAN 2008 molecular spectroscopic database, *J. Quant. Spectrosc. Rad. Trans.*, **110**, 533 (2009).
13. Spectra from NIST Chemistry WebBook (http://webbook.nist.gov/chemistry).
14. Tobias, R. S., Raman spectroscopy in inorganic chemistry. I. Theory, *J. Chem. Educ.*, **44**, 2 (1967).

13 The Quantized Rigid Rotor and the Vib-Rotor

INTRODUCTION

In this chapter, we continue our effort to treat topics in physical chemistry in a way that does not skip over the main details but still tries to simplify the presentation. We need to present the derivation for at least a partial answer if a few students ask, "Where did that come from?" However, let us be clear at the outset of this chapter that the desired goal is for students to understand pure rotational and vib-rotational spectra of molecules and know where the H atom orbital shapes come from. If you follow the derivation with pencil and paper that may help you understand the derivation, but if not at least learn $E_l = \dfrac{l(l+1)\hbar^2}{2\mu r^2}$ and the selection rule $\Delta l = \pm 1$. Here we find a shortcut that makes this problem simpler than the harmonic oscillator. The classic text by Pauling and Wilson [1] is one of the best sources of the full derivation but they still gloss over a few details requiring additional steps from a text on differential equations (Figure 13.1). Even the excellent text by Eyring, Walter, and Kimball [2] does not give the full derivation. Perhaps this is due to the fact that the key equation was worked out by Legendre in 1785 (!) and many previous mathematics texts expand on the Legendre equation. If you follow the derivation in this chapter, it will really pay off in making the next chapter on the H atom easier. Since the H atom solution is the model for the entire periodic chart, it really is essential! In this chapter, quantum chemistry [3–5] is only part of the list of essential topics but we need to solve the rigid rotor problem prior to the H atom solution because the angular wave functions are the same for both problems. We have struggled to make this chapter understandable using only basic calculus but in a correct way [6]. However, a wise student will look beyond the derivation to study the applications to spectroscopy, which probe the quantized behavior of molecules.

In the previous chapter, we solved the problem of the quantized harmonic oscillator and derived key concepts such as the reduced mass and the isotope shift. We were on the verge of treating rotation but you will soon see it is a two-dimensional problem, which needs to be split into two one-dimensional problems. Basically the motion of a gas-phase molecule is translation and free rotation and it takes two coordinates (θ, ϕ) to describe such rotational motion even when we assume constant bond lengths within the molecule. We know from the previous chapter that molecules do vibrate but the motion of the vibrations is much smaller than rotations described by (θ, ϕ). Therefore it is a good approximation to assume constant bond lengths. Thus, we have to solve the Schrödinger equation for a problem in more than one dimension.

THREE-DIMENSIONAL PARTICLE-IN-A-BOX

As we venture into problems with more than one dimension we need to see an overview and so we revert back to the particle-in-a-box (PIB) model in a rectangular container with dimensions ($L_x = a$, $W_y = b$, $H_z = c$). Let $V = 0$ inside the box but the walls are impenetrable as in the one-dimensional case. Thus $V = 0$ for $0 < x < a$, $0 < y < b$ and $0 < z < c$ but $V = \infty$ at all the wall surfaces.

FIGURE 13.1 Linus Carl Pauling (1901–1994) was an American chemist who made substantial contributions to quantum chemistry and molecular biology. He was one of the most influential scientists in the twentieth century and his textbooks remain classics. He is only the second person to have won two Nobel Prizes in two different fields (the Chemistry and the Peace Prize) without sharing the prize with another. Marie Curie won Nobel Prizes in chemistry and physics.

Then we have $\left[\dfrac{-\hbar^2}{2m}\left(\dfrac{\partial^2}{\partial x^2}+\dfrac{\partial^2}{\partial y^2}+\dfrac{\partial^2}{\partial z^2}\right)+0\right]\psi=E\psi$. This example is to teach a method we need very soon for other problems. The trick here is to *assume* that we can write $\psi(x, y, z)=X(x)Y(y)Z(z)$. While this may seem like a bold assumption, the strategy is to try it and maybe it will work? Here that assumption is trivial because (x, y, z) coordinates are clearly independent, but in other systems, the coordinates may mix as we convert from (x, y, z) to say (r, θ, ϕ). Then we can write $\left[\dfrac{-\hbar^2}{2m}\left(\dfrac{\partial^2}{\partial x^2}+\dfrac{\partial^2}{\partial y^2}+\dfrac{\partial^2}{\partial z^2}\right)+0\right]XYZ=[E]XYZ$ and we note that $\dfrac{\partial}{\partial x}$ only operates on X, $\dfrac{\partial}{\partial y}$ only on Y, and $\dfrac{\partial}{\partial z}$ only on Z, and we can also assume that the unknown energy can be broken up into separate values as E_x, E_y, and E_z. When we take the derivatives we find

$$-\frac{\hbar^2}{2m}YZ\left(\frac{\partial^2 X}{\partial x^2}\right)-\frac{\hbar^2}{2m}XZ\left(\frac{\partial^2 Y}{\partial y^2}\right)-\frac{\hbar^2}{2m}XY\left(\frac{\partial^2 Z}{\partial z^2}\right)=(E_x+E_y+E_z)XYZ.$$

The next step is amazing! If we divide the whole equation by XYZ the three-dimensional problem separates into three one-dimensional problems since the variables (x, y, z) are linearly independent and we obtain $\dfrac{\frac{-\hbar^2}{2m}\frac{\partial^2 X}{\partial x^2}}{X}+\dfrac{\frac{-\hbar^2}{2m}\frac{\partial^2 Y}{\partial Y^2}}{Y}+\dfrac{\frac{-\hbar^2}{2m}\frac{\partial^2 Z}{\partial z^2}}{Z}=(E_x+E_y+E_z)$, which is really three separate equations that are identical to the one-dimensional PIB except that a, b, and c are different.

Then $\dfrac{-\hbar^2}{2m}\dfrac{\partial^2 X}{\partial x^2} - E_x X = 0$ and similarly for Y and Z. We have solved this before so we can write

$$XYZ = \sqrt{\dfrac{2}{a}}\sin\left(\dfrac{n_x\pi x}{a}\right)\sqrt{\dfrac{2}{b}}\sin\left(\dfrac{n_y\pi y}{b}\right)\sqrt{\dfrac{2}{c}}\sin\left(\dfrac{n_z\pi z}{c}\right) \text{ and } E_{\text{tot}} = \dfrac{h^2}{8m}\left(\dfrac{n_x^2}{a^2}+\dfrac{n_y^2}{b^2}+\dfrac{n_z^2}{c^2}\right), \text{ Q.E.D.}$$

Note this could be the solution for a freely flying particle in the universe if we knew the dimensions of the universe. Because a, b, and c would be billions of light years squared in the denominator while the triad (n_x, n_y, n_z) are integers, the difference between any two energy levels would be (are) so tiny that translational energy does approach a continuous variable. However, the point of this exercise is to show that *often we can assume a multivariable wave function can be factored into parts, which only depend on one variable at a time.* For future reference, note two parts of the procedure. First, we assumed the wave function could be factored and second, we assumed that the unknown energy could be made up of separate energy values for each coordinate and it worked!

RIGID ROTOR

Here we seek to solve the Schrödinger equation for a rotating molecule. We have seen earlier that the energy levels of a particle in a three-dimensional box are separated by an infinitesimal amount when the size of the box increases to be measured in light years. We also saw in the previous chapter that the quantized harmonic oscillator has specific energy levels that are equally spaced (for low energy) with a sufficient size gap to give rise to absorption bands in the infrared range. We might expect rotational energy-level gaps somewhere between the translation and vibrational ranges since only rotational kinetic energy is involved. We assume there is no potential energy interaction so the usual Schrödinger equation is simply $H_{\text{rot}}\psi_{\text{rot}} = E_{\text{rot}}\psi_{\text{rot}}$ and we can use $I = \mu r^2$ to write

$$\left(\dfrac{-\hbar^2 \nabla^2}{2I}\right)\psi_{\text{rot}} = E_{\text{rot}}\psi_{\text{rot}},$$

where usually $\nabla^2 = \left(\dfrac{\partial^2}{\partial x^2}+\dfrac{\partial^2}{\partial y^2}+\dfrac{\partial^2}{\partial z^2}\right)$ but here we need to represent rotation in polar coordinates. We can use the reduced mass μ for diatomic molecules and this can be generalized for polyatomic cases but we will use diatomic examples here for simplicity. So far in this text we have tried to show only examples that can be worked out cleanly with calculus (and perseverance) but now we come to a situation where it is questionable whether it is worth the time and frustration to get past a big hurdle with full details. The problem is that we have to convert $\nabla^2 (x, y, z)$ to $\nabla^2 (r, \theta, \phi)$. We will sketch out the strategy but this is the sort of thing that a graduate student in physical chemistry or physics needs to check once in their life but here we only need to know the result.

Basically we have to use the Cartesian axes to show the relationships between the Cartesian and polar coordinates and we can write down some facts as formulas based on Figure 13.2. $x = r\sin\theta\cos\phi$, $y = r\sin\theta\sin\phi$ and $z = r\cos\theta$ for $(x, y, z) \rightarrow (r, \theta, \phi)$ as well as $r^2 = x^2 + y^2 + z^2$, $\cos\theta = \dfrac{z}{\sqrt{x^2 + y^2 + z^2}}$ and $\tan\phi = \dfrac{y}{x}$ for $(r, \theta, \phi) \rightarrow (x, y, z)$.

With these relationships, we can set up chain rule derivatives and use trigonometry identities to simplify the results. We note for instance that, in general, the x derivative actually depends on all three polar coordinates as $\left(\dfrac{\partial f}{\partial x}\right)_{y,z} = \left(\dfrac{\partial f}{\partial r}\right)_{\theta,\phi}\left(\dfrac{\partial r}{\partial x}\right)_{y,z} + \left(\dfrac{\partial f}{\partial \theta}\right)_{r,\phi}\left(\dfrac{\partial \theta}{\partial x}\right)_{y,z} + \left(\dfrac{\partial f}{\partial \phi}\right)_{r,\theta}\left(\dfrac{\partial \phi}{\partial x}\right)_{y,z}$. That seems reasonable, so we can work out the partial derivatives and generate the first derivative. That would need to be repeated to obtain the three terms of the gradient of the function as

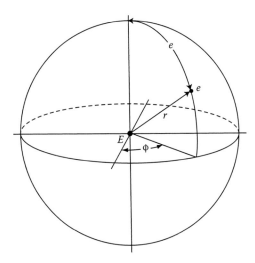

FIGURE 13.2 Spherical coordinate system, E the equatorial x-y plane, ϕ measured from the x-axis in the equatorial plane and θ measured from the vertical z-axis. (Reprinted with permission from White, H.E., *Phys. Rev.*, 37, 1416, 1931. Copyright 1931 by the American Physical Society.)

$\vec{\nabla}f = \left(\dfrac{\partial f}{\partial x}\right)\hat{i} + \left(\dfrac{\partial f}{\partial y}\right)\hat{j} + \left(\dfrac{\partial f}{\partial z}\right)\hat{k}$. This is a vector quantity but if we take the dot product for the second derivative, we will get the desired second derivative as a scalar, that is, $\vec{\nabla} \cdot \vec{\nabla}f = \nabla^2 f$. That is good news because the dot product eliminates any cross derivatives between the Cartesian components. The next step is to find the second derivatives in terms of the chain rules and that is straightforward but tedious! To maintain bookkeeping let us call $\vec{\nabla}f = g_x\,\hat{i} + g_y\,\hat{j} + g_z\,\hat{k}$. In principle, we just repeat the derivative process with the chain rule again for g_x, g_y, and g_z. Then

$$\left(\frac{\partial g_x}{\partial x}\right)_{y,z} = \left(\frac{\partial g_x}{\partial r}\right)_{\theta,\phi}\left(\frac{\partial r}{\partial x}\right)_{y,z} + \left(\frac{\partial g_x}{\partial \theta}\right)_{r,\phi}\left(\frac{\partial \theta}{\partial x}\right)_{y,z} + \left(\frac{\partial g_x}{\partial \phi}\right)_{r,\theta}\left(\frac{\partial \phi}{\partial x}\right)_{y,z} \quad \text{and likewise for } g_y \text{ and } g_z.$$

Then some identities from trigonometry simplify the nine possible terms to yield

$$\nabla^2(r, \theta, \phi) = \frac{1}{r^2}\frac{\partial}{\partial r}\left(r^2\frac{\partial}{\partial r}\right) + \frac{1}{r^2 \sin\theta}\frac{\partial}{\partial \theta}\left(\sin\theta\frac{\partial}{\partial \theta}\right) + \frac{1}{r^2 \sin^2\theta}\frac{\partial^2}{\partial \phi^2}.$$

To ease remembrance we have ordered the terms so that the denominators go as r^2, $r^2 \sin\theta$, and $r^2 \sin^2\theta$ and we see that the third term in ϕ is the simplest in form. We will use this often!

Now we are ready to set up and solve the Schrödinger equation for the rigid rotor problem. First, as with the three-dimensional particle-in-a-box (3D-PIB), we assume the total wave function can be written as a product of three functions with each in only one variable as $\Psi(r, \theta, \phi) = R(r)\,\Theta(\theta)\,\Phi(\phi)$.

Then we have $H\Psi(r, \theta, \phi) = HR(r)\,\Theta(\theta)\,\Phi(\phi) = E_{\text{tot}}\,R(r)\,\Theta(\theta)\,\Phi(\phi) = E_{\text{tot}}\Psi(r, \theta, \phi)$ where

$H = \dfrac{-\hbar^2\nabla^2}{2\mu} + V(r)$ and we expect that E_{tot} can be separated into radial and angular energies.

Thus the Schrödinger equation in polar coordinates can be written as

$$\frac{1}{r^2}\frac{\partial}{\partial r}\left(r^2\frac{\partial\Psi}{\partial r}\right) + \frac{1}{r^2 \sin\theta}\frac{\partial}{\partial \theta}\left(\sin\theta\frac{\partial\Psi}{\partial \theta}\right) + \frac{1}{r^2 \sin^2\theta}\frac{\partial^2\Psi}{\partial \phi^2} + \frac{2\mu}{\hbar^2}[E - V(r)]\Psi = 0.$$

Now divide by $\Psi(r, \theta, \phi) = R(r)\,\Theta(\theta)\,\Phi(\phi)$ just as for the 3D-PIB problem to obtain

$$\frac{1}{R(r)r^2}\frac{d}{dr}\left(r^2\frac{dR}{dr}\right) + \frac{1}{\Theta(\theta)r^2\sin\theta}\frac{d}{d\theta}\left(\sin\theta\frac{d\Theta}{d\theta}\right) + \frac{1}{\Phi(\phi)r^2\sin^2\theta}\frac{\partial^2\Phi}{\partial\phi^2} + \frac{2\mu}{\hbar^2}[E - V(r)] = 0.$$

KEY STEP!

Next multiply through by $r^2\sin^2\theta$ and set $\frac{1}{\Phi}\frac{\partial^2\Phi}{\partial\phi^2} = -m^2$ since that term only depends on ϕ.

$$\frac{\sin^2\theta}{R(r)}\frac{d}{dr}\left(r^2\frac{dR}{dr}\right) + \frac{\sin\theta}{\Theta(\theta)}\frac{d}{d\theta}\left(\sin\theta\frac{d\Theta}{d\theta}\right) + \frac{1}{\Phi(\phi)}\frac{\partial^2\Phi}{\partial\phi^2} + \frac{2\mu}{\hbar^2}[E - V(r)]r^2\sin^2\theta = 0$$

$$\frac{\sin^2\theta}{R(r)}\frac{d}{dr}\left(r^2\frac{dR}{dr}\right) + \frac{\sin\theta}{\Theta(\theta)}\frac{d}{d\theta}\left(\sin\theta\frac{d\Theta}{d\theta}\right) - m^2 + \frac{2\mu}{\hbar^2}[E - V(r)]r^2\sin^2\theta = 0.$$

Now divide by $\sin^2\theta$. Note this leaves r^2 on the last term and $I = \mu r^2$.

$$\frac{1}{R(r)}\frac{d}{dr}\left(r^2\frac{dR}{dr}\right) + \frac{1}{\Theta(\theta)\sin\theta}\frac{d}{d\theta}\left(\sin\theta\frac{d\Theta}{d\theta}\right) - \frac{m^2}{\sin^2\theta} + \frac{2\mu r^2}{\hbar^2}[E - V(r)] = 0.$$

Just as we separated the energy terms in the 3D-PIB, we can split the terms in $R(r)$ from the terms in $\Theta(\theta)$ by setting the separate parts to $\pm\beta$ so as to obtain two separate equations.

$$\frac{1}{R(r)}\frac{d}{dr}\left(r^2\frac{dR}{dr}\right) + \frac{2\mu r^2}{\hbar^2}[E - V(r)] = +\beta \quad \text{and} \quad \frac{1}{\Theta(\theta)\sin\theta}\frac{d}{d\theta}\left(\sin\theta\frac{d\Theta}{d\theta}\right) - \frac{m^2}{\sin^2\theta} = -\beta$$

so that we still have $\beta - \beta = 0$. Then we can rearrange the separate equations. In particular, we multiply the $R(r)$ equation by $\left(\dfrac{R}{r^2}\right)$ and multiply the $\Theta(\theta)$ equation by (Θ) to obtain

$$\frac{1}{r^2}\frac{d}{dr}\left(r^2\frac{dR}{dr}\right) - \frac{\beta}{r^2}R + \frac{2\mu}{\hbar^2}[E - V(r)]R = 0 \quad \text{and} \quad \frac{1}{\sin\theta}\frac{d}{d\theta}\left(\sin\theta\frac{d\Theta}{d\theta}\right) - \frac{m^2}{\sin^2\theta}\Theta + \beta\Theta = 0.$$

We can pause to solve the easy equation for $\Phi(\phi)$ by inspection. $\dfrac{1}{\Phi}\dfrac{\partial^2\Phi}{\partial\phi^2} = -m^2$, and with a little thought we see that $\dfrac{d^2\Phi}{d\phi^2} = -m^2\Phi$ and $\dfrac{d^2\Phi}{d\phi^2} + m^2\Phi = 0$ so $(D + im)(D - im)\Phi = 0$. This is the now familiar POR problem, which has a normalized solution of $\Phi_m(\phi) = \dfrac{1}{\sqrt{2\pi}}e^{\pm im\phi}$. *Note that m is an integer and $m = 0$ is a valid solution!*

It will be important to remember the $R(r)$ equation in the next chapter for the solution to the H atom and the discussion here is very efficient because when we solve the $\Theta(\theta)$ and $\Phi(\phi)$ equations here, we will also have the angular solutions for the H atom. However, here we invoke the conditions of a rigid rotor that has a fixed unchanging bond length between two atoms (temporarily ignoring the fact that we know from the previous chapter that the bond length vibrates!). That condition immediately makes $\dfrac{dR}{dr} = 0$ and if we add the condition that $V(r) = 0$ for free rotation, it greatly simplifies the $R(r)$ equation to $-\dfrac{\beta}{r^2}R + \dfrac{2\mu}{\hbar^2}[E]R = 0$. Multiplying through by r^2 again

produces the moment of inertia $I = \mu r^2$ so that $\left[\dfrac{2IE}{\hbar^2} - \beta\right]R(r) = 0$. Once we find β we will know E!

Let us consider the $\Theta(\theta)$ equation $\dfrac{1}{\sin\theta}\dfrac{d}{d\theta}\left(\sin\theta\dfrac{d\Theta}{d\theta}\right) - \dfrac{m^2}{\sin^2\theta}\Theta + \beta\Theta = 0$. First use the

substitution $z = \cos\theta$ and change the name of the variable $P(z) = \Theta(\theta)$ and note that $\sin^2\theta = 1 - z^2$.

That leads to $\dfrac{d\Theta}{d\theta} = \dfrac{dP}{dz}\dfrac{dz}{d\theta} = \dfrac{dP}{dz}\left(\dfrac{d\cos\theta}{d\theta}\right) = -\dfrac{dP}{dz}\sin\theta$, which transforms our $\Theta(\theta)$ equation to

$\dfrac{d}{dz}\left[(1 - z^2)\dfrac{dP(z)}{dz}\right] + \left[\beta - \dfrac{m^2}{1 - z^2}\right]P(z) = 0$. Here we have used a clever trick in that

$\dfrac{1}{\sin\theta}\dfrac{d}{d\theta} = \dfrac{-d}{d\cos\theta} = \dfrac{-d}{dz}$ so that we have the first term as a function of z as well as the second.

$$\dfrac{-d}{dz}\left[\sin\theta\left(-\sin\theta\dfrac{dP}{dz}\right)\right] = \dfrac{-d}{dz}\left[-\sin^2\theta\dfrac{dP}{dz}\right] = +\dfrac{d}{dz}\left[(1 - \cos^2\theta)\dfrac{dP}{dz}\right] = \dfrac{d}{dz}\left[(1 - z^2)\dfrac{dP}{dz}\right].$$

Thus we have

$$\dfrac{d}{dz}\left[(1 - z^2)\dfrac{dP(z)}{dz}\right] + \left[\beta - \dfrac{m^2}{1 - z^2}\right]P(z) = 0$$

or an alternate form by applying the derivative to the first bracket as

$$(1 - z^2)\dfrac{d^2P(z)}{dz^2} - 2z\dfrac{dP(z)}{dz} + \left[\beta - \dfrac{m^2}{1 - z^2}\right]P(z) = 0.$$

The $m = 0$ Shortcut

At this point we would have to use a difficult "indicial equation" to find β in the general case

because the quantity $\left[\dfrac{m^2}{1 - z^2}\right]$ is undefined (blows up) when $z = \pm 1$. However, when $m = 0$ it leads

to an equation easily solved using a polynomial. That is a special case of an equation solved by
a brilliant French mathematician, A. M. Legendre (1753–1833), who published the work in 1785!
Now we can solve a simpler equation

$$(1 - z^2)\dfrac{d^2P(z)}{dz} - 2z\dfrac{dP(z)}{dz} + \beta P(z) = 0.$$

Let us assume $P(z) = \displaystyle\sum_{l=0} a_l z^l$ so that $\dfrac{dP(z)}{dz} = \displaystyle\sum_{l=0} la_l z^{l-1}$ and $\dfrac{d^2P(z)}{dz^2} = \displaystyle\sum_{l=0} l(l - 1)a_l z^{l-2}$.

Then

$$(1 - z^2)\sum_{l=0} l(l - 1)a_l z^{l-2} - 2z\sum_{l=0} la_l z^{l-1} + \beta\sum_{l=0} a_l z^l = 0$$

or

$$\sum_{l=0} l(l - 1)a_l z^{l-2} - \sum_{l=0} l(l - 1)a_l z^l - 2z\sum_{l=0} la_l z^{l-1} + \beta\sum_{l=0} a_l z^l = 0.$$

Collecting the same powers of z this can be rearranged to

$$\sum_{l=0} l(l-1)a_l z^{l-2} - \sum_{l=0}[l(l-1)+2l-\beta]a_l z^l = \sum_{l=0} l(l-1)a_l z^{l-2} - \sum_{l=0}[l(l+1)-\beta]a_l z^l = 0.$$

The l-index of the first sum must advance to $l+2$ before the power of z is equal to that in the second sum. The l-index steps separately in the two sums and we can set up a recursion ratio as

$$a_{l+2} = \frac{[l(l+1)-\beta]}{(l+2)(l+1)}a_l \quad \text{and} \quad \text{so} \lim_{l\to\infty}\left(\frac{a_{l+2}}{a_l}\right) = \lim_{l\to\infty}\frac{[l(l+1)-\beta]}{(l+2)(l+1)} \to \frac{l^2}{l^2} \to 1.$$

Of course there will be separate even and odd solutions with arbitrary a_0 and a_1 but in the series $P(z) = \sum_{l=0} a_l z^l$ the ratio test shows that far out in the later terms the coefficients do not get smaller so the series will still diverge at $z = \pm 1$! Thus we see that just as we had to terminate the series in the harmonic oscillator solution, we also have to terminate this series to keep it finite. Fortunately we have $[l(l+1)-\beta]$ in the numerator of the ratio test so we can cut off the series when

$\beta = l(l+1)$, $l = 0$, 1, 2, 3...! Further, we now know that $\left[\dfrac{2IE}{\hbar^2} - \beta\right]R(r) = 0$ means that

$E_l = \dfrac{l(l+1)\hbar^2}{2I}$ (at least for $m = 0$)!

Once again the imposition of a relationship to an integer expression leads to a quantized energy! It is not known at what point Schrödinger realized he was dealing with the previously studied Legendre polynomials [7], but the real genius of what Schrödinger did was to recognize the connection between quantization and orthogonal sets of eigenfunctions! Mathematicians had known about Legendre's work since 1785 but it was Schrödinger who made the connection to wave mechanics. While the use of the chain rule to convert the problem to polar coordinates was a necessary chore, this polynomial solution was actually easier than that of the harmonic oscillator due to using the $m = 0$ shortcut! Once we find $\beta = l(l+1)$ and $E_l = \dfrac{l(l+1)\hbar^2}{2I}$, we have what we need for most of the applications in the rest of this chapter. However, we must remember that we are also interested in the angular shapes of the wave functions as used in the solution of the H atom in the next chapter.

RIGID ROTOR WAVE FUNCTIONS

Perhaps Schrödinger recognized this problem as having been thoroughly studied in 1785 by Legendre. A very similar equation was solved by Legendre who knew about the singular points at ± 1 in the general equation and also studied the polynomial solutions on the interval $-1 \le x \le 1$ for the equation $(1-x^2)\dfrac{d^2 P(x)}{dx^2} - 2x\dfrac{dP(x)}{dx} + l(l+1)P(x) = 0$. We can immediately see the analogy to the equation we reached above if $m = 0$. In fact Gatz [8] points this out in an excellent concise appendix. The solutions to Legendre's equation are now known to be

$P_l(x) = \dfrac{1}{2^l l!}\dfrac{d^l}{dx^l}(x^2-1)^l$ and there are also *associated Legendre* functions that are known to be

related to the derivatives as $P_l^{|m|}(x) = (1-x^2)^{|m|/2}\dfrac{d^{|m|}}{dx^{|m|}}P_l(x)$. The Legendre polynomials have been extensively studied by mathematicians over many years, and Anderson [9] gives a useful overview. Some of the first few Legendre polynomials are given in Table 13.1 from [10]. Then some of the associated Legendre polynomials are given in Table 13.2 where $x = \cos(\theta)$ [11]. Note that $\cos(\theta)$ has the perfect range of $-1 \le x \le 1$, so almost all the applications of Legendre polynomials use $x = \cos(\theta)$.

TABLE 13.1
A Short List of $P_l(x)$ Legendre Polynomials

$P_0(x) = 1$

$P_1(x) = x$

$P_2(x) = \dfrac{1}{2}(3x^2 - 1)$

$P_3(x) = \dfrac{1}{2}(5x^3 - 3x)$

$P_4(x) = \dfrac{1}{8}(35x^4 - 30x^2 + 3)$

$P_5(x) = \dfrac{1}{8}(63x^5 - 70x^3 + 15x)$

Source: Lesk, A.M., *Introduction to Physical Chemistry*,
Prentice-Hall, Inc., Englewood Cliffs, NJ, 1982,
pp. 312 and 732; see also Park, D., *Introduction
to Quantum Theory*, 2nd Edn., McGraw-Hill
Book Co., New York, 1974, p. 648.

TABLE 13.2
Selected Associated Legendre Polynomials $P_l^{|m|}(\cos\theta)$

$P_0^0(x) = 1$

$P_1^0(x) = x = \cos(\theta)$

$P_1^1(x) = (1 - x^2)^{1/2} = \sin\theta$

$P_2^0(x) = \dfrac{1}{2}(3x^2 - 1) = \dfrac{1}{2}(3\cos^2\theta - 1)$

$P_2^1(x) = 3x(1 - x^2)^{1/2} = 3\cos\theta\sin\theta$

$P_2^2(x) = 3(1 - x^2) = 3\sin^2\theta$

$P_3^0(x) = \dfrac{1}{2}(5x^2 - 3x)$

$P_3^1(x) = \dfrac{3}{2}(5x^2 - 1)(1 - x^2)^{1/2}$

$P_3^2(x) = 15x(1 - x^2)$

$P_3^3(x) = 15(1 - x^2)^{3/2}$

Source: McQuarrie, D.A., *Quantum Chemistry*, 2nd Edn., University
Science Books, Sausalito, CA, 2008, p. 293.

We can test to show that the Legendre polynomials are solutions to the $m = 0$ equation. Consider $P_2(x) = \dfrac{1}{2}(3x^2 - 1)$ in the equation $(1 - x^2)\dfrac{d^2P(x)}{dx^2} - 2x\dfrac{dP(x)}{dx} + l(l + 1)P(x) = 0$.

$$\dfrac{dP_2(z)}{dz} = 3x \quad \text{and} \quad \dfrac{d^2P_2(z)}{dz^2} = 3 \text{ so } (1 - x^2)(3) - 2x(3x) + 2(3)\left[\dfrac{1}{2}(3x^2 - 1)\right] = 0?$$

Then we find that $3 - 3x^2 - 6x^2 + 9x^2 - 3 = 0$, Q.E.D.!

In order to make our shortcut with $m = 0$ pay off, we need to show that the associated Legendre polynomials satisfy the general Legendre equation. Consider $P_2^2(x) = 3(1 - x^2)$.

$$(1 - z^2)\frac{d^2 P_l^{|m|}(z)}{dz^2} - 2z\frac{dP_l^{|m|}(z)}{dz} + \left[l(l+1) - \frac{m^2}{1-z^2}\right]P_l^{|m|}(z) = 0$$

$\dfrac{dP_2^2(x)}{dx} = -6x$ and $\dfrac{d^2 P_2^2(x)}{dx^2} = -6$ along with $l = 2$ and $m = 2$ in the general Legendre equation.

Thus $(1 - x^2)(-6) - 2z(-6x) + \left[6 - \dfrac{2^2}{1-x^2}\right](3)(1-x^2) = 0$? or $-6 + 6x^2 + 12x^2 + 18(1-x^2) -$

$\dfrac{12(1-x^2)}{(1-x^2)} = 0$? So $-6 + 6x^2 + 12x^2 + 18 - 18x^2 - 12 = 0$ Q.E.D.! The proof due to Anderson [9] is given in Appendix A of this book using just basic calculus. Then considering the odd and even cases of the basic Legendre polynomials, we have in general

$$P_l^m(x) = (-1)^m (1 - x^2)^{\frac{m}{2}} \frac{d^m P_l(x)}{dx^m}.$$

Here we consider it sufficient to have shown that the $P_l^{|m|}(x)$ are related to the derivatives of $P_l(x)$ but in either case $\beta = l(l+1)$. Thus the $P_l^{|m|}(x)$ are the solution to the general Legendre equation as shown in Appendix A and our $m = 0$ shortcut leads to a general solution.

RIGID ROTOR RESULTS

Finally, we recall that the wave function of the rigid rotor is a product of $\Phi(\phi)$ and $\Theta(\theta)$. The normalized product is $Y_l^m(\theta, \phi) = (-1)^{\frac{m+|m|}{2}}\sqrt{\dfrac{(2l+1)(l+|m|)!}{4\pi(l-|m|)!}}P_l^{|m|}(\cos\theta)e^{im\phi}$ [10]. This is the total wave function for the rigid rotor and the energy is given as

$$E_l = \frac{l(l+1)\hbar^2}{2I} = \frac{l(l+1)\hbar^2}{2\mu r^2}, \quad l = 0, 1, 2, 3, 4, \ldots.$$

We also note that m must be an integer and $|m| \leq l$ for a degeneracy of $(2l+1)$ different m values. We show this form here to achieve a combined single eigenfunction, but you can see that the factor of $(-1)^{\frac{m+|m|}{2}}$ causes some tricky sign changes for the various members of the set. This can be avoided by using separate functions for $\Phi(\phi)$ and $\Theta(\theta)$ and we will do that for the H atom. It is worth alerting students that there is a so-called phase problem with wave functions. Note that $\int \psi^*\psi \, d\tau$ is real because of the use of the complex conjugate ψ^*, but we could have also used $\int (-\psi^*)(-\psi)d\tau = \int \psi^*\psi \, d\tau$ so there is an ambiguity of sign and that shows up in the $(-1)^{\frac{m+|m|}{2}}$ factor. This is a well known annoyance and an interested student can check [10] for a discussion of the various options but the thing to remember is that $\int \psi^*\psi \, d\tau$ is a real number by whatever phase convention one uses.

ANGULAR WAVE FUNCTIONS

Although we have solved the problem in principle, the most important result is for students to be able to envision the shape of the functions. In Table 13.3, we provide a few of the normalized functions with shape labels relating to the orbitals that will be found for the H atom in the next chapter, but the energy levels of the rigid rotor are much closer together than the electronic energies of the H atom and we should be aware that the rigid rotor levels can have much higher l values at room temperature. We may see l values of 20 or more in rotational spectra, while for the H atom we will usually deal with l values of 4 or less.

TABLE 13.3
Selected Normalized $Y_l^m(\theta, \phi)$

$Y_l^m(\theta, \phi)$	"Shape-Type"
$Y_0^0 = \dfrac{1}{\sqrt{4\pi}}$	s
$Y_1^0 = \sqrt{\dfrac{3}{4\pi}}\cos\theta$	p
$Y_1^{+1} = -\sqrt{\dfrac{3}{8\pi}}e^{+i\phi}\sin\theta$	p
$Y_1^{-1} = \sqrt{\dfrac{3}{8\pi}}e^{-i\phi}\sin\theta$	p
$Y_2^0 = \sqrt{\dfrac{5}{16\pi}}(3\cos^2\theta - 1)$	d
$Y_2^1 = -\sqrt{\dfrac{15}{8\pi}}e^{+i\phi}\sin\theta\,\cos\theta$	d
$Y_2^{-1} = \sqrt{\dfrac{15}{8\pi}}e^{-i\phi}\sin\theta\,\cos\theta$	d
$Y_2^2 = \sqrt{\dfrac{15}{32\pi}}e^{2i\phi}\sin^2\theta$	d
$Y_2^{-2} = \sqrt{\dfrac{15}{32\pi}}e^{-2i\phi}\sin^2\theta$	d

Source: Lesk, A.M., *Introduction to Physical Chemistry*, Prentice-Hall, Inc., Englewood Cliffs, NJ, 1982, pp. 312 and 732; see also Park, D., *Introduction to Quantum Theory*, 2nd Edn., McGraw-Hill Book Co., New York, 1974, p. 648.

Another point worth noting is that the $\Phi_m(\phi)$ function is complex (involves $\sqrt{-1} \equiv i$) and the normalization coefficient sometimes has a sign change according to $(-1)^{\frac{m+|m|}{2}}$. If one studies the form of the Legendre function in $\Theta(\theta)$, it can be determined that the m can take on values from $-l$ to $+l$ for a total of $(2l + 1)$ values and m can be negative but $|m|$ is always positive. That is probably more interesting for electronic orbitals but here we want to examine what it means in terms of *rotation*! We want to work toward a picture or diagram that tells us how a linear molecule behaving as a rigid rotor can actually rotate. After all, the wave function tells us that only certain shapes are allowed depending on the state function $|l, m\rangle = Y_l^m(\theta, \phi)$ where we start to think of the state of the rigid rotor specified in terms of two integer quantum numbers. Quantum mechanics tells us that if rotation is quantized then *angular momentum* is quantized as well, and we should consider what values are allowed for the angular momentum. Reminding ourselves that this is an undergraduate text, we will just sketch the following treatment of quantized angular momentum.

ANGULAR MOMENTUM

We now return to one of the most difficult aspects of quantum mechanics, the ability to treat angular variables. We had a taste of the problem with the particle-on-a-ring but it referred to a flat ring while we need to describe (θ, ϕ) space. We also recall that momentum can be more fundamental than energy and here we encounter *quantized angular momentum*. First we define what we mean by angular momentum. In a rotating system, the moment of inertia $(I = \mu r^2)$ takes the place of a mass and linear velocity is replaced by angular velocity (ω). Then angular momentum is defined as "L," although some texts use "M," Angular momentum involves a twisting motion and is usually represented as a 3×3 determinant as

$$\vec{L} = I\omega = \vec{r} \times \vec{p} = \begin{vmatrix} \hat{i} & \hat{j} & \hat{k} \\ x & y & z \\ p_x & p_y & p_z \end{vmatrix} = \hat{i}(yp_z - p_yz) + \hat{j}(zp_x - p_zx) + \hat{k}(xp_y - p_xy).$$

Thus, $\vec{L}$ is a vector as the result of a cross product and $\vec{L} = \hat{i}L_x + \hat{j}L_y + \hat{k}L_z$. Then to form a quantum mechanical operator, one just thinks of the (x, y, z) coordinates as operators and substitutes $p_x = \dfrac{h}{i}\dfrac{\partial}{\partial x}, p_y = \dfrac{h}{i}\dfrac{\partial}{\partial y}$ and $p_z = \dfrac{h}{i}\dfrac{\partial}{\partial z}$ (Figure 13.3). That leads to three, relatively simple, operators in the Cartesian representation (note cyclic permutation in alphabetic order, x, y, z, then y, z, x, and finally z, x, y). We want to show the effect of quantized angular momentum in the rigid rotor. There are three operators for the Cartesian components of the angular momentum

$$L_x = \frac{\hbar}{i}\left(y\frac{\partial}{\partial x} - x\frac{\partial}{\partial y}\right), \quad L_y = \frac{\hbar}{i}\left(x\frac{\partial}{\partial z} - z\frac{\partial}{\partial x}\right), \quad \text{and} \quad L_z = \frac{\hbar}{i}\left(x\frac{\partial}{\partial y} - y\frac{\partial}{\partial x}\right).$$

Since we know that $H = \dfrac{-\hbar^2\nabla^2}{2I}$ for the rigid rotor and $E_l = \dfrac{l(l+1)\hbar^2}{2I}$, we deduce from $\dfrac{-\hbar^2\nabla^2}{2I}Y_l^m(\theta, \phi) = \dfrac{\hbar^2 l(l+1)}{2I}Y_l^m(\theta, \phi) = E_l Y_l^m(\theta, \phi)$ that $\dfrac{-\hbar^2\nabla^2}{2I} \rightarrow E$ but in a rotating system $E = \dfrac{L^2}{2I}$ so $L^2 = 2IE$ and we find an important relationship the easy way as

$$L_{op}^2 Y_l^m(\theta, \phi) = l(l+1)\hbar^2 Y_l^m(\theta, \phi).$$

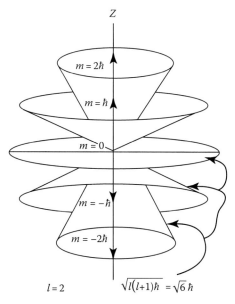

FIGURE 13.3 Cones of definite L_z angular momentum but indeterminant L_x and L_y for $l = 2$.

We would have needed nine partial derivatives for $\nabla^2(r, \theta, \phi)$ but we only need six to convert L_z to polar coordinates:

$$\frac{\partial r}{\partial x} = \frac{x}{r} = \sin(\theta)\cos(\phi), \quad \frac{\partial r}{\partial y} = \frac{y}{r} = \sin(\theta)\sin(\phi), \quad \frac{\partial r}{\partial z} = \cos(\theta),$$

$$\frac{\partial \theta}{\partial x} = \frac{\cos(\theta)\cos(\phi)}{r}, \quad \frac{\partial \theta}{\partial y} = \frac{\cos(\theta)\sin(\phi)}{r}, \quad \frac{\partial \theta}{\partial z} = \frac{-\sin(\theta)}{r},$$

$$\frac{\partial \phi}{\partial x} = \frac{-\sin(\phi)}{r\,\sin(\theta)}, \quad \frac{\partial \phi}{\partial y} = \frac{\cos(\phi)}{r\,\sin(\theta)}, \quad \frac{\partial \phi}{\partial z} = 0.$$

Then we can use the chain rule to convert L_z as $L_z = \frac{\hbar}{i}\left(x\frac{\partial}{\partial y} - y\frac{\partial}{\partial x}\right)$ to

$$L_z = \frac{\hbar}{i}\left\{ r\sin(\theta)\cos(\phi)\left[\left(\frac{\partial r}{\partial y}\right)\frac{\partial}{\partial r} + \left(\frac{\partial \theta}{\partial y}\right)\frac{\partial}{\partial \theta} + \left(\frac{\partial \phi}{\partial y}\right)\frac{\partial}{\partial \phi}\right]\right\}$$
$$- \frac{\hbar}{i}\left\{ r\sin(\theta)\sin(\phi)\left[\left(\frac{\partial r}{\partial x}\right)\frac{\partial}{\partial r} + \left(\frac{\partial \theta}{\partial x}\right)\frac{\partial}{\partial \theta} + \left(\frac{\partial \phi}{\partial x}\right)\frac{\partial}{\partial \phi}\right]\right\}.$$

Amazingly much of this expression cancels out and leaves only $L_z = \frac{\hbar}{i}\frac{\partial}{\partial \phi}$. We leave that proof to the homework problems. We will see that L_z is chosen to be the component of the angular momentum, which commutes with L^2 as in $[L^2, L_z] = 0$ because its polar coordinate form is the simplest of the three components. Then we have a way to "know" the m quantum number of a $Y_l^m(\theta, \phi)$ wave function whether it is for a rigid rotor or an electronic atomic orbital. Since the ϕ dependence of the $Y_l^m(\theta, \phi)$ wave function is just $e^{im\phi}$, $\frac{\partial}{\partial \phi}e^{im\phi} = im$ so that

$$L_z Y_l^m(\theta, \phi) = m\hbar Y_l^m(\theta, \phi).$$

Thus, we can define our wave function with just two quantum numbers as $|l, m\rangle$.

Maybe you thought it is obvious that you can interchange the order of quantum mechanical operators since in real arithmetic $2 \times 3 = 3 \times 2$, but let us try it on the Cartesian forms for L_x, L_y, and L_z. Consider $[L_x, L_y]$ applied to an arbitrary test function "f" where the square brackets refer to *a commutator, the amount by which interchanging the order of the operators makes a difference* as in $[L_x, L_y]f = L_x L_y f - L_y L_x f$.

$$L_x L_y f = (-\hbar^2)\left[\left(y\frac{\partial}{\partial z} - z\frac{\partial}{\partial y}\right)\left(z\frac{\partial}{\partial x} - x\frac{\partial}{\partial z}\right)\right]f$$
$$= (-\hbar^2)\left[\left(y\frac{\partial f}{\partial x}\right) + \left(yz\frac{\partial^2 f}{\partial z \partial x}\right) - \left(yx\frac{\partial^2 f}{\partial z^2}\right) - \left(z^2\frac{\partial^2 f}{\partial y \partial x}\right) + \left(zx\frac{\partial^2 f}{\partial y \partial z}\right)\right],$$

then in reverse

$$L_y L_x f = (-\hbar^2)\left[\left(z\frac{\partial}{\partial x} - x\frac{\partial}{\partial z}\right)\left(y\frac{\partial}{\partial z} - z\frac{\partial}{\partial y}\right)\right]f$$
$$= (-\hbar^2)\left[\left(zy\frac{\partial^2 f}{\partial x \partial z}\right) - \left(z^2\frac{\partial^2 f}{\partial x \partial y}\right) - \left(xy\frac{\partial^2 f}{\partial z^2}\right) + \left(x\frac{\partial f}{\partial y}\right) + \left(xz\frac{\partial^2 f}{\partial z \partial y}\right)\right]$$

and so
$$[L_x, L_y]f = L_xL_yf - L_yL_xf = (-\hbar^2)\left[y\frac{\partial}{\partial x} - x\frac{\partial}{\partial y}\right]f = (i\hbar)(L_z).$$ We could repeat this for the other cases and find

$$[L_x, L_y] = i\hbar L_z, \quad [L_y, L_z] = i\hbar L_x \quad \text{and} \quad [L_z, L_x] = i\hbar L_y.$$

Thus, the angular momentum components DO NOT commute among themselves! What does that mean? Basically it means that if we somehow make a measurement of $\langle L_x\rangle$ followed by a measurement of $\langle L_y\rangle$ we will not get the same answer as if we reversed the order of the measurement! What about the relationship between the individual components of the angular momentum and L^2?

$$[L^2, L_z]f = \left(L_x^2L_z - L_zL_x^2 + L_y^2L_z - L_zL_y^2 + L_z^3 - L_z^3\right)f$$

Clearly $\left(L_zL_z^2 - L_z^2L_z\right)f = 0$ and we can use $[L_x, L_z] = -i\hbar L_y$ as well as $[L_y, L_z] = i\hbar L_x$ to find $\left[L_x^2, L_z\right]f = \{L_x(L_xL_z) - (L_zL_x)L_x\}f$ which becomes $\left[L_x^2, L_z\right]f = \{L_x(L_zL_x - i\hbar L_y) - (L_xL_z + i\hbar L_y)L_x\}f$ and then $\left[L_x^2, L_z\right]f = \{-i\hbar(L_xL_y + L_yL_x)\}f$. Similarly $\left[L_y^2, L_z\right]f = \{L_y(L_yL_z) - (L_zL_y)L_y\}f$ which becomes $\left[L_y^2, L_z\right]f = \{L_y(L_zL_y + i\hbar L_x) - (L_yL_z - i\hbar L_x)L_y\}f$ and then $\left[L_y^2, L_z\right]f = \{+i\hbar(L_yL_x + L_xL_y)\}f$. Finally $[L^2, L_z]f = \{-i\hbar(L_xL_y + L_yL_x)f + i\hbar(L_yL_x + L_xL_y)f + 0\} = 0$ or $[L^2, L_z]f = 0$. (Write it out!)

We could also go through the same exercise above to show that $[L^2, L_x] = 0$ and $[L^2, L_y] = 0$, so all the angular momentum components DO commute with L^2. While this discussion may seem very abstract, we are leading up to an important result that can be used for both the rigid rotor and the H atom. Recall the proof that operators that commute can have the same set of eigenfunctions (Theorem 3, Chapter 11). Here the situation is that each of the angular momentum operators does commute with $L^2 = L_x^2 + L_y^2 + L_z^2$ but the individual components do not commute among themselves. Because we will see that the L_z operator has the simplest form in polar coordinates, it has become a standard *convention* to use $Y_l^m(\theta, \phi)$ as the eigenfunctions of L^2 and L_z. Thus we can "know" (measure) the eigenvalues of these two operators. However, if we use the $Y_l^m(\theta, \phi)$ functions as the eigenfunctions we cannot simultaneously "know" (measure) the eigenvalues of L_x or L_y! The usual way to interpret this situation is to represent the l and m values as a vector with components on an (x, z) plane with $m\hbar$ for the z projection and $\hbar\sqrt{l(l + 1)}$ for the length of the hypotenuse of a triangle with an angle of $\theta = \cos^{-1}\left(\dfrac{m}{\sqrt{l(l + 1)}}\right)$ [12] measured from the z-axis.

Since we cannot "know" $\langle L_x\rangle$ or $\langle L_y\rangle$ by direct measurement or by calculation using the $Y_l^m(\theta, \phi)$ eigenfunctions, we can only say that the x- and y-components are somewhere on a circle around the z-axis forming a cone about the z-axis. If we assume there is an arbitrary direction such as a gravitational or magnetic field ("up"/"down") we now realize that only certain angles of rotation will occur. For high values of l the angles of rotations will sweep out the volume of a sphere but for low values of l not all angles will occur for a given molecule (assuming a reference direction). The bottom line for the two important operators can now be stated as eigenfunction equations.

$$L_zY_l^m(\theta, \phi) = m\hbar Y_l^m(\theta, \phi) \quad \text{and} \quad L^2Y_l^m(\theta, \phi) = l(l + 1)\hbar^2 Y_l^m(\theta, \phi).$$

ROTATIONAL SPECTRUM OF CO

This example takes advantage of the fact that the lines in a pure rotation spectrum are equally spaced. Precise data is available for CO and it is a convention to use "j" for rotational energy levels but retain "l" for electronic orbitals (Figure 13.4). The space between rotational (inverted) peaks is

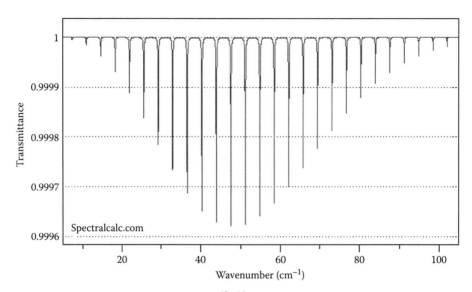

FIGURE 13.4 The pure rotational spectrum for $^{12}C^{16}O$ synthesized from data at the HITRAN site using Spectralcalc software at http://www.spectralcalc.com/calc/spectralcalc.php (From Dr. Keeyoon Sung of the CalTech/Jet Propulsion Laboratory.)

3.8626 cm^{-1} averaged over 16 peaks [13]. For perspective, note that the main absorption peaks are around 50 cm^{-1} and that $E = h\nu = h\dfrac{c}{\lambda} = hc\bar{\nu}$. Then given the Rydberg constant for the H atom as energy $= (109737.31568527$ cm$^{-1})hc = 13.60569193$ eV, we see 1 eV $= (8065.544639$ cm$^{-1})hc$ so that $(50$ cm$^{-1}/8065.544$ cm^{-1}/eV$) = 6.1992 \times 10^{-3}$ eV or roughly 6.2 meV. Just remember that 1 cm$^{-1} = 1.15534 \times 10^{-4}$ eV, which is a very small amount of energy on a macroscopic scale. One wavenumber is a small amount of energy!

Given $E_j = \dfrac{j(j+1)\hbar^2}{2I}$, we can study the pattern of peaks due to $\Delta E = E_{j+1} - E_j = h\nu$.

$\Delta E = [(j+1)(j+2) - j(j+1)]\left(\dfrac{\hbar^2}{2I}\right) = (2j+2)\left(\dfrac{\hbar^2}{2I}\right) = \dfrac{hc}{\lambda} = hc\bar{\nu} = 2(j+1)\left(\dfrac{\hbar^2}{2I}\right)$, but this

depends on knowing the value of j, which may not be easy to assign. However, the spacing between the peaks is very nearly constant so we can use the spacing to find the *bond length* of the molecule!

$$\Delta(\Delta E) = [2(j+1+1) - 2(j+1)]\left(\dfrac{\hbar^2}{\mu r^2}\right) = 2\left(\dfrac{\hbar^2}{2\mu r^2}\right) = h(\nu_{j+1} - \nu_j).$$

Using $[\Delta(\Delta E)]$ eliminates the need to know the actual j value and we only need to know the difference between the corresponding frequencies $(\nu_{j+1} - \nu_j) = \dfrac{h^2}{4\pi^2\mu r^2 h} = \dfrac{h}{4\pi^2\mu r^2}$ and so

$r = \sqrt{\dfrac{h}{(\nu_{j+1} - \nu_j)4\pi^2\mu}}$ but we need the reduced mass μ.

$$\mu = \dfrac{(12.000000 \text{ g/mol})(15.9949146196 \text{ g/mol})}{(27.99491461 \text{ g/mol})(6.02214179 \times 10^{23}/\text{mol})} = 1.138500035 \times 10^{-23} \text{ g}.$$

$$\Delta\bar{\nu} = \nu_{j+1} - \nu_j = c\left(\dfrac{1}{\lambda_{j+1}} - \dfrac{1}{\lambda_j}\right) = c(\Delta\bar{\nu}) = (2.99792458 \times 10^{10} \text{ cm/s})(3.8626/\text{cm})$$

$$= 1.15797834 \times 10^{11}/\text{s}.$$

Then we can find the bond length for $^{12}_{6}C \equiv ^{16}_{8}O$ as

$$r = \sqrt{\frac{6.62606896 \times 10^{-27} \text{ g} \cdot \text{cm}^2/\text{s}}{(1.157978348 \times 10^{11}/\text{s})4\pi^2(1.138500035 \times 10^{-23} \text{ g})}} = 1.128318 \times 10^{-8} \text{ cm} \cong 1.1283 \text{ Å}.$$

The experimental value from the 90th Edn. of the *CRC Handbook* [14] is 1.1283 Å!

Consultation with several molecular spectroscopists reveals that modern Fourier transform infra-red spectrometers are so precise that they usually measure only one or two rotational lines at a time with very high resolution and save the data in a computer file in a data bank such as the HITRAN site [15]. Then at a later time they use a computer program to splice together the various files to make a complete spectrum. The above-mentioned spectrum for CO was constructed in such a manner.

FOURIER TRANSFORM SPECTROMETRY

A major innovation in several forms of spectrometry has occurred in the last 30 years due to the availability of a computer algorithm developed by Cooley of IBM and Tukey of Princeton in 1965 [16]. Although the algorithm was first used by Carl Friedrich Gauss to study the orbits of asteroids in 1805, an efficient application of the method had to await the development of modern programmable computers. The algorithm is known as the Cooley–Tukey fast Fourier transform (FFT) and has found wide applications, especially IR and NMR spectroscopy. For infrared, a modified interferometer design is used in which a plate of 50% transmission and 50% reflection is placed between two mirrors, a bright light source, and some sort of detector. One (or both, see the National Institute of Standards and Technology FT spectrometer at http://physics.nist.gov/Divisions/Div842/Gp1/fts_intro.html) of the reflecting mirrors is methodically moved on a track and the interference transmitted to the detector is a time-dependent series of light and dark waves. A sample can be placed in one of the beams and the detector signal is recorded digitally. FT-IR is in principle a single-beam experiment in which a "spectrum" is recorded for the sample and another spectrum of the "background" is measured and subtracted digitally from the sample spectrum.

This method has several advantages compared to older dispersive spectrometers in which a range of frequencies is sent through a dispersing prism or reflected from a grating through various slits with slow scans of the frequency range. Compared to simple scanning dispersive IR spectrometers of only 30 years ago, the FT-IR spectrometers are a true breakthrough in advanced technology. Specific advantages can be cited [17]:

1. Multiplex (Fellgett) advantage, all frequencies are scanned simultaneously and the signal-to-noise ratio $(S/n) \propto \sqrt{N}$ where N is the number of scans. Also scans are faster so many scans can improve the (S/n).
2. Throughput (Jacquinot) advantage, about 150 times greater throughput than a dispersive instrument that has several slits.
3. Registration (Connes) advantage with internal wavelength reference (He-Ne laser).

Note that in the schematic a He-Ne laser beam can be used for calibration at an exact wavelength (6328 Å) and the monochromatic wave comes through the interferometer at a single frequency while at the same time the broad range of frequencies is represented in the interferogram, which requires use of the Cooley–Tukey FFT to untangle the full spectrum. The raw signal from the detector is the complicated interferogram resulting from the interference of many frequencies all passing through the sample at once. At the "zero position," there are many wave signals present but as the mirror is moved, the phase of the many waves in the light beam with different frequencies is represented in the signal. Once the signal is accumulated and averaged over many repetitive traverses of the moving mirror, a computer is necessary to decompose the signal into the amounts of the various frequencies and this is done using the Cooley–Tukey FFT routine. The fractional amounts of many thousands of waves of different frequencies can then be plotted as a continuous "spectrum" of their wave frequencies.

Although the FFT is done using complex arithmetic $\left(\text{recall that } \cos(x) = \dfrac{e^{ix} + e^{-ix}}{2} \right)$, what is needed is to integrate the product of each desired resolvable frequency from the total by what is essentially a "projection" operation in which the amount (fractional coefficient) of a given frequency is present in the total. Consider a very simple, non-trigonometric, example of the projection of the y-component in the $\hat{j}$ direction from a vector $\vec{r} = 2\hat{i} + 4\hat{j} + 3\hat{k}$. If we take the dot product of the desired component with the total vector as $\hat{j} \cdot \vec{r} = 2\hat{j} \cdot \hat{i} + 4\hat{j} \cdot \hat{j} + 3\hat{j} \cdot \hat{k} = 0 + 4 + 0 = 4$. That is the basic principle of the FFT in which a set of specific frequency $\cos(n\pi cx/\lambda_k)$ functions are integrated as a product with the detector signal and the coefficient of each $\cos(n\pi cx/\lambda_k)$ is extracted from the total signal. The general form of the finite Fourier cosine transform of $F(x)$ (the interferogram) is given as [18]

$$ f_c(n) = \int\limits_0^\pi F(x) \cos(nx)\, dx, \quad n = 0, 1, 2, 3, \ldots $$

Thus you can see the analogy to a projection of a specific cosine component from the interferogram, although the details of the Cooley–Tukey algorithm are much more complicated. If we apply the projection idea to many (thousands) $\cos(nx)$ waves, we can find a fine-grained bar graph that becomes the IR spectrum $\{f_c(n)\} = \left\{ \int_0^\pi F(x) \cos(nx)\, dx \right\}, n = 0, 1, 2, 3, \ldots$

FT-IR IMAGING AND MICROSCOPY

In addition to basic research on molecular species, FT-IR applications have recently been developed for FT imaging of paint [19] and fiber analysis [20] in forensic applications. The main innovation is due to the development of focal plane array detectors, which permit thousands of images to be examined in a microscope with FT-IR analysis of each individual pixel. Further information can be found related to the Varian 610-IR FTIR spectrochemical imaging microscope at http://www.varianinc.com/cgi-bin/nav?products/spectr/ftir/ftir_microscopy/index&cid = LLHQQHINFM (Figures 13.5 through 13.8).

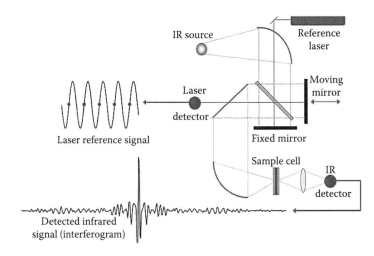

FIGURE 13.5 Diagram from a Varian training presentation by Dr. Ellen Miseo, senior FTIR scientist of the Varian Analytical Division. This shows a typical interferogram from the detector and the use of a secondary signal from a He–Ne laser (6328 Å, red) for internal calibration.

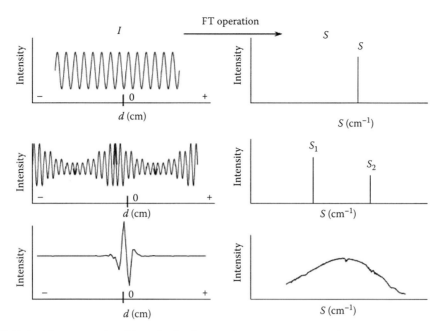

FIGURE 13.6 Three examples of how the Fourier transform converts a wave signal to a representation of the frequencies present in the wave signal. A single frequency produces only one line, a superposition of two waves produces two frequencies and a complicated interferogram produces a frequency spectrum. (From Dr. Ellen Miseo, senior FT-IR scientist, Varian Analytical Division. With permission.)

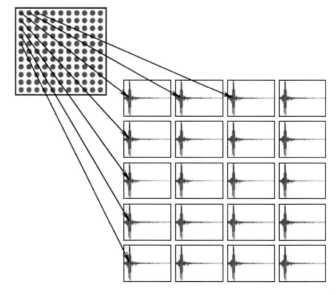

FIGURE 13.7 Up to 4096 pixels (picture elements) with 5.5 micron (1 micron $= 10^{-4}$ cm) resolution can be processed simultaneously in 10 min with a patented Varian Focal Plane Array Detector, U.S. Patent No. 6,141,100. (From John C. Hahn, senior spectroscopy specialist, Varian, Inc. With permission.)

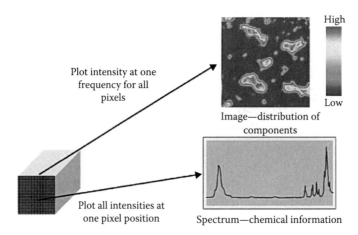

High

Plot intensity at one
frequency for all
pixels

Low

Image—distribution of
components

Plot all intensities at
one pixel position

Spectrum—chemical information

FIGURE 13.8 Schematic of multi-pixel focal plane FT-IR analysis applied to forensic or material science study of surfaces. (From Dr. Ellen Miseo, senior FT-IR scientist, Varian Analytical Division. With permission.)

DIPOLE REQUIREMENT

In this treatment *the molecule must have a dipole moment* that couples with the electric field of the light wave as $\vec{E} \cdot \vec{\mu} = E\mu \cos \theta$. As an aside, there was a controversy regarding the sign of the small dipole moment of CO, which was finally resolved in 1958 by Rosenblum, Nethercot and Townes [21]. The electric dipole of 0.1098 D [14] is attributed to the charges as $C^- O^+$. This somewhat surprising result can be explained qualitatively as due to the greater electronegativity of the O atom pulling the electron density in closer while the electrons near the less electronegative C atom extend further from that end of the molecule. Thus even with more electrons around the O end, the spatially weighted dipole moment $\langle \psi | e_q \vec{r} | \psi \rangle$ turns out to be $C^- O^+$ by a small amount! Here we can use the properties of the $Y_l^m(\theta, \phi)$ wave functions to understand how the electric field of a light wave interacts with the dipole of a molecule. Recall that $Y_l^m(\theta, \phi) = (-1)^{\frac{m+|m|}{2}} \sqrt{\dfrac{(2l+1)(l+|m|)!}{4\pi(l-|m|)!}}$

$P_l^{|m|}(\cos \theta) e^{im\phi}$ so that the θ part consists of an associated Legendre function. Legendre functions have been studied for hundreds of years and it is known that they satisfy a recursion relationship linking P_l^m to P_{l+1}^m and P_{l-1}^m as in the formula $(2l+1)xP_l^{|m|}(x) = (l-|m|+1)P_{l+1}^{|m|}(x) + (l+|m|)P_{l-1}^{|m|}(x)$ [9]. Let us demonstrate this for one case. $P_0^0(x) = 1$, $P_1^0(x) = x = \cos(\theta)$ and $P_2^0(x) = \dfrac{1}{2}(3x^2 - 1) = \dfrac{1}{2}(3\cos^2\theta - 1)$ from Table 13.2 so $(3)(x)P_1^0(x) = (2)P_2^0(x) + (1)P_0^0(x)$ and we find for this simple case that indeed $(3)(x)(x) = (2)[(1/2)(3x^2 - 1)] + (1)(1) = 3x^2 - 1 + 1 = 3x^2$, Q.E.D. This is a very useful formula. We can use this in the dipole transition matrix element $\langle l', m' | \mu \cos \theta | l, m \rangle = \mu \langle l', m' | x | l, m \rangle \cdot (x)P_l^{|m|}(x) = \dfrac{(l-|m|+1)P_{l+1}^{|m|}(x) + (l+|m|)P_{l-1}^{|m|}(x)}{(2l+1)}$ where $x = \cos \theta$, so we substitute this into the matrix element, factor out the integration over ϕ, and use the orthogonality of the $\{P_l^m\}$. $\langle l'm' | x | l, m \rangle = \left\{ \dfrac{(l-|m|+1)\langle l'|l+1\rangle}{(2l+1)} + \dfrac{(l+|m|)\langle l'|l-1\rangle}{(2l+1)} \right\} \int_0^{2\pi} e^{i(|m|-m')} d\phi$. More simply this is $\langle l'm' | x | l, m \rangle = \left\{ \dfrac{(l-|m|+1)\delta_{l',l+1}}{(2l+1)} + \dfrac{(l+|m|)\delta_{l',l-1}}{(2l+1)} \right\} (\delta_{m',|m|})$. So the selection rule is

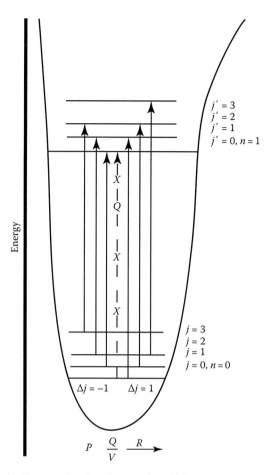

FIGURE 13.9 A schematic diagram showing the way in which the P, Q, and R features of the vib-rotor spectrum occur due to the selection rules for vibrational and rotational transitions. The symbol marked as $\bar{V}$ refers to the wavenumber scale as $\bar{v}$ in cm^{-1}. The Q-branch is missing because the rotational selection rule is that $\Delta j = \pm 1$ but $\Delta j = 0$ for the Q-branch.

$\Delta l = \pm 1$ with $\Delta m = 0$ in order to have a nonzero interaction. This is the main rule we will use to interpret vib-rotor spectra. However, Kauzmann [22] shows a difficult derivation beyond the scope of this text, which proves that $\Delta m = 0, \pm 1$ when one includes the possibility of coupling with the μ_x and μ_y components as well as the simpler derivation for only the μ_z component. This leads to a nonzero Q-branch in nonlinear polyatomic molecules (Figure 13.9). We will define "Q-branch" in the next section and use it for a linear molecule.

VIB-ROTOR INFRARED SPECTROSCOPY

Strictly speaking, we would have to treat the combined wave function for vibration and rotation as $\Psi = \psi_{\text{trans}}(x, y, z)\, \psi_{\text{vib}}(r)\, \psi_{\text{rot}}(\theta, \phi)$, but at this point we know the center of mass is flying about as a 3D-PIB and the molecules are vibrating while they rotate. This is so complicated that we resort to just showing an energy diagram in which a diatomic molecule has one vibrational mode with $E_n = (n + 1/2)h\nu$ levels equally spaced in a harmonic potential well with each n-level having a

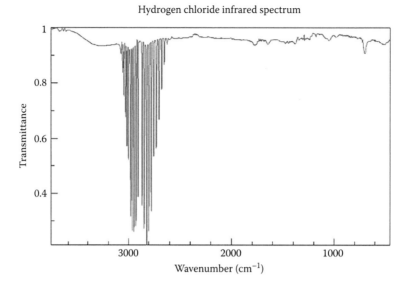

FIGURE 13.10 Low-resolution spectrum of HCl showing some rotational fine structure and the missing Q-branch. (From *NIST Chemistry WebBook*, http://webbook.nist.gov/chemistry).

progression of $E_j = j(j+1)\dfrac{\hbar^2}{2I}$ j-levels more closely spaced than the vibrational levels. Now that we know the vibrational selection rule $\Delta n = \pm 1$ and the main rotational selection rule $\Delta j = \pm 1$ ($j = l$), we can try to interpret some fine details of molecules that vibrate and rotate. We will temporarily ignore the $\Delta m = 0, \pm 1$ rule because the changes in energy due to $\Delta m = \pm 1$ are much smaller than what can be easily resolved in near-infrared spectra so we will assume $\Delta m = 0$ for simplicity at this level.

We previously considered a low-resolution pure rotational spectrum of HCl to make it clear that there are transitions in the far infrared and microwave spectral region for changes in rotation. Now we want to examine the main vibrational bands that have added detail due to rotation (Figure 13.10) [23]. In Figure 13.11, we see a high-resolution spectrum of HCl in the vibrational region. The bond in HCl is sufficiently strong that at room temperature the molecule will be in the lower $n = 0, 1, 2, 3, \ldots$, levels, which are equally spaced so that we only expect a single band, mainly from $n = 0$ to $n = 1$. However, we see a lot of additional details. In addition to mostly $\Delta n = 1$ in absorption, we also have $\Delta n = 1 + \Delta j$; $\Delta j = \pm 1$. Since the j levels are closely spaced, the possible j values are much higher at room temperature. First we notice that there is no transition from $|n = 0, j = 0\rangle \rightarrow |n = 1, j = 0\rangle$.

This is a characteristic of diatomic molecules and is called the "Q" branch at the frequency for $|n = 0\rangle \rightarrow |n = 1\rangle$, which is not there due to the fact that j must change. On the low-energy side of the Q-branch, we have the so-called P-branch of the spectrum for $|n = 0, j\rangle \rightarrow |n = 1, j - 1\rangle$ absorptions and on the high-energy side of the Q-branch we have the so-called R-branch for $|n = 0, j\rangle \rightarrow |n = 1, j + 1\rangle$ absorptions. (Note a memory device: The R-branch *raises* j!) The intensity of the absorptions decreases as one examines higher j values because the Boltzmann principle makes it less probable that the higher energy j values will be occupied at a given temperature. Thus the envelope of the rotational lines is roughly symmetrical on either side of the Q-branch. At low resolution, the overall band shape resembles a slightly distorted "W" due to the Q-branch.

We have chosen a very clean spectrum of HCl, which also shows splitting of the rotational lines due to the presence of two isotopes of chlorine, $^{35}_{17}\mathrm{Cl}$ and $^{37}_{17}\mathrm{Cl}$ (Figures 13.12 and 13.13). The moments of inertia of the two isotopic forms of HCl will be slightly different so the rotational

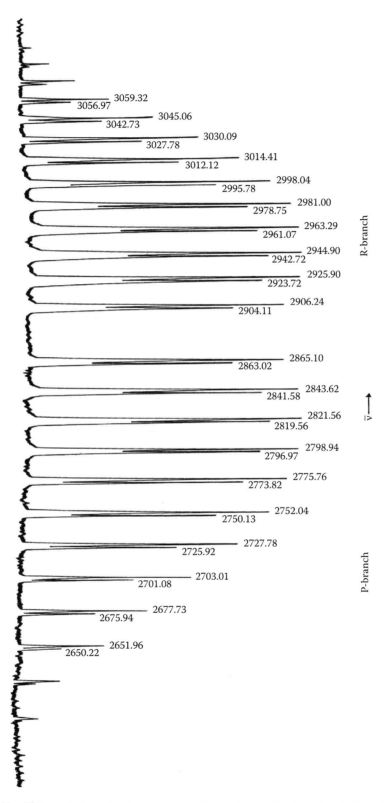

FIGURE 13.11 High-resolution vib-rotor spectrum of HCl showing Cl isotopic splitting. (From Barrow, G.A., *Physical Chemistry*, 5th Edn., McGraw-Hill, New York, 1988, p. 569. With permission.)

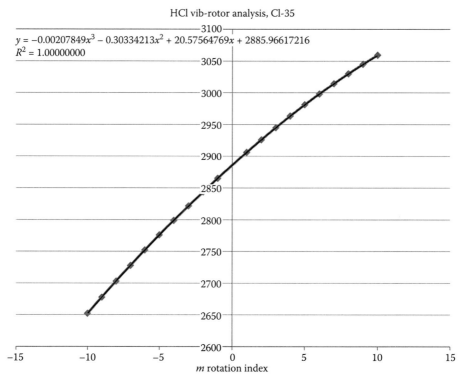

FIGURE 13.12 Combined (P-R, m_j-Index) plot of vib-rotor transitions for $\bar{\nu}_{m_j}\left(\text{H-}^{35}_{17}\text{Cl}\right)$.

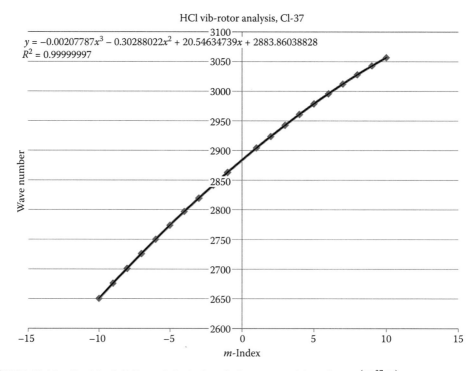

FIGURE 13.13 Combined (P-R, m_j-Index) plot of vib-rotor transitions for $\bar{\nu}_{m_j}\left(\text{H-}^{37}_{17}\text{Cl}\right)$.

lines in the P- and R-branches will be at slightly different energies. The Cl isotope splitting is shown here:

$$\mu_{35} = \frac{m_H m_{35_{Cl}}}{(m_H + m_{35_{Cl}})N_A} = \frac{(1.007825032 \text{ g/mol})(34.96885268 \text{ g/mol})}{(35.97667771 \text{ g/mol})(6.02214179 \times 10^{23}/\text{mol})} = 1.626651403 \times 10^{-24} \text{ g},$$

$$\mu_{37} = \frac{m_H m_{37_{Cl}}}{(m_H + m_{37_{Cl}})N_A} = \frac{(1.007825032 \text{ g/mol})(36.96590259 \text{ g/mol})}{(37.97372762 \text{ g/mol})(6.02214179 \times 10^{23}/\text{mol})} = 1.629116897 \times 10^{-24} \text{ g}.$$

Let us use the experimental bond length of 1.2746 Å [14] with the calculated reduced masses to find the isotope split based on an arbitrary j value

$$(v_{j+1} - v_j)_{35} = \frac{h}{4\pi^2 \mu r^2} = \frac{6.62606896 \times 10^{-27} \text{ g} \times \text{cm}^2/\text{s}}{4\pi^2(1.626651403 \times 10^{-24} \text{ g})(1.2746 \times 10^{-8} \text{ cm})^2} = 6.351172293 \times 10^{11}/\text{s},$$

$$(v_{j+1} - v_j)_{37} = \frac{h}{4\pi^2 \mu r^2} = \frac{6.62606896 \times 10^{-27} \text{ g} \times \text{cm}^2/\text{s}}{4\pi^2(1.629116897 \times 10^{-24} \text{ g})(1.2746 \times 10^{-8} \text{ cm})^2} = 6.341560474 \times 10^{11}/\text{s},$$

$$\Delta(v_{j+1} - v_j) = 9.61181927 \times 10^8/\text{s}; \text{ so } \Delta\bar{v} = \frac{\Delta v}{c} = \frac{9.61181927 \times 10^8/\text{s}}{2.99792458 \times 10^{10} \text{ cm/s}} = 0.032062 \text{ cm}^{-1}.$$

That is a very small number for the isotope splitting, which means that we are looking at a very high-quality spectrum! The last observation is that the $^{37}_{17}Cl$ absorptions are smaller (less intense) than the $^{35}_{17}Cl$ absorptions and this is due to the natural abundance of chlorine on planet Earth (75.76% $^{35}_{17}Cl$, 24.24% $^{37}_{17}Cl$).

With a wealth of precise data from this excellent spectrum, we can see that the spacing in the P and R branches is *not quite the same*. Further, we expect that the bond length will be slightly different in the $n = 0$ vibrational state than in the $n = 1$ state. In Table 12.3, we reported some complicated parameters for a number of gas-phase molecules. We can now show the meaning of those parameters and carry out the analysis given in the laboratory text by Shoemaker, Garland, and Nibler [24] using the excellent data we have for HCl. This analysis has been used in an excellent teaching experiment for many years [24] and is included here to show how the various corrections are applied to the fine details of the spectrum of a diatomic molecule [25]. The numbers we obtain using lower precision in a spectrum from a dispersive spectrometer will lead to some slight discrepancies compared to the most recent FT-IR values [14] but the method is instructive for understanding anharmonicity and vibrational-rotational coupling.

It should be understood that we start from an exact solution to the rigid rotor and the exact solution to the harmonic oscillator for a perfect parabola and then apply common sense corrections to the idealized solutions by fitting a polynomial to experimental data so the corrections are empirical rather than quantum mechanical. We use a formula based on quantum mechanics plus some empirical terms (with Shoemaker–Garland–Nibler notation).

$$\varepsilon(n, J) = \frac{E(n, J)}{hc} = \bar{v}_0(n + 1/2) - \bar{v}_0 x_e(n + 1/2)^2 + B_e J(J + 1) + D_e[J(J + 1)]^2 - \alpha_e(n + 1/2)J(J + 1).$$

Consider the philosophy of this model. The first and third terms are just the idealized quantum mechanical formulas that give the main idea of the solution. The energy of each transition is expressed using the quantity $\varepsilon(n, J)$ directly in wavenumbers. Here $I = \mu r^2$ and we can define $B_c = \frac{h}{8\pi^2 Ic}$ for wavenumbers. The second and fourth terms are corrections in the spirit of a Taylor series expansion. Since what we know about the quantum mechanical solutions is in terms of the $E(n, J)$ quantum numbers, it is natural to express higher-order corrections in terms of the only information we have in terms of the quantum numbers. We use "n" for the vibrational quantum number even though spectroscopists use a script "v" because in our character set, the symbol for frequency is the same, and besides, the quantum number for the pure parabolic case was "n". Thus we have ad hoc second-order (quadratic) corrections as if we were carrying out a Taylor expansion. It is possible to write out a Taylor expansion in two variables, which would involve a term in which the two variables occur as a mixed product of the variables, so the last term has a "to-be-fitted" parameter α_e multiplied by a product of the vibrational and rotational quantum numbers. It is important for your education going forward to understand how such intelligent empirical reasoning is carried out. Now the question is how to fit this model to the data.

The model formula in the preceding text incorporates the ideas of bond stretching and interaction between rotation and vibration so simple treatments trying to relate transitions between two levels either beginning or ending in a common rotational level are shown to be flawed by the observation that *the space between the transition lines is not constant!* First consider the R-branch energies.

$$\varepsilon(1, J+1) = \bar{v}_0(3/2) - \bar{v}_0 x_e(9/4) + B_c(J+1)(J+2) - \alpha_e(3/2)(J+1)(J+2) + D_e(J+1)^2(J+2)^2$$

$$\varepsilon(0, J) = \bar{v}_0(1/2) - \bar{v}_0 x_e(1/4) + B_c(J)(J+1) - \alpha_e(1/2)(J)(J+1) + D_e(J)^2(J+1)^2$$

Then $\bar{v}_R(\Delta n = +1, \Delta J = +1) = \varepsilon(n=1, J'=J+1) - \varepsilon(n=0, J)$

$$\bar{v}_R(J) = \bar{v}_0(1/2) - 2\bar{v}_0 x_e + (2B_c - 3\alpha_e) + (2B_c - 4\alpha_e)J - \alpha_e J^2$$
$$+ D_e\left[(J+1)^2(J+2)^2 - (J)^2(J+1)^2\right]$$

The last term in D_e will not lead to simple analysis in this model [24]. Next consider the P-branch.

$$\varepsilon(1, J-1) = \bar{v}_0(3/2) - \bar{v}_0 x_e(9/4) + B_c(J-1)(J) - \alpha_e(3/2)(J-1)(J) + D_e(J-1)^2(J)^2$$

$$\varepsilon(0, J) = \bar{v}_0(1/2) - \bar{v}_0 x_e(1/4) + B_c(J)(J+1) - \alpha_e(1/2)(J)(J+1) + D_e(J)^2(J+1)^2$$

$$\bar{v}_P(\Delta n = +1, \Delta J = -1) = \varepsilon(n=1, J'=J-1) - \varepsilon(n=0, J)$$

$$\bar{v}_P = \bar{v}_0 - 2\bar{v}_0 x_e - 2B_c J - \alpha_e J^2 + \alpha_e(3/2 + 1/2)J + D_e\left[(J-1)^2 J^2 - J^2(J+1)^2\right]$$

$$\bar{v}_P = \bar{v}_0 - 2\bar{v}_0 x_e - (2B_c - 2\alpha_e)J - \alpha_e J^2 + 4D_e J^3.$$

In this case, the terms in D_e lead to a simple form but actually refer to the lower $n=0$ vibrational state. Then this model [24] assumes the D_e value is the same for both the $n=0$ and $n=1$ states, which makes the formulas simpler and is a good approximation since we expect the D_e correction is very small (it is). The real problem is that we do not actually have a value for the Q-branch ($\bar{v}_e$). The irregular spacing in the P and R branches is noticeably different from the equal spacing of the pure rotational spectra and implies the gap where the Q-branch should be is *not* the midpoint of the gap, but now we come to the most amazing aspect of this model. We have two formulas for the P- and R-branches:

$$\bar{v}_P = \bar{v}_0 - 2\bar{v}_0 x_e - (2B_c - 2\alpha_e)J - \alpha_e J^2 + 4D_e J^3$$

and

$$\bar{v}_R(J) = \bar{v}_0(1/2) - 2\bar{v}_0 x_e + (2B_c - 3\alpha_e) + (2B_c - 4\alpha_e)J - \alpha_e J^2 + D_e\left[(J+1)^2(J+2)^2 - (J)^2(J+1)^2\right].$$

The amazing thing is that we can define $m_j = -J$ for the P-branch so that $J_P = -m_j$ and then for the R-branch we define $m_j = J+1$ or $J_R = m_j - 1$ and get a unified formula for *both* branches as $\bar{v}(m_j) = \bar{v}_e + (2B_c - 2\alpha_e)m_j - \alpha_e m_j^2 - 4D_e m_j^3$. (Here m_j is a counting index used for compatibility

TABLE 13.4
Wavenumber Peaks from Figure 13.11
Using the m_j-Index

$\bar{\nu}_{m_j}$ (H-$^{35}_{17}$Cl)	m_j	$\bar{\nu}_{m_j}$ (H-$^{37}_{17}$Cl)
3059.32	10	3056.97
3045.06	9	3042.73
3030.09	8	3027.78
3014.41	7	3012.12
2998.04	6	2995.78
2981.00	5	2978.75
2963.29	4	2961.07
2844.90	3	2942.72
2925.90	2	2923.72
2906.24	1	2904.11
2865.10	−1	2863.02
2843.62	−2	2841.58
2821.56	−3	2819.56
2798.94	−4	2796.97
2775.76	−5	2773.82
2752.04	−6	2750.13
2727.78	−7	2725.92
2703.01	−8	2701.08
2677.73	−9	2675.94
2651.96	−10	2650.22

Source: Barrow, G.A., *Physical Chemistry*, 5th Edn.,
McGraw-Hill, New York, 1988, page 569.

with Ref. [24] but different from the m-rotational quantum number.) Amazing! Note that we have simply ignored the intractable expression for the D_e part of the R-branch equation and use the simpler D_e formula of the P-branch, which assumes the two are *approximately* the same! The next thing we might notice about the two $(m_j, \bar{\nu})$ plots is that the $m_j = 0$ intercept is not the same, 2885.966 cm^{-1} for $\bar{\nu}_{m_j}$(H-$^{35}_{17}$Cl) and 2883.860 cm^{-1} for $\bar{\nu}_{m_j}$(H-$^{37}_{17}$Cl) (Table 13.4).

The analysis that led to the m_j-index considered the value of $\bar{\nu}_0 = \bar{\nu}_e - 2\bar{\nu}_e x_e$ where x_e is a parameter expressing the anharmonicity of the actual potential well compared to the idealized parabolic well. *Assuming the electronic bond strength is the same in the two isomeric forms*, we can expect the more massive isotope to be more anharmonic than the lighter isotope as can be proven by a complicated argument given by Herzberg [25] according to the relationship $\dfrac{\bar{\nu}_e^* x_e^*}{\bar{\nu}_e x_e} = \dfrac{\mu}{\mu^*}$, and of course we know that the ratio of the ν_e frequencies by themselves satisfy the isotope shift $\dfrac{\bar{\nu}_e^*}{\bar{\nu}_e} = \sqrt{\dfrac{\mu}{\mu^*}}$ so we make the only reasonable assumption [26] to solve two equations in two unknowns using $\mu_{35} = \dfrac{m_H m_{35_{Cl}}}{(m_H + m_{35_{Cl}})N_A} = 1.626651403 \times 10^{-24}$ g and $\mu_{37} = \dfrac{m_H m_{37_{Cl}}}{(m_H + m_{37_{Cl}})N_A} =$ 1.629116897 × 10^{-24} g   so $\left(\dfrac{\mu_{35}}{\mu_{37}}\right) = \dfrac{1.626651403 \times 10^{-24} \text{ g}}{1.629116897 \times 10^{-24} \text{ g}} = 0.998486607$ and

$\left(\dfrac{\mu_{35}}{\mu_{37}}\right)^{1/2} = 0.999243017$. Thus we have

$$\bar{\nu}_0\left(\text{H-}^{35}_{17}\text{Cl}\right) = 2885.966172 = \bar{\nu}_e - 2x_e\bar{\nu}_e \text{ and so } 2885.966172 - \bar{\nu}_e = -2x_e\bar{\nu}_e.$$

We also have

$$\bar{v}_0^*\left(H\text{-}_{17}^{37}Cl\right) = 2883.860388 = \bar{v}_e\left(\frac{\mu}{\mu^*}\right)^{1/2} - 2x_e\bar{v}_e\left(\frac{\mu}{\mu^*}\right).$$

Then substituting $-2\bar{v}_e x_e = \bar{v}_0 - \bar{v}_e$ we have

$$\bar{v}_0^* = \bar{v}_e\sqrt{\frac{\mu_{35}}{\mu_{37}}} + (\bar{v}_0 - \bar{v}_e)\left(\frac{\mu_{35}}{\mu_{37}}\right) \quad \text{and} \quad \bar{v}_0\left(\frac{\mu_{35}}{\mu_{37}}\right) - \bar{v}_0^* = \bar{v}_e\left[\left(\frac{\mu_{35}}{\mu_{37}}\right) - \sqrt{\frac{\mu_{35}}{\mu_{37}}}\right],$$

and finally we have

$$\bar{v}_e = \left\{\frac{\bar{v}_0\left(\frac{\mu_{35}}{\mu_{37}}\right) - \bar{v}_0^*}{\left[\left(\frac{\mu_{35}}{\mu_{37}}\right) - \sqrt{\frac{\mu_{35}}{\mu_{37}}}\right]}\right\} = \left\{\frac{(2885.96617216)(0.998486607) - 2883.86038828}{[0.998486607 - 0.999243017]}\right\} \text{ so we find}$$

$\bar{v}_e = \dfrac{-2.26181701}{-0.00075641} = 2990.199773$, but we have to round to six significant figures because the data is only given to six figures and then we obtain $\bar{v}_e = 2990.20$ cm^{-1}, which applies to both isotopic species. We had to do this tedious process because we observed that the rotational transitions are *not* equally spaced, which led to this use of anharmonicity to find the common $\bar{v}_e$ for both isotopic species.

Next we can solve for $x_e = -(\bar{v}_0 - \bar{v}_e)/2\bar{v}_e = \dfrac{(2990.199 - 2885.966)}{2(2990.199)}$ and we find $x_e = 0.017429$ rounded to six significant figures and $\bar{v}_e x_e = 52.1165$ cm^{-1}.

BOND LENGTH OF H-$_{17}^{35}$Cl

From the Excel polynomial fit for H-$_{17}^{35}$Cl, the coefficient of the fourth term (first term on the graph, the x^3 term) is -0.00207849 so $-4D_e m^4 = -0.00207849 m^4 \Rightarrow D_e = 519.623 \times 10^{-6}$ cm^{-1}. The coefficient of the third term (second term on the graph) is -0.30334213 so we have $\alpha_e = 0.30334213$ cm^{-1} and we can use that in the second term (third term on the graph) where $(2B_e - 2\alpha_e) = 20.57564769 \Rightarrow B_e = [20.57564769 + 2(0.30334213)]/2 = 10.59116598$ cm^{-1}.

Then we can use $B_e = \dfrac{h}{8\pi^2\mu r_e^2 c} = 10.59116598 \Rightarrow r_e = \sqrt{\dfrac{h}{8\pi^2\mu c(10.59116598 \text{ cm}^{-1})}}$ and so

$$r_e = \sqrt{\frac{6.62606896 \times 10^{-27} \text{ erg s}}{8\pi^2(1.626651403 \times 10^{-24} \text{ g})(2.99792459 \times 10^{10} \text{ cm/s})(10.59116598/\text{cm})}}$$

$$= 1.27468 \times 10^{-8}\text{cm}.$$

Thus we find the bond length for H-$_{17}^{35}$Cl $= 1.27468$ Å to six significant figures. Finally we can calculate the force constant from $\bar{v}_e = \dfrac{1}{2\pi c}\sqrt{\dfrac{k}{\mu}} \Rightarrow k = 4\pi^2 c^2 \bar{v}_e^2 \mu$ and we find that

$$k = 4\pi^2 c^2 \bar{v}_e^2 \mu = (4\pi^2)(2.99792458 \times 10^{10} \text{ cm/s})^2(2990.20/\text{cm})^2(1.626651403 \times 10^{-24} \text{ g})$$

$$k = 5.16055 \times 10^5 \text{ g(cm/s)}^2/\text{cm}^2 = 5.16055 \times 10^5 \text{ g(cm/s}^2)/\text{cm} = 5.16055 \text{ N/cm}.$$

TABLE 13.5
Comparison of Vib-Rotor Parameters for H-$^{35}_{17}$Cl

Source	k (N/cm)	ω_e (cm^{-1})	$\omega_e x_e$ (cm^{-1})	B_e (cm^{-1})	α_e (cm^{-1})	D_e (10^{-6} cm^{-1})	r_e (Å)
90th CRC	5.16	2990.95	52.82	10.5934	0.30718	531.94	1.27455
Example	5.16055	2990.20	52.17	10.5912	0.30334	519.62	1.27468

Source: Lide, D. R., *CRC Handbook of Chemistry and Physics*, 90th Edn., 2009–2010, CRC Press, Boca Raton, FL, pp. 9–105.

Now, we can compare our computed values to the values from the 90th Edn. of the *CRC Handbook* as given in Chapter 12.

We see that the agreement is good but not exact. Since we have used a 10-place calculator and carried the Excel polynomial fit to eight places after the decimal, our arithmetic should be within the uncertainty of the six significant figure wavenumbers in the data. It is probable that the discrepancy lies with the use of a more precise vib-rotor spectrum [14]. While the spectrum we used is excellent for a dispersive IR spectrometer, it is probable that a more recent FT-IR spectrum would provide the precision needed to yield the values in the 90th Edn. of the *Chemical Rubber Handbook*. Nevertheless, we have carried out an exercise that you might do in a physical chemistry laboratory course using the text by Shoemaker, Garland, and Nibler [24], and we have benefitted from communication with Prof. Nibler [26] regarding the anharmonicity calculation! For this author, the most interesting aspect of this example is the evaluation of the parameter α_e, which is a measure of the coupling between the vibrational and the rotational quantization. Since α_e is about 0.303 cm^{-1} (roughly ten times the size of the isotope splitting) and α_e indicates that there is some mixing between the vibration and rotation quantization, our idea of a clean separation of vibration and rotation is not quite true. Even so, the coupling is still very small compared to $\bar{\nu}_e = 2990.20$ cm^{-1}.

SUMMARY

1. Well, this really is a mind bending chapter in terms of the mathematics! There is a difficult derivation of the associated Legendre polynomials as solutions to the quantized rigid rotor problem, although the $m = 0$ shortcut made it easier than it might have been. These angular functions also apply to the H atom in the next chapter. A qualitative explanation of recent breakthroughs in Fourier transform technology was given for applications to forensic problems and material science as well as a very precise FT-IR pure rotation spectrum of CO. We also discussed a treatment of the vib-rotor spectrum of HCl to obtain the bond length and to calculate coupling between vibration and rotation in a detailed empirical model. Angular momentum operators were discussed and it was shown that one can only "know" one component of the angular momentum (L_z) with L^2 while the other components (L_x, L_y) are indeterminant. The first main problem is to grapple with the polar coordinate form of the second derivative as

$$\nabla^2(r, \theta, \phi) = \frac{1}{r^2} \frac{\partial}{\partial r} \left(r^2 \frac{\partial}{\partial r} \right) + \frac{1}{r^2 \sin \theta} \frac{\partial}{\partial \theta} \left(\sin \theta \frac{\partial}{\partial \theta} \right) + \frac{1}{r^2 \sin^2 \theta} \frac{\partial^2}{\partial \phi^2}.$$

In this chapter, we held the *r*-coordinate fixed as for a rigid bond to obtain the pure rotational energy levels in terms of (θ, ϕ) coordinates only and found a rotating rigid diatomic has energy levels given by $E_j = j(j+1) \frac{\hbar^2}{2I}$, $j = 0, \pm 1, \pm 2, \pm 3, \ldots$, and $I = \mu r^2$ is the moment of inertia. Note that this system has degeneracies of $g_j = (2j+1)$ as 1, 3, 5, 7, 9, … and the very same angular solutions as the more familiar s, p, d, f, … orbitals of the H atom except that for rotations, the j-quantum number can easily go to much higher values since there is essentially no barrier to the rotation of a molecule in the gas phase. The only condition is that the rotational energy exists in discrete "chunks" or quanta.

2. Two important additional eigenvalue equations were found for the $Y_l^m(\theta, \phi)$ wave functions: $L_z Y_l^m(\theta, \phi) = m\hbar Y_l^m(\theta, \phi)$ and $L^2 Y_l^m(\theta, \phi) = l(l+1)\hbar^2 Y_l^m(\theta, \phi)$.

3. A dipole selection rule exists here as well and is found to be $\Delta j = \pm 1$. This leads to discrete absorption lines in the far infrared (low energy) range of the electromagnetic spectrum. The formula for the transitions is $\Delta E = E_{j+1} - E_j = h\nu = \dfrac{hc}{\lambda} = hc\bar{\nu} = [(j+1)$
$(j+2) - (j)(j+1)]\dfrac{\hbar^2}{2I}$. This leads to $hc\bar{\nu} = [(j+1)(j+2) - (j)(j+1)]\dfrac{\hbar^2}{2I} = 2(j+1)\dfrac{\hbar^2}{2I}$.
This is inconvenient to use since it requires a guess as to a value of "j." However, the spacing between lines can be used in $\Delta(\Delta E) = \Delta E_{j+1} - \Delta E_j = [2(j+2) - 2(j+1)]\dfrac{\hbar^2}{2I} = 2\dfrac{\hbar^2}{2I} = \dfrac{\hbar^2}{\mu r_e^2}$ and this can lead to a value for r_e, the bond length! So pure rotational spectrum is at very low energy in the far infrared region but gives valuable information regarding bond lengths. The lines are equally spaced and there are no gaps in the progression. The heights of the lines are determined by the temperature of the sample according to Boltzmann populations of the individual levels.

4. The vib-rotor spectrum can also be recorded for molecules that are vibrating while they are rotating. Here we have $\Delta n = \pm 1$ for vibration and $\Delta j = \pm 1$ for rotation in the same spectrum. Here the energy levels are not equally spaced, although an indexing scheme (m_j) can be used to unify the sequence of lines around what is a gap due to the forbidden $\Delta j = 0$ transition, which results in the missing line/gap of the Q-branch in the lines of a diatomic molecule. The higher-energy (higher wavenumber) lines form the R-branch and the lower-energy lines form the P-branch sequence. Using a model Hamiltonian, which allows some mixing between rotation and vibration, the lines can be fitted to the spectra and it is found that the coupling between rotation and vibration is very small but not zero, only to the extent of a few wavenumbers compared to several thousand wavenumbers for the vibrational transition. The detailed analysis of the m_j-index model Hamiltonian provides a better understanding of the energetics of gas-phase diatomic molecules and offers a comforting understanding of the vib-rotor interactions including anharmonicity of the vibrations.

5. Two contributions from FT-IR training slide shows from Varian, Inc. provide the basic idea of how FT-IR spectra are obtained and the need for an associated computer to carry out the Cooley–Tukey FFT.

6. The complicated solution to the rigid rotor problem in terms of Legendre polynomials should be read with awe relative to the fact that Legendre derived the main aspects of this set of polynomials in 1785!

PROBLEMS

13.1 Compute the moment of inertia of one CO molecule using the atomic weights of 12.000 for C and 15.9997 for O with a bond length of 1.1283 Å.

13.2 Calculate the reduced mass for $^{12}_{6}C^{18}_{8}O$ and $^{12}_{6}C^{16}_{8}O$ using the mass of $^{18}_{8}O$ as 17.999161 g/mol and $^{16}_{8}O = 15.9949146196$ g/mol with the standard that $^{12}_{6}C = 12.0000000$ and then calculate the isotope splitting due to the oxygen isotopes in the infrared rotational spectrum. $^{18}_{8}O$ has a natural abundance of 0.205% on planet Earth. Use a bond length of 1.1283 Å. The isotope splitting calculated here was not seen in Figure 13.3 because the spectrum was of only the $^{12}C\ ^{16}O$ isotopic species. However would the isotope splitting be observable considering the scale of the wavenumber scale?

13.3 Test the equation $(2l+1)xP_l^{|m|}(x) = (l-|m|+1)P_{l+1}^{|m|}(x) + (l+|m|)P_{l-1}^{|m|}(x)$ using (P_1^0, P_2^0, P_3^0) and again using (P_1^1, P_2^1, P_3^1).

13.4 Calculate the eigenvalues of the following operations:

(a) $L_z Y_3^2 = ?$ (b) $L^2 Y_3^2 = ?$ (c) $L_z Y_2^1 = ?$ (d) $L^2 Y_3^{-1} = ?$ (e) $L_z Y_2^{-1} = ?$

13.5 Hollenberg reported [27] a low-resolution pure rotational spectrum for HCl with six lines spaced at an average of 19.8 cm^{-1}. Use this very rough data to compute the bond length of H-$_{17}^{35}$Cl and compare your value to the modern value of 1.2746 Å given in the 90th Edn. of the *CRC Handbook*. Use the major isotope $_{17}^{35}$Cl = 34.96885269.

13.6 Use the "m_j-polynomial" coefficients for H-$_{17}^{37}$Cl to calculate all the values in Table 13.5 for H-$_{17}^{37}$Cl.

13.7 Calculate the partial derivatives $\dfrac{\partial r}{\partial x}, \dfrac{\partial r}{\partial y}, \dfrac{\partial r}{\partial z}, \dfrac{\partial \theta}{\partial x}, \dfrac{\partial \theta}{\partial y}, \dfrac{\partial \theta}{\partial z}, \dfrac{\partial \phi}{\partial x}, \dfrac{\partial \phi}{\partial y}$, and $\dfrac{\partial \phi}{\partial z}$. You will need to recall that $\dfrac{d}{dx}[\arccos(u)] = \dfrac{-1}{\sqrt{1-u^2}}\dfrac{du}{dx}$ and $\dfrac{d}{dx}[\arctan(u)] = \dfrac{1}{1+u^2}\dfrac{du}{dx}$.

13.8 Prove $L_z = \dfrac{\hbar}{i}\dfrac{\partial}{\partial \phi}$ in spherical polar coordinates.

REFERENCES

1. Pauling, L. and E. B. Wilson, *Introduction to Quantum Mechanics, with Applications to Chemistry*, McGraw-Hill Book Inc., New York, 1935, Chapters III, IV and V. The copyright is now owned by Dover Press, New York and published as an unabridged reprint.
2. Eyring, H., J. Walter, and G. Kimball, *Quantum Chemistry*, John Wiley & Sons, Inc., New York, 1944.
3. Kauzmann, W., *Quantum Chemistry*, Academic Press, New York, 1957.
4. Dicke, R. H. and J. P. Wittke, *Introduction to Quantum Mechanics*, Addison-Wesley Publishing Company, Inc., Reading, MA, 1960.
5. McQuarrie, D. A., *Quantum Chemistry*, University Science Books, Sausalito, CA, 1983.
6. Rainville, E. D., *Elementary Differential Equations*, The Macmillan Company, New York, 1958.
7. Legendre, A. M., Spherical polynomials, *Mem. Math. Phys.* presentes a l'Acad. Sci. par divers savants, **10**, 411 (1785). See http://edocs.ub.uni-frankfurt.de/volltexte/2007/3757/pdf/A009566090.pdf
8. Gatz, C. R., *Introduction to Quantum Chemistry*, Charles E. Merrill Publishing Company, Columbus, OH, 1971, Appendix C, p. 529.
9. Anderson, J. M., Spherical polynomials, *Introduction to Quantum Chemistry*, W. A. Benjamin, Inc., New York, 1969, Appendix 2, p. 322. Also private communication with Prof. Anderson.
10. Lesk, A. M., *Introduction to Physical Chemistry*, Prentice-Hall, Inc., Englewood Cliffs, NJ, 1982, pp. 312 and 732; see also D. Park, *Introduction to Quantum Theory*, 2nd Edn., McGraw-Hill Book Col, New York, 1974, p. 648.
11. McQuarrie, D. A., *Quantum Chemistry*, 2nd Edn., University Science Books, Sausalito, CA, 2008, p. 293.
12. Reference [3, p. 263].
13. Barrow, G. A., *Physical Chemistry*, 6th Edn., The McGraw-Hill Companies, Inc., New York, 1996, p. 575.
14. Lide, D. R., Ed., *CRC Handbook of Chemistry and Physics*, 90th Edn., CRC Press, Boca Raton, FL, 2009–2010.
15. Rothman, L. S., I. E. Gordon, A. Barbe, D. ChrisBenner, P. F. Bernath, M. Birk, V. Boudon et al., The HITRAN 2008 molecular spectroscopic database, *J. Quant. Spectrosc. Radiat. Transf.*, **110**, 533 (2009).
16. Cooley, J. W. and J. W. Tukey, An algorithm for the machine calculation of complex Fourier series, *Math. Comput.* **19**, 297 (1965).
17. Private communication from Dr. Ellen Miseo of the Varian FT-IR Division, an employee of the pioneering firm DigiLab which developed FT-IR spectrometers and she continues with Varian after DigiLab was bought by Varian.
18. Lide, D. R., *CRC Handbook of Chemistry and Physics*, 90th Edn., CRC Press, Taylor & Francis, Boca Raton, FL, 2009–2010, p. A-61.

19. Flynn, K., R. O'Leary, C. Lennard, C. Roux, and B. J. Reedy, Forensic applications of infrared chemical imaging: multi-layered paint chips, *J. Forensic Sci.*, **50**, 832 (2005).
20. Flynn, K., R. O'Leary, C. Roux, and B. J. Reedy, Forensic Analysis of Bicomponent fibers using infrared chemical imaging, *J. Forensic Sci.*, **51**, 586 (2006).
21. Rosenblum, B., A. H. Nethercot, and C. H. Townes, Isotopic mass ratios, magnetic moments and the sign of the lectric dipole moment in carbon monoxide, *Phys. Rev.*, **109**, 400 (1958).
22. Reference [3, p. 659].
23. Spectra from NIST Chemistry WebBook (http://webbook.nist.gov/chemistry).
24. Shoemaker, D. P., C. W. Garland, and J. W. Nibler, *Experiments in Physical Chemistry*, 6th Edn., The McGraw-Hill Companies, Inc., New York, 1996, pp. 398–401.
25. Herzberg, G., *Molecular Spectra and Molecular Structure I: Spectra of Diatomic Molecules*, 2nd Edn., Van Nostrand, Princeton, NJ, 1950, Chapter 11.
26. Nibler, J., Department of Chemistry, Oregon State University, Private communication.
27. Hollenberg, L., Pure rotation spectra of HCl and NH_3: A physical chemistry experiment, *J. Chem. Educ.*, **43**, 9 (1966).

14 The Schrödinger Hydrogen Atom

INTRODUCTION

In this chapter we hope that the previous rigid rotor derivation will pay off. Here we only have to solve the radial equation since we already know the $\Theta_l(\theta)\Phi_m(\phi) = Y_l^m(\theta, \phi)$ solutions. We now have to relax the rigid potential to permit the Coulomb attraction between the electron and the nucleus.

$$\frac{1}{r^2}\frac{d}{dr}\left(r^2\frac{dR}{dr}\right) - \frac{\beta}{r^2}R + \frac{2\mu}{\hbar^2}[E - V(r)]R = 0, \quad \text{where } V(r) = \frac{-Ze^2}{r} \quad \text{is no longer constant.}$$

As a student, your task is to read the method of the solution of this very important problem and be able to sketch/draw functions and work simple problems related to the result. The task of the author is to present a correct description of the derivation with tables that can be referred to later and to present some meaningful applications. That means you can read the highlights now and go to the problems. There are excellent modern treatments by McQuarrie [1] and by Atkins and Friedman [2], but the most complete treatment is still to be found in the 1935 text by Pauling and Wilson [3] (P&W). In our experience, those sources would require working some examples from the later chapters of the differential equations text by Rainville [4]. However, we will attempt to add sufficient verbal description so that you can get the main ideas just using basic calculus. Details may also be found in the text by Kauzmann [5] and the earlier work by Eyring, Walter, and Kimble [6]. We should appreciate the patient brilliance of Schrödinger [7] but we will only sketch the solution. In this author's opinion, this sketch is better than just presenting the results without explanation but it would take almost a semester of mathematics to appreciate all the details of the derivation. However, *the wave mechanical solution of the H atom is the foundation for a comprehensive model of the whole periodic chart* and much can be gained by just studying the results.

STRATEGY TO SOLVE THE PROBLEM

By now you should be familiar with the polynomial method of solving a differential equation.

1. Rearrange the equation using one or more changes in variable to absorb physical constants into an equation with only pure numbers and the new variable(s).
2. Check the equation to see if it "blows up" for some value of the variable. Try to solve a simpler version of the equation, which avoids the undefined behavior. If possible, change the variable to shift the solution function away from the undefined conditions. Factor out the well-behaved form of a partial solution.
3. Attempt to substitute a polynomial with unknown coefficients into the equation hoping boundary conditions will determine the coefficients. Here $L(\rho) = \sum_{m=0} a_m\rho^m$.
4. Check to see if there is a part of the problem that cannot be represented by any combination of integer powers of the variable? In the rigid rotor case, we avoided this problem

using the $m = 0$ equation but here we may (will) have to allow something like $L(\rho) = \rho^s \sum_{m=0} a_m \rho^m$ where ρ^s is some function related to a_0 before the progression of the ρ^m terms in the power series.

5. Check the series for the quantum mechanical requirements that the function be single valued, finite, and continuous. We may need to truncate (cut off) the series to keep the solution finite!

After solving the harmonic oscillator and rigid rotor problems, we should be prepared to carry out the necessary steps. We are aided considerably by following the path laid out by Pauling and Wilson [3] but we will fill in a few details for you.

We return to the $R(r)$ equation we left unsolved in the rigid rotor problem.

$$\frac{1}{r^2} \frac{d}{dr}\left(r^2 \frac{dR}{dr}\right) - \frac{l(l+1)}{r^2} R + \frac{2\mu}{\hbar^2}\left[E + \frac{Ze^2}{r}\right] R = 0, \quad \text{using} \quad V(r) = \frac{-Ze^2}{r} \quad \text{and} \quad \beta = l(l+1)$$

This can be simplified somewhat by defining some constants and changing the variable. We show this step because the definitions of α and λ are crucial at the final step while many of the later steps can be described in words that hide a great deal of intermediate work. Let $\alpha^2 \equiv \frac{-2\mu}{\hbar^2} E, \lambda \equiv \frac{\mu Ze^2}{\hbar^2 \alpha}$ and $\rho = 2\alpha r \Rightarrow r = \frac{\rho}{2\alpha} \Rightarrow dr = \frac{d\rho}{2\alpha}$. Define a new function $S(\rho)$ where $\frac{dR}{dr} = \frac{dS}{d\rho} \frac{d\rho}{dr} = (2\alpha)\frac{dS}{d\rho}$.

There is a subtle trick here in defining α^2 as a negative number, which leads to $\left(\frac{-1}{4}\right)$ in the transformed equation and that leads to a factor of $e^{-\left(\frac{\rho}{2}\right)}$ later.

$$\frac{4\alpha^2}{\rho^2} \frac{2\alpha d}{d\rho}\left(\frac{\rho^2}{4\alpha^2} \frac{2\alpha d}{d\rho} S(\rho)\right) - \frac{l(l+1)4\alpha^2 S(\rho)}{\rho^2} + \left(\frac{2\mu}{\hbar^2}\right)\left[E + \frac{Ze_q^2(2\alpha)}{\rho}\right] S(\rho) = 0$$

$$\div 4\alpha^2 \Rightarrow \frac{1}{\rho^2} \frac{d}{d\rho}\left(\rho^2 \frac{d}{d\rho} S(\rho)\right) - \frac{l(l+1)S(\rho)}{\rho^2} + \left(\frac{2\mu}{4\alpha^2 \hbar^2}\right)\left[E + \frac{Ze_q^2(2\alpha)}{\rho}\right] S(\rho) = 0$$

so

$$\frac{1}{\rho^2} \frac{d}{d\rho}\left(\rho^2 \frac{d}{d\rho} S(\rho)\right) - \frac{l(l+1)S(\rho)}{\rho^2} - \frac{S(\rho)}{4\alpha^2\left(\frac{-\hbar^2}{2\mu E}\right)} + \frac{\mu Ze_q^2}{\alpha \hbar^2 \rho} S(\rho) = 0.$$

Then we have the transformed equation as $\frac{1}{\rho^2} \frac{d}{d\rho}\left(\rho^2 \frac{dS(\rho)}{d\rho}\right) + \left[\frac{-1}{4} - \frac{l(l+1)}{\rho^2} + \frac{\lambda}{\rho}\right] S(\rho) = 0$ but since $0 \le \rho \le \infty$ we have a problem with $\frac{1}{\rho}$ and $\frac{1}{\rho^2}$ when $\rho = 0$. As with the harmonic oscillator solution, we see an asymptotic function when ρ is very large as the solution to the limiting form of the equation $\frac{d^2 S}{d\rho^2} = \left(\frac{1}{4}\right) S$ or as $\left(D + \frac{1}{2}\right)\left(D - \frac{1}{2}\right) S(\rho) = 0$ where $D \equiv \frac{d}{d\rho}$. This has solutions $S = e^{\pm\frac{\rho}{2}}$ but only $S = e^{-\left(\frac{\rho}{2}\right)}$ satisfies the requirement to remain finite. Thus we consider another function $S = e^{-\left(\frac{\rho}{2}\right)} F(\rho)$, apply the derivative (a bit of work) and then divide by $e^{-\left(\frac{\rho}{2}\right)}$ to obtain an equation in $F(\rho)$ as $\frac{d^2 F}{d\rho^2} + \left(\frac{2}{\rho} - 1\right)\frac{dF}{d\rho} + \left[\frac{\lambda}{\rho} - \frac{l(l+1)}{\rho^2} - \frac{1}{\rho}\right] F = 0$, but still $0 \le \rho \le \infty$! Thus we have factored out the long range part of the function but still have a strong singularity at $\rho = 0$;

the equation "blows up" when $\rho = 0$! Let us try to guess a solution that has a factor of ρ^s in the numerator, which might cancel out the problematic ρ^{-2} and ρ^{-1}. Thus we try this as a factor with a further substitution $F(\rho) = \rho^s L(\rho)$ and then use a power series expansion as $L(\rho) = \sum_{m=0} a_m \rho^m$; $a_0 \neq 0$. We can evaluate the derivatives of $F(\rho) = \rho^s L(\rho)$ as $\frac{dF}{d\rho} = s\rho^{s-1}L(\rho) + \rho^s \frac{dL}{d\rho}$ and for the second derivative $\frac{d^2F}{d\rho^2} = s(s-1)\rho^{s-2}L + 2s\rho^{s-1}\frac{dL}{d\rho} + \rho^s\frac{d^2L}{d\rho^2}$ so we can substitute these into the equation, multiply through by ρ^2 and collect terms to find a lot of terms (another chore). This is quite an exercise in algebra but we want to find out the value of s for the a_0 term to determine the ρ^s part of the function, which cannot be represented by an integer power of ρ^m in the series expansion. After collecting terms, Pauling and Wilson [3] isolate the a_0 term so as to factor $F(\rho) = \rho^s L$, which will only have the terms in $(\rho^s a_0)$ as

$$\rho^{s+2}\frac{d^2L}{d\rho^2} + 2s\rho^{s+1}\frac{dL}{d\rho} + s(s-1)\rho^s L + 2\rho^{s+1}\frac{dL}{d\rho} + 2s\rho^s L - \rho^{s+2}\frac{dL}{d\rho}$$

$$-s\rho^{s+1}L + (\lambda - 1)\rho^{s+1}L - l(l+1)\rho^s L = 0.$$

Looking at the terms that have $\rho^s L = \rho^s \sum_{m=0} a_m \rho^m$, we can collect only those terms that have $(\rho^s a_0)$ for the $m = 0$ terms of the series that have no ρ in the expression. This leads to $[s(s-1) + 2s - l(l+1)]a_0 = 0$. Since each of the a_m terms has to be equal to zero independently and yet $a_0 \neq 0$, we have $[s(s-1) + 2s - l(l+1)] = [s(s+1) - l(l+1)] = 0$. By inspection we see that there are two solutions: one is $s = +l$ that leads to $\rho^s = \rho^l$, which is not singular at $\rho = 0$ and so is acceptable. On the other hand, the solution of $s = -(l+1)$ leads to $\rho^{-(l+1)}$, which is unacceptable since it is still singular (blows up) at $\rho = 0$. Therefore, we use $F(\rho) = \rho^l L(\rho) = \rho^l \sum_{m=0} a_m \rho^m$. Next we examine the terms in the equation that did not vanish for $s = l$ with a_n for $n > 0$ and divide them all by ρ^{l+1} to find that we can combine several terms as

$$\rho\left(\frac{d^2L}{d\rho^2}\right) + [2(l+1) - \rho]\left(\frac{dL}{d\rho}\right) + (\lambda - l - 1)L = 0,$$

where all the a_0 terms are now zero due to the value of $s = +l$ so the power series begins with a_1 but we can still equate each ρ^m coefficient to zero. Thus we have

$$(\lambda - l - 1)a_0 + 2(l+1)a_1 = 0$$

$$(\lambda - l - 1 - 1)a_1 + [2 \cdot 2(l+1) + 1 \cdot 2]a_2 = 0$$

$$(\lambda - l - 1 - 2)a_2 + [3 \cdot 2(l+1) + 2 \cdot 3]a_3 = 0$$

and in general we find that

$$(\lambda - l - 1 - m)a_m + [(m+1) \cdot 2(l+1) + m(m+1)]a_{m+1} = 0.$$

The ratio test for this series is $\lim_{m \to \infty} \frac{a_{m+1}}{a_m} = \lim_{m \to \infty}\left[\frac{-(\lambda - l - 1 - m)}{2(l+1)(m+1) + m(m+1)}\right] \to \left(\frac{1}{m}\right)$, which looks like it converges to some limit. However, let us consider the Taylor series for e^ρ expanded about the point $\rho = 0$

$$e^\rho = \sum_{m=0}^{\infty} \frac{\rho^m \left[\frac{d^m e^\rho}{d\rho^m}\right]_{\rho=0}}{m!} = 1 + \rho + \frac{\rho^2}{2} + \frac{\rho^3}{3!} + \cdots + \frac{\rho^m}{m!} + \frac{\rho^{m+1}}{(m+1)!} + \cdots$$

The ratio test for this series is also $\lim_{m \to \infty} \dfrac{a_{m+1}}{a_m} \to \dfrac{1}{m}$, so both series converge to something, but in the case of the series for e^ρ, the function will behave as e^ρ, which is not finite over the full range of $0 \le \rho \le \infty$. So yes, the $F(p) = \rho L(\rho) = \rho^l \sum_{m=0}^{\infty} a_m \rho^m$ series converges, but by comparing it to e^ρ we see that it will not remain finite as $\rho \to \infty$! So once again after all that work we have a polynomial that does not remain finite! However, we know what to do this time. We see that the variable λ contains the E value and that there is a minus sign in the recursion relationship so we set $(\lambda - l - 1 - m) = 0$. *That can only be true if λ is an integer n since the other values are integers!*

Recall $\lambda = \dfrac{\mu Z e^2}{\hbar^2 \alpha}$ and $\alpha = \sqrt{\dfrac{-2\mu E_n}{\hbar^2}}$ so $\lambda = n = \dfrac{\mu Z e^2}{\hbar^2 \sqrt{\dfrac{-2\mu E_n}{\hbar^2}}}$ and we can square that to find

$E_n = -\dfrac{\mu Z^2 e^4}{2n^2 \hbar^2}, n = 1, 2, 3 \ldots$. From $(\lambda - l - 1 - m) = 0$ we see that $n > 0$ even if $(l + m) = 0$. *Wait a minute, that looks like the same formula as for the Bohr atom!* Yes, the s-orbital energies $(l = 0)$ have the same formula as the Bohr pancake orbitals but now the Schrödinger wave solution introduces the $Y_l^m(\theta, \phi)$ shapes and generalizes the s-orbitals with $l = 0$ to spheres in three dimensions. Note that for the special case of only one electron, all the orbitals in a given shell have the same energy (degenerate levels). In multi-electron atoms the m-sub-shells will have different energies due to electron–electron interactions but the orbitals are energy degenerate within the m-sub-shell. Note the energy formula has only an n-quantum number, l or m do not occur in the one-electron H atom energy formula.

ASSOCIATED LAGUERRE POLYNOMIALS

As before with the $\Theta_l(\theta)$ functions, Schrödinger probably recognized the radial polynomials as associated Laguerre polynomials [7] after he had gone through the tedious process earlier. The associated Laguerre polynomials had been developed earlier by a French mathematician, Edmond Laguerre (1834–1886), so they were available to Schrödinger in 1926. The Laguerre polynomials are defined in several ways but one is by $L_r(\rho) = e^\rho \dfrac{d^r}{d\rho^r}(\rho^r e^{-\rho})$, and the associated Laguerre polynomials are the sth derivatives $L_r^s(\rho) = \dfrac{d^s}{s\rho^s} L_r(\rho)$. Maybe now you understand why we used the symbol L in the derivation! To a large extent Schrödinger's discovery was the application of families of orthogonal polynomials to quantized systems as is evident from the title of his 1926 paper "Quantisierung als Eigenvertproblem" (Quantization as an Eigenvalue Problem). To make connection to the Laguerre Polynomials, we note that they satisfy the equation [3] $\rho \dfrac{d^2 L_r(\rho)}{d\rho^2} + (1 - \rho) \dfrac{dL_r(\rho)}{d\rho} + r L_r(\rho) = 0$, and the sth derivative of the rth Laguerre polynomial $L_r^s(\rho) = \dfrac{d^s L_r(\rho)}{d\rho^s}$ satisfies $\rho \dfrac{d^2 L_r^s}{d\rho^2} + (s + 1 - \rho) \dfrac{dL_r^s}{d\rho} + (r - s) L_r^s = 0$. It is pointed out in the Pauling and Wilson [3] text that if you replace the integer r by $(n + l)$ and s by $(2l + 1)$, you find the generalized equation for the associated Laguerre polynomials as

$$\rho \frac{d^2 L_{n+l}^{2l+1}(\rho)}{d\rho^2} + [2(l + 1) - \rho] \frac{dL_{n+l}^{2l+1}(\rho)}{d\rho} + (n - l - 1) L_{n+l}^{2l+1}(\rho) = 0.$$

Very few texts list the associated Laguerre polynomials but Pauling and Wilson [3] give the general expression as

$$L_{n+l}^{2l+1}(\rho) = \sum_{k=0}^{n-l-1} (-1)^{k+1} \left\{ \frac{[(n + l)!]^2}{(n - l - 1 - k)!(2l + 1 + k)!k!} \right\} \rho^k.$$

We can use this to find the associated Laguerre polynomial for $n = 4$ and $l = 1$ (4p orbital).

$$L_5^3(\rho) = \frac{-(5!)^2\rho^0}{(4-1-1-0)!(3+0)!(0)!} + \frac{(5!)^2\rho^1}{(4-1-1-1)!(3+1)!(1)} - \frac{(5!)^2\rho^2}{(4-1-1-2)!(3+2)!(2)}$$

or $L_5^3(\rho) = -1200 + 600\rho - 60\rho^2$. Then $\dfrac{dL_5^3(\rho)}{d\rho} = 600 - 120\rho$ and $\dfrac{d^2L_5^3(\rho)}{d\rho^2} = -120$. Inserting these expressions into the general equation produces

$$\rho(-120) + [2(2) - \rho](600 - 120\rho) + (4-2)[-1200 + 600\rho - 60\rho^2]$$
$$= -120\rho + 2400 - 480\rho - 600\rho + 120\rho^2 - 2400 + 1200\rho - 120\rho^2 = 0, \quad \text{Q.E.D.!}$$

Thus we have demonstrated how the $L_{n+l}^{2l+1}(\rho)$ polynomials can be generated and that they do satisfy the general associated Laguerre polynomial equation. Schrödinger worked out the Hydrogen orbitals from these functions in his third revolutionary paper [7] and perhaps we can appreciate the patience required to carry the derivation to useful results!

This sketched derivation is difficult but we want to introduce these ideas to you so that at some later time you will have a starting point for serious study. In the opinion of this author, there is no better place to turn to than the P&W text [3], which is still available as a Dover reprint for a very reasonable price. There are two reasons why P&W is so good. First, it was written in English in 1935 at a time when many scientists were struggling with the original concepts in Schrödinger's papers that were in German and, second, those authors must have proofread the manuscript very thoroughly because it is exceptionally clean as text books go!

By comparing known properties and the equation for the associated Laguerre polynomials, it is likely that Schrödinger was pleased to find that the $L_r(\rho)$ he derived from the radial equation for the H atom satisfies the equation for the previously studied associated Laguerre polynomials. However, there was still a tedious normalization integral that needed to be worked out as $\int_0^\infty e^{-\rho}\rho^{2l}\left[L_{n+l}^{2l+1}(\rho)\right]^2\rho^2 d\rho = \dfrac{2n[(n+l)!]^3}{(n-l-1)!}$ where the ρ^2 in the integrand is the radial coordinate part of the volume element. With that normalization factor we can write $R_{nl}(r) = -\left[\left(\dfrac{2Z}{na_0}\right)^3\dfrac{(n-l-1)!}{2n[(n+l)!]^3}\right]^{1/2} e^{-\left(\frac{\rho}{2}\right)}\rho^l L_{n+l}^{2l+1}(\rho)$ in terms of the original r coordinate using $\rho = 2\alpha r = \dfrac{2\mu Ze_q^2}{n\hbar^2}r = \dfrac{2Z}{na_0}r$ and $a_0 = \dfrac{\hbar^2}{\mu e_q^2}$ is the Bohr radius. Although the P&W text uses a minus sign for R_{nl}, we know from previous discussion that the sign is arbitrary due to the phase problem.

INTERPRETATION

Despite the tricky polynomial analysis, we need to make some key observations here. First, it is important to notice that the r coordinate is scaled according to Z and n as $\left(\dfrac{Z}{n}\right)$, so even though the functions are given in terms of ρ, *the effective size of the model will contract as Z increases and swells as n increases* mainly due to the factor of $e^{-\frac{\rho}{2}} = e^{-\frac{Zr}{na_0}}$. Larger Z causes the exponential to decrease more rapidly and so "shrinks" the radial scale while the inverse n in the exponent causes the radial orbital to decrease less rapidly for larger values and expand for outer shells. Both considerations are relative to a_0, the Bohr radius (0.52917720859 Å)! Note we have used e_q to denote the charge on an electron while $e = 2.718281828459045\ldots$ is the base of natural logarithms. Due to the concept of Z being the atomic number, *the H atom model proves to be a major rationale for the arrangement of atoms in the periodic chart!*

The quantum number n is called the "principal quantum number" or shell number. Within each shell there are different orbital shapes according to the l quantum number with a nomenclature that is left over from lines of atomic emission spectra, $0 \Rightarrow s$, $1 \Rightarrow p$, $2 \Rightarrow d$, $3 \Rightarrow f, \ldots$ from "sharp," "principal," "diffuse," "fine," etc. There is no true correspondence between appearance of spectral lines and the Schrödinger quantum numbers. The labels are just historical vestiges of a former technology. Note that each radial function has the strongest ρ^l dependence as ρ^{n-1} while the lesser powers of ρ merely serve to provide shapes to maintain orthogonality among the various eigenfunctions so that the 2s function has a little part like 1s in it along with an outer part of opposite sign, the 3s function has a little part like 2s and another little part like 1s with appropriate signs so that the integrated product will sum to zero, that is, be orthogonal! However, the main meaning of the radial functions comes from $R_{nl} \sim \rho^{n-1} e^{-\frac{\rho}{2}}$; this observation will be important in constructing electronic orbitals for molecules.

Another simple observation is that the number of nodes in the radial function is given by # $nodes = n - l - 1$. *If we assume the derivation given earlier is merely to explain where the energy and wave functions come from, a student only needs to be able to qualitatively sketch the radial functions with the proper number of nodes and to understand the radial orthogonality.* An observant student should also remember that the $e^{-\left(\frac{\rho}{2}\right)}$ part of the wave function means that the long-range "tail" of the radial function tapers asymptotically to zero.

An important question often asked by students is, "How can orbitals of an atom be orthogonal to each other if they exist in roughly the same volume of space?" We are used to thinking in Cartesian (x, y, z) space where the coordinates are clearly in linearly independent directions. Quantum mechanics uses a "linear vector space" in which the components are orthogonal polynomials (as Schrödinger found and developed) instead of functions that "point" in different directions. The polynomials achieve orthogonality by virtue of sign changes so that the idea of a dot product in Cartesian space is generalized to an integrated "inner product" as in the case of

$$[\vec{a} \cdot \vec{b}] = |ab| \cos \theta_{ab} \rightarrow \int \psi_a^* \psi_b d\tau = \langle a|b \rangle.$$ Thus ψ_a and ψ_b can occupy the same general volume

but if the integral product $\int \psi_a^* \psi_b d\tau = \langle a|b \rangle = 0$, the functions are said to be orthogonal. This is most easily seen in Figure 14.1 where the total (normalized) 1s, 2s, and 3s wave functions are shown on the same scale. Clearly, the 1s function inhabits the space in close to the nucleus but the 2s and 3s wave functions achieve orthogonality in the sense that the $\int \psi_{1s}^* \psi_{2s} d\tau = \langle 1s|2s \rangle = 0$ by including

an amount of the positive portion that will overlap with the 1s function but have an equal minus amount where the 1s function is essentially zero. The 3s wave function then includes a portion like both the 1s and 2s functions with two sign changes in such a way that the integrated products are again zero $\langle 1s|2s \rangle = \langle 2s|3s \rangle = \langle 1s|3s \rangle = 0$. However, the separate functions do emphasize separate regions. This same effect occurs when the angular integration is included but is more complicated so we only show the simple case of the radial distributions for the s-orbitals in Figure 14.2. However, note the physical meaning is due to $|\psi^* \psi|$, the electron density, which forms "rings" since the square of the function is always positive.

Next we come to the angular eigenfunctions to provide the three-dimensional shapes of the H atom orbitals as well as model the orbitals for the entire periodic chart! We have previously solved the rigid rotor problem to obtain the $Y_l^m(\theta, \phi) = \Theta_l(\theta)\Phi_m(\phi)$ solution as a normalized product [8] (Table 14.1). However, we have noted there are some annoying sign changes with those functions, although they are correct. To try and make the topic simpler, we present here mathematically equivalent forms in terms of trigonometric functions. A student should be aware that many problems can be treated in the $Y_l^m(\theta, \phi)$ set of functions but perhaps it is also good to know that there are other ways to represent the angular functions. Then in Table 14.2 we show the equivalent real trigonometric forms only for the problematical $\Phi_m(\phi)$ part of the total wave function.

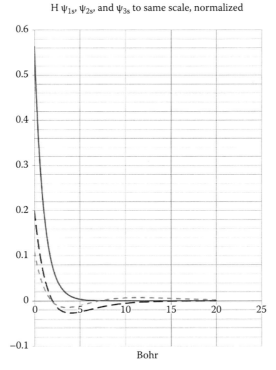

H ψ_{1s}, ψ_{2s}, and ψ_{3s} to same scale, normalized

FIGURE 14.1 Normalized 1s, 2s, and 3s total wave functions for the H atom showing how outer orbitals are orthogonal to inner orbitals by means of sign changes. $(n - l - 1) = \#$ nodes.

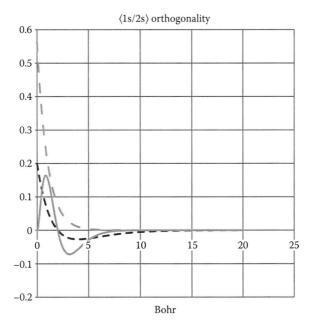

$\langle 1s/2s \rangle$ orthogonality

FIGURE 14.2 The solid curve beginning at (0, 0) is the product of $4\pi\psi_{1s}\,\psi_{2s}\,r^2$ showing both positive and negative values, which would be summed in the inner product integral $\int \psi_{1s}\psi_{2s}r^2 dr = 0$ showing the orthogonality over the radial space.

TABLE 14.1
Selected Normalized $Y_l^m(\theta, \phi)$

l	m	$Y_l^m(\theta, \phi)$	"Shape-Type"
0	0	$Y_0^0 = \dfrac{1}{\sqrt{4\pi}}$	s
1	0	$Y_1^0 = \left(\dfrac{3}{4\pi}\right)^{1/2} \cos\theta$	p
1	±1	$Y_1^{\pm 1} = \mp\left(\dfrac{3}{8\pi}\right)^{1/2} e^{\pm i\phi} \sin\theta$	p
2	0	$Y_2^0 = \left(\dfrac{5}{16\pi}\right)^{1/2} (3\cos^2\theta - 1)$	d
2	±1	$Y_2^{\pm 1} = \mp\left(\dfrac{15}{8\pi}\right)^{1/2} e^{\pm i\phi} \sin\theta \cos\theta$	d
2	±2	$Y_2^{\pm 2} = \left(\dfrac{15}{32\pi}\right)^{1/2} e^{\pm 2i\phi} \sin^2\theta$	d
3	0	$Y_3^0(\theta, \phi) = \left(\dfrac{7}{16\pi}\right)^{1/2} (5\cos^3\theta - 3\cos\theta)$	f
3	±1	$Y_3^{\pm 1}(\theta, \phi) = \mp\left(\dfrac{21}{64\pi}\right)^{1/2} e^{\pm i\phi} \sin\theta(5\cos^2\theta - 1)$	f
3	±2	$Y_3^{\pm 2}(\theta, \phi) = \left(\dfrac{105}{32\pi}\right)^{1/2} e^{\pm 2i\phi} \sin^2\theta \cos\theta$	f
3	±3	$Y_3^{\pm 3}(\theta, \phi) = \mp\left(\dfrac{35}{64\pi}\right)^{1/2} e^{\pm 3i\phi} \sin^3\theta \cos\theta$	f

etc.

Source: Lesk, A.M., *Introduction to Physical Chemistry*, Prentice-Hall, Inc., Englewood Cliffs, NJ, 1982, pp. 312 and 732.

The shapes are difficult to visualize because the $\Phi_m(\phi)$ function is complex (involves $\sqrt{-1} \equiv i$). Most chemistry texts get around this problem by using the so-called real spherical harmonics with $R_l^m(\theta, \phi) \equiv \dfrac{Y_l^m + (-1)^m Y_l^{-m}}{\sqrt{2}}$ for $m > 0$ and $R_l^{\bar{m}}(\theta, \phi) \equiv \dfrac{Y_l^m - (-1)^m Y_l^{-m}}{i\sqrt{2}}$, for $m < 0$ using $\bar{m}$ and $R_l^0(\theta, \phi) = Y_l^0$. This uses the identities of $\cos\theta = \dfrac{e^{i\theta} + e^{-i\theta}}{2}$ and $\sin\theta = \dfrac{e^{i\theta} - e^{-i\theta}}{2i}$ to produce real functions in terms of (θ, ϕ). However, Slater [9] has made these even simpler using the Cartesian equivalent formulas.

The functions in Table 14.3 are still subject to the normalization over the 4π solid angle of (θ, ϕ) and radial dependence but it is comforting to see the shape dependence in more familiar Cartesian coordinates.

Another important observation is that for each principal shell number n there are $l = n$ orbital shape types and for each l-shell there are $(2l + 1)$ m sublevels. The $Y_l^m(\theta, \phi)$ eigenfunctions are the same for the H atom as for the rigid rotor. The l-values in the H atom correspond to j-values in the

TABLE 14.2
$\Phi_m(\phi)$ Eigenfunctions of the H Atom [1]
in Trigonometric Form

m	$e^{im\phi}$	$\sin(m\theta)$ or $\cos(m\theta)$
0	$\dfrac{1}{\sqrt{2\pi}}$	$\dfrac{1}{\sqrt{2\pi}}$
1	$\dfrac{e^{i\phi}}{\sqrt{2\pi}}$	$\dfrac{\cos\phi}{\sqrt{\pi}}$
-1	$\dfrac{e^{i\phi}}{\sqrt{2\pi}}$	$\dfrac{\sin\phi}{\sqrt{\pi}}$
2	$\dfrac{e^{2i\phi}}{\sqrt{2\pi}}$	$\dfrac{\cos 2\phi}{\sqrt{\pi}}$
-2	$\dfrac{e^{-2i\phi}}{\sqrt{2\pi}}$	$\dfrac{\sin 2\phi}{\sqrt{\pi}}$
etc.		

Note: $\sin\phi = \dfrac{e^{i\phi} - e^{-i\phi}}{2i}$ and $\cos\phi = \dfrac{e^{i\phi} + e^{-i\phi}}{2}$.

TABLE 14.3
Real Angular Wave Functions for s, p, and d Orbitals

Type: m	Formula	Common Name
s:$m = 0$	1	s
p:$\|m\| = 0$	$\sqrt{3}\left(\dfrac{z}{r}\right)$	p_z
p:$\|m\| = 1$	$\sqrt{3}\left(\dfrac{x}{r}\right)$, $\sqrt{3}\left(\dfrac{y}{r}\right)$	p_x, p_y
d:$\|m\| = 0$	$\sqrt{5}\left[\dfrac{z^2 - 1/2(x^2 + y^2)}{r^2}\right]$	d_{z^2}
d:$\|m\| = 1$	$\sqrt{15}\left(\dfrac{xz}{r^2}\right)$, $\sqrt{15}\left(\dfrac{yz}{r^2}\right)$	d_{xz}, d_{yz}
d:$\|m\| = 2$	$\sqrt{\dfrac{15}{4}}\left(\dfrac{x^2 - y^2}{r^2}\right)$, $\sqrt{15}\left(\dfrac{xy}{r^2}\right)$	$d_{x^2-y^2}, d_{xy}$

Source: Slater, J.C., *Quantum Theory of Matter*, 2nd Edn., McGraw-Hill
Book Co., New York, 1968, p. 127.
$x = r\sin\theta\cos\phi$, $y = r\sin\theta\sin\phi$, $z = r\cos\theta$.

rigid rotor but the *m*-values have the same meaning in both cases. Most atoms can be treated with $l \leq 5$, even allowing for excited states. However, the *j*-values of the rigid rotor often reach much higher values in rotation than the *l*-values reach in electronic states. You can visualize the H atom as a two-particle rotor with the electron end much less massive than the proton end but the mathematics for the $Y_l^m(\theta, \phi)$ part of the problem is identical in both cases!

Fortunately for us the Pauling and Wilson text [3] has saved us a lot of trouble and combined the three solutions into the total wave function for most of the lower states of the H atom as

$$\psi_{nlm}(r, \theta, \phi) = R_{nl}(r)\Theta_{lm}(\theta)\Phi_m(\phi) \text{ with}$$

$$\Phi_m(\phi) = \frac{1}{\sqrt{2\pi}}e^{im\phi} \text{ (but in the trigonometric form),}$$

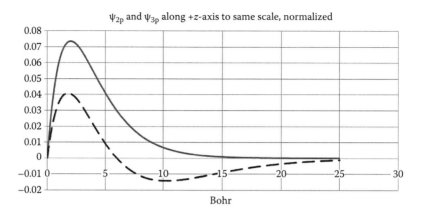

FIGURE 14.3 The positive lobe of the normalized 2p (solid) and 3p (dashed) H orbitals along the x-axis showing how $3p_x$ is orthogonal to $2p_x$ and $(n - l - 1) = 1$ node.

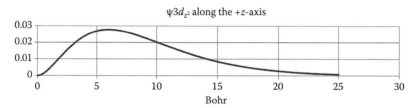

FIGURE 14.4 The positive lobe of the tall $3d_{z^2}$ H orbital along the z-axis showing there is no node in the radial part of the wave function and that the main "fat" region of the orbital is in the same region as the $3p_z$ orbital but the p-orbital is an "odd" function with a sign change above and below the x–y plane while the $3d_{z^2}$ wave function is an "even" function with the same sign above and below the x–y plane. Thus the orthogonality is achieved by the $(3\cos^2 \theta - 1)$ angular part of the $3d_{z^2}$ orbital compared to the $\cos \theta$ angular part of the $2p_z$ and $3p_z$ orbitals.

$$\Theta_{lm}(\theta) = \left[\frac{(2l + 1)(l - |m|)!}{2(l + |m|!)} \right]^{1/2} P_l^{|m|}(\cos \theta)$$

and

$$R_{nl}(r) = -\left[\left(\frac{2Z}{na_0} \right)^3 \frac{(n - l - 1)!}{2n|(n + 1)!|^3} \right]^{1/2} e^{-\frac{\rho}{2}} \rho^l L_{n+l}^{2l+1}(\rho).$$

It is important to note that $\rho = \left(\frac{2Z}{na_0} \right) r$ so the "scale" of the coordinate ρ depends on $\left(\frac{Z}{n} \right)$; later this is clarified using $\sigma = \left(\frac{Z}{a_0} \right)$ so that we can see the effect of the n-shell number making the exponential factor more diffuse for higher n values and canceling the 2 from the $\rho/2$ exponent. The best news for students is that the P&W text has simplified all these problems (particularly the conversion to real functions) and we present their final orbital formulas here. The only hint of the conversion from complex to real functions is the $\pm m$ quantum number (Figures 14.3 and 14.4).

PICTURES OF ANGULAR ORBITALS

A student who is eager for a picture instead of polynomial equations should find Figure 14.5 a delight. The illustration is from a 1932 paper by H. E. White [11] written at a time when scientists were also curious as to what the Schrödinger wave equation for the H atom looked like. Not only is

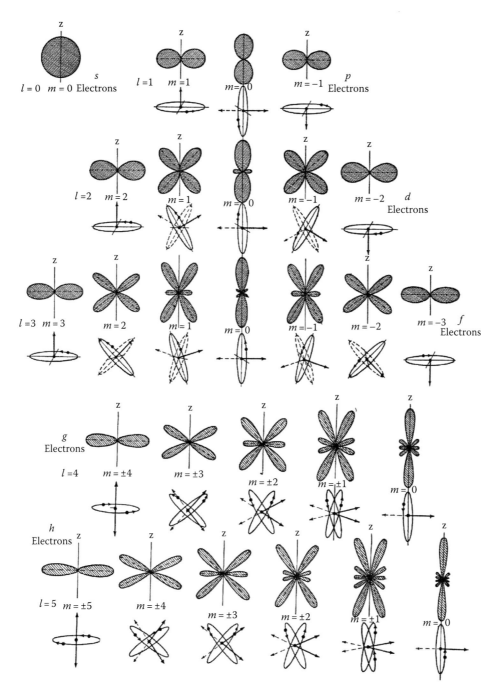

FIGURE 14.5 Angular shapes of the H orbitals according to both Schrödinger and Bohr. (Reprinted with permission from White, H.E., *Phys. Rev.*, 37, 1416, 1931. Copyright 1931 by the American Physical Society.)

this an excellent illustration of the orbital shapes but it shows the corresponding interpretation of the Bohr atom. In 1931, there were still advocates of the Bohr H atom solution and it is interesting to see the three p-orbitals represented by three circular orbits relative to three different axes. Similarly the d-orbitals are represented by circular orbits canted at various angles for the $(2l + 1)$ m-sublevels. That would have made sense at the time relative to the structure of atoms, but today we realize that

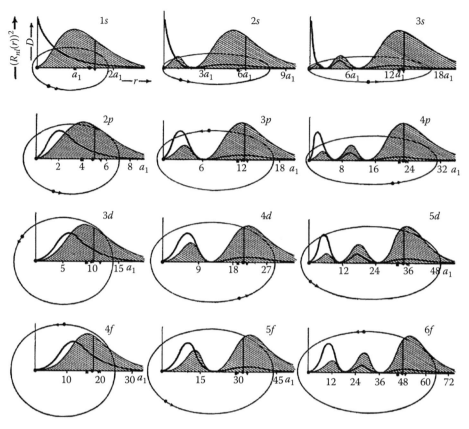

FIGURE 14.6 Radial probability density of the H atom from both Schrödinger and Bohr. Note that the scale of the x-axis changes so that the size of the figures is compressed horizontally as the n-quantum number increases. The dark line is the square of the radial function $(R_{nl}(r))^2$. The shaded area is the radial probability density as $4\pi r^2 (R_{nl}(r))^2$. The radial nodes in the orbitals show up here as probability ring separations because the square of the wave functions is always positive compared to the sign changes in the orbitals themselves. (Reprinted with permission from White, H.E., *Phys. Rev.*, 37, 1416, 1931. Copyright 1931 by the American Physical Society.)

canted circular orbits do not give a good picture of how chemical bonds are formed in molecules. Even so, it is interesting to see that by canting the sublevel orbits relative to the z-axis, the Bohr model was capable of extension to higher orbits.

Figure 14.6 is also very complete in that it shows the square of the radial functions $[R_{nl}(r)]^2$ represented as a dark curve and the *radial probability of the wave function* $4\pi r^2 [R_{nl}(r)]^2$ with a shaded area. The functions used by White are essentially identical to the functional forms found in Pauling and Wilson [3] except that White uses a factor of $(\sin^m \theta)$ in place of what we have shown as $(-1)^{\frac{m+|m|}{2}}$. Of more interest is the shaded area showing $[R_{nl}(r)]^2 (4\pi r^2)$, which is the probability density over the surface of a sphere and gives higher weight to values at larger radius. Once again the corresponding size of the Bohr orbit is given and the dark vertical line shows the value of the Bohr radius.

The most important conclusion from the appearance of the radial orbitals is a rationale for how valence shell effects can have an effect on "chemical shifts" in nuclear magnetic resonance (NMR), which is after all a measurement of a nuclear spin state (to be discussed later). We can see from Figure 14.6 that while the main probability of an outer 3s orbital could be involved in a valence bonding situation, there are also inner portions of that same orbital near the nucleus to cause transmission of valence effects to the electronic environment close to the nucleus. While more modern pictures have been developed for molecular orbitals, these pictures of atomic orbitals are as good today as they were in 1931!

An added explanation for the d-orbitals in Figure 14.6 is that the $m = 2$ and $m = -2$ views are from the side of the d_{xy} and $d_{x^2-y^2}$ orbitals that lie in the x–y plane. They actually have the same "four-leaf-clover" shape with alternating lobe signs as the d_{xz} and d_{yz} $m = +1$ and $m = -1$ orbitals. The $d_{x^2-y^2}$ orbital has lobes along the x–y axes while the d_{xy} orbital has lobes "between" the x–y axes rotated at a 45° angle relative to the $d_{x^2-y^2}$ orbital. In Figure 14.6, the $m = 0$ d-orbital is the so-called d_{z^2} orbital, which looks like it just has two small "ears" at the midpoint, but a three-dimensional rendering would show that feature as a "tire" ring around the middle with both large lobes positive and the tire negative since the orbital really should be labeled as $d_{2z^2-x^2-y^2}$ and can actually be synthesized from a renormalized combination of $(d_{xz} + d_{yz})$. Perhaps a better way to describe the d_{z^2} orbital might be as $2d_{z^2} - d_{(x^2+y^2)}$, which emphasizes the circular nature of the $(x^2 + y^2)$ "tire" as we recall the equation for a circle is $r^2 = x^2 + y^2$.

POWELL EQUIVALENT d-Orbitals

Once a student becomes accustomed to realizing that what counts is the probability density of orbitals as positive numbers, it should be possible to make simple freehand drawings of at least 1s, 2s, $2p_x$, $2p_y$, $2p_z$, 3s, $3p_x$, $3p_y$, $3p_z$, $3d_{z^2}$, $3d_{xz}$, $3d_{yz}$, $3d_{xy}$, and $3d_{x^2-y^2}$ orbitals. That is what a student could do after skipping over the tedious polynomial derivations! After doing those sketches, students may wonder why there are three equivalent shapes for the p-orbitals but the d-orbitals require a different shape for the d_z^2 orbital. This is due to the fact that there are actually six ways to draw the four-leaf-clover shape of the d-orbitals in Cartesian representation but only $(2l + 1) = 5$ orbitals for the $l = 2$ shell in the (θ, ϕ) representation. However, one can form a d_z^2 orbital by combining two four-leaf clovers as $d_{z^2} = N(d_{xz} + d_{yz})$ and renormalizing to find N. Thus the usual Cartesian d-orbitals given by Slater in Table 14.4 can all be made from the basic four-leaf-clover shape. You can visualize that the double counting of the z-part of the d_z^2 makes the orbital taller along the z-axis and the presence of both x and y parts leads to a circular "tire" around the middle of the orbital in the x–y plane. Thus the basic building block of the d-shell is the use of the six Cartesian orbitals as five linear combinations. Basically, nature (or at least the mathematics of nature) is telling us that while we may want to think in (x, y, z) space, the natural orbital shapes are better represented in polar coordinates! (Figures 14.7 through 14.9)

Now that you have pondered over how six four-leaf clovers can be made into five orbitals, we can ask you to wonder how we are going to describe the bonding in ferrocene. Ferrocene is an organometallic "sandwich" compound with an iron atom (Fe^{+2}) between two planar cyclopenta-diene anions $(C_5H_5)^-$ and it is possible to take linear combinations of (3d, 4s, 4p) orbitals on the iron atom to make hybrid linear combinations that rationalize the bonding between the Fe^{+2} and the (staggered) cyclopentadienide rings. However, that exercise is very challenging for the human mind while nature does it easily! In 1968, Powell [10] showed that it is possible to take linear combinations of the usual (canonical, standard) d-orbital shapes and form five equivalent d-orbitals that are tilted from the z-axis by about 48°, which form a set of five equivalent orbitals rotated about the z-axis by $\left(\dfrac{2\pi}{5}\right) = 72°$ in what is called a C_5 rotation. In a later chapter, we will use this idea of rotation by a C_n operation as rotation around a principal axis by $\left(\dfrac{2\pi}{n}\right)$. Here $n = 5$ and the principal axis is the z-axis, a convention as for polar coordinates.

We need to start using matrix algebra so we introduce the idea of a matrix as an array of numbers in a table surrounded by [ ] or in some cases by ( ) and the dimension of the matrix will be

TABLE 14.4
Final Combined Forms of the H Atom Wave Functions

n	l	m	Type	$\psi_{nlm}(r, \theta, \phi)$
1	0	0	1s	$\dfrac{1}{\sqrt{\pi}}\left(\dfrac{Z}{a_0}\right)^{3/2} e^{-\sigma}$
2	0	0	2s	$\dfrac{1}{4\sqrt{2\pi}}\left(\dfrac{Z}{a_0}\right)^{3/2}(2-\sigma)e^{-\frac{\sigma}{2}}$
2	1	0	$2p_z$	$\dfrac{1}{4\sqrt{2\pi}}\left(\dfrac{Z}{a_0}\right)^{3/2}\sigma e^{-\frac{\sigma}{2}}\cos\theta$
2	1	± 1	$2p_x$	$\dfrac{1}{4\sqrt{2\pi}}\left(\dfrac{Z}{a_0}\right)^{3/2}\sigma e^{-\frac{\sigma}{2}}\sin\theta\cos\phi$
2	1	± 1	$2p_y$	$\dfrac{1}{4\sqrt{2\pi}}\left(\dfrac{Z}{a_0}\right)^{3/2}\sigma e^{-\frac{\sigma}{2}}\sin\theta\sin\phi$
3	0	0	3s	$\dfrac{1}{81\sqrt{3\pi}}\left(\dfrac{Z}{a_0}\right)^{3/2}(27-18\sigma+2\sigma^2)\sigma e^{-\frac{\sigma}{3}}$
3	1	0	$3p_z$	$\dfrac{\sqrt{2}}{81\sqrt{\pi}}\left(\dfrac{Z}{a_0}\right)^{3/2}(6-\sigma)\sigma e^{-\frac{\sigma}{3}}\cos\theta$
3	1	± 1	$3p_x$	$\dfrac{\sqrt{2}}{81\sqrt{\pi}}\left(\dfrac{Z}{a_0}\right)^{3/2}(6-\sigma)\sigma e^{-\frac{\sigma}{3}}\sin\theta\cos\phi$
3	1	± 1	$3p_y$	$\dfrac{\sqrt{2}}{81\sqrt{\pi}}\left(\dfrac{Z}{a_0}\right)^{3/2}(6-\sigma)\sigma e^{-\frac{\sigma}{3}}\sin\theta\sin\phi$
3	2	0	$3d_{z^2}$	$\dfrac{1}{81\sqrt{6\pi}}\left(\dfrac{Z}{a_0}\right)^{3/2}\sigma^2 e^{-\frac{\sigma}{3}}(3\cos^2\theta-1)$
3	2	± 1	$3d_{xz}$	$\dfrac{\sqrt{2}}{81\sqrt{\pi}}\left(\dfrac{Z}{a_0}\right)^{3/2}\sigma^2 e^{-\frac{\sigma}{3}}\sin\theta\cos\theta\cos\phi$
3	2	± 1	$3d_{yz}$	$\dfrac{\sqrt{2}}{81\sqrt{\pi}}\left(\dfrac{Z}{a_0}\right)^{3/2}\sigma^2 e^{-\frac{\sigma}{3}}\sin\theta\cos\theta\sin\phi$
3	2	± 2	$3d_{xy}$	$\dfrac{1}{81\sqrt{2\pi}}\left(\dfrac{Z}{a_0}\right)^{3/2}\sigma^2 e^{-\frac{\sigma}{3}}\sin^2\theta\cos 2\phi$
3	2	± 2	$3d_{x^2-y^2}$	$\dfrac{1}{81\sqrt{2\pi}}\left(\dfrac{Z}{a_0}\right)^{3/2}\sigma^2 e^{-\frac{\sigma}{3}}\sin^2\theta\sin 2\phi$

Source: Pauling, L. and Wilson, E.B. *Introduction to Quantum Mechanics, with Applications to Chemistry*, McGraw-Hill Book Inc., New York, 1935, Chaps III–V. The copyright is now owned by Dover Press, New York, and is published as an unabridged reprint.

Note: $\sigma = \dfrac{Zr}{a_0}$.

(n-rows) × (m-columns). Linear algebra defines matrix multiplication as an operation where each element of an (n-row) is sequentially multiplied across by the sequential elements downward in the (m-column) of a matrix or column to the right and then the sum of the multiplications is placed in the (n, m) entry of a product matrix or column. For example, Powell's matrix is given as a 5×5 array in the following and the element for $d_0 = \left(\sqrt{\dfrac{1}{5}}\right)d_{z^2} + \left(\sqrt{\dfrac{2}{5}}\right)d_{xz} + (0)d_{yz} + \left(\sqrt{\dfrac{2}{5}}\right)d_{x^2-y^2} + (0)d_{xy}$,

multiplying and summing across the (5×5) for the first element in the (1×5) column. In linear algebra, multiplication is only defined for $[(n \times k) \times (k \times m)] = (n \times m)$ dimensions when the inner dimension "k" is the same. Here we see $(5 \times 1) = [(5 \times 5) \times (5 \times 1)]$. While we reserve the symbol $| \; |$

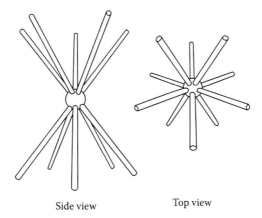

Side view   Top view

FIGURE 14.7 Stick models of the staggered arrangement of five-equivalent d-orbital bond directions. The application to the bonding in ferrocene seems obvious! (From Powell, R.E., *J. Chem. Educ.*, 45, 45, 1968. With permission.)

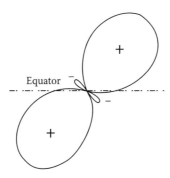

FIGURE 14.8 The angular shape of one of the Powell equivalent d-orbitals. The overall shape is not cylindrical but somewhat flattened and the major axis of the orbital is tipped up from the *x–y* plane by about 48.2°. (From Powell, R.E., *J. Chem. Educ.*, 45, 45, 1968. With permission.)

for a determinant (a determinant is a single number) and for absolute values, matrices are arrays of values in rows and columns subject to the rules of multiplication shown earlier. There really is no matrix division defined in linear algebra but it is possible to find the inverse of a matrix M^{-1} and then use that in multiplication. Addition or subtraction of two matrices is carried out by simple addition or subtraction of each element of one matrix with the corresponding element of another. Addition or subtraction of matrices is only defined for matrices of the same dimensions. You may or may not have had formal coursework in linear algebra but this author learned operational use of linear algebra (which is essential for computer applications of quantum mechanics) from some informal notes. One benefit of this latter part of the book is that you should end up with an operational understanding of linear algebra!

Numerically, the elements in Powell's (5×5) matrix form what is called a *unitary transformation matrix*. The main property of a unitary matrix is that it can scramble the relative amounts of the components but it does not change the overall "length" (magnitude) of the vector result. This can be shown by evaluating the determinant of the matrix, which is exactly 1.0, but in this case evaluating a (5×5) determinant is a tedious process!

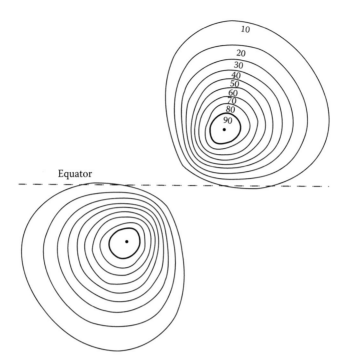

FIGURE 14.9 The probability contours of the $|\psi^* \psi|$ of one of the Powell equivalent d-orbital. (From Powell, R.E., *J. Chem. Educ.*, 45, 45, 1968. With permission.)

$$
\begin{bmatrix} d_0 \\ d_{1/5} \\ d_{2/5} \\ d_{3/5} \\ d_{4/5} \end{bmatrix} = \begin{bmatrix} \dfrac{1}{\sqrt{5}} & \sqrt{\dfrac{2}{5}} & 0 & \sqrt{\dfrac{2}{5}} & 0 \\[2ex] \dfrac{1}{\sqrt{5}} & \sqrt{\dfrac{2}{5}}\cos\dfrac{2\pi}{5} & \sqrt{\dfrac{2}{5}}\sin\dfrac{2\pi}{5} & -\sqrt{\dfrac{2}{5}}\cos\dfrac{2\pi}{5} & \sqrt{\dfrac{2}{5}}\sin\dfrac{2\pi}{5} \\[2ex] \dfrac{1}{\sqrt{5}} & -\sqrt{\dfrac{2}{5}}\cos\dfrac{2\pi}{5} & \sqrt{\dfrac{2}{5}}\sin\dfrac{2\pi}{5} & \sqrt{\dfrac{2}{5}}\cos\dfrac{2\pi}{5} & -\sqrt{\dfrac{2}{5}}\sin\dfrac{2\pi}{5} \\[2ex] \dfrac{1}{\sqrt{5}} & -\sqrt{\dfrac{2}{5}}\cos\dfrac{2\pi}{5} & -\sqrt{\dfrac{2}{5}}\sin\dfrac{2\pi}{5} & \sqrt{\dfrac{2}{5}}\cos\dfrac{2\pi}{5} & \sqrt{\dfrac{2}{5}}\sin\dfrac{2\pi}{5} \\[2ex] \dfrac{1}{\sqrt{5}} & \sqrt{\dfrac{2}{5}}\cos\dfrac{2\pi}{5} & -\sqrt{\dfrac{2}{5}}\sin\dfrac{2\pi}{5} & -\sqrt{\dfrac{2}{5}}\cos\dfrac{2\pi}{5} & -\sqrt{\dfrac{2}{5}}\sin\dfrac{2\pi}{5} \end{bmatrix} \begin{bmatrix} d_{z^2} \\ d_{xz} \\ d_{yz} \\ d_{x^2-y^2} \\ d_{xy} \end{bmatrix}
$$

Powell's equivalent d-orbitals point their maximum electron density lobes directly at the pi-orbitals in the cyclopentadienide rings providing an easy explanation of the bonding in ferrocene based on orbital overlap. No doubt hybridization in the $2sp$ shell of the carbon atom is familiar from courses in organic chemistry where linear sp^1, trigonal sp^2 and tetrahedral sp^3 orbitals were discussed and here we see a form of hybridization within the 3d shell. It is probably worth mentioning that the canted Bohr orbits would not have provided any easy model for the bonding in ferrocene. Thus we see that once bonding geometry is in question, the Bohr orbitals are of little interpretive value. A pancake orbital canted at an angle is still a pancake and the Bohr orbits fail to provide a mechanism for rationalizing directional bonding. One further development was that while it might be assumed from Powell's pictures that the equivalent d-orbitals are cylindrical, a paper by Pauling and McClure in 1970 [12] examined the cross section of the main lobes of the d-orbitals and found them to be more nearly oval than circular.

TABLE 14.5
Unsöld's Constant for Selected l-Shells
of the H Atom

Type	l	$\pm m$	$\sum_{m=-l}^{+l}[\Theta_{lm}(\theta)]^2$
s	0	± 0	$1/2$
p	1	± 1	$3/2$
d	2	± 2	$5/2$
f	3	± 3	$7/2$

Source: White, H.E., *Phys. Rev.*, 37, 1416, 1931.

UNSÖLD'S THEOREM

Students may wonder whether atoms are truly spherical as monatomic gases. There is an answer according to work by Unsöld. Pauling and Wilson [3] mention the theorem, and the paper by White [11] for the orbital diagrams also gives results for Unsöld's theorem with values for various shells. The idea is that if one sums over all the m-sublevels of a given *l*-shell, the result is a constant independent of (θ, ϕ). This implies that a completely filled *l*-shell is spherical. The calculation is for $\sum_{m=-l}^{+l} \Theta_{lm}^*(\theta)\Phi_m^*(\phi)\Theta_{lm}(\theta)\Phi_m(\phi) = C$ where C is a constant. White [11] gives the values of the constant for several shells shown in Table 14.5.

White quotes only $\sum_{m=-l}^{+l} [\Theta_{lm}(\theta)]^2 = C$ because the exponents cancel when $\Phi_m(\phi) = \dfrac{e^{im\phi}}{\sqrt{2\pi}}$ is multiplied by its complex conjugate leaving only the normalization constant. The value of the constant is not important, the significance is that there is a constant independent of (θ, ϕ) so that means that the filled shell rare gas atoms are surely spherical and other atoms will spin freely unless there is some strong reference direction to a magnetic field.

AUFBAU PRINCIPLE AND THE SCALED H ATOM

Now that we have the H atom orbitals, we want to emphasize the role played by $\left(\dfrac{Z}{n}\right)$ in their formula. Extending Z to higher atomic numbers allows us to "build up" ("aufbau" translates as "build") other atoms by adding electrons to more and more orbitals. The trends in the periodic chart can be seen to be related to the number of electrons filling the orbitals of an H atom model whose inner shells contract drastically as Z increases, but when the inner electrons shield some of the nuclear charge, the outer electrons still behave as if they are in hydrogenic orbitals. Shielding of part of the nuclear charge allows the outer "valence shell" electrons to behave as if they were in the electrostatic field of a lesser Z value charge, although the effective value of Z may no longer be an integer! However, as more and more electrons are added to this model, the small amount of electron–electron repulsion accumulates until the combination of a non-integer, effective nuclear charge, and the mutual electron interaction distorts the order of the energy levels of the H atom formula even using an effective nuclear charge. As the number of electrons increases, another effect becomes more important. Coupling of the small magnetic fields generated by swirling electron orbits with the spin magnetic moments distorts the simple single electron model of the H atom. This effect becomes dominant somewhere near $Z = 29$ (Cu) or $Z = 30$ (Zn) using the (L,S) coupling scheme discussed in the following but the shape of the orbitals remains roughly hydrogen-like well past $Z = 92$ (U). The order of the energy levels can be further explained using a revised way of counting the angular momentum with $|J, J_z\rangle$ coupling where $J = L + S$ so the H atom model is a very good way to explain the shell filling of 2, 8, 8, 18, 18, etc., observed in the earlier Bohr model.

The H atom model provides a particularly satisfying interpretation of the relative unreactivity of the rare gases as having completely filled shells. Unsöld's theorem together with the known near inertness of the rare gases calls attention to their completely filled spherical shells. This observation can be combined with known valence tendencies of Group I metals to donate an outer electron to Halogen atoms to postulate that the rare gas structures are so stable electronically that there is a chemical principal at work relative to electronic structure. *The main principle of chemistry is that atoms in molecules trade or share electrons in such a way as to approach filled shells!* The electron donor–acceptor concept can be extended to the whole periodic chart to designate "metals" as donors, "nonmetals" as acceptors and the relatively inert rare gases.

TERM SYMBOLS AND SPIN ANGULAR MOMENTUM

During the Aufbau building-up principle, the first 30 elements can be described pretty well by a labeling scheme that could be described by the notation shown for Zn ($Z = 30$) in terms of H orbitals as $(1s^2)(2s^2\, 2p^6)(3s^2\, 3p^6\, d^{10})(4s^2)$ and it is easily rationalized to explain the common charge of Zn^{+2} when the outer two electrons are lost. The H atom model "explains" the stability of He as $(1s^2)$, Ne as $(1s^2)(2s^2\, 2p^6)$, Ar as $(1s^2)(2s^2\, 2p^6)(3s^2\, 3p^6)$, etc. This can be augmented by the observation that "the state of maximum multiplicity lies lowest!" according to Hund's rule to understand the magnetism of the half-filled 3d^5 shell of Mn and even the half-filled 4f^7 shell of Gd. All of these freshman chemistry concepts hang comfortably on the shoulders of the H atom model so we see why we have spent time on the details of the derivation!

We have avoided the topic of "spin" for quite a while in this text but now we have to consider what it means. The easiest explanation is the most mysterious. We can say that electrons are particles that have an *intrinsic* amount of angular momentum. We know energy comes in quantum chunks and that kinetic energy is made from momentum so we can extend that idea to say that electrons just have $\alpha = +\dfrac{\hbar}{2}$ or $\beta = -\dfrac{\hbar}{2}$ as small amounts of angular momentum. In effect, they cannot stop rotating although the amount of rotation is tiny. Even a tiny amount of rotation of a charged particle generates a tiny magnetic field so that relative to some reference field (magnetic field of the Earth) one can observe $|\alpha\rangle$ as a "spin up" state and $|\beta\rangle$ as a "spin down" state. It is not so much that the electrons are spinning wildly, but rather that they have a certain (small) amount of angular momentum and really just cannot stop; the spin is an intrinsic property of the particle. We know that other particles can have spin and that is used in NMR just as electron spin can be observed using electron spin resonance (ESR). Actually, there is a higher-order treatment of the H atom due to P. Dirac, which treats electron spin within the Hamiltonian of the H atom but it is too difficult for an undergraduate treatment. In the Dirac H atom, the $|\alpha\rangle$ and $|\beta\rangle$ electron orbitals are slightly different spatially. In the usual "nonrelativistic" Schrödinger treatment, we blithely assign two electrons to each ψ_{nlm} spatial orbital and make the reasonable assumption that somehow the opposite magnetic moments of the two electrons in each spatial orbital tend to form a stable "electron pair." In fact, we know from chemical principles that a great deal of chemical bonding can be explained on the basis of electron pairs so there is something to that idea. However, the idea of intrinsic angular momentum is very profound and requires delving into the theory of elementary particles.

The most famous experiment related to spin was the "Stern-Gerlach" experiment [13] in 1922. In that experiment, silver was vaporized in an oven and allowed to exit as a beam, which traveled between a long pair of magnet poles each machined to a knife edge. From chemistry, we know Ag has one outer electron and the Schrödinger model would describe the electron orbital occupancy as $(1s^2)(2s^2\, 2p^6)(3s^2\, 3p^6\, 3d^{10})(4s^2\, 4p^6\, 4d^{10})(5s^1)$. As the Ag atoms traveled through the long path, the interaction of the outer electron with the magnetic field caused the Ag beam to separate into two beams that were deposited on a glass plate at the end of the apparatus. At first, they thought the experiment was inconclusive but Professor Stern examined the glass plate while smoking a cigar and the cigar smoke had enough H_2S to precipitate dark brown Ag_2S on the glass plate and two images were "developed" by the cigar smoke! Analysis of the beam conditions and the separation of the two images led to the conclusion that neutral Ag atoms have a tiny magnetic moment of $\pm\dfrac{\hbar}{2}$ due to the outer electron. Note that also implies that the spin angular momentum magnetic moments of the other 46 electrons in the Ag atom cancel to zero!

HUND'S RULE

Friedrich Hund (1896–1997) was a German physicist who was an assistant to N. Bohr and studied many atomic spectra so as to conclude his famous "Hund's rule" for the electronic structure of the lowest energy of atoms (ground state): *the ground state is the state of maximum spin multiplicity.* Using the Aufbau principle, Hund observed that the electrons successively fill sub-shells of the H atom model in a certain way. In succession, every orbital in a sub-shell is filled with just one electron before any is doubly occupied and all singly occupied orbitals have the same spin. Hund's rule is a conclusion based on observation. Further, the ground state of

$$
\begin{array}{cccc}
\alpha\beta & \alpha & \alpha & 2p^4 \\
m=+1 & 0 & -1 &
\end{array}
\qquad
\begin{array}{ccc}
\alpha & \alpha & 2p^2 \\
m=+1 & 0 & -1
\end{array}
$$

$\alpha\beta 2s^2$
$m=0$

$\alpha\beta 2s^2$
$m=0$

$\alpha\beta 1s^2$
$m=0$ Oxygen $={}^3P$

$\alpha\beta 1s^2$
$m=0$ Carbon $={}^3P$

an atom can be described by a "term symbol" such as $^{(2s+1)}T\left(\sum_i m_i\right)$. Here T is a generic symbol for the total sum of the m-quantum numbers of the occupied orbitals: S(0), P(1), D(2), F(3), G(4), H(5), etc. Consider the example of the ground state of an oxygen atom $1s^2\,2s^2\,2p^4$ compared to the carbon atom $1s^2 2s^2 2p^2$. Both atoms have two unpaired electrons. We show the 1s orbital energy as the lowest with an implied vertical energy scale and it is found in atoms that the 2s energy is slightly lower than the 2p set of three energy- degenerate 2p orbitals. According to the Aufbau principle we fill the orbitals as α, β, α, β, α,.... until we come to a degenerate shell as the 2p orbitals. Then we fill all the degenerate levels first with alpha electrons starting with the highest m value and then come back and add beta electrons until all are assigned, and here that is α, α, α, β in the 2p shell. Now we have the sum of m_i for O as $\sum_i m_i = (0+0)_{1s} + (0+0)_{2s} + (1+0-1+1)_{2p} = 1 \Rightarrow P$. Then we add up the spins in $\hbar$ units as $S = \sum_i s_i = \left(\dfrac{1}{2}-\dfrac{1}{2}\right)_{1s} + \left(\dfrac{1}{2}-\dfrac{1}{2}\right)_{2s} + \left(\dfrac{1}{2}-\dfrac{1}{2}+\dfrac{1}{2}+\dfrac{1}{2}\right)_{2p} = 1 \Rightarrow (2S+1) = 3$, so the term symbol is ^3P, pronounced "triplet P." Why is the state a "triplet"? In a magnetic field (of the Earth or a laboratory magnet), the possible spin states [5] for two electrons can be as follows (quantized along a reference z-axis):

$$|\alpha\alpha\rangle, \quad S_z = 1$$

$\dfrac{(|\alpha\beta\rangle + |\beta\alpha\rangle)}{\sqrt{2}}$, $S_z = 0$ implies three states in a magnetic field, that is, a "triplet."

$$|\beta\beta\rangle, \quad S_z = -1$$

The $S_z = 0$ state is somewhat unexpected but we must realize that electrons are indistinguishable so we have to consider both ways we can have one spin up and the other spin down. In fact, just as you may have learned in counting NMR peaks the total number of spin states for N unpaired electrons is $(N + 1)$. The most complicated case in the first row of elements is for nitrogen as for the electron configuration $1s^2\, 2s^2\, 2p^3$.

$$\begin{array}{cccc} \alpha & \alpha & \alpha & 2p^3 \\ m = +1 & 0 & -1 & \end{array}$$

$$\begin{array}{l} \alpha\beta 2s^2 \\ m = 0 \end{array}$$

$$\begin{array}{ll} \alpha\beta 1s^2 & \\ m = 0 & \text{Nitrogen} = {}^4S \end{array}$$

$\sum_i m_i = 0, \Rightarrow S$ and $\sum_i s_i = \dfrac{3}{2}, \Rightarrow \left[2\left(\dfrac{3}{2}\right) + 1 \right] = 4$, so the term symbol is ^4S, quartet S.

Why do we care about the term symbols? First, it tells us about the magnetic properties of atoms due to unpaired spins. In addition, you will need to know the term symbols in upper level courses in inorganic chemistry because the overall value of the $\sum_i m_i$ as S, P, D, etc., gives spatial information as to what orbitals form molecular orbitals in octahedral metal complexes. A large part of physical chemistry is preparation for other aspects of chemistry! For now just use the term symbol to relate to magnetic aspects of atoms such as Mn $(1s^2\, 2s^2\, 2p^6\, 3s^6\, 3p^6\, 3d^5\, 4s^2) \Rightarrow {}^6S$ with five unpaired electrons! For Mn, Hund's rule confirms the stability of the half-filled 3d shell. In fact, planar pentagonal Mn_5 has been reported to have 25 unpaired electrons [14] when trapped in a frozen Argon matrix!

$|L, S_z\rangle$ VERSUS $|J, J_z\rangle$ COUPLING

We mentioned earlier that the Aufbau Principle can be used along with Hund's rule to simply lay out a diagram of the H atom energy levels and populate them with electrons starting with the lowest energy $1s$ orbital. That works pretty well up to about $Z = 30$ (Zn) and we know that angular momentum is quantized so the momentum of each orbital can be described by the quantized vector of $\vec{L} + \vec{S}_z$. For light elements up to $Z = 30$ the small coupling between the two forms of angular momentum allows them to be considered separately. However, due to *spin–orbit coupling*, there is an energy effect given by $\xi \vec{L} \cdot \vec{S}$ as the interaction of the two types of angular momentum and ξ is a measureable quantity that increases with Z. As the number of electrons increases with atomic number, there is more of both $\vec{L}$ and $\vec{S}$ and their coupling becomes comparable to electron repulsion effects. Then the basic H atom energy level scheme is distorted to the point that another way has to be used to explain the quantization of the angular momentum past $Z = 30$ with the so-called $|J, J_z\rangle$ scheme where $\vec{J} = \vec{L} + \vec{S}_z$. Past $Z = 30$, the order of the levels is modified due to spin–orbit interactions so it is better to use the $|J, J_z\rangle$ scheme.

Up to this point, we have simply placed electron pairs into hydrogenic orbitals but the electrons do have a small amount of angular momentum measured in units of $\pm \dfrac{\hbar}{2}$. As such the spins also obey momentum quantization, which gives rise to another set of equations.

$$S^2 \psi_s = s(s + 1)\hbar^2 \psi_s \quad \text{and} \quad S_z \psi_s = s_z \hbar \psi_s.$$

At this point, we are flirting with the more difficult relativistic Dirac equation for the H atom. That method involves representing the spin function, ψ_s, as 2×2 Pauli matrices. Not only is that method more complicated, it is so correct that it predicts the existence of antimatter electrons called positrons. At the level we are using here, we just designate the spin functions as α (spin "up") or β (spin "down") for electrons but students should be aware that spin can be treated in a more detailed way in advanced texts.

$$S^2\alpha = \frac{1}{2}\left(\frac{1}{2}+1\right)\alpha \quad \text{and} \quad S^2\beta = \frac{1}{2}\left(\frac{1}{2}+1\right)\beta \quad \text{with} \quad S_z\alpha = \frac{1}{2}\alpha \quad \text{and} \quad S_z\beta = -\frac{1}{2}\beta \text{ (in units of } \hbar).$$

In this "semiclassical" (not quite full Dirac treatment) formalism, we still need to use

$$\int \alpha^*\alpha d\tau = 1, \quad \int \beta^*\beta d\tau = 1, \quad \text{and} \quad \int \alpha^*\beta d\tau = 0.$$

Continuing with this approximation, we limit the discussion to the H atom. It would seem that we need to add a term to the Hamiltonian to account for the interaction of the magnetic moment of the electron spin and the orbital angular momentum. That extra term in the Hamiltonian is of the form

$$H_{so} = \sum_i \xi_i \vec{L}_i \cdot \vec{S}_i.$$

For the single electron in the H atom or a hydrogenic ion, the value was worked out by Thomas [15] using the first-order terms of a relativistic expansion [15–19].

$$\xi = \frac{1}{2m^2c^2}\left(\frac{1}{r}\frac{dV}{dr}\right).$$

The dot product of the two forms of angular momentum can be found by squaring the total momentum as

$$\vec{J}^2 = (\vec{L}+\vec{S})^2 = \vec{L}^2 + 2\vec{L}\cdot\vec{S} + \vec{S}^2 \Rightarrow \vec{L}\cdot\vec{S} = (1/2)(\vec{J}^2 - \vec{L}^2 - \vec{S}^2).$$

Then, since we know the $Y_l^m(\theta, \phi)$ functions are simultaneously the eigenfunctions of all the angular momentum operators and we know the rules for the spin eigenfunctions, we find

$$\vec{L}\cdot\vec{S} = (1/2)[j(j+1) - l(l+1) - s(s+1)]\hbar^2.$$

The details are given in Appendix D but the final formula can be derived cleanly to within first order of a relativistic formula for the electron velocity as

$$\xi_{nl} = \frac{e^2\hbar^2}{2m^2c^2a_0^3} \frac{Z^4}{n^3l(l+1/2)(l+1)}.$$

Thus very good approximations can be made for the exact energy levels of the hydrogen and hydrogenic ions using $H_{so} = \sum_i \xi_i \vec{L}_i \cdot \vec{S}_i$. While this is a mathematically clean process for one electron, the process gets very complicated for many-electron systems. We can write

$$\vec{L}\cdot\vec{S} = \sum_i \sum_j L_i \cdot S_j$$

but this introduces a great deal of complexity into the calculation of energy levels. However, it is worth observing the dependence on the fourth power of the atomic number (Z^4) in the numerator. Even if it is difficult to calculate the spin–orbit splitting exactly, we can now understand why the simple idea of just putting (α, β) spin pairs into orbitals starts to fail for heavier (higher Z values) elements. Students interested in a more detailed explanation can consult Appendix D but the main message is that as the atomic number (Z) increases, the effect of spin–orbit coupling rapidly increases. At about $Z = 30$ (Zn) the simple idea of keeping $\vec{L}$ and $\vec{S}$ separate as if they do not interact fails. Then one must use $\vec{J} = (\vec{L} + \vec{S})$ in a new scheme called $|\vec{J}, J_z\rangle$ coupling. This explains the need for a more sophisticated approach with heavier elements. Even so, the simpler $|\vec{L}, S_z\rangle$ scheme will be sufficient to treat compounds formed from light elements up to Zn. A lot of Chemistry can be explained using the $|L, S_z\rangle$ scheme including most of organic chemistry, but the inorganic chemistry of heavier elements is necessarily described in terms of the $|J, J_z\rangle$ coupling scheme. However, even for higher Z values the shape of the orbitals remains basically as described by the Schrödinger model. Spin–orbit coupling mainly affects the order of the energy levels, not the shape of the orbitals. We leave $|J, J_z\rangle$ coupling and relativistic effects in heavier elements to more advanced texts.

SUMMARY

Chapters 12 through 14 have been the most challenging so far in terms of mathematical analysis and perhaps now you understand why only a very few problems have an exact solution for the Schrödinger equation. Although we only sketched the tedious polynomial analysis, we found the Schrödinger H atom yields exactly the same energy for the 1s level as the earlier Bohr theory, but the orbital solutions are more spatially detailed than the flat Bohr orbitals. Schrödinger orbitals can hybridize to form directional bonds as in the case of the Powell d-orbitals as well as the familiar sp hybrid orbitals of carbon. The Aufbau principle and the H atom model along with Hund's rule and a simplified treatment of electron spin help us organize the periodic chart. This provides a strong framework for understanding how elements trade electrons in compounds so as to more nearly approach the rare gas electronic structures. Metals release electrons to nonmetals and covalent compounds share electrons generally in a way that leads to electronic structures more like rare gas–filled shells, a fundamental principle of chemistry all "explained" on the basis of the H atom model!

1. There is a polynomial solution for the radial part of the H atom solution of the Schrödinger equation. The functions are related to previously studied Laguerre polynomials. The total solution for the H atom is the product of the rigid rotor (Θ, Φ) wave functions with the Laguerre radial functions. The eigenvalues for the orbitals are exactly the same as for the Bohr model, $E_n = -\dfrac{\mu Z^2 e^4}{2\hbar^2 n^2}$, $n = 1, 2, 3, \ldots$ but the shapes of the Schrödinger orbitals are three dimensional. The actual probability density of electrons is given by $\psi^*\psi$.

2. The H atom solutions have a principal shell number "n," and each shell has n different l shapes. Within each l shape there are $(2l + 1)$ sublevels with "m" quantum numbers as for 4f-orbitals ($n = 4$, $l = 3$) with $m = -3, -2, -1, 0, +1, +2, +3$. While some orbital shapes have been standardized as "canonical" Cartesian (x, y, z) shapes in textbooks, later work by Powell shows that in any l-shell an equivalent shape can be found for all $(2l + 1)$ members of the shell. In practice this is obvious for p-orbitals and occasionally useful for five equivalent d-orbitals or even seven equivalent f-orbitals.

3. Although a proper description of electron spin requires the higher-order Dirac equation, the Schrödinger orbitals of the H atom are useful to rationalize the orbital occupancy of atoms in the entire periodic chart! The simple Aufbau (building) principle works well with simple spin pairing up to $Z = 30$ (Zn) but the orbitals can still be used in a modified (J, J_z) spin coupling scheme to explain spectra of all known elements.

4. The Laguerre radial solutions form an orthonormal set of functions and the orthogonality of the various members of the set is accomplished using a system of $(n - l - 1)$ radial nodes (zero crossings in the r-coordinate). The 2s orbital has a small part like the 1s in it

with a new part which is negative, the 3s orbital has parts like the 1s and 2s shapes with alternating sign, and so forth so that when any $\langle R_n \mid R_m \rangle = \delta_{nm}$ orbital product is integrated over $0, \infty$ the result is zero and they are orthogonal as well as normalized (orthonormal).

5. A simple description of the magnetic state can be given by term symbols, which provide a compact way to summarize the spin and angular momentum of a given orbital occupancy of an atom; this also forms the foundation for similar descriptors of molecular electronic states.

6. The simple $|L, S_z\rangle$ model falls victim to the Z^4 increase of the $\vec{L} \cdot \vec{S}$ coupling between electron spin ($\vec{S}$) and the angular momentum ($\vec{L}$) of the orbitals in many-electron atoms at about $Z = 30$. Heavier elements need to be described by the $|J, J_z\rangle$ scheme where $\vec{J} = \vec{L} + \vec{S}$.

PROBLEMS

14.1 Sketch the shapes of the 1s, 2s, 3s, 4s, 2p, 3p, 4p, 3d, and 4d orbitals both as the radial dependence with the correct number of nodes with outer tails that taper asymptotically to zero and the angular shapes. Sketch the radial and angular functions separately. Sketch all the m-suborbitals for the s, p, and d shapes. Compare a sketch of the Powel equivalent 3d orbitals to the canonical shapes.

14.2 Show that $\lim\limits_{m \to \infty} \dfrac{a_{m+1}}{a_m} = \lim\limits_{m \to \infty} \left[\dfrac{-(\lambda - l - 1 - m)}{2(m+1)(l+1) + m(m+1)} \right] \to \left(\dfrac{1}{m} \right)$

14.3 Work out the term symbol for Cu ($Z = 29$), consider both possibilities remembering that a filled shell tends to be more stable than a partially filled shell.

14.4 Work out the term symbol for manganese, $Z = 25$.

14.5 Look up the structure of ferrocene and sketch the organic representation of the $2p_z$ orbitals of the cyclopentadienide ($C_5H_5^-$) anion. Then sketch a picture of how the Powell 3d orbitals on Fe^{+2} would overlap the $2p_x$ orbitals on both anions when the rings are staggered. Would five canted Bohr orbitals explain the bonding?

14.6 Work out the term symbols for the ground state of H, He, Li, Be, B, C, N, O, F, and Ne.

14.7 Work out the associated Laguerre polynomial for the 3s orbital ($n = 3$, $l = 0$).

14.8 Show that your polynomial from Problem 14.7 is a solution to the associated Laguerre equation.

14.9 On graph paper draw energy-level lines for the 1s, 2s, 2p, 3s, 3p, 3d, and 4s orbitals of the H atom on a vertical scale to show the gap between levels decreases as n increases.

14.10 Look up the history of the discovery of ferrocene and write a chemical reaction for its synthesis.

REFERENCES

1. McQuarrie, D. A., *Quantum Chemistry*, 2nd Edn., University Science Books, Sausalito, CA, 2008.
2. Atkins, P. W. and R. S. Friedman, *Molecular Quantum Mechanics*, 3rd Edn., Oxford University Press, New York, 1997.
3. Pauling, L. and E. B. Wilson, *Introduction to Quantum Mechanics, with Applications to Chemistry*, McGraw-Hill Book Inc., New York, 1935, Chaps III–V. The copyright is now owned by Dover Press, New York, and is published as an unabridged reprint.
4. Rainville, E. D., *Elementary Differential Equations*, The Macmillan Company, New York, 1958.
5. Kauzmann, W., *Quantum Chemistry*, Academic Press, New York, 1957.
6. Eyring, H., J. Walter, and G. Kimball, *Quantum Chemistry*, John Wiley & Sons, Inc., New York, 1944.
7. Schrödinger, E., An undulatory theory of the mechanics of atoms and molecules, *Phys. Rev.*, **28**, 1049 (1926).
8. Lesk, A. M., *Introduction to Physical Chemistry*, Prentice-Hall, Inc., Englewood Cliffs, NJ, 1982, pp. 312 and 732.
9. Slater, J. C., *Quantum Theory of Matter*, 2nd Edn., McGraw-Hill Book Co., New York, 1968, p. 127.

10. Powell, R. E., The five equivalent d orbitals, *J. Chem. Educ.*, **45**, 45 (1968).

11. White, H. E., Pictorial representations of the electron cloud for hydrogen-like atoms, *Phys. Rev.*, **37**, 1416 (1931).

12. Pauling, L. and V. McClure, Five equivalent d orbitals, *J. Chem. Educ.*, **47**, 15 (1970).

13. Gerlach, W. and O. Stern, Das magnetische Moment des Silberatoms, *Z. Phys.*, **9**, 353 (1922).

14. Shillady, D. D., P. Jena, B. K. Rao, and M. R. Press, A theoretical study of the geometry of Mn5, *Int. J. Quantum Chem.*, **S22**, 231 (1988).

15. Thomas, L. H., The motion of the spinning electron, *Nature*, **117**, 514 (1926).

16. Murrell, J. N., S. F. A. Kettle, and J. M. Tedder, *Valence Theory*, 2nd. Edn., John Wiley & Sons, London, U. K., 1964, p. 118.

17. Pilar, F. L., *Elementary Quantum Chemistry*, The McGraw-Hill Book Co., New York, 1968, p. 301.

18. McGlynn, S. P., T. Azumi, and M. Kinoshita, *Molecular Spectroscopy of the Triplet State*, Prentice-Hall, Englewood Cliffs, NJ, 1969, p. 188.

19. P. W. Atkins and R. S. Friedman, *Molecular Quantum Mechanics*, 3rd Edn., Oxford University Press, Oxford, 1997, p. 208.

15 Quantum Thermodynamics

INTRODUCTION

In terms of "essentials" we only have a few more topics. Here we will give an introduction to a field called "statistical thermodynamics," which depends on an understanding of quantized energy but then extends the statistical approach pioneered by Boltzmann to macroscopic Thermodynamics. Our intention here is to just show the "tip-of-an-iceberg" view of statistical thermodynamics, enough to "explain" the Boltzmann energy averaging and show that the statistical approach gives high-temperature limiting forms of the energy partition concept. There are large and complete texts on statistical thermodynamics available for graduate study but here we only attempt to justify the Boltzmann principle used in earlier chapters. The brevity of this chapter is based on our experience of what can be treated in a two-semester course along with other essential topics. This might be the last chapter in a two-semester sequence. The direct calculation of A and G from spectroscopic data is left to more advanced texts but a student should be aware that those calculations are possible, at least for small molecules in the gas phase. What we are calling "quantum thermodynamics" to emphasize the connection between the quantized energies and thermodynamic quantities would be called a "microcanonical ensemble" in statistical thermodynamics, A microcanonical ensemble assumes there are many (mole quantities) identical systems (molecules) that can be treated statistically, each with exactly the same energy, volume, and number of particles. There are whole texts [1,2] suitable for undergraduates and graduate courses in statistical thermodynamics but here we only give the "essential" ideas. We worked hard to derive the energy levels of several molecular systems and now we can apply them to thermodynamic ideas.

$$\text{Particle-in-a-box translation:} \quad E_n = \frac{n^2 h^2}{8mL^2}, \quad n = 1, 2, 3, \ldots$$

$$\text{Free rotation (diatomic):} \quad E_J = j(j+1)\frac{\hbar^2}{2\mu r^2}, \quad j = 0, 1, 2, 3, \ldots$$

$$\text{Harmonic vibration:} \quad E_n = (n + 1/2)h\nu, \quad n = 0, 1, 2, 3, \ldots$$

Now recall the simple but important example of averaging student grades as $\overline{G} = \frac{\sum_i n_i G_i}{\sum_i n_i}$. We went over Boltzmann averaging in an earlier chapter as one of the most profound principles in science so we want to use it for the quantized forms of energy. Here we use G_i to mean the individual "grade" value because Boltzmann averaging can be generalized using the letter "g" to mean the degeneracy of a given level. In Boltzmann averaging we can write the number of atoms or molecules in a given quantized level as $n_i = g_i e^{-\left(\frac{E_n}{kT}\right)}$, which corresponds to the number of students in the weighted average that have a certain grade. From now on in this chapter, g_i means the degeneracy of the ith energy level.

The next derivation is the link between Boltzmann's famous equation $S = k \ln \Omega$ and the macroscopic world. Think back to the statistics we did for the entropy of mixing derivation. The following steps have always been amazing to this author in the sense that we can relate tedious quantities to enormous numbers of tiny molecular species and get macroscopic thermodynamic quantities. Consider N *distinguishable* particles (N is on the order of Avagadro's mole number) where small groups of the total are in specific energy states that may be degenerate. Then we have

$\Omega = \dfrac{(N_{\text{tot}}!) \prod_i g_i^{N_i}}{(N_1!)(N_2!) \dots (N_n!)} = \dfrac{(N_{\text{tot}}!)}{\prod_i (N_i!)} \left\{ \prod_i \left(g_i^{N_i} \right) \right\}$. While this looks complicated, it uses the same reasoning we encountered with the egg crate and poker chips except that now we have to include the possible degeneracies as well $\prod_i (N_i!) \equiv (N_1!)(N_2!)(N_3!) \dots$ introduces the product symbol $\prod_i n_i$. We first consider the very large number of molecules as N as if they were distinguishable and then divide out the partial distinguishability due to the fact that N_i molecules are in a given energy level. In each of the i possible levels there are g_i degenerate levels so the possible ways for N_i to be in one of a possible g_i levels is $g_i^{N_i}$ but when there are i levels, we end up with the factor $\left\{ \prod_i \left(g_i^{N_i} \right) \right\}$ in the numerator. Well, that is complicated but we can think it through till we see that it is correct.

Next we consider that S is entropy, which always tends to a maximum, so "all we have to do" is take the derivative of S and maximize it! That is a real problem but we can just as well maximize $\ln(\Omega)$ noting the logarithmic dependence, $\Omega = e^{\left(\frac{S}{k} \right)}$. Fortunately, we know Stirling's approximation works very well for large numbers and we recall $\ln N! \cong N \ln N - N$. Next we come to the clever trick of using Lagrangian multipliers. This sometimes mysterious process is really very simple if we briefly consider its principle. We want to let S maximize and so maximize $\ln \Omega$. The simplest way to do this is to merely set $\Omega = \infty$ but maybe that is not what we really want to do. We want to maximize S but at the same time keep the number of particles constant and the total energy constant. Thus, what we really want to maximize is $f = \left[\ln \Omega - \alpha \sum_i N_i - \beta \sum_i (N_i \varepsilon_i) \right]$ where we maintain $N_{\text{tot}} = N_0 + N_1 + N_2 = N_3 + \cdots$ and also $E_{\text{tot}} = N_0 \varepsilon_0 + N_1 \varepsilon_1 + N_2 \varepsilon_2 + \cdots$. *That is, we want to maximize* $\ln \Omega$ *but remove/subtract from the variation anything that would change either the total number of particles or the total energy.* Perhaps you can think of this as specifically sorting through a list of all possible N_i occupation numbers but discarding any of those that might alter the total energy or the total number of particles, this is what the Lagrangian multipliers do. Some texts use a plus sign for the Lagrangian multipliers leaving the determination of the sign until the final step but we offer the idea that the Lagrangian multipliers should initially be thought of as minus to remove any changes that violate the desired constraints. Thanks to Stirling's approximation, the derivative process is greatly simplified. When we use the Lagrangian multipliers α and β, they allow the process of maximization to be carried out in a way that "removes" any variation that would change the total number of particles or the total energy, hence the minus signs. Now we set

$$\frac{\partial f(N_i)}{\partial N_i} = 0, \quad \Rightarrow \frac{\partial}{\partial N_i} \left[\ln \Omega - \alpha \sum_i N_i - \beta \sum_i (N_i \varepsilon_i) \right] = 0.$$

The first step is easy and we obtain $\dfrac{\partial \ln \Omega}{\partial N_i} - \alpha - \beta \varepsilon_i = 0$ where ε_i is the energy of the arbitrary ith level with an occupation of N_i. Now consider the first term alone $\dfrac{\partial \ln \Omega}{\partial N_i}$ using Stirling's approximation carefully and often!

$$\frac{\partial \ln \Omega}{\partial N_i} = \frac{\partial}{\partial N_i} \{ \ln (N_{\text{tot}}!) + N_1 \ln g_1 + N_2 \ln g_2 + \cdots - N_1 \ln N_1 + N_1 - N_2 \ln N_2 + N_2 + \cdots \}$$

and further $\dfrac{\partial \ln \Omega}{\partial N_i} = \dfrac{\partial}{\partial N_i} \left\{ N_1 \left(1 + \ln \dfrac{g_1}{N_1} \right) + N_2 \left(1 + \ln \dfrac{g_2}{N_2} \right) + \cdots + \ln (N_{\text{tot}}!) \right\}$;   there will only be one N_i term and N_{tot} is constant so $\dfrac{\partial \ln \Omega}{\partial N_i} = \dfrac{\partial}{\partial N_i} \left[N_i \left(1 + \ln \dfrac{g_i}{N_i} \right) \right] =$

$\left(1 + \ln \dfrac{g_i}{N_i} \right) + N_i \left[\dfrac{1}{\left(\frac{g_i}{N_i} \right)} \left(\dfrac{-g_i}{N_i^2} \right) \right] = \ln \left(\dfrac{g_i}{N_i} \right)$. Thus we find that what seemed so complicated turns out to be $\ln\left(\dfrac{g_i}{N_i}\right) - \alpha - \beta\varepsilon_i = 0$, $i = 1, 2, 3, \ldots$ or $\dfrac{g_i}{N_i} = e^{\alpha+\beta\varepsilon_i}$. We can rearrange this to obtain the population in the ith level as $N_i = g_i e^{-\alpha-\beta\varepsilon_i}$. It turns out that it is not necessary to evaluate α except to say it is there and it maintains the number of particles as a constant. We can eliminate α by taking the ratio of two populations as $\dfrac{N_i}{N_0} = \left(\dfrac{g_i}{g_0}\right)e^{-\beta(\varepsilon_i-\varepsilon_0)}$ where the usual convention is to relate the energy to the population in the lowest energy level. While $\varepsilon_0 = 0$ in most cases, the case of molecular (harmonic) vibrations will have $\varepsilon_0 = (h\nu/2)$. What is β? Recall our four thermodynamic equations, in particular where U is the internal energy E. We have for an open system with chemical potential μ, $dU = TdS - PdV + \mu dN$ (Note restricting N to N_{tot} makes the problem a closed system).

Thus relating U to the internal energy E we have $dS = \dfrac{dU}{T} + \dfrac{PdV}{T} - \dfrac{\mu dN}{T}$ or $dS = \dfrac{d(E - E_0)}{T} + \dfrac{PdV}{T} - \dfrac{\mu dN}{T}$ so that $\left(\dfrac{\partial S}{\partial E}\right)_{V,N} = \dfrac{1}{T}$.

Now recall $\Omega = \dfrac{(N_{tot}!)}{\prod_i (N_i!)}\left\{\prod_i \left(g_i^{N_i}\right)\right\}$ and $S = k \ln \Omega$ where we can use $N_i = g_i e^{-\alpha-\beta\varepsilon_i}$ and $\dfrac{g_i}{N_i} = e^{+\alpha+\beta\varepsilon_i}$ in $S = k\left\{N_{tot} \ln N_{tot} - N_{tot} - \sum_i N_i \ln N_i + \sum_i N_i\right\} + k\sum_i N_i \ln g_i$ to find that $S = k\left\{N_{tot} \ln N_{tot} + \sum_i N_i \ln\left(\dfrac{g_i}{N_i}\right)\right\} = k\left\{N_{tot} \ln N_{tot} + \sum_i N_i \ln\left(e^{+\alpha+\beta\varepsilon_i}\right)\right\}$. Then further $S = k\left\{N_{tot} \ln N_{tot} + \alpha \sum_i N_i + \beta \sum_i N_i\varepsilon_i\right\}$. Noting that $\sum_i N_i\varepsilon_i = E$ and the first two terms are constants, we find that $\left(\dfrac{\partial S}{\partial E}\right)_{V,N} = k\beta = \dfrac{1}{T} \Rightarrow \beta = \dfrac{1}{kT}$. Wow, we just derived the "Boltzmann principle"! Finally the working equation is

$$N_i = N_0 g_i e^{-\left(\frac{\varepsilon_i-\varepsilon_0}{kT}\right)}$$

Here we assume the ground state has $g_0 = 1$, which is usually the case, but the exponential term in $(\varepsilon_i - \varepsilon_0)$ will cause a problem with the harmonic oscillator $\varepsilon_0 = \dfrac{h\nu}{2}$ zero point energy. However, it turns out that you really cannot physically get rid of the zero point energy even at absolute zero Kelvin so in this text we will define our statistical energies as $U \equiv (E - E_0)$, relative to the lowest physically attainable energy.

(ENERGY) PARTITION FUNCTION

Now we come to a major shortcut in obtaining the average values of various forms of quantized energy. We define a quantity called the *partition function* $Q \equiv \sum_i g_i e^{-\left(\frac{\varepsilon_i-\varepsilon_0}{kT}\right)}$ where we remind ourselves that $(\varepsilon_i - \varepsilon_0)$ is the quantized energy above the lowest state ε_0 of the quantized system with energy levels. This will only be important for the harmonic oscillator. We stand on the shoulders of intellectual giants after they used much scratch paper but we can see from calculus that there is a relationship between the partition function and the way we do Boltzmann averaging.

Suppose we did not know $\beta = \dfrac{1}{kT}$, then $Q = \sum_{i=0}^{\infty} g_i e^{-\beta\varepsilon_i}$ and $\dfrac{dQ}{d\beta} = -\sum_{i=0}^{\infty} \varepsilon_i g_i e^{-\beta\varepsilon_i}$, which resembles the numerator of our Boltzmann average formula. If we divide that by Q we will have

$\dfrac{dQ/d\beta}{Q} = \dfrac{-\sum_{i=0}^{\infty} \varepsilon_i g_i e^{-\beta\varepsilon_i}}{\sum_{i=0}^{\infty} g_i e^{-\beta\varepsilon_i}} = -\langle E \rangle$ or $-\dfrac{d \ln Q}{d\beta} = \langle E \rangle$. Since $\beta = \dfrac{1}{kT} \Rightarrow d\beta = \dfrac{-1}{kT^2}dT$, we find

$$\langle E \rangle = -(-1)kT^2 \frac{d \ln Q}{dT} = +kT^2 \frac{d \ln Q}{dT}.$$

So all we have to do to get an average energy is set up the formula for Q, take the derivative of the ln Q with respect to T and multiply by kT^2!

AVERAGE TRANSLATION ENERGY IN ONE DIMENSION

We know the particle-in-a-box energy levels are very close together when the dimension "L" of the box is large compared to the size of the particle. Thus we use $\sum_{n=0}^{\infty} \rightarrow \int_0^{\infty} (\ )dn$ with calculus integration. Thus, we form $Q_{\text{trans}} = \int_0^{\infty} e^{-\left(\frac{n^2 h^2}{8mL^2 kT}\right)} dn = \frac{\sqrt{\pi}}{2\left(\sqrt{\frac{h^2}{8mL^2 kT}}\right)} = \sqrt{\frac{2\pi mkT}{h^2}} L$ where we have

used $\int_0^{\infty} x^{2p} e^{-a^2 x^2} dx = \frac{1 \cdot 3 \cdot 5 \cdots (2p-1)\sqrt{\pi}}{2^{p+1} a^{2p+1}} \Rightarrow \int_0^{\infty} e^{-a^2 x^2} dx = \frac{1}{2}\sqrt{\frac{\pi}{a}}$, $p = 0$. So then we use our

Q formula as $\langle \varepsilon_{\text{trans}} \rangle = kT^2 \frac{d \ln Q}{dT} = kT^2 \frac{dQ/dT}{Q} = kT^2 \frac{\left(\frac{1}{2\sqrt{T}}\right)\sqrt{\frac{2\pi mk}{h^2}} L}{\sqrt{\frac{2\pi mkT}{h^2}} L} = kT^2\left(\frac{1}{2T}\right) = \frac{kT}{2}.$ An

observant student will notice that we integrated over the variable n as a continuous variable while we know it came from the quantized formula where n was definitely an integer. This approximation works here because the energy levels are so close together that the energy is approaching the limit of a continuous function even though we know its quantization would be more noticeable in a length L, which is small compared to the size of the particle. The next step is to extrapolate this to three dimensions and for Avagadro's number of particles where $Q_{\text{trans}}(x, y, z) = \left(\frac{2\pi mkT}{h^2}\right)^{3/2} L_x L_y L_z = \left(\frac{2\pi mkT}{h^2}\right)^{3/2} V$ so that in three dimensions we obtain $\langle E_{\text{trans}}(x, y, z) \rangle = \frac{3kT}{2}(N_{\text{Avagadro}}) = \frac{3RT}{2}$ as before from the Boltzmann averaging process, except that this time we used quantum mechanics and simple statistics! It should be obvious that it is also easy to calculate the constant volume heat capacity as $C_V = \left(\frac{\partial U}{\partial T}\right)_V = \left(\frac{\partial E}{\partial T}\right)_V = \frac{3R}{2}.$

AVERAGE ROTATIONAL ENERGY OF A DIATOMIC MOLECULE

Since we have a large intellectual investment in the quantized rigid rotor, let us apply our quantized energy formula with our new partition function formula using Q. Perhaps we should ask, "What does Q mean?" It is a mathematical description of how the energy is "partitioned" among the quantized levels allowing for the Boltzmann factor. Once again the rotational levels are close together, although not as close as those for the translational case. Even so it is still a good approximation to use $\sum_{j=0}^{\infty} \rightarrow \int_0^{\infty} (\ )dj$ so $Q_{\text{rot}} = \int_0^{\infty} (2j+1)e^{-\left(\frac{j(j+1)h^2}{8\pi^2 IkT}\right)} dj$. Let $x = j(j+1)$ so that

$dx = (2j+1)dj$, then we have an easy integral of the form $\int_0^{\infty} e^{-ax} dx = \frac{e^{-ax}|_0^{\infty}}{-a} = \frac{0-1}{-a} = \frac{1}{a}$ so

$Q_{\text{rot}} = \int_0^{\infty} (2j+1)e^{-\left(\frac{j(j+1)h^2}{8\pi^2 IkT}\right)} dj = \int_0^{\infty} e^{-\left(\frac{h^2}{8\pi^2 IkT}\right)x} dx = \frac{e^{-\left(\frac{h^2}{8\pi^2 IkT}\right)x}\Big|_0^{\infty}}{\left(\frac{-h^2}{8\pi^2 IkT}\right)} = \left(\frac{8\pi^2 IkT}{h^2}\right).$ Then we use

the formula for energy as $\langle \varepsilon_{\text{rot}} \rangle = kT^2 \frac{d \ln Q}{dT} = kT^2 \left[\frac{\frac{8\pi^2 Ik}{h^2}}{\frac{8\pi^2 IkT}{h^2}}\right] = kT$ or with $R = kN_{\text{Avogadro}}$,

$E = RT/mole$. Going back to the law of equipartition of energy we see that a diatomic molecule can only rotate in two ways in (θ, φ) space to produce $E_{rot} = 2\left(\dfrac{RT}{2}\right) = RT$. Interestingly, there is no way to mark a diatomic molecule on one end with a dot of green paint to detect rotation about the bond axis, so physically, the rule of "what you see is what you get" (WYSIWYG) really works since there is no rotational degree of freedom about the bond line. Without doing a proof we can extrapolate the rule of $E = \dfrac{RT}{2}$ for each degree of freedom to three dimensions for a nonlinear polyatomic molecule to $E = \dfrac{3RT}{2}$ for rotation. Once again the heat capacity is easy to find as $\left(\dfrac{\partial U}{\partial T}\right)_V = \left(\dfrac{\partial E}{\partial T}\right)_V = R = C_V$ for diatomic rotation, $\dfrac{3R}{2}$ for polyatomic molecules.

AVERAGE VIBRATIONAL ENERGY

Next we come to the case of vibrational energy. At this point the size of the vibrational quanta is too large to use the continuous integration method and we must use a discrete summation. In addition, we remind ourselves that we are counting the energy starting above $\varepsilon_0 = \dfrac{h\nu}{2}$. The discrete summation of the terms in Q is essentially the same as Planck used for the treatment of blackbody radiation. $Q_{vib} = \sum_{i=0}^{\infty} e^{-\left(\frac{nh\nu}{kT}\right)} = 1 + e^{-\left(\frac{h\nu}{kT}\right)} + e^{-\left(\frac{2h\nu}{kT}\right)} + e^{-\left(\frac{3h\nu}{kT}\right)} + \cdots$ Let $y = e^{-\left(\frac{h\nu}{kT}\right)}$

so that $\qquad Q = 1 + y + y^2 + y^3 + y^4 + \cdots$, where $y < 1 \Rightarrow \lim_{n \to \infty} y^n = 0$

Then $\qquad \underline{(-1)[yQ = y + y^2 + y^3 + y^4 + \cdots]}$

Since $y < 1$, $\quad (Q - yQ) = Q(1 - y) = 1$ and we find that $Q_{vib} = \dfrac{1}{(1 - y)} = \dfrac{1}{\left(1 - e^{-\left(\frac{h\nu}{kT}\right)}\right)}$.

Then $\quad \langle \varepsilon - \varepsilon_0 \rangle = kT^2 \dfrac{d \ln Q}{dt} = kT^2 \left[\dfrac{-\left(1 - e^{-\frac{h\nu}{kT}}\right)^{-2}\left(-e^{-\frac{h\nu}{kT}}\right)\left(\frac{+h\nu}{kT^2}\right)}{\left(1 - e^{-\frac{h\nu}{kT}}\right)^{-1}}\right] = \left[\dfrac{h\nu e^{-\frac{h\nu}{kT}}}{\left(1 - e^{-\frac{h\nu}{kT}}\right)}\right] = \dfrac{h\nu}{\left(e^{\frac{h\nu}{kT}} - 1\right)}$.

Make note that this is the average energy above the zero-point energy. In this case, C_V is not easy to calculate. Define $x \equiv \left(\dfrac{h\nu}{kT}\right) = \left(\dfrac{hc\bar{\nu}}{kT}\right)$ and assume one mole of oscillators so that $(N_{Ava})k = R$.

Then $\quad U = \langle E - E_0 \rangle = (N_{Ava})\left[\dfrac{(h\nu)}{(e^x - 1)}\right]\left(\dfrac{kT}{kT}\right) = \left[\dfrac{(N_{Ava}k)T\left(\frac{h\nu}{kT}\right)}{(e^x - 1)}\right] = \dfrac{RTx}{(e^x - 1)}$ so we can

calculate $\left.\dfrac{\partial U}{\partial T}\right)_V$ as $C_V = \left.\dfrac{\partial}{\partial T}\right)_V\left[\dfrac{RTx}{(e^x - 1)}\right] = \dfrac{Rx}{(e^x - 1)} + \dfrac{RT\left(\frac{dx}{dT}\right)}{(e^x - 1)} + \dfrac{RTx(-1)e^x\left(\frac{dx}{dT}\right)}{(e^x - 1)^2}$ and

$\left(\dfrac{dx}{dT}\right) = -\dfrac{h\nu}{kT^2} = -\dfrac{x}{T}$.

Then $C_V = \left.\dfrac{\partial}{\partial T}\right)_V\left[\dfrac{RTx}{(e^x - 1)}\right] = \dfrac{Rx}{(e^x - 1)} + \dfrac{RT\left(\frac{-x}{T}\right)}{(e^x - 1)} + \dfrac{RTx(-1)e^x\left(\frac{-x}{T}\right)}{(e^x - 1)^2} = \dfrac{Rx^2e^x}{(e^x - 1)^2}$. According

to the law of equipartition of energy there should be $\left(\dfrac{2}{2}\right)RT$ energy per vibration and the heat

capacity should be $\left(\dfrac{2}{2}\right)R$ but instead we find $C_V = \dfrac{Rx^2e^x}{(e^x - 1)^2}$. Let us take a look at the vibrational heat capacity, C_V, for the HCl molecule, which we found has a fundamental frequency of

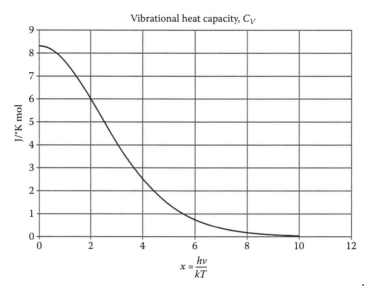

FIGURE 15.1 The vibrational heat capacity of a quantized oscillator as a function of $x = \dfrac{h\nu}{kT}$. Note that the high temperature limit will occur at $x = 0$.

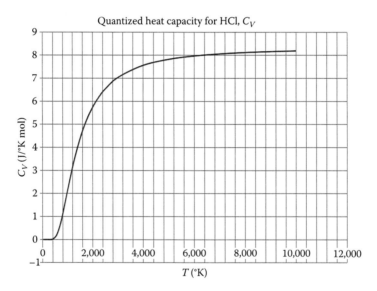

FIGURE 15.2 A graph of the heat capacity, C_V, for HCl using a fundamental of 2990.2 cm^{-1}.

2990.2 cm^{-1} (Figures 15.1 and 15.2). Here we first plot the heat capacity for any quantized harmonic oscillator against x. This graph is difficult to understand because T is in the denominator of x, but we see that at very high temperatures, $\lim\limits_{T \to \infty} \left(\dfrac{h\nu}{kT}\right) = 0$ so the left edge of the graph is the high-temperature limit, which is seen to be 8.314 J/K mol $= R$. So the quantized formula gives the expected answer only at very high temperature. On the other hand, we see that for very low temperature, large x, the heat capacity is asymptotic to zero, meaning that a quantized oscillator can be "frozen" down to only the zero-point energy $h\nu/2$.

If we plot the heat capacity of HCl, which we found to have a fundamental frequency of 2990.2 cm^{-1}, we can use $x = \left(\dfrac{hc\bar{\nu}}{kT}\right)$ to plot the heat capacity against temperature, T.

The vibrational heat capacity for the HCl molecule does taper off asymptotically to 8.314 J/°K mol, but only at quite high temperatures. Near room temperature the vibrational heat capacity is very small.

Let us evaluate C_V for HCl at 25°C.

$$x = \left(\frac{hc\bar{v}}{kT}\right) = \frac{(6.62606896 \times 10^{-27} \text{ erg s})(2.99792458 \times 10^{10} \text{ cm/s})(2990.2 \text{ cm}^{-1})}{(1.3806504 \times 10^{-16} \text{ erg/°})(298.15 \text{ °K})} \quad \text{or}$$

$x = 14.42973498$ (unitless). Then we can insert the numbers into the formula to find

$$C_V = \frac{(8.314 \text{J/°K mol})x^2 e^x}{(e^x - 1)^2} = \frac{(8.314 \text{J/°K mol})(14.42973498)^2 \, e^{14.42973498}}{(e^{14.42973498} - 1)^2} \cong 9.366405 \times 10^{-4} \text{ J/°K}$$

mol. This is negligible at room temperature for this molecule due to the strong H–Cl bond.

As an atmospheric contaminant, HCl cannot hold much heat in vibrations and so should not be considered a green house gas. Conversely, molecules that have many vibrational modes or some modes that have low vibrational frequencies can get higher up on the C_V curve at room temperature and those molecules are the green house gasses. For instance CO_2 has four $(3N - 5)$ vibrational modes of which two are low-frequency bending modes, so it is one of the main green house gases. Methane is a gas molecule that has $15 - 6 = 9$ vibrational modes, although they are not particularly low frequencies, so CH_4 is a green house gas by virtue of having many ways to absorb and hold heat in the form of vibrational energy.

HIGH-TEMPERATURE LIMIT FOR VIBRATIONAL HEAT CAPACITY

We can see the vibrational heat capacity approaches R asymptotically but the example for HCl shows that at room temperature, the actual heat capacity can be quite low. What does the formula tell us? Consider again that $\lim_{T \to \infty} x = \lim_{T \to \infty} \left(\frac{hv}{kT}\right) \to 0$ and apply L'Hopital's rule *twice*!

$$\lim_{T \to \infty} C_V = \lim_{x \to 0} \frac{Rx^2 e^x}{(e^x - 1)^2} = \lim_{x \to 0} R\left[\frac{2xe^x + x^2 e^x}{2(e^x - 1)e^x}\right] = \lim_{x \to 0} R\left[\frac{2 + 2x}{2e^x}\right] = R.$$

Thus we see that mathematically the vibrational heat capacity does tend toward R at high temperatures but if the bond is very stiff the value can be considerably smaller than R at room temperature.

HEAT CAPACITY OF A POLYATOMIC SPECIES: WATER

Next, let us evaluate the heat capacity of a polyatomic molecule, H_2O. From the $3N - 6$ rule for a nonlinear molecule there are three normal modes of vibration (Figure 15.3). However, a gas molecule will be rapidly rotating and a "movie" of one molecule would show a blur to the

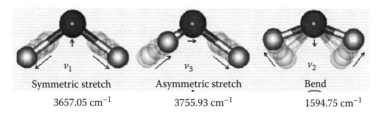

v_1 v_3 v_2

Symmetric stretch   Asymmetric stretch   Bend

3657.05 cm^{-1} 3755.93 cm^{-1} 1594.75 cm^{-1}

FIGURE 15.3 The three $(3N - 6 = 3)$ normal mode motions of H_2O with the frequencies for the $^{16}_8 O$ isotopic species. By permission from Prof. Martin Chaplin of the London South Bank University as shown at http://www1.lsbu.ac.uk/water/vibrat.html which includes real time animation of the motion of the atoms in these modes. Interested students are encouraged to visit Prof. Chaplin's site and gain appreciation of the dynamic action of the normal mode motions.

human eye. The normal modes are those linearly independent motions which are the fundamental components of the vibrations. Prof. Martin Chaplin has set up an Internet site where the normal modes are simulated. Since this chapter is about quantum thermodynamics, let us calculate the heat capacity of steam at 500°K (226.85°C) so we can be sure we are treating a gas. We will use Prof. Chaplin's six significant figure vibrational frequencies. Then we can calculate the C_V heat capacity for a mole of water molecules *in the gas phase* at 500°K.

$$C_V = \left(\frac{3R}{2}\right)_{trans} + \left(\frac{3R}{2}\right)_{rot} + \frac{Rx_1^2 e^{x_1}}{(e^{x_1}-1)^2} + \frac{Rx_2^2 e^{x_2}}{(e^{x_2}-1)^2} + \frac{Rx_3^2 e^{x_3}}{(e^{x_3}-1)^2}.$$

First we calculate x_1, x_2 and x_3.

$$x_1 = \left(\frac{hc\bar{v}}{kT}\right) = \frac{(6.62606896 \times 10^{-27}\ \text{erg s})(2.99792458 \times 10^{10}\ \text{cm/s})(3657.05\ \text{cm}^{-1})}{(1.3806504 \times 10^{-16}\ \text{erg/°})(500°K)}$$

or

$$x_1 = \left(\frac{hc\bar{v}}{kT}\right) = (2.87755032 \times 10^{-3}\ \text{cm})(3657.05\ \text{cm}^{-1}) = 10.52334540,$$

$$x_2 = (2.87755032 \times 10^{-3})\,(1594.75) = 4.588973373,$$

and

$$x_3 = (2.87755032 \times 10^{-3})\,(3755.93) = 10.80787757.$$

Then we can calculate the C_V terms.

$$C_V(x_1) = \frac{Rx_1^2 e^{x_1}}{(e^{x_1}-1)^2} = (8.314472\ \text{J/mol °K}) \left[\frac{10.52334540}{e^{10.52334540}-1}\right]^2 e^{10.52334540} = 0.0244770504\ \text{J/mol °K}.$$

$$C_V(x_2) = \frac{Rx_2^2 e^{x_2}}{(e^{x_2}-1)^2} = (8.314472\ \text{J/mol °K}) \left[\frac{4.588973373}{e^{4.588973373}-1}\right]^2 e^{4.588973373} = 1.816238266\ \text{J/mol °K}.$$

$$C_V(x_3) = \frac{Rx_3^2 e^{x_3}}{(e^{x_3}-1)^2} = (8.314472\ \text{J/mol °K}) \left[\frac{10.80787757}{e^{10.80787757}-1}\right]^2 e^{10.80787757} = 0.019657645\ \text{J/mol °K}.$$

We will round to five sig. figs. at the end since the experimental data is only given to five sig. fig. Then we add R to get C_P using $C_P = C_V + R$, so with the translational and rotational terms we obtain

$$C_P = C_V + R = \left(\frac{3R}{2}\right)_{trans} + \left(\frac{3R}{2}\right)_{rot} + \frac{Rx_1^2 e^{x_1}}{(e^{x_1}-1)^2} + \frac{Rx_2^2 e^{x_2}}{(e^{x_2}-1)^2} + \frac{Rx_3^2 e^{x_3}}{(e^{x_3}-1)^2} + R.$$ Numerically we find

$$C_P = [4.0(8.314472) + 0.0244770504 + 1.816238266 + 0.019657645]\ \text{J/mol °K}$$

$$C_P = 35.11826096\ \text{J/mol °K} \cong 8.393\ \text{cal/°K mol using "quantum thermodynamics."}$$

The 90th *CRC Handbook* gives a value [3] directly for C_P of steam at 1 bar and 500°K as 35.259 J/mol °K.

Quantum thermodynamics: C_P (1 bar, 500°K) = 35.118 J/mol °K
Experimental (CRC, 90th Edn.): C_P (1 bar, 500°K) = 35.259 J/mol °K

These values are sufficiently close together (within 0.4%) to give us confidence in the result from the statistical thermodynamic formula. Today there are computer programs that can routinely compute such thermodynamic quantities using the quantum formulas for molecules up to the size of benzene but for larger molecules polynomial fits to heat capacity data are still very useful. Another recent innovation is the ability to calculate theoretical vibrational frequencies from quantum chemistry programs for large polyatomic molecules. These calculated frequencies are often too high by 10% or so but their values can be corrected and used to obtain qualitative thermodynamic quantities. This offers future improvement using more accurate vibrational frequencies within known statistical thermodynamic formulas.

COMBINING PARTITION FUNCTIONS

We saw in Chapter 13 that the energy levels of the vib-rotor result from both vibrational and rotational levels. There we ignored the fact that the gas molecules also have translational energies and that within the structure of the molecule there are electronic and even nuclear energy levels. It is possible to combine the separate partition functions into one total partition function. We can illustrate this by considering only the combined levels of the vib-rotor in a single partition function

$$Q_{\text{vib-rot}} = \sum_{n=0}^{\infty} \sum_{j=0}^{\infty} (2j+1)\, e^{-\left[nh(v-v_0)+j(j+1)\hbar^2/2I\right]/kT}$$

$$= \left(\sum_{n=0}^{\infty} e^{-nh(v-v_0)/kT} \right) \left(\sum_{j=0}^{\infty} (2j+1)\, e^{-j(j+1)\hbar^2/2IkT} \right) = Q_{\text{vib}} Q_{\text{rot}}.$$

Thus we see that we can easily write a total partition function as a product of each separate quantized subsystem of energy levels as

$$Q_{\text{tot}} = Q_{\text{nuc}} Q_{\text{elec}} Q_{\text{vib}} Q_{\text{rot}} Q_{\text{trans}} \cong 1 \cdot 1 \cdot Q_{\text{vib}} Q_{\text{rot}} Q_{\text{trans}}.$$

For most cases in the range of macroscopic interest, the ambient energy is too small to consider higher energy levels of electronic or nuclear states, although in the case of photochemical reactions Q_{elec} might have to be specified to include low-lying electronic excited states. That is also the case for molecules such as NO and O_2, which have unpaired electrons in the ground state that are easily excited. There might also be some cases involving nuclear spin systems that require a specific summation for Q_{nuc}. However, for most cases, we can consider $Q_{\text{nuc}} Q_{\text{elec}} = 1$. The product form is also convenient when using

$$\ln Q_{\text{tot}} = \ln Q_{\text{nuc}} + \ln Q_{\text{elec}} + \ln Q_{\text{vib}} + \ln Q_{\text{rot}} + \ln Q_{\text{trans}} \quad \text{and} \quad \ln(1) = 0.$$

Now that we see that we can combine partition functions for all the quantized energy systems into a total partition function, we can think of other ways to use the quantized energy formulas. There is a curious history for this approach. We can see above that Q_{vib} is an important part of the total partition function and yet for many years low-resolution infrared spectra blurred many of the $3N-6$ vibrational modes of molecules typically larger than benzene. Thus the equations for "quantum thermodynamics" were known before 1940 but could only be applied to cases of small molecules in the gas phase using experimental vibrational frequencies. Since about 1985, quantum chemistry programs have included the calculation of vibrational frequencies with some correction factors that now make it possible to write down the full partition function by including theoretical vibrational frequencies. Today, it is routine for large quantum chemistry programs to end a calculation of vibrational frequencies with the full set of thermodynamic quantities.

STATISTICAL FORMULAS FOR OTHER THERMODYNAMIC FUNCTIONS

Now that we can calculate $\langle E \rangle = (U - U_0)$ using quantum statistics, it is a simple matter to use $(H - E_0) = (U - U_0) + RT$ for an ideal gases or for real gases at least up to about 10 atmospheres and perhaps 1000°K. As when we discovered S from the Carnot Cycle, the key to extending these thermodynamic concepts is to find the entropy expression for $S(T)$. Although other texts show explicit formulas for the Helmholtz free energy, A, and the Gibbs free energy, G, it is sufficient here to recall that $A = U - TS$ and $G = H - TS$. In principle, once we can calculate a value for $(U - U_0)$ and $S(T)$, we can find A and/or G using a calculator. We leave analytical calculations of A and G to specialized texts with just a comment that $(U - U_0)$ and $S(T)$ are the key quantities for the other quantities linking quantum statistics and thermodynamics.

There is a small trick at this point, which eliminates pages of discussion. Barrow [4] shows this method and expands it over three chapters. We form the ratio of the population in one given quantum level, N_i, to the total number of molecules in the sample to obtain a useful formula.

$$\left(\frac{N_i}{N_{tot}}\right) = \left(\frac{g_i e^{-(\varepsilon_i - \varepsilon_0)/kT}}{\sum_i^{\infty} g_i e^{-(\varepsilon_i - \varepsilon_0)/kT}}\right) = \frac{g_i e^{-(\varepsilon_i - \varepsilon_0)/kT}}{Q} \quad \text{which we rearrange to} \quad \frac{N_i}{g_i} = \left(\frac{N_{tot}}{Q}\right) e^{-(\varepsilon_i - \varepsilon_0)/kT}.$$

The main trick is to invert this equation as $\dfrac{g_i}{N_i} = \dfrac{Q}{N_{tot}} e^{+(\varepsilon_i - \varepsilon_0)/kT}$ and take the natural logarithm of the whole equation, which produces $\ln\left(\dfrac{g_i}{N_i}\right) = \dfrac{(\varepsilon_i - \varepsilon_0)}{kT} + \ln\left(\dfrac{Q}{N_{tot}}\right).$

STATISTICAL FORMULA FOR S(T)

Now consider Boltzmann's definition of entropy as $S = k \ln \Omega$ and then enumerate Ω as before but for *indistinguishable* particles, which removes the factor of $N_{tot}!$ from Ω. Then

$$\Omega = \frac{g_1^{N_1} g_2^{N_2} g_3^{N_3} \cdots}{N_1! N_2! N_3! \cdots} = \frac{\prod_i g_i^{N_i}}{\prod_i N_i!} \quad \text{so that} \quad \ln \Omega = \sum_i N_i \ln g_i - \sum_i \ln N_i! = \sum_i N_i \ln g_i - \sum_i (N_i \ln N_i - N_i)$$

which leads to $\ln \Omega = \sum_i N_i \left(1 + \ln \dfrac{g_i}{N_i}\right)$ and then to insertion into $S = k \ln \Omega$. We find that

$$S = k\left\{\sum_i N_i\left[1 + \left(\frac{\varepsilon_i - \varepsilon_0}{kT}\right) + \ln\left(\frac{Q}{N_A}\right)\right]\right\} = \left(\frac{1}{T}\right)\left[\sum_i N_i(\varepsilon_i - \varepsilon_0)\right] + kN_A\left(1 + \ln\frac{Q}{N_A}\right) \quad \text{where}$$

$\sum_i N_i = N_A$ is Avogadro's number for 1 mol. This shortcut step depends on the fact that

$$k\sum_i N_i\left[\ln\left(\frac{Q}{N_A}\right)\right] = k\left[\ln\left(\frac{Q}{N_A}\right)\right]\left(\sum_i N_i\right) = k\left[\ln\left(\frac{Q}{N_A}\right)\right]N_A = R\ln\left(\frac{Q}{N_A}\right). \quad \text{Thus, for 1 mol we}$$

find the formula

$$S = \left(\frac{U - U_0}{T}\right) + R\ln\left(\frac{Q}{N_A}\right) + R$$

since $(U - U_0) = \langle E \rangle$ is the internal energy for one mole and $kN_A = R$. We can also represent this in another way using the statistical formula for $\langle E \rangle = (U - U_0)$ as

$$S = RT\left(\frac{d \ln Q}{dT}\right) + R\ln\left(\frac{Q}{N_A}\right) + R.$$

This can be expanded to $S = RT \left(\dfrac{d \ln Q_{\text{trans}} Q_{\text{rot}} Q_{\text{vib}}}{dT} \right) + R \ln \left(\dfrac{Q_{\text{trans}} Q_{\text{rot}} Q_{\text{vib}}}{N_A} \right) + R$ and usually the last term of R is calculated along with N_A in the translational entropy for a gas so that we can itemize the total as the famous Sackur–Tetrode Equation

$$S_{\text{tot}} = S_{\text{trans}} + S_{\text{rot}} + S_{\text{vib}}.$$

$$S_{\text{trans}} = RT \left(\frac{d \ln Q_{\text{trans}}}{dT} \right) + R \ln \left(\frac{Q_{\text{trans}}}{N_A} \right) + R \quad \text{or} \quad S_{\text{trans}} = \frac{(U - U_0)_{\text{trans}}}{T} + R \ln \left(\frac{Q_{\text{trans}}}{N_A} \right) + R.$$

$$S_{\text{rot}} = RT \left(\frac{d \ln Q_{\text{rot}}}{dT} \right) + R \ln (Q_{\text{rot}}) \quad \text{or} \quad S_{\text{rot}} = \frac{(U - U_0)_{\text{rot}}}{T} + R \ln (Q_{\text{rot}}),$$

$$S_{\text{vib}} = RT \left(\frac{d \ln Q_{\text{vib}}}{dT} \right) + R \ln (Q_{\text{vib}}) \quad \text{or} \quad S_{\text{vib}} = \frac{(U - U_0)_{\text{vib}}}{T} + R \ln (Q_{\text{vib}}).$$

SAKUR–TETRODE FORMULA FOR ABSOLUTE ENTROPY OF A GAS

The Sackur–Tetrode equation was derived independently by Otto Sackur (1880–1914), a German physical chemist, and Hugo Martin Tetrode (1895–1931), a Dutch physicist, in 1912. While it is possible to calculate the absolute entropy of molecules including polyatomic gases with their multiple vibrations, we will only give a brief illustration for diatomic CO gas that uses the energy formulas we have previously derived. Consider $^{12}C \equiv {}^{16}O$ gas at 298.15°K. First we consider the translational entropy as

$$S_{\text{trans}}(298.15°\text{K}) = \frac{3R}{2} + R \ln \left[\left(\frac{2\pi mkT}{h^2} \right)^{\frac{3}{2}} \frac{V}{N_A} \right] + R = \frac{5R}{2} + R \ln \left[\left(\frac{2\pi mkT}{h^2} \right)^{\frac{3}{2}} \frac{V}{N_A} \right].$$

At 25°C and 1 bar pressure $V = \left(\dfrac{nRT}{P} \right) = \dfrac{1\,\text{mol}(0.08314\,L\,\text{bar}/\text{K}°\text{mol})(298.15\,\text{K}°)}{(1.0\,\text{bar})} = 24.789L =$ 24789 cm³. Here we have $m_{CO} = \dfrac{(27.9994\,\text{g/mol})}{(6.02214179 \times 10^{23}/\text{mol})} = 4.649408961 \times 10^{-23}$ g

$$\left[\frac{2\pi(4.649408961 \times 10^{-23}\,\text{g})(1.3806504 \times 10^{-16}\,\text{erg}/°\text{K})(298.15\,°\text{K})}{(6.62606896 \times 10^{-27}\,\text{erg s})^2} \right]^{\frac{3}{2}} = 1.433429624 \times 10^{26}/\text{cm}^3$$

$$S_{\text{trans}} = (2.5)(8.314472) + (8.314472) \ln \left[\frac{(1.433429624 \times 10^{26}/\text{cm}^3)(24789\,\text{cm}^3/\text{mol})}{6.02214179 \times 10^{23}/\text{mol}} \right]$$

$$S_{\text{trans}}(298.15\,°\text{K}) = 20.78618 + 129.6270875 = 150.4132675 \,(\text{J/mol}\,°\text{K})$$

Next the rotational entropy is $S_{\text{rot}}(298.15°\text{K}) = R + R \ln \left[\dfrac{8\pi^2 IkT}{h^2} \right] = R \left[1 + \ln \left(\dfrac{2IkT}{\hbar^2} \right) \right]$

$$\mu = \frac{(12.000000\,\text{g/mol})(15.99491462\,\text{g/mol})}{(27.99491462\,\text{g/mol})(6.02214179 \times 10^{23}/\text{mol})} = 1.138500035 \times 10^{-23}\,\text{g}$$

$$r = \sqrt{\frac{6.62606896 \times 10^{-27}\,\text{g} \cdot \text{cm}^2/\text{s}}{(1.157978348 \times 10^{11}/\text{s})4\pi^2(1.138500031 \times 10^{-23}\,\text{g})}} = 1.128317603 \times 10^{-8}\,\text{cm}$$

$$\cong 1.1283\,\text{Å}$$

$$I = \mu r^2 = (1.138500035 \times 10^{-23} \text{ g})(1.128317603 \times 10^{-8} \text{ cm})^2 = 1.449425093 \times 10^{-39} \text{ g cm}^2$$

$$\left[\frac{2(1.449426113 \times 10^{-39} \text{ g cm}^2)(1.3806504 \times 10^{-16} \text{ erg/}^\circ\text{K})(298.15^\circ\text{K})}{(1.054571628 \times 10^{-27} \text{ erg s})^2} \right] = 107.2981273$$

$$S_{\text{rot}}(298.15^\circ\text{K}) = 8.314472[1.0 + \ln(107.2981273)] = 47.18971037 \text{ (J/mol }^\circ\text{K)}$$

and then the vibrational entropy is

$$S_{\text{vib}}(298.15\,^\circ\text{K}) = \frac{Rx}{(e^x - 1)} + R \ln \left[\frac{1}{(1 - e^{-x})} \right] = R \left[\frac{x}{(e^x - 1)} - \ln(1 - e^{-x}) \right]$$

$\bar{\nu} = 2169.81/\text{cm}$ for CO (see Chapter 12) and so

$$x = \frac{hc\bar{\nu}}{kT} = \frac{(6.62606896 \times 10^{-27} \text{ erg s})(2.99792458 \times 10^{10} \text{ cm/s})(2169.81/\text{cm})}{(1.3806504 \times 10^{-16} \text{ erg/}^\circ\text{K})(298.15\,^\circ\text{K})} = 10.47079903$$

$$S_{\text{vib}}(298.15^\circ\text{K}) = (8.314472) \left[\left(\frac{10.47079903}{35269.38821} \right) - \ln(0.999971647) \right]$$

$$S_{\text{vib}}(298.15\,^\circ\text{K}) = (8.314472)[2.968806538 \times 10^{-4} + 2.835279694 \times 10^{-5}]$$

$$S_{\text{vib}}(298.15\,^\circ\text{K}) = 2.70414442 \times 10^{-3} \text{(J/mol }^\circ\text{K)}.$$

Thus we have the total entropy as

$$S_{\text{tot}}(298.15^\circ\text{K}) = S_{\text{trans}} + S_{\text{rot}} + S_{\text{vib}} = 197.605682 \cong 197.61 \text{ (J/mol }^\circ\text{K)}.$$

The entropy value given in the 90th Edn. of the *CRC Handbook* for the standard state of 1 bar pressure and 298.15°K is $S(298.15^\circ\text{K}) = 197.7$ J/mol °K [5] so the theoretical calculations are within $(0.1/197.7) \times 100 = 0.05\%$ of the standard value. We have used the latest values of the constants from the 90th Edn. of the *CRC Handbook* [5] to be as accurate as is possible with just a 10 place calculator and then rounded the answer to 5 places at the end since the temperature is only given to five significant figures.

There is reason for some discussion of this entropy calculation. First we see that most of the entropy comes from the translation component, which is understandable considering the "random" stipulation in the Boltzmann kinetic theory of gases and the actual motion of gas molecules. Next we see that the rotational entropy is surprisingly large for such a simple molecule. Of course there is no potential hindering rotation and we expect that rather high l-values occur in gas rotation so there are plenty of low-lying energy states to populate in a Boltzmann distribution and the energy levels are close together. We previously saw the actual rotational spectrum of CO in Chapter 12 so we know many rotational states are populated even at only 25°C. Next we come to the entropy due to populating vibrational states but we have previously noted that the CO molecule has a strong bond with a high vibrational frequency. Thus the spacing between vibrational levels is larger than for most molecules and we see that there is little Boltzmann population of the upper vibrational levels. This is a good example of what we have said before in that on planet Earth the temperature is such that most molecules are in their $n = 0$ vibrational level. This is certainly true here for CO and we see that there is almost no contribution to the entropy from vibration at 25°C. Other molecules with lower vibrational frequencies will have larger entropy contributions from vibration.

SUMMARY

The mathematical foundation for statistical thermodynamics was set by Ludwig Boltzmann in the late 1800s and has been developed to a useful technology for small molecules. Other aspects of thermodynamics still need use of empirically fitted polynomials, but as computers continue to improve, statistical thermodynamics offers a detailed treatment to connect quantized energy levels of molecules to macroscopic thermal properties. There are extensive treatises on statistical thermodynamics but here we have tried to give just the basic "essentials" as a way to use some of the quantized energy-level formulas we have obtained from quantum mechanics.

1. Boltzmann's $S = k \ln \Omega$ can be used by maximizing S subject to maintaining constant energy and the number of particles in order to "derive" the $e^{-\frac{\varepsilon}{kT}}$ part of the Boltzmann probability with the help of Stirling's approximation $\ln N! \cong N \ln N - N$. This "proves" the $(1/kT)$ part of the Boltzmann principle.

2. A special generalized "partition function," $Q \equiv \sum_i g_i \, e^{-\left(\frac{\varepsilon_i}{kT}\right)}$ was derived/discovered so that one can find the Boltzmann-weighted energy of a quantized system from the partition function as $\langle E \rangle = +kT^2 \dfrac{d \ln Q}{dT}$. When this is applied to the particle-in-a-box problem with $E_n = \dfrac{n^2 h^2}{8mL^2}$, the usual average translational energy of $\left(\dfrac{kT}{2}\right)$ is obtained for each direction and for the rotational energy of a diatomic molecule with $E_j = j(j+1)\dfrac{\hbar^2}{2I}$ the expected value of kT is found. The translation and rotation energy levels are close together allowing a classical interpretation and indicating a high temperature limit of the quantized system is reached easily at room temperature. However, with more widely spaced vibrational quantum levels, the partition function approach leads to $U = \langle E - E_0 \rangle = \dfrac{RTx}{(e^x - 1)}$ where $x = \dfrac{hv}{kT}$ and the thermodynamic U is measured above the zero-point energy. The heat capacity for a given vibrational mode is the derivative of the energy with respect to temperature, which turns out to be $C_V = \dfrac{Rx^2 e^x}{(e^x - 1)^2}$. This is the main departure from the classical thermodynamic result of R/mode. However, is it shown that at very high temperature $\lim\limits_{T \to \infty} C_V(T) = R$. We can form a table that shows that the classical results from the kinetic theory of gases are realized at high temperature. The vibrational formulas can be applied to each individual vibrational mode in a molecule (Table 15.1).

TABLE 15.1
Selected Statistical Thermodynamic Formulas

Degree of Freedom	$U(T)$	C_V	High Temp. C_V
Translation	$\dfrac{3RT}{2}$	$\dfrac{3R}{2}$	$\dfrac{3R}{2}$
Rotation (diatomic)	$\dfrac{2RT}{2}$	$\dfrac{2R}{2}$	$\dfrac{2R}{2}$
Rotation (polyatomic)	$\dfrac{3RT}{2}$	$\dfrac{3R}{2}$	$\dfrac{3R}{2}$
Vibration/mode	$\dfrac{RTx}{(e^x - 1)}$	$\dfrac{Rx^2 e^x}{(e^x - 1)^2}$	R

where $x = \left(\dfrac{hv}{kT}\right)$

---

3. It was shown that by using some data for the bond length and vibrational frequency the absolute entropy can be calculated with the Sakur–Tetrode equation. While this was only demonstrated for a diatomic molecule (CO), the method can be applied to polyatomic molecules if the molecular formula, vibrational frequencies, and molecular geometry are known.

PROBLEMS

15.1 Using $\varepsilon_j = j\,(j+1)\,\hbar^2/2I$, derive the average rotational energy per mole for a diatomic molecule using a partition function. How fast can you do this derivation if it is a test question?

15.2 Use $\varepsilon_n = \dfrac{n^2 h^2}{8mL^2}$ with a partition function to derive the average translational energy of a particle in one dimension using a partition function. How fast can you do this if it is a test question? Have you memorized the odd and even cases of a Gaussian integral?

15.3 Using the vibrational frequency of HCl as 2990.2 cm^{-1} calculate the energy $(E - E_0)$ of 1 mol of HCl at 1000°K including translational, rotational, and vibrational energy.

15.4 Using the vibrational frequency of HCl as 2990.2 cm^{-1} calculate the total heat capacity as C_V and as C_P at 1500°K including translational, rotational, and vibrational terms.

15.5 Sometimes the variable of interest in the formulas derived with Boltzmann statistics is expressed as a "characteristic temperature" θ where $\theta = \left(\dfrac{h\nu}{k}\right)$ since the exponent should be unitless and the expressions have a Boltzmann exponent of $\left(\dfrac{h\nu}{kT}\right) = \dfrac{\theta}{T}$. Given the characteristic temperature for Cl$_2$ is 810°K, calculate C_P and C_V for Cl$_2$ at 100°C including translation, rotation, and vibration.

15.6 Using $\theta = 810$°K for Cl$_2$, calculate $(E - E_0)$ at 500°C including translation, rotation, and vibration.

15.7 Given the vibrational fundamental of O$_2$ is at 1556 cm^{-1}, calculate the heat capacities C_P and C_V for O$_2$ at 800°K including translation, rotation, and vibration.

15.8 Using the same data that the fundamental for O$_2$ is at 1556 cm^{-1}, calculate $(E - E_0)$ for 1 mol of O$_2$ at 1000°K, include the translation, rotational, and vibrational energy.

15.9 Use the Sakur–Tetrode equation to calculate the absolute entropy of gas phase BH at 298.15°K and 1 bar pressure using a vibrational frequency of 2366.9/cm, the atomic weights of 1.00794 g/mol for H g/mol and 10.811 g/mol for B with a bond length of 1.2324 Å. Compare your answer to the CRC value of 171.8 (J/mol °K) [6].

REFERENCES

1. Rushbrook, G. S., *Introduction to Statistical Mechanics*, Oxford University Press, Oxford, U.K., 1962.
2. Hill, T. L., *An Introduction to Statistical Thermodynamics*, Addison-Wesley Publishing Co., Inc. Reading, MA, 1960.
3. Lide, D. R., Ed., *CRC Handbook of Chemistry and Physics*, 90th Edn., CRC Press, Boca Raton, FL, 2009–2010, pp. 5–57.
4. Barrow, G. M., *Physical Chemistry*, 6th Edn., The McGraw-Hill Companies, Inc., New York, 1996, Chaps 4–6.
5. Lide, D. R., Ed., *CRC Handbook of Chemistry and Physics*, 90th Edn., CRC Press, Boca Raton, FL, 2009–2010, pp. 5–46.
6. ibid. pp. 5–6.

16 Approximate Methods and Linear Algebra

INTRODUCTION

In terms of "essentials" we only have a few more topics. Specifically we have four goals:

1. To describe and illustrate the simplest form of first-order perturbation theory
2. To describe and illustrate the variation principle (H 1s orbital)
3. To explain the secular equation which results from the LCAO approximation
4. To describe and illustrate the technique of matrix diagonalization

While this may seem to be intense ad hoc mathematics, it represents what took this author at least 5 years to locate and assemble into the necessary tools to function using the linear combination of atomic orbitals and a sense of how to approach seemingly impossible problems in quantum chemistry. We offer it here as a shortcut for interested undergraduates and auxiliary mathematics as an abbreviated course in linear algebra, which can be useful in a number of areas of chemistry. We will mention that this chapter is usually the end of a nine week (six credit, two semester) course and example 2 is almost always treated and tested.

Now that we have some better understanding of where the H atom orbitals come from, the next topic should be the electronic structure of molecules and ways to treat problems for which we are unable to solve the Schrödinger equation exactly. Recall the difficulty of solving the Schrödinger equation for just one electron in the H atom. Then perhaps you may faint when you consider the notion of how one might treat the electronic structure of benzene with 12 atoms and 42 electrons! Well, there is no known exact solution for even the He atom with only two electrons so do not faint but continue to wonder about how we are going to treat the multi-electron case for molecules. There are two main methods: the variation method and perturbation theory. In this chapter, we will emphasize the variation method, which is the most powerful mathematical approach, and give a few key examples. However, we will first mention the basic approach of perturbation theory but without much elaboration since it is the weaker of the two methods.

SIMPLE FIRST-ORDER PERTURBATION THEORY

Undergraduates should be aware that the concept of perturbation theory can be carried to very high order, up to eighth order in some cases, but the Variation Principle is much more powerful as a first resort so we will only show perturbation theory to first order to illustrate the principle. Imagine we have a problem for which we seek the solution to the Schrödinger equation but we cannot solve the equation. Then suppose we could solve it except for one annoying part of the problem (usually some detail in the potential energy term). Since we know that the true solutions of the Schrödinger equation have eigenfunctions that form complete sets (usually polynomials), maybe we could express the part of the problem we cannot treat in terms of the complete set of solutions for the problem which is nearly like the one we want to solve [1].

Let $H = H + \lambda V = H_0^{(0)} + \lambda H^{(1)}$ where $\lim_{\lambda \to 0} H = H^{(0)}$ and we know $H^{(0)}\psi_n^{(0)} = E_n^{(0)}\psi_n^{(0)}$ so we assume $\{\psi_n^{(0)}\}$ is a complete set under the boundary conditions of the original problem. Then we use expansions

$$\psi_n = \psi_n^{(0)} + \lambda\psi_n^{(1)} + \lambda^2\psi_n^{(2)} + \lambda^3\psi_n^{(3)} + \cdots \text{ where } \lim_{\lambda \to 0}\psi_n = \psi_n^{(0)} \quad \text{and}$$

$$E_n = E_n^{(0)} + \lambda E_n^{(1)} + \lambda^2 E_n^{(2)} + \lambda^3 E_n^{(3)} + \cdots \text{ where again } \lim_{\lambda \to 0}E_n = E_n^{(0)}.$$

Now we set up the equation we want to solve as $H\psi_n = H_n E_n$ We can multiply out the terms and separate them according to the powers of λ as

$$\left(H^{(0)}\psi_n^{(0)} - E_n^{(0)}\psi_n^{(0)}\right)\lambda^0 + \left(H^{(0)}\psi_n^{(1)} + V\psi_n^{(0)} - E_n^{(0)}\psi_n^{(1)} - E_n^{(1)}\psi_n^{(0)}\right)\lambda$$
$$+ \left(H^{(0)}\psi_n^{(2)} + V\psi_n^{(1)} - E_n^{(0)}\psi_n^{(2)} - E_n^{(1)}\psi_n^{(1)} - E_n^{(2)}\psi_n^{(0)}\right)\lambda^2 + \cdots = 0$$

The first term with λ^0 is the zeroth-order term from the exactly solvable problem. For the equation to be true, each power of λ^n must be zero individually and we can see how higher-order treatments might be possible using the terms with higher powers of λ. Thus we obtain several equations as

$$\left(H^{(0)}\psi_n^{(0)} - E_n^{(0)}\psi_n^{(0)}\right) = 0 \quad \text{(zeroth order, must be solved for } \{\psi_n^{(0)}, E_n^{(0)}\}.\text{)}$$
$$\left(H^{(0)}\psi_n^{(1)} + V\psi_n^{(0)} - E_n^{(0)}\psi_n^{(1)} - E_n^{(1)}\psi_n^{(0)}\right) = 0 \quad \text{(first-order terms)}$$
$$\left(H^{(0)}\psi_n^{(2)} + V\psi_n^{(1)} - E_n^{(0)}\psi_n^{(2)} - E_n^{(1)}\psi_n^{(1)} - E_n^{(2)}\psi_n^{(0)}\right) = 0, \quad \text{etc.}$$

Then we use the complete set $\{\psi_n^{(0)}\}$ to expand the unknown first-order wave function with as-yet-unknown coefficients a_{nk} as

$\psi_n^{(1)} = \sum_{k \neq n} a_{nk}\psi_k^{(0)}$ so that $\left(H^{(0)} - E_n^{(0)}\right)\sum_{k \neq n} a_{nk}\psi_k^{(0)} = (E_n^{(1)} - V)\psi_n^{(0)}$, but $H^0\psi_k^{(0)} = E_k^{(0)}\psi_k^{(0)}$ in the terms on the left so we obtain $\sum_{k \neq n}\left(E_k^{(0)} - E_n^{(0)}\right)a_{nk}\psi_k^{(0)} = \left(E_n^{(1)} - V\right)\psi_n^{(0)}$. Next we multiply the equation from the left by $\psi_n^{(0)}$ and integrate both sides of the equation and recall that the members of the set $\{\psi_n^{(0)}\}$ are all orthogonal to each other so that we find that

$$\sum_{k \neq n}\left(E_k^{(0)} - E_n^{(0)}\right)a_{nk}\langle\psi_n^{(0)}|\psi_k^{(0)}\rangle = 0, \quad n \neq k \Rightarrow 0 = E_n^{(1)}\langle\psi_n^{(0)}|\psi_n^{(0)}\rangle - \langle\psi_n^{(0)}|V|\psi_n^{(0)}\rangle.$$

Thus we find a simple result $E_n^{(1)} = \langle\psi_n^{(0)}|V|\psi_n^{(0)}\rangle$. In other words, *the first-order correction to the energy due to the perturbation is simply the expectation value of the perturbation*! What about the wave function? We can multiply the first-order terms by $\psi_k^{(0)}$ from the left and integrate to obtain

$a_{nk}\left(E_k^{(0)} - E_n^{(0)}\right) = -\langle\psi_k^{(0)}|V|\psi_n^{(0)}\rangle$ and so $a_{nk} = \dfrac{-\langle\psi_k^{(0)}|V|\psi_n^{(0)}\rangle}{E_k^{(0)} - E_n^{(0)}}$, $k \neq n$. In summary

$$E_n^{(1)} = \langle\psi_n^{(0)}|V|\psi_n^{(0)}\rangle \quad \text{and} \quad \psi_n^{(1)} = \psi_n^{(0)} - \left\{\sum_{k \neq n}\frac{\langle\psi_k^{(0)}|V|\psi_n^{(0)}\rangle}{E_k^{(0)} - E_n^{(0)}}|\psi_k^{(0)}\rangle\right\}.$$

Note that all the information needed to calculate the first-order energy and wave function is available from the zeroth order problem. For each higher order, the corrections to the energy and wave function only depend on the $(n-1)$th order information so, although tedious, each order of corrections can be obtained by a sort of "mathematical bootstrap" process. Perhaps perturbation theory is conceptually useful in that we now know that we can look at an unsolved problem and by mental modeling see that the answer is like a solvable problem with some modification.

Example of Anharmonicity in a Harmonic Oscillator

Given $\psi_0^{(0)} = \left(\dfrac{k}{hv\pi}\right)^{1/4} e^{-\left(\frac{kx^2}{2hv}\right)}$, check the normalization of this function for the $n=0$ level of the

harmonic oscillator and then calculate the *total* energy in the presence of a perturbation given by $H^{(1)} = ax^3 + bx^4$. First we integrate the square of the function since it is not complex.

$$\int_{-\infty}^{+\infty} \left(\frac{k}{hv\pi}\right)^{1/2} e^{-\left(\frac{2kx^2}{2hv}\right)} dx = \left(\frac{k}{hv\pi}\right)^{1/2} 2\int_0^{+\infty} e^{-\left(\frac{kx^2}{hv}\right)} dx = \left(\frac{k}{hv\pi}\right)^{1/2}\frac{2}{2}\sqrt{\frac{\pi}{\left(\frac{k}{hv}\right)}} = 1 \quad \text{Q.E.D.}$$

Next we evaluate the first-order correction $E_n^{(1)} = \langle \psi_n^{(0)} | V | \psi_n^{(0)} \rangle$.

$$E_0^{(1)} \int_{-\infty}^{+\infty} \left[\left(\frac{k}{hv\pi}\right)^{1/4} e^{-\left(\frac{kx^2}{2hv}\right)}\right](ax^3 + bx^4)\left[\left(\frac{k}{hv\pi}\right)^{1/4} e^{-\left(\frac{kx^2}{2hv}\right)}\right] dx$$

$$= \left(\frac{k}{hv\pi}\right)^{1/2} \int_{-\infty}^{+\infty} e^{-\left(\frac{kx^2}{hv}\right)}(ax^3 + bx^4)\, dx.$$

This is a very interesting problem when you notice the limits on the integral are $(-\infty, +\infty)$. The part of the integrand that is ax^3 is an odd function, which changes sign at $x=0$ while the $e^{-\left(\frac{kx^2}{2hv}\right)}$ part of the integrand is an even function, which does not change sign! Over the full range of the integration, the ax^3 term will integrate to zero. Thus we only have to evaluate the bx^4 term using $\int_0^\infty e^{-a^2x^2} x^{(2p)} dx = \dfrac{1 \cdot 3 \cdots (2p-1)\sqrt{\pi}}{2^{p+1} a^{2p+1}}$ and multiply the integral by 2 for the full $(-\infty, +\infty)$ range.

$$E_0^{(1)} = \left(\frac{k}{hv\pi}\right)^{1/2}\int_{-\infty}^{+\infty} e^{-\left(\frac{kx^2}{hv}\right)}(bx^4)\, dx = \left(\frac{k}{hv\pi}\right)^{1/2} 2b\int_0^{+\infty} e^{-\left(\frac{kx^2}{hv}\right)} x^4 dx = \left(\frac{k}{hv\pi}\right)^{1/2}\frac{2b\sqrt{\pi}(1\cdot 3)}{2^3\left(\frac{k}{hv}\right)^{5/2}}, \text{ so}$$

we find $E_0 = \dfrac{hv}{2} + \dfrac{3b}{4}\left(\dfrac{hv}{k}\right)^2$ and the ax^3 term contributes nothing to the first-order perturbation!

However, "a" might be important in a higher-order treatment. In practice, additional data would be needed to fit the value of "b" to a given spectrum. We now have a way to approximately correct the harmonic potential for anharmonicity.

PRINCIPLES OF PERTURBATION THEORY

One major flaw in perturbation theory is that there is no bound. The correction may be too large, too small, or even alternate as high, low, high, low in successive orders. At one time, it was hoped that perturbation theory could be used to higher order with the corrections forming a converging oscillation as is sometimes observed with successive orders. The fact that the wave function of a given order is used for the next higher-order energy correction implies a sort of progressive bootstrap procedure that might lead to the correct answer at some higher order; however, tests of this idea have shown it is not valid. All that can be said about a given level of perturbation is that "you get what you get"! Another principle worth noting is that the success of perturbation theory depends a lot on the size of the perturbation relative to the original zeroth order problem $E_0^{(0)}$. One good example is where ligand effects in lanthanide compounds are applied to the relatively shielded inner 4f shell of electrons [2]. There the effect is small and first-order perturbation works very well for "Crystal Field Theory." If the same approach is used for transition metal complexes and applied to the d-orbital valence shell where the ligand orbitals overlap directly with the metal orbitals, the success of a "Ligand Field theory" [3] is good but less precise and hints strongly that the zeroth-order Hamiltonian needs to be expanded to include delocalized electrons on the metal and the ligands. Obviously, if the perturbation "V" approaches 50% of the original $E_0^{(0)}$, first-order perturbation will be inconclusive, but if "V" is only 1% of $E_0^{(0)}$ there is a good chance for semiquantitative accuracy!

The expression for the first-order perturbed wave function is often useful in organic chemistry (especially π-orbitals) to predict which excited states may affect the ground state the most in a reaction.

$$\psi_n^{(1)} = \psi_n^{(0)} - \left\{ \sum_{k \neq n} \frac{\langle \psi_k^{(0)} | V | \psi_n^{(0)} \rangle}{E_k^{(0)} - E_n^{(0)}} | \psi_k^{(0)} \rangle \right\}.$$

We can see two effects in the expression for the perturbed orbital. First, the effect will be greatest when $\left(E_k^{(0)} - E_n^{(0)} \right)$ is small in the denominator as for the HOMO–LUMO energy difference. Effects from higher excited states will decrease rapidly as the energy difference increases. Second, even if the energy gap is small, the $\langle \psi_k^{(0)} | V | \psi_n^{(0)} \rangle$ integral needs to be large. Usually $V = \dfrac{e^2}{r_{12}}$ is the perturbation linking the ground state orbitals with unoccupied orbitals, but other perturbations can be used.

VARIATION METHOD

Once again, much of what we show here is given in the text by Pauling and Wilson [4]. The variation method depends on principles we have shown now for the particle-in-a-box, the particle-on-a-ring, the harmonic oscillator, the rigid rotor and the H atom. Think back and recall that in each case we found there are a number of eigenvalues for the energies and a corresponding orthonormal set of wave eigenfunctions. Now suppose we want to solve a new problem and find the energy as $\langle E \rangle = \int \psi^* H \psi d\tau$. We need to be able to write a correct Hamiltonian operator, H, which includes all the interactions we want to use in the problem. Then we want to solve the Schrödinger equation $H\psi = E\psi$. (The time-independent equation; we probably already have enough difficulty without worrying over time dependence!) After a period of frustration, we might do what many students do, we will *guess* a functional form which seems to have some properties we need (finite, single-valued, and continuous plus whatever other features you think are important). Let our guess function be ϕ_g, chosen so that using calculus or a computer program we can evaluate $E_g = \int \phi_g^* H \phi_g d\tau$. Now it is time to recall the characteristics of the problems that have exact solutions. They each had a set of eigenvalues $\{E_n\}$. Although there might have been degeneracies in the energy values (all p-orbitals, all d-orbitals have the same energy, etc.), we recall that there was one lowest energy, which we can call E_0. The other energy values followed $E_n \leq E_{n+1}$ (allowing for degeneracies). Further, the set of eigenfunctions $\{\psi_n\}$ had several properties such as $H\psi_n = E_n\psi_n$ and $\int \psi_n^* \psi_m d\tau = \delta_{nm}$. Then depending on the work of mathematicians such as Hermite, Legendre, and Laguerre the eigenfunctions can be shown to form a *complete* set. That means that within the same boundary conditions, there is no other function that cannot be described by some linear combination of the orthonormal set of eigenfunctions, the set is *complete*! Thus in principle, our guess-function can be represented by a linear combination of the true eigenfunctions $\{\psi_n\}$ (which we do not know) using weighting coefficients c_n (which we do not know) as $\phi_g = \sum_n c_n \psi_n$. So then we try to evaluate the energy with our guessed function and the Hamiltonian, as long as we can do the integral correctly, and operate inside the integral using $H\psi_m = E_m\psi_m$ to find

$$E_g = \int \phi_g^* H \phi_g d\tau = \sum_n \sum_m c_n^* c_m \int \psi_n^* H \psi_n d\tau = \sum_n \sum_m c_n^* c_m E_m \int \psi_n^* \psi_m d\tau = \sum_n \sum_m c_n^* c_m E_m \delta_{nm}.$$

Because the true (but unknown) eigenfunctions are orthonormal, the Kronecker delta δ_{nm} eliminates all the terms except when $n = m$, so we obtain $E_g = \sum_n |c_n^2| E_n$ although we do not really know the set of eigenvalues $\{E_n\}$ or the coefficients, c_n! How does this help us? Well, we know that one of the true (unknown) eigenvalues is the lowest energy and we have seen in the problems we have solved that $E_n \leq E_{n+1}$ so let us subtract that lowest (unknown) eigenvalue from both sides of our equation as $E_g - E_0 = \sum_n |c_n^2| E_n - (1)E_0$. We emphasize the factor of (1) because we are only subtracting E_0 "once" *and* because we need to normalize our guess function as follows

$$1 = \int \phi^* \phi d\tau = \sum_n \sum_m c_n^* c_m \int \psi_n^* \psi_m d\tau = \sum_n \sum_m c_n^* c_m \delta_{nm} = \sum_n |c_n^2| = 1.$$ Then we can write

$$E_g - E_0 = \sum_n |c_n^2| E_n - (1)E_0 = \sum_n |c_n^2| E_n - \sum_n |c_n^2| E_0 = \sum_n |c_n^2|(E_n - E_0)$$ and now we are

ready for the main conclusion. If $E_n \leq E_{n+1}$ then $\sum_n |c_n^2|(E_n - E_0) \geq 0$ on the right side of the equation so $E_g - E_0 \geq 0$ or in conclusion we find that

$$E_g \geq E_0!$$

What does that mean? Well it means that we can guess whatever we want (intelligently) for ϕ_g and as long as we evaluate the integral $E_g = \int \phi_g^* H \phi_g d\tau$, either analytically or by using a computer, the answer will either be correct (E_0) or higher than the true value. How can we use this? We can put a lot of parameters in the guess function, take the derivative with respect to each parameter, set it to zero, and then use the value of that parameter at the minimum to find the minimum energy for that function (which may still be higher than the true minimum energy). *The key point is that we know E_g cannot go below the true energy as long as we do the integral correctly!* Pauling and Wilson credit this method to Eckart [5] and it is indeed a powerful concept!

Example

Starting with a variational trial function $\phi_g \sim e^{-\alpha r}$ for the 1s H atom solution, show in detail that $\alpha_{\min}$ yields the *exact* nonrelativistic Schrödinger–Bohr energy. (A frequent examination question worth learning thoroughly!) First we need to normalize the trial-guess. Our function ϕ_g is finite, single-valued, continuous, and physically models the electron wave function near the nucleus with less probability for large r. You need to recall an integral formula we derived in an earlier chapter as

$$\int_0^\infty x^n e^{-ax} dx = \frac{n!}{a^{n+1}}.$$

$$1 = 4\pi \int_0^\infty (Ne^{-ar})^*(Ne^{-ar}) r^2 dr = 4\pi N^2 \int_0^\infty e^{-2\alpha r} r^2 dr = 4\pi N^2 \frac{2!}{(2\alpha)^3} \Rightarrow N = \sqrt{\frac{\alpha^3}{\pi}}$$

so

$$\phi_g = \sqrt{\frac{\alpha^3}{\pi}} e^{-\alpha r}.$$

The Hamiltonian is $H = \frac{\hbar^2}{2m} \nabla^2 - \frac{Ze_q^2}{r}$. This looks like a very difficult problem but the trial wave function has no dependence on (θ, ϕ) so we have a great simplification here.

$$\frac{1}{r^2} \frac{\partial}{\partial r}\left(r^2 \frac{\partial}{\partial r}\right) + \frac{1}{r^2 \sin\theta} \frac{\partial}{\partial \theta}\left(\sin\theta \frac{\partial}{\partial \theta}\right) + \frac{1}{r^2 \sin^2\theta} \frac{\partial^2}{\partial \phi^2} = \nabla^2(r, \theta, \phi)$$ becomes much simpler.

$$\nabla^2(r, \theta, \phi) \rightarrow \frac{1}{r^2} \frac{\partial}{\partial r}\left(r^2 \frac{\partial}{\partial r}\right) = \nabla^2(r)$$ because there is no part of ϕ_g depending on (θ, ϕ)!

So $\frac{1}{r^2}\frac{\partial}{\partial r}\left(r^2\frac{\partial}{\partial r}\right)e^{-\alpha r} = \frac{1}{r^2}\frac{\partial}{\partial r}[-\alpha r^2 e^{-\alpha r}] = \frac{1}{r^2}[-2\alpha r + \alpha^2 r^2]e^{-\alpha r}$. We then insert this into the averaging integral to find the expectation value of the Hamiltonian.

$$E_g\int\phi_g^*H\phi_g d\tau = 4\pi\left(\frac{\alpha^3}{\pi}\right)^{\frac{2}{2}}\left\{\int_0^\infty e^{-\alpha r}\left(\frac{-\hbar^2}{2m}\right)\left[\frac{-2\alpha r + \alpha^2 r^2}{r^2}\right]e^{-\alpha r}r^2 dr - Ze_q^2\int_0^\infty e^{-2\alpha r}\left(\frac{1}{r}\right)r^2 dr\right\}$$

$$E_g\int\phi_g^*H\phi_g d\tau = 4\alpha^3\left\{\left(\frac{-\hbar^2}{2m}\right)\left[\frac{-2\alpha}{(2\alpha)^2} + \frac{2\alpha^2}{(2\alpha)^3}\right] - \frac{Ze_q^2}{(2\alpha)^2}\right\} = \left(\frac{-\hbar^2}{2m}\right)[-2\alpha^2 + \alpha^2] - Ze_q^2\alpha$$, and so

$$E_g\int\phi_g^*H\phi_g d\tau = +\frac{\hbar^2\alpha^2}{2m} - Ze_q^2\alpha = E_g(\alpha).$$ Now we have the energy of the guessed function as a function of the parameter α and we could insert numerical values into the constants, write a small computer program and print out the energy for various values of α from $\alpha = 0.001$ to $\alpha = 1000$ in increments of 0.001 and look for the lowest value of the energy from the guessed function. That would work but be very tedious! It is worth noting that the kinetic energy term is positive and the potential energy term is negative so there must be some value of α at which a minimum occurs. How do we know there is a minimum? The variation theorem we proved earlier tells us that $E_g \geq E_0$. As long as we computed the energy correctly from the guessed trial function, the answer $cannot$ go $lower$ $than$ E_0!

Why not use calculus to find the minimum? $\frac{dE_g(\alpha)}{d\alpha} = 2\frac{\hbar^2\alpha}{2m} - Ze_q^2 = 0 \Rightarrow \alpha_{min} = \frac{Ze_q^2 m}{\hbar^2}$ so we know

α_{min}! Then the minimum energy is $E_g(\alpha_{min}) = \left(\frac{\hbar^2}{2m}\right)\left(\frac{Ze_q^2 m}{\hbar^2}\right)^2 - Ze_q^2\left(\frac{Ze_q^2 m}{\hbar^2}\right) = -\frac{Z^2 me_q^4}{2\hbar^2}$. Hey,

wait a minute, $that$ is the $Schrödinger–Bohr$ H $1s$ $energy!$ This is a fortuitous case where the value of the energy from the guessed trial function is exactly the minimum energy. This occurred because we made an excellent guess which is of course the correct functional form for a H 1s orbital, but we did not know the value of α_{min}. This example shows us that an energy from a guessed trial function approaches the true energy from above and that if you have the correct functional form, the guess will give the correct energy. This is very rare and most of the time the calculated energy will not be the exact energy but will be above the true value. In the 1950–1970 era, computer programs were being written and tested to compute the energy of molecules and research papers were published with ever lower energies tending to the exact ground state energy as a measure of the accuracy of the calculations. Note that this process only works for the lowest energy of a given symmetry and so is usually only applied to the ground state of atoms and molecules.

MOLECULAR ORBITALS AND THE SECULAR EQUATION

One of the main problems in the extension of the H atom orbital ideas to molecules is the fact that atomic functions are centered on a nuclear position as a centro-symmetric coordinate system but a molecule involves more than one center. As soon as a second nucleus is introduced as with H_2, HCl, H_2O, NH_3, or CH_4, we have a problem relating atomic orbitals centered on one nucleus to that of a neighbor. Not only that but we have good reason to believe on chemical grounds that in covalent compounds there will be considerable electron sharing. Practice using Lewis electron-dot structures in freshman chemistry reveals many cases of covalent sharing. In fact the simplest ideas of bonding imply that there needs to be $orbital$ $overlap$ for bonding to occur except in the most extreme ionic electron-transfer bonding. Thus the simplest ideas of molecular orbitals recognize the concept of overlap between two orbitals on different centers. Traditionally $S_{ij} = \langle i|j\rangle = \int\phi_{ia}^*\phi_{jb}d\tau$ where the two orbitals are on different atoms or maybe in orbitals on the same center that are not mutually orthogonal because they are trial functions.

From the example above we now know that we could put a lot of parameters into a trial wave function and then minimize the energy of the molecule with respect to the parameters. We also appreciate the fact that a good guess will produce a good result. This thought process has led to the linear combination of atomic orbitals (LCAO) approximation. The idea is that when an atom is in a molecule, electrons close to an atom will likely be in orbitals close to what they would be for that atom by itself. However, if/when the (indistinguishable) electrons move from the neighborhood of one atom and wander to another nearby atom (and so on throughout the whole molecule) they may be in orbitals, which are partly like both atoms. Thus, the LCAO orbital is taken to be a linear combination of orbitals that have been obtained from variationally optimized orbitals for individual atoms.

Using $\psi = \sum_i c_i \phi_i$, the ϕ_i basis functions are assumed to be fixed functions and only *the coefficients c_i are treated as variational parameters*. Thus, we need a way to simultaneously vary many c_i values to reach an energy minimum. Since we will be interested in the interaction between electrons in orbitals ϕ_i and ϕ_j, we can see that we will have a set of equations in an (i, j) representation. Thus we set up the Schrödinger equation in terms of the LCAO function within a *basis set* of atomic orbitals as a matrix equation $[H][C] = [S][C][E]$. Here, $[H]$ is a square matrix with matrix elements $H_{i,j} = \int \phi_i^* H \phi_j d\tau$, $[S]$ is a square matrix with matrix elements $S_{ij} = \int \phi_i^* \phi_j d\tau$, and $[C]$ is a square matrix whose *columns are the molecular orbital coefficients, c_i*. If there are N ϕ_i functions in the linear combination, there will also be N molecular orbitals, but maybe only a few of the lowest energy orbitals will be occupied by an extension of the Aufbau principle idea. This model also incorporates what is called a "single electron model" in the sense that we seek linear combinations of atomic orbitals to be normalized descriptions of single electrons, but of course we fill them with two electrons as an $\alpha\beta$ spin-pair.

In order to evaluate the energy as a function of the unknown coefficients c_i we set up the expectation process for the energy, but actually we will find N different energies, ε_i, which are the one-electron orbital energies. In this simple model the total energy will be the simple sum of the orbital energies but in more sophisticated treatments this is not exactly true. For this model

$$\frac{\sum_i \sum_j c_i c_j \int \phi_i^* H \phi_j d\tau}{\sum_i \sum_j c_i c_j \int \phi_i^* \phi_j d\tau} = \langle E \rangle$$ where we have to make sure the linear combinations are normalized.

Then $\sum_i \sum_j c_i c_j \int \phi_i^* H \phi_j d\tau = (E) \sum_i \sum_j c_i c_j \int \phi_i^* \phi_j d\tau$ and next we take the derivative with respect to any specific c_j coefficient. This derivative will be made easier with an example aside.

Suppose we seek the derivative of an expression of weighted elements of a Hermitian matrix, where $A_{ij}^* = A_{ji}$ (all real) with respect to one coefficient, say c_2, where A_{ij} could be H_{ij} or S_{ij}.

$$\frac{\partial}{\partial c_2} \sum_{i=1}^3 \sum_{j=1}^3 c_i c_j A_{ij} = \frac{\partial}{\partial c_2} \left[c_1^2 A_{11} + c_1 c_2 A_{12} + c_1 c_3 A_{13} + c_2 c_1 A_{21} + c_2^2 A_{22} + c_2 c_3 A_{23} + c_3 c_1 A_{31} \right.$$
$$\left. + c_3 c_2 A_{32} + c_3^2 A_{33} \right]$$

so we find $\frac{\partial}{\partial c_2} \sum_{i=1}^3 \sum_{j=1}^3 c_i c_j A_{ij} = c_1 A_{12} + c_1 A_{21} + 2c_2 A_{22} + c_3 A_{23} + c_3 A_{32} = 2 \sum_{i=1}^3 c_i A_{2i}$ or the derivative of the double sum is just twice the single sum: $\frac{\partial}{\partial c_2} \sum_{i=1}^3 \sum_{j=1}^3 c_i c_j A_{ij} = 2 \sum_{i=1}^3 c_i A_{2i} = 2 \sum_i^3 c_i A_{i2}$.

Now we can take the derivative as

$$2 \sum_i c_i H_{ij} = \left(\frac{\partial E}{\partial c_j} \right) \sum_i \sum_j c_i c_j \int \phi_i^* \phi_j d\tau + 2E \sum_i c_i S_{ij}.$$

Then we can use the idea that at the minimum $\left(\dfrac{\partial E}{\partial c_j}\right) = 0$. Finally, we can cancel the factor of 2 on each side of the equation to produce $\sum_i c_i H_{ij} = E \sum_i c_i S_{ij}$ or $\sum_i (H_{ij} - ES_{ij})c_i = 0$. However, there are "$n$" such equations for n basis orbitals in the LCAO, one for each c_j, which leads to a system of equations all $= 0$.

$$(H_{11} - ES_{11})c_1 + (H_{12} - ES_{12})c_2 + (H_{13} - ES_{13})c_3 + \cdots (H_{1n} - ES_{1n})c_n = 0$$
$$(H_{21} - ES_{21})c_1 + (H_{22} - ES_{22})c_2 + (H_{23} - ES_{23})c_3 + \cdots (H_{2n} - ES_{2n})c_n = 0$$
$$\cdots\cdots\cdots\cdots\cdots\cdots\cdots\cdots\cdots\cdots\cdots\cdots\cdots\cdots\cdots\cdots\cdots\cdots\cdots\cdots\cdots\cdots = 0$$
$$(H_{n1} - ES_{n1})c_1 + (H_{n2} - ES_{n2})c_2 + (H_{n3} - ES_{n3})c_3 + \cdots (H_{nn} - ES_{nn})c_n = 0$$

This is a linear system of equations in n variables where the unknown variables are the c_i coefficients. Formally, this calls into effect the Cayley–Hamilton theorem [6] because the right-hand sides of the equations are all zero. The Cayley–Hamilton theorem [6] states that a square matrix, A, satisfies its characteristic equation and if we have a characteristic polynomial of the eigenvalues of the matrix $\{\lambda_i\}$, which are the roots of the polynomial $\lambda^n + c_{n-1}\lambda^{n-1} + \cdots + c_0 = 0$, then the matrix also satisfies $A^n + c_{n-1}A^{n-1} + \cdots + c_0 = 0$. Later, we will see that if a square matrix can be transformed to "diagonal form," the Cayley–Hamilton theorem becomes obvious. This can be explained using an example of two equations in two unknowns solved by Cramer's rule using determinants.

Given two equations in two unknowns as:

$$3x + 2y = 4$$
$$2x + 4y = 7$$

we would normally solve this by setting up a denominator determinant with the coefficients of the unknowns. Then we would set up the numerator determinant using the constants in place of the unknown coefficients. The answer is then the quotient of the two determinants!

$$x = \frac{\begin{vmatrix} 4 & 2 \\ 7 & 4 \end{vmatrix}}{\begin{vmatrix} 3 & 2 \\ 2 & 4 \end{vmatrix}} = \frac{16 - 14}{12 - 4} = \frac{2}{8} = \frac{1}{4} \quad \text{and} \quad y = \frac{\begin{vmatrix} 3 & 4 \\ 2 & 7 \end{vmatrix}}{\begin{vmatrix} 3 & 2 \\ 2 & 4 \end{vmatrix}} = \frac{21 - 8}{12 - 8} = \frac{13}{8}$$

$$\text{Check: } 3\left(\frac{1}{4}\right) + 2\left(\frac{13}{8}\right) = \frac{3 + 13}{4} = 4 \quad \text{and} \quad 2\left(\frac{1}{4}\right) + 4\left(\frac{13}{8}\right) = \frac{1 + 13}{2} = 7 \quad \text{Q.E.D.}$$

That works fine and can be extended to N unknowns as long as the right sides of the equation are nonzero. In the variational molecular orbital problem, all the right sides of the equation are zero and if we use Cramer's rule we only get the trivial solution with all values equal to zero. The Cayley–Hamilton theorem implies that if all the equations equal zero, you can still get a nonzero solution by forcing the denominator determinant to be zero, that is, by solving for the roots of the corresponding polynomial. In effect, this forces the equations to $\left(\dfrac{0}{0}\right)$, which is not zero; it is undefined! This leads to the so-called secular determinant as $|H - E \cdot S| = 0$.

CHEMICAL BONDS OF ETHYLENE

In the 1930s there were no computers and severe approximations were often made to facilitate paper-and-pencil calculations. A simple method due to Hückel [7] formed a foundation for the notion of molecular orbitals which stretched out across the whole molecule, but the method was limited mostly to the $2p_z\pi$ orbitals of aromatic hydrocarbons and other $2p_z\pi$ heteroaromatic

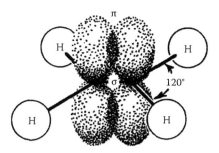

FIGURE 16.1   Diagram of π-bonding in ethylene. (From Roberts, J.D., *Notes on Molecular Orbital Calculations* Benjamin Press, New York, 1961. With permission.) The whole text can be downloaded from http:// caltechbook.library.caltech.edu/23/. This sort of diagram was a very exciting connection between Organic Chemistry and Quantum Chemistry in the 1960s. Prof. Roberts notes in his text that the HCH angle is closer to 117° than the 120° shown here but only the $2p_z\pi$ orbitals enter into the calculation.

hydrocarbons (Figure 16.1). Historically, this method was brought to the forefront of organic chemistry by Roberts [7] in a small book that went through 14 printings in the 1960s through 1978 (Figure 16.2). Today, these methods have been totally eclipsed by very accurate all-electron computer calculations, but the Hückel orbitals are still useful for an introduction to the molecular orbital concept. The main assumptions are that we base all parameters on $2p_z\pi - sp^2$ carbon

FIGURE 16.2   John D. Roberts (born 1918) is an American chemist who has contributed greatly to the integration of organic chemistry, physical chemistry, and spectroscopy. Roberts received the Priestly Medal in 1987, the National Medal of Science in 1990, the Glenn T. Seaborg Medal in 1991, and the NAS Award in Chemical Sciences in 1999. He is credited with bringing the first female graduate student into the Caltech Chemistry department when he moved from MIT. (Gift of John D. Roberts, Chemical Heritage Foundation Collections.)

compounds and it is found that even with that restriction a surprising number of electronic properties can be parameterized with $H_{ii} = \alpha$, $H_{ij} = \beta$ (nearest neighbors only), $S_{ii} = 1$ and $S_{ii} = 0$. Consider the ethylene molecule using this scheme. The parameters α and β are in units of electronic energy in the one-electron sense but two spins per orbital will be used. There is a conflict in standard notation in that Hückel and Roberts use α and β for electronic energy and then use $\uparrow$ and $\downarrow$ for the spins. We consider only the two $2p_z\pi$ orbitals as our basis set and set up a 2×2 secular determinant from a

matrix equation as $\begin{bmatrix} \alpha & \beta \\ \beta & \alpha \end{bmatrix} \begin{bmatrix} c_1 \\ c_2 \end{bmatrix} = \begin{bmatrix} E & 0 \\ 0 & E \end{bmatrix} \begin{bmatrix} c_1 \\ c_2 \end{bmatrix}$ to $\begin{bmatrix} \alpha - E & \beta \\ \beta & \alpha - E \end{bmatrix} \begin{bmatrix} c_1 \\ c_2 \end{bmatrix} = 0$ and a system of

equations as a determinant $\begin{vmatrix} (\alpha - E) & \beta \\ \beta & (\alpha - E) \end{vmatrix} = 0$.

We introduced the basic ideas of matrix arithmetic in Chapter 14 relative to the Powell d-orbitals and we expand on matrix use here. The determinant $|f(E)|$ is a *single number* as a function of E in this case and this can be solved for E as $(\alpha - E)^2 - \beta^2 = 0$. It will be important for later discussion to see that this equation came from the matrix equation $[H][C] = [S][C][E]$ as $H\psi(c_i)$ $E\psi(c_i)$. A matrix is an array which contains many relationships as coefficients placed in a [ ] or

$\begin{bmatrix} \alpha & \beta \\ \beta & \alpha \end{bmatrix} \begin{bmatrix} c_1 \\ c_2 \end{bmatrix} = \begin{bmatrix} 1 & 0 \\ 0 & 1 \end{bmatrix} \begin{bmatrix} c_1 \\ c_2 \end{bmatrix} \begin{bmatrix} E & 0 \\ 0 & E \end{bmatrix}$ using $[S] = \begin{bmatrix} 1 & 0 \\ 0 & 1 \end{bmatrix}$ but the whole process can be stream-

lined by dividing the Cayley–Hamilton determinant by β and defining $x \equiv \dfrac{(\alpha - E)}{\beta}$. Then

$\begin{vmatrix} (\alpha - E)/\beta & \beta/\beta \\ \beta/\beta & (\alpha - E)/\beta \end{vmatrix} = 0$ leads to $\begin{vmatrix} x & 1 \\ 1 & x \end{vmatrix} = 0$ and so $x^2 - 1 = 0$ and the solutions $x = 1$ and $x = -1$.

The determinant uses straight vertical bars and is a single number while a matrix has little right-angle extensions on the vertical bars and represents values in a linear system of equations. Note that a Cayley–Hamilton matrix led to a determinant due to the zeros on the right side of the secular equations. We can use these roots to find the coefficients of the $2p_z\pi$ molecular orbital coefficients. When $x = -1$ we can go back to the original system of equations but still use the determinant with x so we have $xc_1 + 1c_2 = 0$ or $-c_1 + c_2 = 0$ and $c_1 = c_2$. Note that we could use either equation and we will get the same result! However, we need to normalize the molecular orbitals so we set up the

integral $1 = \int (c_1\phi_1 + c_2\phi_2)^*(c_1\phi_1 + c_2\phi_2)d\tau = c_1^2 + 0 + 0 + c_2^2 = 2c_1^2 = 1$ so for the solution of

$\left(\dfrac{\alpha - E}{\beta}\right) = -1$, we have $E = \alpha + \beta$ and $\psi_\pi = \dfrac{\phi_1 + \phi_2}{\sqrt{2}}$. When $x = +1$ we see $xc_1 + 1c_2 = 0$ so

$c_1 = -c_2$ but normalization still produces $2c_1^2 = 1$ or $c_2 = \dfrac{-1}{\sqrt{2}} = -c_1$. Then $\left(\dfrac{\alpha - E}{\beta}\right) = +1$ or

$E = \alpha - \beta$.

In summary: $E = \alpha - \beta$ when $\psi_{\pi*} = \dfrac{\phi_1 - \phi_2}{\sqrt{2}}$ (antibonding orbital)

$E = \alpha + \beta$ when $\psi_\pi = \dfrac{\phi_1 + \phi_2}{\sqrt{2}}$ (bonding orbital)

We see that the bonding orbital has the lower energy since both α and β are apparently negative by nature of their definitions as matrix elements of the Hamiltonian for "bound orbitals." If α and β were not negative (bound), the electrons in those orbitals would fly away! The excited orbital has a node (sign change) that makes it antibonding while the bonding orbital has a plus sign for both $2p_z\pi$ components. The transition between the two levels is 2β and calibration of electronic excitations for many compounds leads to a value of -2.71 eV. This predicts an excitation of

about $\lambda_{\pi \to \pi*} = \dfrac{12398}{2(2.71 \text{ eV})} = 2287 \text{ Å} \cong 229$ nm, which is to the red of the main experimental

---

TABLE 16.1
Hückel Heteroatom Parameters

$h_{O-} = 2.0$	$k_{C-O} = 0.8$
$h_{O=} = 1.0$	$k_{C=O} = 1.0$
$h_{N-} = 0.5$	$k_{C-N} = 0.8$
$h_{N=} = 1.5$	$h_{N=} = 1.5$

Sources: Streitwieser, A., *Molecular Orbital Theory for Organic Chemists*, John Wiley & Sons LTD., London, U.K., 1961, Chap. 4; Murrell, J.N. et al., *Valence Theory*, 2nd Edn., John Wiley & Sons, LTD., London, U.K., 1970, p. 296.

---

absorption at about 160 nm, but the value of the calculation is mainly in providing a model that is qualitatively accurate.

It is possible to extend the simple Hückel method to hetero-atom cases in organic pi-electron systems by adjusting the α and β values in terms of the basic carbon parameters. Two additional parameters are introduced as $\alpha_X = \alpha_C + h_X \beta_{C-C}$ and $\beta_{C-X} = k_{C-X} \beta_{C-C}$ (Table 16.1).

A second example for the pi-orbitals of formaldehyde will suffice to illustrate the hetero-pi method. Once again we only have two orbitals, $2p_z C$ and a $2p_z O$. Note this method ignores the nonbonded orbitals on oxygen and the C–H sigma bonds as well as the two inner shell 1s orbitals; we are only treating the pi-orbitals. This time the Hückel determinant becomes $\begin{bmatrix} (\alpha + \beta - E) & \beta \\ \beta & (\alpha - E) \end{bmatrix} \begin{bmatrix} c_O \\ c_C \end{bmatrix} = 0, \Rightarrow \begin{vmatrix} (x+1) & 1 \\ 1 & x \end{vmatrix} = 0$ so we get $x(x+1) - 1 = 0$ and we need to use the quadratic equation to find $x \dfrac{-1 \pm \sqrt{1+4}}{2} = \dfrac{-1 \pm \sqrt{5}}{2}$ so $x = 0.618$ and $x = -1.618$. When $x = 0.618 = \dfrac{(\alpha - E)}{\beta}$, $E = \alpha - 0.618\beta$, then $1.618c_O + 1c_C = 0$ and $1.618c_O = -1c_C$. Now $1^2 c_O^2 + (-1.618)^2 c_O^2 = 3.618 c_O^2 = 1$ so $\psi_{\pi*} = 0.526\phi_O - 0.851\phi_C$ (anti-bonding). When $x = -1.618 \left(\dfrac{\alpha - E}{\beta} \right)$, $E = \alpha + 1.618\beta$ then $-0.618c_O + 1c_C = 0$ and $0.618c_O = c_C$. Now $1^2 c_O^2 + (0.618)^2 c_O^2 = 1.382 c_O^2 = 1$ so $\psi_\pi = 0.851\phi_O + 0.526\phi_C$ (bonding). Once again the excited state has a node but the weighting is unsymmetrical.

The Hückel pi-electron method has been applied to numerous compounds and Lowe [9] has documented two general solutions for pi-hydrocarbons using a graphical argument.

1. Linear polyene molecules

$$x = -2\cos[k\pi/(n+1)], \quad k = 1, 2, \ldots, n \quad \text{and} \quad c_{lk} = \sqrt{2/(n+1)} \sin[kl\pi/(n+1)]$$

where $l = $ atom index and $k = $ MO index

2. Cyclic $2p\pi$ molecules

$$x_k = -2\cos(2\pi k/n), \ k = 0, 1, 2, \ldots, (n-1) \quad \text{with} \quad c_{lk} = \left(1/\sqrt{n} \right) \exp[2\pi i k(l-1)/n]; \quad i = \sqrt{-1}$$

The Hückel–Roberts π-electron model at least brings with it (1) the use of hydrogenic orbitals, (2) the "overlap-bonding" relationship, and (3) the application of the secular equation. Such orbitals follow the amplitude of the particle-in-a-box in Chapter 12, and now one can even sketch a simple representation of the molecular orbitals since they are linear combinations of all the $2p_z$ atomic orbitals. However, to critique this model, one can ask what the inner 1s orbitals and the valence sigma orbitals are doing. Further, the model only applies to delocalized pi systems. Still this is a credible result for the orbital model of molecules.

This type of calculation is very approximate and its only present value is to illustrate the basic concepts, although historically it stimulated great interest in molecular orbital theory [9,10]. It is especially noteworthy to mention the effect Roberts' text [8] had on physical organic chemistry in the early 1960s. That text went through 14 printings and convinced many young chemistry students, including this author, that orbital theory and even some point group theory was useful in organic chemistry. Ironically, Roberts' treatment used many of the ideas from the 1944 text "Quantum Chemistry" by Eyring, Walter, and Kimble [11] which was considered too difficult by many chemists. Roberts has coined a term in his memoirs as the abbreviation ATRPATRT, and his enthusiastic "Notes..." text proves the principle of being "at the right place at the right time"! A very interesting history by Roberts is available at http://www.quantum-chemistry-history.com/Roberts1.htm.

ELEMENTARY LINEAR ALGEBRA

It is debatable whether we should have presented this section before we talked about the Hückel–Roberts pi-electron method. Nevertheless, it is the opinion of this author that you would not have read this section as a topic presented without an application. Perhaps you now want to know what this matrix business is all about. If we want to proceed with a few modern computational applications, we need a major concept that may not be on your transcript. This author never took a course in linear algebra but a working understanding of the principles of linear algebra has proved to be indispensible in research, easily in the list of "essentials." Further, if you have taken a course in linear algebra but never developed the skill to diagonalize or invert a small matrix on scratch paper, you still need to learn these skills. Actually the author's understanding is based mainly on a short six-page set of notes from Prof. Walter Kauzmann at Princeton. So here is a short course in linear algebra.

A matrix can be an operator or result of an operation in terms of basis set coordinates. $\begin{bmatrix} 3 & 1 \\ 1 & 2 \end{bmatrix} \begin{bmatrix} x \\ y \end{bmatrix} = \begin{bmatrix} (3x + 1y) \\ (1x + 2y) \end{bmatrix}$. This matrix operation changed a vector $\vec{r} = x\hat{i} + y\hat{j}$ to another vector $\vec{r} = x'\hat{i} + y'\hat{j}$ where the new coordinate is $x' = 3x + 1y$ and $y' = 1x + 2y$. That changed the direction of the vector and its length which is a general and severe thing to do, but possible under matrix multiplication. Operationally matrix multiplication is carried out using the "row × column" multiplication across a row and down a column, summing the individual products. Here

$$[A][B] = [C] \Rightarrow c_{ij} = \sum_k (a_{ik}b_{kj}).$$

A more gentle matrix operation is to just change the direction of the (column) vector by rotating it through an angle θ as shown by the operation

$$\begin{bmatrix} \cos\theta & \sin\theta \\ -\sin\theta & \cos\theta \end{bmatrix} \begin{bmatrix} x \\ y \end{bmatrix} = \begin{bmatrix} (x\cos\theta + y\sin\theta) \\ (-x\sin\theta + y\cos\theta) \end{bmatrix} = \begin{bmatrix} x' \\ y' \end{bmatrix}.$$

This operation is ambiguous as to whether the vector rotated by θ or the axis system rotated but generally we assume the vector was rotated by the matrix "operator" (using the "row × column" multiplication). This matrix operation is *unitary* because it maintained the length of the vector as constant during the rotation. The Pythagorean sum of squares of the components remains the same. Matrices can do other "symmetry operations" such as inversion via $\begin{bmatrix} -1 & 0 & 0 \\ 0 & -1 & 0 \\ 0 & 0 & -1 \end{bmatrix} \begin{bmatrix} x \\ y \\ z \end{bmatrix} = \begin{bmatrix} -x \\ -y \\ -z \end{bmatrix}$, again using the multiplication element by element (row $\times$

column). You can see that this matrix "inverts" all the coordinates to their negative values. Next

consider $\begin{bmatrix} 1 & 0 & 0 \\ 0 & 1 & 0 \\ 0 & 0 & -1 \end{bmatrix} \begin{bmatrix} x \\ y \\ z \end{bmatrix} = \begin{bmatrix} x \\ y \\ -z \end{bmatrix}$. This operation causes a "reflection" in the $x-y$ plane and

is called a C_s symmetry operation. Again, $\begin{bmatrix} 0 & -1 & 0 \\ 1 & 0 & 0 \\ 0 & 0 & 1 \end{bmatrix} \begin{bmatrix} x \\ y \\ z \end{bmatrix} = \begin{bmatrix} -y \\ x \\ z \end{bmatrix}$ is a 90° or C_4 rotation

about the z-axis. That matrix is one example of a more general case of a C_n rotation matrix for

a rotation about the z-axis by an angle of $\left(\dfrac{2\pi}{n}\right)$ as $C_n = \begin{bmatrix} \cos\left(\frac{2\pi}{n}\right) & \sin\left(\frac{2\pi}{n}\right) & 0 \\ -\sin\left(\frac{2\pi}{n}\right) & \cos\left(\frac{2\pi}{n}\right) & 0 \\ 0 & 0 & 1 \end{bmatrix}$. We previ-

ously used a C_5 operation to generate the members of the Powell 5-equivalent d-orbitals for the H atom. Those are a few symmetry operations that have the effect of moving a given Cartesian point to another place without changing the length of the position vector, they are *unitary* operations. We have seen how analytical functions can satisfy eigenvalue equations and now we will show that matrices can satisfy the same condition. In quantum mechanics, we know $\psi^*\psi = \psi\psi^*$ so it has been found that when matrix operators are used in quantum mechanics they are of a slightly restricted type called Hermitian matrices, which means the operators are square $(n \times n)$ matrices in which $a_{ji}^* = a_{ij}$. This only matters when the operator being represented by the matrix has some imaginary part (such as angular momentum) but usually $a_{ji} = a_{ij}$ in most cases. However, not all matrices are square and a single row or column can be a matrix. Note for matrix multiplication to be defined, the inner dimension of the two matrices in the product must be the same. For instance, when we multiply $(m \times n) \times (n \times k) = (m \times k)$ the important thing is that the n-dimension matches. For

example $[c_1\ c_2] \begin{bmatrix} c_1 \\ c_2 \end{bmatrix} = c_1^2 + c_2^2$ is a scalar, a single number, because a $(1 \times 2) \times (2 \times 1) =$

$(1 \times 1) = $ a number.

Now we come to the main useful mystery of linear algebra, the *diagonalization* process. We have shown earlier that matrices can change the directions of vectors so it is possible to do generalized rotations that result in zeros everywhere in the matrix except on the diagonal from the upper left corner to the lower right corner.

Example

Find the eigenvalues and eigenvectors of the matrix equation given and show the matrix of the column eigenvectors diagonalizes the original matrix.

Given: $\begin{bmatrix} 3 & 1 \\ 1 & 3 \end{bmatrix} = [M]$ where $[M][C] = [\lambda][C]$ as $\begin{bmatrix} 3 & 1 \\ 1 & 3 \end{bmatrix}\begin{bmatrix} c_1 \\ c_2 \end{bmatrix} = \begin{bmatrix} \lambda & 0 \\ 0 & \lambda \end{bmatrix}\begin{bmatrix} c_1 \\ c_2 \end{bmatrix}$, we can see

that this is an eigenvalue equation where the unknown coefficients are the eigenvector coefficients and there will be two values of λ that are the eigenvalues in this (2×2) space. It is important to note for future reference that while we are used to writing $H\psi = E\psi$ with the eigenvalue first that will lead to problems if the overlap matrix is other than a unit matrix. We should realize we could as well write $[M][C] = [C][\lambda]$, which turns out to be correct if the overlap is not a unit matrix but here it does not matter because the overlap is a unit matrix. We can learn a lot about linear algebra from just this simple case, which is similar to the Hückel pi-electron problem for ethylene. Note that this is a symmetric (Hermitian) matrix because $a_{ji} = a_{ij}$. Further if the off-diagonal values of 1 were 0 we would have a system where both eigenvalues would obviously be 3 and 3, which would be a case of degenerate eigenvalues. However, we will see that the off-diagonal values of 1 will split the degeneracy and that is a lesson for cases where the matrix is an energy matrix.

Off-diagonal interactions will split degenerate energy levels. Last, for future comparison note that the sum of the diagonal elements is $(3+3) = 6$. The sum of the diagonal elements is called the "trace" in English but in German papers it is the "spur," which relates to the English word for the trail of an animal as a "spoor" and that helps understand the translation as a trace or trail. Many of the early scientific papers on quantum mechanics were formulated in "matrix mechanics" and published in German.

Next, we combine both sides of the equation element by element to set up the Cayley–Hamilton condition. In matrix addition or subtraction $[A] + [B] = [C] \Rightarrow c_{ij} = a_{ij} + b_{ij}$, so $\begin{bmatrix} (3-\lambda) & 1 \\ 1 & (3-\lambda) \end{bmatrix} \begin{bmatrix} c_1 \\ c_2 \end{bmatrix} = 0$, which corresponds to a linear system of (two) equations as

$$(3-\lambda)c_1 + c_2 = 0,$$

$$c_1 + (3-\lambda)c_2 = 0.$$

Since the right side constants are 0, we have to invoke the Cayley–Hamilton theorem just as we did with the pi-electron treatment of ethylene. We force the determinant of the coefficients to be zero! Then the eigenvalues of the matrix will be the roots of the polynomial in λ, which fulfills the Cayley–Hamilton Theorem!

$$\begin{vmatrix} (3-\lambda) & 1 \\ 1 & (3-\lambda) \end{vmatrix} = 0 \Rightarrow (3-\lambda)^2 - 1 = \lambda^2 - 6\lambda + 8 = (\lambda - 4)(\lambda - 2) = 0.$$

Thus $\lambda = 4$ and 2 which are the roots of the Cayley–Hamilton polynomial as well as the eigenvalues of the matrix! When $\lambda = 4$ we can insert this root back into either equation to find the corresponding eigenvector.

Arbitrarily choose (either equation works!) $(3-4)c_1 + 1c_2 = 0 \Rightarrow -c_1 + c_2 = 0 \Rightarrow c_1 = c_2$. Now just as with the ethylene pi-electron problem, we need to find eigenvectors that are normalized to 1 because we are going to perform a generalized coordinate rotation to a new orientation but we want to do it in a unitary way; do not change the length of the vector, just rotate it. So we set $[c_1 \ c_1] \begin{bmatrix} c_1 \\ c_1 \end{bmatrix} = c_1^2 + c_1^2 = 2c_1^2 = 1 \Leftrightarrow c_1 = \dfrac{1}{\sqrt{2}} = c_2$. Let us check to see if this forms an eigenvector. $\begin{bmatrix} 3 & 1 \\ 1 & 3 \end{bmatrix} \begin{bmatrix} 1/\sqrt{2} \\ 1/\sqrt{2} \end{bmatrix} = \begin{bmatrix} (3/\sqrt{2} + 1/\sqrt{2}) \\ (1/\sqrt{2} + 3/\sqrt{2}) \end{bmatrix} = 4 \begin{bmatrix} 1/\sqrt{2} \\ 1/\sqrt{2} \end{bmatrix}$, Q.E.D. it is an eigenvector with $\lambda = 4$.

Next find the eigenvector for $\lambda = 2$. Again $(3-2)c_1 + 1c_2 = 0 \Rightarrow c_1 = -c_2$ and again we require $[c_1 \ -c_1] \begin{bmatrix} c_1 \\ -c_1 \end{bmatrix} = c_1^2 + c_1^2 = 2c_1^2 = 1 \Leftrightarrow c_1 = \dfrac{1}{\sqrt{2}} = -c_2$. Let us check this eigenvector also.

$\begin{bmatrix} 3 & 1 \\ 1 & 3 \end{bmatrix} \begin{bmatrix} 1/\sqrt{2} \\ -1/\sqrt{2} \end{bmatrix} = \begin{bmatrix} (3/\sqrt{2} - 1/\sqrt{2}) \\ (1/\sqrt{2} - 3/\sqrt{2}) \end{bmatrix} = 2 \begin{bmatrix} 1/\sqrt{2} \\ -1/\sqrt{2} \end{bmatrix}$, Q.E.D. it is an eigenvector with $\lambda = 2$.

The next steps are almost magical and well worth the preceding effort. Form a matrix called T (for Transformation) by lining up the eigenvectors as columns. $[T] = \begin{bmatrix} 1/\sqrt{2} & 1/\sqrt{2} \\ 1/\sqrt{2} & -1/\sqrt{2} \end{bmatrix}$. It turns out you can put the eigenvector columns into T in any order as long as they are columns and are normalized. It is easy to show that the eigenvectors are orthogonal as well as normalized to 1, $\begin{bmatrix} 1/\sqrt{2} & 1/\sqrt{2} \end{bmatrix} \begin{bmatrix} 1/\sqrt{2} \\ -1/\sqrt{2} \end{bmatrix} = \dfrac{1}{2} - \dfrac{1}{2} = 0$, so they are orthonormal!

Students who have not taken any linear algebra often ask, "What is going on here, how does this work?" A very elegant proof/demonstration is given by Williams in his text [12]. Actually the theorem Williams gives is more general than what we need for symmetric Hermitian matrices but it illustrates another very important truth that with diagonal representations, you can perform ordinary arithmetic on the diagonal elements and in this case diagonal matrices commute since real numbers commute. The whole proof/principle of diagonalization depends on the key step of commuting the diagonal matrix of eigenvalues to the right side of the eigenvector matrix. We have paraphrased Williams' theorem in the conversational style of this text but this author greatly appreciates Williams' formal description.

UNITARY SIMILARITY DIAGONALIZATION OF A SQUARE HERMITIAN MATRIX [12]

Let $[A]$ be an $n \times n$ matrix (and Hermitian symmetric for quantum mechanics) with eigenvectors V_1, V_2, $V_3, \ldots$, V_n and corresponding eigenvalues λ_1, λ_2, $\lambda_3, \ldots$, λ_n. Let $[T] = \begin{bmatrix} V_1 & V_2 \ldots & V_n \\ \ldots & \ldots & \ldots \\ \ldots & \ldots & \ldots \end{bmatrix}$, a matrix with the eigenvectors of $[A]$ as columns in any order. Then

$$[A][T] = \begin{bmatrix} AV_1 & AV_2 \ldots & AV_n \\ \ldots & \ldots & \ldots \\ \ldots & \ldots & \ldots \end{bmatrix} = \begin{bmatrix} \lambda_1 V_1 & \lambda_2 V_2 \ldots & \lambda_n V_n \\ \ldots & \ldots & \ldots \\ \ldots & \ldots & \ldots \end{bmatrix} = [T] \begin{bmatrix} \lambda_1 & 0 & 0 \\ 0 & \ldots & 0 \\ 0 & 0 & \lambda_n \end{bmatrix}.$$ That is a

key step because real numbers commute and we have factored out the eigenvalue matrix to the right of $[T]$. Then it is a simple matter to multiply from the left by $[T]^{-1}$ to prove that a unitary similarity transformation can produce a diagonal form of the original matrix $[A]$ since $[T]^{-1}[T] = 1_n$, a unit matrix of dimension n.

$$[T]^{-1}[A][T] = [T]^{-1}[T]\begin{bmatrix} \lambda_1 & 0 & 0 \\ 0 & \ldots & 0 \\ 0 & 0 & \lambda_n \end{bmatrix} = \begin{bmatrix} \lambda_1 & 0 & 0 \\ 0 & \ldots & 0 \\ 0 & 0 & \lambda_n \end{bmatrix} = [A]_{\text{diag}}, \quad \text{Q.E.D.}$$

From this demonstration, you can see that we need to find a way to "divide" matrices as in "$[T] \div [T]$" $= [T]^{-1}[T] = [T][T]^{-1}$. Well, there is no such thing as matrix division although there is a way to get an inverse as T^{-1}. A student should know/learn that there is a tedious way to find the inverse of a matrix as $a_{ij}^{-1} = \dfrac{(-1)^{(i+j)}|a_{ji}|}{|A|}$ where this means that each a_{ij}^{-1} element of the matrix is the sub-determinant of the original matrix of the a_{ji} element divided by the determinant of the whole matrix and multiplied by a minus sign alternating across a row or column. For this (2×2) case this is easy, but for larger matrices, the sub-determinant is actually a small determinant of dimension only one less than the original matrix, which produces a very lengthy process! We will see that this method is seldom used but you ought to know it exists.

Example

Find the inverse of $[A] = \begin{bmatrix} 3 & 1 \\ 1 & 2 \end{bmatrix}$. The determinant of $[A]$ is $\begin{vmatrix} 3 & 1 \\ 1 & 2 \end{vmatrix} = (3 \times 2) - (1 \times 1) = 5$. Thus

$[A]^{-1} = \begin{bmatrix} 2/5 & -1/5 \\ -1/5 & 3/5 \end{bmatrix}$. So this tedious formula can be used to form the inverse of $[A]$ and then

$[A][A]^{-1} = \begin{bmatrix} 3 & 1 \\ 1 & 1 \end{bmatrix}\begin{bmatrix} 2/5 & -1/5 \\ -1/5 & 3/5 \end{bmatrix} = \begin{bmatrix} (6/5 - 1/5) & (-3/5 + 3/5) \\ (2/5 - 2/5) & (-1/5 + 6/5) \end{bmatrix} = \begin{bmatrix} 1 & 0 \\ 0 & 1 \end{bmatrix},$ Q.E.D.!

Fortunately *there is a short cut* we can use most of the time when $[T]$ is unitary (formed from orthonormal columns) as here $[T] = \begin{bmatrix} 1/\sqrt{2} & 1/\sqrt{2} \\ 1/\sqrt{2} & -1/\sqrt{2} \end{bmatrix}$ and we see that $|T| = \dfrac{-1}{2} - \dfrac{1}{2} = -1$, of magnitude 1 with a negative phase, but we know the phase of a wave function is arbitrary.

Then according to the formula $[T]^{-1} = \begin{bmatrix} \left(\frac{(-1/\sqrt{2})}{-1} \right) & \left(\frac{(-1/\sqrt{2})}{-1} \right) \\ \left(\frac{(-1/\sqrt{2})}{-1} \right) & \left(\frac{(+1/\sqrt{2})}{-1} \right) \end{bmatrix} = \begin{bmatrix} 1/\sqrt{2} & 1/\sqrt{2} \\ 1/\sqrt{2} & -1/\sqrt{2} \end{bmatrix} = [T],$

so $[T]^{-1} = [T]^{\dagger}$, i.e., $t_{ji} = t_{ij}$ as long as $[T]$ is unitary as here for this special case, which is used every time we do this unitary transformation $[T]^{-1}[M][T] = [M]_{\text{diag}}$. Thus when $[T]$ is formed from columns of orthonormal eigenvectors, you can just "flip" the elements around the diagonal to get the inverse with $t_{ij} = t_{ji}$. In our example, the minus sign was in the (2, 2) position so flipping it leaves it unchanged. If we had arbitrarily set up the transformation as $[T] = \begin{bmatrix} 1/\sqrt{2} & 1/\sqrt{2} \\ -1/\sqrt{2} & 1/\sqrt{2} \end{bmatrix}$, then the

inverse is $[T]^{-1} = \begin{bmatrix} 1/\sqrt{2} & -1/\sqrt{2} \\ 1/\sqrt{2} & 1/\sqrt{2} \end{bmatrix}$.

We are taking a lot of space here to illustrate the very important technique of diagonalizing a matrix with a simple (2×2) example but there should be no doubt that this is one of the "essentials" of modern physical chemistry. Modern quantum chemistry programs use this concept hundreds of times in every calculation, often for matrices of dimension (100×100) or more! Well then let us see it! We need to multiply out $[T]^{-1}[M][T] = [M]_{\text{diag}}$. Note that if we were to multiply in the order $[M][T][T]^{-1}$ with $[T]^{-1}$ on the right side, $[M][T][T]^{-1} = [M]$ and nothing happens to $[M]$ since it is just multiplying a unit matrix. Now perform $[T]^{-1}[M][T]$ as

$\begin{bmatrix} 1/\sqrt{2} & 1/\sqrt{2} \\ 1/\sqrt{2} & -1/\sqrt{2} \end{bmatrix} \begin{bmatrix} 3 & 1 \\ 1 & 3 \end{bmatrix} \begin{bmatrix} 1/\sqrt{2} & 1/\sqrt{2} \\ 1/\sqrt{2} & -1/\sqrt{2} \end{bmatrix} = \begin{bmatrix} 1/\sqrt{2} & 1/\sqrt{2} \\ 1/\sqrt{2} & -1/\sqrt{2} \end{bmatrix} \begin{bmatrix} (4/\sqrt{2}) & (2/\sqrt{2}) \\ (4/\sqrt{2}) & (-2/\sqrt{2}) \end{bmatrix}$. Note

that the intermediate matrix on the right clearly demonstrates Williams' proof showing each eigenvector is multiplied by its eigenvalue! Using Williams' proof [12], we could factor

out the eigenvalues to the right as $\begin{bmatrix} 1/\sqrt{2} & 1/\sqrt{2} \\ 1/\sqrt{2} & -1/\sqrt{2} \end{bmatrix} \begin{bmatrix} (4/\sqrt{2}) & (2/\sqrt{2}) \\ (4/\sqrt{2}) & (-2/\sqrt{2}) \end{bmatrix} =$

$\begin{bmatrix} 1/\sqrt{2} & 1/\sqrt{2} \\ 1/\sqrt{2} & -1/\sqrt{2} \end{bmatrix} \begin{bmatrix} 1/\sqrt{2} & 1/\sqrt{2} \\ 1/\sqrt{2} & -1/\sqrt{2} \end{bmatrix} \begin{bmatrix} 4 & 0 \\ 0 & 2 \end{bmatrix} = \begin{bmatrix} 4 & 0 \\ 0 & 2 \end{bmatrix}$ or we can do it the hard way and

just complete the second multiplication. $\begin{bmatrix} 1/\sqrt{2} & 1/\sqrt{2} \\ 1/\sqrt{2} & -1/\sqrt{2} \end{bmatrix} \begin{bmatrix} (4/\sqrt{2}) & (2/\sqrt{2}) \\ (4/\sqrt{2}) & (-2/\sqrt{2}) \end{bmatrix} =$

$\begin{bmatrix} (4/2 + 4/2) & (2/2 - 2/2) \\ (4/2 - 4/2) & (2/2 + 2/2) \end{bmatrix} = \begin{bmatrix} 4 & 0 \\ 0 & 2 \end{bmatrix}$, wow! While this is not very impressive for the (2×2) case, if we did this for a (100×100) matrix (operator) and it went from a bewildering array of 10,000 numbers to just 100 numbers along the diagonal, we would really appreciate the greater simplicity of the diagonal form where we can just read the eigenvalues off the diagonal list. But wait! The trace of the original matrix was 6 and the trace of the new diagonal matrix is still 6! *The trace of a Hermitian matrix remains invariant under a unitary transformation!* So not only is the diagonal matrix simpler, its effect is still the same and besides we now know the pure orthonormal eigenvectors that satisfy the eigenvalue equation $[A][c] = \lambda[c]$ for each individual column vector! Was this example worth the effort? You better believe it! Without the technique of diagonalization the matrix form of quantum mechanics would be a hopeless set of arrays sprinkled densely with off-diagonal numbers that could not be separated!

One final word is to look back at how the off-diagonal "1" elements split the degenerate set of "3" values on the diagonal of the original matrix. The interaction with the off-diagonal elements changed (3, 3) to (4, 2) so one diagonal element increased while the other decreased and the amount of the increase/decrease is the off-diagonal amount! When this happens with degenerate energy levels in a matrix like the Hamiltonian energy matrix, it is said to be the "non-crossing rule" where two energy levels (of the same symmetry) interact, one is pushed up and the other is pushed down by the interaction, a very general principle.

JACOBI ALGORITHM FOR DIAGONALIZATION USING A COMPUTER

What we have shown earlier for a simple (2×2) case is solvable with paper-and-pencil work and we believe we have packed a lot of principles in that one example so it is worth reading several times. Nevertheless, modern computation has developed several computer code methods to diagonalize Hermitian matrices since it is such a key step. Some of these algorithms are optimized for speed in use of very large matrices, but are complicated gems of computer science. The simplest computer code is called the Jacobi method originally written by F. J. Corbato and M. Merwin at the Massachusetts Institute of Technology and graciously documented by Offenhartz [13]. In the Jacobi method, we seek a solution to $[H][C] = [E][C]$. Given that $[H]$ is a square, Hermitian ($n \times n$) matrix, we expect to find "n" eigenvectors in $[C]$ with the normalized eigenvectors in either rows or, preferably, the columns of $[C]$. Initially, $[C]$ is set up as an ($n \times n$) unit matrix with all zeros except for 1 along the diagonal. Then an initial array is also set up as a unit matrix for $[T_0]$ and examination of the upper triangle values of $[H]$ is carried out since $H_{ij} = H_{ji}$. Then the key equation is $\tan(2\theta) = \dfrac{2H_{ij}}{H_{jj} - H_{22}}$ with θ restricted to the first and fourth quadrants and if $H_{ii} = H_{jj}$ then $\theta = (\pi/4) = 45°$. With this angle, a new $[T_1]$ matrix is generated with 1 along the diagonal except for the (i, j) subspace, which contains the (2×2) rotation matrix $\begin{bmatrix} \cos\theta & \sin\theta \\ -\sin\theta & \cos\theta \end{bmatrix}$; the $[T_1]^{-1}$ matrix is just "flipped" about the diagonal and a new $[H_1] = [T_1][H_0][T_1]$ is computed, which is only partially diagonalized. Next another H_{ij} is selected for use of a different $[T_2]$ and a new $[H_2] = [T_2]^{-1}[T_1]^{-1}[H_0][T_1][T_2]$ is computed. Each time a new $[T_n]$ is formed, it is cumulatively applied to the initial unit matrix for $[C]$. Thus the computer program sweeps over a double loop of i and j with $j > i$ since we are only interested in making the off-diagonal elements zero and $H_{ij} = H_{ji}$. On each step, it computes a new $[T_n]$ and uses that for $[H_n] = [T_n]^{-1}[H_{n-1}][T_n]$ and $[C_n] = [T_n][C_{n-1}]$. The Corbato–Merwin routine scans through the off-diagonal elements and chooses the largest off-diagonal element to treat because the diagonalization of one (2×2) subspace may make another subspace less diagonal. Even so a simple-minded routine could just treat every (i, j) pair in the list and run through the list maybe ten times or until the largest H_{ij} is less than some desired threshold. While such a procedure is unthinkable for pencil-and-paper work, the miracle of computers makes this task take less than a second for matrices up to about (35×35). One warning is that the Jacobi method does sag under the computational burden somewhere around (35×35) and other methods are faster for larger matrices, although (100×100) matrices can be diagonalized on modern PCs in a few seconds. A typical transformation matrix for the $H(2, 5)$ element in a (6×6) space would look like

$$[T_n] = \begin{bmatrix} 1 & 0 & 0 & 0 & 0 & 0 \\ 0 & \cos\theta & 0 & 0 & \sin\theta & 0 \\ 0 & 0 & 1 & 0 & 0 & 0 \\ 0 & 0 & 0 & 1 & 0 & 0 \\ 0 & -\sin\theta & 0 & 0 & \cos\theta & 0 \\ 0 & 0 & 0 & 0 & 0 & 1 \end{bmatrix}.$$

ORDER MATTERS!

In order to warn undergraduates of a pitfall in linear algebra, let us consider the important case of a matrix eigenvalue case where *the basis set is not orthogonal* as in

$$HC = SCE = (?)ESC.$$

While we have become accustomed to writing $H\psi = E\psi$, we need to be very careful about the meaning of the indices of the matrices we use and we actually have to "think" about what we are doing rather than just rearranging matrices. What follows is edited from a set of notes by Prof. James Harrison of the Chemistry Department of Michigan State University. Let us define a basis expansion for ψ_i in terms of a (probably incomplete) set of basis functions $\{\chi_\mu\}$ so that we have

$$\psi_{i\text{-orbital}} = \sum_{\mu\text{-basis}} c_{\mu\text{-basis},i\text{-orbital}}(\chi_{\mu\text{-basis}}) = \sum_{\mu b} c_{\mu b,io}(\chi_{\mu b}) = \psi_{io}.$$

Here we have used abbreviations io for the orbital index and μb for the basis member index. Let us assume we can write our problem as $HC = ESC$ and focus on just one typical matrix element using inner-index sums for the multiplication steps using "row-column" multiplication.

$$(HC)_{\mu b,io} = \sum_\sigma H_{\mu b,\sigma} C_{\sigma,io} = (ESC)_{\mu b,io} = \sum_\sigma \sum_\lambda E_{\mu b,\sigma} S_{\sigma,\lambda} C_{\lambda,io} = \sum_\sigma \sum_\lambda \varepsilon_{\mu b} \delta_{\mu b,\sigma} S_{\sigma,\lambda} C_{\lambda,io}.$$

We seek an answer where E is a diagonal matrix with all off-diagonal elements equal to zero, which is accomplished by the use of the Kroneker delta $\delta_{\mu b,\sigma} = 1(\mu b = \sigma); = 0(\mu b \neq \sigma)$. Then

$$(HC)_{\mu b,io} = \sum_\sigma H_{\mu b,\sigma} C_{\sigma,io} = (ESC)_{\mu b,io} = \sum_\sigma \sum_\lambda E_{\mu b,\sigma} S_{\sigma,\lambda} C_{\lambda,io} = \varepsilon_{\mu b} \sum_\lambda S_{\mu b,\lambda} C_{\lambda,io}.$$

That results in $(HC)_{\mu b,io} = \varepsilon_{\mu b} \sum_\lambda S_{\mu b,\lambda} C_{\lambda,io}$ and this does not make any useful sense because the energy is $\varepsilon_{\mu b}$ and would refer to some energy of a basis function, not an orbital. Thus we see that using the order $HC = ESC$ does not lead to a useful result. Now consider $HC = SCE$.

$$(HC)_{\mu b,io} = \sum_\sigma H_{\mu b,\sigma} C_{\sigma,io} = (SCE)_{\mu b,io} = \sum_\sigma \sum_\lambda S_{\mu b,\sigma} C_{\sigma,\lambda} E_{\lambda,io} = \sum_\sigma \sum_\lambda S_{\mu b,\sigma} C_{\sigma,\lambda} \delta_{\lambda,io} \varepsilon_{io}.$$

That results in $(HC)_{\mu b,io} = \varepsilon_{io} \sum_\sigma S_{\mu b,\sigma} C_{\sigma,io}$. Now we have "orbital energy" ε_{io} and we can see that the summation over σ runs across $S_{\mu b,\sigma}$ and down a column (second index of $C_{\sigma,io}$) so the orbital energy refers to the eigenvalue of an eigenvector column in the $C_{\mu b,io}$ matrix, which is what we want. Sometimes we have to be very careful about the meaning of the indices of the matrices we use. The simple Roberts–Hückel case gives us some basic concepts but its simplicity disguises the pitfall we see in the preceding text for the case of a nonorthogonal basis set.

SUMMARY

There are two main ways to approximate the Schrödinger equation solution for "unsolvable problems": perturbation theory and the variation principle. Perturbation theory is not bounded but can be useful when the perturbation is small. The variation principle is much more powerful and can be used to approach accurate energy values for the lowest state of a given symmetry from above with a guaranteed lower bound. The variation principle inspires the linear combination of atomic orbitals (LCAO) approximation with fixed atomic orbitals and parametric coefficients. We found a way to represent the pi-orbitals of ethylene in terms of specific $2p_z$ atomic orbitals. Hückel pi-electron theory was implemented in a particularly effective way by J.D. Roberts and we can use pi-electron equations to learn a working knowledge of linear algebra, including the Cayley–Hamilton theorem, matrix inversion, and especially the very valuable technique of matrix diagonalization. Modern computer programs routinely diagonalize very large matrices which simplifies quantum chemistry problems from "unimaginable" to merely "difficult." A demonstration was given that the order of the nonorthogonal eigenvalue problem must be $HC = SCE$. While this material looks difficult, you can teach yourself enough linear algebra to understand the essentials of modern quantum chemistry in a few hours! The linear algebra section was presented after the Hückel–Roberts pi-electron method challenged us to learn some key concepts needed in quantum chemistry/mechanics.

PROBLEMS

16.1 Show that $\psi_0^{(0)} = \left(\dfrac{k}{h\nu\pi}\right)^{1/4} e^{-\left(\frac{kx^2}{2h\nu}\right)}$ is an eigenfunction of the harmonic oscillator Hamiltonian

operator $H = \dfrac{-\hbar^2}{2m}\dfrac{d^2}{dx^2} + \dfrac{kx^2}{2}$. What is the eigenvalue? Hint: use $\nu = \dfrac{1}{2\pi}\sqrt{\dfrac{k}{m}}$.

16.2 Find the inverse to the matrix $[M] = \begin{bmatrix} 1 & 2 & 3 \\ 2 & 4 & 5 \\ 3 & 5 & 6 \end{bmatrix}$ and prove that $[M][M]^{-1} = \begin{bmatrix} 1 & 0 & 0 \\ 0 & 1 & 0 \\ 0 & 0 & 1 \end{bmatrix}$.

16.3 Diagonalize the matrix $[M] = \begin{bmatrix} 2 & 0 & 1 \\ 0 & 3 & 0 \\ 1 & 0 & 2 \end{bmatrix}$ by noting the M_{22} element is already a diagonal

value of 3 and so the eigenvector can be obtained by inspection as $\begin{bmatrix} 0 \\ 1 \\ 0 \end{bmatrix}$. Then take the matrix

apart and diagonalize the (1, 3) subspace, which is now only $\begin{bmatrix} 2 & 1 \\ 1 & 2 \end{bmatrix}$. When you find the

eigenvalues and eigenvectors of this (2 × 2) subspace, reassemble the transformation matrix

$[T] = \begin{bmatrix} ? & 0 & ? \\ 0 & 1 & 0 \\ ? & 0 & ? \end{bmatrix}$ and flip it to find $[T]^{-1}$. Then show explicitly that $[T]^{-1}[M][T] = [M]_{\text{diag}}$.

Hint, this question frequently appears on final examinations.

16.4 Use the Jacobi algorithm to find the eigenvalues and eigenvectors of the matrix $[M]$ in Problem 16.3.

TESTING, GRADING, AND LEARNING?

Realistically, this may be the end of the course although sometimes the lectures can be extended to introductory group theory given in Chapter 18 with applications to Raman spectroscopy. Thus we offer here a final examination given in 2008 at Virginia Commonwealth University as a continuation of the 2008 examinations from the first semester. This written examination was only part of the final examination and the first 110 min consisted of a standardized test in physical chemistry from the American Chemical Society. There was a break after the ACS examination and another 90 min allowed for this written "Part II" examination. These questions are easy and designed only to determine whether the students read Chapters 10 through 16 with comprehension. The ACS test grade must be made a part of the final average so that the students will actually try their best and not just fill in the bubble sheets randomly. They are encouraged to write more derivations after answering the eight questions and can score more than the total of 100 points on Part II.

CHEM 304   Final Exam (Part II) Summer 2008   D. Shillady, Professor
(points)   (Attempt all problems!)   (90 min)

(15) 1. Given $\psi_0 = \left(\dfrac{k}{h\nu\pi}\right)^{1/4} e^{-\left(\frac{kx^2}{2h\nu}\right)}$ for the $n = 0$ level of a harmonic oscillator, calculate the total

energy in the presence of a perturbation $H' = ax^3 + bx^4$. $\left(\text{Answer: } E_{tot} = \left(\dfrac{h\nu}{2}\right) + \dfrac{3b}{4}\left(\dfrac{h\nu}{k}\right)^2\right)$

(15) 2. Given the vibrational fundamental of O_2 is 1556 cm^{-1}, calculate $(E - E_0)$ for 1 mol of O_2 gas in calories and joules at 1000 K (23.0 kJ/mol = 5.498 kcal/mol).

(10) 3. Draw a schematic of a 20 keV electron microscope with component labels. Show how Auger spectra could be used to detect Sn ($Z = 50$) and calculate the (2S->1S) Bohr wavelength ($\lambda \cong 0.486$ Å according to the Bohr Model).

(10) 4. Compute the De Broglie wavelength of an electron ejected from Pt, whose work function is 5.0 eV, by the combined energy of a *two photon* absorption from an intense N_2 laser pulse with a wavelength of 3371 Å ($\lambda_{matter} \cong 7.98$ Å).

(10) 5. Sketch the 1S, 2S, 3S, 2P, 3P, and 3D angular and radial functions for the H atom. show the "standard" 3D orbitals and Powell/s five-equivalent-D orbitals (see Chapter 14).

(15) 6. Given the vibrational fundamental of O_2 (again) is 1556 cm^{-1}; this time calculate the total heat capacities C_P and C_V for O_2 at 800 K including translation and rotation ($C_V = 6.042$ cal/mol °K; $C_P = 8.029$ cal/mol °K).

(15) 7. Using $\varepsilon_j = j(j+1)\left(\dfrac{\hbar^2}{2I}\right)$, calculate the average rotational energy per mole for a diatomic molecule using a partition function $\left(q_{rot} = \left(\dfrac{2IkT}{\hbar^2}\right), E_{rot} = RT \right)$.

(10) 8. Calculate the x-ray wavelength of the (2S->1S) Auger wavelength emission from O ($Z = 8$) in the $(SiO_2)_x$ glass of a TV screen using 30 keV electrons (cathode ray tube). Estimate using the Bohr model ($\lambda \cong 18.985$ Å; soft x-ray).

What follows is a more difficult final examination from Summer 2009; it still followed an ACS Standardized Examination of 110 min and there were some very high percentile scores. Students were allowed to start this part if they finished the ACS test early.

CHEM 304   Final Exam (Part II) Summer 2009   D. Shillady, Professor
(points)   (Attempt all problems!)   (90 min)

(10) 1. Derive the Michaelis–Menten equation for enzyme kinetics in the case of competitive inhibition by species "I"; define K_m and show a graph (see Chapter 8).

(10) 2. Crystalline NaCl has a simple cubic structure with an NaCl distance of 2.76 Å. Calculate the Bragg *angle* of the $n = 2$ LEED pattern for diffraction of 150 eV electrons by this NaCl distance $\left(\sin(\theta) = \dfrac{nh}{mv2d} = 0.3625 \text{ } (ans \text{ } \theta = 21.25°) \right)$.

(10) 3. Adjust the length "L" of the Particle-in-a-Box" model to fit the first $\pi \rightarrow \pi^*$ transition of all-*trans*-hexatriene to the experimentally observed wavelength of 2510 Å ($L \cong 7.29$ Å).

(20) 4. Normalize the variational trial function $\Phi = Ne^{-\alpha r}$ for the 1S state of the H atom and show $\langle E \rangle$ is exact (non-relativistic) compared to the Bohr energy.

$$\left(\Phi = \sqrt{\frac{\alpha^3}{\pi}}e^{-\alpha r}, \quad \nabla^2 = \frac{1}{r^2}\frac{\partial}{\partial r}\left(r^2 \frac{\partial}{\partial r}\right) + 0 + 0, \quad \langle E \rangle_{guess} = \frac{-Z^2 me^4}{2\hbar^2} = E(Z, n = 1)_{Bohr} \right)$$

(10) 5. Use the Eyring Transition State theory to compute $\Delta S^{\pm}$ and $\Delta H^{\pm}$ at 25°C for N_2O_5 decomposition if $k(25°C) = 3.46 \times 10^{-5}$/min and $k(35°C) = 13.5 \times 10^{-5}$/min ($\Delta H^{\pm} = 24.246$ kcal/mol, $\Delta S^{\pm} = -5.76$ cal/mole °, note sign).

(10) 6. In the far IR region the pure rotation spectrum of HCl shows adjacent lines/peaks at 286.1 and 306.1 cm^{-1}. Use the rigid rotor model to estimate the bond length of HCl; use $M_H = 1.008$ g/mol and $M_{Cl} = 35.00$ for the $^{35}_{17}$Cl isotope ($\mu = 1.629 \times 10^{-24}$ g, $r_e = 1.31$ Å).

(20) 7. The invariance of symmetry matrices under unitary transformation is the main principle of point group theory. Show $T^{-1}MT = M_{diag}$, $T^{-1}T = [1]$ and trace invariance of

$$M = \begin{bmatrix} 2 & 0 & 0 \\ 0 & 2 & 1 \\ 0 & 1 & 2 \end{bmatrix} \left(T = \begin{bmatrix} 1 & 0 & 0 \\ 0 & 1/\sqrt{2} & 1/\sqrt{2} \\ 0 & 1/\sqrt{2} & -1/\sqrt{2} \end{bmatrix}, \quad M_{diag} = \begin{bmatrix} 2 & 0 & 0 \\ 0 & 3 & 0 \\ 0 & 0 & 1 \end{bmatrix}, \quad \text{Trace } M = 6 \right)$$

(10) 8. Qualitatively sketch the radial orbitals and angular shapes of the H atom 1S, 2S, 3S, 2P, 3P and 3D orbitals. Compare the canonical 3D shapes to the Powell equivalent 3D orbitals.

STUDY, TEST, AND LEARN?

Next, we give the final examination for the CHEM 312 course at Randolph Macon College, which we previously showed as RMC-Quiz #1. Again the final examination was cumulative but there was also a special project to use PCLOBE to calculate the vibrational modes of a water molecule H-bonded with a molecule of formic acid. This was of interest to the class because of ongoing research in the Department regarding organic molecules trapped in ice and experimental IR spectra was available. The final grade in this course was again dependent on a student's average with the final examination or the score on the final examination alone (scaled to 100), whichever is higher. There was no ACS test on this occasion and the time limit was 180 min. The grades ranged from A$^+$ to C$^+$ so the test was a challenge to some students, although the A$^+$ student had a perfect score. The Randolph Macon College CHEM 311–312 courses are analogous to the VCU CHEM 303–304 but only chemistry majors took the 312 course while other majors took only the 311 course so this is similar to what will happen under the new ACS requirements with a smaller class taking the second semester.

Chemistry 312   Randolph Macon College   Spring 2009   D. Shillady, Professor
(points)   Final Examination (Attempt all problems!)   (180 min)

(30) 1. Normalize the variational wave function $\psi = Ne^{-\alpha r}$ for the 1s state of the H atom and show $\langle E \rangle_{min}$ is exact! (See Chapter 16)

(30) 2. In the far IR region the pure rotation spectrum of HCl shows lines/peaks at 206.2 and 226.7 cm^{-1} (among others). Use the rigid rotor approximation to estimate the bond length of HCl using $M_H = 1.0079$ and $M_H = 35.0$ g/mol. Assume the lines/peaks are adjacent ($r_e = 1.295$ Å).

(25) 3. Compute the De Broglie matter wavelength of an electron ejected from Zn ($W_f = 3.60$ eV) by a two-photon absorption of 4880 Å light from an Ar$^+$ laser ($\lambda_{DB} \cong 10.07$ Å).

(30) 4. Derive the formula for the energy of a "particle-on-a-ring," normalize the wave function and apply the "Perimeter Model" to anthracene ($C_{14}H_{10}$) to estimate the wavelength λ of the first $\pi \rightarrow \pi^*$ transition (453 nm).

(25) 5. Assuming the bond force constant $k = 4.8 \times 10^5$ dynes/cm is the same for H^{35}Cl and H^{37}Cl, calculate $\Delta \bar{v} = (\Delta v/c)$ for the splitting in cm^{-1} of their vibrational frequencies IR spectra. Use $M_H = 1.0079$, $M_{Cl-35} = 35.0$ and $M_{Cl-37} = 37.0$ g/mol ($\Delta \bar{v} = 2.1915$ cm^{-1}).

(25) 6. Qualitatively sketch the radial orbitals and orbital shapes of the H atom 1s, 2s, 3s, 2p, 3p, and 3d orbitals. Compare canonical and Powell equivalent 3d orbitals.

(25) 7. Use the Bohr model to estimate the Auger x-ray wavelength from Si ($Z = 14$) on the inside of a TV-cathode-ray-tube due to a 10,000 V electron beam ($\lambda_{x\text{-ray}} \cong 6.2$ Å).

(10) 8. If two eigenfunctions of the same Hermitian operator have different eigenvalues, prove the eigenvalues are real numbers and the eigenfunctions are orthogonal (See Chapter 11).

REFERENCES

1. Pauling, L. and E. B. Wilson, *Introduction to Quantum Mechanics with Applications to Chemistry*, McGraw-Hill Book Co, Inc. New York, 1935. The copyright is now owned by Dover Press, New York, and the unabridged reprint is available from Dover., see p. 156.
2. Wybourne, G. G., *Spectroscopic Properties of Rare Earths*, John Wiley & Sons, Inc. New York, 1965.
3. Ballhausen, C. J., *Introduction to Ligand Field Theory*, McGraw-Hill Book Co., New York, 1962.
4. Pauling, L. and E. B. Wilson, *Introduction to Quantum Mechanics with Applications to Chemistry*, McGraw-Hill Book Co, Inc., New York, 1935, p. 180.
5. Eckart, C., The theory and calculation of screening constants, *Phys. Rev.*, **36**, 878 (1930).
6. Ayres, F., *Schaum's Outline of Theory and Problems of Matrices*, Schaum, New York, 1962, p. 181.
7. Hückel, E., Quantentheoretische Beiträge zum Benzolproblem I. Die Elektronenkonfiguration des Benzols und verwandter Verbindungen, *Z. Phys.* **70**, 204(1931); Hückel, E., Quanstentheoretische Beiträge zum Benzolproblem II. Quantentheorie der induzierten Polaritäten, *Z. Phys.* **72**, 310(1931); Hückel, E., Quantentheoretische Beiträge zum Problem der aromatischen und ungesättigten Verbindungen. III, *Z. Phys.* **76**, 628(1932).
8. Roberts, J. D., *Notes on Molecular Orbital Calculations*, W. A. Benjamin Press, New York, 1961; This text is now available as a single-user download from http://caltechbook.library.caltech.edu/23/
9. Lowe, J. P. and K. Petersen, *Quantum Chemistry*, 3rd. Edn., Elsevier, Academic Press, Boston, MA, 2005.
10. Murrell, J. N., S. F. A. Kettle, and J. M. Tedder, *Valence Theory*, 2nd Edn., John Wiley & Sons, LTD., London, U.K., 1970, p. 296.
11. Eyring, H., J. Walter, and G. E. Kimball, *Quantum Chemistry*, John Wiley & Sons, New York, 1944.
12. Williams, G., *Linear Algebra with Applications*, 4th Edn., Jones and Bartlett Publishers, Sudbury, MA, 2001, p. 283.
13. Offenhartz, P. O., *Atomic and Molecular Orbital Theory*, McGraw-Hill Book Co., New York, 1970, Appendix 2, p. 338.

17 Electronic Structure of Molecules

INTRODUCTION

In terms of "essentials" this topic may seem advanced but with the availability of fast personal computers and some 50 years of development, quantum chemistry is now a standard aspect of chemistry. As such, undergraduates should have some awareness of the standard methods and we will use a program set up for personal computers (PCLOBE). There are a number of excellent modern large scale quantum chemistry programs that will run on personal computers but they are expensive and/or complicated to install. The main advantage of PCLOBE is that, while much of it is obsolete for research, it is easy to install on Microsoft systems and very easy to use. Now that we have a better understanding of where the H-atom orbitals come from, the next topic should be the electronic structure of molecules. From Chapter 16, you should now have a better understanding of the use of linear algebra for orbital calculations. This might have been an advanced topic 10 years ago but today Hartree–Fock–Roothaan calculations are the foundation of more exotic methods; so undergraduates need to be aware of what is now a basic method.

It is traditional to spend a lot of time on the H_2 molecule, but it is our experience that class time is limited and students want to understand how orbital calculations are used for molecules with more structure than H_2. On the other hand, we need to choose a small example to be able to fit the program details into this text. Thus, we are going to briefly mention the Li atom and proceed directly to the BH molecule which permits full printout of the results. We are going to take what would have been the high road of "ab initio" methods 30 years ago. However, students may come across vestiges of the "semiempirical" struggle to make progress in molecular quantum mechanics when only limited computer resources were available (Figure 17.1). Today, personal computers are more powerful than mainframe computers of the 1970s but a niche still lingers for the application of approximate methods to very large molecules. Actually, a great deal of modern research was guided and supplemented by semiempirical quantum chemistry prior to the availability of computers fast enough to make "ab initio" methods feasible. This was particularly true in the use of pharmaceutical drug design. So let's jump in to the deep water and grab hold of some of the invaluable writings by Slater [1] as intellectual life savers. While Roberts molecular orbital book [2] lured us to the surf, we hope that we can navigate deeper explanations with help from Slater and Pople [3] when we try to understand the amazing derivation of the Roothaan-self-consistent-field calculations (SCF) orbitals.

HARTREE–FOCK–ROOTHAN LCAO CALCULATIONS

Looking back at the pi-electron calculations, we motivated the linear combination of atomic orbitals (LCAO) concept for electron orbitals, which spread over a whole molecule. We also have in hand the polynomial functions for the radial functions of the H-atom. In the previous example for pi-electrons, we used the variational principle to optimize the coefficients of the atomic orbitals in the linear combinations. We have also observed that the H-atom model can be modified by varying the value of the effective nuclear charge, Z, to model heavier elements. In 1930, Slater [4] proposed deleting the radial nodes and small parts of the H atom orbitals to use functions of the

FIGURE 17.1 While semiempiricists struggled to catch up using spectroscopic data in approximate calculations, they were chasing these "ab initio" leaders in 1970. Here is Prof. Hugh Kelly (on the left), Dr. Enrico Clementi (then at IBM) at the board and Prof. Per O. Lowdin (on the right) of Uppsala University and Organizer of the Quantum Theory Project at the University of Florida in a 1970 Sanibel Meeting at a spartan, but tropical, meeting place called Casa Ybel. Prof. Kelly is still known for one of the most accurate treatments of the Be atom (four electrons). Dr. Clementi was the main author of IBMOL which was the first widely distributed ab initio computer program for molecular quantum mechanics. Prof. Lowdin is famous for his Lowdin Orthogonalization procedure but more so for organizing the annual Sanibel Meetings and Summer Schools at which more than 5000 Physicists and Chemists have been trained in Quantum Chemistry. Prof. Lowdin was awarded the Lavoisier Medal in Gold by the French Academy of Sciences in 1981, Chevalier of the Legion of Honor in 1982, Niels Bohr Medal in 1987, and the Oscar Carlson Medal in Gold of the Swedish Chemical Society in 1993. (By permission from the Quantum Theory Project of the University of Florida at Gainesville, FL.)

form $R_{nl}(r) = Nr^l e^{-\zeta r} Y_{lm}(\theta, \phi)$, which is only the largest part of the radial polynomial. In molecules, the smaller parts of the H-atom radial orbitals are modified by the need to be orthogonal among themselves in the presence of other atoms, so only the largest component of the H-atom orbitals retain the main nature of the atomic environment. Slater proposed his own values for ζ which were based on atomic scattering of x-rays; but they were only approximate. After computers became available, a valuable paper by Clementi and Raimondi [5] used Slater's functional form and revised the ζ values by carrying out self-consistent-field calculations (SCF) on atoms in which they optimized the ζ values. *The main idea here is that if you are going to use the LCAO method for molecules, the functions describing the atoms should be optimized first and then you build molecular orbitals from the linear combinations of the "best" atomic functions.* More recently, the idea of atom-optimized basis sets has been extended to other functional forms and we will consider examples using Gaussian orbitals in PCLOBE.

CHEMICAL EFFECTS IN ORBITAL SCREENING

A variational calculation can be done for H_2 using a small computer program and it produces one very interesting result in that the best value of ζ (the parametric form of the σ exponent in the hydrogenic radial function) is greater than the nuclear charge of H which is 1.0! The 1s exponent on H can vary from 1.17 in H_2 to more than 1.4 in highly acidic compounds. H is the only element in which the 1s exponent is greater than the nuclear charge and the orbital acts as if it has been robbed of electron density leaving only a core-spike of what had been the H1s orbital. While it might seem that the H atom has more attraction for a shared electron, the situation is actually that the neighboring atom in the bond pair has pulled electron density away from the H atom and so what is left is described by a tight spike for the 1s orbital. In the case of acid compounds, you might interpret the high screening constant for H as a "skinny" remainder of the 1s orbital after the anion part of the compound has drained away a lot of the electron density from the H atom. That is why the compound is acidic since there is less electron density shared by the H atom and so it leaves the compound easily as a separate proton. If the $H\zeta_{1s}$ value is say 1.3 or 1.4 in a compound, it really means that H atom has lost a lot of its electron density. In Slater's early rules for such exponents, every other element has the 1s screening constant of $\zeta_{1s} = Z - 0.3$ while in the Clementi–Raimondi [5] values in Table 16.1 we see that ζ_{1s} varies from 1.6875 in He to 17.5075 in Ar ($Z = 18$). Thus Slater's rule of a constant screening value of 0.3 does not hold exactly but the ζ_{1s} value is less than the Z value for all elements other than H (Table 17.1).

The 1s screening effect is caused by one of the electrons hiding or screening the full electrostatic view of the positive nucleus by the other 1s electron. As the nuclear charge Z increases in heavier elements the 1s orbitals are pulled in closer to the nucleus and the ability of one electron to screen out the other electron is greater in a smaller confined space of the 1s orbital region. Thus we see the screening increases in the Clementi–Raimondi 1s values as Z increases and the ζ_{1s} value decreases.

Since there is a lot of interest in the C atom for organic molecules we should note that the best single ζ value for the 2s orbital (1.6083) is higher than the value for the 2p orbitals (1.5679). This means that the energies of the 2s and 2p orbitals are not degenerate in the C atom (the 2p set is still degenerate) and in particular the lower exponent of the 2p orbitals means the 2p orbitals extend a little further out from the nucleus compared to the 2s orbital which is slightly more compact.

TABLE 17.1
Selected Clementi–Raimondi Best-Single Zeta (ζ) Values for Atomic Orbitals

Atom	Z	ζ_{1s}	ζ_{2s}	ζ_{2p}	ζ_{3s}	ζ_{3p}
H	1	1.24[a]				
He	2	1.6875				
Li	3	2.6906	0.6396			
Be	4	3.6848	0.9560			
B	5	4.6795	1.2881	1.2107		
C	6	5.6722	1.6083	1.5679		
N	7	6.6651	1.9237	1.9170		
O	8	7.6579	2.2458	2.2266		
F	9	8.6501	2.5638	2.5500		
Ne	10	9.6421	2.8792	2.8792		
Na	11	10.6259	3.2857	3.4009	0.8358	
Mg	12	11.6089	3.6960	3.9129	1.1025	
Al	13	12.5910	4.1068	4.4817	1.3724	1.3552
Si	14	13.5745	4.5100	4.9725	1.6344	1.4284
P	15	14.5578	4.9125	5.4806	1.8806	1.6288
S	16	15.5409	5.3144	5.9885	2.1223	1.8273
Cl	17	16.5239	5.7152	6.4966	2.3561	2.0387
Ar	18	17.5075	6.1152	7.0041	2.5856	2.2547

Source: Clementi, E. and Raimondi, D. L., *J. Chem. Phys.*, 38, 2686, 1963.

[a] The ζ value for H is an optimized average for the H_2 molecule and other hydrides.

This lends added credibility to the concept that 2p pi-orbitals are held less tightly than sigma orbitals and that the pi-electrons can delocalize more easily.

MANY-ELECTRON WAVE FUNCTIONS

During the 1930s, a great deal of effort was put into what was called "Atomic Physics," culminating in the nuclear age of the 1940s. Along the way there was a lot of analysis of the emission spectra of atoms. A typical (and very valuable) text of that era was by Condon and Shortley [6] which tabulated and explained ways to carry out calculations on atoms to assign the spectral lines. The equations developed in that text provide an excellent description of how complicated matrix elements over large determinants can be reduced to specific integrals (see Appendix B). Slater made another very important contribution in the 1930s when it was noted that wave functions describing electrons in atoms needed to be "antisymmetric" with respect to interchanging any two electrons. It was found that correct assignments for atomic spectral calculations had to involve antisymmetric wave functions. Slater invented a compact way to write such a wave function as a determinant of spin-orbital functions. In the LCAO formalism, one writes spatial orbitals as one of a set of atomic orbitals multiplied by either an α or a β spin function and then places these "spin-orbitals" into a determinant which has been multiplied by a normalization constant. You need to remember that a determinant is a single number as different from a matrix which is an array of numbers. When such a Slater-determinant is built up from functions that depend on coordinates like (x, y, z) the whole determinant still has only one value for whatever (x, y, z) coordinates are used.

DETERMINANTAL WAVE FUNCTIONS FOR MANY-ELECTRON SYSTEMS

You can see that as the number of electrons increases the determinantal wave function will become difficult to actually write on a page. However there is a simple pattern there.

$$\Psi(1,2,3,\ldots,n) = N \begin{vmatrix} \psi_1(1)\alpha(1) & \psi_1(1)\beta(1) & \psi_2(1)\alpha(1)\ldots & \psi_n(1)\beta(1) \\ \psi_1(2)\alpha(2) & \psi_1(2)\beta(2) & \psi_2(2)\alpha(2)\ldots & \psi_n(2)\beta(2) \\ \ldots & \ldots & \ldots & \ldots \\ \psi_1(n)\alpha(n) & \psi_1(n)\beta(n) & \psi_2(n)\alpha(n)\ldots & \psi_n(n)\beta(n) \end{vmatrix}$$

Note that you just write all the functions (spin-orbitals) in the wave function across a row with the same electron but repeat the same orbital down a column with the different electrons. The rows and columns can be interchanged but your choice must continue through the calculation. If any two rows or columns are interchanged the determinant changes sign which satisfies the antisymmetry requirement. The form of the determinant above is seldom shown in other texts or even research papers because it is too big to write but you need to see it here to understand what follows. That big determinant is too difficult to write for 42 electrons in benzene or 10 electrons in the water molecule and just barely imaginable for the six electrons in BH which we are going to consider! Instead, most texts write the condensed form recognizing that every term in the expanded determinant will consist of a product of all the functions in the basis set used.

$$\Psi(1,2,3,\ldots n) = N \sum_P (-1)^P P|\psi_1(1)\alpha(1)\psi_1(2)\beta(2)\ldots\psi_n(n)\beta(n)|.$$

Again N is the normalization constant. On first encounter, one may ask, just what is "P"? The operation denoted by P is a "permutation" of the order of the orbitals in the basis set performed in the mind of the reader. There is no "calculus operator P"—it is just the operation you use in your mind when you write down all the possible permutations that evolve from expanding the determinant. With this form of the wave function one can proceed to set up the expectation integral for the energy and it usually takes a whole lecture to slowly write the matrix elements on a (large) black/white board. The details are given by Condon and Shortley [6], by Pople and Beveridge [3], by Trindle and Shillady [7] and here in Appendix B.

Example

Suppose we expand the determinant for three electrons in a Li atom using a basis of only $1s\alpha$, $1s\beta$, and $2s\alpha$.

$$\Psi(1,2,3) = \frac{1}{\sqrt{3!}} \begin{vmatrix} 1s\alpha(1) & 1s\beta(1) & 2s\alpha(1) \\ 1s\alpha(2) & 1s\beta(2) & 2s\alpha(2) \\ 1s\alpha(3) & 1s\beta(3) & 2s\alpha(3) \end{vmatrix},$$

$$\Psi(1,2,3) = \frac{1}{\sqrt{3!}} \{ [1s\alpha(1)1s\beta(2)2s\alpha(3)] - [1s\alpha(1)2s\alpha(2)1s\beta(3)]$$

$$- [1s\beta(1)1s\alpha(2)2s\alpha(3)] + [1s\beta(1)2s\alpha(2)1s\alpha(3)]$$

$$+ [2s\alpha(1)1s\alpha(2)1s\beta(3)] - [2s\alpha(1)1s\beta(2)1s\alpha(3)] \}.$$

Here the (1, 2, 3) order has been maintained and the alternating order of determinant expansion has been used.

$$\begin{vmatrix} a & b & c \\ d & e & f \\ g & h & i \end{vmatrix} = a \begin{vmatrix} e & f \\ h & i \end{vmatrix} - b \begin{vmatrix} d & f \\ g & i \end{vmatrix} + c \begin{vmatrix} d & e \\ g & h \end{vmatrix} = a(ei - hf) - b(di - gf) + c(dh - ge).$$

ATOMIC UNITS

We noted in Chapter 9 that when computer programs were first being written for electronic structure problems in the 1960s, the physical constants were being updated rapidly. The Bohr radius was believed to be 0.529167 Å in the 1960s, then 0.529172 Å in the 1970s, and more recently 0.529177 Å in the 1980s; so those changes effected published results in the scientific literature. The latest value [8] is 0.52917720859 Å. A new system of units called "atomic units" was adopted to avoid this problem. In atomic units $c = e_q = \hbar = m_e = 1$; so in these units the energy of a H atom is

$$E(n = 1, Z = 1) = -\frac{Z^2 m_e e_q^4}{2n^2 \hbar^2} \to -\frac{1}{2} = -0.5 \text{ units and the energy units were called hartrees in}$$

honor of a team [9] that carried out many of the early SCF calculations on atoms in the 1930s. Today we use 1 hartree = 27.21138386 eV [8]. Thus the Hamiltonian in atomic units (au) emphasizes the mathematics, is independent of the latest values of the physical constants, and R_{ab} is now measured in bohrs. To make further progress we note

$$H = \sum_i^n \left(\frac{-\nabla_i^2}{2} \right) - \sum_i^n \sum_K^N \left(\frac{Z_K}{r_{iK}} \right) + \sum_{i>}^n \sum_j^n \frac{1}{r_{ij}} + \sum_{K>}^N \sum_L^N \frac{Z_K Z_L}{R_{KL}}$$

$$= \sum_i^n f_i + \sum_{i>}^n \sum_j^n g_{ij} + \sum_{K>}^N \sum_L^N \frac{Z_K Z_L}{R_{KL}}$$

that the kinetic energy terms and the nuclear attraction terms are called "one-electron operators" as denoted by $\sum_i^n f_i$ in the Hamiltonian for "n" electrons and "N" nuclei; these interactions turn out to be easy to compute (see Appendix C). On the other hand, all the interactions (and there are many!) between two electrons, $\sum_{i>} \sum_j g_{ij}$, are called "two-electron operators" and they are difficult to compute (see Appendix C) as well as being very numerous! Thus in atomic units and noting the types of operators, we have

$$H = \sum_i^n f_i + \sum_{i>}^n \sum_j^n g_{ij} + \sum_{K>}^N \sum_L^N \frac{Z_K Z_L}{R_{KL}}.$$

For a water molecule we have 10 electrons and 3 nuclei but the number of two-electron interactions is $(n)(n-1)/2 = (10)(9)/2 = 45$ since each of the 10 electrons can interact with 9 other electrons but each interaction is only counted as one interaction! Similarly, the repulsions between a nucleus with charge Z_N and another nucleus with charge Z_M separated by distance R_{KL} are only counted once since two interacting nuclei only have one interaction but those terms are very easy to compute!

ROOTHAAN'S LCAO HARTREE–FOCK EQUATION

In the late 1920s and early 1930s, a team led by Hartree [9] formulated a "self-consistent-field" iterative numerical process to treat atoms. In 1930, Fock [10] noted that the Hartree-SCF method needed a correction due to electron "exchange" and the combined method was known as the Hartree–Fock SCF method. It was not until 1951 that a molecular form of the LCAO-SCF method was derived by Roothaan [11] as given in Appendix B but we can give a brief outline here. The Roothaan method allows the LCAO to be used for more than one atomic center and so the path was open to treat molecules! Now, "all we have to do" is to carry out the integral for the expectation value of the energy as

$$\langle E \rangle = \int \ldots \int \Psi^*(1,2,3,\ldots,n) H \Psi(1,2,3,\ldots,n)\, d\tau_1 d\tau_2 \ldots d\tau_n ds_1 ds_2 \ldots ds_n.$$

This turns out to be a tremendous problem keeping track of all the terms but it has been done [1,3] in many texts which we will only summarize here. Note in particular that we have to integrate over the spin functions as well as the spatial coordinates using simple rules.

$$\int \alpha^* \alpha\, ds = \int \beta^* \beta\, ds = 1 \quad \text{and} \quad \int \alpha^* \beta\, ds = \int \beta^* \alpha\, ds = 0$$

Among the many permutations of the $\Psi^*(1,2,3,\ldots,n)$ and the $\Psi(1,2,3,\ldots,n)$ there will be many integrals in the long summation (there may be thousands of terms) of the general form

$$\int ds_1 \int ds_2 \int d\tau_1 \int d\tau_2 [\psi_1(1)\alpha(1)\psi_1(2)\beta(2)\ldots]^*\, H[\psi_1(1)\alpha(1)\psi_1(2)\beta(2)\ldots]$$

In many of these integrals there will be some misalignment of the order of the orbitals between $\Psi^*(1,2,3,\ldots,n)$ and $\Psi(1,2,3,\ldots,n)$ for that particular term of the two determinants; so the integral will be zero by orthogonality, assuming the $\{\psi_i\}$ are orthogonal. However, if the order in the left and right part of the integral match exactly or differ by only one orbital and the spins match, the result will be $\int \phi_\mu^* f_i \phi_\nu d\tau_i = \langle \mu | f_i | \nu \rangle$. Again when the order matches for all but two orbitals in $\Psi^*(1,2,3,\ldots,n)$ and $\Psi(1,2,3,\ldots,n)$ for a particular set of cross terms there can be four possible integrals which are coupled by the g_{ij} two-electron operators resulting from the form of the $\left(\frac{1}{r_{12}}\right)$ operator.

Much of the mathematics for the SCF process was derived for atoms using texts like Condon and Shortly [6] but in 1951 Roothaan [11] solved the problem for molecules using the LCAO philosophy. Instead of using just single orbitals such as $1s\alpha$ in our example above for the

Li atom, the LCAO concept says that $\Psi(1, 2, \ldots, n)$ is built from many spin orbitals ψ_i which are linear combinations of basis functions ϕ_μ. Thus $\psi_i = \sum_\mu^M c_{\mu i}\phi_\mu$. The original paper by Roothaan [11] is a masterpiece but not easy reading, as is often the case with descriptions of new advances. However, *there is a succinct explanation given by Slater in a later text* [1]. This is a difficult topic but we will follow Slater's description in the faint hope that it is understandable with the calculus we have used so far. Since we will discuss the final equation in Poples [3] notation we will write Slater's derivation using Greek letters for basis functions and charge-density "$(1 \times 1/2 \times 2)$" rounded parenthesis notation for the two electron integrals to merge with that notation as

$$(\mu\nu/\lambda\sigma) \equiv \iint \phi_\mu^*(1)\phi_\nu(1)\frac{1}{r_{12}}\phi_\lambda^*(2)\phi_\sigma(2)d\tau_1 d\tau_2.$$

After collecting all the non-zero terms in the calculation of $\langle E \rangle = \int \Psi^* H \Psi d\tau$, we find for just the electrons (fixed nuclei)

$$\langle E \rangle = \sum_i^n \sum_\mu^M \sum_\nu^M c_{\mu i}^* c_{\nu i} \langle \mu|f_i|\nu \rangle + \sum_{i>}^n \sum_j^n \sum_\mu^M \sum_\nu^M \sum_\lambda^M \sum_\sigma^M c_{\mu i}^* c_{\nu i} c_{\lambda j}^* c_{\sigma j} \{(\mu\nu/\lambda\sigma) - (\mu\lambda/\nu\sigma)\},$$

where we now have the energy in terms of the $\psi_i = \sum_\mu^M c_{i\mu}\phi_\mu$ LCAO expansion of the LCAO basis $\{\phi_\mu\}$. Here we are using $(11/22)$ notation for the g_{12} operator for N electrons and M basis functions. The alternative $\langle 12|12 \rangle$ notation would mean

$$\langle nk|pq \rangle = \iint \phi_n^*(1)\phi_k^*(2)\frac{1}{r_{12}}\phi_p(1)\phi_q(2)d\tau_1 ds_1 d\tau_2 ds_2,$$

where the complex conjugate orbitals are together on the left and the real function on the right. Since some research papers use a $\langle 12|12 \rangle$ notation you always need to look at the definition a given author uses! We also have to remind ourselves to integrate over the spins. The effect of the permutations in the g_{12} two-electron terms is to produce a "coulomb term" $(\mu\nu/\lambda\sigma)$ as well as the "exchange term" from the permutation as $(\mu\lambda/\nu\sigma)$. These integrals have not yet been integrated over the spins which will lead to the four cases treated in Ref. [3].

ELECTRON EXCHANGE ENERGY

Pople and Beveridge [3] give a beautiful analysis of exchange, seldom written anywhere else, but we show the details in Appendix B. In the grand scheme of the detailed derivation you will be looking at energy terms like

$$\iint \Psi^*(1, 2, 3, \ldots)\left[\frac{1}{r_{ij}}\right]\Psi(1, 2, 3, \ldots)d\tau_1 d\tau_2 ds_1 ds_2.$$

Remember that $\Psi(1, 2, 3, \ldots)$ is a big determinant on both sides of the operator. All the terms for a minimum basis treatment of the Li atom with only three electrons will fill some nine pages for the open-shell (one unpaired electron) case. The smallest closed shell general case would be for the four electron ground state of the Be atom. If this is taught in a classroom it might be possible to write out just a few terms of the general case for something like the six electrons in the BH molecule on a wide black/white board. You can easily write out all the terms for the two electrons in He but unfortunately that is not the general case. With more than four electrons you could see the four exchange terms shown in Appendix B as well as orbital products which do/do not match for the other electrons. Assuming all the $\{\psi_i\}$ can be made orthonormal (they can by choice of the expansion coefficients in $\psi_i = \sum_\mu^M c_{\mu i}\phi_\mu$) all the parts of the wave functions for electrons which are not i or j will just integrate to factors of 1 when the g_{ij} part is nonzero, but you will be left with the four possibilities shown in Appendix B that occur due to the permutations of the determinants subject to the spin integration.

As a result of spin orthogonality, only $(1/2)$ of the exchange terms are nonzero but they are there! This was first pointed out by Fock [10] and was added as a correction to the method then developed by Hartree [9]. Today this method is almost obsolete and modern methods tend to be some form of "density functional theory, DFT" where the whole program is made much faster by using numerical tricks and replacing the two-electron operators by carefully designed functions of the electron density. *However, after more than 20 years of research on DFT methods, the most accurate calculations have been found to require at least some fractional amount of this so-called "Hartree–Fock Exchange" [12,13]; it is still a part of modern research!* Further, the exchange terms enter into the energy with a negative sign, so the phenomenon of exchange lowers the energy and is necessary for the most accurate energy calculations.

THE HARTREE–FOCK–ROOTHAAN EQUATIONS FOR $2n$ ELECTRONS

As with the pi-electron model we want to treat the orbital coefficients as variation variables and use something like the Clementi–Raimondi–Slater atomic orbitals for the basis functions, or at least something like them which are easy to integrate. We want to minimize the energy by varying the values of the c_{ni} but we also want to maintain the orthonormality of the linear combination of basis functions as orthonormal one-electron orbitals. They are formed from linear combinations of basis functions $\{\phi_i\}$ which may have nonzero overlap $\int \phi_\mu^* \phi_\nu d\tau = S_{\mu\nu}$.

$$\int \psi_i^* \psi_j d\tau = \sum_\mu^M \sum_\nu^M c_{\mu i} c_{\nu j} \int \phi_\mu^* \phi_\nu d\tau = \sum_\mu^M \sum_\nu^M c_{\mu i} c_{\nu j} S_{\mu\nu} = \delta_{ij}.$$

Thus we use the idea of Lagrangian multipliers to remove from the minimization procedure that part which might change the orthonormal nature of the one-electron orbitals $\{\psi_i\}$. We therefore need the derivative of $\dfrac{\partial}{\partial c_{\rho i}^*}\left[\langle E \rangle - \sum_i \sum_j \lambda_{ij} \sum_\mu^M \sum_\nu^M c_{\mu i}^* c_{\nu j} S_{\mu\nu}\right] = 0$ where λ_{ij} are the Lagrangian multipliers that prevent loss of orthonomality of the one-electron orbitals. We might as well take the derivative with respect to $c_{\rho i}^*$ to get rid of one of the potentially complex numbers. Remember, as for the pi-orbital coefficients in Roberts treatment [2], the $c_{\rho i}$ coefficients are the variational variables, so we can take derivatives with respect to their value. Thus the constrained derivative of the energy expression $\dfrac{\partial}{\partial c_{\rho i}^*}\left[\langle E \rangle - \sum_i \sum_j \lambda_{ij} \sum_\mu^M \sum_\nu^M c_{\mu i}^* c_{\nu j} S_{\mu\nu}\right] = 0$ uses

$$\langle E \rangle = \sum_i^n \sum_\mu^M \sum_\nu^M c_{\mu i}^* c_{\nu i}(\mu/f_1/\nu) + \sum_{i>}^n \sum_j^n \sum_\mu^M \sum_\nu^M \sum_\lambda^M \sum_\sigma^M c_{\mu i}^* c_{\nu i} c_{\lambda j}^* c_{\sigma j}\{(\mu\nu/\lambda\sigma) - (\mu\lambda/\nu\sigma)\}$$

to find

$$\sum_\nu^M c_{\nu j}\left\{(\mu/f_1/\nu) + \sum_j^n \sum_\lambda^M \sum_\sigma^M c_{\lambda j}^* c_{\sigma j}[(\mu\nu/\lambda\sigma) - (\mu\lambda/\nu\sigma)]\right\} - \sum_j^n \sum_k^M c_{\nu j} \lambda_{ij} S_{\mu\nu} = 0.$$

Although the process is for an arbitrary $c_{\rho i}^*$, in effect it removes the summation over μ and summations associated with the index for electron i. As with the Hückel pi-electron derivation, *this is the minimization requirement for just one $(c_{\rho i})$ coefficient for the orbital of just one electron (i) and just one equation in the rows of equations that form the system that leads to the Cayley–Hamilton situation.* From the pi-electron case we can see that this can be put into a matrix equation where we will need to use diagonalization.

There remains one very mysterious step. Note the subscripts on λ_{ij} which implies that each orbital ψ_i (which we do not know yet) is dependent somehow on all the other ψ_j to maintain orthogonality to them as well as normalization. Recall that the big wave function $\Psi(1, 2, 3, \ldots)$ is a determinant which is a single number at each (x, y, z) point in space and we know some sort of unitary transformation $[T]^{-1}[\lambda_{ij}][T] = [\lambda_{ii}]_{diag}$ could be applied to the orbitals ψ_i (if we knew them!) which would not change the value of the overall wave function, but only make the λ_{ij} interactions diagonal. If we assume this has been done and solve the equation subject to that constraint we will make it true! Thus we end up with an equation for all the electrons in a given basis set as a Cayley–Hamilton problem to find the coefficients c_{ni} and from them find the energy once the coefficients are known. A diagonal λ_{ii} eliminates the $\sum_j^n \lambda_{ij}$ for the overlap term and leads to

$$\sum_v^M c_{vi} \left\{ (\mu/f_1/v) + \sum_j^n \sum_\lambda^M \sum_\sigma^M c_{\lambda j}^* c_{\sigma j} [(\mu v/\lambda \sigma) - (\mu \lambda/v \sigma)] \right\} - \sum_v^M c_{vi} \lambda_{ii} S_{\mu v} = 0, \quad \text{for basis } \phi_\mu,$$

but the summation over j remains on the two-electron terms. *Note that here we use (enforce mathematically) $\lambda_{ii} = \varepsilon_i$ as the now diagonal Lagrangian multiplier in an energy equation so that these will be assumed to be the one-electron orbital energies!* So far we have not integrated over the spins.

Before we go further, note that the secular equation to solve for the $c_{\mu i}$ coefficients involves the $c_{\mu i}$ coefficients; the answer depends on the answer! We will need to make a good guess at the orbital coefficients and then iterate the calculation in some way to find convergence!

After integrating over the spins we need to vary the energy with respect to c_{vi} (and c_{vi}^*) and define a population matrix, $P_{\lambda \sigma} \equiv 2 \sum_j^{occ.} c_{\lambda j} c_{\sigma j}$ using the Aufbau Principle of filling the lowest energy orbitals with spin-pairs up to the highest occupied orbital. As shown in Appendix B, we find the working equations $\sum_v (F_{\mu v} - \varepsilon_i S_{\mu v}) c_{vi} = 0$ for each c_{vi} leading to the Cayley-Hamilton secular equation where $F_{\mu v} = H_{\mu v} + \sum_\lambda \sum_\sigma P_{\lambda \sigma} [(\mu v/\lambda \sigma) - 1/2(\mu \lambda/v \sigma)]$. We can interpret this as a one-electron equation for each electron where $H_{\mu v}$ is the kinetic energy of that electron and the attractive potential energy due to the positive nuclei and then $P_{\lambda \sigma}$ tells us how that electron is repelled by and exchanges with other electrons.

PRACTICAL IMPLEMENTATION AND EXAMPLES

Quantum chemistry calculations are currently quite common and we recall the excitement this author experienced when first reading Roberts "Notes on Molecular Orbital Theory"[2] in the early 1960s. Therefore, we include some modern but simple examples here in the hope that the amazement factor is still possible for undergraduates eager to learn up-to-date material. First we can write down the main Hartree–Fock–Roothaan energy operator and at least interpret the various terms. We have used Slater's derivation [1] of the Roothaan LCAO form of the Hartree–Fock equations but prefer Pople's implementation [3] for computer code. First, the one-electron operator in Pople's notation is $H_{\mu v} = \langle \mu | -\dfrac{\nabla^2}{2} - \sum_k^{Nuc} \left(\dfrac{Z_k}{r_k} \right) |v\rangle$ is called the core-Hamiltonian, as a one-electron matrix. This part of the calculation is relatively simple and results in about 80% or more of the total electronic energy. It is easy to compute and is a fixed matrix for a given framework of frozen positions of the nuclei. The other 20% of the electronic energy comes from the wild scramble of electrons trying to get into orbitals as attracted by the terms of the core-Hamiltonian but repelling each other and exchanging with each other! Note that Pople [3] uses (11/22) notation as $(\mu v/\lambda \sigma) \equiv$

$$\int \int \phi_\mu^*(1)\phi_v(1) \frac{1}{r_{12}} \phi_\lambda^*(2)\phi_\sigma(2) d\tau_1 d\tau_2.$$ To some people this makes sense as a charge density $(\phi^*\phi)_1$

is interacting with another charge density $(\phi^*\phi)_2$ via Coulomb's Law as $\dfrac{(\phi_\mu^*\phi_v)_1 (\phi_\lambda^*\phi_\sigma)_2}{r_{12}}$.

We noted above that the answer depends on the answer; so Pople defines a "population matrix" (related to the first-order density matrix) as $P_{\mu\nu} = \sum_i^{occ} 2c_{\mu i}c_{\nu i}$ based on the Aufbau Principle where the orbitals are ordered by lowest energy first and *all the integrals are integrated over the spins so that the spinless form* of the Hartree–Fock–Roothaan equation is

$$FC = SCE \quad \text{where } F_{\mu\nu} = H_{\mu\nu} + \sum_\lambda \sum_\sigma P_{\lambda\sigma}[(\mu\nu/\lambda\sigma) - 0.5(\mu\lambda/\nu\sigma)] \quad \text{and} \quad S_{\mu\nu} = \int \phi_\mu^* \phi_\nu d\tau.$$

This adds meaning to the concept that we stand on the shoulders of intellectual giants! We have tried to present an understandable derivation of Roothaan's equation in Appendix B leaning on Slater's short form [1], and there is the final SCF equation (in Pople's notation [3])!

Let us pause and discuss the meaning of the equation. We see a matrix form of the Schrödinger equation in the nonorthogonal representation. The $H_{\mu\nu}$ matrix is actually about 80% of the problem and contains the kinetic energy of each electron and the attraction of each electron to all of the nuclear charges. This is a one-electron model in terms of the weighting coefficients to be found in the eigenvector columns of the C matrix. This problem can be solved easily by ignoring the two electron terms for an initial guess of the final solution to begin iteration; usually on the values of the calculated energy of each iteration. Now the main focus should be on what is happening in the two-electron terms of the equation. The population matrix shown here as $P_{\lambda,\sigma}$ is restricted to an even number of electrons populating the lowest energy orbitals two at a time, assuming integration over the spin integrals; so this equation is spinless at this point. This is the so-called "restricted Hartree–Fock" (RHF) method (used in about 95% of all ground state calculations!). The matrix element for the charge density $(\mu\nu]$ is then integrated with charge density $[\lambda\sigma)$ as weighted by $P_{\lambda,\sigma}$ in a standard Coulomb interaction. That would be all that is needed for a Hartree-SCF [9] calculation, but when we add Fock's [10] exchange term we have to include the second "exchange integral" which is shown in Appendix B to be negative and only half as large as the coulomb term. In a computer program most of the details are centered on the data processing of how to form this two-electron part of the $F_{\mu\nu}$ matrix. Then we only have to worry about how to solve the overall $FC = SCE$ equation!

We know how to diagonalize a matrix now but the nonzero S matrix makes this more difficult.

$$FC = SCE$$

Note that C is not symmetric and E must be last here when we have S as a nonunit matrix.

The solution to this is to find a matrix $S^{-1/2}$ and then define a new $C' = S^{1/2}C$ matrix so that $C = S^{-1/2}C'$. Then $FS^{-1/2}C' = SS^{-1/2}C'E$. Next multiply by $S^{-1/2}$ from the left to obtain

$$S^{-1/2}FS^{-1/2}C' = S^{-1/2}SS^{-1/2}C'E \Rightarrow F'C' = C'E \quad \text{where } F' = S^{-1/2}FS^{-1/2}.$$

It should be noted that this leaves F' symmetric and eligible for diagonalization. The strategy here is to set up the product $S^{-1/2}SS^{-1/2} = 1$ and once S is simplified we just use some computer program like the Jacobi algorithm to solve $F'C' = C'E$ to find the eigenvalues and the eigenvectors in C'. The desired eigenvectors are found from $C = S^{-1/2}C'$. How do we get $S^{-1/2}$? Recall that scalar arithmetic applies to a diagonal matrix, so diagonalize the overlap matrix as $[T]^{-1}[S][T] = [S]_{diag}$.

Change each $s_{ii} \rightarrow \dfrac{1}{\sqrt{s_{ii}}}$ and then reverse the transformation to go back to the original basis as

$[T]\left[\dfrac{1}{\sqrt{s_{ii}}}\right]_{diag}[T]^{-1} = [S]^{-1/2}$. We anticipate iteration but this operation only needs to be done once and the $[S]^{-1/2}$ matrix can be saved in computer memory for frequent use.

The traditional way to treat the long list of $(\mu\nu/\lambda\sigma)$ integrals is to compute them once and save them in a (large) file. There are a number of techniques to store and retrieve these integrals which has a maximum number of almost M^4 for M basis functions but usually the number is approximately M^3 depending on the shape of the molecule and the number of basis functions, if there is a threshold for neglecting small integrals. PCLOBE neglects two-electron integrals less than 10^{-6} hartrees for minimum basis sets but has a tighter threshold of 10^{-12} hartrees for the more accurate split basis sets such as 6-31G**.

SCF ITERATION

For the first iteration we use $H_{\mu\nu} = \langle\mu| -\frac{\nabla^2}{2} - \sum_k^{Nuc} \left(\frac{Z_k}{r_k}\right) |\nu\rangle$ and choose an initial guess for the orbitals by solving $HC = SCE$. Then we form $P_{\mu\nu} = 2\sum_i^{occ} c_{\mu i}^* c_{\nu i}$ for the "population matrix" where integration over the spins leaves identical spatial orbital with two electrons and only the lowest energy orbitals are "occupied." The solution to the $HC = SCE$ leaves the orbitals sorted with the lowest energy first. Next we form the first Fock matrix using the formula $F_{\mu\nu} = H_{\mu\nu} + \sum_\lambda \sum_\sigma P_{\lambda\sigma}[(\mu\nu/\lambda\sigma) - 0.5(\mu\nu/\lambda\sigma)]$. This usually requires some complicated computer science algorithm to read the $(\mu\nu/\lambda\sigma)$ values from the large file and fill them in to the Fock matrix along with the proper population $P_{\lambda\sigma}$ value. When the Fock matrix $(F_{\mu\nu})$ is completed we solve $FC = SCE$ for a new C matrix and compute the electronic energy according to the spin integrated energy formula where $(1/2)$ is used to only count the $i-j$ interaction once.

$$\langle E \rangle_{electronic} = \sum_\mu \sum_\nu P_{\mu\nu} H_{\mu\nu} + (1/2)\sum_\mu \sum_\nu \sum_\lambda \sum_\sigma P_{\mu\nu} P_{\lambda\sigma}[(\mu\nu/\lambda\sigma) - (1/2)(\mu\lambda/\nu\sigma)]$$

Fortunately this can also be computed as $\langle E \rangle_{electronic} = (1/2)\sum_\mu \sum_\nu P_{\mu\nu}[H_{\mu\nu} + F_{\mu\nu}]$.

Thus an iterative scheme is set up as $H \rightarrow C \rightarrow P \rightarrow F \rightarrow E_1 \rightarrow C \rightarrow P \rightarrow F \rightarrow E_2 \rightarrow C\ldots$

The iteration may require a number of special techniques to achieve convergence and may take anywhere from 10 to 100 iterations depending on the convergence technique and the initial guess. Upon electronic convergence the nuclear repulsion energy is added to the total.

GAUSSIAN BASIS SETS

Although many very talented people attempted to develop formulas for $(\mu\nu/\lambda\sigma)$ using Slater type orbitals (STO) optimized by Clementi and Raimondi [5], the results were discouraging and many investigators turned their attention to a now famous paper by Boys [14] in 1950 (Figures 17.2 through 17.4). The formulas Boys derived can be written on a $3'' \times 5''$ card but require one special function as we see in Appendix C. They have been tabulated by Shavitt [15] and take up less than a page to write. The basic idea is to use simple Gaussian functions $Ne^{-\alpha(r-R)^2}$ where $\vec{R} = (x, y, z)$ is the center of the Gaussian orbital. We can see from Figure 17.5 that a single Gaussian is very poor near the nucleus ($r = 0$), matches well near 1 bohr which is near a typical bond length and then is poor again at larger values of r. Thus Gaussian orbitals are very poor representatives of the true exponential close to a nucleus or far away but the integrals are easy to do for polyatomic molecules and the description of electron density at nominal bond lengths is good.

The next improvement is to use a linear combination of Gaussians fitted to the STO shape, the so-called STO-NG orbitals [16] where N is the number of Gaussian terms in the fitted linear combination. Since 1951 there have been hundreds of research papers investigating various ways to optimize linear combinations of Gaussians to atomic orbitals. Figure 17.6 shows an early variational fit of three Gaussians to a H1s orbital and we can see that the fit is greatly improved over just one Gaussian. Later orbitals were fit in a least-squares sense to the Clementi–Raimondi orbitals or optimized variationally and today there is a whole technology for basis sets of s, p, d, f, g, and even higher angular momentum orbitals in some cases.

Variationally (Chapter 16) it has been found that lower (better) energies are obtained if so-called "split basis sets" are used in which the outer component of the linear combination is allowed to be optimized separately from the inner components which remain in fixed ratios as fitted to atoms. The outer components can adjust variationally to a molecular environment. In PCLOBE one of the

FIGURE 17.2 Prof. John Clarke Slater (1900–1976) was a noted American physicist and theoretical chemist. He is known for his definition of Slater-type-orbitals which led to the STO-NG Gaussian basis sets and the use of determinants to guarantee antisymmetric wave functions for electrons. His writing is especially lucid and among his many books are *Quantum Theory of Matter (2nd Edn.)* and *Quantum Theory of Molecules and Solids*, Vols. 1, 2, 3, 4 published by McGraw-Hill. Here he is giving a lecture at an early Sanibel Meeting. Note his famous $\rho^{4/3}$ formula on the board, a formula which encouraged development of modern density functional theory. (Photo by Prof. Sam Trickey of the Quantum Theory Project at the University of Florida in Gainesville.)

best basis sets stored in the program is the 6-31G** basis set. The inner 1s orbital is composed of six Gaussians while the 2sp shell is made from four Gaussian components in which the inner three are held fixed but the outer component adjusts to the molecular environment. One "*" indicates an additional "polarization" basis that consists of an optimized set of 3d orbitals on the first row atoms. The second "*" indicates that for H atoms there will be an optimized set of 2p polarization functions. For very fast runs there is a very robust STO-3/2G basis which is the smallest Gaussian basis that gives reliable molecular geometries although the energies it produces are poor (high).

One peculiarity of PCLOBE is that it uses an old technique which makes the evaluation of the integrals simple (Appendix C). In PCLOBE the p-orbitals are made by putting Gaussian spheres on opposite sides of the origin with opposite signs and d-orbitals are made by putting off-center Gaussian spheres in cloverleaf-patterns. This greatly simplifies the integral evaluation but the orbital shapes are not quite perfect fits to the spherical harmonic shapes. However, in a variational sense the orbitals adjust quite well to molecular environments and the energies with these orbitals are essentially equal to energies with spherical harmonic functions to within 10^{-6} hartrees. At one time these orbitals were criticized as being the incorrect shape but we have refined the fits and PCLOBE is probably the most refined program of its type relative to the basis functions.

FIGURE 17.3 This a photograph of the 1962 participants in one of the first Sanibel Meetings. Front (left to right): Arthur Frost, Roy McWeeny, Donald S. McClure, Art Freeman, Eli Burstein and Per O. Lowdin (the Organizer). Back (left to right): Stig Flodmark, Gilda Loew, Frank Harris, Reuben Pauncz, Joop de Heer, Harrison Shull, Inga Fischer-Hjalmars and John Pople. At the time of this picture John Pople may have been best known for his work elucidating the theory of nuclear magnetic resonance in a now classic 1959 text with W. G. Schneider and H. J. Bernstein but he later became Sir John Anthony Pople, KBE, FRS (1925–2004), an English mathematician who came to the United States in 1964. Although he regarded himself as a mathematician, theoretical chemists considered him a leader among them and he received the Nobel Prize in chemistry in 1998 for his contributions to theoretical chemistry. Probably because he was British, his scientific papers were written in excellent English and this author found his papers far and away the most easily understandable. This author's advice to a graduate student is to read Pople's papers on a given topic and if you do not understand that, other papers will be more difficult! (By permission of the Quantum Theory Project at the University of Florida in Gainesville, FL.)

In Figure 17.6, we see an improved linear combination of three Gaussian orbitals fitted variationally to the H atom. A linear combination of Gaussians will always have difficulty representing a 1s orbital at $(r=0)$ because the bell-shape of a Gaussian is flat at the top. Sambe [17] and others have investigated the accuracy of up to 10 Gaussians in the linear combination and the Gaussian exponents have been more thoroughly optimized. Here we show graphs of $\psi_{1s} = (0.15583039)e^{-(4.45)r^2} + (0.21749787)e^{-(0.676)r^2}(0.11140671)e^{-(0.151)r^2}$, with $\zeta = 1.0$ and $\psi_{B2s} = (0.19822186)e^{-(0.180)r^2} + (0.19822186)e^{-(8.99760584)r^2}$ as scaled to $\zeta = 1.5$ for boron.

The energy of this three-Gaussian 1s mimic in Figure 17.6 is $E = 0.496979$ hartrees [17]. In the related graph in Figure 17.7 we show a nodeless 2s-orbital from the BH calculation developed to make PCLOBE faster. It uses the largest component of Sambe's [17] 2s orbital ($\alpha = 0.020$ for $\zeta_{H2s} = 0.5$), which had a node in it due to the variational optimization. It is canceled at $(r=0)$ by a second Gaussian with equal weight and the exponent optimized ($\alpha = 0.997339820289$ for $\zeta_{H2s} = 0.5$) to provide the same maxima as a STO 2s-orbital. Using this 2G-2S function reaps the benefit of a good fit to the Sambe hydrogenic 2S in the important outer tail of the function. Also, the SCF process and the $S^{-1/2}$ orthogonalization will use the 1S function in the calculation to improve the inner part of the 2S function while at the same time we can use known optimized scaling for STO-2S orbitals. Finally, using fewer Gaussians for the 2S orbital really speeds up the calculation!

Scaling the Gaussians from the variational fit to the H atom is easy in the lobe basis. Once the exponent is modified, the function is renormalized by dividing the coefficients by the square root of the overlap of the total orbital with itself.

FIGURE 17.4 Prof. Henry F. Schaefer III, Director of the Center for Computational Chemistry at the University of Georgia. Professor Schaefer's major awards include the American Chemical Society Award in Pure Chemistry (1979), the American Chemical Society Leo Hendrik Baekeland Award (1983); the Schrödinger Medal (1990); the Centenary Medal of the Royal Society of Chemistry (London, 1992); the American Chemical Society Award in Theoretical Chemistry (2003). In 2003, he also received the annual American Chemical Society Ira Remsen Award. The *Journal of Physical Chemistry* published a special issue in honor of Dr. Schaefer on April 15, 2004. In 2009, the journal *Molecular Physics* published five consecutive issues in honor of Professor Schaefer. He was elected a Fellow of the American Academy of Arts and Sciences in 2004. He was the recipient of the prestigious Joseph O. Hirschfelder Prize of the University of Wisconsin for the academic year 2005–2006. He became a Fellow of the Royal Society of Chemistry (London) in 2005. He was also among the inaugural class of Fellows of the American Chemical Society, chosen in 2009. (With the permission of Prof. Schaefer.)

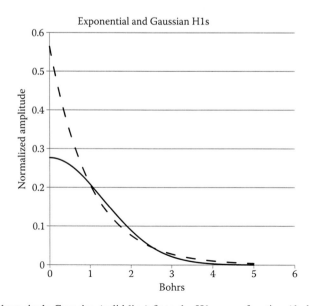

FIGURE 17.5 The best single Gaussian (solid line) fit to the H1s wave function (dashed line).

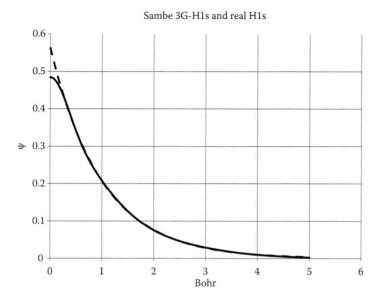

FIGURE 17.6 Variational fit of three Gaussians (solid line) to the H1s wave function (dashed line), from the parameters given by [17].

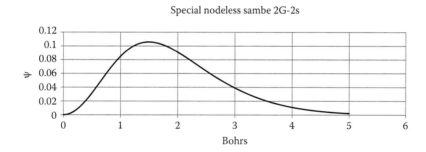

FIGURE 17.7 Special 2G-2s orbital optimized using the outer tail of the Sambe 4G-2s with just one additional inner Gaussian optimized to produce the same radial maximum as a Slater-Type 2s function. Here the orbital is scaled to $\zeta_{2s} = 1.5$ for the B2s orbital. Note vertical scale is much smaller for the 2s than the 1s in Figure 17.6.

Example

For the first term of the B2s orbital we assume that the Sambe exponent of 0.02 is for the H atom orbital with an exponent of 0.5. Since the variable r is squared in the Gaussian exponent we "scale the r-coordinate" by $\left(\dfrac{1.5}{0.5}\right)^2 (0.02) = 0.180$ and for the special exponent fitted to guarantee the same radial maximum as a STO-2S function $\left(\dfrac{1.5}{0.5}\right)^2 (0.997339820) = 8.97605838$.

In a molecular environment the orbitals will orthogonalize due to $S^{1/2}$ and the diagonalization in $FC = SCE$. Nodes will form in the molecular orbitals as needed and each orbital is normalized within PCLOBE.

Example, BH

Let us set up a calculation for a molecule slightly larger than H_2. We select BH because it is more covalent than LiH, may involve p-orbitals and, most importantly, the resulting matrices fit the page size here! Experts will recognize this calculation as a very low level example by present standards and yet it illustrates the full application of the Hartree–Fock–Roothaan equation we described above. As complicated as this technology is, such a calculation is trivial in modern quantum chemistry; so we have included it in the list of "Essential Physical Chemistry"! Although BH is not a household item, it is a real molecule; so we start with a small STO-3/2G basis and the geometry that is stored within the standard library accompanying the program. A "minimum basis set" only uses the orbitals required to accommodate all the electrons. In this case we need B1s, B2s, B2px, B2py, B2pz, and H1s so that our matrices will be (6×6) and we use orbital ζ values suggested for this basis by Pople's group [16]. We only use a 2-Gaussian B2s and a 2G (four sphere) B2p set so that this is really a tiny basis set requiring only 20 Gaussian spheres at carefully selected positions. The output has been edited and spaced for this text but we can see how the 20 spheres are "contracted" into six linear combinations that mimic STO orbital shapes. We see the program report on blocks of two-electron integrals having been completed and the eigenvectors with eigenvalues of the initial guess H-core matrix.

We have deleted many of the SCF iterations because this was not an easy path to convergence. In the initial guess orbitals, we see that orbital columns No. 3 and 4 are a degenerate pair of 2p-pi orbitals on the boron atom while what would/should be the H1s hydride orbital is not filled because it is No. 5 in the initial guess order. Fortunately, the SCF process wandered slowly to a situation where the sigma orbital (column) with a large H1s coefficient is lower than the degenerate pi-orbitals and the process converges to place six electrons in the first three sigma orbitals with the empty 2p-pi orbitals left over as "virtual orbitals". The virtual orbitals are not true excited states, they are left over "junk" since the optimization focused only on the occupied orbitals. However, the virtual orbitals have to be orthogonal to the occupied orbitals by virtue of the diagonalization process so in fact they do resemble excited orbitals, at least in symmetry. We will see later that calculating excited states accurately is even more complicated and involves more determinants than just the one optimized in the SCF equations but qualitatively the virtual orbitals look like excited states. As a result of the diagonal λ_{ii} in the SCF derivation, the one-electron energies are approximately the energy required to remove an electron from its orbital (Koopman's Theorem, see Ref. [1]). Even then this is only a one-determinant result and better values for the ionization potential of a molecule require more elaborate calculations. However, this one-electron SCF model is of higher credibility than the simpler Hückel pi-electron model [2].

In the calculation shown the bond length is optimized and the final geometry is treated to a numerical estimate of the second derivative with respect to nuclear displacements (hessian), then the hessian is mass-weighted according to the atomic weight of the atoms to obtain the multidimensional force constant matrix using the method of E. B. Wilson shown in Figure 17.8 with other now-famous chemists. The eigenvalues of the diagonalized force constant matrix are the vibrational frequencies [18] along with small or imaginary frequencies. The imaginary frequencies are due to the square root of a negative force constant. Usually this is due to small numerical errors in the hessian but it is possible for a saddle-point potential to occur which produces imaginary frequencies. For N atoms there are $(3N - 6 \text{ or } 3N - 5)$ vibrational motions (1 for BH) but the force constant matrix will be $N \times N$ so the remaining eigenvalues and eigenvectors represent impure vibrations mixed with translation and rotation. Possible errors are due to the numerical second derivatives in the force constant matrix and the fact that the calculation is done at not quite the true minimum in the energy surface.

In PCLOBE an optimization threshold of "3" or an energy gradient of 10^{-3} hartrees/bohr only optimizes the geometry, a "4" for a 10^{-4} gradient does the IR calculation but "5" for 10^{-5} hartrees/bohr is usually better. There is only one vibrational mode here.

The eigenvectors of the force constant matrix are the motions of the atoms in the normal mode [18]. Due to the diagonalization process, the normal modes are linearly independent normalized motions. Here we separate the (x, y, z) motions of each atom in the eigenvector column. If this were a larger molecule we could examine the (x, y, z) motions of each atom in the various modes, all we see here is

FIGURE 17.8 A relatively rare photograph of E. Bright Wilson attending the opening of the new Technion Chemistry Building in Israel in 1962, from left to right: Sir Alexander Todd (1957 Nobel Prize in chemistry for research on nucleotides, coenzymes, and vitamins), Robert Burns Woodward (reaching for his lighter) (1965 Nobel Prize in chemistry for many complex organic syntheses including cephalosporin and vitamin B12), E. Bright Wilson (Harvard coauthor with L. Pauling of *Introduction to Quantum Mechanics* and coauthor of *Molecular Vibrations* with J. C. Decius and P.C. Cross. This scholarly text contains the Wilson FG-matrix method), and Sir Christopher Kelk Ingold, FRS (noted for introduction of modern electronic structure rules into organic chemistry including mechanisms and is known for terms such as nucleophile, electrophile, inductive and resonance effects, and such symbols as S_N1, S_N2, E1, and E2. He is regarded as one of chief pioneers of physical organic chemistry). (Gift of John D. Roberts, Chemical Heritage Foundation Collections.)

that the atoms move back and forth along the bond line, but the motion is greater for the less massive H atom. The numerical evaluation of the second derivative is of low quality here and larger programs have complicated analytical formulas for the second derivative but it was expedient to add the numerical derivatives as a way to obtain qualitative vibrational frequencies for their educational value. It is well known that a small basis set makes the energy potential wells too narrow [19] (there are insufficient basis functions to describe small features of the overall wave function) and that makes the calculated frequencies too high. A set of correction factors is available for various basis sets and here we use the value of 0.89 which is an average value for the STO-3G basis.

The dipole moment is computed along with estimated charges using a formula developed by Mulliken [20] for what is called "Mulliken population analysis" for bond orders and partial charges on atoms. The dipole moment is a bona fide quantum mechanical expectation value as calculated from this wave function, although SCF dipoles tend to be about 10% higher from a single determinant wave function, and can be improved using a multideterminant wave function. If the calculated partial atomic charges are placed at the positions of the atoms, the dipole moment calculated from these point charges is usually much higher than the quantum mechanical expectation value. The Mulliken analysis [20] is an ad hoc formula based on the orbital overlap and usually gives a *noninteger number qualitatively proportional to the bond strength*. Because the Mulliken analysis is based on dividing the overlap exactly in half it works well for covalent bonds, but strains credibility for ionic situations. The Wiberg–Mayer [21,22] bond order is usually a number much closer to what might be expected for a single or multiple bonds and the value of 0.9909 here is very close to the expected value of integer 1 for a single bond. A similar calculation for the H_2 molecule produces a Wiberg–Mayer bond order of 1.000 and a calculation for N_2 produces a bond order of 3.000 so we favor the Wiberg–Mayer bond orders over the Mulliken values (Table 17.2).

TABLE 17.2
Edited Output from PCLOBE for the BH Molecule

PCLOBE	PCLOBE	PCLOBE	PCLOBE	PCLOBE	PCLOBE	PCLOBE	PCLOBE	PCLOBE

Gaussian-Lobe Program for Organic Molecules
Adapted to Personal Computers by Don Shillady
Virginia Commonwealth University
Richmond, Virginia
1978, 30 May, 2003

Historical Foundations of Gaussian-Lobe Basis Sets

1. S. F. Boys, *Proc. Roy. Soc.* A200, 542 (1950).
2. H. Preuss, *Z. Naturf. A*, 11, 823 (1956).
3. J. L. Whitten, *J. Chem. Phys.*, 39, 349 (1963).
4. I. Shavitt, *Methods Comp. Phys.*, 2, 1 (1963).
5. H. Preuss, *Mol. Phys.*, 8, 157 (1964).
6. H. Sambe, *J. Chem. Phys.*, 42, 1732 (1965).
7. H. Preuss and G. Diercksen, *Int. J. Quantum Chem.*, I, 605 (1967).
8. J. F. Harrison and L.C. Allen, *Mol. Spectrosc.*, 29, 432 (1969).
9. F. Dreisler and R. Ahlrichs, *Chem. Phys. Lett.*, 23, 571 (1973).
10. H. Le Rouzo and B. Silvi, *Int. J. Quantum Chem.*, XIII, 297 (1978).
11. S. Y. Leu and C.Y. Mou, *J. Chem. Phys.*, 101, 5910 (1994).
12. P. Otto, H. Reif, and A. Hernandez-Laguna, *J. Mol. Struct. (Theochem)*, 340, 51 (1995).

BH.XYZ

THE AVAILABLE BASIS SETS ARE:

ibs = 0, STO-3/22G with Sambe 2P for H-Ne; STO-3/33* for Na-Ar

Nuclear Coordinates (Angstroms) from Input

```
Atomic Core    X             Y             Z           Basis

    5.       0.000000      0.000000      0.000000        0
         Z1s =  4.680     Z2s =  1.500    Z2p =  1.500


    1.       1.225554      0.000000      0.000000        0
         Z1s =  1.240     Z2s =  0.000    Z2p =  0.000
```

Basis Size = 6 and Number of Spheres = 20 for 6 Electrons

Distance Matrix in Angstroms

```
        B        H
B    0.0000   1.2256
H    1.2256   0.0000
```

TABLE 17.2 (continued)
Edited Output from PCLOBE for the BH Molecule

Spherical Gaussian Basis Set

```
No.  1 alpha = 0.10445304E+03 at X =  0.000000 Y =  0.000000 Z =  0.000000 a.u.
No.  2 alpha = 0.15696662E+02 at X =  0.000000 Y =  0.000000 Z =  0.000000 a.u.
No.  3 alpha = 0.33783755E+01 at X =  0.000000 Y =  0.000000 Z =  0.000000 a.u.
No.  4 alpha = 0.18000000E+00 at X =  0.000000 Y =  0.000000 Z =  0.000000 a.u.
No.  5 alpha = 0.89760584E+01 at X =  0.000000 Y =  0.000000 Z =  0.000000 a.u.
No.  6 alpha = 0.29430000E+00 at X =  0.110667 Y =  0.000000 Z =  0.000000 a.u.
No.  7 alpha = 0.12690000E+01 at X =  0.110667 Y =  0.000000 Z =  0.000000 a.u.
No.  8 alpha = 0.29430000E+00 at X = -0.110667 Y =  0.000000 Z =  0.000000 a.u.
No.  9 alpha = 0.12690000E+01 at X = -0.110667 Y =  0.000000 Z =  0.000000 a.u.
No. 10 alpha = 0.29430000E+00 at X =  0.000000 Y =  0.110667 Z =  0.000000 a.u.
No. 11 alpha = 0.12690000E+01 at X =  0.000000 Y =  0.110667 Z =  0.000000 a.u.
No. 12 alpha = 0.29430000E+00 at X =  0.000000 Y = -0.110667 Z =  0.000000 a.u.
No. 13 alpha = 0.12690000E+01 at X =  0.000000 Y = -0.110667 Z =  0.000000 a.u.
No. 14 alpha = 0.29430000E+00 at X =  0.000000 Y =  0.000000 Z =  0.110667 a.u.
No. 15 alpha = 0.12690000E+01 at X =  0.000000 Y =  0.000000 Z =  0.110667 a.u.
No. 16 alpha = 0.29430000E+00 at X =  0.000000 Y =  0.000000 Z = -0.110667 a.u.
No. 17 alpha = 0.12690000E+01 at X =  0.000000 Y =  0.000000 Z = -0.110667 a.u.
No. 18 alpha = 0.73328489E+01 at X =  2.315961 Y =  0.000000 Z =  0.000000 a.u.
No. 19 alpha = 0.11019426E+01 at X =  2.315961 Y =  0.000000 Z =  0.000000 a.u.
No. 20 alpha = 0.23716991E+00 at X =  2.315961 Y =  0.000000 Z =  0.000000 a.u.
```

Contracted Orbital No. 1

```
1.60447595*(1)   2.20815846*(2)   1.18220933*(3)
```

Contracted Orbital No. 2

```
0.19822186*(4)   -0.19822186*(5)
```

Contracted Orbital No. 3

```
1.87064030*(6)   1.11225426*(7)   -1.87064030*(8)   -1.11225426*(9)
```

Contracted Orbital No. 4

```
1.87064030*(10)   1.11225426*(11)   -1.87064030*(12)   -1.11225426*(13)
```

Contracted Orbital No. 5

```
1.87064030*(14)   1.11225426*(15)   -1.87064030*(16)   -1.11225426*(17)
```

Contracted Orbital No. 6

```
0.21882501*(18)   0.30115771*(19)   0.16123456*(20)
```

***** Nuclear Repulsion Energy in au =   2.15893061757377 *****

(continued)

TABLE 17.2 (continued)
Edited Output from PCLOBE for the BH Molecule

Initial-Guess-Eigenvectors by Column

The solution to the secular equations is a modified form
of the routines HDIAG and EIGEN given in the text by
P.O. Offenhartz in ''Atomic and Molecular Orbital Theory'',
1970, originally written by F. J. Corbato and M. Merwin at
M.I.T. We have modified the routines to use less memory
via the symmetric nature of the H-matrix.

#At-Orb	1	2	3	4	5	6
1 B 1s	0.970	−0.011	0.000	0.000	−0.254	−0.174
2 B 2s	0.142	−0.028	0.000	0.000	0.754	0.944
3 B 2px	0.027	0.861	0.000	0.000	−0.339	0.752
4 B 2py	0.000	0.000	−0.981	−0.196	0.000	0.000
5 B 2pz	0.000	0.000	0.196	−0.981	0.000	0.000
6 H 1s	−0.060	0.247	0.000	0.000	0.409	−1.276

One-Electron Energy Levels

$E(1) =$ −9.722037579344
$E(2) =$ −2.354649254845
$E(3) =$ −2.229474929327
$E(4) =$ −2.229474929327
$E(5) =$ −1.910839316872
$E(6) =$ −1.048994859272

(1, 6/ 6, 6) are done.
(2, 6/ 6, 6) are done.
(3, 6/ 6, 6) are done.
(4, 6/ 6, 6) are done.
(5, 6/ 6, 6) are done.
(6, 6/ 6, 6) are done.

Block No.   1 Transferred to Disk/Memory
The Two-Electron Integrals Have Been Computed

The construction of the Fock matrix will use the
algorithm of Preuss and Diercksen, Int. J. Quant. Chem.
I, p605, 1967 as modified here to merge with the other
routines of the PCLOBE program.

Electronic Energy =   −25.5720869058 a.u., Dif. =   25.572086905840
Electronic Energy =   −25.5777791182 a.u., Dif. =   0.005692212328
Electronic Energy =   −26.0140297726 a.u., Dif. =   0.436250654482
Electronic Energy =   −26.0164143999 a.u., Dif. =   0.002384627271
Electronic Energy =   −27.0336901449 a.u., Dif. =   0.000000000001
Iteration No.145 Energy Second Derivative = −0.00000000000036
alpha = 0.900000
Electronic Energy =   −27.0336901449 a.u., Dif. =   0.000000000000

TABLE 17.2 (continued)
Edited Output from PCLOBE for the BH Molecule

```
Iteration No.146 Energy Second Derivative = −0.00000000000016
alpha = 0.900000
Electronic Energy =      −27.0336901449 a.u., Dif. =      0.000000000000
Iteration No.147 Energy Second Derivative = −0.00000000000009
alpha = 0.900000
Electronic Energy =      −27.0336901449 a.u., Dif. =      0.000000000000
Iteration No.148 Energy Second Derivative = −0.00000000000005
alpha = 0.900000
Electronic Energy =      −27.0336901449 a.u., Dif. =      0.000000000000
```

```
      ##### Total Energy =      −24.8747595273 hartrees #####

         Virial Ratio = −<E>/<T> = 0.985615

            One-Electron Energy Levels

               E(1) =    −7.530946193339
               E(2) =    −0.597487711549
               E(3) =    −0.281512715999
               E(4) =     0.243234519411
               E(5) =     0.243234519411
               E(6) =     0.666414452060
```

```
SCF spin-Orbitals for 3 Filled Orbitals by Column


#At-Orb    1        2        3        4        5        6

1 B 1s    0.993    0.164    0.136    0.000    0.000   −0.070
2 B 2s    0.037   −0.603   −0.780    0.000    0.000    0.713
3 B 2px   0.007   −0.231    0.561    0.000    0.000    1.027
4 B 2py   0.000    0.000    0.000   −0.977   −0.213    0.000
5 B 2pz   0.000    0.000    0.000    0.213   −0.977    0.000
6 H 1s   −0.009   −0.471    0.431    0.000    0.000   −1.205
```

```
   Using Schlegel-Gradient Equations, see:
   Pople, Krishnan, Schlegel and Binkley,
   I.J.Q.C. Symp. 13, p225 (1979); equation (21).
```

```
Varying six coordinates by 0.000100 until RMS-norm = 0.0001000 hartrees/bohr
```

```
   The construction of the Fock matrix will use the
   algorithm of Preuss and Diercksen, Int. J. Quant. Chem.
   I, p605, 1967 as modified here to merge with the other
   routines of the PCLOBE program.
```

(continued)

TABLE 17.2 (continued)
Edited Output from PCLOBE for the BH Molecule

```
##### Total Energy =    -24.8747595273 hartrees #####

        Virial Ratio = -<E>/<T> = 0.985615


Atom  X(Ang.)   Y(Ang.)   Z(Ang.)    dE/Dx     dE/dY     dE/dZ

1    0.000000  0.000000  0.000000  -0.004641  0.000002  0.000002
2    1.225554  0.000000  0.000000   0.004646  0.000002  0.000002


   Minimization No. 1 RMS Gradient = 0.0124093 hartrees/bohr
```

```
                            .
                            .
                            .

   ##### Total Energy =    -24.8748901469 hartrees #####

        Virial Ratio = -<E>/<T> = 0.985374


Atom   X(Ang.)    Y(Ang.)    Z(Ang.)    dE/dX      dE/dY     dE/dZ

1     0.007725  -0.000004  -0.000013  -0.000029  0.000002  0.000001
2     1.217794  -0.000016  -0.000007   0.000035  0.000001  0.000002


   Minimization No. 12 RMS Gradient = 0.0000855 hartrees/bohr
```

```
   ##### Total Energy =    -24.8748901504 hartrees #####

        Virial Ratio = -<E>/<T> = 0.985374


              GEOMETRY CONVERGED

   CHEMSITE RESTART COORDINATES, COPY AND PASTE INTO INPUT FILE *.CTA


B  0.00774  -0.00001  -0.00001
H  1.21777  -0.00002  -0.00001


   MOLUCAD RESTART COORDINATES, COPY AND PASTE INTO INPUT FILE *.XYZ


B  0.007742  -0.000006  -0.000014
H  1.217773  -0.000016  -0.000008


              Distance Matrix in Angstroms


        B        H

B   0.0000   1.2100
H   1.2100   0.0000
```

TABLE 17.2 (continued)
Edited Output from PCLOBE for the BH Molecule

~~~~~~
~~~~~~
~~~~~~

VIBRATIONAL ANALYSIS

~~~~~~
~~~~~~
~~~~~~

Approximate Energy Second-Derivative Matrix

	1	2	3	4	5	6
1	0.307E+00	0.638E-04	0.682E-04	−0.308E+00	0.692E-04	0.650E-04
2	0.638E-04	0.851E-03	0.711E-08	0.693E-04	0.897E-03	0.266E-06
3	0.682E-04	0.711E-08	0.864E-03	0.651E-04	0.238E-06	0.846E-03
4	−0.308E+00	0.693E-04	0.651E-04	0.307E+00	−0.690E-04	−0.644E-04
5	0.692E-04	0.897E-03	0.238E-06	−0.690E-04	0.851E-03	0.149E-06
6	0.650E-04	0.266E-06	0.846E-03	−0.644E-04	0.149E-06	0.864E-03

Mass-Weighted Force Constant Matrix

	1	2	3	4	5	6
1	0.284E-01	0.591E-05	0.631E-05	−0.933E-01	0.210E-04	0.197E-04
2	0.591E-05	0.787E-04	0.657E-09	0.210E-04	0.272E-03	0.807E-07
3	0.631E-05	0.657E-09	0.799E-04	0.197E-04	0.721E-07	0.256E-03
4	−0.933E-01	0.210E-04	0.197E-04	0.305E+00	−0.685E-04	−0.639E-04
5	0.210E-04	0.272E-03	0.721E-07	−0.685E-04	0.844E-03	0.148E-06
6	0.197E-04	0.807E-07	0.256E-03	−0.639E-04	0.148E-06	0.857E-03

Normal Modes Relative to Optimized Geometry

Ordered Atom-1(x,y,z),Atom-2(x,y,z), ...
Including Rotational and Translational False Modes

	1	2	3	4	5	6
1	0.293	0.003	−0.003	0.100	−0.090	0.947
2	0.000	0.021	−0.303	0.090	−0.943	−0.100
3	0.000	0.287	0.019	0.948	0.100	−0.092
4	−0.956	0.001	−0.001	0.030	−0.027	0.290
5	0.000	0.064	−0.951	−0.029	0.301	0.028
6	0.000	0.956	0.065	−0.285	−0.030	0.024

(continued)

TABLE 17.2 (continued)
Edited Output from PCLOBE for the BH Molecule

Normal Mode Eigenvalues (hartrees/(amu*bohr^2))

0.333573   0.000934   0.000931   0.000004   −0.000007   −0.000115

Harmonic Vibrational Frequencies and Corrections

J.A. Pople et. al. Int. J. Quantum Chem. S15, 269 (1981)
We apply 3-21G factor of 0.89 to any basis set initially.

(K Newtons/cm;   frequency 1/cm   ;   0.89 × 1/cm)
(5.193368   ;   2968.93   ;   2642.34)
(0.014537   ;   157.08   ;   139.80)
(0.014496   ;   156.85   ;   139.60)
(0.000066   ;   10.57   ;   9.41)
(−0.000106   ; Imag   13.39   ; Imag   11.92)
(−0.001788   ; Imag   55.10   ; Imag   49.04)

Imaginary frequencies (Imag) indicate either an
insufficient optimization or a less complete basis set.
Try further optimization and/or a larger basis set.

Total Energy =   −24.8748901504 hartrees

Virial Ratio = −<E>/<T> = 0.985374

Reference State Orbitals for 3 Filled Orbitals by Column

# At-Orb	1	2	3	4	5	6
1 B 1s	0.993	0.163	−0.136	0.000	0.000	−0.070
2 B 2s	0.038	−0.595	0.785	0.000	0.000	0.730
3 B 2px	0.007	−0.234	−0.561	0.000	0.000	1.036
4 B 2py	0.000	0.000	0.000	0.775	0.632	0.000
5 B 2pz	0.000	0.000	0.000	0.632	−0.775	0.000
6 H 1s	−0.009	−0.474	−0.426	0.000	0.000	−1.224

Dipole Moment Components in Debyes

Dx = 1.247953   Dy = −0.000005   Dz = −0.000002

Resultant Dipole Moment in Debyes = 1.247953

Computed Atom Charges

Q(1) = 0.096   Q(2) = −0.096

TABLE 17.2 (continued)
Edited Output from PCLOBE for the BH Molecule

```
                Orbital Charges

 1.998510  1.830296  1.075670  0.000000  0.000000  1.095525
```

```
              Mulliken Overlap Populations

    #At-Orb        1       2       3       4       5       6

    1 B 1s       2.063  −0.061   0.000   0.000   0.000  −0.003
    2 B 2s      −0.061   1.943   0.000   0.000   0.000  −0.052
    3 B 2px      0.000   0.000   0.738   0.000   0.000   0.337
    4 B 2py      0.000   0.000   0.000   0.000   0.000   0.000
    5 B 2pz      0.000   0.000   0.000   0.000   0.000   0.000
    6 H 1s      −0.003  −0.052   0.337   0.000   0.000   0.813
```

```
          Total Overlap Populations by Atoms

        B        H
B     4.6222   0.2823
H     0.2823   0.8133
```

```
              Wiberg-Mayer W-matrix

    #At-Orb        1       2       3       4       5       6

    1 B 1s       1.999   0.016   0.062   0.000   0.000  −0.052
    2 B 2s       0.016   1.830  −0.652   0.000   0.000   0.540
    3 B 2px      0.022  −0.241   1.076   0.000   0.000   0.766
    4 B 2py      0.000   0.000   0.000   0.000   0.000   0.000
    5 B 2pz      0.000   0.000   0.000   0.000   0.000   0.000
    6 H 1s      −0.026   0.284   1.092   0.000   0.000   1.096
```

```
        Wiberg-Mayer Bond Orders Between Atoms
        (See Chem. Phys. Letters, v97, p270 (1983))


        B        H
B     8.8181   0.9909
H     0.9909   1.2002
```

```
        ********* NORMAL FINISH OF PCLOBE *********
        (Use file lobe.xyz for RasMol or Viewerlite display.)
```

At present PCLOBE only has limited graphical output for RASMOL [23] so we present some results from the SPARTAN program [24] which provides excellent graphical output. The results shown are from a calculation using a slightly bigger basis set, the so-called 3-21G basis and a much bigger basis called 6-31G(d,p). The 6-31G(d,p) basis is very interesting in that it includes a set of 3d orbitals on the B atom and a set of 2p orbitals on the H atom! These are called "polarization functions" which allow the SCF process to refine the shape of the wave function by incorporating small amounts of the additional shapes into the ground state description. These basis uses functions of the form $Nx^l y^m z^n e^{-\alpha r^2}$, the so-called Cartesian Gaussian type orbitals (GTO). There is similarity in that both the Gaussian lobe orbital (GLO) basis and the GTO basis only use a single 3G-1s Gaussian expansion for the H atom as well as for the B1s orbital in the 3-21 set but a 6G 1s representation in the 6-31G(d,p) representation. While the GLO basis uses four off-axis Gaussian spheres for the 2p orbitals, the GTO basis uses a 2sp shell of a fixed set of 2 Gaussians with a third outer tail Gaussian free to be optimized for both the 2s and 2p orbitals on the Boron atom. In principle, the 3-21G basis should produce better results than the PCLOBE 3/2G basis because the 2-1G part of the 3-21G basis has more functions and more variational flexibility. In Figures 17.9 (a–h) we show graphical results only for the better 6-31G** = 6-31G(d,p) basis set.

DIPOLE MOMENT OF BH

One comment can be made regarding the dipole moment we have calculated here as 1.247953 Debye using the tiny 3/2G basis in PCLOBE. We do not expect the calculated properties to be very accurate in this tiny basis and the energy is lower (better) with the 3-21G basis. The 3-21G and 6-31(d,p) dipoles are higher as well. The 3G1s orbitals are essentially the same in the PCLOBE 3/2G basis and the 3-21G basis, so the lower energy of the 3-21G calculation is due to more Gaussians in the 2s orbital and the variational flexibility of the 2-1 split. While there is no known experimental value for the dipole moment of BH, a thorough calculation by Halkier, Klopper, Helgaker, and Jorgensen found a dipole moment of 1.27 ± 0.21 D [25]. This sort of comparison of results should converge to what is called the Hartree–Fock limit when very large basis sets are used. However, the "SCF model" assumes each electron moves in the "frozen average field" of all the electrons and there is an energy error of about 1% due to the fact that the motion of the electrons is dynamic and more complicated, that is, "correlated." Dipole moments are also higher by about 10% in the Hartree–Fock limit of very large basis sets. This is what has led to other improved methods and the modern use of the DFT which tries to correct these small errors. Although qualitative trends from SCF calculations are very useful, absolute thermodynamic energies are difficult to obtain due to the correlation error and research continues to try to solve this problem. We note a severe numerical condition here

$$1 \text{ hartree} = 27.21138386 \text{ eV}(1.602176487 \times 10^{-19}\text{C})(6.02214170^{23}/\text{mol})/(4.184 \text{ J/cal})$$

so 1 *hartree* = 627.509 *kcal/mol* and the reciprocal produces 1 *kcal/mol* $\cong$ 1.5936×10^{-3} *hartrees* = 0.0015936 *hartrees* $\leq$ 0.002 *hartrees*. As most thermochemical measurements can be measured to within 1 kcal/mol, this means that in order to reach "chemical accuracy" we need to be sure of the calculated energy to within 0.001 hartrees!

This level of accuracy is the desired goal in quantum chemistry but you can now realize it is a very demanding goal which is seldom reached! (Table 17.3).

EXCITED STATES OF BH

The vast majority of quantum mechanical calculations are on the ground state of molecules. Calculations of excited states of molecules are still a research frontier but for small molecules very complicated multideterminantal wave functions can be used to obtain accurate calculations of the energies of

TABLE 17.3
Electronic Energy and Dipole Moment of BH for Selected Basis Sets

Basis Set	Energy (Hartrees)	Dipole (Debyes)	Bond Length (Å)
PCLOBE 3/2G	−24.874890	1.248	1.2100
GAUSSIAN09 3-21G [26]	−24.976803	1.595	1.2287
GAUSSIAN09 6-31G(d,p)	−25.119238	1.495	1.2271

excited states as a function of bond stretching to dissociation into atoms. Early *ab initio* calculations by Harrison and Allen on BH [27], triplet-CH_2 [28] and the spin resonance of triplet-CH_2 [29] added credibility to quantum chemistry calculations and recently Harrison has updated the BH calculation with improved MCSCF results using MOLPRO [30]. The method is described as a multi-configurational-self-consistent-field (MCSCF) in a complete-active-space (CASSCF) which means that the orbitals have been optimized for all of the determinants simultaneously. The "complete active space" means that all possible permutations of orbital occupancy have been included from a selected few low-lying excited configurations. Another awesome possibility of a "complete configuration interaction" is inclusion of all possible orbital occupancies in all orbitals which is seldom carried out due to the size and difficulty of the problem. When you say "Complete CI" it should be in an echo chamber! (Figure 17.9).

We cannot give the details here but we mention this to indicate that in spite of the tedious derivation for optimizing just one determinant, it should be clear that even more sophisticated methods are needed for excited states. Since the virtual (empty) orbitals from the SCF process are orthogonal to the occupied orbitals, other determinants can be built by deleting one or more orbitals from the SCF determinant and replacing them by virtual orbitals to make new determinants. Then a further variational Cayley–Hamilton calculation is carried out for $HC = SCE$ where *now H is the Hamiltonian expressed in a linear combination of determinants and the eigenvector coefficients in C are the weights of the added determinants*. PCLOBE can do this "Configuration Interaction" if the bubble marked Boys–Reeves Configuration Interaction is clicked to use up to 200 spin-projected configurations for arbitrary spin multiplets. However, the next question is whether the orbitals are optimum for the excited determinants and PCLOBE only does a limited optimization. Some minor improvements can be obtained if the "Natural Orbitals" are iterated 2 or 3 times but that is not a true MCSCF optimization. Natural orbitals are those orbitals which make the one-electron population matrix (in the molecular orbital basis) diagonal but this does not include the important two electron terms. The MCSCF results from Harrison using the MOLPRO [30] program include full optimization of the orbitals as well as the weighting of the added determinants (Figures 17.10 and 17.11).

MESOIONIC BOND ORDERS

One of the most interesting conjectures posed in Coulson's "Valence" [31] concerned the dilemma of how to describe the bonding in 3-methyl sydnone and related mesoionic compounds. Often, synthetic chemists bypass this dilemma and simply draw a ring in the heterocyclic structure with a plus sign and then indicate a negative carbonyl oxygen atom. Some valence bond diagrams can be drawn but they also use separated charges (Figure 17.12). That rationalizes the high dipole moment of these compounds but is an oversimplification of the bonding. This question has been addressed [32] using methods more sophisticated than the single determinantal method used here but without the benefit of the Wiberg–Mayer Bond Order analysis. We can now do a single determinantal treatment using PCLOBE as a homework problem with PCLOBE.

We will see a larger-than-usual C–N bond across the ring indicating a latent three-membered ring. Coulson considered this bond but rejected it as unlikely due to its length but later experiments added

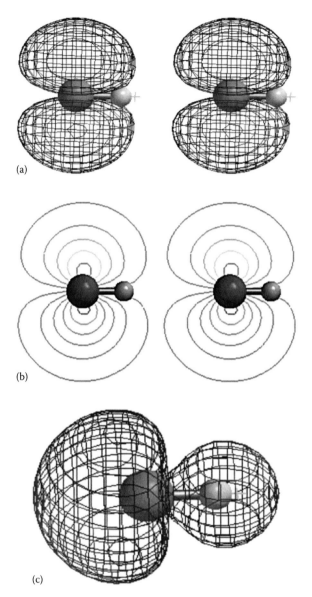

(a)

(b)

(c)

FIGURE 17.9 (a) A wireframe surface for the LUMO orbitals using the GTO 6-31G(d,p) basis in the SPARTAN program for the surface of the wave function of 0.032. There are actually two of these π-orbitals perpendicular to the B–H bond and they are identical but in different planes. Note there are 2p orbitals on the H atom in this basis set so the LUMO extends slightly toward the H atom but the orbital is mainly represented by 2p orbitals on the boron atom. What looks like a plus sign to the right of the H atom is the positive end of a dipole moment arrow depicted by the SPARTAN program. All the 17-9 plots were graciously provided by Prof. Carl Trindle of the Chemistry Department of the University of Virginia. (b) Both the degenerate unoccupied LUMO 1π orbitals are identical except one is in the xz plane and the other is in the yz plane with the BH bond along the x-axis. This is the contour diagram from a higher quality 6-31G(d,p) basis set using SPARTAN which includes 3d orbitals on the B atom and 2p orbitals on the H atom. We might expect to see some small contribution of the H2p orbitals but the contours seem to indicate almost total B2p orbitals. (c) This is the wireframe surface at 0.032 total wave function for the HOMO 3σ.

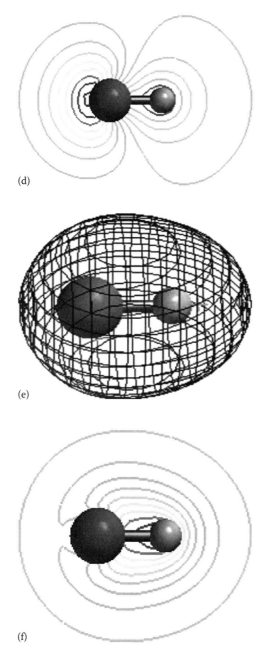

(d)

(e)

(f)

FIGURE 17.9 (continued) (d) This is the contour diagram for the 3σ HOMO orbital using the better 6-31G (d,p) basis set with the SPARTAN PROGRAM. The tendency for the H atom to be like a hydride ion is more noticeable here. (e) This is the wireframe surface of the 2σ orbital at a value of 0.032 using the 6-31G(d,p) basis set with the SPARTAN program. Here we see the wave function shifted more to the H atom. This is mainly a linear combination of the B2s and H1s orbitals. (f) This is the contour diagram for the 2σ orbital which shows a shift of the wave function toward a higher contribution of the H 1s orbital but there is no node between B and H; so this is a covalent sharing situation.

(continued)

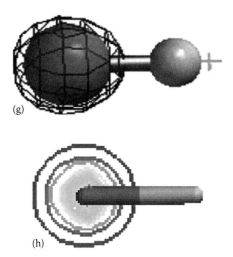

(g)

(h)

FIGURE 17.9 (continued) (g) Here we see a very tight orbital which is almost 100% 1s on the boron atom. The wireframe surface value is 0.032 and the results are from the 6-31G(d,p) calculation using the SPARTAN program. A slight distortion toward the H atom is barely noticeable and once again the plus tail of the dipole moment arrow definitely shows that most of the electron is associated with the B atom. (h) This is the contour diagram for the 1σ orbital which is almost entirely B1s with no evidence of any participation by the H1s orbital. This is from the 6-31G(d,p) calculation using the SPARTAN program and the B1s orbital is made up of six Gaussian components for a better representation of an exponential 1s function. The H1s function is a linear combination of only three Gaussians and the nuclear charge of +5 on B compared to only +1 on H results in this lower energy orbital being essentially the atomic B1s orbital. (All the 17-9 plots were graciously provided by Prof. Carl Trindle of the Chemistry Department of the University of Virginia.)

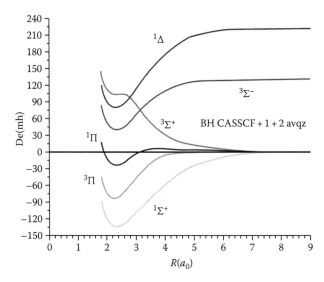

FIGURE 17.10 Excited states of BH calculated using a multi-configurational-self-consistent-field (MCSCF) complete-active-space method using the MOLPRO program, the state-of-the-art!

FIGURE 17.11 Prof. James F. Harrison of the Michigan State University Chemistry Department carried out early calculations on the excited states of molecules using accurate configuration interaction methods, particularly BH [27] under the direction of his advisor Prof. Leland C. Allen at Princeton University. He was one of the first to use accurate computational methods to study the triplet methylene molecule · CH_2 · which is very important in synthetic organic chemistry. In particular, his calculation of the electron spin resonance parameters of CH_2 was a rare case of computational theory guiding experimental spectroscopy at a time in the history of quantum chemistry that established the credibility of such calculations. He is also a contributor to Chapters 16 and 17.

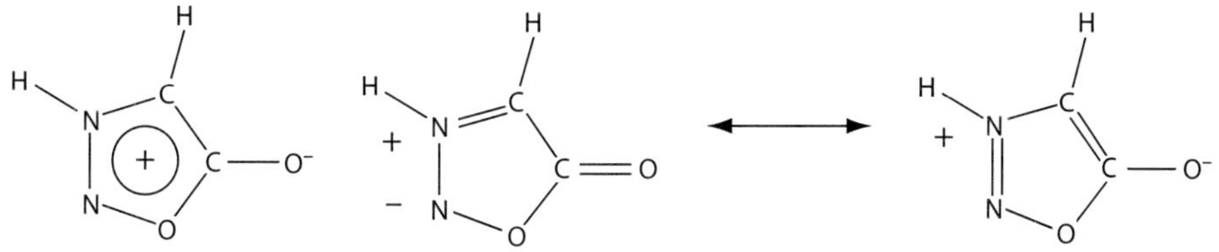

FIGURE 17.12 Proposed structures for bonding in sydnone.

credibility to the idea. We also see the expected weakening of the carbonyl bond to less than 2 but at about 1.84 it is still essentially a double bond. Thus we can draw the structure as a single covalent compound if we include the C–N bond across the ring and achieve "normal" valence rules for all the atoms in the molecule (4 for C, 3 for N, 2 for O, and 1 for H) with an explanation of the large dipole moment as due to the lone-pair of the ring N and the lone pairs of the furan-like O for the negative end and the opposite side of the molecule as the positive end of the dipole (Table 17.4).

While we might claim that the Wiberg–Mayer bond orders allow a single structure to be drawn with "normal" valence values, it is clear that there are many contributing ionic structures which can be drawn if one allows formal charges to be drawn on some atoms. This is a very interesting molecule from the point of view of bonding. One important piece of experimental evidence for a bond across the ring is that there is a patent on a photochemical device [33] which produces the three-membered ring as the end product of a photochemical reaction and another paper in which the three-membered ring has actually been trapped [34] and a crystal structure established. That information was not available to Coulson when he wrote Valence (Figures 17.13 and 17.14).

If your copy of PCLOBE that came from a CD in your book does not have the coordinates for this molecule in the file PCLOBE.DATA you can copy the file below and paste it or type it into a new file

TABLE 17.4
Selected SCF Results for Bonding in 3-Methyl Sydnone Using PCLOBE

Basis Set	C–N Wiberg-Mayer Bond	C=O Wiberg-Mayer Bond	Dipole Moment (D)
STO-3/2G	0.3129	1.8366	6.94700
STO-4G	0.2984	1.8448	6.8812

FIGURE 17.13 Prof. Robert Sanderson Mulliken (1896–1986), on the left, was an American physicist and chemist who received the Nobel Prize in 1966 for developing molecular orbital theory and the Priestly Medal in 1983. Dr. C. C. J. Roothaan was his student. On the right is Prof. Charles Alfred Coulson, FRS (1910–1974) who was a prominent English theoretical chemist and the author of the classic text *Valence* which contains many chemical insights. The meeting was in Prof. Coulson's office at Oxford University in 1953. (Gift of John D. Roberts, Chemical Heritage Foundation Collections.)

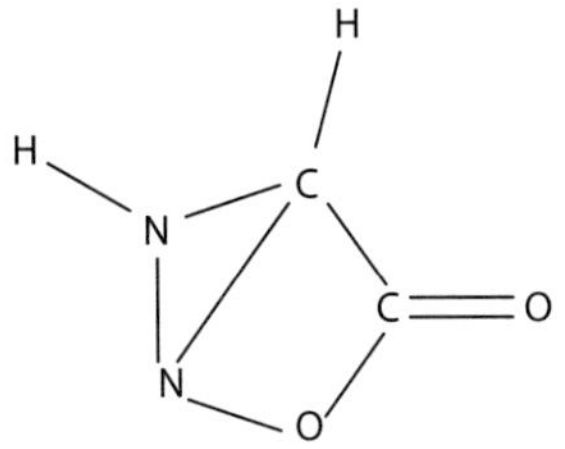

FIGURE 17.14 The prototype structure for the bond across the ring satisfies "normal" valence rules and does not require structures with charge separation. However, the bond across the ring is weak and the carbonyl stretching is redshifted in the infrared spectrum indicating less than a strong double bond, so this "normal" structure does not tell the whole story either. It is however more amenable to a description using only one determinant. There should be no doubt the mesoionic compounds provide challenges to bonding theory and Coulson called attention to this interesting case in "Valence". References [33,34] give experimental support for the cross-ring bond.

in PCLOBE.DATA. If you are using WINDOWS-VISTA or WINDOWS-7 you will have to run the program in "Administrator" mode but PCLOBE will run "as-is" with WINDOWS-XP or earlier versions of WINDOWS. The data file must end immediately after the last character of the last atom card-image, in this case a "0". If you press "ENTER" after the last digit while editing the input file, the program will look for another line of data but it will not be there and the program will stop. The data set of coordinates given below has already been extensively optimized in the STO-3/2G basis to save you the time for the preliminary molecular geometry. Note that on the BH output there is a place after convergence is obtained to copy the last coordinates for input back into the PCLOBE program file PCLOBE.DATA or use as input to either the MOLUCAD or CHEMSITE modeling programs.

PCLOBE also provides an output data file "lobe.xyz" which can be used to visualize the molecule with either RASMOL or Viewerlite modeling programs. Once you achieve a result for 3-methyl sydnone you should be able to do calculations easily for molecules up to the size of substituted benzene compounds in a few minutes using the fast STO-3/2 basis and experiment using other basis sets.

At this time PCLOBE can be downloaded from http://www.crcpress.com/product/isbn9780849384066 by clicking on the "updates" tab if the CD from your book is lost.

N	0.000000	0.000000	0.000000
N	−0.443824	−1.282047	0.000000
O	−1.820544	−1.165865	0.000000
C	−0.948648	0.952358	0.000000
H	−0.737145	2.028028	0.000000
C	−2.222417	0.185178	0.000000
O	−3.390925	0.503277	0.000000
C	1.480578	0.228700	0.000000
H	1.901626	−0.238412	0.907403
H	1.901626	−0.238412	−0.907403
H	1.650401	1.317754	0.000000

SUMMARY

This is a long chapter, but really a short introduction to quantum chemistry. The length of the chapter was increased due to the nearly complete output of an SCF calculation for the BH molecule. The difficult derivation of the Roothaan equation is given in Appendix B for those who are interested and the lobe basis integrals are given in Appendix C. It would be good to try and use PCLOBE to see what it does, particularly for the interesting sydnone molecule. Perhaps your teacher will not include this chapter in your course or you will run out of time in the second semester, but this material is here for a few interested students. In every class there will be a few who really want to know what is going on in modern physical chemistry, and this is the area of research most familiar to this author. It is almost a certainty that there will be one or two students out of a class of 40 who will be interested in this chapter, and for those students it is here if you want it!

PROBLEMS

17.1 Following the procedure used for the variational treatment of the H1s orbital in the previous chapter, assume another trial function of the form $\phi_{guess} = Ne^{-\alpha r^2}$. Normalize this function to find N and evaluate the energy, noting that once again there is no angular dependence in the function. Use $H = \dfrac{-\hbar^2}{2m}\nabla^2(r, \theta, \phi) - \dfrac{e_q^2}{r}$. Minimize the energy to find α_{min} and insert α_{min} in the energy expression to find the minimum energy. Compare this "best single Gaussian" energy to −0.5 au for the Schrödinger–Bohr H1s orbital. This would be a difficult question in a final exam (where it has often appeared) unless you prepare for it by doing it over several times until you "know" it. Hundreds of other students have "learned/memorized" this problem in the past and it should provide a fundamental understanding of the use of Gaussian orbitals in modern calculations.

17.2 Run PCLOBE for H_2 and N_2 confirm the Mayer–Wiberg bond orders and then run PCLOBE for CO and BF to compare to the ideal bond orders. You might want to look up the Lewis Electron models for complete octets in CO and BF in your freshman chemistry text.

17.3 Run PCLOBE for the water molecule using the input file H2OSYM along with the constraint of C_{2V} and set the optimization to "4" for a gradient of 0.0001 hartrees/bohr. Make a table of the vibrational frequencies (using the scaling factor) for STO-3G, STO-4G, "Slater-Transform-Preuss" STP-544* and 6-311G** basis sets along with the values of the frequencies given in Chapter 15. Compare the total energies of the STP-544* basis calculation with that of the 6-31G**. The STP-544* basis is built from a finite set of Gaussians fitted to orbital shapes which represent "infinite-zeta" basis sets [35–37] rather than single-zeta or double-zeta functions. If you only had access to PCLOBE would you use STP-544* or 6-31G** basis sets for calculation of vibrational frequencies? What about the run times? This problem is intended to show a need to use "Occams Razor," that is, only use what is needed to obtain the results you need. In addition we see that modern calculations need to use huge basis sets and exotic methods to approach thermodynamic or spectroscopic accuracy in quantum chemistry calculations.

17.4 Optimize the geometry of ethylene using PCLOBE to a gradient of 0.001 hartrees/bohr in the STO-4G basis set with C_S symmetry. Identify the π and π^* orbitals as the molecular orbitals which only have nonzero coefficients for the 2Pz orbitals. Compare the signs of the orbital coefficients to the Roberts–Hückel diagram in the previous chapter and make a sketch of the 2Pz-π and 2Pz-π^* orbitals as Roberts has drawn them and superimpose a bar graph of the coefficients from the SCF orbitals.

Note, it might be a good idea to look at the symmetry chart in the next chapter for the following problems. We are trying to motivate you to pay attention to symmetry while you play around with PCLOBE!

17.5 Optimize the geometry of *trans*-1,4 butadiene in C_S symmetry using PCLOBE and sketch the coefficients of the 2Pz atomic orbital coefficients on a line and compare the pattern to the first four levels of the Particle-in-a-Box.

17.6 Use the C6H6 data file in PCLOBE.DATA to run (without optimization) PCLOBE for benzene in either C_S or D_{6h} symmetry with the STO-3G basis to identify the six pi-orbitals of benzene within the SCF C-matrix. Why does the SCF solution only have two degenerate orbital pairs compared to the infinite number of degenerate orbital pairs in the Particle-on-a-Ring? Hint: how many 2Pz orbitals are included in the basis set? Note that the D_{6h} point group symmetry includes the single mirror plane of the C_S point group.

17.7 Use the coordinates provided for 3-methyl sydnone to calculate the Mulliken atom charges and the Mayer–Wiberg bond orders with PCLOBE using the STO-4G or STO-3G basis. Draw a large picture of the molecule on a sheet of paper and write the bond orders on each bond and the atom charges near each atom. Include a dashed line across the ring where the C–N bond occurs.

17.8 Optimize the bond length of BH in the 6-31G** basis to a gradient of 0.001 hartrees bohr (type a "3" in the optimization panel) and then use the Configuration Interaction (CI) option in PCLOBE for more than one determinant by clicking on the Boys–Reeves bubble in the Post-Hartree–Fock panel for BH. Compare the energies of the excited states to Harrison's Multi-reference CI diagram at the minimum energy bond length. Identify the excited states by State number to Harrison's symmetry labels for the three lowest states.

17.9 Optimize the geometry of the NH_3 molecule to a gradient of 0.001 hartrees/bohr using PCLOBE and C_1 (i.e., no symmetry) by putting N on the z-axis. Make a rough note of the time and examine the symmetry of the final geometry, is it truly C_{3V}? Then run PCLOBE using the input file NH3SYM using enforced C_{3V}. Does the final structure have an exact threefold symmetry about the z-axis? What does this say about odd-fold rotational axes compared to symmetry features which are easily represented in right-angle (x, y, z) coordinates? Hypothesis: Odd-fold axes do not easily resolve components cleanly in (x, y, z) Cartesian coordinates?!

17.10 Using Windows-VISTA or Windows-7 CLOSE ALL FILES and *right-click* on the PCLOBE icon to select "Run as Administrator". Click on the "browse" button at the top of the panel and select "C6H6SYM" as the input file. Then select the point group D_{6h} in the lower left corner of the input panel, set the gradient tolerance to "3" before clicking "Launch". Use the simple STO-3G basis option. Make a rough note of the run time. Next, rerun the calculation for the same input file "C6H6SYM" but use the C_1 point group in the lower left corner of the input panel with a gradient tolerance of "3" and set the number of iterations to "99" (which will really allow 300 iterations). Let it run overnight if necessary or just note that the run without use of symmetry takes a long time. Note the input file includes the coordinates for all 12 atoms in benzene but when D_{6h} symmetry is used the program only uses the coordinates of the unique C-H pair of atoms first in the list. When using the symmetry option the unique atoms must start with the repeating unit along the *x*-axis relative to the central *z*-axis. See the documentation for the program. Are you a believer in the importance of symmetry yet?

REFERENCES

1. Slater, J. C., *Quantum Theory of Molecules and Solids, Vol. 1, Electronic Structure of Molecules*, McGraw-Hill Book Co. Inc., New York, 1963, Appendix 7, p. 277.
2. Roberts, J. D., *Notes on Molecular Orbital Calculations*, W.A. Benjamin Press, New York, 1961.
3. Pople, J. A. and D. L. Beveridge, *Approximate Molecular Orbital Theory*, McGraw-Hill Book Co., New York, 1970.
4. Slater, J. C., Atomic shielding constants, *Phys. Rev.*, **36**, 57 (1930).
5. Clementi, E. and D. L. Raimondi, Atomic screening constants from SCF functions, *J. Chem. Phys.*, **38**, 2686 (1963).
6. Condon, E. U. and G. H. Shortley, *The Theory of Atomic Spectra*, Cambridge University Press, Cambridge, U.K., 1935.
7. Trindle, C. and D. Shillady, *Electronic Structure Modeling, Connections between Theory and Software*, CRC Press, Boca Raton, FL, 2008, Chap. 3.
8. Lide, D. R., *CRC Handbook of Chemistry and Physics*, 90th Edn., CRC Press, Taylor and Francis, Boca Raton, FL, 2009–2010, Chap. 3.
9. Hartree, D. R., The wave mechanics of an atom with a non-coulomb central field. Part III. Term values and intensities in series in optical spectra, *Proc. Cambridge Phil. Soc.*, **24**, 426 (1928).
10. Fock, V., Self-consistent field mit Austausch für Natrium. [Self-consistent field with exchange for Sodium], *Z. Physik.*, **62**, 795 (1930).
11. Roothaan, C. C. J., New developments in molecular orbital theory, *Rev. Mod. Phys.*, **23**, 69 (1951).
12. Reference [7], Chap. 11.
13. Jaramillo, J., G. E. Scuseria, and M. Ernzerhof, Local hybrid functional, *J. Chem. Phys.*, **118**, 1068 (2003).
14. Boys, S. F., Electronic wave functions I. A general method of calculation for the stationary states of any molecular system, *Proc. Roy. Soc.*, **A200**, 542 (1950).
15. Shavitt, I., The Gaussian function in calculations of statistical mechanics and quantum mechanics. In *Methods in Computational Physics, Advances in Research and Applications*, B. Alder, S. Fernbach, and M. Rotenberg, Eds., Vol. 2, Academic Press, New York, 1963.
16. Hehre, W. H., R. F. Stewart, and J. A. Pople, Self-consistent molecular-orbital methods. I. Use of Gaussian expansions of slater-type atomic orbitals, *J. Chem. Phys.*, **51**, 2657 (1969).
17. Sambe, H., Use of 1s Gaussian wave functions for molecular calculations. I. The hydrogen atom and the hydrogen molecule ion, *J. Chem. Phys.*, **42**, 1732 (1965).
18. Wilson, E. B., J. S. Decius, and P. C. Cross, *Molecular Vibrations: The Theory of Infrared and Raman Vibrational Spectra*, Dover Reprint, New York, 1980.
19. Irikura, K. K., R. D. Johnson III, and R. N. Kacker, Uncertainties in scaling factors for ab initio vibrational frequencies, *J. Phys. Chem. A.*, **109**, 8430 (2005).
20. Mulliken, R. S., Electronic population analysis on LCAO-MO molecular wave functions. I, *J. Chem. Phys.*, **23**, 1833 (1955).
21. Wiberg, K., Application of the Pople-Santry-Segal CNDO method to the cyclopropylcarbinyl and cyclobutyl cation and to bicyclobutane, *Tetrahedron*, **24**, 1083 (1968).

22. Mayer, I., Charge, bond order and valence in the AB initio SCF theory, *Chem. Phys. Lett.*, **97**, 270 (1983).

23. Mueller, A. and H. Berstein, http://rasmol.org/, 1998; see open source GNU program manual by Sayles, R., at http://rasmol.org/doc/rasmol.html

24. Hehre, W., *SPARTAN*, Wavefunction, Inc., Irvine, CA.

25. Halkier, A., W. Klopper, T. Helgaker, and P. Jrgensen, Basis-set convergence of the molecular electric dipole moment, *J. Chem. Phys.*, **111**, 4424 (1999).

26. Foresman, J. B. and A. Frisch, *Exploring Chemistry with Electronic Structure Methds*, 2nd Ed., Gaussian, Inc., Pittsburg, PA, 1996.

27. Harrison, J. F. and L. Allen, The electronic structure and molecular properties of Boron Hydride in its ground and excited states, *Mol. Spectrosc.*, **29**, 432 (1969).

28. Harrison, J. F. and L. C. Allen, Electronic structure of methylene, *J. Am. Chem. Soc.*, **91**, 807 (1969).

29. Harrison, J. F., An ab initio study of the zero field splitting parameters of $^3B_1$ methylene, *J. Chem. Phys.*, **54**, 5415 (1971).

30. Werner, H.-J. and P. J. Knowles, MOLPRO, maintained at the University of Cardiff and Universitat Stuttgart, see http://www.molpro.net/info/users

31. Coulson, C. A., *Valence*, 2nd Edn., Oxford University Press, Oxford, Great Britain, U.K., 1961.

32. Shillady, D. D., S. Cutler, L. F. Jones, and L. B. Kier, A molecular orbital valence bond study of 3-methyl sydnone and 3-methyl pseudosydnone, *Int. J. Quant. Chem. Symp.*, **24**, 153 (1990).

33. Nespurek, S. and M. Sorm, Sydnones and their photochromic properties, *Coll., Czech. Chem. Commun.*, **42**, 811 (1977).

34. Gotthardt, H. and F. Reiter, A. Gieren, and V. Lamm, A new unusual photoisomerization of sydnones, *Tetrahedron Lett.*, **30**, 2331 (1978).

35. Shillady, D. D., Slater transform functions. Application to the Helium atom ground state, *Chem. Phys. Lett.*, **3**, 104 (1969).

36. Yurtsever, E. and D. D. Shillady, A slater-transform-preuss (STP) wave-function for the ground state of Be, *Chem. Phys. Lett.*, **40**, 447 (1976).

37. Yurtsever, E., W. W. Scholler, and D. D. Shillady, Improved Gaussian basis sets for molecular calculations. Applications to GAUS70, *Chim. Acta Turcica*, **10**, 165 (1982).

18 Point Group Theory and Electrospray Mass Spectrometry

INTRODUCTION

One of the homework problems (17.10) in the previous chapter should have convinced you that symmetry considerations can be quite useful in reducing computation and simplifying spectroscopic analysis. Undergraduates should have some introduction to the use of symmetry and how the techniques are applied in spectroscopy because the principles offer shortcuts in interpreting spectra. Texts by Eyring, Walter, and Kimble [1] (EWK), Cotton [2] and Harris and Bertolucci [3] are recommended. The Harris–Bertolucci text has the most complete collection of character tables. Here we will study the water molecule as a simple example, which still uses all the features of symmetry analysis for a brief introduction to this topic. At the end of this chapter, we will switch gears to show how a new method to study high molecular weight biological samples has recently been developed. This is an exercise to introduce forensic and pre-health science students to a new technique that is sure to become more important in the future. At the very least, spectroscopic applications of point group theory, resonance Raman spectroscopy and developments in electrospray mass spectrometry are introductions to future research and can serve as the beginning point for a special project/term paper by an upper-level undergraduate.

BASIC POINT GROUP THEORY

Let us be clear that we are trying to give a basic introduction to the use point group theory rather than a mathematical seminar. This is a low level "how to do it" introduction!

First, we introduce the flowchart that is used to select the point group to apply to a given molecule, as shown in Figure 18.1. All of the symmetry operations refer to a common "point" origin within the molecule, usually an obvious "center" point as the origin. *The symmetry operations are those that can be performed on a given molecule, which leave the molecule unchanged!* For the water molecule shown in Figure 18.1, the oxygen atom is placed on the z-axis with the O–H bonds in the xz plane. Let us introduce what is called the C_{2V} character table, which contains the "characters" of the simplest possible matrices that satisfies the four symmetry operations in the collection of possible matrices in the C_{2V} point group. The characters are actually the "trace" or "spur" (spoor) sum-of-diagonal-elements of the symmetry operation matrices. These characters are invariant under a unitary transformation as was demonstrated in Chapter 16. As long as the symmetry operation leaves the molecule unchanged, the traces are invariant.

In the study of crystals there is a similar system of mathematics for what is called "space group theory," but here we are concerned with the symmetry properties of individual molecules relative to some point within the molecule. This body of mathematics is called "point group theory."

Using the diagram in Figure 18.2 or a model of the water molecule you can ascertain that you can rotate the model by 180° twice about an axis through the O atom bisecting the HOH angle and each time the model will look the same so that is a $((2\pi/\pi) = 2)$ C_2 rotation. Next you can actually check with a small handheld mirror that there are two mirror planes that will leave the molecule

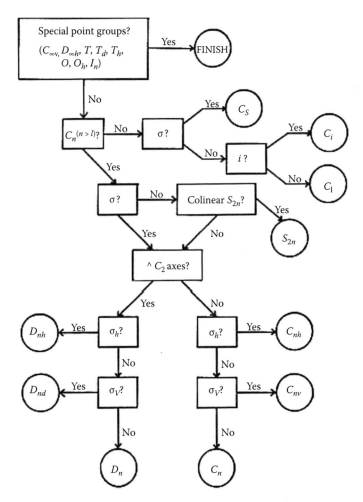

FIGURE 18.1 Flowchart for selection of the "character table" appropriate for a molecule. (From Harris, D.C. and Bertolucci, M.D., *Symmetry and Spectroscopy, An Introduction to Vibrational and Electronic Spectroscopy*, Oxford University Press, New York, 1978. With permission.)

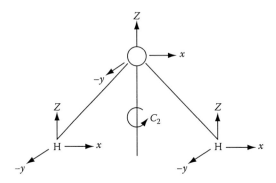

FIGURE 18.2 The Cartesian "arrow" representation of the atoms in the water molecule. Note that we have maintained a right-handed coordinate system with the positive y-axis pointing to the back, but what really counts is that the y-coordinate changes sign when the image is reflected in the plane of the drawing.

unchanged, one in the HOH plane and the other perpendicular to that. The symbol for a mirror plane in point group theory is σ, and when the reflection contains a rotation axis (as here) it is called σ_V, if the mirror plane is perpendicular to the principle axis it is σ_h, and finally if the mirror plan bisects the angle of two lesser rotation axes it is called a σ_d plane. Another symmetry operation that is not present here is the inversion operation, i, which inverts all points through the origin. There is some ambiguity in this particular (low symmetry) point group as to which is the σ_V plane and which is the σ'_V plane so we will designate them with a further "xz" or "yz" label. A further convention in point group theory is that the highest order rotation axis (here C_2) is chosen to be the z-axis. Finally, there is one trivial operation that results in the molecule looking the same: If we do nothing and just leave the molecule unchanged it will appear unchanged! That operation is called the identity operation and is labeled as "I" here but some other texts use the symbol "E" for the identity operation.

Once you can identify the specific symmetry operations a given molecule possesses you can then follow the flowchart to identify the proper name of the point group, here it is C_{2V}. For this point group, there are four symmetry operations and we can state the mathematical definition of a point group while you visualize the operations in your mind. Chemists usually have developed "minds-eye" visualization due to studying organic chemistry and point group symmetry operations should be easy for chemists.

What is a "group"? A point group is a collection of symmetry operations (four in the water example), which satisfy certain mathematical properties, but before we go on let us appreciate the main principle operating here. We emphasize again that all of these symmetry operations that result in a molecule looking exactly alike are automatically *unitary transformations*! As such we know a matrix describing the operation will have an invariant (unchanging) trace (sum of diagonal elements) which we call the "character" in point group theory. Thus "trace," "spur" (German, spoor), and "character" all mean the same thing in the applied linear algebra we will use. The wonderful thing about a matrix trace is that it is independent of any arbitrary orientation of a molecule. In fact, in some more difficult cases of point group theory (the T_d or tetrahedral point group for CH_4 or CCl_4), one can visualize the molecule in one orientation to see a reflection plane but choose another orientation to see a C_3 axis because *we know all the operations are unitary transformations and their trace/character is independent of orientation*! A further implication from our previous excursions into linear algebra is that the transformations are formed from normalized *orthogonal* eigenvectors, which we will see implies a higher order orthogonality on the transformation matrices as well.

Loosely paraphrasing the Cotton text [2] we can define the properties of a point group:

1. *The product of any two (symmetry) elements in the group and the square of each element must result in an element of the group.* This is easily demonstrated using the four elements of the C_{2V} symmetry of the water molecule, since the successive application of any two symmetry operations results in leaving the molecule look the same, all the symmetry elements share the results of the identity element I.
2. *One element of the group must commute with all the members and leave them unchanged.* Obviously, the identity (I) element satisfies this requirement since a unit matrix will commute with any other matrix of the same dimension from either the right or left.
3. *The simple associative law of multiplication must hold, $A(BC) = (AB)C$.* Although this is a formal requirement, we have already used this idea when we demonstrated matrix diagonalization in a previous chapter. We could remind the reader that diagonal matrices obey ordinary scalar arithmetic, so we could prove this requirement by transforming the representation to the diagonal form of the particular matrix we want to commute to the right or left using scalar arithmetic as in $2 \times 3 = 3 \times 2 = 6$.
4. *Every (symmetry) element must have a reciprocal, which is also a member of the set.* Again, this is almost obvious since every symmetry operation can be applied in reverse to achieve the identity (I) result, but it does guarantee that rotations about an axis are

continuous across a total of 2π rotation. This condition can be demonstrated using linear algebra equations, but that is not necessary if one pictures the symmetry operations mentally and then considers the inverse operation as a reversal of that operation. Because the symmetry matrices are unitary, their inverse is equal to their transpose as we have seen in the diagonalization process. The transpose matrix reverses the operation. For instance, a rotation in a plane can be reversed by rotating back using the transpose of the original operations as shown by a (2×2).

$$\begin{bmatrix} \cos\theta & \sin\theta \\ -\sin\theta & \cos\theta \end{bmatrix}^{-1} = \begin{bmatrix} \cos\theta & -\sin\theta \\ \sin\theta & \cos\theta \end{bmatrix}$$

Next we come to a mysterious aspect of point group theory, which is in the realm of PhD theses in mathematics. The question is how does one determine the "irreducible representations"? While there are formal ways in mathematics to determine the very simplest matrices, which obey the same rules as the set of operations in the group, we will accept the character tables as a given. Often the smallest matrix will be "1" with the property that it can also be "−1" but for applications we are dependent on "character tables" that are tabulated in books. Let us not forget that the numbers in the character tables are the traces of the smallest irreducible representations in the smallest possible matrix dimension.

But wait! We are humans who tend to think in Cartesian (x, y, z) coordinates so the procedure is to mock up the symmetry operations in Cartesian representations and then try to "reduce" them. In the following text, we see the (9×9) Cartesian representations for what might be called the "Cartesian arrow representation," which is obtained by placing the Cartesian axes on each atom as small arrows and analyzing how the $(\hat{i}, \hat{j}, \hat{k})$ "arrows" transform under the symmetry operations. These four matrices will carry out the symmetry operations of the C_{2V} point group as seen by a human who wants to think in Cartesian (x, y, z) space. We have placed "0" in the subblocks where all nine entries in that subblock are zero. *If you multiply a Cartesian column vector by one of the matrices from the left, the result will be the new positions of the coordinate "arrows" after the operation!* However, these matrices are "reducible," that is, they have (9×9) dimensionality that is greater than the collection of the smallest possible matrices, which can also satisfy the definitions of the C_{2V} point group. Although we may not be able to visualize the irreducible matrices, mathematicians assure us the "characters" of the irreducible representations are given by the numbers in the character tables! In C_{2V}, they are all $+1$ or -1. By convention, the A representations are those that are "even" with respect to the highest order rotation axis (C_2 here) and the B representations are "odd." Note that there are four rows for the four "representations" or ways of using the small matrices as A1, A2, B1, or B2 and by convention we use "h" as the order of the group. Here $h = 4$ for four symmetry operations.

$$\begin{bmatrix} 1 & 0 & 0 & & & \\ 0 & 1 & 0 & & 0 & \\ 0 & 0 & 1 & & & \\ & & & 1 & 0 & 0 & \\ & 0 & & 0 & 1 & 0 & & 0 \\ & & & 0 & 0 & 1 & \\ & & & & & & 1 & 0 & 0 \\ & & 0 & & & 0 & 0 & 1 & 0 \\ & & & & & & 0 & 0 & 1 \end{bmatrix} \begin{bmatrix} O_x \\ O_y \\ O_z \\ H_x \\ H_y \\ H_z \\ H'_x \\ H'_y \\ H'_z \end{bmatrix} \quad \begin{bmatrix} 1 & 0 & 0 & & & \\ 0 & -1 & 0 & & 0 & \\ 0 & 0 & 1 & & & \\ & & & 1 & 0 & 0 & \\ & 0 & & 0 & -1 & 0 & & 0 \\ & & & 0 & 0 & 1 & \\ & & & & & & 1 & 0 & 0 \\ & & 0 & & & 0 & 0 & -1 & 0 \\ & & & & & & 0 & 0 & 1 \end{bmatrix} \begin{bmatrix} O_x \\ O_y \\ O_z \\ H_x \\ H_y \\ H_z \\ H'_x \\ H'_y \\ H'_z \end{bmatrix}$$

$$\qquad\qquad\qquad I \qquad\qquad\qquad\qquad\qquad\qquad\qquad\qquad\qquad \sigma_V(xz)$$

$$\begin{bmatrix} \begin{matrix} -1 & 0 & 0 \\ 0 & -1 & 0 \\ 0 & 0 & 1 \end{matrix} & 0 & 0 \\ 0 & \begin{matrix} -1 & 0 & 0 \\ 0 & -1 & 0 \\ 0 & 0 & 1 \end{matrix} & 0 \\ 0 & 0 & \begin{matrix} -1 & 0 & 0 \\ 0 & -1 & 0 \\ 0 & 0 & 1 \end{matrix} \end{bmatrix} \begin{bmatrix} O_x \\ O_y \\ O_z \\ H_x \\ H_y \\ H_z \\ H'_x \\ H'_y \\ H'_z \end{bmatrix} \quad \begin{bmatrix} \begin{matrix} -1 & 0 & 0 \\ 0 & 1 & 0 \\ 0 & 0 & 1 \end{matrix} & 0 & 0 \\ 0 & \begin{matrix} -1 & 0 & 0 \\ 0 & 1 & 0 \\ 0 & 0 & 1 \end{matrix} & 0 \\ 0 & 0 & \begin{matrix} -1 & 0 & 0 \\ 0 & 1 & 0 \\ 0 & 0 & 1 \end{matrix} \end{bmatrix} \begin{bmatrix} O_x \\ O_y \\ O_z \\ H_x \\ H_y \\ H_z \\ H'_x \\ H'_y \\ H'_z \end{bmatrix}$$

$$C_2 \qquad\qquad\qquad\qquad\qquad\qquad\qquad \sigma_V(yz)$$

After you use this type of table it becomes familiar, but we only have space to use it once here so we need to pay attention to the details (Table 18.1). It is easily shown that the dot product of the various representations is zero so the representations are orthogonal and linearly independent. This super hierarchy of orthogonality of the matrices themselves as well as the orthogonality of the eigenvector columns in the matrices is due to "the great orthogonality theorem." The proof is given on p. 371 of EWK [1] and it is difficult, so following Cotton [2] we will just state the theorem and in our opinion you will believe it after we do some demonstrations of its use.

$$\sum_R [\Gamma_i(R)_{mn}][\Gamma_j(R)_{m'n'}] = \frac{h}{\sqrt{l_i l_j}} \delta_{ij} \delta_{mm'} \delta_{nn'}$$

Here $(R)_{mn}$ is a symmetry matrix of the representation Γ_i for the symmetry operation "R." This implies the "dot-product" of two one-dimensional representations is zero when $m = m' = n = n' = 1$, they are orthogonal! However, the theorem also applies to individual (m, n) and (m', n') elements of (2×2) representations of odd-fold rotations such as C_{3V} or C_{5V}.

Example

For the C_{2V} representation all the irreducible matrices are (1×1) so $m = m' = n = n' = 1$. Thus

$$\sum_R [\Gamma_{A1}][\Gamma_{B1}] = (1 \cdot 1) + (1 \cdot (-1)) + (1 \cdot 1) + (1 \cdot (-1)) = 0.$$

Similarly, $\sum_R [\Gamma_{B1}][\Gamma_{B1}] = (1 \cdot 1) + ((-1) \cdot (-1)) + (1 \cdot 1) + ((-1) \cdot (-1)) = 4 = h.$

TABLE 18.1

C_{2V} Point Group Character Table

C_{2V}	I	C_2	$\sigma_V(xz)$	$\sigma_{V'}(yz)$	(1)	(2)	(3)
A1	1	1	1	1	z	x^2, y^2, z^2	$z^3, z(x^2 - y^2)$
A2	1	1	-1	-1	R_z	xy	xyz
B1	1	-1	1	-1	x, R_y	xz	$xz^2, x(x^2 - 3y^2)$
B2	1	-1	-1	1	y, R_x	yz	$yz^2, y(3x^2 - y^2)$
Γ_{xyz}	9	-1	3	1			

Source: Harris, D.C. and Bertolucci, M.D., *Symmetry and Spectroscopy, an Introduction to Vibrational and Electronic Spectroscopy*, Oxford University Press, New York, 1978.

This is relatively easy to see when the characters are one-dimensional (as many are) but the theorem holds even for odd-fold rotations that reduce to two-dimensional irreducible matrices because the characters are invariant to orientation transformations. In summary, the orthogonality of the symmetry operations extends to their representations in what is a fortuitous relationship that is very useful.

Now go back to the idea of a "projection operation" we have used before. Suppose we want to know how much of a vector $\vec{r} = 3\hat{i} + 2\hat{j} + 5\hat{k}$ is in the $\hat{j}$ direction. We can find out by "projecting" the $\hat{j}$ component using a dot product with $\hat{j}$ itself as $\hat{j} \cdot \vec{r}$ where we have $\hat{j} \cdot \vec{r} = \hat{j} \cdot (3\hat{i} + 2\hat{j} + 5\hat{k}) = 0 + 2 + 0 = 2$. That worked because of the mutual orthogonality of the unit vectors $(\hat{i}, \hat{j}, \hat{k})$ so we can use the orthogonality of the symmetry representations to project the amounts of the various irreducible representations that are in our human Cartesian matrices. First we have to determine the characters (traces) of our (9×9) Cartesian arrow matrices. We have written a row of numbers as Γ_{xyz} along the bottom of the character table as the trace of the Cartesian symmetry matrices. This is the part you have to do yourself. You have to look at the picture of the water molecule with the Cartesian arrows and imagine how those arrows will change under the symmetry matrices. Fortunately one observation is very helpful: *if an atom moves during the symmetry operation, the symmetry matrix will have the block of its coordinates in an off-diagonal position and so the arrows of that atom cannot contribute to the trace!* For atoms that stay in place, we merely have to add up the sum of the diagonal elements. In the case of an odd-fold axis such as C_3 in the NH_3 case with a

C_{3V} point group, the C_3 matrix would contain
$$\begin{bmatrix} \cos\left(\frac{2\pi}{3}\right) & -\sin\left(\frac{2\pi}{3}\right) & 0 \\ \sin\left(\frac{2\pi}{3}\right) & \cos\left(\frac{2\pi}{3}\right) & 0 \\ 0 & 0 & 1 \end{bmatrix} = \begin{bmatrix} -\frac{1}{2} & \frac{-\sqrt{3}}{2} & 0 \\ \frac{\sqrt{3}}{2} & -\frac{1}{2} & 0 \\ 0 & 0 & 1 \end{bmatrix} \text{ so}$$

the trace (character) $= 0$.

This type of symmetry operation, which is intrinsically two-dimensional in the x–y subblock, is called an E representation (some texts also use E for the identity operation). The A and B representations are always one-dimensional. Otherwise, if an atom moves during a symmetry operation it contributes nothing to the character of that matrix. In the case of the I operation, the trace is the same as the dimension of the space, which is 9 here. Thus we can find out how much of each representation is in our human Cartesian matrices but to normalize to 1 we have to divide by "h." We use the formula (# irreducible representations) $= \left(\frac{1}{h}\right) \sum_i \left[\chi_i(irred.) \cdot \chi_{\Gamma_{xyz}}(red.)\right]$ where χ are the characters.

$$(\#A1) = \left(\frac{1}{4}\right)[(1 \cdot 9) + (1 \cdot (-1)) + (1 \cdot 3) + (1 \cdot 1)] = \frac{12}{4} = 3A1$$

$$(\#A2) = \left(\frac{1}{4}\right)[(1 \cdot 9) + (1 \cdot (-1)) + (-1 \cdot 3) + (-1 \cdot 1)] = \frac{4}{4} = 1A2$$

$$(\#B1) = \left(\frac{1}{4}\right)[(1 \cdot 9) + (-1 \cdot (-1)) + (1 \cdot 3) + (-1 \cdot 1)] = \frac{12}{4} = 3B1$$

$$(\#B1) = \left(\frac{1}{4}\right)[(1 \cdot 9) + (-1 \cdot (-1)) + (-1 \cdot 3) + (1 \cdot 1)] = \frac{8}{4} = 2B2$$

Thus we find that $\Gamma_{xyz} = 3A1 + 1A2 + 3B1 + 2B2 = (9 \times 9)$ and the individual dimensions of the irreducible representations add up to 9, which is the dimension of our human Cartesian matrices.

We are about to reap great benefit from other considerations provided by the character tables. On the right side of the character table is a list of Cartesian coordinates in first, second, and third order combinations (only Harris and Bertolucci [3] show the third order). The (x, y, z) entries there serve several purposes. They show how atomic orbitals (using the l-quantum number: 1 for p-orbitals, 2 for d-orbitals, 3 for f-orbitals, etc.) transform under the symmetry operations of the group and which representations they belong to. In addition, we are interested here in how the dipole and polarizability transitions transform under the symmetry of the point group. The entries marked R_x, R_y, and R_z are the row representations in which *rotations* would belong, along with the first order properties that depend on x, y, or z such as (x, y, z) *translations* and including properties like the *dipole moment* and the *dipole selection rule* for absorption spectra. The second order set of Cartesian products refer to what might occur in an *electron polarizability tensor* such as $\alpha = \begin{bmatrix} \alpha_{xx} & \alpha_{xy} & \alpha_{xz} \\ \alpha_{yx} & \alpha_{yy} & \alpha_{yz} \\ \alpha_{zx} & \alpha_{zy} & \alpha_{zz} \end{bmatrix}$.

The Raman effect is a secondary vibrational-electronic effect in which a molecule is excited from a given vibrational state in the electronic ground state up to some higher electronic state and then falls down with emission of light in the visible range but containing infrared information, as shown in Chapter 12. There (Chapter 12) we stressed how the transition selection rule requires a dipole moment in the molecule to provide a mechanism for absorption of light energy. In the case of homonuclear diatomics or other symmetric molecules such as CO_2, the dipole moment is zero leading to a blank IR spectrum. However, the polarizability tensor tells us how the electron distribution changes as a molecular geometry changes shape during a vibrational mode. The polarizability tensor also tells us that small components of an "induced dipole moment" can be caused by the electric field $\vec{E}$ of a light wave.

$$\vec{\mu} = \alpha\vec{E} = \begin{bmatrix} \alpha_{xx} & \alpha_{xy} & \alpha_{xz} \\ \alpha_{yx} & \alpha_{yy} & \alpha_{yz} \\ \alpha_{zx} & \alpha_{zy} & \alpha_{zz} \end{bmatrix} \begin{bmatrix} E_x \\ E_y \\ E_z \end{bmatrix} = \begin{bmatrix} \mu_x \\ \mu_y \\ \mu_z \end{bmatrix}.$$

Thus light of a particular frequency can simultaneously induce a dipole moment in a molecule and then couple with the dipole components to result in light absorption! Raman spectra are observed within the spectrum of light scattered from an intense source. Induced vibrational transitions are observed with a dispersive device (monochrometer) and some sort of electronic detection (in the visible range) *at 90° from the light source (laser) beam*. Remarkably, C. V. Raman first observed this effect with a handheld spectroscope in 1928 for which he received the Nobel Prize in 1930. Thus we can examine the symmetry properties of second-order combinations of the Cartesian coordinates (in column "2") and use them to indicate a yes/no answer as to whether a given molecular vibration will occur in Raman spectroscopy.

First let us sort out the Γ_{xyz} components using columns "1" and "2."

$$\Gamma_{xyz} = 3A1 + 1A2 + 3B1 + 2B2$$

Rotation $= A2 + B1 + B2$
Translation $= A1 + B1 + B2$
Vibration $= \Gamma_{xyz} - A1 - A2 - 2B1 - 2B2 = 2A1 + B1 = 3$ vibrations, $3N - 6 = 3$, Q.E.D.

Thus using symmetry analysis we find the expected three vibrations but now we know their symmetry as well! We can assign the symmetric O–H stretch mode to $A1$ ($\sim 3700/cm$) and the symmetric V-shape wagging motion also as $A1$ ($\sim 1600/cm$) so the $B1$ mode must be the asymmetric out-of-phase O–H stretch ($\sim 3800/cm$).

CALCULATION OF MOLECULAR VIBRATIONS

This analysis can also be carried out using the PCLOBE program using the file "H2OSYM" with selection of C_{2V} in the lower left corner of the setup panel and a gradient tolerance of "$n = 4$" for 10^{-4} hartrees/bohr. You will have to run PCLOBE in "Administrator Mode" if your PC is running WINDOWS VISTA or WINDOWS-7. We can see from the PCLOBE results that although the frequencies are high due to the use of a small basis set, the largest "false frequency" is only about 86 cm^{-1}. That is small compared to the smallest true frequency (Mode No. 3). The main cause of the false frequencies is that the geometry may not be at the true energy minimum and the numerical second derivatives are then evaluated "on the side of the energy bowl" instead of at the minimum. The imaginary frequencies result from "saddle-points" in the energy hypersurface and are also much smaller than lowest real vibrational frequencies. By comparison to the Roberts–Hückel pi-electron treatment in Chapter 16, we see that the all-electron PCLOBE treatment of the water molecule offers far more details even if there are still some numerical limitations.

Mode No. 3 is the easiest mode eigenvector to visualize. It shows the O atom moving up in the z direction with the two H atoms moving down and moving closer together if you note the signs of the x-coordinate and the original positions. This is a symmetric "wagging" motion. Mode No. 2 is more difficult to visualize but we see the displacements are slightly larger and still symmetrical when you take into account that one of the H atoms has a negative x-coordinate at the minimum. Mode No. 1 is easier to understand since the x-coordinate of the O atom moves to $+x$ so we see that the motion is asymmetric, a sort of "rocking" motion. You can also see from the table of force constants that Mode. No. 3 is a "soft, mushy motion" while the higher-frequency modes have much higher force constants. Although the PCLOBE results are only qualitative, there is enough information there to show you what to look for in a more sophisticated quantum chemistry program (Table 18.2).

We have already given a phenomenological discussion of Raman spectroscopy in Chapter 12. Here we can compare the experimental frequencies of the Raman bands of the water molecule to the calculated frequencies from PCLOBE. As expected, a small basis set of the minimum number of orbitals can only produce results for limited volumes of electron density in the calculations. There are no diffuse functions in the basis set to allow electrons to roam in regions of low probability. The electrons are tightly restricted to the valence orbital regions, which is good enough for perhaps 90% of the true wave function but is not a complete description of the electron density. In effect, the small basis set constrains the electron density to narrow potential wells and pushes the vibrational levels higher. The net result is that the frequencies are too high. This is a common characteristic of limited basis sets in quantum chemistry calculations. The problem can be partially solved by adding more functions to the orbital basis set but even the most accurate calculations are seldom within 10 cm^{-1} of the experimental values. Exact vibrational frequencies are not easy to compute! In the PCLOBE example, we apply a reduction factor of 0.89 to the frequencies. Extensive research has found that there is no single value that will correct all the frequencies, so 0.89 is an average correction. However, the corrected pattern of (2057, 3856, and 4007 cm^{-1}) should enable us to match the bands to the experimental pattern. Modeling the molecular orbitals with PCLOBE or some other quantum chemistry program is useful for interpretation of the experimental spectra. Thus the main use of the PCLOBE results is to match the motions of the atoms with symmetry labels for each vibrational mode.

The Raman spectrum is observed by using an intense monochromatic source applied to a sample and *the scattered light is gathered at 90° from the incident beam*. There is considerable intensity from the incident beam and the scattered light at the wavelengths of interest on either side of the monochromatic source are much less intense. Thus the exciting wavelength may obscure the Raman wavelengths and care is required to design the optical system. Here we show in Figure 18.3 a custom spectrometer constructed by Prof. James Terner at Virginia Commonwealth University. This spectrometer is optimized to gather very-low-intensity emissions in the Stokes region of light scattered from laser excitation. The main object in Raman spectroscopy is to record

TABLE 18.2
Selected Vibrational Output from PCLOBE for H_2O Using a STO-3G Basis Set

Molucad Restart Coordinates, Copy and Paste into Input File *.XYZ

O	0.000000	0.000000	0.000000
H	0.740353	0.000000	0.654266
H	−0.740353	0.000000	0.654266

Distance matrix in Å

O	H	H
O	0.0000	0.9880	0.9880
H	0.9880	0.0000	1.4807
H	0.9880	1.4807	0.0000

A-B-C arcs in degrees for three atoms

arc(1, 2, 3) = 41.47
arc(1, 3, 2) = 41.47
arc(2, 1, 3) = 97.06
arc(2, 3, 1) = 41.47
arc(3, 1, 2) = 97.06
arc(3, 2, 1) = 41.47

Normal modes relative to optimized geometry

Ordered atom-1 (x, y, z), atom-2 (x, y, z), ... by column
Including rotational and translational false modes

Mode	1	2	3
Atom			
1	0.257	0.001	0.000
2	0.000	0.000	0.000
3	−0.001	0.221	0.252
4	−0.510	−0.533	0.467
5	0.000	0.000	−0.001
6	−0.451	−0.441	−0.501
7	−0.514	0.529	−0.466
8	0.000	0.000	−0.001
9	0.454	−0.438	−0.500

Harmonic vibrational frequencies and corrections [4]
We apply 3-21G factor of 0.89 to any basis set initially

(K N/cm;	Frequency 1/cm;	$0.89 \times 1/cm$)
(11.943648;	4502.39;	4007.13)
(11.058170;	4332.28;	3855.73)
(3.147630;	2311.36;	2057.11)

The following are translation + rotation frequencies

(0.004361;	86.04;	76.57)
(0.002026;	58.64;	52.19)
(0.000713;	34.79;	30.96)
(−0.000940;	Imag 39.95;	Imag 35.55)
(−0.003225;	Imag 73.99;	Imag 65.85)
(−0.024098;	Imag 202.24;	Imag 179.99)

Imaginary frequencies (Imag) indicate either an insufficient optimization or a less complete basis set
Try further optimization and/or a larger basis set
Fortunately, imaginary frequencies are usually in the range where they are not easily measured below 400 cm^{-1}
C_{2v} symmetry w.r.t. atom-1 at (0, 0, 0)

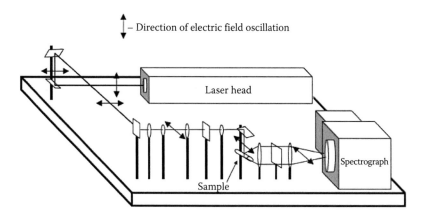

– Direction of electric field oscillation

Laser head

Spectrograph

Sample

FIGURE 18.3   The custom Raman spectrometer designed by Prof. Jim Terner at Virginia Commonwealth University with emphasis on orientation of the polarization of the excitation beam which is important in some cases. Note the right angle between excitation and emission and the square cutoff filter between the sample and the spectrograph.

very-low-intensity spectra in the presence of a very intense exciting light source so the emphasis is on the Stokes lines to the neglect of the lower intensity anti-Stokes lines. That emphasis permits the use of an optical "cutoff filter," which is a plate of absorbing material that is selected to block the light from the exciting line but permit the light of the Stokes lines to proceed to the monochrometer. More advanced applications also involve analysis of the polarization of the emitted light, so this spectrometer has been constructed with careful attention to the plane of the electric field polarization of the emitted light.

The spectrometer in Figure 18.3 is optimized to gather intensity of the Stokes lines for "resonance Raman" [5] spectra of heme, porphyrin, and other large compounds in which exciting light is used with a wavelength close to or within the electronic transition band of the main compound to observe very-low-intensity Raman wavelengths close to the wavelength of the exciting light source. Thus in recent years, there have been two main advances in Raman spectroscopy. First, the intensity of a monochromatic (single-wavelength) polarized laser beam is ideal for Raman excitation. Second, the development of the resonance Raman application to sites involving a chromophore offers increased sensitivity for localized mechanistic studies of catalytic sites as for instance metal ions in porphyrin/heme compounds. You can appreciate the difficulty in observing a weak emission in the presence of a very intensive light source. In such cases, a specific Raman spectrometer will be optimized to gather light from weak signals and yet use a dispersing device (grating monochrometer) capable of high resolution. This is shown in Figure 18.4 with the Raman spectra of H_2O and D_2O where the optical strategy pays off with observation of the very-low-intensity band near 1600 cm^{-1}, which others have found difficult to see. Thus, great care is used to employ optical cutoff filters and use of a high-resolution monochrometer to shut out all light from the excitation wavelength and to the blue of that line.

FUTURE DEVELOPMENT OF ELECTROSPRAY MASS SPECTROMETRY?

In our attempt to treat the "essential" topics of physical chemistry, we take this opportunity to look at a new development, which may soon become essential. We consider it "essential" for students to understand the *discovery and innovation process*. The invention of polaroid sunglasses and instant photography by Edwin Land in the 1950s and the early formation of the Microsoft Corporation by William Gates and Paul Allen from 1975–1980 are amazing stories that may not be fully appreciated by students of today, so we highlight a more recent innovation. The 2002 Nobel Prize awarded to Prof. John Fenn at Virginia Commonwealth University generated interest in the field of electrospray

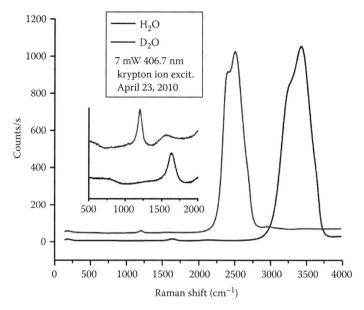

FIGURE 18.4 Raman spectra of H_2O (lower) and D_2O (upper) contributed by Prof. Jim Terner of the Chemistry Department of Virginia Commonwealth University. The optical path included a 600 lines/mm grating in a single 0.5 m Spex monochromator with a 60 micron slit width; a configuration optimized for light gathering ability to record low intensity spectra. While the smaller band near 1620 cm^{-1} has been reported as too weak to observe in some of the literature, Prof. Terner has no difficulty recording that band with this optical configuration and a Kr$^+$ laser. The inset shows amplification of the very small bands in the main spectra. Here we also see the isotope shift in the redshifted D_2O spectrum, the 1620 cm^{-1} A1 vibration and more intense, but barely resolved, A1 and B1 bands of H_2O near 3600 cm^{-1}.

mass spectrometry and students often ask questions about this new field. Since the applications are important to large biological molecules, there are potential uses to be developed in medical research and forensic science.

Mass spectrometry is a very old technique going back to J. J. Thompson in 1897 to measure the (m/e) ratio for the electron. Early mass spectrometers used the "magnetic sector" design based on the behavior of charged particles in a magnetic field using the familiar $\vec{F} = q(\vec{V} \times \vec{B}) = m\vec{a}$ response (Figure 18.5). This led to the use of large magnets to deflect a beam of charged particles by mass using slow scans of either the ion-accelerating voltage or the field of an electromagnet.

Most modern mass spectrometers use a quadrupole field device that is smaller and more sophisticated [6]. The quadrupole electrical field is time dependent and charged particles travel in a beam between the cylindrical poles (Figure 18.6). When the time-dependent field is tuned to a certain radio frequency, most of the charged particles are deflected away from whatever electrical detector is at the end of the quadrupole rods, while a few particles arrive at the detector according to their (mass/charge) ratio. Other particles spiral away from the detector and are swept out by powerful pumps, which maintain a high vacuum in the analyzer chamber. Ideally, the cross section of the metal rods should be hyperbolic facing each other, but cylindrical rods are often used for ease of fabrication. This design has been so successful that in 1978, it was extended to three such quadrupoles in succession for ultimate mass resolution by Yost and Enke at Michigan State University [7]. While the quadrupole design is much more compact than the magnetic sector design, the range of its effectiveness is between $1 < (m/z^+) < 5000$ at most, even though much effort was made to extend the range to higher masses.

Recent advances in biological science bring chemical questions about enzymes, DNA, RNA, cell membrane structures, etc., all of which are high-molecular-weight materials. In addition, future

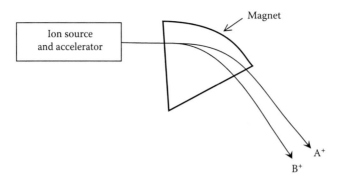

FIGURE 18.5 Schematic of a magnetic sector mass spectrometer. Here the magnetic field lines are perpendicular to the plane of the image going into the image so that for positive particles the right-hand-rule applies but the deflection depends on the mass of the article; here the B^+ ion has less mass than the A^+ ion. (From Prof. Scott E. Van Bramer, Widener University, http://science.widener.edu/~svanbram/ With permission.)

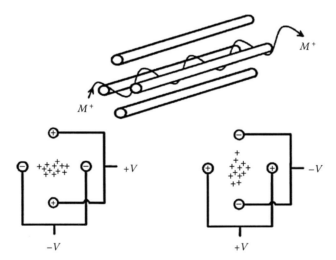

FIGURE 18.6 The principle of the quadrupole mass analyzer. The voltage applied to the pairs of conducting metal rods alternates rapidly using a radio frequency matched to the velocity of charged ions and only ions with a specific (m/z^+) ratio stay in the center of the rods and find their way to the detector. His web site gives an excellent overview of mass spectrometry. (From Prof. Scott E. Van Bramer, Widener University, http://science. widener.edu/~svanbram/ With permission.)

forensic applications may include identification of biological samples of high molecular weight. Nature keeps challenging chemistry with more macroscopic structures whose function depends on coarse-grain structure beyond that of atoms and small molecules and yet has the specificity of organic reactions. Clearly, there was motivation to develop a technique that offered separation and identification of these large biomolecular species—but how?

"MAKING ELEPHANTS FLY"

This is a quote from John Fenn's presentations explaining his 2002 Nobel Prize in chemistry, which he shared with Koichi Tanaka and Kurt Wuthrich. Electrospray ionization is a new way to produce intact ions in a vacuum from large and often fragile biological/chemical species in solution (Figure 18.7). While solution methods of collection and separation of large biological molecules has been in

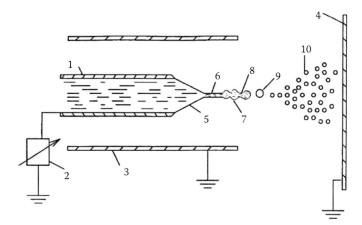

FIGURE 18.7 An enlarged schematic of the basic idea of "electrospray" generation of charged droplets containing dissolved high molecular weight materials. Item (1) is a hypodermic needle or a thin metal tube, (2) is a high voltage source capable of several thousand volts DC relative to the ground (3) with (4) as the target ground for positive ions and (5 through 10) is a sequence of droplet formation that spreads out in a conical beam on its way toward the final ground. Considerable discussion in the literature ensued regarding solvent evaporation since this apparatus is in a high vacuum, although dry nitrogen can be infused at the needle. The hypothesis is that the solvent droplets evaporate until the accumulated positive charges (H^+) on the sample are concentrated to the point where their mutual repulsion exceeds the surface tension of the drop and the droplet disintegrates to release individual molecules which are still highly charged. In samples such as large peptides (enzymes) there are many places where H^+ ions can bind temporarily, long enough to travel through the quadrupole mass analyzer system to a sensitive galvanometer detector. (Reproduced from Fenn, J.B., *J. Biomol. Tech.*, 13, 101, 2002. With permission of the Association of Biomolecular Resource Facilities.)

use for a long time, the ability to transfer these species to a vacuum is a valuable new development. Proteins up to a molecular weight of 130,000 have been observed as mass spectra but there does not seem to be an upper limit in molecular weight for applications of this technique. The main characteristic of this method is that some ions are multiply charged and mass peaks are observed for ions differing by one charge in a series of peaks.

To give a short overview, this research is sufficiently recent to show how such developments are brought to a practical reality. There were actually a number of people who realized that there was a need to use mass spectrometry for high-molecular-weight materials but the way research unfolds is still through applied experimentation. Prof. Fenn acknowledged that his early interest in what became electrospray ionization was largely due to a paper by Dole et al. [8] in 1968 (Figure 18.8).

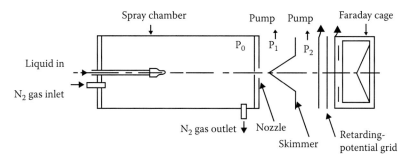

FIGURE 18.8 A schematic of Dole's original electrospray nozzle. (Reproduced from Fenn, J.B., *J. Biomol. Tech.*, 13, 101, 2002. With permission of the Association of Biomolecular Resource Facilities.)

FIGURE 18.9 John Bennet Fenn (1917–2010) was an American scientist who shared the 2002 Nobel Prize in chemistry for his work in developing electrospray mass spectrometry. A professor of chemistry at Virginia Commonwealth University at the time of his Nobel award, he has been on the faculty at Yale University and Princeton University.

Then improvements in tandem quadrupole mass analyzers occurred in 1978 due to the work of Yost and Enke [7], and others realized there was a challenge to extend applications of mass spectrometry to fragile biological compounds of high molecular weight. There were papers on polyethylene glycol [9] by Fenn and his associates to prove the method, but great interest was not aroused until a key paper appeared in *Science* in 1989 [10] showing applications to biological samples. The sequence of developments was later reviewed in 2002 by Prof. Fenn [11] (Figure 18.9).

In this introductory presentation, there are two main points. First, an astute student should readjust his/her value judgments to note that interest in technological advances is heightened in the case of biomedical applications. However, the foremost lesson here is that if there is a device that can accurately "weigh molecules" up to a value of (m/z^+) of 4000 or 5000 atomic mass units (amu) for charges of $z = +1$, then increasing the value of the total charge to higher z^+ numbers will allow the (m/z^+) ratio to remain in the useable range of the mass analyzer! Suppose a given biological molecule travels through the mass analyzer associated with 50 H^+, then the upper limit of the molecular weight the quadrupole analyzer can handle is about $(50)(4,000) = 200,000$! In actual practice, even higher molecular weights have been reported up to 5 million [10] due to the number of positive ions (H^+) that can associate with a large polymeric molecule.

The initial "nozzle" design of Dole was improved in Prof. Fenn's laboratory and the lesson here is that careful work can often succeed where an initial idea can be greatly improved by a few key technical features (Figure 18.10). Fenn's nozzle design went through several stages of improvement with the end goal of obtaining clean mass spectra of high-molecular-weight materials. A particular concern was the safety/danger associated with the high voltage of the source needle and the use of wet chemicals in a laboratory environment. This problem was solved in a clever way by using a capillary that is metalized on the ends to attach leads to create an electrical potential well entirely within the cylindrical volume of the capillary while the electrospray needle

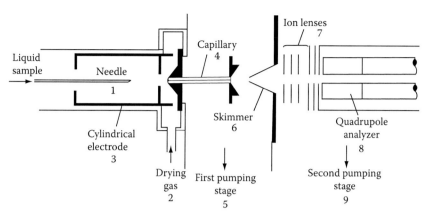

FIGURE 18.10 The most recent nozzle design from Prof. Fenn's laboratory showing several refinements in design including a safer electrode using a metalized capillary (4) as a restrictive skimmer to select only those ions moving in a very straight path toward the quadrupole analyzer with a set of staged-potential plates to serve as ion lenses focusing the beam toward the entrance to the quadrupole analyzer. This design is safer because the needle is at ground and the high voltage potential well is between the metalized ends of the capillary. The ionization process is entirely in the capillary and the needle is only used to form the initial stream of droplets. The drying gas is a stream of nitrogen which carries away solvent vapor before it can enter into the ionizing chamber in the capillary. This design is a careful tradeoff between sweeping away solvent vapor but allowing a short distance for the heavier droplets to make their way into the ionizing capillary. It is important to have multiple vacuum pumps to "pull" the ion stream through the apparatus unimpeded by collisions in the gas phase. (Reproduced from Fenn, J.B., *J. Biomol. Tech.*, 13, 101, 2002. With permission of the Association of Biomolecular Resource Facilities.)

remains at ground and only serves to create the liquid droplets. Another innovation was to provide a countercurrent flow of dry nitrogen within the electrospray chamber to sweep away solvent vapor before the droplets entered the capillary.

So what are the results? Well the results are just short of fantastic! We show only a few of the amazing spectra of biological molecules obtained by Prof. Fenn and his students. The first spectra in Figure 18.11 show examples of molecular weights that are higher than the nominal capability of a quadrupole mass analyzer because the (m/z^+) ratio has been reduced by the presence of many H^+ ions associated with the large molecules. The individual peaks differ by only one H^+ but there are

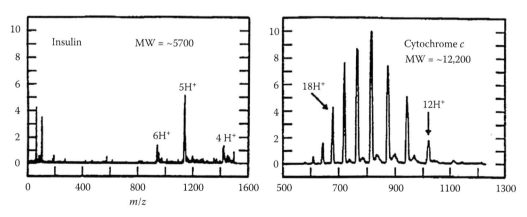

FIGURE 18.11 Electrospray ionization mass spectra (ESI-MS) of low molecular weight samples. (Reproduced from Fenn, J.B., *J. Biomol. Tech.*, 13, 101, 2002. With permission of the Association of Biomolecular Resource Facilities.)

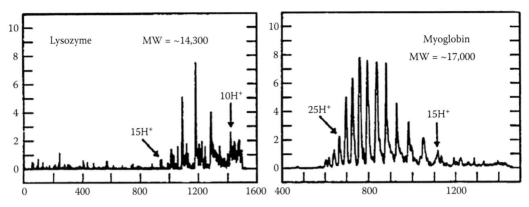

FIGURE 18.12 ESI-MS spectra for medium molecular weight peptides but well beyond the nominal capability of a quadrupole mass analyzer if $z^+ = 1$. This is made possible by the careful design of the electrospray nozzle and the attached H^+ ions which brings the (m/z^+) ration into range of the quadrupole mass analyzer. (Reproduced from Fenn, J.B., *J. Biomol. Tech.*, 13, 101, 2002. With permission of the Association of Biomolecular Resource Facilities.)

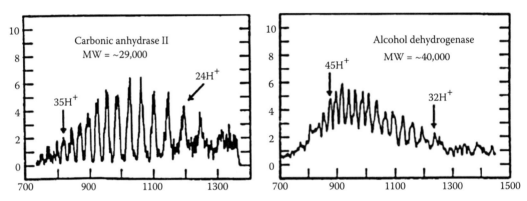

FIGURE 18.13 Examples of ESI-MS spectra of enzymes that would be entirely out of range for the quadrupole mass analyzer for $z = +1$. (Reproduced from Fenn, J.B., *J. Biomol. Tech.*, 13, 101, 2002. With permission of the Association of Biomolecular Resource Facilities.)

many different species. *The difference in mass between two peaks gives the molecular weight of the molecule (plus 1 amu for the added H^+).* In Figure 18.12, we see more nice clean spectra of what might be termed "medium molecular weight" peptide samples. Then in Figure 18.13 we can clearly see mass spectra resolved for higher-molecular-weight enzymes. Note that as the sample molecule gets larger, the number of H^+ ions that accompany it can be larger. We can see 45 H^+ associated with the alcohol dehydrogenase enzyme and a molecular weight of about 40,000 amu. This new technique provides a fast way to determine the molecular weight of biological materials, which would ordinarily be much more complicated. Interested students will find a tutorial on electrospray mass spectrometry at http://www.ionsource.com/tutorial/spectut/spec1.htm

When faced with the raw data of this type of spectra, one only knows that the peaks represent ions with progressively increasing numbers of H^+ ions attached. You do need to know the amu values for certain peaks. The equation that must be fitted to the data is of the form

$$X(\text{amu}) = \frac{(M + nH^+)}{nH^+} = \frac{(M + n)}{n}, \quad M = \text{molecular weight}$$

It helps to know an approximate molecular weight to begin the fitting but basically you just have to keep assigning the value of n to a given peak until it agrees with a calculated value of another peak with a different value of n. In Figures 18.12 and 18.13, this has already been done but when you obtain the raw data you need to carry out the fitting process.

SUMMARY

This chapter has been added to introduce more sophisticated spectroscopic techniques with a minimum of numerical problems. We make no apology for discussing instrumentation in this text because physical chemistry is an experimental science as well as a collection of mathematical descriptions of physical phenomena. Students in physical chemistry should be interested in how instrumentation is used as well as adept in the mathematical description of an experiment. This may be the end of a long semester but this material is intended to inspire students to develop new applications of spectroscopic methods to forensic analysis and characterization of molecules. Raman spectroscopy brings with it demonstrations of the use of point group theory for a few applications and students should anticipate further encounters with group theory as a normal part of physical chemistry. The main advances in Raman spectroscopy will likely be in the use of "resonance Raman" spectra where the excitation wavelength is chosen to electronically excite a chromophore near the site of interest within the molecular application. Finally, the amazing story of electrospray mass spectrometry is probably in the early chapters of development for use in aspects of biomolecular science [12]. There are calls from within the American Chemical Society for education in the discovery process. Thus it is valuable to show students the development of a recent innovation in the form of electrospray mass spectrometry [10]. The topics of resonance Raman spectra and electrospray mass spectrometry offer a high probability of becoming "essential physical chemistry" applications in the very near future, and at the very least it is an essential part of science education to illustrate the process of discovery and development of new applications of basic principles.

PROBLEMS

18.1 Given the C_{3V} character table work out the symmetry types of the vibrations of NH_3. Which vibrations will be "allowed" in infrared absorption and which in Raman spectroscopy?

C_{3V}	I	$2C_3$	$3\sigma_V$	(1)	(2)
A1	1	1	1	z	$x^2 + y^2, z^2$
A2	1	1	-1	R_z	
E	2	-1	0	$(x, y), (R_x, R_y)$	$(x^2 - y^2, xy)(xz, yz)$
Γ_{xyz}	3	0	1		
Γ_{NH_3}	?	?	?		

Note that there are two C_3 rotations and three σ_V operations when you do the projection decomposition. You only have to "see" one of each to get the character and then multiply by the number of equivalent operations. For instance, when you can visualize one of the σ_V operations and get the character of its "arrow" representation, just multiply that by 3 for all of the σ_V operations. The E representation is two dimensional and will require a rotation matrix for the C_3 operation. Harris and Bertolucci also give a big hint with the Γ_{xyz} numbers; just multiply those numbers by the number of atoms that do not move under the symmetry operation to obtain the Cartesian "arrow" characters. To check your work you could run PCLOBE using the NH3SYM file and the C_{3V} point group in the lower left corner of the PCLOBE input panel; use the STO-3G basis set with a gradient setting of "4."

18.2 Given the T_d character table work out the symmetry types of the vibrations of CCl_4. Which vibrations will be "allowed" in infrared absorption and which in Raman spectroscopy? Compare the number and type of your vibrations to the Raman Stokes lines in Figure 12.23.

T_d	I	$8C_3$	$3C_2$	$6S_4$	$6\sigma_d$	(1)	(2)
$A1$	1	1	1	1	1		$(x^2 + y^2 + z^2)$
$A2$	1	1	1	−1	−1		
E	2	−1	2	0	0		$(2z^2 - x^2 - y^2, x^2 - y^2)$
T_1	3	0	−1	1	−1	(R_x, R_y, R_z)	
T_2	3	0	−1	−1	1	(x, y, z)	(xy, xz, yz)
Γ_{xyz}	3	0	−1	−1	1		
Γ_{CH_4}	?	?	?	?	?		

This is one of the more difficult examples for small molecules like CH_4 and CCl_4. Perhaps this should be a group project with the teacher assisting. Make a molecular model out of Tinker Toys or Dreiding models and remember that under unitary transformations you can change the way you hold the model to look for the symmetry elements. Harris and Bertolucci [3] also give a great help with the Γ_{xyz} characters, just multiply those numbers by the number of atoms that are stationary for each symmetry element and that gives you the Cartesian characters. Also if you can see one of the multiple symmetry elements of a given type that is all you need because the characters will all be the same for all members even if you cannot "see" all of them in your mind's eye. So hold the model anyway you need to and find just one operation of each type and then multiply the number of stationary atoms under that symmetry operation by the number in the Γ_{xyz} row and that is the character for your molecule! Once you have the Cartesian characters you can use the projection operation as we did for C_{2v}. The T representations are intrinsically three-dimensional and the S_4 operations are rotations by $(2\pi/4)$ followed by a mirror reflection in a plane perpendicular to the rotations axis (allene has an S_4 axis along the $C=C=C$ bond line). The E representation is two-dimensional and may lead to a degenerate/near degenerate band. While you are there you should run PCLOBE using the coordinates in the file CH4SYM using the STO-3G basis and the T_d point group in the lower left corner of the PCLOBE panel and a gradient setting of "4." The type and number of bands should be the same for any T_d molecule. With some thought and help from a teacher in a group project you should be able to assign each of the STO-3G PCLOBE modes a symmetry label according to which representation each belongs after you decompose the reducible representation. You should be able to make the CCl_4 band assignments in Figure 12.23 using the analogy to the calculated CH_4 bands.

18.3 In Figure 18.13, it appears that the "$36H^+$" peak is very close to the x-axis marker of 800 amu and the peak for $24H^+$ is very close to the x-axis marker for 1200 amu. Using those two estimates, calculate the average molecular weight of carbonic anhydrase and compare to the value of the inset in the figure.

18.4 In Figure 18.13, it appears that the "$44H^+$" peak is very close to the x-axis marker of 900 amu and the peak for $33H^+$ is very close to the x-axis marker for 1200 amu. Using those two estimates, calculate the average molecular weight of alcohol dehydrogenase and compare to the value of the inset in the figure.

REFERENCES

1. Eyring, H., J. Walter, and G. E. Kimball, *Quantum Chemistry*, John Wiley & Sons, Inc., New York, 1944.
2. Cotton, F. A., *Chemical Applications of Group Theory*, 2nd Edn., Wiley-Interscience, New York, 1971.
3. Harris, D. C. and M. D. Bertolucci, *Symmetry and Spectroscopy, an Introduction to Vibrational and Electronic Spectroscopy*, Oxford University Press, New York, 1978; available as a Dover Reprint.
4. Pople, J. A., H. B. Schlegel, R. Krishnan, D. J. Defrees, J. S. Binkley, M. J. Frisch, R. A. Whiteside, R. F. Houk, and W. J. Hehre, *Int. J. Quantum Chem.*, **S15**, 269 (1981).

5. Ward, K. R., R. W. Barbee, P. S. Reynolds, I. P. Torres Filho, M. H. Tiba, L. Torres, R. N. Pittman, and J. Terner, Oxygenation monitoring of tissue vasculature by resonance Raman spectroscopy, *Anal. Chem.*, **79**, 1514 (2007).

6. Paul, W. and H. Steinwedel, US patent 2939952 Apparatus For Separating Charged Particles Of Different Specific Charges, June 1960.

7. Yost, R. A. and C. G. Enke, Selected ion fragmentation with a tandem quadrupole mass spectrometer, *J. Am. Chem. Soc.*, **100**, 2274 (1978).

8. Dole, M., L. L. Mach, R. L. Hines, R. C. Mobley, L. P. Ferguson, and M. B. Alice, Molecular beams of macroions, *J. Chem. Phys.*, **49**, 2240 (1968).

9. Wong, S. F., C. K. Meng, and J. B. Fenn, Multiple charging in electrospray ionization of poly(ethylene glycols), *J. Phys Chem.*, **92**, 546 (1988).

10. Fenn, J. B., M. Mann, C. K. Meng, S. F. Wong, and C. M. Whitehouse, Electrospray ionization for mass spectrometry of large biomolecules, *Science*, **246**, 64 (1989).

11. Fenn, J. B., Electrospray ionization mass spectrometry: How it all began, *J. Biomol. Tech.*, **13**, 101 (2002).

12. Cole, R. B., Ed., *Electrospray Ionization Mass Spectrometry, Fundamentals, Instrumentation, Instrumentation and Applications*, John Wiley & Sons, Inc. New York, 1997.

19 Essentials of Nuclear Magnetic Resonance

INTRODUCTION

This topic is beyond what is normally covered in a 9 week summer course but is certainly an "essential topic." The development of magnetic resonance imaging (MRI) in health science continues and is an outgrowth of nuclear magnetic resonance, which has had its own modern developments. In fact, NMR research has grown into a sophisticated specialty field and we will only attempt to show the basics here but in such a way as to form a foundation for further study. In modern chemistry curricula, applications of NMR are usually given in organic chemistry but without much of the underlying mathematics. We will attempt to fill in some of the mathematics here as well as give specific examples from several Internet tutorials. Evaluation of the available technology related to pulsed NMR spectroscopy indicates it has become a very complex technology requiring continued study including advanced quantum mechanics, electrical engineering and a lot of advanced physics best acquired in graduate study. However, for pre-medical students and forensic majors we present a few examples which should be considered introductory material designed to whet the appetite for further study.

EARLY NMR SPECTROMETERS

The earliest experiments in NMR go back to 1945 with work by Bloch [1] at Stanford University and his contemporary Purcell at Harvard University for which they shared a Nobel Prize in Physics in 1952. In 1959, a very important text was published by Pople, Schneider, and Bernstein (PSB) [2] with the title "High resolution nuclear magnetic resonance." At that time many NMR spectrometers used permanent magnets of only 10,000 gauss (1 Tesla, 1 T) with a proton resonance of 42.5775 MHz. That seems like a very low frequency by today's standards although some of Purcell's work used a magnetic field of only 7000 gauss. Even so, the PSB text is still amazingly up to date for the phenomenolgical mathematics it contains and it is still a desirable book to have on your shelf if you are a person using NMR. Jumping to the present, if you can afford only one other text for NMR it should be "NMR: The Toolkit" [3] by Hore, Jones, and Wimperis. Usually the manual accompanying the spectrometer at your site will have extensive information and, for instance, the Bruker manual is very complete. There are many tutorials on the Internet and other excellent texts but "The Toolkit" says it better with fewer words than most other texts; it is to "Pulsed-NMR" what Roberts' MO book was to organic chemistry in the 1960s.

A student may wonder what a gauss or Tesla really is. Probably all students have "felt" the magnetic *force* from a small magnet as an attraction or repulsion *relative to motion* of a magnetic material like iron or another magnet. The definition of a magnetic field is intimately tied to the connection between moving electrical charges and a magnetic field. A charged particle will experience a force moving in a magnetic field and a moving charged particle will generate a magnetic field! The definition in the SI unit system is given by

$$1\ \text{T} \equiv 1\frac{\text{W}}{\text{m}^2} \equiv 1\frac{\text{N}}{\text{C}\left(\frac{\text{m}}{\text{s}}\right)} = 1\frac{\text{kg}\left(\frac{\text{m}}{\text{s}^2}\right)}{\text{C}\left(\frac{\text{m}}{\text{s}}\right)} \equiv 10^4\ \text{gauss}$$

This is a difficult unit to visualize, but we can see that it relates force and a moving charge. The recent development of superconductivity allows large currents to move in a circle in what is called "persistent mode" with *zero resistance* as long as the temperature is kept below some temperature threshold. Modern superconducting magnets typically operate below the boiling point of liquid He at 4.2°K and are further insulated with an outer shell of liquid N_2 at 77°K. Such coils of superconductors in persistent mode can generate very high magnetic fields to allow NMR spectrometers to use fields of 18 T or more for routine use, as long as the magnet coils are kept below the superconducting threshold temperature. A superconducting magnet can be severely damaged if the temperature rises above the threshold temperature while the magnet is operating in persistent mode.

NMR SPIN HAMILTONIAN

If for no other reason, the exercise shown in the following is *the only chance we have in this undergraduate text to show use of the Pauli spin operators*, which apply to spin-1/2 nuclei as well as electrons. While it has been sufficient to describe electron spins as either α or β, we now need to consider nuclear spin in more detail in a quantized "spin space" based on angular momentum. There is a corresponding form of spectroscopy for electrons called "electron spin resonance" (ESR) but here we are emphasizing NMR because it yields information about geometrical molecular structure while ESR only gives information about electronic structure. Because most nuclear spins of interest are spin-1/2, the mathematics for spin-1/2 electrons is similar in NMR and ESR, but NMR is more general and extends to nuclei of other spin types. In the 1960s, NMR spectrometers used a large magnet with a sample in a region of high field homogeneity and scanned the sample with a radio frequency (RF) to record frequencies where energy absorption occurred due to "spin flips." At first, analog recording was used only for proton resonance but it was known that many other nuclei could be studied at other frequencies.

In the 1980s, two main advances occurred in NMR spectroscopy. First, superconducting magnets became available with much higher magnetic fields so resolution increased dramatically. Second, RF pulse technology made use of the Cooley–Tukey fast Fourier transform (FFT) algorithm possible on readily available minicomputers, which could be dedicated to a specific spectrometer. Let us go back to the early form of the phenomological "spin Hamiltonian" to show the quantum mechanics involved for a simple case of two spin-1/2 nuclei before we move on to the Bloch equations and pulsed-NMR with FFT.

Let us take a look at what is the simplest AB NMR pattern of proton NMR in terms of a spin Hamiltonian and a wave function built from a linear combination of spin eigenfunctions as given by PSB [3]. For two nonequivalent nuclei, the possible spin states would be $|spin_A \, spin_B\rangle$ in four ways $|\alpha\alpha\rangle$, $|\alpha\beta\rangle$, $|\beta\alpha\rangle$, and $|\beta\beta\rangle$. This same model can be used in our later discussion of coherent spectroscopy (COSY) so it is worth understanding. In principle, we could use the HF molecule as our example since ^{19}F is a spin-1/2 nucleus as is ^1H, but we will use a more familiar organic example, not to mention difficulty in handling HF to get a spectrum! A better example is the case of 2-bromo-5-chloro thiophene, which has one proton on the 3-carbon and another on the 4-carbon that are not equivalent due to the halide substituents (Figure 19.1). This will be basically a matrix perturbation treatment like the LCAO formulation for electrons. We will set up a (4 × 4) Hamiltonian matrix and solve the matrix form of the eigenvalue problem $HC = EC$ in a (4 × 4) space, which leads to the usual secular determinant $|H_{mn} - E\delta_{mn}| = 0$. The normal resonance of the two protons will be modified by chemical shift values as $H_A = H_0(1 - \sigma_A)$ and $H_B = H_0(1 - \sigma_B)$ and the relative chemical shift is given by $(\sigma_B - \sigma_A)$. The Hamiltonian matrix is a (4 × 4) [3,4] as $HC = CE$ where

$$H = \begin{bmatrix} v_0(1 - 1/2\sigma_A - 1/2\sigma_B) + 1/4J & 0 & 0 & 0 \\ 0 & v_0(-1/2\sigma_A + 1/2\sigma_B) - 1/4J & 1/2J & 0 \\ 0 & 1/2J & v_0(1/2\sigma_A - 1/2\sigma_B) - 1/4J & 0 \\ 0 & 0 & 0 & v_0(-1 + 1/2\sigma_A + 1/2\sigma_B) + 1/4J \end{bmatrix}$$

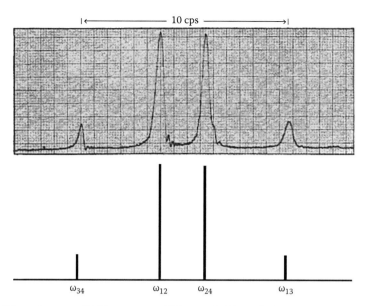

FIGURE 19.1   The early proton NMR spectrum of 2-bromo-5-chlorothiophene. The stick figure spectrum was fitted using $v_0(\sigma_B - \sigma_A) \cong 4.7$ cps and $|J| \cong 3.9$ cps. (Reprinted with permission from Anderson, W., *Phys. Rev.*, 102, 151, 1956. Copyright 1956, America Institute of Physics.)

Here we have used $v_0 = \dfrac{\gamma H_0}{2\pi}$ where γ is the gyromagnetic ratio for the nucleus in question (^{1}H here) and H_0 is the external magnetic field in the notation of PSB [3]. The gyromagnetic ratio is characteristic of each nucleus and is the ratio of its magnetic moment to its angular momentum. The Hamiltonian is given by spin operators I made up of

$$H = \frac{1}{2\pi} \sum_i \gamma_i H_i I_z(i) + \sum_i \sum_{<j} J_{ij} I(i) \cdot I(j)$$

with (2×2) Pauli spin operators. $I_x = 1/2 \begin{bmatrix} 0 & 1 \\ 1 & 0 \end{bmatrix}$, $I_y = 1/2 \begin{bmatrix} 0 & i \\ -i & 0 \end{bmatrix}$, and $I_z = 1/2 \begin{bmatrix} 1 & 0 \\ 0 & -1 \end{bmatrix}$.

The problem is reduced from a (4×4) diagonalization to only a (2×2) problem since we can use the theorem that if two eigenfunctions have different eigenvalues for the same operator they are orthogonal. Thus $|\alpha\alpha\rangle$ has an eigenvalue of $+1$ and $|\beta\beta\rangle$ and eigenvalue of -1 so they will not mix with each other or $|\alpha\beta\rangle$ and $|\beta\alpha\rangle$, which have an eigenvalue of 0. However $|\alpha\beta\rangle$ and $|\beta\alpha\rangle$ will mix and do have an off-diagonal element. Let us define how the Pauli spin operators operate on the spin vectors represented as (2×1) columns where $\alpha = \begin{bmatrix} 1 \\ 0 \end{bmatrix}$ and $\beta = \begin{bmatrix} 0 \\ 1 \end{bmatrix}$ in "spin space" (energy in $\hbar$ units). For the diagonal elements we have J_{12} coupling which can be worked out using

$$\langle\alpha|I_x|\alpha\rangle = [1 \quad 0]1/2 \begin{bmatrix} 0 & 1 \\ 1 & 0 \end{bmatrix} \begin{bmatrix} 1 \\ 0 \end{bmatrix} = [1 \quad 0]1/2 \begin{bmatrix} 0 \\ 1 \end{bmatrix} = 0,$$

$$\langle\alpha|I_y|\alpha\rangle = [1 \quad 0]1/2 \begin{bmatrix} 0 & i \\ -i & 0 \end{bmatrix} \begin{bmatrix} 1 \\ 0 \end{bmatrix} = [1 \quad 0]1/2 \begin{bmatrix} 0 \\ -i \end{bmatrix} = 0,$$

$$\langle\alpha|I_z|\alpha\rangle = [1 \quad 0]1/2 \begin{bmatrix} 1 & 0 \\ 0 & -1 \end{bmatrix} \begin{bmatrix} 1 \\ 0 \end{bmatrix} = [1 \quad 0]1/2 \begin{bmatrix} 1 \\ 0 \end{bmatrix} = \frac{1}{2},$$

$$\langle\alpha|I_x|\beta\rangle = [1 \quad 0]1/2\begin{bmatrix} 0 & 1 \\ 1 & 0 \end{bmatrix}\begin{bmatrix} 0 \\ 1 \end{bmatrix} = [1 \quad 0]1/2\begin{bmatrix} 1 \\ 0 \end{bmatrix} = \frac{1}{2},$$

$$\langle\alpha|I_y|\beta\rangle = [1 \quad 0]1/2\begin{bmatrix} 0 & i \\ -i & 0 \end{bmatrix}\begin{bmatrix} 0 \\ 1 \end{bmatrix} = [1 \quad 0]1/2\begin{bmatrix} i \\ 0 \end{bmatrix} = \frac{i}{2}, \quad \langle\beta|I_y|\alpha\rangle = \frac{-i}{2},$$

$$\langle\alpha|I_z|\beta\rangle = [1 \quad 0]1/2\begin{bmatrix} 1 & 0 \\ 0 & -1 \end{bmatrix}\begin{bmatrix} 0 \\ 1 \end{bmatrix} = [1 \quad 0]1/2\begin{bmatrix} 0 \\ -1 \end{bmatrix} = 0,$$

$$\langle\beta|I_x|\beta\rangle = [0 \quad 1]1/2\begin{bmatrix} 0 & 1 \\ 1 & 0 \end{bmatrix}\begin{bmatrix} 0 \\ 1 \end{bmatrix} = [0 \quad 1]1/2\begin{bmatrix} 1 \\ 0 \end{bmatrix} = 0,$$

$$\langle\beta|I_y|\beta\rangle = [0 \quad 1]1/2\begin{bmatrix} 0 & i \\ -i & 0 \end{bmatrix}\begin{bmatrix} 0 \\ 1 \end{bmatrix} = [0 \quad 1]1/2\begin{bmatrix} i \\ 0 \end{bmatrix} = 0,$$

and finally

$$\langle\beta|I_z|\beta\rangle = [0 \quad 1]1/2\begin{bmatrix} 1 & 0 \\ 0 & -1 \end{bmatrix}\begin{bmatrix} 0 \\ 1 \end{bmatrix} = [0 \quad 1]1/2\begin{bmatrix} 0 \\ -1 \end{bmatrix} = \frac{-1}{2},$$

all in units of $\hbar$. Note $\langle\beta|I_y|\alpha\rangle = \frac{-i}{2}$ has an opposite sign to $\langle\alpha|I_y|\beta\rangle = \frac{i}{2}$ as a Hermitian requirement! Then we can use these matrix elements for $\langle\alpha\alpha|J_{12}\,I(1) \cdot I(2)|\alpha\alpha\rangle$ and $\langle\alpha\beta|J_{12}\,I(1) \cdot I(2)|\alpha\beta\rangle$ in the diagonal elements.

$$\langle\alpha\alpha|J_{12}I(1) \cdot I(2)|\alpha\alpha\rangle = J_{12}\left[\langle\alpha|I_x|\alpha\rangle^2 + \langle\alpha|I_y|\alpha\rangle^2 + \langle\alpha|I_z|\alpha\rangle^2\right] = 1/4J_{12}$$

and

$$\langle\alpha\beta|J_{12}I(1) \cdot I(2)|\alpha\beta\rangle = J_{12}\left[\langle\alpha|I_x|\alpha\rangle\langle\beta|I_x|\beta\rangle + \langle\alpha|I_y|\alpha\rangle\langle\beta|I_y|\beta\rangle + \langle\alpha|I_z|\alpha\rangle\langle\beta|I_z|\beta\rangle\right]$$
$$= J_{12}[0 + 0 + (1/2)(-1/2)]$$
$$= \frac{-J_{12}}{4}.$$

Similar expressions are found for $\langle\beta\beta|J_{12}I(1) \cdot I(2)|\beta\beta\rangle = 1/4J_{12}$ and $\langle\beta\alpha|J_{12}I(1) \cdot I(2)|\beta\alpha\rangle = \frac{-J_{12}}{4}$. Next we need

$$\langle\alpha\beta|J_{12}I(1) \cdot I(2)|\beta\alpha\rangle = J_{12}\left[\langle\alpha|I_x|\beta\rangle\langle\beta|I_x|\alpha\rangle + \langle\alpha|I_y|\beta\rangle\langle\beta|I_y|\alpha\rangle + \langle\alpha|I_z|\beta\rangle\langle\beta|I_z|\alpha\rangle\right]$$
$$= J_{12}\left[\left(\frac{1}{2}\right)\left(\frac{1}{2}\right) + \left(\frac{i}{2}\right)\left(\frac{-i}{2}\right) + (0)(0)\right] = \frac{J_{12}}{2}.$$

The rest of the diagonal elements can be evaluated by assuming the field strength H_0 is constant even though some early spectrometers did sweep over a small range of field strength. Next we can evaluate the unperturbed resonant frequencies in terms of the relative chemical shift difference and the diagonal angular momentum with $+1/2$ for α and $-1/2$ for β as

---

TABLE 19.1
**Basis Spin States for Two Protons
with (a, b) as Normalized Mixing
Coefficients**

1.	$|\alpha\alpha\rangle$
2.	$a|\alpha\beta\rangle + b|\beta\alpha\rangle$
3.	$a|\alpha\beta\rangle - b|\beta\alpha\rangle$
4.	$|\beta\beta\rangle$

$$\langle\alpha\alpha|H^0|\alpha\alpha\rangle = \nu_0(1/2 + 1/2 - \sigma_A/2 - \sigma_B/2),$$

$$\langle\beta\beta|H^0|\beta\beta\rangle = \nu_0(-1/2 + -1/2 + \sigma_A/2 + \sigma_B/2),$$

$$\langle\alpha\beta|H^0|\alpha\beta\rangle = \nu_0(+1/2 - 1/2 - \sigma_A/2 + \sigma_B/2) = \nu_0(-\sigma_A/2 + \sigma_B/2)$$

$$\langle\beta\alpha|H^0|\beta\alpha\rangle = \nu_0(-1/2 + 1/2 + \sigma_A/2 - \sigma_B/2) = \nu_0(+\sigma_A/2 - \sigma_B/2).$$

Although this is a set of phenomenological equations only good to first order, it has been found to work very well. Depending on the value of J_{12}, simulated spectra can be drawn using the idea that there will only be transitions in the spectrum for states that differ by a spin on a given nucleus. That condition is worth remembering because even though we are treating more than one nucleus, the implication is that we are exciting/flipping each one individually and that may not be the case later if we are concerned with spectra from a special sequence of RF pulses. Here we have transitions between states for which there is only a difference for a certain nucleus and for the resonant frequency to cause a change of $+1$ or -1. For instance, suppose the value of J_{12} is such that after diagonalization we have four states for two nuclei where states 2 and 3 have mixed due to the off-diagonal term. The normalized mixing coefficients (a, b) come from solving the $(2, 3)$ subspace of the Hamiltonian. Only transitions $(1->2)$, $(1->3)$, $(4->2)$ and $(4->3)$ will occur for a total of four lines and their relative peak intensities will be determined by the numerical coefficients (a, b) (Table 19.1).

But wait! What is J_{12}? Therein lies a dilemma in that a lot of effort was required to fit the value of J_{12} to measured spectra and that is only for two nuclei. It turns out that a least squares fitting procedure can be used easily up to about seven nuclei (protons) [5] but larger molecules present a challenge. At first, increasing the field strength offered a way to spread out a given spectrum but for things like DNA fragments and enzymes this approach is hopeless!

FORENSIC APPLICATION OF 1D-NMR

Benzoylecgonine (BEG) is one of the main metabolites of cocaine as formed by hydrolysis of cocaine in the liver and is mainly excreted in urine. Therefore BEG in a urine sample is a telltale sign of cocaine and "crack" use (Figure 19.2). Although there are other methods of detection such as gas chromatography/mass spectrometry, two-dimensional NMR (2D-NMR) [6] has recently been developed as a screening technique since sample size is not critical in this case. While detection of cocaine use is important in criminal cases, NMR detection is relatively expensive and unlikely to become a routine method for this analysis. However, we will see that the simplest form of such analysis might be proton NMR using an inexpensive low-field spectrometer. The crack/rock form of cocaine can rapidly lead to addiction with a single use followed by devastating

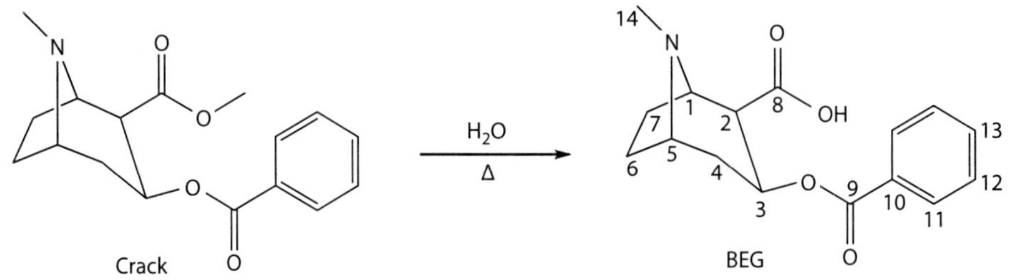

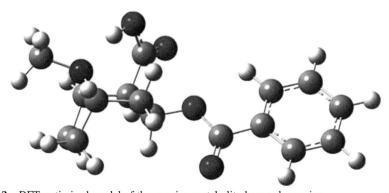

FIGURE 19.2   Research preparation of BEG by mild hydrolysis of "crack" cocaine.

personal and social consequences; death has occurred from use of crack (do an Internet search on "Len Bias," a tragic story). The key point here is that Pedersoli et al. [6] made an all-out study of the NMR spectra of BEG using 2D techniques and while we might find simple proton NMR to be sufficient for forensic use, physical chemists should have a working understanding of 2D-NMR, possibly from graduate courses.

Pedersoli et al. [6] started from a sample of "crack" cocaine and hydrolyzed the methyl ester using hot water. The metabolite was synthesized and characterized by 2D-NMR using three advanced techniques for a very thorough study of BEG by NMR (Figure 19.3). Details of the study were kindly supplied by Prof. Roberto Rittner of The Physical Organic Chemistry Laboratory, Chemistry Institute, State University of Campinas, Campinas, SP, Brazil.

Spectra: ^{1}H and ^{13}C NMR spectra run in a VARIAN INOVA 500 (499.88 MHz for ^{1}H and 125.70 MHz for ^{13}C), temperature 300°K, in CD_3CN and TMS (Figure 19.4).

Structure: It was optimized at B3LYP/cc-pVDZ level of theory.

The modeling program is a modern density functional theory (DFT) [7] molecular orbital method, which uses an empirically adjusted exchange potential and a "valence double zeta" (VDZ) basis set of orbitals (a more refined form of the Slater X α exchange discussed in Chapter 9, and additional orbitals are included in a calculation beyond what would be needed for the minimum description), which is modeling including electronic structure effects and this method is regarded as capable of producing reasonable bond lengths and angles by minimizing the conformational energy (Table 19.2).

After this analysis of the overall NMR spectra of BEG, the forensic application depends on the resonance of the protons (No. 14 on the structural formula) at 2.54 ppm due to the *N*-methyl substituent. That peak is strong enough and in a sufficiently unique region to be observable from a urine extract sample or a suitable bowel sample from a newborn infant. One interesting structural observation from the paper is that the carboxyl H is attracted to the N lone pair forming an internal

FIGURE 19.3   DFT-optimized model of the cocaine metabolite benzoylecgonine.

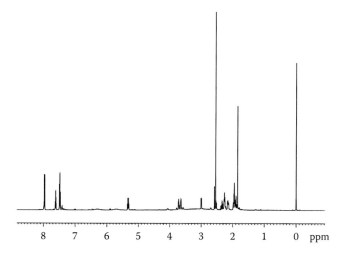

FIGURE 19.4 The 500 MHz proton NMR spectrum of the benzoylecgonine (BEG) metabolite. (From Pedersoli, S. et al., *Spectrosc. Lett.*, 41, 101, 2008. With permission; Spectrum courtesy of Prof. Rittner of The Physical Organic Chemistry Laboratory, Chemistry Institute, State University of Campinas, Campinas, SP, Brazil.)

TABLE 19.2
Summary of the BEG Cocaine Metabolite NMR Data

C	δC (ppm)	H	δH (ppm)	Multiplicity	Coupling Constants (Hz)	gHMBC	gCOSY	gHSQC
1	65.3	1	3.71(1H)	Dd	$J_{1,2}=3.1; J_{1,7\alpha}=6.1$	C-2, 3, 5, 6, 14	$H_2, H_{7\alpha}, H_{7\beta}$	H_1
2	49.3	2	2.99(1H)	Dd	$J_{2,3}=6.5; J_{2,1}=3.1$	C-3, 4, 8	H_1, H_3	H_2
3	66.3	3	5.32(1H)	Dt	$J_{3,4\alpha}=11.3;$ $J_{3,4\beta}=J_{3,2}=6.5$	C-2, 4, 8, 9	$H_2, H_{4\alpha}, H_{4\beta}$	H_3
4	34.6	4_α	2.33(1H)	Ddd	$J_{4\alpha,4\beta}=13.5; J_{4\alpha,3}=11.3;$ $J_{4\alpha,5}=3.1$	C-3, 5, 6	$H_3, H_{4\beta}, H_5$	$H_{4\alpha}, H_{4\beta}$
4		4_β	2.14(1H)	Ddd	$J_{4\beta,4\alpha}=13.5; J_{4\beta,3}=6.5;$ $J_{4\beta,5}=3.1$	C-2, 3	$H_3, H_{4\alpha}, H_5$	
5	61.9	5	3.64(1H)	Dt	$J_{5,6\alpha}=6.4;$ $J_{5,4\alpha}=J_{5,4\beta}=3.1$	C-3, 7	$H_{6\alpha}, H_{4\alpha}, H_{4\beta}$	H_5
6	25.7	6_α	2.26(1H)	Dd	$J_{6\alpha,6\beta}\approx9.8; J_{6\alpha,5}=6.4$	C-1, 4, 5, 7	$H_5, H_{6\alpha}, H_{7\beta}$	$H_{6\alpha}, H_{6\beta}$
6		6_β	1.96(1H)	T	$J_{6\beta,7\beta}=J_{6\beta,6\alpha}=9.8$	C-1, 4, 5, 7	$H_5, H_{6\beta}, H_{7\alpha}$	
7	24.6	7_α	2.26(1H)	Dd	$J_{7\alpha,7\beta}=9.8; J_{7\alpha,1}=6.1$	C-1, 2, 5, 6	$H_1, H_{6\beta}, H_{7\beta}$	$H_{7\alpha}, H_{7\beta}$
7		7_β	1.89(1H)	T	$J_{7\beta,7\alpha}=J_{7\beta,6\beta}=9.8$	C-1, 2, 5, 6	$H_1, H_{6\alpha}, H_{6\beta}, H_{7\alpha}$	
8	174.0	8	—	—	—	—	—	—
9	166.4	9	—	—	—	—	—	—
10	131.2	10	—	—	—	—	—	—
11	130.4	11	7.96(2H)	Dd	$J_{11,12}=8.5; J_{11,3}=1.4$	C-9, 10, 12, 13	H_{12}, H_{13}	H_{11}
12	129.6	12	7.48(2H)	Dd	$J_{12,11}=8.5; J_{12,13}=7.3$	C-9, 10, 11, 13	H_{11}, H_{13}	H_{12}
13	134.2	13	7.61(1H)	Tt	$J_{13,12}=7.3; J_{13,11}=1.4$	C-11	H_{11}, H_{12}	H_{13}
14	38.5	14	2.54(3H)	S	—	C-1, 5	—	H_{14}

Source: Pedersoli, S. et al., *Spectrosc. Lett.*, 41, 101, 2008. With permission.

H-bond which makes the structure more rigid but this may not persist in solution. It is likely that the main forensic application will depend on the large methyl-proton resonance at 2.54 ppm, which can be observed using less expensive spectrometers. In the research of the NMR spectra of the BEG metabolite, all the assignments had to be made and the 2D spectra was extremely helpful in assigning the resonance peaks. However, one should appreciate the excellent resolution provided by the high field in the 1D spectrum.

NUCLEAR MAGNETIC RESONANCE: PULSE ANALYSIS

It is interesting that the key paper by Bloch [1] was published in 1946 and even the 1959 printing of the PSB text includes the description of the Bloch equations in the early chapters, but it was not until about the 1980s that it was applied to a sophisticated sequence of RF pulses. Early scanning NMR spectrometers used 10,000 gauss or 10 kilogauss. Today, superconducting magnets are rated in Tesla $= 1$ W/m$^2 = 10$ kilogauss $= 10,000$ gauss and typical superconducting magnets are rated at 6 T or much more. A 4 T magnet can pull a loose (steel) screwdriver off a nearby table and cause a lot of damage in a laboratory. A typical laboratory Alnico magnet can reach about 1 T in a 1 mm gap. Today, spectrometers are available with 18.7 T fields to measure 800 MHz proton spectra, and "whole body" magnetic resonance imaging (MRI) solenoids typically operate at 1.5 T! There is even one NMR spectrometer in Lyon, which can measure 1.1 GHz $= 1100$ MHz proton spectra! The point here is that the magnetic fields are so strong that precise control and analysis of spin alignment is now possible with brief intense pulses of RF energy perpendicular to the main field.

One of the main innovations in NMR spectroscopy became a practical reality around 1980 when spectrometers changed from scanning spectrometers to "pulse-analyzing spectrometers."

Going back to the idea of representing a function by a Fourier series [8] as

$$f(x) = \frac{a_0}{2} + \sum_{n=1}^{\infty} [a_n \cos(nx) + b_n \sin(nx)]$$

one can find the coefficients from $a_n = \dfrac{1}{\pi} \displaystyle\int_{-\pi}^{+\pi} f(x) \cos(nx) dx$ and $b_n = \dfrac{1}{\pi} \displaystyle\int_{-\pi}^{+\pi} f(x) \sin(nx) dx$ using the idea of component projection we discussed for FT-IR. However, if $f(x)$ is a rectangular "pulse" such as $[f(x) = 0, x < 0; f(x) = 1, 0 \geq x \leq a; f(x) = 0, x > a]$, it will be found that the fit requires essentially $n \to \infty$ to get a good fit, especially at the corners of the pulse (Figure 19.5). On the one hand that means that it takes an infinite number of waves with a Fourier series to fit a square wave pulse, but on the other hand it implies that if you do apply a brief square-wave pulse of a RF centered near where the absorptions of a nuclear resonance occurs, *the pulse will approximate a simultaneous irradiation by an infinite number of frequencies!* In practice, a pulse of 10–200 W with a fixed frequency in the range of the desired resonance is applied for a short time of about 1–10 μs. Although the actual short pulses may not be perfectly rectangular they still represent a wide range of frequencies in the Fourier sense.

In Chapter 9, we emphasized the use of electromagnetic radiation to characterize quantized energy levels. Figure 9.1 shows that a light (radio) wave has a magnetic field as well as an electric field. Thus there is a mechanism by which a magnetic field can be set up by a brief but intense RF wave *across the main field direction* and cause a tendency for the magnetic moments to precess about the direction of the RF beam and tip/rotate many of the spin moments toward the opposite quantized position. Basically, the NMR probe measures magnetization so a model has been developed called "the vector model" which represents the total sum of the magnetic moments in the sample. For a two-state spin system (proton spin "up" and spin "down") relative to the field

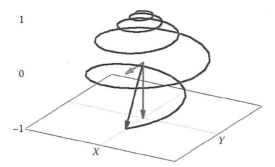

FIGURE 19.5 A schematic of how the tip of the magnetic moment of the spin-1/2 nucleus is "flipped" from one quantized state to the other by a 180° side magnetic field. The $+1, 0, -1$ refer to energy in energy units of $\hbar/2$. (Drawing by Dr. Walter Scott, University of British Columbia.)

direction, there will be a state where the magnetic field of a proton will be "parallel" (p) with the field which is a favored orientation [magnet-N (N-spin-S) S-magnet] with energy $-\dfrac{\gamma\hbar}{2}B_0$. There will also be a less stable orientation of the spin "antiparallel" (ap) to the field [magnet-N (S-spin-N) S-magnet] with energy $+\dfrac{\gamma\hbar}{2}B_0$ [4]. Thus $\Delta E = \gamma\hbar B_0$. The two-state model also applies to ^{13}C and ^{15}N nuclei although the resonant frequencies are different.

At thermal equilibrium in a strong magnetic field, the energy difference between the two orientations is large at a magnetic field of 18.7 T (800 MHz for protons) but even then at 300°K the population in the upper energy level is quite large. This is the main reason that NMR is less sensitive than other spectroscopic methods and requires a fairly large sample. Here γ is the gyromagnetic ratio which is the ratio of the magnetic dipole moment to its angular momentum with SI units of radian per second per tesla ($\text{s}^{-1} \cdot \text{T}^{-1}$). For nuclei the ratio is expressed relative to the nuclear magnetron $\gamma = \dfrac{+e}{2m_p}g = g\dfrac{\mu_N}{\hbar}$. The gyromagnetic ratio for a proton is measured to be $\gamma_p = 2.675222099 \times 10^8/\text{T}$ s. Thus we have $\dfrac{N_{ap}}{N_p} = e^{-\left(\frac{\Delta E}{kT}\right)} = \exp\left[-\left(\dfrac{+\gamma\hbar B_0}{kT}\right)\right]$ according to the Boltzmann Principle. Numerically that is $\dfrac{N_{ap}}{N_p} = 1.273715856 \times 10^{-4}$. Assume the sample is about 0.001 M in protons or about 6×10^{20} total protons. Then $N_{ap} \cong 1.273716 \times 10^{-4}N_p$ and $N_p + N_{ap} = (1 + 1.273716 \times 10^{-4})N_{ap} = 6 \times 10^{20}$ so we find $N_p = 5.999236 \times 10^{20}$ and $N_{ap} = 7.641323 \times 10^{16}$. It might seem from the powers that there are only about ten thousand more parallel than antiparallel spins but N_p is so large that there are plenty of spins to flip from the low level. Staying with the two-state model for spin (1/2) nuclei, we note that the difference in energy $(+\hbar/2) - (-\hbar/2) = \hbar$ so the resonant frequency is due to $\Delta E = h\nu = (h\nu/2\pi)$ for spin 1/2 nuclei. All of this is complicated by the fact that different nuclei can have $+/-$ gyromagnetic ratios (Table 19.3), but for undergraduates the main fact is that the parallel orientation is the lower energy [4]. We also note the spins will be precessing at the Larmor frequency $\omega_0 = -\gamma B_0$ (note the minus sign, see Figure 19.9) about the field z-direction. The precession is the same phenomenon we saw with the angular momentum of the d-orbital states in the H atom where the z-direction is defined but the x- and y-angular momentum is undefined. The Larmor frequency is a real frequency given by the product of the gyromagnetic ratio and the field strength $\omega_0 = \gamma B_0$. For protons, this is $(2.675222099 \times 10^8/\text{T s})(18.7 \text{ T})/(2\pi) \cong 796.2 \times 10^6/\text{s}$ in the assumed 800 MHz spectrometer. Thus you need to visualize a milli-mole sample with at least 6×10^{20} protons precessing in the field multiplied by the number of protons in each molecule and the precession cones are either spin-up or spin-down (Table 19.3).

TABLE 19.3
Selected Two-State Nuclear Spin (1/2) Properties

Z	Isotope	Abundance%	I	$\nu/1\ T = \gamma_p/2\pi$	μ/μ_N
1	^{1}H	99.9885	1/2	42.5775	+2.792847337
6	^{13}C	1.07	1/2	10.7084	+0.7024118
7	^{15}N	0.364	1/2	4.3173	−0.2831888
9	^{19}F	100	1/2	40.0776	+6.628868
14	^{29}Si	4.685	1/2	8.4655	−0.55529
15	^{31}P	100	1/2	17.2515	+1.13160

Source: Lide, D.R., *CRC Handbook of Chemistry and Physics*, 90th Edn., CRC Press,
Taylor & Francis, Boca Raton, FL, 2009–2010, pp. 1–5.
The nuclear magnetron $\mu_N = \left(\dfrac{e\hbar}{2m_p}\right) = 5.05078324 \times 10^{-27}$ J/T [9].

ROTATING COORDINATE SYSTEM

Now we come to the most important consideration of this discussion, *the rotating coordinate system*. A simple analogy is to imagine you observe children playing on a "foot-scooter-driven" merry-go-round at a playground. Standing away from the rotating platform you see it rotate just as you would see a precessing magnetic moment in the laboratory frame. However, if you step onto the rotating platform, the objects on the platform appear to be still while the outside "world" appears to rotate! If the rotation frequency is the same as the precession frequency, the precession will appear to be at a standstill! Thus it is easier to analyze the effects of the RF magnetic pulse in a rotating coordinate system. If the rotating frequency (f) is not quite the same as the precessing frequency (Ω) the precession will appear to proceed slowly with an angular "offset frequency" which is the difference between the RF frequency and the precession frequency as ($\Omega - f$). Usually, the main magnetic field direction is assumed to be in the z-direction with x- and y-directions perpendicular to the main field in what is called the (x, y, z) laboratory framework. Since the RF pulses are controlled by any one of several programs in the control computer, the signal is analyzed as to when a major magnetization signal such as the tetramethysilane (TMS) decreases to zero, goes negative, and comes back to zero as establishing the net magnetization cycle of 360°. That time can then be divided by 2 and 4 to obtain the times for so-called 90°, 180°, and 270° pulse durations. It is important to understand that these degree angles refer to the net magnetization of the whole sample (at one particular frequency) and not the precession phase of any given nucleus. We see in the following a simulation of the magnetization vector undergoing a change in direction due to an "180°" RF pulse.

The use of the vector model can be illustrated by an early technique used to intensify or "refocus" weak signals. The technique was first documented by Hahn [10] but the results were of marginal use. Carr and Purcell [11] published an improvement later that showed stronger signal enhancement. Today this "Spin Echo" technique is less important than other pulse techniques but it is a good educational exercise (Figure 19.6). The diagram uses rotating coordinate axes of (x', y', z') with $z' = z$ of the laboratory frame.

In Figure 19.6A we see the net magnetization pointing along the main field z-axis. In Figure 19.6B there is a 90° pulse, that is a pulse long enough to nullify the z-magnetization to zero as shown in Figure 19.6C with the net vector in the x–y plane of the rotating system and a time of τ per rotation. Then a suitable time delay passes in which the individual spins start to fan out in the x–y plane due to their different chemical shift effective fields. In Figure 19.6E, an RF field is applied for a time that is twice as long as the original 90° pulse resulting in a 180° pulse and full inversion of the magnetization in Figure 19.6F. However, the individual chemical shifts now tend to coalesce in

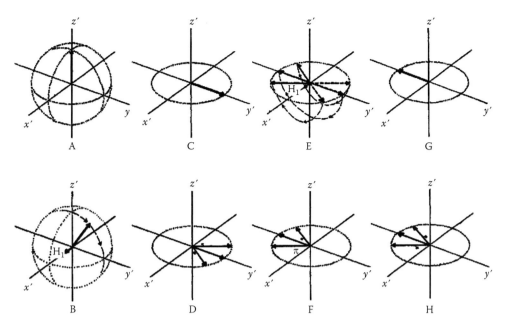

FIGURE 19.6 The classic "spin echo" pulse sequence in the rotating coordinate system from an early paper by Carr and Purcell. (Reprinted with permission from Carr, H.Y. and Purcell, E.M., *Phys. Rev.*, 94, 630, 1954. Copyright 1954 by the American Physical Society.)

Figure 19.6G to form an "echo" of the original signal at a time of 2τ and finally the differences in chemical shifts leads to a fan out of the individual magnetization vectors again. This effect was an important milestone in the development of pulse-NMR but now it serves mainly to illustrate the ideas of the vector model. Unfortunately, the vector model is only an introduction to pulse analysis and is insufficient to explain "COSY." Coherent analysis is very important for direct measurement of coupling constants but at the very least involves a higher description of "product operators" in which the basis represents coherent excitation of two nuclei at once. We will give examples in the following text to motivate students for further study, but the topic requires a separate course for proper understanding.

DETECTION OF MAGNETIC FIELDS

Before we go on with more sophisticated pulse sequences, we need to take a long pause and find out how the Fourier transform is used and what signal is analyzed. Thus we need to take a look at the Bloch equations and we summarize the Internet tutorial of Prof. Suzana Straus of the University of British Columbia available at http://www.chem.ubc.ca/faculty/straus/Nlecture1.pdf. Basically, an NMR spectrometer measures a *time-dependent magnetization* (Figure 19.7). An electrical coil in the x–y plane can detect changes in the magnetic field passing through the coil, which induce electrical currents. According to work by Pierre Curie with magnetism prior to the work on isolating radium, magnetization is the sum of the magnetic moments per unit volume as $\vec{M} = \left(\sum_i \mu_i / V \right)$. At relatively high temperature (room temperature) and assuming the z is the field direction, Curie's Law reduces to $M_0 = M_z = \dfrac{N(\gamma\hbar)^2 I(I+1)B_0}{3kT} = \dfrac{N(\gamma\hbar)^2 B_0}{4kT}$; $I = 1/2$. The Curie Law is the appropriate expression at 300°K, but the derivation is not trivial. However this is what an NMR spectrometer measures at 300°K. Note that due to thermal equilibrium, the magnetization is only along the z-axis direction of the field since the x and y components of the field average to zero by virtue of the precession.

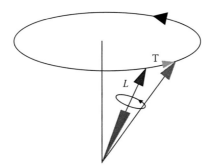

FIGURE 19.7 An illustration of how quantized angular momentum of a magnetic dipole will precess around a magnetic field. (Drawing by permission of Prof. Suzana Straus, of the Chemistry Dept. at the University of Vancouver, British Columbia, http://www.chem.ubc.ca/faculty/straus/Nlecture1.pdf)

BLOCH EQUATIONS

The 1952 Nobel Prize in Physics was awarded to Felix Bloch for results that were published in 1946 [1]. The implementation of what are now the famous "Bloch Equations" led to a major advance in NMR spectroscopy and "magnetic resonance imaging" (MRI) which is now in common use in hospitals. Magnetically, $\vec{\mu} = \gamma \vec{L}$ but the Larmor frequency is given by $\omega_0 = -\dfrac{\mu B_0}{L}$ with a minus sign. The absolute sign is determined by the magnetic moment of a given nucleus, which can be either sign. Prof. Straus goes all the way back to defining the gyromagnetic ratio but we have already defined that as $\vec{\mu} = \gamma \vec{L}$ or $\gamma = \dfrac{\mu}{L}$. Our interpretation depends on following the precession of the spin moments, although we remind ourselves that due to chemical shifts, not all the spin-1/2 particles are precessing at exactly the same frequency.

The first step, which is unfamiliar to chemists, is that $\dfrac{d\vec{L}}{dt} = \vec{T}$ but we should recall Newton's second law in words as "a change in momentum is a force" and torque is a force. Then a precession is caused by a torque (T) on $\vec{\mu}$ as $\vec{T} = \vec{\mu} \times \vec{B}_0$. That is a hard thing to understand but it is common experience when we recall the way in which a toy top precesses while spinning in a gravitational field. Multiply the equation by $\left(\dfrac{\gamma}{V}\right)$ on both sides to obtain the time dependence of the magnetization, the key equation $\dfrac{d\vec{M}}{dt} = \gamma \vec{M} \times \vec{B}_0$, which is a vector equation with a cross-product. Here is where we see the key steps in the derivation of the Bloch equations when we expand the cross-product:

$$\frac{dM_x}{dt} = \gamma(M_y B_z - M_z B_y)$$

$$\frac{dM_y}{dt} = \gamma(M_z B_x - M_x B_z)$$

$$\frac{dM_z}{dt} = \gamma(M_x B_y - M_y B_x)$$

Now we assume $B_x = B_y = 0$ (field in the z-direction) and $B_z = B_0$ with $M_y(t=0) = 0$ as a future condition. Those conditions knock out parts of the cross-product as we see in

$$\frac{dM_x}{dt} = \gamma(M_y B_0)$$

$$\frac{dM_y}{dt} = -\gamma(M_x B_0)$$

$$\frac{dM_z}{dt} = 0$$

So simply using the conditions of the magnet geometry leads us to the key step of the derivation. Next, take the second derivative of the M_y component, *this is the key step!*

$$\frac{d^2 M_y}{dt^2} = -\gamma B_0 \left[\frac{dM_x}{dt}\right] = -(\gamma B_0)^2 M_y$$

and this can be factored as

$$\left(\frac{d}{dt} + i\gamma B_0\right)\left(\frac{d}{dt} - i\gamma B_0\right) M_y = 0.$$

So $M_y = C_1 \cos(\gamma B_0 t) + C_2 \sin(\gamma B_0 t) = 0$ at $t = 0 \Leftrightarrow C_1 = 0$ and so $M_y = C_2 \sin(\gamma B_0 t)$.

Here we have used the sin and cos solution form with as-yet-unknown coefficients.

Then $\frac{dM_y}{dt} = C_2 \frac{d}{dt}[\sin(\gamma B_0 t)] = C_2 \gamma B_0 \cos(\gamma B_0 t) = -\gamma B_0 M_x \Rightarrow M_x = M_x(0)\cos(\gamma B_0 t)$ where we have chosen to call $M_x(0) = -C_2$ since it is arbitrary, as from a RF pulse, but with that choice $C_2 = -M_x(0)$. Then $M_y = C_2 \sin(\gamma B_0 t) = -M_x(0)\sin(\gamma B_0 t)$. Further, we use $\gamma B_0 = \omega_0$ to find

$$M_x(t) = M_x(0)\cos(\omega_0 t),$$

$$M_y(t) = -M_x(0)\sin(\omega_0 t),$$

$$M_z(t) = M_z(0).$$

However, we come to a convention here that requires thought. The sign of the gyromagnetic ratio determines the sense of rotation and, therefore, whether or not the right-hand rule applies. A positive ratio implies the right-hand rule and a negative ratio (^{15}N, e.g.) the opposite. You can see this in the torque equation: $\vec{T} = \vec{\mu} \times \vec{B}_0 = \gamma(\vec{L} \times \vec{B}_0)$. The direction of the torque (and therefore the direction of precession) depends on the sign of gamma. Thus, if the RF pulse is along the x-axis from $+x$ to $-x$ the magnetization will tend to rotate clockwise rather than the usual calculus counterclockwise positive angle. Actually, the Bloch equations are telling us that the M_y component wants to go opposite, to the minus direction. Thus in the diagrams that follow, we will use the right-hand rule for rotation of the magnetization and maintain the usual right-handed coordinate axes system (Figures 19.8 and 19.9). Several NMR books emphasize this convention, which we illustrate in Figure 19.9. *We say again, an RF pulse along the x-axis from $+x$ to $-x$ rotates the magnetization vector clockwise for nuclei with a positive gyromagnetic ratio.*

According to this, the rotation coordinate system would rotate indefinitely about a constant $M_z(0)$ component. However, two effects intervene. First the $M_z(t)$ loses coherence due to physical effects in the sample such as wall collisions and Brownian motion in the sample so Bloch modeled this empirically as $\frac{dM_z}{dt} = -\frac{(M_z - M_0)}{T_1}$ so that as T_1 (the *spin-lattice* relaxation time) increases $M_z \to M_0$ from below. M_z changes dramatically when so many spins are flipped but gradually thermal equilibrium is restored and $M_z \to M_0$. Second, the $\sin(\omega_0 t)$ and $\cos(\omega_0 t)$ waves will gradually decrease in amplitude as the spin–spin interactions of nuclei in slightly different chemical shift frequencies interfere with each other in what is called the *spin–spin relaxation time, T_2.*

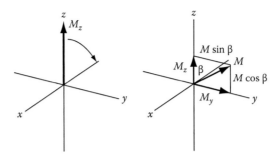

FIGURE 19.8 Definition of a right-hand-rule rotation of the magnetization by an angle $+\beta$ due to a magnetic pulse from the $+x$ direction toward the negative $(-x)$ direction. (From Dr. Keith Brown of the Chemistry Department of the University of Saskatchewan, Canada, at http://chem4823.usask.ca/nmr/practical_nmr.html)

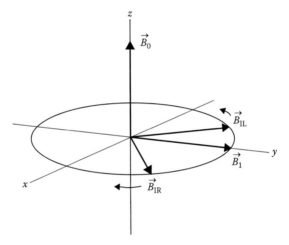

FIGURE 19.9 Resolution of a *RF* pulse plane wave into two $\pm$ circular waves. The circular wave opposite to precession can be neglected as "off-resonance," and when the resonant wave is tuned to the precession it becomes motionless in the rotating system. (From Dr. Keith Brown of the Chemistry Department of the University of Saskatchewan, Canada, at http://chem4823.usask.ca/nmr/practical_nmr.html)

This damping effect is applied as a simple exponential factor $e^{-\left(\frac{t}{T_2}\right)}$. Thus we come to the final (empirical) Bloch equations (Figure 19.10).

$$M_x(t) = M_x(0) \cos(\omega_0 t) e^{-\left(\frac{t}{T_2}\right)} \qquad \frac{dM_x}{dt} = \gamma B_0 M_y - \frac{M_x}{T_2}$$

$$M_y(t) = M_x(0) \sin(\omega_0 t) e^{-\left(\frac{t}{T_2}\right)} \qquad \frac{dM_y}{dt} = -\gamma B_0 M_x - \frac{M_y}{T_2}$$

$$\frac{dM_z}{dt} = \frac{-(M_z - M_0)}{T_1}$$

Thus the wave signals will fade away in time T_2 and the whole effect can be written in a general

equation as $\dfrac{d\vec{M}}{dt} = \gamma \vec{M} \times \vec{B}_0 - R[\vec{M} - \vec{M}_0]$ with $R = \begin{bmatrix} \frac{1}{T_2} & 0 & 0 \\ 0 & \frac{1}{T_2} & 0 \\ 0 & 0 & \frac{1}{T_1} \end{bmatrix}$.

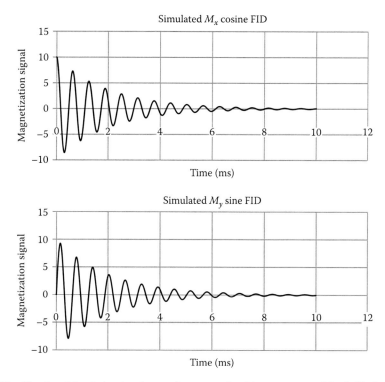

FIGURE 19.10 Simulated free induction decay of a pure cosine M_x component of the field after the 90° pulse from the side. The M_y component is 90° different as a sine wave while the magnetization rotates in the X–Y plane. For a real sample these would not be clean waves but would contain the information from many different chemical shift environments. The time scale could also be as much as 10 times longer in MRI. The two components are available for Fourier transformation with M_x as the real part and M_y as the imaginary part of the complex transform integral.

The Bloch equations are like a mixture of classical electricity and magnetism with some quantum mechanics to determine the energy levels. Note there are two time intervals here. T_1 is the so-called spin–lattice time constant, which is a characteristic time for the loss of the magnetization for a given cycle due to bumping and jostling of the sample molecules in the solvent, the container, etc., the mechanical conditions of the sample. On the other hand, the T_2 time interval is due to the loss of magnetization coherence due to spin–spin interactions (de-phasing) in the sample. The fact that different nuclei are in differing chemical shift environments leads to what is called "free induction decay" (FID) and the coherence decays.

In MRI the signal is primarily due to ^1H resonances from H_2O and $-CH_2-$, which are about 220 cps apart in a typical 1.5 T "body magnet." Different organs and tissue vary in proportions of water and fat content but in general $T_2 < T_1$ with T_1 on the order of 1 s and T_2 on the order of 50–100 ms. For imaging, so-called contrast agents can be infused as water-soluble salts of Mn^{+2} with five unpaired electrons or Gd^{+3} with seven unpaired electrons to shorten T_2 due to paramagnetic interaction with protons. If the salts go preferentially to one site, the part of the image due to T_2 can be changed. Usually, T_1 and T_2 are longer in MRI than in chemical NMR because O_2 is also paramagnetic with two unpaired electrons so dissolved O_2 in chemical preparations usually causes a shorter T_2 for organic chemistry samples. However, the chemical shift information is in the FID signal, and the Cooley–Tukey FFT process projects the spectrum out of the FID signal and then another cycle proceeds, and again for many cycles to improve the average signal (Figure 19.11).

FIGURE 19.11 Felix Bloch (1905–1983) was a Swiss-born physicist who emigrated to the United States in1933 after education in Zurich where he studied under Schrodinger, Heisenberg, Pauli and others who developed quantum mechanics. His early research was on measuring the magnetic moment of the neutron but after WWII he published work on measurement of proton magnetic moments. This work formed the foundation for modern nuclear magnetic resonance and even magnetic resonance imaging. He was awarded the Nobel Prize in Physics in 1952. Here he is speaking at a banquet as President of the AIP in 1965. (Photograph by Robert M. St. John, courtesy AIP Emilio Segre Visual Archives, Physics Today Collection.)

COMPLEX FOURIER TRANSFORM

In the previous use of a Fourier transform for infrared spectra, we only needed the real part but the complex form of the Fourier transform is ideally suited for two dimensions that are orthogonal by using M_x as the real part and plotting M_y as the imaginary part just to indicate the orthogonality of M_x and M_y vector components.

$$f(\omega) = \int_{-\infty}^{+\infty} f(t)e^{i\omega t}dt = \int_{-\infty}^{+\infty} f(t)[\cos(\omega t) - i\sin(\omega t)]dt$$

With very fast computer control, the complex FID can be sampled and the transform calculated in real time for several reasons using the trick of sampling the FID at time delays that correspond to a phase shift to obtain the $\sin(\omega t)$ sample as $\sin(\omega t) = \cos(\omega t + \pi/2)$. Another feature of the pulse acquisition is that it is repeated many times to average the results. Thus we have a situation where a computer controls the timing of RF pulses perpendicular to the z-axis of the main magnet and many programs are available to carry out a number of experiments with the same spectrometer. As such modern NMR spectrometers are "programmable experiments."

2D-NMR COSY

Now we come to the part we have been working toward ever since we noted the spin-Hamiltonian approach needed a way to find the coupling constants. COSY has revolutionized the use of NMR because it is no longer necessary to worry about the values of the coupling constants and there is

TABLE 19.4
Product Operators of RF Pulses in the Rotating Coordinate System

	IS	$1/2E$	S_x	S_y	S_z
	$1/2E$	$1/2E$	S_x	S_y	S_z
	I_x	I_x	$2I_xS_x$	$2I_xS_y$	$2I_xS_z$
$2x$					
	I_y	I_y	$2I_yS_x$	$2I_yS_y$	$2I_yS_z$
	I_z	I_z	$2I_zS_x$	$2I_zS_y$	$2I_zS_z$

Source: Hore, P.J. et al., *NMR: The Tookit*, Oxford University Press, Oxford, U.K., 2000.
For two spins there are 16 possible product operators:
I_z, S_z: z component of magnetization (unobservable).
I_x, I_y, S_x, S_y: x and y components of magnetization (observable).
$2I_zS_z$: longitudinal two-spin order (unobservable).
$2I_xS_z$, $2I_yS_z$, $2I_zS_x$, $2I_zS_y$: antiphase magnetization (observable).
$2I_xS_x$, $2I_xS_y$, $2I_yS_x$, $2I_yS_y$: two-spin coherence (unobservable).
$E/2$: unity operator.
The factors of 2 are for normalization.

a way to use product operators to directly measure and plot the coupling constants! First we need to set out a carefully prepared table of the product operators where the operators that only operate on one nucleus only have one letter but the operators with two letters operate successively on two nuclei. It is a historical convention to treat the system of two nuclei with the symbols I for the "insensitive" nucleus and S for the "sensitive" nucleus, but modern convention usually assigns $I = {}^1H$ and $S = {}^{13}C$ or ${}^{15}N$. However, this notation corresponds to the labels of A and B in our spin-Hamiltonian example at the beginning of this chapter and the analysis is based on the same four energy levels we found for the AB case (Table 19.4).

Note especially in what follows that certain operations are not observable because only operations that cause a change in magnetization flux through the x–y plane receiver will be recorded in the detector. This helps simplify the operator analysis by neglecting the unobservable changes.

COHERENT SPECTROSCOPY

Now we come to an interesting consideration. The main magnetic field is so strong and so homogeneous all in one very well-defined direction that projection of vectors into sine and cosine components is very precise. The text by Hore, Jones, and Wimperis [3] gives geometric diagrams to show the effects of the product operators, but a tutorial by Dr. Keith Brown at the University of Saskatchewan in Canada has carefully worked out the product string for four repeated events in the COSY pulse sequence (http://chem4823.usask.ca/nmr/practical_nmr.html) in a direct use of repeated application of trigonometric resolution of the vector components (Figure 19.12).

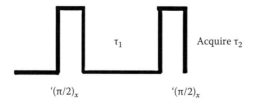

$'(\pi/2)_x$ $'(\pi/2)_x$

FIGURE 19.12   The COSY pulse sequence.

Note in particular that the two time variables between pulses are variable, which is critical to the final process. The two 90° pulses are bursts of RF energy perpendicular to the main field for a time equal to (1/4) of the time it takes the total magnetization signal to go from zero to a maximum, to zero again, to a minimum, and then back to zero magnetization signal again. That is the way in which the spectrometer is calibrated to establish (90/360) degrees of the magnetization signal. The final equations obtained by Dr. Brown are complicated but can still be deciphered.

First, we need to define some experimental parameters relative to phase angles, times, and chemical shifts in both Greek symbols and the abbreviations used in Dr. Brown's program

Ω_I is the chemical shift of nucleus I (could be nucleus A) = wI
Ω_S is the chemical shift of nucleus S (could be nucleus B) = wS
$t_1 = \tau_1$ is the variable time delay between the two 90° pulses (computer controlled)
$t_2 = \tau_2$ is the variable time delay after the second pulse (computer controlled)
J is the coupling constant between the two nuclei (what we seek!)
$\pi = 180° = p$

PRODUCT OPERATOR COSY ANALYSIS USING DR. BROWN'S AUTOMATED SOFTWARE

"For two spins, we start with equilibrium I magnetization and follow the evolution of things as pulses and time delays are applied. Spins I and S are, in this case, homonuclear in opposition to the general literature convention of using them as heteronuclear spin symbols. One could perhaps use I1 and I2 but I find that with all of the terms generated it makes things clearer to use I and S."

$$+ \mathbf{Iz}$$

$$|$$

$$\text{p/2 pulse on I, +x phase}$$

$$|$$

$$\backslash/$$

$$- \mathbf{Iy}$$

$$|$$

$$\text{chemical shift time: t1}$$

$$|$$

$$\backslash/$$

$$- \mathbf{Iy}\text{*cos(wIt1)} + \mathbf{Ix}\text{*sin(wIt1)}$$

$$|$$

$$\text{coupling evolution time: t1}$$

$$|$$

$$\backslash/$$

$$- \mathbf{Iy}\text{*cos(pJt1)*cos(wit1)} + \mathbf{2IxSz}\text{*sin(pJt1)*cos(wit1)}$$

$$+ \mathbf{Ix}\text{*cos(pJt1)*sin(wit1)} + \mathbf{2IySz}\text{*sin(pJt1)*sin(wit1)}$$

|

p/2 pulse on I, +x phase

|

\/

$- \mathbf{Iz}*\cos(pJt1)*\cos(wIt1) + \mathbf{2IxSz}*\sin(pJt1)*\cos(wIt1)$

$+\mathbf{Ix}*\cos(pJt1)*\sin(wIt1) + \mathbf{2IzSz}*\sin(pJt1)*\sin(wIt1)$

|

p/2 pulse on S, +x phase

|

\/

$- \mathbf{Iz}*\cos(pJt1)*\cos(wIt1) - \mathbf{2IxSy}*\sin(pJt1)*\cos(wIt1)$

$+\mathbf{Ix}*\cos(pJt1)*\sin(wIt1) - \mathbf{2IzSy}*\sin(pJt1)*\sin(wIt1)$

|

Removing Unobservables

|

\/

$+\mathbf{Ix}*\cos(pJt1)*\sin(wIt1) - \mathbf{2IzSy}*\sin(pJt1)*\sin(wIt1)$

|

chemical shift time: t2

|

\/

$+\mathbf{Ix}*\cos(wIt2)*\cos(pJt1)*\sin(wIt1)$

$+\mathbf{Iy}*\sin(wIt2)*\cos(pJt1)*\sin(wIt1)$

$-\mathbf{2IzSy}*\cos(wIt2)*\sin(pJt1)*\sin(wIt1)$

$+\mathbf{2IzSx}*\sin(wIt2)*\sin(pJt1)*\sin(wIt1)$

|

coupling evolution time: t2

|

\/

$+\mathbf{Ix}*\cos(pJt2)*\cos(wIt2)*\cos(pJt1)*\sin(wIt1)$

$+\mathbf{2IySz}*\sin(pJt2)*\cos(wIt2)*\cos(pJt1)*\sin(wIt1)$

$+\mathbf{Iy}*\cos(pJt2)*\sin(wIt2)*\cos(pJt1)*\sin(wIt1)$

$-\mathbf{2IxSz}*\sin(pJt2)*\sin(wIt2)*\cos(pJt1)*\sin(wIt1)$

$-\mathbf{2IzSy}*\cos(pJt2)*\cos(wSt2)*\sin(pJt1)*\sin(wIt1)$

$+\mathbf{Sx}*\sin(pJt2)*\cos(wSt2)*\sin(PiJt1)*\sin(wIt1)$

$+\mathbf{2IzSx}*\cos(pJt2)*\sin(wSt2)*\sin(pJt1)*\sin(wIt1)$

$+\mathbf{Sy}*\sin(pJt2)*\sin(wSt2)*\sin(pJt1)*\sin(wIt1)$

The analysis of the (complicated) final expression is fairly straightforward. First, looking at the chemical shift modulation terms for I_x, I_y, S_x, and S_y, we see that the I terms are modulated only by w_I during both the t_1 and t_2 delay times but the S terms are modulated by both w_I in t_1 and w_S in t_2. In other words, the I terms correspond to on-diagonal peaks in the spectrum and the S terms to off-diagonal peaks or cross-peaks in NMR speak. The $2I_xS_z$ and $2I_yS_z$ terms are modulated by w_I and the $2I_zS_x$ and $2I_zS_y$ terms are modulated by w_S and correspond to antiphase peaks close to the diagonal. If we were to start with S_x equilibrium magnetization instead, the result would be similar except that I and S would be exchanged. The actual result of the pulse sequence using Greek symbols is

$$+ I_X \cos(\pi J t_2) \cos(\Omega_I t_2) \cos(\pi J t_1) \sin(\Omega_I t_1) + 2I_Y S_Z \sin(\pi J t_2) \cos(\Omega_I t_2) \cos(\pi J t_1) \sin(\Omega_I t_1)$$

$$+ I_Y \cos(\pi J t_2) \sin(\Omega_I t_2) \cos(\pi J t_1) \sin(\Omega_I t_1) - 2I_X S_Z \sin(\pi J t_2) \sin(\Omega_I t_2) \cos(\pi J t) \sin(\Omega_I t_1)$$

$$- 2I_Z S_Y \cos(\pi J t_2) \cos(\Omega_S t_2) \sin(\pi J t_1) \sin(\Omega_I t_1) + S_X \sin(\pi J t_2) \cos(\Omega_S t_2) \sin(\pi J t_1) \sin(\Omega_I t_1)$$

$$+ 2I_Z S_X \cos(\pi J t_2) \sin(\Omega_S t_2) \sin(\pi J t_1) \sin(\Omega_I t_1) + S_Y \sin(\pi J t_2) \sin(\Omega_S t_2) \sin(\pi J t_1) \sin(\Omega_I t_1)$$

The last four terms have the property that $(\Omega_I t_1)$ occur together and $(\Omega_S t_2)$ occur together so that a grid of values can be computed under computer control to vary τ_1 in small increments while holding τ_2 constant and then varying them in reverse order.

We should have gained an appreciation for the way in which the magnetic moments are snapped into new orientations in a rapid crisp manner in such a high field that the RF pulses perform sudden precise operations. Thus the various sine and cosine terms are also very precise even though they

FIGURE 19.13 Dr. Keith Brown of the Chemistry Department at the University of Saskatchewan in Canada has coded a program for automatic generation of results for the product operators of various RF pulse sequences in NMR which solves a problem of tremendous tedium and allows a creative level of higher order planning and creating unique pulse sequences as well as understanding standard manufacturer pulse sequences. As research in extending pulsed-NMR to ever more detailed understanding of molecular structure is pursued, this automatic pulse-design software offers a key to creative applications without the tedium of mental imagery. His web site is at http://chem4823.usask.ca/nmr/practical_nmr.html

occur in such a complicated formula. *Finally*, we can see how varying (t_1, t_2) over a grid of values solves the problem of determining the chemical shift coupling constant J directly! We will see in the following that manufacturers have installed even more complicated pulse sequences in the data acquisition part of the cycle for the purposes of creating gradients that sharpen the image of the various 2D plots. But we are indebted to Dr. Brown for his analysis of the COSY pulse sequence to explain the main feature of the 2D plots. Both Dr. Brown at the University of Saskatchewan and Prof. Hoffman use a Bruker spectrometer similar to the one at Virginia Commonwealth University but other manufacturers (Varian and Jeol) may use slightly different sequences in the data acquisition portion of the pulse train. Here we should look again at Figure 19.12 and understand that the (τ_1, τ_2) grid is the basis of the 2D COSY plot (Figure 19.13).

ANATOMY OF A 2D EXPERIMENT

The construction of a 2D experiment is simple: In addition to preparation and detection, which are already known from 1D experiments, the 2D experiment has an indirect evolution time t_1 and a mixing sequence. This scheme, documented below by Prof. Roy Hoffman, can be seen in Figure 19.14.

Let us look at a few examples of the COSY-NMR technique as given on the Internet site at the Hebrew University. First we can see from the COSY spectrum of ethyl benzene that when the ^{1}H 1D spectrum is displayed with a right angle display against itself there is a graphical way to display the spin–spin interactions on the plot and *where the off-diagonal resonances occur we can immediately read off the coupling constant*s in a group when the cross-peak is resolved (Figure 19.15). We can immediately obtain the value of the coupling constants between the viscinal protons on carbons 1 and 2 by reading the scale in ppm where the horizontal and vertical lines connect the resonances for protons on the 1 and 2 carbon atoms and obtain the coupling constants from the fine details of the cross-peak.

This seems too easy! It really is easy for small organic molecules but we will show a few more complicated examples. However, *the COSY plot totally eliminates the need for a computer program to calculate/simulate the spectrum to obtain the coupling constants*! Further, the main interest in the coupling constants was to determine the molecular structure or at least the connection diagram for the molecule. All we really want to know is what atoms are connected and this graph tells us which atoms are close enough to show significant coupling.

But wait! We are not quite finished. The desired payoff of all this complicated analysis is to make the answer simple, right? Early in this chapter we pointed out the difficulty in fitting the coupling constant between neighboring spins, J_{ab}. In the spectrum mentioned earlier from Prof. Hoffman's COSY spectrum of ethyl benzene, we realize the peaks along the diagonal are just what we would

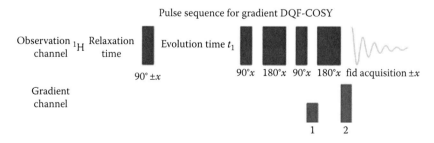

FIGURE 19.14 Typical pulse sequence for coherent detection of NMR showing the points at which additional sensitivity is gained through use of gradient detection. (From Prof. Roy Hoffman, Chemistry Department, Hebrew University. http://chem.ch.huji.ac.il/nmr/techniques/newassignment.htm With permission. Site contains an excellent NMR tutorial.)

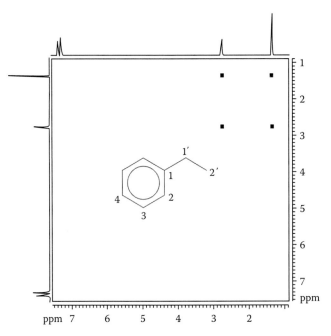

FIGURE 19.15 COSY spectrum plot for ethyl benzene from the NMR tutorial maintained at http://chem.ch. huji.ac.il/nmr/techniques/newassignment.htm. (From Prof. Roy Hoffman, Hebrew University Chemistry Department. With permission.)

see in a 1D NMR spectrum. *It is the off-diagonal cross-peaks that give us the coupling information.* In Prof. Hoffman's spectrum the cross-peaks due to the interaction between the $-CH_3$ protons and the $-CH_2-$ protons are not resolved as shown. However, it is possible to enlarge the region of the cross-peaks to finally achieve the desired J_{ab} coupling constants directly in a graphical way without any computation! In Figure 19.16, we present a schematic of the resolved peaks within the single dot shown on Prof. Hoffman's COSY spectrum. In the completely resolved cross-peak, we can directly measure the J_{ab} value graphically!

Prof. Hofmann's tutorial shows that even more can be done if all we want is to understand which atoms are coupled in a molecular structure. We can do more by remembering the square wave pulse can include a wide range of frequencies including the resonances of other nuclei. Thus it is possible to configure the pulse sequence to obtain another type of 2D plot, the so-called HMBC (hetero-nuclear multiple bond correlation) plot in which we can determine which ^1H resonances are coupled to ^{13}C resonances (both are spin-1/2 nuclei) even though the relative abundance of ^{13}C is only about 1.08%! In the next panel we see the HMBC plot (Figure 19.17) for ethyl benzene where more wonderful information is easily obtained from plotting the ^{13}C 1D spectrum against the ^1H 1D spectrum along with the spin–spin coupling in the off-diagonal space. It is possible to reduce the sensitivity by an alternate pulse sequence to only show correlations one bond apart, but to save space we show the full correlation pattern here. This is amazing! It is now possible to fill in the complete connection table shown in Figure 19.17!

Next, with a sense of overconfidence, we can look at a more complex molecule and illustrate the power of the HSQC (heteronuclear single quantum correlation) 2D-NMR plot. This is perhaps the most useful type of NMR experiment because it enables direct deduction of the nearest neighbor connection table of the molecular structure and it can be used for N–H as well as C–H coupling. The 1D NMR plot for the ^1H resonances is plotted horizontally at the top of the display with the ^{13}C 1D resonance spectrum displayed along the left edge but the experiment has measured the coupling between the nearest neighbors so we can identify the bonding connection table even though

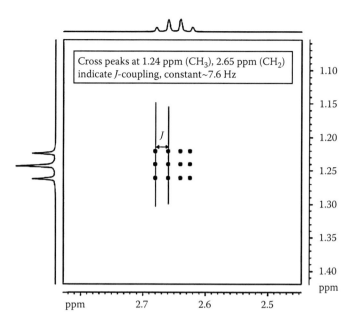

FIGURE 19.16 A schematic of the fully resolved cross peak for the alkyl protons in ethyl benzene. (Special thanks are due to Dr. Yun Qu, director of the NMR Center of the Virginia Commonwealth University Chemistry Department for this drawing.)

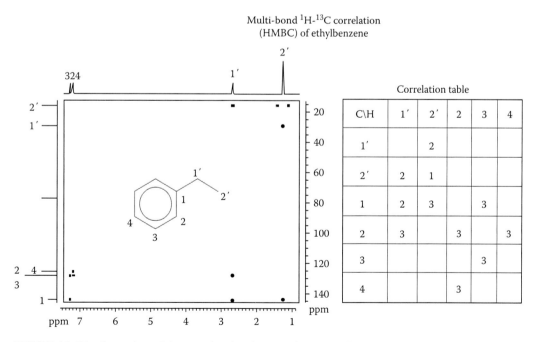

FIGURE 19.17 Generation of the complete bond-connection table of ethyl benzene using the full HMBC diagram. Usually this would be preceded by a less sensitive HMBC plot in which only the immediate closest neighbor correlations would be displayed. This plot is rather sparse for a simple molecule but such diagrams can be more densely populated for larger molecules. (From Prof. Roy Hoffman, Hebrew University Chemistry Department. With permission.)

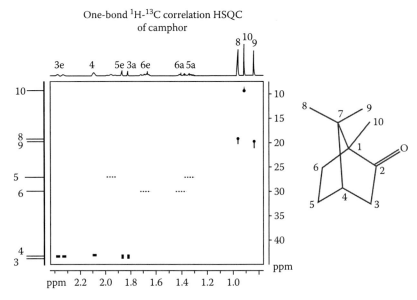

FIGURE 19.18 HSQC NMR chart showing the one-bond connection pattern of C–H links. (From Prof. Roy Hoffman, Chemistry Department, Hebrew University. http://chem.ch.huji.ac.il/nmr/techniques/newassignment.htm With permission.)

camphor has an unusual bridgehead structure! Note a characteristic of the ^{13}C spectrum is that there is essentially no splitting due to spin coupling between the various C atoms so we only see unique single lines for each C atom. In the ^{1}H spectrum, we see the usual complications due to spin–spin coupling but the off-diagonal cross-peak resonances tell us right away which protons are connected to a C atom! Fine details of the spin–spin coupling can be examined by zooming in on specific off-diagonal cross-peak resonances and the structure can be deduced from the bond-connection diagram (Figure 19.18).

In these 2D-NMR plots, the diagonal signals are due to spins that have not changed during the side pulses while those spins that interact and exchange magnetization during the mixing time of the pulse sequence show up in the off-diagonal cross-peaks of the plot. These cross-peak signals contain the really important information about the near-neighbor nuclei. This is why the pulse excitation and use of FFT detection with the Bloch equations simplifies NMR analysis—*COSY NMR directly displays the J_{AB} values!* When you consider the difficulty in least-squares fitting the many coupling constants in the 1970 treatment of 1D-NMR spectra, the use of the 2D-NMR analysis is an amazing simplification! The use of these sophisticated features makes 2D-NMR a powerful tool for structural elucidation in chemistry including inorganic chemistry [4] using other NMR-active nuclei. None of these conveniences would be possible without careful design by electrical engineers and physicists, but chemists have learned to use the power of modern NMR spectroscopy for structural elucidation so that it is the predominant form of spectroscopy in chemistry. While organic chemists are the main beneficiaries of this capability, the physics content in NMR spectroscopy makes pulse-NMR one of the "essential" parts of education in physical chemistry!

There are some complications of course, as in the seven-membered ring of the BEG metabolite where the vicinal coupling of the protons on the 3-4-5 portion of the ring led to very similar off-diagonal coupling in the 2D map, but the simple Karplus equations [12] helped sort that out. There are several other types of 2D-NMR techniques that have been developed, which can even elucidate the structure of polypeptide conformations, and the NMR field is so highly specialized that some fields of organic chemistry and biochemistry depend almost entirely on just a laboratory for wet chemistry preparations and a modern NMR spectrometer! Probably physicists will continue to develop more sophisticated NMR spectrometers and organic/biochemists will be the main users but today it is "essential" that physical chemists are aware of these amazing new developments in NMR spectroscopy! One definite conclusion is that modern NMR spectrometers are really devices capable of many different experiments. There probably remain a lot of new experiments that can be carried out with more complicated pulse sequences. In particular, a Web site maintained by Prof. Malcolm Levitt of the chemistry department of Southampton, UK at http://www.mhl.soton.ac.uk/public/Main/index.html, documents many other pulse simulations.

SUMMARY

This is an optional, difficult chapter that might be the basis of a special topic project for an upper-level student with an interest in NMR or perhaps a way to discuss the dangers of cocaine with undergraduates. Premedical students need some basic introduction to the mechanisms in MRI, and perhaps forensic students will be interested in the possible use of NMR spectroscopy in drug analysis. In some drug crime cases, it is necessary to prove the structure of an evidentiary sample is indeed the isomer specified in a law so there may be occasions to use NMR spectroscopy for a complete structural analysis of a compound.

PROBLEMS

19.1 Given the data: [11] in ν/MHz at 1 T for $^1H = 42.5775$, $^{13}C = 10.7084$, $^{15}N = 4.3173$, $^{19}F = 40.0776$, $^{29}Si = 8.4655$ and $^{31}P = 17.2515$ calculate the magnetic field needed for ^1H resonance at 600 MHz and then recalculate the resonant frequencies of the other spin-1/2 nuclei at that field strength.

19.2 Given the Pauli spin matrices $I_x = 1/2 \begin{bmatrix} 0 & 1 \\ 1 & 0 \end{bmatrix}$, $I_y = 1/2 \begin{bmatrix} 0 & i \\ -i & 0 \end{bmatrix}$, and $I_z = 1/2 \begin{bmatrix} 1 & 0 \\ 0 & -1 \end{bmatrix}$, work out the commutators $[I_x, I_y] = ?$, $[I_x, I_z] = ?$ and $[I_y, I_z] = ?$ Are those results compatible with the z-axis as the primary axis of quantization with the x-axis and y-axis undergoing precession?

19.3 The most educational thing a student can do with this coupling of chemical reasoning and graphical interpretation of 2D plots, is to work through the excellent online tutorial by Prof. Roy Hoffman at http://chem.ch.huji.ac.il/nmr/techniques/newassignment.htm

19.4 Work out the details of the $H(2, 2)$ and $H(2, 3)$ matrix elements of the AB spin-Hamiltonian.

19.5 Use Figures 19.14 through 19.16 to estimate the coupling constant between the protons on the $-CH_3$ and $-CH_2-$ groups of ethyl benzene.

REFERENCES

1. Bloch, F., Nuclear induction, *Phys. Rev.* **70**, 460–474 (1946).
2. Pople, J. A., W. G. Schneider, and H. J. Bernstein, *High-Resolution Nuclear Magnetic Resonance*, McGraw-Hill Book Company, New York, 1959.
3. Hore, P. J., J. A. Jones, and S. Wimperis, *NMR: The Toolkit*, Oxford University Press, Oxford, U.K., 2000.
4. Drago, R. S., *Physical Methods in Chemistry*, W. B. Saunders Company, Philadelphia, PA, 1977, Chap. 9.
5. Wiberg, K. B., *Computer Programming for Chemists*, W. A. Benjamin, Inc., New York, 1965, p. 196.
6. Pedersoli, S., L. Lombardi, N. F. Ho, and R. Rittner, Assignments of 1H and 13C NMR spectral data for benzoylecgonine, a cocaine metabolite, *Spectrosc. Lett.*, **41**, 101 (2008).

7. Trindle, C. and D. Shillady, *Electronic Structure Modeling, Connections between Theory and Software*, CRC Press, Boca Raton, FL, 2008, Chap. 11.

8. Churchill, R. V., *Fourier Series and Boundary Value Problems*, McGraw-Hill Book Company, Inc., New York, 1941, Chap. IV.

9. Lide, D. R., *CRC Handbook of Chemistry and Physics*, 90th Edn., CRC Press, Taylor & Francis, Boca Raton, FL, 2009–2010, pp. 1–5.

10. Hahn, E. L., *Phys. Rev.*, **80**, 580 (1950).

11. Carr, H. Y. and E. M. Purcell, *Phys. Rev.* **94**, 630 (1954).

12. Karplus, M. J., *J. Am. Chem. Soc.*, **85**, 2870 (1963).

Appendix A: Relation between Legendre and Associated Legendre Polynomials

The purpose of this exercise is to show that the relatively simple solution of the $m=0$ case of the Legendre equation can be extended to the general case for nonzero m values. The proof is given by Anderson [1] in two steps. First, define a function $\left[f(x) \equiv (1-x^2)^{\frac{m}{2}}g(x)\right]$ and substitute it into the associated equation for nonzero values of m.

$$\frac{d}{dx}\left[(1-x^2)\frac{df}{dx}\right] + \left[n(n+1) - \frac{m^2}{(1-x^2)}\right]f = 0,$$

$$\frac{d}{dx}\left[(1-x^2)\frac{d}{dx}(1-x^2)^{\frac{m}{2}}g\right] + \left[n(n+1) - \frac{m^2}{1-x^2}\right](1-x^2)^{\frac{m}{2}}g = 0.$$

Then

$$\frac{df}{dx} = \frac{m}{2}(1-x^2)^{\left(\frac{m}{2}-1\right)}(-2x)g + (1-x^2)^{\frac{m}{2}}\frac{dg}{dx}$$

and

$$(1-x^2)\frac{df}{dx} = -m(1-x^2)^{\left(\frac{m}{2}\right)}(x)g + (1-x^2)^{\left(\frac{m}{2}+1\right)}\frac{dg}{dx},$$

so then we can calculate the first part of the general equation as $\dfrac{d}{dx}\left[(1-x^2)\dfrac{df}{dx}\right]$

$$= -\frac{m^2}{2}(1-x^2)^{\left(\frac{m}{2}-1\right)}(-2x)(x)g - m(1-x^2)^{\frac{m}{2}}g - m(1-x^2)^{\frac{m}{2}}(x)\frac{dg}{dx}$$

$$+ \left(\frac{m}{2}+1\right)(1-x^2)^{\frac{m}{2}}(-2x)\frac{dg}{dx} + (1-x^2)^{\left(\frac{m}{2}+1\right)}\frac{d^2g}{dx^2}.$$

Then we insert this expression into the general equation as

$$(1-x^2)^{\left(\frac{m}{2}+1\right)}\frac{d^2g}{dx^2} + \left[(1-x^2)^{\frac{m}{2}}\left(\frac{m}{2}+1\right)(-2x) - m(x)(1-x^2)^{\frac{m}{2}}\right]\frac{dg}{dx}$$

$$+ \left[\left(\frac{-m^2}{2}\right)(1-x^2)^{\left(\frac{m}{2}-1\right)}(-2x^2) - m(1-x^2)^{\frac{m}{2}}\right]g$$

$$+ \left[n(n+1) - \frac{m^2}{1-x^2}\right](1-x^2)^{\frac{m}{2}}g = 0$$

Now   divide   the   whole   equation   by $(1-x^2)^{\frac{m}{2}}$ to   obtain

$$(1-x^2)\frac{d^2g}{dx^2} + \left[\left(\frac{m}{2}+1\right)(-2x)-mx\right]\frac{dg}{dx} + \left[\frac{-m^2(-2x^2)}{2(1-x^2)}-m+n(n+1)-\frac{m^2}{1-x^2}\right]g = 0.$$ Then

$$(1-x^2)\frac{d^2g}{dx^2} + [-mx-2x-mx]\frac{dg}{dx} + \left[\frac{m^2x^2-m^2}{(1-x^2)}-m+n(n+1)\right]g = 0$$ and   further   to

$$(1-x^2)\frac{d^2g}{dx^2} - 2(m+1)x\frac{dg}{dx} + \left[\frac{-m^2(1-x^2)}{(1-x^2)}-m+n(n+1)\right]g = 0.$$ Canceling $(1-x^2)$ we finally

obtain $(1-x^2)\dfrac{d^2g}{dx^2} - 2(m+1)x\dfrac{dg}{dx} + [n(n+1)-m(m+1)]g = 0$. The general form is

$$(1-x^2)\frac{d^{m+2}f}{dx^{m+2}} - 2(m+1)x\frac{d^{m+1}f}{dx^{m+1}} + [n(n+1)-m(m+1)]\frac{d^mf}{dx^m} = 0.$$

For comparison, we differentiate the entire $m=0$ equation for $m=1$ to obtain the derivative of

$$(1-x^2)\frac{d^2P(x)}{dx^2} - 2x\frac{dP(x)}{dx} + l(l+1)P(x) = 0 \text{ (at the bottom of page 284) as}$$

$$(1-x^2)\frac{d^3P_l}{dx^3} - 2x\frac{d^2P_l}{dx^2} - 2\frac{dP_l}{dx} - 2x\frac{d^2P_l}{dx^2} + l(l+1)\frac{dP_l}{dx} = 0 \text{ using the product rule several times.}$$

Collecting terms we get

$$(1-x^2)\frac{d^3P_l}{dx^3} - 4x\frac{d^2P_l}{dx^2} + [l(l+1)-2]\frac{dP_l}{dx} = 0.$$

Anderson [1] notes that the result of successive differentiation is to produce an additional

$(-2x)\dfrac{d^{m+1}f}{dx^{m+1}}$ and $-m(m+1)\dfrac{d^mf}{dx^m}$. As shown for $m=1$, this is compatible with

$$(1-x^2)\frac{d^{|m|+2}P_l}{dx^{|m|+2}} - 2(|m|+1)x\frac{d^{|m|+1}P_l}{dx^{|m|+1}} + [l(l+1)-|m|(|m|+1)]\frac{d^{|m|}P_l}{dx^{|m|}} = 0.$$

Compare to

$$(1-x^2)\frac{d^2g}{dx^2} - 2(m+1)x\frac{dg}{dx} + [n(n+1)-m(m+1)]g = 0 \Rightarrow g = \frac{d^mf}{dx^m} = \frac{d^mP_l}{dx^m}. \quad \text{Q.E.D.!}$$

Thus, we have in general $P_l^m(x) = (-1)^m(1-x^2)^{\frac{m}{2}}\dfrac{d^mP_l(x)}{dx^m}$ considering the odd and even cases of the
basic Legendre polynomials.

REFERENCE

1. Anderson, J. M., *Introduction to Quantum Chemistry*, W. A. Benjamin, Inc., New York, 1969, pp. 321–322.

Appendix B: The Hartree–Fock–Roothaan SCF Equation

ROOTHAAN'S LCAO HARTREE–FOCK EQUATION

At this point, we are about to follow a derivation of the LCAO-SCF equation first derived by Roothaan [1], streamlined in Slater's text [2], and presented finally in Pople's [3] clear notation. Here is our version of Roothaan's breakthrough derivation (Figure B.1).

"All we have to do" is to carry out the integral for the expectation value of the energy as

$$\langle E \rangle = \int \ldots \int \Psi*(1, 2, 3, \ldots n) H \Psi(1, 2, 3, \ldots n)\, d\tau_1 d\tau_2 \ldots d\tau_n ds_1 ds_2 \ldots ds_n \ldots$$

Note in particular that we have to integrate over the spin functions as well as the spatial coordinates using simple rules:

$$\int \alpha*\alpha\; ds = \int \beta*\beta\; ds = 1 \quad \text{and} \quad \int \alpha*\beta\; ds = \int \beta*\alpha\; ds = 0.$$

Among the many permutations of the $\Psi*(1, 2, 3, \ldots, n)$ and the $\Psi(1, 2, 3, \ldots, n)$ there will be many integrals in the long summation, which may have thousands of terms of the general form

$$\int ds_1 \int ds_2 \int d\tau_1 \int d\tau_2 [\psi_1(1)\alpha(1)\psi_1(2)\beta(2)\ldots]* \; H[\psi_1(1)\alpha(1)\psi_1(2)\beta(2)\ldots]$$

due to the fact that there are many terms in the determinants [4] and the Hamiltonian

$$H = \sum_i^n \left(\frac{-\nabla_i^2}{2}\right) - \sum_i^n \sum_K^N \left(\frac{Z_K}{r_{iK}}\right) + \sum_{i>}^n \sum_j^n \frac{1}{r_{ij}} + \sum_{K>}^N \sum_L^N \frac{Z_K Z_L}{R_{KL}}$$

$$= \sum_i^n f_i + \sum_{i>}^n \sum_j^n g_{ij} + \sum_{K>}^N \sum_L^N \frac{Z_K Z_L}{R_{KL}}.$$

In many of these integrals, there will be some misalignment of the order of the orbitals between $\Psi*(1, 2, 3, \ldots, n)$ and $\Psi(1, 2, 3, \ldots, n)$ for that particular term of the two determinants; so the integral will be zero by orthogonality, assuming the $\{\psi_i\}$ are orthogonal. However, if the order in the left and right parts of the integral match exactly or differ by only one orbital and the spins match, the result will be $\int \phi_\mu^* f_i \phi_\nu d\tau_i = \langle \mu | f_i | \nu \rangle$. Again when the order matches for all but two orbitals in $\Psi*(1, 2, 3, \ldots, n)$ and $\Psi(1, 2, 3, \ldots, n)$ for a particular set of cross terms, there can be four possible integrals that are coupled by the g_{ij} two-electron operators resulting from the form of the $\left(\dfrac{1}{r_{12}}\right)$ operator.

FIGURE B.1 Prof. Clemens Roothaan (1918) was born in the Netherlands but came to the University of Chicago in the United States in 1949. His 1951 paper in *Rev. Mod. Phys.* **23**, 69–89 (1951) deriving the self-consistent-field equations in a LCAO basis set was the most highly cited paper between 1945 and 1954. He was director of the Computation Center of the University of Chicago from 1962 to 1968; this photo was taken in October of 1962. In the 1970s, he remained far ahead of others in writing an efficient MCSCF (multi-configurational-SCF; simultaneously optimizing more than one determinant) program with his student John Detrich and other students so that his group was at least 10 years ahead of other developments. Prof. Roothaans's many honors include a Guggenheim Fellowship in Physics (Cambridge University, 1957); election to The International Academy of Quantum Molecular Science in 1967; named the Louis Block Professor of Physics and Chemistry, the University of Chicago (still Emeritus Title); and honored by a special session in honor of Clemens' 70th birthday in 1988 at the International Symposium on the Electronic Structures and Properties of Molecules and Crystals, Rudor Boskovic Institute Zagreb, Cavtat, Yugoslavia. Today he continues to serve as a consultant to IBM in the development of the Itanium processor. See also http://pubs. acs.org/cen/science/science2001.html for an interview.

Much of the mathematics for the SCF process was derived for atoms using texts like Condon and Shortly [5], but in 1951, Roothaan [1] solved the problem for molecules using the LCAO philosophy. Instead of using just single orbitals such as $1s\alpha$ in our Li example on page 371, the LCAO concept says that $\Psi(1, 2, \ldots, n)$ is built from many spin orbitals ψ_i that are linear combinations of basis functions ϕ_μ. Thus, $\psi_i = \sum_\mu^M c_{\mu i}\phi_\mu$. The original paper by Roothaan [1] is a masterpiece but not easy reading, as is often the case with descriptions of new advances. However, there is a succinct explanation given by J.C. Slater in a later text [2]. This is a difficult topic for an undergraduate text but we will follow Slater's description in the faint hope that it is understandable with the calculus we have used so far. Since we will discuss the final equation in Pople's [3] notation, we will write Slater's derivation using Greek letters for basis functions and charge-density "(1*1/2*2)" rounded parenthesis notation for the two electron integrals to merge with that notation as $(\mu\nu/\lambda\sigma) \equiv \int\int \phi_\mu^*(1)\phi_\nu(1)\frac{1}{r_{12}}\phi_\lambda^*(2)\phi_\sigma(2)d\tau_1 d\tau_2$.

After collecting all the nonzero terms in the calculation of $\langle E \rangle = \int \Psi^* H \Psi d\tau$, we find for just the electrons (fixed nuclei)

$$\langle E \rangle = \sum_i^n \sum_\mu^M \sum_\nu^M c_{\mu i}^* c_{\nu i}\langle\mu|f_i|\nu\rangle + \sum_{i>}^n \sum_j^n \sum_\mu^M \sum_\nu^M \sum_\lambda^M \sum_\sigma^M c_{\mu i}^* c_{\nu i}c_{\lambda j}^* c_{\sigma j}\{(\mu\nu/\lambda\sigma) - (\mu\lambda/\nu\sigma)\},$$

where we now have the energy in terms of the $\psi_i = \sum_\mu^M c_{i\mu}\phi_\mu$ LCAO expansion of the LCAO basis $\{\phi_\mu\}$. Here we are using (11/22) notation for the g_{12} operator for N electrons and M basis functions. The alternative $\langle 12/12\rangle$ notation would mean

$$\langle nk/pq\rangle = \int\int \phi_n^*(1)\,\phi_k^*(2)\frac{1}{r_{12}}\,\phi_p(1)\phi_q(2)d\tau_1 ds_1 d\tau_2 ds_2,$$

where the complex conjugate orbitals are together on the left and the real function on the right. Since some research papers use a $\langle 12/12\rangle$ notation, you always need to look at the definition a given author uses! We also have to remind ourselves to integrate over the spins. The effect of the permutations in the g_{12} two-electron terms is to produce a "coulomb term" $(\mu\nu/\lambda\sigma)$ as well as the "exchange term" from the permutation as $(\mu\lambda/\nu\sigma)$. These integrals have not yet been integrated over the spins, which will lead to the four cases treated in Ref. [3].

ELECTRON EXCHANGE ENERGY

Pople and Beveridge [3] give a beautiful analysis of exchange, seldom written anywhere else, but we take the liberty of rewriting their example in $(1^*, 1/2^*, 2)$ charge density notation. Here we see the effect of the permutations on the order of the basis orbitals in a given term in the product of Ψ^* $(1,2,\dots)$ and $\Psi(1,2,\dots)$ and the spin alternations. Let us examine the possible integrals that occur for the exchange term:

$$\int\int \phi_\mu^*(1)\alpha^*(1)\phi_\nu(1)\alpha(1)\frac{1}{r_{12}}\phi_\nu^*(2)\alpha^*(2)\phi_\mu(2)\alpha(2)d\tau_1 d\tau_2 ds_1 ds_2 \neq 0,$$

$$\int\int \phi_\mu^*(1)\alpha^*(1)\phi_\nu(1)\beta(1)\frac{1}{r_{12}}\phi_\nu^*(2)\beta^*(2)\phi_\mu(2)\alpha(2)d\tau_1 d\tau_2 ds_1 ds_2 = 0,$$

$$\int\int \phi_\mu^*(1)\beta^*(1)\phi_\nu(1)\alpha(1)\frac{1}{r_{12}}\phi_\nu^*(2)\alpha^*(2)\phi_\mu(2)\beta(2)d\tau_1 d\tau_2 ds_1 ds_2 = 0,$$

$$\int\int \phi_\mu^*(1)\beta^*(1)\phi_\nu(1)\beta(1)\frac{1}{r_{12}}\phi_\nu^*(2)\beta^*(2)\phi_\sigma(2)\beta(2)d\tau_1 d\tau_2 ds_1 ds_2 \neq 0.$$

Note that $\int \alpha^*\alpha\ ds = \int \beta^*\beta\ ds = 1$ and $\int \alpha^*\beta\ ds = \int \beta^*\alpha\ ds = 0$.

In the grand scheme of the detailed derivation, you will be looking at energy terms like $\int\int \Psi^*(1,2,3,\ \dots)\left[\frac{1}{r_{ij}}\right]\Psi(1,2,3,\ \dots)d\tau_1 d\tau_2 ds_1 ds_2$. Remember that $\Psi(1,2,3,\dots)$ is a big determinant on both sides of the operator. We can see from the four integrals above that the second and third are zero by spin-orthogonality. You can easily write out all the terms for the two electrons in He but unfortunately that is not the general case. With more than four-electrons you could see the four exchange terms above as well as orbital products that do/do not match for the other electrons. Assuming all the $\{\psi_i\}$ can be made orthonormal (they can by choice of the expansion coefficients in $\psi_i = \sum_\mu^M c_{\mu i}\phi_\mu$), all the parts of the wave functions for electrons which are not i or j will just integrate to factors of 1 when the g_{ij} part is nonzero, but you will be left with the four possibilities shown above that occur due to the permutations of the determinants subject to the spin integration.

As a result of spin orthogonality, only $(1/2)$ of the exchange terms are nonzero but they are there! This was first pointed out by Fock [6] and was added as a correction to the method then developed by Hartree [7]. The combined method is now called the Hartree–Fock method if tabulated numerical orbitals are used, but the Hartree–Fock–Roothaan method in an LCAO basis. Today this method is almost obsolete and modern methods tend to be some form of "density functional theory (DFT)" where the whole program is made much faster by using numerical tricks and replacing the two-electron operators by carefully designed functions of the electron density. However, the discussion above is as important today as it was in 1930 because after more than 20 years of research on DFT methods, the most accurate calculations have been found to require at least some fractional amount of this so-called "Hartree–Fock Exchange" [8]; it is still a part of modern research! Further, the exchange terms enter into the energy with a negative sign so the phenomenon of exchange lowers the energy and is necessary for the most accurate energy calculations.

THE HARTREE–FOCK–ROOTHAAN EQUATIONS FOR 2N ELECTRONS

As with the pi-electron model, we want to treat the orbital coefficients as variation variables and use something like the Clementi–Raimondi–Slater [9] atomic orbitals for the basis functions, or at least something like them that are easy to integrate. We want to minimize the energy by varying the values of the c_m but we also want to maintain the orthonormality of the linear combination of basis functions as orthonormal one-electron orbitals. They are formed from linear combinations of basis functions $\{\phi_i\}$ that may have nonzero overlap $\int \phi_\mu^* \phi_\nu d\tau = S_{\mu\nu}$.

$$\int \psi_i^* \psi_j d\tau = \sum_\mu^M \sum_\nu^M c_{\mu i} c_{\nu j} \int \phi_\mu^* \phi_\nu d\tau = \sum_\mu^M \sum_\nu^M c_{\mu i} c_{\nu j} S_{\mu\nu} = \delta_{ij}.$$

Thus, we use the idea of Lagrangian multipliers to remove from the minimization procedure that part which might change the orthonormal nature of the one-electron orbitals ψ_i. We therefore need the derivative of $\frac{\partial}{\partial c_{\rho i}^*}\left[\langle E\rangle - \sum_i \sum_j \lambda_{ij} \sum_\mu^M \sum_\nu^M c_{\mu i}^* c_{\nu j} S_{\mu\nu}\right] = 0$ where λ_{ij} are the Lagrangian multipliers that prevent loss of orthonormality of the one-electron orbitals. We might as well take the derivative with respect to c_{ri}^* to get rid of one of the potentially complex numbers. Slater shows [2] that if you take the derivative with respect to c_{ri}, you will also get the alternate derivative in terms of c_{ri}^*. When you set the derivatives to zero, you find that since one is real and the other is complex you can set either one separately to zero and end up with an equivalent equation for either one. We will take the derivative with respect to c_{ri}^* so that we end up with an equation for c_{ri}. Remember, as for the pi-orbital coefficients in Roberts' treatment in Chapter 16, the c_{ri} coefficients are the variational variables so we can take derivatives with respect to their value. Thus, the constrained derivative of the energy expression $\frac{\partial}{\partial c_{\rho i}^*}\left[\langle E\rangle - \sum_i \sum_j \lambda_{ij} \sum_\mu^M \sum_\nu^M c_{\mu i}^* c_{\nu j} S_{\mu\nu}\right] = 0$ uses

$$\langle E\rangle = \sum_i^n \sum_\mu^M \sum_\nu^M c_{\mu i}^* c_{\nu i}(\mu/f_1/\nu) + \sum_{i>}^n \sum_j^n \sum_\mu^M \sum_\nu^M \sum_\lambda^M \sum_\sigma^M c_{\mu i}^* c_{\nu i} c_{\lambda j} c_{\sigma j}\{(\mu\nu/\lambda\sigma) - (\mu\lambda/\nu\sigma)\}$$

to find

$$\sum_\nu^M c_{\nu j}\left\{(\mu/f_1/\nu) + \sum_j^n \sum_\lambda^M \sum_\sigma^M c_{\lambda j}^* c_{\sigma j}[(\mu\nu/\lambda\sigma) - (\mu\lambda/\nu\sigma)]\right\} - \sum_j^n \sum_\nu^M c_{\nu j} \lambda_{ij} S_{\mu\nu} = 0.$$

Although the process is for an arbitrary c_{ri}^*, in effect it removes the summation over μ and summations associated with the index for electron i. As with the Hückel pi-electron derivation, this is the minimization requirement for just one (c_{ri}) coefficient for the orbital of just one electron (i) and just one equation in the rows of equations that form the system that leads to the Cayley–Hamilton situation. From the pi-electron case, we can see that this can be put into a matrix equation where we will need to use diagonalization.

This is a difficult derivation, but, to summarize, we now have just one of the equations that will be part of our secular equation for the Cayley–Hamilton solution to find the molecular orbital coefficients. There remains one very mysterious step. Note the subscripts on λ_{ij}, which implies that each orbital ψ_i (which we do not know yet) is dependent somehow on all the other ψ_j to maintain orthogonality to them as well as normalization. Recall the big wave function $\Psi(1, 2, 3, \ldots)$ is a determinant that is a single number at each (x, y, z) point in space, and we know some sort of unitary transformation $[T]^{-1}[\lambda_{ij}][T] = [\lambda_{ij}]_{diag}$ could be applied to the orbitals ψ_i (if we knew them!), which would not change the value of the overall wave function, just make the λ_{ij} interactions diagonal. If we assume this has been done and solve the equation subject to that constraint we will make it true! Thus, we end up with an equation for all the electrons in a given basis set as a Cayley–Hamilton problem to find the coefficients c_{ni} and from them find the energy once the coefficients are known. A diagonal λ_{ij} eliminates the $\sum_j^n \lambda_{ij}$ for the overlap term and leads to

$$\sum_\nu^M c_{\nu i}\left\{(\mu/f_1/\nu) + \sum_j^n \sum_\lambda^M \sum_\sigma^M c_{\lambda j}^* c_{\sigma j}[(\mu\nu/\lambda\sigma) - (\mu\lambda/\nu\sigma)]\right\} - \sum_\nu^M c_{\nu i}\lambda_{ii}S_{\mu\nu} = 0, \text{ for basis } f_m,$$

but it remains on the two-electron terms. Note that here we use (enforce mathematically) $\lambda_{ii} = \varepsilon_i$ as the now diagonal Lagrangian multiplier in an energy equation so these will be assumed to be the one-electron orbital energies! So far we have not integrated over the spins.

Integrating over spins and using the Aufbau Principle to put two electrons as spin pairs in the lowest energy orbitals, we find that from the four coulomb (not shown) and two exchange integrals above we will obtain $2[2(\mu\nu/\lambda\sigma) - (\mu\lambda/\nu\sigma)]$ with $2H_{\mu\nu}$ and $2S_{\mu\nu}$ from the one-electron operators. Within the two-electron integrals the four exchange integrals above are the result of a 2×2 determinant space for two orbitals which do not match in the big determinant terms. That leads to a minus sign for the exchange terms. Recall the factor of $(-1)^P$ in the determinantal wavefunction due to the way in which the determinants are filled with electrons as $\alpha\beta\alpha\beta\alpha\beta \ldots$. Pair occupation produces a factor of 2 which is there on all the terms but which then cancels.

$$\sum_\nu^M c_{\mu i}\left\{2(\mu/f_1/\nu) + \sum_j^n \sum_\lambda^M \sum_\sigma^M c_{\lambda j}^* c_{\sigma j}[4(\mu\nu/\lambda\sigma) - 2(\mu\lambda/\nu\sigma)]\right\} - 2\sum_\nu^M \lambda_{ii}S_{\mu\nu} = 0$$

Now we understand that the coefficients $c_{\mu i}$ refer to doubly-occupied orbitals. After cancelling the factor of 2 we have

$$\sum_\nu^M c_{\mu i}\left\{(\mu/f_1/\nu) + \sum_j^n \sum_\lambda^M \sum_\sigma^M c_{\lambda j}^* c_{\sigma j}[2(\mu\nu/\lambda\sigma) - (\mu\lambda/\nu\sigma)]\right\} - \sum_\nu^M \lambda_{ii}S_{\mu\nu} = 0.$$

Then we can define $P_{\lambda\sigma} \equiv 2\sum_j^{occ.} c_{\lambda j}c_{\sigma j}$ which is the "population matrix" due to summing over the lowest energy orbitals which are occupied by ($\alpha\beta$) spin-pairs. This leads to the final equations suitable for computer code as

$$FC = SC\varepsilon \quad \text{where} \quad F_{\mu\nu} = H_{\mu\nu} + \sum_\lambda \sum_\sigma P_{\lambda\sigma}[(\mu\nu/\lambda\sigma) - 0.5(\mu\lambda/\nu\sigma)].$$

Despite our attempt at a verbal description here, none of the three best references given [1–3] actually show the explicit details. In the final analysis a skeptical student will have to work out the full problem for the Be atom with four electrons to see the explicit details. Nevertheless, our final formula has been verified many times and we should be in awe of Prof. Roothaan for establishing that formula! Note that the secular equation to solve for the $c_{\mu i}$ coefficients involves the $c_{\mu i}$ coefficients; so the answer depends on the answer! We will need to make a good guess at the orbital coefficients and then iterate the calculation in some way to find convergence!

REFERENCES

1. Roothaan, C. C. J., New developments in molecular orbital theory, *Rev. Mod. Phys.*, **23**, 69 (1951).
2. Slater, J. C., *Quantum Theory of Molecules and Solids, Vol. 1, Electronic Structure of Molecules*, McGraw-Hill Book Co. Inc., New York, 1963, Appendix 7, p. 277, also Appendix 4, p. 256.
3. Pople, J. A. and D. L. Beveridge, *Approximate Molecular Orbital Theory*, McGraw-Hill Book Co., New York, 1970.
4. Slater, J. C., The electronic structure of atoms—The Hartree-Fock Method and correlation, *Rev. Mod. Phys.* **35**, 484 (1963).
5. Condon, E. U. and G. H. Shortley, *The Theory of Atomic Spectra*, Cambridge University Press, Cambridge, U.K., 1935.
6. Fock, V., Self-consistent field mit Austausch für Natrium. [Self-consistent field with exchange for Sodium], *Z. Physik.*, **61**, 795 (1930).
7. Hartree, D. R., The wave mechanics of an atom with a non-Coulomb central field. Part III. Term values and intensities in series in optical spectra, *Proc. Cambridge Phil. Soc.*, **24**, 426 (1928).
8. Jaramillo, J., G. E. Scuseria, and M. Ernzerhof, Local hybrid functionals, *J. Chem. Phys.*, **118**, 1068 (2003).
9. Clementi, E. and D. L. Raimondi, Atomic screening constants from SCF functions, *J. Chem. Phys.*, **38**, 2686 (1963).

Appendix C: Gaussian Lobe Basis Integrals

INTEGRALS: THE INEVITABLE DETAILS

Ultimately, the practical reality of using the Roothaan equation for an actual calculation comes down to evaluating some integrals. Eventually, devotees of Slater-type orbitals (STO) gave in to using linear combinations of Gaussian-type orbitals (GTO) based on formulas derived by Boys [1]. The first time this author saw these formulas was when Prof. Jerry Whitten, then a post-doc at Princeton, gave a seminar and passed around a single 3×5 in. card with the formulas! That was an amazing breakthrough to the way in which to actually carry out the Roothaan equations. To save space but provide information for the interested student, we present a very short list of the integral formulas as given by Shavitt [2]. These formulas are for individual un-normalized Gaussian spheres and, in the case of off-axis Gaussian lobe orbitals (GLO) used in PCLOBE, orbital mimics require many of these integrals nested inside linear expansions. These are the fundamental formulas that are implemented in PCLOBE [3].

$$G(\alpha, r_A) \equiv \exp\left(-\alpha r_A^2\right) = e^{-\alpha\left[(x-A_x)^2+(y-A_y)^2+(z-A_x)^2\right]}; \quad r_A = R(A_x, A_y, A_z)$$

$$(aA/bB) = \left(\frac{\pi}{a+b}\right)^{3/2} \exp\left(-\frac{ab}{a+b}(\overline{A}\,\overline{B})^2\right),$$

$$\left(aA\left|-\frac{\nabla^2}{2}\right|bB\right) = \frac{ab}{a+b}\left(3 - \frac{2ab(\overline{A}\,\overline{B})^2}{a+b}\right)(aA/bB),$$

$$\left(aA\left|\frac{1}{r_c}\right|bB\right) = \frac{2\pi}{a+b}F_0\left[(a+b)(\overline{C}\,\overline{P})^2\right]\exp\left(-\frac{ab}{a+b}(\overline{A}\,\overline{B})^2\right)$$

and then for (1*, 1/2*, 2) we have

$$\left[aA, bB\left|\frac{1}{r_{12}}\right|cC, dD\right] = \frac{2\pi^{5/2}}{(a+b)(c+d)\sqrt{(a+b+c+d)}}$$

$$\times F_0\left[\frac{(a+b)(c+d)}{(a+b+c+d)}\overline{PQ}^2\right]\exp\left[-\left(\frac{ab}{a+b}\right)(\overline{A}\,\overline{B})^2 - \left(\frac{cd}{c+d}\right)(\overline{C}\,\overline{D})^2\right].$$

One special function is required but most computer language compilers have an intrinsic function for the error function "$erf(x)$," which can be used directly or a table can be generated within a program with a fine grid and interpolation methods used to speed up the process [4].

$$F_0(t) = \frac{1}{\sqrt{t}}\int_0^{\sqrt{t}} e^{-v^2}\,dv = \left(\frac{1}{2}\sqrt{\frac{\pi}{t}}\right)erf\left(\sqrt{t}\right) \quad \text{or} \quad F_{m+1}(t) = -\frac{d}{dt}F_m(t)$$

and

$$F_m(t) = \frac{1}{2m+1}[2tf_{m+1} + e^{-t}].$$

The key relationship is that the product of two Gaussians is still a third Gaussian. Thus, if

$$G(\alpha, r_A) \equiv \exp(-\alpha r_A^2), \quad \text{then} \quad G(\alpha_1, r_A)G(\alpha_2, r_B) = \exp\left(-\frac{\alpha_1 \alpha_2}{(\alpha_1 + \alpha_2)}(\overline{A\,B})^2\right)G((\alpha_1 + \alpha_2), r_c)$$

where $C_x = \dfrac{\alpha_1 A_x + \alpha_2 B_x}{\alpha_1 + \alpha_2}$, $C_y = \dfrac{\alpha_1 A_y + \alpha_2 B_y}{\alpha_1 + \alpha_2}$, and $C_z = \dfrac{\alpha_1 A_z + \alpha_2 B_z}{\alpha_1 + \alpha_2}$ and that process can be

used still further to define the points that are on a line between the (aA/bB) Gaussian product and the (cC/dD) Gaussian product to form yet another Gaussian for the two-electron integral. Thus, we

can calculate $P_x = \dfrac{aA_x + bB_x}{a+b}$ and $Q_x = \dfrac{cC_x + dD_x}{c+d}$, etc., to obtain $(\overline{P\,Q})^2$, which is just the square of

the distance between the Gaussian centroids of the (aA/bB) and (cC/dD) Gaussian products. While this may look complicated at first encounter, it is far simpler than integrals required in a GTO basis which uses $x^l y^m z^n \exp(-\alpha r^2)$. The mere fact that we can write these formulas in a page or so is amazing after one has delved into formulas for other basis sets, and the GLO basis is a good starting point to learn these ideas. Of course, it takes many spheres with optimized weighting coefficients to form accurate mimics of H-atom-like orbitals, but it can be done if a sufficient number of optimized spheres are used.

Thus, we have $\psi_i = \sum_\mu C_{\mu i} \phi_\mu$ where each molecular orbital is expanded in a basis $\{\phi_\mu\}$ and we also have an expansion of each basis function in terms of Gaussian spheres $\phi = \sum_j c_{j\mu} g(\alpha_j, \vec{A}_j)$. Finally, in PCLOBE we can form integrals over mimics of atomic orbitals, which form basis functions for the Roothaan molecular orbitals. For the one-electron integrals (overlap, kinetic energy, and nuclear attraction, f_1), we have (using $C_{\mu i}$ for orbitals and $c_{j\mu}$ for Gaussians) $(\phi_\mu|f_1|\phi_\nu) = \sum_j \sum_k c_{j\mu} c_{k\nu} (g_j|f_1|g_k)$ and for the two-electron integrals, we have $(\phi_\mu, \phi_\nu|1/r_{12}|\phi_\lambda^*, \phi_\sigma) = \sum_j \sum_k \sum_m \sum_n c_{j\mu} c_{k\nu} c_{m\lambda} c_{n\sigma} (g_j, g_k|1/r_{12}|g_m g_n)$. The $c_{j\mu}$, α_j and $\vec{A}$ are printed out at the beginning of each PCLOBE run for documentation.

Other integrals for dipole moment, quadrupole moment, and even spin-orbital coupling have also been worked out and are available in closed form as given in a recent monograph [3].

REFERENCES

1. Boys, S. F., Electronic Wave Functions I. A general method of calculation for the stationary states of any molecular system, *Proc. Roy. Soc.*, **A200**, 542 (1950).
2. Shavitt, I., The Gaussian function in calculations of statistical mechanics and quantum mechanics. In *Methods in Computational Physics, Advances in Research and Applications*, B. Alder, S. Fernbach and M. Rotenberg, Eds., Vol. 2. Academic Press, New York, 1963.
3. Trindle, C. and D. Shillady, *Electronic Structure Modeling: Connections between Theory and Software*, CRC Press, Boca Raton, FL, 2008, p. 66.
4. Mosier, C. and D. D. Shillady, A fast, accurate approximation for $F_0(z)$ occurring in Gaussian lobe basis electronic calculations, *Math. of Comp.*, **26**, 1022 (1972).

Appendix D: Spin-Orbit Coupling in the H Atom

The inclusion of spin-orbit coupling [1] can be worked out cleanly for the one-electron case of the H atom. It provides a model for an effect that becomes more important in heavy elements [1]. This has been developed and implemented by Prof. Balasubramanian (Figure D.1), Prof. Bloor (Figure D.2), and their students. The extra term in the Hamiltonian is of the form

$$H_{so} = \sum_i \xi_i \vec{L}_i \cdot \vec{S}_i.$$

For the single electron in the H atom or a hydrogenic ion the value was worked out by Thomas [2] using the first order terms of a relativistic expansion. According to the Dirac theory of electrons, the magnetic moment of the electron is given by

$$\vec{\mu} = \frac{e}{mc} \vec{S}$$

and the energy of interaction of the magnetic moment with a magnetic field $\vec{H}$ is

$$E_{so} = -\vec{H} \cdot \vec{\mu}.$$

The magnetic field generated by an electron moving around a nucleus is then

$$\vec{H} = \left(\frac{1}{c}\right)\vec{E} \times \vec{v} = \left(\frac{1}{mc}\right)\vec{E} \times \vec{p}.$$

where $\vec{E}$ is the electric field of the charged nucleus. The electron moves in the centrosymmetric electric field $\vec{E}$ of the nucleus, which is due to the gradient of the electrostatic potential of the nucleus.

$$\vec{E} = -\left(\frac{\vec{r}}{r}\right)\left(\frac{dV}{dr}\right).$$

Thus, the magnetic field $\vec{H}$ generated by the electron is given by the motion of the electron in the electric field gradient of the nucleus

$$\vec{H} = -\frac{1}{mcr}\frac{dV}{dr}(\vec{r} \times \vec{p}) = -\frac{1}{mcr}\frac{dV}{dr}\vec{L}.$$

The electric potential of the nuclear charge is $V = \dfrac{Ze^2}{r}$ so that $\dfrac{dV}{dr} = -\dfrac{Ze^2}{r^2}$ and

$$\vec{H} = \frac{Ze^2}{mcr^3}\vec{L}.$$

In 1926, Thomas [2] carried out the relativistic derivation of this interaction and found the factor of $\left(\dfrac{1}{2mc}\right)$ using the leading term of a relativistic expansion of the velocity.

$$\xi = \frac{1}{2m^2c^2}\left(\frac{-1}{r}\frac{dV(r)}{dr}\right) \quad \text{where} \quad \frac{dV}{dr} = \frac{d}{dr}\left(\frac{Ze^2}{r}\right) = \frac{-Ze^2}{r^2} \quad \text{so} \quad \xi = \frac{Ze^2}{2m^2c^2}\left(\frac{1}{r^3}\right).$$

Then the interaction with the spin of the electron is

$$E_{so} = \langle n, l, m|\left[\frac{Ze^2}{2m^2c^2}\right]\left(\frac{\vec{L}\cdot\vec{S}}{r^3}\right)|n, l, m\rangle.$$

You can anticipate an additional factor of Z^3 from the normalization constant of the radial polynomial and the angular parts over (θ, ϕ) will integrate to 1 since the matrix element only involves radial dependence as $\left\langle\frac{1}{r^3}\right\rangle$

$$R_{nl}(r) = -\left[\left(\frac{2Z}{na_0}\right)^3\frac{(n-l-1)!}{2n|(n+l)!|^3}\right]^{\frac{1}{2}}e^{-\frac{\rho}{2}}\rho^l L_{n+l}^{2l+1}(\rho).$$

The radial integration is over the product of two polynomials but selected powers of r are given in the text by Lesk [3] along with a recursion relationship due to Kramer for other powers of r (Table D.1).

However, we need $\left\langle\frac{1}{r^3}\right\rangle$ so we can use that recursion relationship found by Kramer as given in Lesk's text [3].

FIGURE D.1 Professor Krishnan Balasubramanian, emeritus professor at Arizona State University (presently at Lawrence Livermore, Berkeley Labs, and California State University, Eastbay) and recipient of the 2003 Robert S. Mulliken Award presented at the University Georgia, a Fullbright Scholarship, the Lawrence Livermore Distinguished Service Award, the Camille and Henry Dreyfus Teacher-Scholar Award, and many other awards. Professor Balasubramanian has published over 600 scientific papers to date and is recognized as a pioneer in extending quantum chemistry to relativistic treatment of molecules containing heavy elements [1].

TABLE D.1
Expectation Values of Powers of r with Hydrogenic Orbitals for Nuclear Charge Ze

$$\langle r \rangle_{nlm} = \frac{a_0}{2Z}\left[3n^2 - l(l+1)\right]$$

$$\langle r^2 \rangle_{nlm} = \frac{n^2 a_0^2}{2Z^2}\left[5n^2 + 1 - 3l(l+1)\right]$$

$$\langle r^{-1} \rangle_{nlm} = \frac{Z}{a_0 n^2}$$

$$\langle r^{-2} \rangle_{nlm} = \frac{Z^2}{a_0^2 n^3 [l + 1/2]}$$

Source: Lesk, A.M., *Introduction to Physical Chemistry*, Prentice-Hall, Inc., Englewood Cliffs, NJ, 1982, p. 320.

FIGURE D.2 Prof. John E. Bloor, wrote his PhD thesis in organic chemistry at the University of Manchester on the synthesis and UV spectra of disubstituted benzenes and aza-merocyanine dyes but Prof. Coulson was appointed as chair of his Thesis Committee when his advisor, Prof. Burawoy, passed away suddenly. Although Dr. Bloor's thesis was approved as a PhD in organic chemistry, he responded with a career in theoretical/quantum chemistry due to his contact with Prof. Coulson. He used the Pariser-Parr-Pople Pi-electron SCF method to study UV spectra of conjugated hydrocarbons and their derivatives containing oxygen, nitrogen, and sulfur, and was one of the first to apply the all valence CNDO/2 method to organic molecules. Later in his career at the University of Tennessee, he used relativistic ab initio computer programs, especially the one developed by Balasubramanian [4–7], which included spin-orbit coupling in the valence shell to interpret the photoelectron and electronic spectra of molecules containing heavy transition metals, rare earth metals, and transuranium elements.

$$Z^2\left(\frac{k+1}{n^2}\right)\left\langle\left(\frac{r}{a_0}\right)^k\right\rangle - (2k+1)Z\left\langle\left(\frac{r}{a_0}\right)^{k-1}\right\rangle + k\left(l+\frac{1}{2}+\frac{k}{2}\right)\left(l+\frac{1}{2}-\frac{k}{2}\right)\left\langle\left(\frac{r}{a_0}\right)^{k-2}\right\rangle = 0$$

If we choose $k = -1$, we find $0 - (-1)Z\left\langle\left(\frac{r}{a_0}\right)^{-2}\right\rangle - \left(l+\frac{1}{2}-\frac{1}{2}\right)\left(l+\frac{1}{2}+\frac{1}{2}\right)\left\langle\left(\frac{r}{a_0}\right)^{-3}\right\rangle = 0.$

Using $\langle r^{-2}\rangle_{nlm} = \dfrac{Z^2}{a_0^2 n^3\left[l+\frac{1}{2}\right]}$ we find $\left\langle\left(\frac{r}{a_0}\right)^{-2}\right\rangle = \langle a_0^2 r^{-2}\rangle = \dfrac{a_0^2 Z^2}{a_0^2 n^3\left(l+\frac{1}{2}\right)} = \dfrac{Z^2}{n^3\left(l+\frac{1}{2}\right)}.$ Then

$$-(-1)Z\left[\frac{Z^2}{n^3\left(l+\frac{1}{2}\right)}\right] = (l+0)(l+1)\left\langle\left(\frac{r}{a_0}\right)^{-3}\right\rangle = l(l+1)\left\langle\frac{a_0^3}{r^3}\right\rangle$$

and so we have

$$\left\langle\frac{1}{r^3}\right\rangle = \frac{Z^3}{n^3 a_0^3 l\left(l+\frac{1}{2}\right)(l+1)}.$$

Finally, we can use this in the spin-orbit expression to find

$$E_{so} = \langle n, l, m|\left[\frac{Ze^2}{2m^2c^2}\right]\left(\frac{\vec{L}\cdot\vec{S}}{r^3}\right)|n, l, m\rangle = \left[\frac{Ze^2}{2m^2c^2}\right]\left\{\frac{Z^3[j(j+1)-l(l+1)-s(s+1)]\hbar^2}{n^3 a_0^3 (l)\left(l+\frac{1}{2}\right)(l+1)}\right\}$$

and we see that

$$\xi_{nl} = \left(\frac{e^2\hbar^2}{2m^2c^2 a_0^3}\right)\frac{Z^4}{n^3 l\left(l+\frac{1}{2}\right)(l+1)}.$$

Note the dependence on the fourth power of Z which indicates the importance of this effect for heavier elements!

Although we used non-SI units we have reached agreement with the formula for ξ_{nl} given in a revered text on Quantum Mechanics by Schiff [8] which predates SI units. The excellent text by McGlynn, Azumi and Kinoshita [9] also gives the same expression related to the spin-orbit coupling in triplet states. The important thing for undergraduates to learn from this is the fourth power dependence on the nuclear charge Z. It is also an interesting opportunity to use what we have learned about the eigenvalues of the $J = L + S$ total angular momentum operator. Modern treatments by McQuarrie [10] and by Townsend [11] derive the shift of the energy levels of the H atom

due to spin-orbit coupling as $\Delta E_{so} = \dfrac{m_e c^2 Z^4 \alpha^4}{4n^3 l(l+1/2)(l+1)}\left\{\begin{array}{l} l, j = l+1/2 \\ -(l+1), j = l-1/2 \end{array}\right\}$ where

$\alpha = \dfrac{e^2}{4\pi\varepsilon_0\hbar c}$ in SI units. That predicts the splitting between $j = l \pm 1/2$ is only

$\Delta(\Delta E_{so}) = \dfrac{5.8437 Z^4 \text{ cm}^{-1}}{n^3 l(l+1)}.$ The same wave number value is obtained using cgs constants with the a_0 value [9,12]. While this is important in astronomy due to the abundance of H in space, this difference is tiny relative to most chemical reactions. However past $Z = 30$ for Zn, heavier elements have values of ξ_{nl} in thousands of cm^{-1}.

REFERENCES

1. Balasubramanian, K., *Relativistic Effects in Chemistry*, Parts A & B, Wiley Interscience, New York, 1997.
2. Thomas, L. H., The Motion of the spinning electron, *Nature*, **117**, 514 (1926).
3. Lesk, A. M., *Introduction to Physical Chemistry*, Prentice-Hall, Inc., Englewood Cliffs, NJ, 1982, p. 320.
4. Balasubramanian, K., N. Tanpipat, and J. E. Bloor, Spectroscopic properties of SbH, *J. Mol. Spectroc.*, **124**, 458 (1987).
5. Tanpipat, N., Ab initio relativistic effective potential and relativistic configuration interaction calculations on diatomic molecules containing heavy elements (Lutecium and Lawrencium), PhD thesis, University of Tennessee, 1987.
6. Balasubramanian, K. and Pitzer, K. S. Relativistic quantum chemistry, in *Advances in Chemical Physics: Ab Initio Methods in Quantum Chemistry Part I*, Vol. 67 (ed K. Lawley), John Wiley & Sons, Inc., Hoboken, NJ, 2007.
7. Balasubramanian, K., Relativistic calculations of electronic states and potential energy surfaces of Sn_3, *J. Chem. Phys.*, **85**, 3401 (1986).
8. Schiff, L. I., *Quantum Mechanics*, 2nd Edn., McGraw-Hill Book Co., New York, 1955. p. 286.
9. McGlynn, S. P., T. Azumi, and M. Kinoshita, *Molecular Spectroscopy of the Triplet State*, Prentice Hall Inc., Englewood Cliffs, NJ, 1969, pp. 188–189.
10. McQuarrie, D. A., *Quantum Chemistry*, 2nd Edn., University Science Books, Sauslito, CA, 2008, p. 364.
11. Townsend, J. S., *A Modern Approach to Quantum Mechanics*, University Science Books, Sausalito, CA, 2000.
12. Shillady, D. D., Some electronic effects of covalency, PhD thesis, University of Virginia, Charlottesville, VA, 1970, p. 185.

Index

Use of PCLOBE

I. Copying files

1. Insert the PCLOBE compact disk into your CD drive and locate its three files. One is a zipped file of illustrative examples and another is the "setup.exe" file.
2. Transfer the zipped file to your computer and extract the compressed files into a new folder, which you could name "PCLOBEXAMPLES."
3. Double click on "setup.exe" on the CD—this will install PCLOBE on your own drive. You will find the program in your programs list; you may construct a shortcut so to start PCLOBE from your desktop.

II. Running PCLOBE

1. Start (click on) PCLOBE either from the shortcut or from the programs list. If all goes well, you will see a Visual Basic display with selection buttons to define PCLOBE calculations.
2. Define a job, using structures supplied in PCLOBEXAMPLES:
   a. Click the Browse button at the top left of the display panel.
   b. Find and click the "name.XYZ" file of your choice (open it).
   c. Select the desired basis from the list at center left.
   d. Enter an integer to define the gradient threshold ("3" $= 10^{-3}$). Selecting the default "0" defines a single-point energy calculation. This is required for certain tasks, and enforced by the interface. A gradient threshold of 4 or 5 will invoke calculation of vibrational frequencies. The default of 30 steepest descent iterations is best reset to "99" for geometry optimization.
3. Click on "Launch PCLOBE" to start the program. You should immediately see a black foreground output window showing progress through the task.
4. When the run finishes, click on the "View Output" button at the lower right part of the Visual Basic panel to look at the total output. If you wish to retain the output, save it with a unique file name.

The WINDOWS-VISTA or WINDOWS-7 operating systems can be used to run PCLOBE in Administrator Mode by right-clicking on the PCLOBE icon, but all other files and applications should be closed before running in Administrator mode.

SPECIAL REQUIREMENTS FOR INPUT FILES

Some options require special care that input files be prepared in accord with certain conventions. For instance, the point group symmetry option requires that the input structure possess the specified symmetry, and requires that the high order symmetry axis coincide with the "z" Cartesian direction. The examples for this input with names xxxsym.XYZ obey these requirements. Consult the "READpclobe.txt" file for more detail and an exception with the nonconventional Cnvx symmetry.

Use of the Rydberg basis requires that the file "rydberg.dat" contain a nonzero value of scaling for each atom to which a 3sp shell is to be added; remember to reset the value to "0.0" for the next run if the Rydberg orbitals are not desired (applies only to CIS in a STO-4G basis).

PRESERVING DATA WRITTEN TO PLACEHOLDER FILES

From some tasks, PCLOBE produces data files labeled "uvspectra.txt," cdspectra.txt," "mcdspectra.txt," and "ordspectra.txt." These are intended for use by graphics programs. The file labeled "lobe.xyz" is intended to provide input data to an external molecular modeler such as RASMOL. The file "lobe.draw" is intended for the external plotting program MOLEKEL. If information in these files is to be retained, they should be saved with more informative names. The original READpclobe file included in the CD of the text *Electronic Structure Modeling, Connections between Theory and Modeling*, CRC Press (2008) by C. Trindle and D. Shillady is on the present CD along with a revised file READPCLOBEV2 which has been contributed by a user (Dr. Charles Castevens). The revised file includes the perspective of a user of the PCLOBE program and gives additional information related to using PCLOBE in Windows-7 which was not available in the 2008 text.

PGSTL